Biotechnology
Volume 6a

Biotechnology

A Comprehensive Treatise in 8 Volumes

edited by H.-J. Rehm and G. Reed

Volume 1
Microbial Fundamentals

Volume 2
Fundamentals of Biochemical Engineering

Volume 3
Biomass, Microorganisms for Special Applications, Microbial Products I, Energy from Renewable Resources

Volume 4
Microbial Products II

Volume 5
Food and Feed Production with Microorganisms

Volume 6a
Biotransformations

Volume 6b
Special Microbial Processes

Volume 7
Enzymes in Biotechnology

Volume 8
Microbial Degradations

Weinheim · Deerfield Beach, Florida · Basel

Biotechnology

A Comprehensive Treatise
in 8 Volumes
edited by H.-J. Rehm and G. Reed

Volume 6a

Biotransformations

Volume Editor: K. Kieslich

Weinheim · Deerfield Beach, Florida · Basel

Prof. Dr. H.-J. Rehm
Institut für Mikrobiologie
der Universität
Correnssstraße 3
D-4400 Münster
Federal Republic of Germany

Dr. G. Reed
Universal Foods Corp.
Technical Center
6143 N 60th Street
Milwaukee, Wisconsin 53218
U.S.A.

Prof. Dr. K. Kieslich
Gesellschaft für Biotechnologische
Forschung mbH
Mascheroder Weg 1
D-3300 Braunschweig-Stöckheim

Publisher's Editor: Dr. Hans F. Ebel
Copy Editor: Christa Maria Schultz
Production Manager: Peter J. Biel

This book contains 232 figures and 100 tables

Deutsche Bibliothek Cataloguing-in-Publication Data

Biotechnology: a comprehensive treatise in 8 vol. / ed. by H.-J. Rehm and G. Reed. –
Weinheim; Deerfield Beach, Florida; Basel: Verlag Chemie
NE: Rehm, Hans-Jürgen [Hrsg.]
Vol. 6a. Biotransformations. – 1984

Biotransformations/vol. ed.: K. Kieslich. –
Weinheim; Deerfield Beach, Florida; Basel: Verlag Chemie, 1984.
(Biotechnology; Vol. 6a)
ISBN 3-527-25768-3 (Weinheim, Basel)
ISBN 0-89573-046-4 (Deerfield Beach, Florida)
NE: Kieslich, Klaus [Hrsg.]

Composition and Printer: Zechnersche Buchdruckerei, D-6720 Speyer
Bookbinder: Klambt-Druck GmbH, D-6720 Speyer
Printed in the Federal Republic of Germany

Preface

Volume 6a of "Biotechnology" describes biotransformations, i.e., enzymatic conversions of natural and chemically synthesized products, into substances having specifically modified structures.

This vast topic is dealt with systematically according to the class of chemical substrate. The authors are all experts and draw upon many years of experience and a profound understanding of their fields.

Differences in the importance of the types of substrates and differences in the state-of-the-art technology required different structures for the individual chapters. The topics covered range from model substrates to laboratory preparative methods and to industrial processes.

The multitude of known biotransformations and the indication of their advantages should stimulate further intensive research in this area. It should also be the starting point for the discovery of other less thoroughly investigated enzymatic reactions. Some areas with good prospects for wider application of such methods are:

- the investigation of further interesting substrate structures and extension to additional classes of substrates,
- the detailed study of previously known or novel enzymatic reactions for preparative purposes,
- the conversion of low cost natural products into useful partial structures by controlled enzymatic partial degradation,
- and, in particular, the use of the biosynthetic bond forming potential for coupling artificial partial structures or for intramolecular cyclizations.

In the final analysis the use of biotransformations for preparative purposes and their economic advantages requires, as a basis, further process development such as the optimization of conventional batch culture methods with charges of cells and enzymes, the immobilization of enzymes on a carrier, and the extended use of membrane reactors.

My hope is that this book, which describes the present state-of-the-art, will elucidate these additional possibilities.

I extend my sincere thanks to the authors for accepting our invitation to contribute to this volume, for their committed and intensive dedication, for their considerable competence, and for their excellent collaborative efforts.

I also wish to thank the scientific editor of Verlag Chemie, Dr. H. F. Ebel, for his consistent support, Christa Schultz for meticulously editing the manuscript, and the series editors, Prof. H.-J. Rehm and Dr. G. Reed, for their helpful comments and critical reading of the text.

Last but not least, I wish to thank my secretary, Ursula Reichel, for handling the enormous amount of correspondence and for her additional collaboration.

Braunschweig, February 1984

Klaus Kieslich

Editorial Advisory Board

of "Biotechnology"

List of Contributors

of Volume 6a

Dr. Matthias Bühler

Henkel KGaA
Abt. Biotechnologie
D-4000 Düsseldorf 1
Federal Republic of Germany

Dr. Peter Egerer

Bayer AG
Verfahrensentwicklung Biochemie
D-5600 Wuppertal 1
Federal Republic of Germany

Dr. Anneliese Crueger

Bayer AG
Verfahrensentwicklung Biochemie
D-5600 Wuppertal 1
Federal Republic of Germany

Dr. Gabriele Engelhardt

Bayerische Landesanstalt für
Bodenkultur und Gartenbau
Abt. Pflanzenschutz
D-8000 München 19
Federal Republic of Germany

Dr. Wulf Crueger

Bayer AG
Biotechnikum Mikrobiologie
D-5600 Wuppertal 1
Federal Republic of Germany

Prof. Dr. Alain Kergomard

Laboratoire de Chimie Organique
Biologique
Université de Clermont-Ferrand II
F-63170 Aubière
France

Prof. Dr. Patrick J. Davis

The University of Texas at Austin
College of Pharmacy
Austin, Texas 78712
U.S.A.

Prof. Dr. Klaus Kieslich

Gesellschaft für Biotechnologische
Forschung mbH
D-3300 Braunschweig-Stöckheim
Federal Republic of Germany

Dr. Victor Krasnobajew

Givaudan Research Company Ltd.
CH-8600 Dübendorf/Zürich
Switzerland

Prof. Dr. Günter Schmidt-Kastner

Bayer AG
Verfahrensentwicklung Biochemie
D-5600 Wuppertal 1
Federal Republic of Germany

Dr. Hans Georg W. Leuenberger

Hoffmann-La Roche & Co AG
Central Research Units
CH-4002 Basel
Switzerland

Dr. Oldrich K. Sebek

The Upjohn Company
Infectious Diseases Research
Kalamazoo, Michigan 49001
U.S.A.

Dr. Christoph K. A. Martin

BASF Aktiengesellschaft
WHZ 89 (Hauptlabor)
D-6700 Ludwigshafen
Federal Republic of Germany

Prof. Dr. Leland L. Smith

University of Texas Medical Branch
Division of Biochemistry
Galveston, Texas 77550
U.S.A.

Dr. Joachim Schindler

Henkel KGaA
Abt. Biotechnologie
D-4000 Düsseldorf 1
Federal Republic of Germany

Prof. Dr. Peter R. Wallnöfer

Bayerische Landesanstalt
für Bodenkultur und Gartenbau
Abt. Pflanzenschutz
D-8000 München 19
Federal Republic of Germany

Contents

Chapter 0

Introduction 1

by *Klaus Kieslich*

Chapter 1

Methodology 5

by *Hans Georg W. Leuenberger*

Chapter 2

Steroids 31

by *Leland L. Smith*

Chapter 3

Sterols 79

by *Christoph K. A. Martin*

Chapter 4

Terpenoids 97

by *Victor Krasnobajew*

Chapter 5

Alicyclic and Heteroalicyclic Compounds 127

by *Alain Kergomard*

Chapter 6

Natural and Semi-synthetic Alkaloids 207

by *Patrick J. Davis*

Chapter 7

Antibiotics 239

by *Oldrich K. Sebek*

Chapter 8

Aromatic and Heterocyclic Structures 277

by *Peter R. Wallnöfer* and *Gabriele Engelhardt*

Chapter 9

Aliphatic Hydrocarbons 329

by *Matthias Bühler* and *Joachim Schindler*

Chapter 10

Amino Acids and Peptides 387

by *Günter Schmidt-Kastner* and *Peter Egerer*

Chapter 11

Carbohydrates 421

by *Anneliese Crueger* and *Wulf Crueger*

Index 469

Chapter 0
Introduction

Klaus Kieslich

Gesellschaft für Biotechnologische Forschung
Braunschweig-Stöckheim, Federal Republic of Germany

I. *What are Bio-Transformations?*

Biotechnology employs microorganisms as well as higher cells and their active principles with the aim of achieving desirable conversions of various substrates. Physiology, biochemistry, molecular biology, and genetics enable us to reveal and comprehend the characteristics of the systems involved. The engineering sciences finally prepare the ground to technically apply these systems, through the development of equipment and technological procedures. Other contributing disciplines, boardering on the aforementioned fields of research, contribute significantly to progress in these conversions.

The desirable products of biotechnology are either pure compounds, mixtures, cell fractions, or biomass. These homogeneous or heterogeneous chemical structures result from *de novo* formation, transformation, or degradation, by means of a vast array of enzymatic reactions, but in some cases merely by a single reaction or a short sequence of reactions. This accounts for the essentially chemical nature of biotechnology, which holds true especially for the biotransformations considered in this volume of 'Biotechnology'. The term 'chemical reactions by microorganisms or enzymes' briefly describes the essence of biotransformations.

Organic compounds of interest generally are obtained by the following procedures:

1. extraction of plants or organs of animals
2. total chemosynthesis
3. partial chemosynthesis starting from natural products
4. native or controlled biosynthesis
5. enzymatic transformation of natural products or chemically synthesized substrates.

The first-mentioned procedure has been the only feasible approach yielding desirable products in past centuries. Since slightly more than a century, however, there has been an build-up of an enormous chemical industry based on chemosynthesis (points 2 and 3, above).

With the exception of a few traditional processes, high yields of biosynthetic products from the primary and secondary metabolism of microorganisms or higher cells by large-scale technologies have been possible only for approximately the last 40 years.

It is important to note, however, that certain fermentations of substrate mixtures have been carried out for centuries utilizing the effect of enzymes from the primary metabolism of microorganisms. Multi-stage reaction pathways are involved in the degradation of biopolymers or macromolecules of complex structures, for which multi-enzyme systems, derived from suitable biological systems, have long been empirically employed (SEBEK, 1982).

In contrast to the aforementioned procedures, biotransformations should be considered as selective, enzymatic modifications of defined pure compounds into defined final products.

In order to differentiate between the three types of procedures, we furthermore want to distinguish: bioconversions, biodegradations, and biotransformations; though, in some cases, it might not be possible to draw a clear line between the individual categories.

Biotransformations are carried out either with pure cultures of microorganisms, with plant cell cultures, or with purified enzymes. The first processes which fulfill these requirements were:

- The oxidation of alcohol to acetic acid by *Bacterium xylinum* (PASTEUR, 1862; BROWN, 1886).
- The oxidation of glucose to gluconic acid by *Acetobacter aceti* (BOUTROUX, 1880).
- The oxidation of sorbit to sorbose by *Acetobacter* sp. (BERTRAND, 1896).
- Acyloin-formation from benzaldehyde and acetaldehyde through yeasts (NEUBERG, 1921).

Although these early procedures were later developed into large-scale technologies, it was only much later that, biotransformations gained their present significance through the discovery of microbial transformations of steroids:

- Reduction of androstenedione to testosterone through yeasts (MAMOLI and VERCELLONE, 1937).
- 11α-Hydroxylation of progesterone by *Rhizopus arrhizus* (PETERSON and MURRAY, 1952).
- Introduction of a Δ^1-double bond into 3-keto-4-enesteroids by *Corynebacterium simplex* (NOBILE and CHARNEY, 1955).
- 11β-Hydroxylation of C_{21}-steroids by *Curvularia lunata* (SHULL and KITA, 1955).
- 16α-Hydroxylation of 9α-fluorhydrocortisone by *Streptomyces roseochromogenes* (THOMA and FRIED, 1957).

In the aftermath, diverse transformations of substrates from many different groups and types of compounds were found, some of which were useful in chemosyntheses.

Today, the general goals of biotransformations may be considered to be the following:

- specific modification of the substrate structure via selective transformation reactions;
- partial degradation of substrates into desirable metabolites by means of controlled microbial reactions or reaction pathways;
- extension of the substrate structure by the use of biosynthetic reactions to artificial structures.

While a variety of reactions leading to specific modifications are known, just as there are various examples of the partial degradation of substrates, the general possibilities for sequential biosynthesis via individual enzymatic reactions have been employed only rarely.

II. Short Description of Contents

The following chapters inform the reader of the current state of knowledge of biotransformations; the subject has been organized according to families of chemical substrates.

An introduction into the methodology and available technological procedures of biotransformations is the concern of Chapter 1, followed by a survey of the most important microbial transformations of steroids in Chapter 2. In awareness of the enormous range of the current knowledge in this field – most interestingly the stereospecific hydroxylations –, the chapter focuses on enzymatic reactions presently employed in industrial technology.

The sterols, described in Chapter 3, have gained economic significance only within the past 15 years, through such procedures as the controlled side chain degradation into suitable substrates for chemosynthesis of steroids. The description of the individual processes emphasizes the successful development of bacterial mutants.

In contrast to such easily crystallizing substrates and products as above, biotransformations of small, highly volatile, and liquid terpenoids require different fermentation and isolation procedures, as delineated in Chapter 4. The author emphasizes industrially employed technologies in the production of aroma compounds.

A choice of alicyclic and heteroalicyclic substrates forms the basis of systematic model investigations of sterical behavior during hydroxylations and keto-reductions, as presented in Chapter 5, along with synthetically important applications.

Biotransformations of alkaloids unfortunately have not as yet been applied on a technical scale, despite the many known reactions of this important group of compounds. Nevertheless, regiospecific enzymatic methylether-cleavage reactions and *N*-demethylations have been shown to yield advantages over analogous chemical reactions, as underlined in Chapter 6.

Antibiotics, with their manifold structures, have been subjected to many different oxidations and reductions as well as to various types of partial oxidative degradations. Hydrolytic reactions and reversed hydrolyses of penicillins have gained technical significance; immobilized systems have been developed and are used industrially. Chapter 7 reports on the details of these procedures.

Research on biotransformations of aromatic and heterocyclic substrates – the subject of Chapter 8 – has been aimed at the elucidation of their enzyme-induced catabolism. Only few examples are known to have been applied for preparative purposes in the laboratory.

Aliphatic substrates with simple hydrocarbon structures can be oxidized to various alcohols and acids, which can also be further degraded enzymatically. Chapter 9 reports the results of detailed investigations of the enzymatic reaction mechanisms.

In the production of amino acids, biotransformations allow enzymatic reactions with highly selective stereospecificity. Examples are the production of amino acids from racemic mixtures and the reductive amination of α-ketones. Technical procedures often employ purified enzymes or immobilized systems. Peptide synthesis – by reversed hydrolysis reactions – is an example for selective biosynthesis by means of biotransformation procedures (Chapter 10).

Carbohydrate substrates (Chapter 11) are the classical example for traditionally applied biotransformations, based on the regiospecificity of the involved enzymatic oxidations and reductions. Next to the significance of obtaining monosaccharides from corresponding biopolymers, the isomerization of aldoses to ketoses has become an important biotechnological method. Today, these reactions are also carried out by immobilized systems.

III. References

BERTRAND, G. (1896). C. R. Acad. Sci. *122*, 900.

BOUTROUX, L. (1880). C. R. Acad. Sci. *91*, 236.

BROWN, A. J. (1886). J. Chem. Soc. *49*, 172.

DEXTERSON, D. H., and MURRAY, H. C. (1952). J. Am. Chem. Soc. *74*, 1871.

MAMOLI, C., and VERCELLONE, A. (1938). Ber. Dtsch. Chem. Ges. *71*, 1686.

NEUBERG, C., and HIRSCH, J. (1921). Biochem. Z. *115*, 282.

NOBILE, A., CHARNEY, W., PERLMAN, P. L., HERZOG, H. L., PAYNE, C. C., TULLY, M. E., JERNIK, M. A., and HERSHBERG, E. B. (1955). J. Am. Chem. Soc. *77*, 4184.

PASTEUR, L. (1862). C. R. Acad. Sci. *55*, 28.

SEBEK, O. K. (1982). "Historical development of microbial transformations". In "Microbial Transformations of Bioactive Compounds" (J. P. ROSAZZA, ed.), Vol. I, p. 1. CRC-Press Boca Raton, Florida.

SHULL, G. M., and KITA, D. A. (1955). J. Am. Chem. Soc. *77*, 763.

THOMA, R. W., FRIED, J., BONNANO, S., and GRABOWICH, P. (1955). J. Am. Chem. Soc. *77*, 763.

Chapter 1

Methodology

Hans Georg W. Leuenberger

Central Research Units, Hoffmann-La Roche & Co. Ltd.
Basel, Switzerland

I. Introduction
II. Properties of Biotransformations and Fields of Application
  1. Functionalization of a certain non-activated carbon
  2. Selective conversion of a functional group among several groups with similar reactivities
  3. Resolution of racemates
  4. Introduction of a chiral center
  5. Mutational biosynthesis
III. Design of Biotransformation Processes
  A. The Substrate
  B. Selection of Microorganisms
  C. Methods for Carrying Out Biotransformation Processes
    1. Biotransformations with growing cultures
    2. Biotransformations with previously grown cells
    3. Biotransformations with immobilized cells
    4. Biotransformations with purified enzymes
    5. Biotransformations with multiphase systems
    6. Combination of two sequential biotransformation steps catalyzed by different microorganisms
IV. Improvement of Biotransformation Processes
  A. Optimization of Environmental Conditions
  B. Genetic Strain Improvement
  C. Elimination of Side Reactions
V. Product Isolation
VI. References

I. Introduction

Microorganisms employ both constitutive and inducible enzymes to degrade and synthesize a great variety of chemical compounds, not only for their viability and reproduction, but also for their secondary metabolism. Each reaction is catalyzed by a particular enzyme in a highly complex and well coordinated metabolic pathway. In addition to their usual substrates, many of these enzymes accept other structurally related compounds and thus catalyze "unnatural" reactions when foreign substrates are added to the medium. Reaction products which are not further degraded can usually be isolated from the fermentation medium. Such chemical reactions mediated by microorganisms or their enzyme preparations are called biotransformations. The scope of reactions which can be catalyzed by microbial enzymes (see Table 1) covers nearly all types of chemical reactions (KIESLICH, 1976). In many cases, biotransformations have been proven to be beneficial for biotechnological applications. Reviews on the methodology involved have been published by PERLMAN (1976) and recently by GOODHUE (1982). This chapter is restricted to methods which are typical for biotransformations. The reader is referred to Volumes 1 and 2 of this Handbook for general microbiological techniques such as isolation, characterization, growth and maintenance of microorganisms as well as general fermentation techniques and reactor design.

Table 1. Reaction Types Mediated by Microorganisms

Oxidations	Hydroxylation, epoxidation, dehydrogenation of C—C bonds. Oxidation of alcohols and aldehydes, oxidative degradation of alkyl, carboxyalkyl or ketoalkyl chains, oxidative removal of substituents, oxidative deamination, oxidation of hetero-functions, oxidative ring fission
Reductions	Reduction of organic acids, aldehydes, ketones and hydrogenation of C=C bonds, reduction of hetero-functions, dehydroxylation, reductive elimination of substituents
Hydrolysis	Hydrolysis of esters, amines, amides, lactones, ethers, lactams etc. Hydration of C=C bonds and epoxides
Condensation	Dehydration, *O*- and *N*-acylation, glycosidation, esterification, lactonization, amination
Isomerization	Migration of double bonds or oxygen functions, racemization, rearrangements
Formation of C—C bonds or hetero-atom bonds	

II. Properties of Biotransformations and Fields of Application

Biotransformations have the following characteristics typical for enzyme reactions:

Reaction specificity. The catalytical activity is usually restricted to a single reaction type. This means that side reactions are not expected as long as one enzyme is involved in a biotransformation.

Regiospecificity. The substrate molecule is usually attacked at the same site, even if several groups of equivalent or similar reactivity are present.

Stereospecificity. The reactive center of an enzyme provides an asymmetrical environment which allows to distinguish between the enantiomers of a racemic substrate. Therefore, only one or at least preferentially one of the enantiomers is attacked. On the other hand, if an enzyme reaction produces a new asymmetric center, usually only one of the possible enantiomers is formed and the product is therefore optically active.

Mild reaction conditions. Activation energy of chemical reactions is distinctly lowered by the interaction of substrate and enzyme, and thus biotransformations take place under mild conditions (temperature below 40 °C, pH near neutrality, normal pressure). Therefore even labile molecules can be converted using low energy consumption without undesired decomposition or isomerization.

Because of these properties, biotransformation provides a method to carry out reaction steps which can hardly be accomplished by chemical methods. This is illustrated by the following examples which are of industrial interest:

1. Functionalization of a certain non-activated carbon

Microbial reactions can selectively introduce functional groups at a certain non-activated position in a molecule, which is not attacked by chemical reagents.

For example, the antiinflammatory activity of corticosteroids depends on an oxygen function at position 11 of the steroid nucleus. Initially, cortisone-acetate was prepared at a very low yield from deoxycholic acid by a chemical synthesis requiring 31 steps. Nine steps were necessary to convert the 12-hydroxy group to the 11-keto group (SARETT, 1946). Several years later, more economical and much shorter pathways were made available by direct microbial 11-hydroxylation of precursors readily available by chemical degradation of inexpensive diosgenin or stigmasterol, namely the 11-α-hydroxylation of progesterone with *Rhizopus nigricans* or *Rhizopus arrhizus* (PETERSON and MURRAY, 1952) and the 11-β-hydroxylation of Reichstein's Compound S with *Curvularia lunata* (SHULL and KITA, 1955).

In the meantime, nearly every carbon in the steroid skeleton can selectively be hydroxylated, as long as the appropriate microorganism is selected (see Chaps. 2 and 3 of this Volume).

2. Selective conversion of a functional group among several groups with similar reactivities

Due to their characteristic regioselectivity, biotransformations can, in contrast to chemical methods, differentiate between several functional groups of similar reactivity within the same molecule. An example is the second step of commercially synthesized ascorbic acid. Here D-sorbitol is converted to L-sorbose by highly selective dehydrogenation with *Acetobacter suboxydans* (REICHSTEIN and GRÜSSNER, 1934).
Among the 6 hydroxy groups of D-sorbitol, microbial oxidation takes exclusively place at position 2, producing L-sorbose as the only product in high yields (see Chap. 11).

3. Resolution of racemates

Biotransformation provides also the possibility to modify only one enantiomer of a racemic mixture. For example several L-amino acids have been produced from readily available racemic precursors as acyl-DL-amino acids. L-Aminoacylase from *Aspergillus oryzae* hydrolyzes exclusively the acyl-L-amino acid of the racemic substrate leaving the D-enantiomer untouched (CHIBATA et al., 1972; see also Chap. 10).

4. Introduction of a chiral center

Chiral products can be obtained by subjecting a prochiral substrate to a biotransformation reaction. Examples are the asymmetric addition of NH_3 by *Escherichia coli* or H_2O by *Brevibacterium flavum* to the double bond of fumaric acid yielding L-aspartic acid (SATO et al., 1979) or L-malic acid (TAKATA et al., 1980), respectively.

5. Mutational biosynthesis

Mutational biosynthesis can be considered a borderline case between biotransformation and secondary metabolite production. This term is used for microbial production of semi-synthetic metabolites by incorporation of an artificial precursor.

A microorganism is mutagenized such, that a certain precursor is required for further secondary metabolite production. If artificial analogues of this precursor are added to the culture, they may be incorporated and thus give rise to a new semi-synthetic product. This technique has been used for the production of new semi-synthetic aminoglycoside antibiotics. For example, mutants of the neomycin producer *Streptomyces fradiae* which depend on 2-deoxystreptamine to produce neomycin are grown in the presence of analogous compounds, streptamine or 2-epi-streptamine, to convert them to a new class of semi-synthetic antibiotics, the so-called hybrimycines (NARA, 1977).

III. Design of Biotransformation Processes

A. The Substrate

Biotransformations are employed when a given reaction step is not easily accomplished by chemical methods. In this case, a readily available substrate A can be converted into a target compound B, the desired end product or an intermediate which can be subjected to further chemical modification. Another aim of biotransformations is to obtain any modification of a relatively complex molecule (i. e., an antibiotic or another drug) by specific reactions. Drug analogues prepared in this way are compared to the original compound with respect to their activity, pharmacokinetics, toxicity etc. Furthermore, their structures may indicate the metabolic fate of drugs in mammalian organisms (SMITH and ROSAZZA, 1975).

For a successful biotransformation, it is absolutely necessary that the substrate molecule comes into contact with the enzyme such that the catalytic capability of the microorganism is not inactivated by the substrate or its product. Therefore, the ideal substrate should be soluble in the fermentation medium and able to pass the cell membrane without being toxic to the microorganism. Such a substrate is usually supplemented to a growing culture as a sterile concentrated solution. Sterilization can be omitted as long as the substrate is added to a resting culture depleted of at least one essential nutrient. The substrate concentration as well as the time of addition allowing optimum conversion have to be determined experimentally.

If the substrate is not very soluble, it is recommended to dissolve it in a relatively non-toxic water miscible solvent (e. g., ethanol, acetone, propyleneglycol, dimethylsulfoxide, dimethylformamide, ethyleneglycol monomethylether) prior to addition to the medium. Otherwise solubility can be enhanced by adding an emulsifying agent (e. g., Tweens) or by grinding crystalline substrates to micron size particles. For example, *Arthrobacter simplex* converts finely powdered cortisol in concentrations up to 50% to crystalline prednisolone with good yields (KONDO and MASUO, 1961). This process is called pseudo-crystallofermentation, a technique which has also been applied to dehydrogenate cortisol using *A. simplex* immobilized on calcium alginate (OHLSON et al., 1979). Immobilized cells can be reactivated between transformations by incubation in presence of nutrients and small amounts of cortisol.

Biphasic reaction systems using a water immiscible organic solvent or a solid support as a reservoir for substrate and product will be discussed later in this chapter (Sect. III. C. 5.).

Substrates which affect the viability and reproduction of cells are best dispensed after completion of growth at a defined rate (semi-continuously in small portions or continuously at a low rate) such that the cells can convert the substrate before it becomes toxic.

If it turns out that the substrate cannot pass the cell membrane, then the cells must be treated by chemical or physical methods in order to make the substrate accessible to the enzyme or the biotransformation carried out by a cell-free enzyme preparation or the purified enzyme. If biotransformation experiments with a given compound do not give satisfactory results, the investigator should consider a chemically modified substrate, e. g., addition, variation, or removal of a protecting group or by changing the oxydation state of a functional group. Such measures might make the molecule more amenable to the active site of the enzyme. Furthermore, these modifications could possibly improve a substrate's properties such as solubility and toxicity not to mention its possible protection against undesired side reactions or degradation. For example, side hydroxylations in positions 7α, 9α, and 14α were responsible for moderate yields in the 11-β-hydroxylation of Reichstein's Compound S with *Curvularia lunata* (SHULL and KITA, 1955). These undesired hydroxylations at the α-side of the molecule could be avoided by using the 17-acetate of Reichstein's Compound S instead of the unprotected compound (DE FLINES and WAARD, 1967) or other substitutions shielding the α-side (KIESLICH, 1980 a, b). – Designing an optically active C_5-precursor for the synthesis of natural α-tocopherol by enantioselective hydrogenation of a double bond with baker's yeast or *Geotrichum candidum* (LEUENBERGER et al., 1979), more than 50 bifunctional, unsaturated, and methyl-branched compounds have been investigated which differed in functional groups (hydroxymethyl, formyl, carboxy) and in their protecting groups (different ester, acetal, or ether groups). Considerable differences in yields of reduction products have been observed with these closely related compounds. Only two of these substrates could be reduced in high concentrations and good yields, while others gave only moderate yields or were not attacked at all. The successful strategy of this work was to find the most suitable substrate for two selected microorganisms known for their ability to hydrogenate double bonds stereoselectively, instead of screening for a microorganism which efficiently converts a given substrate.

B. Selection of Microorganisms

The first task in designing a biotransformation process is to find the microorganism which catalyzes the reaction of interest with the highest possible yield. Usually, it is necessary to perform an extensive screening with a great number of pure cultures which can be obtained from a public culture collection or isolated from natural sources (see Vol. 1 of this Handbook, Chaps. 6a and 6b). In order to eliminate empiricism of a fully random screening, it is advisable to test microorganisms which have been assigned according to prior observations to mediate the desired reaction with structurally related compounds. In addition to the reviews in this volume, valuable sources to identify such microorganisms have been compiled for non-steroid cyclic compounds (KIESLICH, 1976), alkanes, alicyclics, and terpenes (SEBEK and KIESLICH, 1977), steroids (KIESLICH, 1980a; CHARNEY and HERZOG, 1980; IIZUKA and NAITO, 1981), various compounds including steroids (LASKIN and LECHEVALIER, 1974), antibiotics (SEBEK, 1980), alkaloids and nitrogenous xenobiotics (VINING, 1980), bioactive compounds including prostaglandins (ROSAZZA, 1982), and for microbial oxydations (FONKEN and JOHNSON, 1972; JOHNSON, 1978).

For screening experiments, microorganisms are grown in an appropriate medium in flasks which are shaken. The compound to be transformed is added during or after completion of growth. After incubation for 1–7 days, culture samples are removed, the biomass is concentrated by centrifugation (filtration), and the supernatant (filtrate) is prepared for product analysis. Depending on the properties of substrate and product, the analysis of the filtrate is carried out by chromatographic methods (thinlayer chromatography, gaschromatography, high performance liquid chromatography, ion-exchange chromatography) or by the measurement of absorption with a photospectrometer.

Sometimes it is desired or necessary to identify new strains of microorganisms which are capable to carry out a certain biotransformation. Soil samples containing a mixed microbial population are most frequently used as a rich source of test organisms. Since these microorganisms are involved in degradation of complex organic materials, some of them may have the potential to metabolize or modify the desired compound. Such microorganisms can be identified by the so-called soil enrichment technique as follows: Soil samples are collected and incubated in a medium which contains the substrate to be converted as the major or sole carbon source (or nitrogen source if the molecule contains nitrogen). Strains which have the capability to metabolize the substrate will be enriched and selected in this medium. Analysis of the fermentation broth will determine if the desired product is formed at the expense of the substrate. If this is not the case, however, the substrate disappears, then the target compound may still be an intermediate of the substrate metabolism. The enriched microorganisms are mutagenized and the survivors selected (e. g., by replica plating) for mutants which still grow on common carbon sources (e. g., glucose) but have lost the ability to grow on the substrate. Such mutants are blocked in the degradative substrate metabolism and may accumulate the desired biotransformation product as an intermediate (CARGILE and MC CHESNEY, 1974).

Recently, SAWADA et al. (1983) isolated bacteria from soil, capable of performing the optical resolution of α-hydroxy acids. Soil microorganisms growing on agar containing DL-2-hydroxy-4-methyl-pentanoic acid (HMPA) as the sole carbon source, but not an agar containing either enantiomer of this hydroxy acid alone were selected and further investigated for their capability of resolving racemic HMPA in a liquid medium containing glucose. Strains of the genera *Bacillus, Pseudomonas, Streptomyces* as well as coryneform bacteria were identified which selectively degrade one enantiomer of HMPA. Accumulation of 2-keto acid indicated that under the experimental conditions used, the selection was based on enantiospecific dehydrogenation at the C_2 position. Substrate specificity of this reaction proved to be rather broad. Thus besides HMPA many other racemic hydroxy acids could be enantioselectively dehydrogenated by the same microorganisms. The best strains (D-transformer: *Bacillus freudenreichii*; L-transformer: *Brevibacterium albidum, Brevibacterium tegmenticola, Bacillus* sp.) were able to resolve 30 g/L of racemic HMPA within 5–7 days and yielded nearly optically pure L- or D-HMPA, respectively. The accumulated keto acid could be reduced by catalytic hydrogenation to DL-HMPA and recycled.

C. Methods for Carrying Out Biotransformation Processes

1. Biotransformation with growing cultures

The substrate is added to the fermentation medium at the time of inoculation or during a later phase of microbial growth. The optimum time of addition of the substrate must be determined, whereafter incubation is continued until maximum yield of transformation has been reached. The level of the enzyme responsible for the desired biotransformation may be enhanced by induction, if the substrate is added during active growth of microorganisms. On the other hand, if the substrate inhibits cell growth, substrate addition has to be delayed until maximum cell mass is obtained.

Direct substrate addition to the fermentation broth is widely used because of its ease, especially in screening experiments. Total time of incubation is relatively short, because growth and biotransformation take place simultaneously; however, usual measures to avoid contamination by foreign microorganisms have to be taken.

2. Biotransformation with previously grown cells

At first, the microorganism is cultivated under conditions allowing optimum growth whereafter the biomass is harvested by centrifugation or filtration. After resuspending the cells in a simple "transformation medium" (e. g., water or a buffer solution with optimized pH value containing the substrate) the biotransformation is performed in a second step. In some instances it is advisable to supplement the transformation medium with an easily metabolizable nutrient like glucose. This prolongs viability and biotransformation activity of the cells as well as possible cofactor regeneration. The cell suspension is incubated until maximum product yield is obtained.

The strict separation of microbial growth and biotransformation offers the following advantages: (1) Each step can individually be optimized and a negative influence of the substrate or its product on growth is excluded. (2) The cell density yielding optimum conversion rate can easily be determined. (3) Due to the medium composition, the biotransformation step is usually not susceptible to infection by contaminant microorganisms and can therefore be conducted under non-sterile conditions. (4) The product isolation is easier since the transformation medium is less complex.

Biotransformations which belong to this category are those catalyzed by commercially available baker's yeast. One such example from our own laboratories, is the stereoselective hydrogenation of ketoisophorone by baker's yeast. This procedure yields (6*R*)-2,2,6-trimethyl-1,4-cyclohexanedione, a useful precursor for the synthesis of optically active carotenoids (LEUENBERGER et al., 1976). Here 10 kg of baker's yeast are suspended in a fermentor containing a nonsterile solution with 5 kg of sugar in 200 L of deionized water aerated and agitated at 20 °C. After 30 minutes 2 kg of ketoisophorone are added together with 20 mL of polyethyleneglycol monobutylether as an antifoam-agent. The biotransformation is followed by gaschromatographic analysis. During the following days, each kg of transformed ketoisophorone is replaced by adding new substrate, so that its concentration is always kept below the toxic limit of 12 g/L. Furthermore, 2 kg of sugar are added at 3 days intervals. After 17 days of incubation, 13 kg of ketoisophorone are nearly quantitatively converted to the desired product, which is then extracted and recrystallized in chemically and optically pure form at yields higher than 80%.

A special case of biotransformations involving previously grown cells is the use of spores as biocatalysts. Microorganisms are cultivated under conditions supporting high spore formation. Subsequently, spores are separated from the mycelium which then can be used for biotransformations after resuspension in a transformation medium containing the substrate, as described above. Spores may form the same products as the mycelium, and often are more active on a dry weight basis. In some cases germination of spores is necessary in order to obtain maximum biotransformation activity.

The fundamental advantage of spores is their high stability. Large amounts can be prepared and stored for a long time at −20 °C as a paste without losing their activity. Thus, spores can be handled like a chemical catalyst. The procedures of spore propagation, as well as many applications concerning the biotransformation of steroids, fatty acids, triglycerides, penicillin V and carbohydrates have been reviewed by VEZINA et al. (1968).

3. Biotransformation with immobilized cells

Much attention has been given to the immobilization of microorganisms during the past 10 years and recent reviews on immobilization techniques and their applications are available (JACK and ZAJIC, 1977; ABBOTT, 1978; DURAND and NAVARRO, 1978; CHIBATA et al., 1979; MESSING, 1980; FUKUI and TANAKA, 1982).

Microorganisms are grown in an appropriate medium, whereafter they are collected and immobilized by similar methods as enzymes are. The following basic techniques of cell immobilization have been described in literature (see reviews cited above):

1. entrapment in a polymerous porous network (polyacrylamide, κ-carrageenan, alginate, chitosan, collagen, agar, cellulose, urethane);
2. surface adsorption to a water-insoluble, solid support (e. g., DEAE-cellulose, concanavalin A, ion-exchange resins, metal oxide);
3. covalent attachment to a carrier material (e. g., carboxymethyl-cellulose),
4. induction of cell aggregation by physical or chemical cross-linking (e. g., with glutaraldehyde).

Biotransformation with immobilized cells is attractive, because in general, the immobilized biocatalyst shows a higher operational stability than cells free in solution. Moreover, immobilized cells are easily removed from the reaction mixture and can be used repeatedly. If the immobilized biomass is retained within the reaction vessel, the biotransformation can be continuously operated, at a high cell population where a favorable substrate concentration can be maintained for a long time. Furthermore, the product formation rate is high and the inhibitory influences are minimal, thus providing maximum efficiency and stability of the biocatalyst.

On the other hand, the catalytic activity of immobilized cells is generally reduced when compared to the same amount of cells in solution. This is due to damage occurring during the immobilization procedure and to the additional permeability barrier introduced by the carrier material. This loss of catalytic activity can be compensated by increasing the cell density.

Compared to an immobilized enzyme, whole cell immobilization has the following advantages: (1) the often tedious task of enzyme isolation is not necessary; (2) the enzymes remain in their natural environment within the cell and are therefore often more stable than in their isolated form; (3) if the biotransformation involved requires a coenzyme, it is supplied and regenerated within the intact cell. On the other hand, however, purified and immobilized enzymes are favored, if other enzymes contained in the intact cell catalyze undesirable reactions or if the cell membrane prevents permeation of substrates and products into and out of cells.

Among the various methods of cell immobilization, entrapment into polymerous matrix materials (e. g., polyacrylamide gel or natural polymers such as carrageenan, alginate, collagen, cellulose, and agar) have been most extensively applied for both, small-scale laboratory procedures as well as for large-scale industrial applications. Other immobilization techniques (e. g., adsorption, covalent binding and cell aggregation) are not as practical (due to desorption of cells, mechanical weakness etc.) and thus have so far been excluded from industrial application.

In Japan since 1973 technical application of entrapped cells has been employed. *Escherichia coli* cells with high aspartase activity have been immobilized in polyacrylamide gel (SATO et al., 1975) or in κ-carrageenan (SATO et al., 1979), a polysaccharide isolated from seaweeds. This process is applied for the commercial production of L-aspartic acid from fumaric acid. It will briefly be discussed in order to illustrate the powerful method of cell immobilization by entrapment.

Escherichia coli cells are cultivated in a medium containing ammonium fumarate in order to induce maximum aspartase activity. Cells are collected by centrifugation, washed and resuspended in physiological saline (1 g packed cells in 4 mL). Acrylamide monomer (0.75 g) is added to the cell suspension together with the bifunctional crosslinking reagent *N,N'*-methylene bisacrylamide (40 mg) whereafter a 5% solution of β-dimethylaminopropionitrile (0.5 mL) is added as an accelerator of polymerization. Polymerization is initiated by adding a 2.5% solution of potassium persulfate (0.5 mL). The resulting gel is ground to particles of about 4 mm diameter, activated by incubation with ammonium fumarate and packed into a column. A solution of 1 M ammonium fumarate supplemented with Mg^{2+} ions (1 mM), which protect the enzyme activity against thermal inactivation, is pumped through the column at a flow rate allowing complete substrate conversion to L-aspartic acid. Under optimized conditions, the column maintains 50% of its initial reaction rate after 120 days of operation. This continuous process of L-aspartic

Table 2. Immobilization Techniques and Selected Applications

Method of Cell Immobilization	Applications: Biotransformation Process (Microorganism)	Reference
Entrapment with polyacrylamid gel	Reichstein's Compound S ⟶ cortisol *(Curvularia lunata)*	MOSBACH and LARSSON (1970)
	L-Arginine ⟶ L-citrulline *(Pseudomonas putida)*	YAMAMOTO et al. (1974a)
	L-Histidine ⟶ urocanic acid *(Achromobacter liquidum)*	YAMAMOTO et al. (1974b)
	Penicillin G ⟶ 6-APA *(Escherichia coli)*	SATO et al. (1976)
Entrapment with κ-carrageenan	Fumaric acid ⟶ L-aspartic acid *(Escherichia coli)*	SATO et al. (1979)
	Fumaric acid ⟶ L-malic acid *(Brevibacterium flavum)*	TAKATA et al. (1980)
	L-Aspartic acid ⟶ L-alanine *(Pseudomonas dacunhae)*	YAMAMOTO et al. (1980)
Entrapment with alginate	Cortisol ⟶ prednisolone *(Arthrobacter simplex)*	OHLSON et al. (1979)
Entrapment with photo-crosslinkable resin prepolymers or urethane prepolymers	Bioconversion of steroids *(Nocardia rhodochrous)*, Stereoselective hydrolysis of (±)-menthylester *(Rhodotorula minuta* var. *texensis)*	FUKUI and TANAKA (1981)
Entrapment in agar	Sucrose inversion *(Saccharomyces pastorianus)*	TODA and SHODA (1975)
Entrapment in a reconstituted collagen membrane	Glucose isomerization *(Streptomyces venezuelae* or *Bacillus* sp.)	VIETH and VENKATASUBRAMANIAN (1976)
	Cortisol ⟶ prednisolone *(Corynebacterium simplex)*	CONSTANTINIDES (1980)
Entrapment within α-cellulose	Glucose isomerization of whey hydrolyzate *(Actinoplanes missouriensis)*	LINKO et al. (1979)
Entrapment in a chitosan matrix	Penicillin V ⟶ 6-APA *(Pleurotus ostreatus)*	KLUGE et al. (1982)
Entrapment in an epoxy matrix	Penicillin G ⟶ 6-APA *(Escherichia coli)*	KLEIN and WAGNER (1980)
Adsorption to DEAE-cellulose	Cholesterol ⟶ cholest-4-ene-3-one *(Nocardia erythropolis* JMET 7185)	ATRAT et al. (1980)

acid resulted in a reduction of production costs by 40% as compared to the conventional batch procedure with suspended cells (SATO et al., 1975).

A few years later, the same group observed even higher enzyme activity and cell stability when κ-carrageenan was used as a matrix for cell immobilization (SATO et al., 1979). *Escherichia coli* is suspended in physiological saline (8 g in 8 mL) and mixed with a solution of κ-carrageenan in the same saline (2.07 g in 45 mL). Gelation occurs by cooling to 10 °C and gel is strengthened by soaking in a cold solution of 0.3 M KCl. After this treatment, the gel is granulated to a suitable particle size. The operational stability of the particles can be increased by a treatment with hardening agents. Best results with respect to enzyme activity and stability are obtained after treatment with glutaraldehyde and hexamethylenediamine (both 85 mM). The immobilized cell preparation is packed into a column under continuous flow with a 1 M or 1.5 M solution of ammonium fumarate containing 1 mM Mg^{2+}. Under optimized conditions, a half-life of 693 days has been reported for such an immobilized cell reactor. The total production of L-aspartic acid is higher than with other cell preparations. Therefore, the carrageenan immobilization procedure is today the preferred method for industrial production of L-aspartic acid (SATO et al., 1979). L-Aspartic acid can be decarboxylated to L-alanine by *Pseudomonas dacunhae* (CHIBATA et al., 1965). A continuous process has been developed by YAMAMOTO et al. (1980) with cells of the same microorganism immobilized in κ-carrageenan. Recently TAKAMATSU et al. (1982) reported an efficient production of L-alanine from ammonium fumarate by using a mixture of *Escherichia coli* and *Pseudomonas dacunhae,* each immobilized in κ-carrageenan. Thus, the aspartase activity of the former microorganism and the L-aspartate-β-decarboxylase activity of the latter mediate the two-step conversion from fumarate to L-alanine in the same reactor.

WADA et al. (1979) have shown that cells immobilized in κ-carrageenan are still viable. As long as required nutrients are fed, a dense layer of cells is formed during incubation on the gel surface where the cells find the most suitable growth conditions. Enzyme activity of such immobilized cell preparations is enhanced not only due to the increased cell density, but also because the cell layer is near the gel surface, where substrate availability is particularly high. OHLSON et al. (1979) investigated the Δ^1-dehydrogenation of cortisol by *Arthrobacter simplex* immobilized in calcium alginate and observed a 10-fold increase of dehydrogenase activity after incubation of the immobilized cells in nutrient media. This activation is proportional to the increase of cell number during incubation and is thus attributed to microbial growth in the carrier material. Such immobilized cell preparations are promising for mediating single enzyme reactions, as well as complicated multienzyme conversions or even secondary metabolite production.

Table 2 shows a list of successful immobilization techniques and their applications.

4. Biotransformations with purified enzymes

In some cases it may be advantageous or necessary to carry out a biotransformation with a purified enzyme. This is true:

- if the membrane of intact cells prevents proper substrate or product permeation,
- if further product degradation or undesirable side reactions take place due to the presence of other enzyme systems,
- if the enzyme of interest is excreted by the cell and can easily be purified from the medium after biomass removal,
- if an enzyme from animal or plant tissues has to be used,
- if the enzyme of interest is already commercially available.

Purification of enzymes is often tedious, time consuming, and therefore expensive. A variety of basic methods are available which involve the following steps: cell disintegration, extraction, concentration, fractionation, precipitation, drying. For detailed techniques of enzyme purification the reader is referred to Volume 7 of this Handbook and to an earlier compilation of relevant methods edited by JAKOBY (1971). Depending on the problem, thorough enzyme purification is not always necessary, where in some cases biotransformations have been carried out with preparations of low purity or even with crude cell extracts.

Techniques for enzyme immobilization are well established and permit (as in the case of immobilized cells) the repeated use of the biocatalyst and the opportunity to run continuous reactors (see Vol. 7 of this Handbook and the reviews by WINGARD et al., 1976; BRODELIUS, 1978; and BARKER, 1980).

Until now industrial application of immobilized enzymes has largely been restricted to simple reactions which do not require cofactors, such as hydrolysis or isomerization (e. g., penicillin acylase, aminoacylase, lactase, glucose isomerase, glucoamylase). For example, L-aminoacylase from *Aspergillus oryzae* is adsorbed to DEAE-sephadex at a 1000 L scale for continuous production of L-alanine, L-methionine, L-phenylalanine, L-tryptophan and L-valine from acyl-DL-amino acids by enantioselective hydrolysis. This process has been used since 1969 and is said to be the first industrial application of an immobilized enzyme (CHIBATA et al., 1972). The stability of the immobilized enzyme reactor is excellent since at least 60% of activity is maintained after 32 days of operation whereafter the original activity is reconstituted by adding new enzyme. Acyl-D-amino acids remaining are racemized and recycled.

Furthermore, immobilized enzyme technology is commercially used for lactose hydrolysis in milk by lactase from *Kluyveromyces* sp. entrapped in cellulose triacetate fibers. Another example is the hydrolysis of penicillin G to generate 6-aminopenicillanic acid, an intermediate of semi-synthetic penicillins, with penicillin acylase from *Escherichia coli* or *Bacillus megaterium*. By far the largest industrial application is, however, the isomerization of glucose from hydrolyzed corn starch to yield fructose, a sugar with increased sweetness. Design and analysis of immobilized enzyme reactors have been reviewed by VIETH et al. (1976).

An alternative approach to enzyme immobilization has been described by WANDREY and FLASCHEL (1979) where a membrane reactor containing soluble L-aminoacylase was employed economically to produce L-methionine which has also entered industrial use.

Many useful enzymes, such as oxidoreductases and kinases, display their activity only in the presence of a coenzyme (e. g. NAD(H), NADP(H), FAD(H_2), ADP/ATP). The economic feasibility of such enzymes depends on an efficient coenzyme regenerating system. Recently several methods involving enzymatic, chemical or electrochemical regeneration of NAD(H) or NADP(H) have been reported (WANG and KING, 1979).

WICHMANN et al. (1981) used NADH dependent L-leucine dehydrogenase as a catalyst for the reductive amination of α-ketoisocaproate to L-leucine while the coenzyme was oxidized to NAD. In the same reactor, formiate dehydrogenase is used to mediate the NAD requiring oxydation of formiate to CO_2 during which NAD is reduced to NADH. The substrate of the auxiliary NADH regenerating reaction (formiate) is inexpensive and the by-product (CO_2) is easily separated from the reaction mixture which does not inhibit the major enzyme reaction. NAD(H) is covalently linked to polyethylene glycol, to increase the molecular weight where it is retained with both enzymes in a membrane reactor such that L-leucine can be continuously produced from α-ketoisocaproate for 48 hours with a maximum conversion of 99.7% and a yield of 42.5 g/L per day. These results demonstrate the potential of the NAD(H) enzymatic regeneration system.

Another enzymatic method to efficiently regenerate cofactors uses hydrogenase containing microorganisms to catalyze the reduction of oxidized electron carriers with molecular hydrogen (SIMON et al., 1974; KLIBANOV and PUGLISI, 1980). For example, KLIBANOV and PUGLISI employed immobilized whole cells of *Alcaligenes eutrophus* in presence of hydrogen for the regeneration of nicotinamide and flavin coenzymes as well as for the reduction of artificial cofactors as phenazine methosulfate, Janus green, methylene blue and 2,6-dichlorophenol-indophenol.

A combined electrochemical/enzymatic approach to regenerate NADH has recently been described. The authors were able to continuously reduce methylviologen at an electrode in presence of whole cells of Clostridia (SIMON et al., 1981) or yeast (GÜNTHER et al., 1983). Both microorganisms mediated enzymatic formation of NADH from NAD using electrochemically reduced methylviologen as an electron donor. Simultaneously, NADH serves as a cofactor for a reductive biotransformation and the circle starts again by reducing oxidized methylviologen at the electrode. Depending on whether whole cells or the isolated enzyme (enoate reductase from Clostridia) was used, the authors called such a process electro-microbial or electro-enzymatic reduction, respectively.

WANG and KING (1979) have reviewed further methods of NAD(H) regeneration including chemical approaches, which until now have not been very successful on the preparative scale.

With respect to regeneration of ATP from ADP or AMP presently most efficient methods involve the use of immobilized adenylate kinase and acetate kinase (WHITESIDES et al., 1976) or immobilized cells, e. g., heat treated or acetone dried yeast (FUKUI and TANAKA, 1982).

5. Biotransformation with multiphase systems

Since cells grow in aqueous media and their enzymes act on dissolved substrates, the biotransformation of lipophilic compounds (e. g., steroids or terpenoids) is limited by their solubility in water. This situation can be improved to a certain extent by adding water-miscible solvents or detergents (see Sect. III. A.). However, higher concentrations of solvents as well as substrate/product may progressively inhibit the activity of the biotransformation. These problems have been circumvented by adding a further liquid or solid phase to serve as a reservoir to take up the lipophilic substrate and also the reaction product. The microorganism remains in the aqueous phase whereas the reaction can take place either in this phase where the substrate availability depends on its limited water-solubility or at the interface between the two phases. Biotransformations of substrates with low solubility in water have been successfully applied with two-liquid phase systems consisting of water and a poorly water miscible solvent (ANTONINI et al., 1981; LILLY, 1982). For this purpose, an ideal organic solvent should be practically immiscible in water, however, exhibit high solubility for the substrate and its product. Furthermore it should not exert an inhibitory effect on the biocatalyst. For reasons of safety, non-flammable solvents are preferred, where organic solvents such as *n*-alkanes, cyclohexane, toluene, benzene, carbontetrachloride, chloroform, methylene chloride, ethyl or butyl acetate, diethyl or dibutyl ether have been used for biotransformations with intact cells or enzymes. Biotransformations in which water-immiscible solvents have been successfully applied are summarized in Table 3. In all cases, the reaction rate was increased and the product yield distinctly improved by using this technique.

KLIBANOV et al. (1977) have reported another important application of two-liquid phase systems where the water-immiscible solvent is in great excess. In such systems, the equilibrium of reactions catalyzed by hydrolytic enzymes (e. g., α-chymotrypsin) is reversed from hydrolysis to the synthesis of organic compounds, accompanied by water formation (syntheses of esters, amides, peptides; polymerization of sugars, nucleotides etc.) (KLIBANOV et al., 1977; SEMENOV et al., 1981).

Table 3. Biotransformations With Water-Immiscible Organic Solvents

Biocatalyst	Biotransformation	Organic Solvent	Effect of multi-phase system	Reference
Nocardia sp. wet biomass suspended in solvent	cholesterol ⟶ cholestenone	carbontetrachloride	100-fold increase in reaction rate	BUCKLAND et. al. (1975)
Nocardia erythropolis immobilized by adsorption to DEAE-cellulose	cholesterol ⟶ cholestenone	carbontetrachloride or toluene	improved reaction rate	ATRAT et al. (1980)
Nocardia rhodochrous gel-entrapped cells	various steroid transformations	benzene, chloroform, *n*-heptane (mixed in different ratios)	markedly increased reaction rate and yield obtained in combination with the properties of the gel selected for cell immobilization	FUKUI and TANAKA (1981)
Pseudomonas oleovorans	1.7-octadiene ⟶ 7.8-epoxy-1-octene + 1,2 – 7,8-diepoxyoctane	cyclohexane (20–50%)	5-fold increase of yield (90%), reduction of toxicity exerted by epoxide products	SCHWARTZ and McCOY (1977)
Rhodotorula minuta var. *texensis*: gel-entrapped cells	stereoselective hydrolysis of (±)menthyl-ester	*n*-heptane	increased reaction rate, yield (86%) and catalyst stability observed in combination with gel entrapment	FUKUI and TANAKA (1981)
immobilized lipase	regiospecific interesterification of triglyceride in organic solvent	*n*-hexane	entrapment markedly enhanced the operational stability of lipase	YOKOZEKI et al. (1982)

Solid lipophilic adsorbents [Amberlite XAD-2 or XAD-4 and poly-(dimethyl-syloxane), respectively] have been used as a reservoir for steroid conversions (MARTIN and WAGNER, 1976; BHASIN et al., 1976).

The advantages of liquid-liquid or solid-liquid multiphase systems are the following: (1) High concentrations of the lipophilic substrate can be maintained in the reactor due to high solubility in the non-aqueous phase. (2) The substrate is continuously released to the aqueous phase where it is converted to the end product which is withdrawn by the organic phase. Substrate and product concentrations in the aqueous phase are low; thus possible inhibitory effects on the biocatalyst are minimized. (3) The phases are easily separated where the product is recovered from the organic phase, while the aqueous phase contains the biocatalyst which can be used again. – On the other hand it is possible that the catalyst can be inhibited by the organic phase. The technique of multiphase fermentations is still in its infancy, but will certainly attract much attention in the future.

6. Combination of two sequential biotransformation steps catalyzed by different microorganisms

Biotransformations involving two (or several) sequential steps, which are catalyzed by different microorganisms, are favorably carried out as a one-stage process using a mixture of both microorganisms in a single reactor. If the two microorganisms require different fermentation conditions, a two-stage fermentation can be considered, where the broth of the first step is added to the second fermentation without isolation of the intermediate product. The following examples are of industrial interest:

a) Continuous production of L-alanine from ammonium fumarate (intermediate product: L-aspartic acid) mediated by aspartase from immobilized *Escherichia coli* and L-aspartate-β-decarboxylase from immobilized *Pseudomonas dacunhae* in the same reactor (TAKAMATSU et al., 1982).

b) Complete conversion of DL-α-amino-ε-caprolactam (up to 100 g/L) into L-lysine by enantioselective hydrolysis of the racemic compound with *Cryptococcus laurentii* and simultaneous racemization of the D-enantiomer with *Achromobacter obae* (FUKUMURA, 1977).

c) Ethanol formation from cellulose by simultaneous hydrolysis of cellulose with *Trichoderma reesei* and fermentation of formed glucose with *Saccharomyces cerevisiae* (GOSH et al., 1982). Cellulose hydrolysis rate was enhanced in the combined reaction process by 13–30%, because the inhibition of cellulases by ethanol was less severe than that by the same concentration of glucose or cellobiose.

d) Two-stage production of 2-keto-L-gulonate (a precursor of L-ascorbic acid) from D-glucose on the 10 m^3 scale: in a first fermentor D-glucose is transformed to 2,5-diketo-D-gluconate by *Erwinia* sp. After treatment with dodecyl sulfate, the broth is added directly to the second fermentor, where conversion to 2-keto-L-gulonate is catalyzed by pregrown *Corynebacterium* sp. After the first fermentation step, neither cell removal, nor intermediate product concentration was necessary, and a 84.6% yield was achieved (SONOYAMA et al., 1982).

IV. Improvement of Biotransformation Processes

As soon as a technically feasible biotransformation process has been established, major interest is to improve the yield and the reaction rate. The efficiency of a biotransformation process with a selected microorganism is maximum, if optimal environmental conditions for highest possible enzyme formation and maximum enzyme activity are established. Furthermore biological prerequisites for enzyme overproduction may be established by genetic modification. If side reactions are observed, they can be eliminated by physical, chemical or mutagenic treatment to improve the biotransformation.

A. Optimization of Environmental Conditions

Two aspects have to be considered: firstly optimization of environmental conditions should optimize the biomass such that the level of the enzyme is at a maximum. Secondly, the environment should provide conditions under which the microorganisms containing this enzyme (or the purified enzyme) display maximum biotransformation activity. Both, enzyme formation and enzyme activity, are heavily influenced by environmental conditions, such as medium composition (kind and concentration of nutrients; substrate/product concentration), temperature, pH, and dissolved oxygen (depending on aeration and agitation). As the enzyme content may vary considerably during the different growth phases, inoculum size and fermentation time suitable for substrate addition (in biotransformations with growing cultures) or biomass harvest (for biotransformations with resting cultures) are further important criteria. At first, the optimal physical and chemical conditions have to be determined by variation of the above mentioned parameters. Besides the empirical approach, several mathematical models have been applied (DOBRY and JOST, 1977; WEIGAND, 1978).

If the conditions for growth and associated enzyme synthesis are not optimal for the biotransformation, both processes must be separated and optimized individually (biotransformation with resting cells). For industrial processes, economic aspects (e. g., cheap raw materials, short reaction time, increased substrate concentration, low energy consumption) have to be considered. Furthermore, technical development requires the elaboration of appropriate reactor design, process control (see Vol. 2 of this Handbook) and scaling-up to industrial levels (BANKS, 1979).

During microbial growth many enzymes are synthesized independently from medium composition. These enzymes are called constitutive enzymes, in contrast to inducible and/or repressible enzymes. Inducible enzymes are formed in response to the presence of an inducing agent, which may be the substrate itself or a structurally related compound, while the formation of repressible enzymes is prevented in the presence of a repressing compound, e. g., the end product of a metabolic pathway (end product repression) or catabolites of a rapidly metabolizable carbon source such as glucose (catabolite repression). Thus, gene expression of an enzyme may be controlled by mechanisms such as induction or repression. Overproduction of an enzyme can be obtained, if an appropriate inducer initiates or increases enzyme production in the absence of repressing compounds. Many biotransformations have been improved considerably by adding small amounts of the substrate or an analogous compound to the growth medium. On the other hand, if a substance is identified as a repressor of enzyme synthesis, its presence in the medium or its formation during microbial growth should be avoided as much as possible. For example, β-galactosidase activity has been increased in *Escherichia coli* 1000-fold by induction with certain glycosides and kynureinase by the same level in *Pseudomonas fluorescens* after addition of kynureine or its precursor tryptophan to the medium (see reviews by PARDEE, 1969, and DEMAIN, 1971). In all strains of *Trichoderma reesei* cellulase formation has to be induced by cellulose or certain metabolites (e. g., lactose and sophorose) whereas it is repressed by glucose (MANDELS, 1982). Usually, steroid biotransformation activities are markedly enhanced by substrate induction. Recent examples of process development by induction are listed in Table 4.

Table 4. Recent Examples of Process Improvement by Enzyme Induction

Microorganism	Biotransformation	Enzyme	Inducer (Concentration)	Increase of Enzyme Activity by Induction	Reference
Pseudomonas sp.	glutaryl-7-ACA[a] ⟶ 7-ACA	glutaryl-7-ACA-acylase	glutaric acid (0.1%)	5-fold	SHIBUYA et al. (1981)
Pseudomonas dacunhae	L-aspartic acid ⟶ L-alanine	L-aspartate β-decarboxylase	Na-L-glutamate (2%)	2-fold	SHIBATANI et al. (1979)
Rhodotorula gracilis	deamination of D-amino acids	D-amino acid oxydase	D-alanine (10 mM)	4-fold	SIMONETTA et al. (1982)
Cryptococcus laurentii	L-α-amino-ε-caprolactam ⟶ L-lysine	L-α-amino-ε-caprolactamase	L-α-amino-ε-caprolactam (1%)	5-fold	FUKUMURA (1976)
Streptomyces hydrogenans	reduction of ketosteroids	$3\beta,17\beta$-hydroxysteroid- and $3\alpha,20\beta$-hydroxysteroid dehydrogenase	various steroids (0.34 mM)	5- to 34-fold	BAUER and TRÄGER (1982)

[a] ACA Amino cephalosporanic acid

B. Genetic Strain Improvement

The aim of genetic strain development is the amplification of beneficial properties and/or the elimination of detrimental properties of the microorganism involved in a biotransformation. This can be accomplished by inducing genetic modifications and selecting for mutants which improve the biotransformation. One target may be to obtain mutants which are constitutive with respect to the enzyme mediating the considered biotransformation. Due to inactivation of regulatory genes, enzyme formation with these mutants will proceed at the same or even higher rate as with a completely induced strain independent of an inducer and not affected by repressors. Enzyme overproduction can also result from an increased number of gene copies responsible for enzyme formation. Other targets of genetic development may be enhancement of substrate/product tolerance, increase of membrane permeability facilitating substrate transport into the cell and product efflux from the cell, as well as inactivation of enzymes catalyzing side reactions or product degradation. The latter aspect will be discussed separately in the next section. The traditional method of strain development is mutagenization by commonly used agents (e. g., UV light or chemical mutagens) and subsequent selection for superior strains among survivors. Recently in our laboratories (BECHER et al., 1981), we have isolated a *Bacillus* sp. from soil which is able to hydrolyze selectively the *R*-enantiomer of racemic 3-acetoxy-4-oxo-β-ionone. This optical resolution yields an optically active precursor of natural astaxanthin. We were able to improve this process 12-fold after UV treatment and mutant selection combined with optimization of the fermentation parameters. Up to now empirical methods have predominantly been used because very little is known about the genetics of most microorganisms involved in biotransformation processes. However, screening efficiency can be considerably enhanced by adopting a screening strategy which selects superior strains that give higher yields of the desired enzyme. DEMAIN (1971) has reviewed several target-directed strategies for the selection of constitutive mutants.

The most suitable screening technique must be chosen for each biotransformation system. For instance, an efficient procedure has been employed by ICHIKAWA et al. (1981) for the development of *Pseudomonas* sp. able to deacylate glutaryl-7-aminocephalosporanic acid (GL-7-ACA). The enzyme involved (GL-7-ACA acylase) was found to hydrolyze also glutarylanilide into glutaric acid and aniline. However, the liberated aniline showed a strong growth inhibitory effect. Thus, after mutagenic treatment survivors were selected on an agar medium containing glutarylanilide and a subtoxic level of aniline. As expected, mutants with high GL-7-ACA activity failed to grow on these plates. Among these mutants, a strain with 5-fold GL-7-ACA acylase activity was isolated. Like the parent strain, this mutant requires glutaric acid as an inducer. Since higher concentrations of glutaric acid inhibited growth and associated acylase production, even higher acylase levels were expected with constitutive mutants. Therefore, after repeated mutagenization, colonies which grew on an agar medium without glutaric acid were analyzed for their acylase activity. By this approach isolates with an additional 2.4-fold acylase production over induced strains were obtained. Thus, the total improvement of 12-fold was achieved by genetic manipulation.

Recombinant DNA technology provides new prospects of strain improvement. Plasmid or phage vectors can transfer any DNA fragment from eukaryotic, prokaryotic or even synthetic origin into a suitable recipient microorganism. This technology has been described by PÜHLER and HEUMANN (1981) in Volume 1 of this Handbook and has also been reviewed extensively by MALIK (1981). Briefly, recombinant DNA technology involves the following basic steps:

a) Fractionation of foreign DNA with endonuclease, identification and isolation of fragments containing the desired genetic information. Alternatively, genes may be chemically synthesized from amino acid sequences of known proteins or by reverse transcription of isolated mRNAs into complementary DNAs (cDNA).

b) Enzymatic ligation of these DNA fragments with a suitable plasmid or phage vector to form the so-called, "recombinant DNA".

c) Insertion of the recombinant DNA into the recipient host organism.

d) Selection of recombinant DNA clones, which express the wanted genetic information. Easy selection is possible, if two different antibiotic resistance genes are present in the original vector plasmid. If one resistance gene is inactivated by inserting the foreign DNA fragment within the antibiotic resistance site, recombinant clones can be detected, as they cannot grow in the presence of both antibiotics.

Since plasmids replicate independently from chromosomal DNA, frequently they exist in many copies per cell and thus a high gene dosage is obtained.

Potential benefits of recombinant DNA technology for development of biotransformation processes are manifold: (1) increased enzyme production due to amplified gene dosage, (2) transfer of biotransformation ability into a well growing microorganism which is easier to handle, (3) construction of microorganisms which can carry out several sequential reaction steps.

Prerequisite of successful gene cloning is detailed knowledge of the genetics of the recipient microorganism as well as the appropriate host-vector system. To date, these properties have only been fulfilled by a few microorganisms, such as *Escherichia coli, Bacillus subtilis* and *Saccharomyces cerevisiae* and therefore, applications in the field of biotransformations are few so far. One example of technical use is given by the construction of a new *E.coli* strain via recombinant DNA producing a 7 times increased level of penicillin acylase with constitutive enzyme formation (MAYER et al., 1980). There is no doubt that future development of biotransformation processes will take increasingly advantage of recombinant DNA technology.

C. Elimination of Side Reactions

Side reactions in a biotransformation process are observed, if a substrate or its product is attacked by undesired enzyme activities present in the cell. Suppression of such side reactions is desired in order to increase the yield of the target product and to simplify its isolation. In order to eliminate side reactions selectively, cells have to be treated under conditions suppressing the activities of undesired enzymes, while maintaining the desired biotransformation activity. Physical (heat) and/or chemical treatment of biomass (pH shift with acid or alkaline solutions, addition of detergents or organic solvents) can cause selective inactivation of enzymes which are responsible for side reactions. Examples are shown in Table 5.

A more laborious approach is mutagenization and screening of mutants which have lost the ability to form enzyme systems responsible for side reactions. For example, only moderate yields of conversion of *cis*-tetrahydro-2-oxo-4*n*-pentylthieno-(3,4-α)-imidazoline to biotinol and biotin with *Corynebacterium* sp. were observed because of subsequent side chain degradation through the β-oxidation pathway. Large amounts of biotin could be produced, however, by selecting mutants with repressed β-oxidation among survivors after *N*-methyl-*N*-nitro-*N'*-nitrosoguanidine (NTG) treatment (OGINO et al., 1974).

SHIBUYA et al. (1981) isolated a *Pseudomonas* sp. capable of deacylating glutaryl-7-aminocephalosporanic acid, in order to produce 7-aminocephalosporanic acid as a starting material for semi-synthetic cephalosporins. This strain, however, contained also β-lactamase activity which distroyed the antibiotic. After mutagenic treatment of the original strain with NTG, this problem was avoided by selecting a β-lactamase-deficient mutant with high acylase activity.

A further possibility to protect a molecule from side reactions, is to chemically modify it in such a way that it is still ac-

Table 5. Elimination of Side Reactions by Physical and/or Chemical Treatment of Biomass

Microorganism	Desired Biotransformation (Enzyme)	Side Reaction (Enzyme)	Method of Selective Elimination of the Side Reaction	Reference
Escherichia coli	fumaric acid ⟶ L-aspartic acid (aspartase)	fumaric acid ⟶ L-malic acid (fumarase)	treatment of culture broth at pH 5 and 45 °C for 1 h	TAKAMATSU et al. (1982)
Pseudomonas dacunhae	L-aspartic acid ⟶ L-alanine (L-aspartate-β-decarboxylase)	L-alanine ⟶ D-alanine (alanine racemase)	treatment of culture broth at pH 4.75 and 30 °C for 1 h	TAKAMATSU et al. (1982)
Brevibacterium flavum *Brevibacterium ammoniagenes*	fumaric acid ⟶ L-malic acid (fumarase)	fumaric acid ⟶ succinic acid (succinate dehydrogenase)	treatment of immobilized cells with 0.6% bile extract[a]	TAKATA et al. (1980) YAMAMOTO et al. (1976)
Achromobacter liquidum	L-histidine ⟶ urocanic acid (L-histidine ammonia-lyase)	urocanic acid ⟶ imidazolone propionic acid (urocanase)	heat treatment of cells at 70 °C for 30 min	YAMAMOTO et al. (1974b)
Nocardia sp.	β-sitosterol ⟶ 17-ketosteroids (multistep conversion)	steroid decomposition by ring cleavage	biotransformation in presence of 0.3 mM α,α′-dipyridyl	MARTIN and WAGNER (1976)

[a] Bile acid or deoxycholic acid inactivated also succinate dehydrogenase selectively, but bile extract was chosen for ecconomic reasons. Methanol or *n*-butanol suppressed side reaction but decreased also fumarase activity.

cepted by the target enzyme but not by enzymes responsible for the side reactions.

Thus, by using Reichstein S-17-acetate instead of the unprotected compound for the 11-β-hydroxylation with *Curvularia lunata,* improved yields of hydrocortisone were obtained (DE FLINES and WAARD, 1967). The ester group prevents enzymic attack at the α-side of the molecule and thus formation of undesired 7α, 9α and 14α-hydroxylated products. Changing the sterical shape of the substrate by chemical substitution or variation can generally guide the direction of hydroxylations to altered positions (KIESLICH, 1980 a, b; JONES, 1973).

V. Product Isolation

Usually biotransformation products are extracellular compounds of low or medium molecular weight. They are dissolved or suspended in the fermentation medium and can be isolated from the whole broth or, if preferable, from the supernatant after removal of the biomass by filtration (if necessary with a filter aid) or centrifugation. Larger particles as mycelial pellets or immobilized cells may be separated by sieving or sedimentation. Cell rupture is not necessary, however, thorough washing of the separated biomass with a buffer or water is advisable because a certain amount of product might be contained in the solid phase.

Product isolation involves subsequent concentration, fractionation, and final purification. The methods chosen depend on the various components present in the broth, physical as well as chemical properties of the product to be isolated (polarity and solubility in different solvents, volatility, electric charge, diffusion coefficient, affinity to other materials, crystallizability, sensitivity to temperature, pH, and auxiliary chemicals, molecular weight etc.). Useful reviews have been published by EDWARDS (1969), BELTER (1979), SCHMIDT-KASTNER and GÖLKER (1982), and VOSER (1982). A generalized scheme of product isolation is shown in Table 6. If necessary, after the solids are separated, the desired product is concentrated and fractionated from other compounds.

Lipophilic products are usually separated by extraction. A water immiscible organic solvent is added and agitated with the broth, where the lipophilic products are enriched in the organic phase and concen-

Table 6. Basic Operations and Common Methods for the Isolation of Biotransformation Products

	Basic Operation	Method
fermentation broth ↓	removal of biomass and other solids (if necessary)	centrifugation filtration sedimentation sieving
	concentration, fractionation	extraction ion exchange adsorption precipitation distillation
raw product ↓	final purification	chromatography distillation decolorization crystallization drying
final product		

trated by solvent removal in a vacuum evaporator. Since the distribution coefficient of lipophilic products may be dependent on pH, salt concentration, temperature etc., the conditions for most efficient product extraction have to be determined for each product. The theoretical basis and the technical devices employed for liquid-liquid extraction have been described by SCHREINER (1967, 1969).

Due to their electric charge or surface binding by van der Waals forces, hydrophilic products not extractible by organic solvents, can be isolated either on solid or liquid ion exchangers (KUNIN and WINGER, 1962) or by selective adsorption to polymeric resins (e.g., Amberlite XAD-resins, Rohm and Haas Co., Philadelphia, USA). Resins of a wide range of polarity are available, so that also lipophilic compounds can be processed by this method (VOSER, 1982). Elution of the products is accomplished by suitable liquids.

Products which form an insoluble derivative can be precipitated and recovered from the solid phase. Volatile compounds are purified by distillation. Further methods of product removal are dialysis, ultrafiltration, gelfiltration, and evaporation of solvent with subsequent crystallization.

Final purification is accomplished in most cases either by crystallization, drying, fractionated distillation, or column chromatography (adsorption chromatography, ion exchange chromatography, affinity chromatography, reversed phase chromatography, gel chromatography). If necessary, the color of the product can be removed by treatment with active charcoal. Preparative gas chromatography and high performance liquid chromatography are powerful tools for the separation of small product quantities.

In recent years, the method of liquid or supercritical carbon dioxide extraction has attracted much attention. This method has been applied to extract natural products from plants and is, for instance, commercially used for the decaffeination of coffee. Aldehydes, ketones, esters, and alcohols of low molecular weight are very soluble in liquid CO_2, while sugars, salts, organic acids, and proteins are insoluble (GRIMMETT, 1981). CALAME and STEINER (1982) used supercritical CO_2 extraction for the isolation of flavors and perfumes from plants and fruits. Furthermore, these authors applied the same method to successfully extract a fermentation broth wich produced heavy emulsions with all other solvents. In many cases this technique provides a powerful method of extraction under extremely mild conditions without formation of emulsions. The solvent is easily removed without evaporation, an energy consuming step. CO_2 is non-toxic, non-flammable, chemically inert and inexpensive. In the future, it is anticipated that extraction with supercritical gases will be further developed for its application in fermentation industry.

VI. References

ABBOTT, B. J. (1978). "Immobilized cells". Annu. Rep. Ferment. Proc. *2*, 91–123.

ANTONINI, E., CARREA, C., and CREMONESI, P. (1981). "Enzyme catalyzed reactions in water-organic solvent two-phase systems." Enzyme Microb. Technol. *3*, 291–296.

ATRAT, P., HÜLLER, E., and HÖRHOLD, C. (1980). "Steroid-Transformationen mit immobilisierten Mikroorganismen. I. Transformation von Cholesterin zu Cholestenon in organischen Lösungsmitteln." Z. Allg. Mikrobiol. *20*, 79–84.

BANKS, G. T. (1979). "Scale-up of fermentation processes." In "Topics in Enzyme and Fermentation Biotechnology" (A. WISEMAN, ed.), Vol. 3; pp. 170–266. Ellis Horwood Ltd., Chichester.

BARKER, S. A. (1980). "Immobilized enzymes." In "Economic Microbiology," Vol. 5 (A. H. ROSE, ed.), pp. 331–367. Academic Press, London, New York.

BAUER, B., and TRÄGER, L. (1982). "Enzyme induction in *Streptomyces hydrogenans*." Z. Allg. Mikrobiol. *22*, 287–292.

BECHER, E., ALBRECHT, R., BERNHARD, K., LEUENBERGER, H. G. W., MAYER, H., MÜLLER, R. K., SCHÜEP, W., and WAGNER, H. P. (1981). "Synthese von Astaxanthin aus β-Ionon. I. Erschließung der enantiomeren C_{15}-

Wittigsalze durch chemische und mikrobiologische Racematspaltung von 3-Acetoxy-4-oxo-β-ionon." Helv. Chim. Acta *64*, 2419–2435.

Belter, P. A. (1979). "General procedures for isolation of fermentation products." In "Microbial Technology," Vol. II (H. J. Peppler and D. Perlman, eds.), pp. 403–432. Academic Press New York, San Francisco, London.

Bhasin, D. P., Gryte, C. C., and Studebaker, J. F. (1976). "A silicone polymer as a steroid reservoir for enzyme-catalyzed steroid reactions." Biotechnol. Bioeng. *18*, 1777–1792.

Brodelius, P. (1978). "Industrial application of immobilized biocatalysts." Adv. Biochem. Eng. *10*, 75–129.

Buckland, B. C., Dunnill, P., and Lilly, M. D. (1975). "The enzymatic transportation of water insoluble reactants in non-aqueous solvents. Conversion of cholesterol to cholest-4-ene-3-one by a *Nocardia* sp." Biotechnol. Bioeng. *17*, 815–826.

Calame, J. P., and Steiner, R. (1982)." CO_2 extraction in the flavour and perfumery industry." Chem. Ind. London *1982*, 399–402.

Carglie, N. L. and Mc Chesney, J. D. (1974). "Microbiological steroid conversions: Utilization of selected mutants." Appl. Microbiol, *27*, 991–994.

Charney, W., and Herzog, H. L. (1967). "Microbial Transformations of Steroids. A Handbook." 2nd Ed. Academic Press, New York.

Chibata, I., Kakimoto, T., and Kato, J. (1965). "Enzymatic production of L-alanine by *Pseudomonas dacunhae*." J. Appl. Microbiol. *13*, 638–645.

Chibata, I., Tosa, T., Sato, T., Mori, T., and Matsuo, Y. (1972). "Preparation and industrial application of immobilized aminoacylase." In "Fermentation Technology Today" (G. Terui, ed.). Proc. IVth Ferment. Symp., Soc. Ferment. Technol. Osaka, pp. 383–389.

Chibata, I., Tosa, T., and Sato, T. (1979). "Use of immobilized cell systems to prepare fine chemicals". In "Microbial Technology" (H. J. Peppler and D. Perlman, eds.), Vol. 2, pp. 433–461. Academic Press, New York.

Constantinides, A. (1980). "Steroid transformation at high substrate concentrations using immobilized *Corynebacterium simplex* cells". Biotechnol. Bioeng. *22*, 119–136.

Demain, A. L. (1971). "Increasing enzyme production by genetic and environmental manipulations." Meth. Enzymol. *22*, 86–95.

Dobry, D. D., and Jost, J. L. (1977). "Computer applications to fermentation operations." Annu. Rep. Ferment. Proc. *1*, 95–114.

Durand, G., and Navarro, J. M. (1978). "Immobilized microbial cells." Process Biochem. *1978*, 14–23.

Edwards, V. H. (1969). "Separation techniques for the recovery of materials from aqueous solutions." In "Fermentation Advances" (D. Perlman, ed.), pp. 273–298. Academic Press, New York, London.

de Flines, J., and Waard v. d., F. (1967). Netherlands Patent Nr. 6 605 514.

Fonken, G. S., and Johnson, R. A. (1972). "Chemical Oxydation with Microorganisms." Marcel Dekker, New York.

Fukui, S., and Tanaka, A. (1981). "Bioconversions of lipophilic compounds by immobilized microbial cells in organic solvents." Acta Biotechnol. *1*, 339–350.

Fukui, S., and Tanaka, A. (1982). "Immobilized microbial cells." Annu. Rev. Microbiol. *36*, 145–172.

Fukumura, T. (1967). "Hydrolysis of L-α-amino-ε-caprolactam by yeasts." Agric. Biol. Chem. *40*, 1695–1698.

Fukumura, T. (1977). "Conversion of D- and DL-α-amino-ε-caprolactam into L-lysine using both yeast cells and bacterial cells." Agric. Biol. Chem. *41*, 1327–1330.

Ghommidh, C., Navarro, J. M., and Durand, G. (1982). "A study of acetic acid production by immobilized *Acetobacter* cells." Biotechnol. Bioeng. *24*, 605–617.

Goodhue Ch. T. (1982). "The methodology of microbial transformations of organic compounds." In "Microbial Transformations of Bioactive Compounds," Vol. I. (J. P. Rosazza, ed.), pp. 9–44. CRC-Press, Boca Raton, Florida.

Gosh, P., Pamment, N. B., and Martin, W. R. B. (1982). "Simultaneous saccharification and fermentation of cellulose: effect of β-D-glucosidase activity and ethanol inhibition of cellulases." Enzyme Microbiol. Technol. *4*, 425–430.

Grimmett, Ch. (1981). "The use of liquid carbon dioxide for extracting natural products." Chem. Ind. London *1981*, 359–362.

Günther, H., Frank, C., Schuetz, H. J., Bader, J., and Simon, H. (1983). "Elektromikrobielle Reduktion mit Hefen." Angew. Chem. *95*, 325–326.

Ichikawa, S., Murai, Y., Yamamoto, S., Shibuya, Y., Fujii, T., Komatsu, D., and Kodaira, R. (1981). "Isolation and properties of *Pseudomonas* mutants with an enhanced productivity of 7-β-(4-carboxybutan-amido)-cephalosporanic acid acylase." Agric. Biol. Chem. *45*, 2225–2229.

IIZUKA, H., and NAITO, A. (1981). "Microbial Transformations of Steroids and Alkaloids". University of Tokyo Press, Tokyo.

JACK, T. R., and ZAJIC, J. E. (1977). "The immobilization of whole cells". Adv. Biochem. Eng. *5*, 125–145.

JAKOBY, W. B. (editor) (1971). "Enzyme Purification and Related Techniques." Methods in Enzymology, Vol. *22*. Academic Press, New York, London.

JOHNSON, R. A. (1978). "Oxygenation with microorganisms." In "Oxidation in Organic Chemistry," Part C, pp. 131–210. Academic Press, New York.

JONES, E. R. H. (1973). "Microbiological hydroxylation of steroids and related compounds." Pure Appl. Chem. *33*, 39–52.

KIESLICH, K. (1976). "Microbial Transformations of Non-steroid Cyclic Compounds." Georg Thieme Verlag, Stuttgart.

KIESLICH, K. (1980 a). "Steroid conversions." In "Economic Microbiology", Vol. *5*. (A. H. ROSE, ed.), pp. 369–465. Academic Press, London, New York.

KIESLICH K. (1980 b). "New examples of microbial transformations in pharmaceutical chemistry". Bull. Soc. Chim. France Nr. 1, 2 II., 9–17.

KLEIN, J., and WAGNER, F. (1980). "Immobilization of whole microbial cells for the production of 6-aminopenicillanic acid." In "Enzyme Engineering", Vol. *5* (H. WEETALL and G. ROYER, eds.), pp. 335–345. Plenum Press, New York, London.

KLIBANOV, A. M., and PUGLISI, A. V. (1980). "The regeneration of coenzymes using immobilized hydrogenase." Biotechnol. Lett. *2*, 445–450.

KLIBANOV, A. M., SAMOKHIN, G. P., MARTINEK, K., and BEREZIN, I. V. (1977). "A new approach to preparative enzymatic synthesis." Biotechnol. Bioeng *19*, 1351–1361.

KLUGE, M., KLEIN, J., and WAGNER, F. (1982). "Production of 4-aminopenicillanic acid by immobilized *Pleurotus ostreatus*." Biotechnol. Lett. *4*, 293–296.

KONDO, E., and MASUO, E. (1961). "Pseudocrystallofermentation of steroid. A new process for preparing prednisolone by a microorganism." J. Gen. Appl. Microbiol. *7*, 113–117.

KUNIN, R., and WINGER, A. G., (1962). "Technologie der flüssigen Ionenaustauscher". Chem. Ing. Tech. *34*, 461–467.

LASKIN, A. I., and LECHEVALIER, H. A. (1974). "CRC Handbook of Microbiology," Vol. IV. CRC Press, Cleveland, Ohio.

LEUENBERGER, H. G. W., BOGUTH, W., WIDMER, E., and ZELL, R. (1976). "Synthese von optisch aktiven Carotinoiden und strukturell verwandten Naturprodukten I." Helv. Chim. Acta *59*, 1832–1849.

LEUENBERGER, H. G. W., BOGUTH, W., BARNER, R., SCHMID, M., and ZELL, R. (1979). "Totalsynthese von natürlichem α-Tocopherol. I. Herstellung bifunktioneller, optisch aktiver Synthesebausteine für die Seitenkette mit Hilfe mikrobiologischer Umwandlungen." Helv. Chim. Acta *62*, 455–463.

LILLY, M. D. (1982). "Two-liquid-phase biocatalytic reactions." J. Chem. Tech. Biotechnol. *32*, 162–169.

LINKO, Y., POUTANEN, K., WECKSTRÖM, L., and LINKO, P. (1979). "Cellulose bead entrapped microbial cells for biotechnological applications." Enzyme Microbiol. Technol. *1*, 26–30.

MALIK, V. S. (1981). "Recombinant DNA technology." Adv. Appl. Microbiol. *27*, 1–84.

MANDELS, M. (1982). "Cellulases." Annu. Rep. Ferment. Proc. *5*, 35–78.

MARTIN, C. K. A., and WAGNER, F. (1976). "Microbial transformations of β-sitosterol by *Nocardia* sp. M 29." Eur. J. Appl. Microbiol. *2*, 243–255.

MAYER, H., COLLINS, J., and WAGNER, F. (1980)." Cloning of penicillin-G-acylase gene of *E.coli* ATCC 11 105 on multicopy plasmids." In "Enzyme Engineering", Vol. 5 (H. WEETALL and G. P. ROYER, eds.), pp. 61–69. Plenum Press, New York - London.

MESSING, R. A. (1980). "Immobilized microbes". Annu. Rep. Ferment. Proc. *4*, 105–121.

MOSBACH, K., and LARSSON, P. O. (1970). "Preparation and application of polymer entrapped enzymes and microorganisms in microbial transformation processes with special reference to steroid 11-β-hydroxylation and Δ^1-dehydrogenation." Biotechnol. Bioeng. *12*, 19–27.

NARA, T. (1977). "Aminoglycoside antibiotics." Annu. Rep. Ferment. Proc. *1*, 299–326.

OGINO, S., FUJIMOTO, S., and AOKI, Y. (1974). "Production of biotin by microbial transformation of dl-*cis*-tetrahydro-2-oxo-4*n*-pentylthieno-(3,4-d)-imidazoline." Agric. Biol. Chem. *38*, 707–712.

OHLSON, S., LARSON, P. O., and MOSBACH, K. (1979). "Steroid transformation by living cells immobilized in calcium alginate." Eur. J. Appl. Microbiol. Biotechnol. *7*, 103–110.

PARDEE, A. B. (1969). "Enzyme production in bacteria." In "Fermentation Advances" (D. PERLMAN, ed.), pp. 3–14. Academic Press, New York, London.

PERLMAN, D. (1976). "Procedures useful in studying microbial transformations." In "Application of Biochemical Systems in Organic Chemistry" (J. B. JONES, C. J. SIH, D. PERLMAN, eds.), Part I, pp. 47-68. John Wiley, New York.

PETERSON, D. H., and MURRAY, H. C. (1952). "Microbial oxygenation of steroids at carbon 11". J. Am. Chem. Soc. *74*, 1871-1872.

PÜHLER, A., and HEUMANN, W. (1981). "Genetic engineering." In Biotechnology - a Comprehensive Treatise in 8 Volumes" (H. J. REHM and G. REED, eds.), Vol. 1, pp. 331-354. Verlag Chemie, Weinheim - Deerfield Beach, Florida - Basel.

REICHSTEIN, T., and GRÜSSNER, A. (1934). "Eine ergiebige Synthese der L-Ascorbinsäure (Vitamin C)." Helv. Chim. Acta *17*, 311-328.

ROSAZZA, J. P. (1982). "Microbial Transformations of Bioactive Compounds." CRC-Press, Boca Raton, Florida.

SARETT, L. (1946). "The structure of some derivatives of 3-α-hydroxy-$\Delta^{9,11}$-cholenic acid." J. Biol. Chem. *162*, 591-610.

SATO, T., MORI, T., TOSA, T., CHIBATA, I., FURUI, M., YAMASHITA, K., and SUMI, A. (1975). "Engineering analysis of continuous production of L-aspartic acid by immobilized *Escherichia coli* cells in fixed beds." Biotechnol. Bioeng. *17*, 1797-1804.

SATO, T., TOSA, T., and CHIBATA, I. (1976). "Continuous production of 6-aminopenicillanic acid from penicillin by immobilized microbial cells." Eur. J. Appl. Microbiol. *2*, 153-160.

SATO, T., NISHIDA, Y., TOSA, T., and CHIBATA, I. (1979). "Immobilization of *Escherichia coli* cells containing aspartase activity with κ-carrageenan: Enzymic properties and application for L-aspartic acid production." Biochim. Biophys. Acta *570*, 179-186.

SAWADA, T., OGAWA, M., NINOMIYA, R., YOKOSE, K., FUJIU, M., WATANABE, K., SUHARA, Y., and MARUYAMA, H. B. (1983). "Microbial resolution of α-hydroxy acids by enantiospecifically dehydrogenating bacteria from soil." Appl. Environ. Microbiol. *45*, 884-891.

SCHMIDT-KASTNER, G., and GÖLKER, CH. (1982). "Aufarbeitung in der Biotechnologie." In Handbuch der Biotechnologie" (P. PRÄVE, U. FAUST, W. SITTIG, and D. A. SUKATSCH, eds.), pp. 215-249. Akademische Verlagsgesellschaft, Wiesbaden.

SCHREINER, H. (1967). "Flüssig-flüssig Extraktion. I. Berechnungsgrundlagen." Chem. Ztg. *91*, 667-676.

SCHREINER, H. (1969). "Flüssig-flüssig Extraktion. Extraktionsapparate." Chem. Ztg. *93*, 971-982.

SCHWARTZ, R. D., and MCCOY, C. I. (1977). "Epoxidation of 1,7-octadiene by *Pseudomonas oleovorans:* Fermentation in the presence of cyclohexane." Appl. Environ. Microbiol. *34*, 47-49.

SEBEK, O. K. (1980). "Microbial transformations of antibiotics." In "Economic Microbiology," Vol. *5* (A. H. ROSE, ed.), pp. 575-612. Academic Press, London, New York.

SEBEK, O. K., and KIESLICH, K. (1977). "Microbial transformation of organic compounds." "Annu. Rep. Ferment. Proc. *1*, 267-297.

SEMENOV, A. N., BEREZIN, I. V., and MARTINEK, K. (1981). "Peptide synthesis enzymatically catalyzed in a biphasic system: water-water-immiscible organic solvent." Biotechnol. Bioeng. *23*, 355-360.

SHIBATANI, T., KAKIMOTO, T., and CHIBATA, I. (1979). "Stimulation of L-aspartate β-decarboxylase formation by L-glutamate in *Pseudomonas dacunhae* and improved production of L-alanine." Appl. Environ. Microbiol. *38*, 359-364.

SHIBUYA, Y., MATSUMOTO, K., and FUJII, T. (1981). "Isolation and properties of 7-β-(4-carboxybutanamido)cephalosporanic acid acylase producing bacteria." Agric. Biol. Chem. *45*, 1561-1567.

SHULL, G. M., and KITA, D. A. (1955). "Microbial conversion of steroids. I. Introduction of the 11-β-hydroxyl group into C_{21} steroids." J. Am. Chem. Soc. *77*, 763-764.

SIMON, H., RAMBECK, B., HASHIMOTO, H., GÜNTHER, H., NOHYNEK, H., and NEUMANN, H. (1974)." Stereospezifische Hydrierungen mit Wasserstoffgas und Mikroorganismen als Katalysatoren." Angew. Chem. *86*, 675-676.

SIMON, H., GÜNTHER, H., BADER, J., and TISCHER, W. (1981). "Electro-enzymatic and electro-microbial stereospecific reductions." Angew. Chem. Int. Ed. *20*, 861-863.

SIMONETTA, M. P., VANONI, M. A., and CURTI, B. (1982). D-Amino acid oxydase activity in the yeast *Rhodotorula gracilis*." FEMS Microbiol. Lett. *15*, 27-31.

SMITH, R. V., and ROSAZZA, J. P. (1975)." Microbial systems for study of the biotransformation of drugs." Biotechnol. Bioeng. *17*, 785-814.

SONOYAMA, T., TANI, H., MATSUDA, K., KAGEYAMA, B., TANIMOTO, M., KOBAYASHI, K., YAGI, S., KYOTANI, H., and MITSUHIMA, K. (1982). "Production of 2-keto-L-gulonic acid from D-glucose by two-stage fermentation." Appl. Environ. Microbiol. *43,* 1064–1069.

TAKAMATSU, S., UMEMURA, I., YAMAMOTO, K., SATO, T., TOSA, T., and CHIBATA, I. (1982)." Production of L-alanine from ammonium fumarate using two immobilized microorganisms." Eur. J. Appl. Microbiol. Biotechnol. *15,* 147–152.

TAKATA, I., YAMAMOTO, K., TOSA, T., and CHIBATA, I. (1980). "Immobilization of *Bacterium flavum* with carrageenan and its application for continuous production of L-malic acid." Enzyme Microbiol. Technol. *2,* 30–36.

TODA, K., and SHODA, M. (1975). "Sucrose inversion by immobilized yeast cells in a complete mixing reactor." Biotechnol. Bioeng. *17,* 481–497.

VEZINA, C., SHEGAL, S. N., and SINGH, K. (1968). "Transformation of organic compounds by fungal spores." Adv. Appl. Microbiol. *10,* 211–268.

VIETH, W. R., and VENKATASUBRAMANIAN, K. (1976). Process engineering with glucose isomerization by collagen-immobilized whole microbial cells. Meth. Enzymol. *44,* 768–776.

VIETH, W. R., VENKATASUBRAMANIAN, K., CONSTANTINIDES, A., and DAVIDSON, B. (1976). "Design and analysis of immobilized enzyme flow reactors." In "Applied Biochemistry and Bioengineering," Vol. *1* (L. B. WINGARD, E. KATCHALSKI-KAZIR, and L. A. GOLDSTEIN, eds.), pp. 222–327. Academic Press, New York.

VINING, L. C. (1980). "Conversion of alkaloids and nitrogenous xenobiotics." In "Economic Microbiology," Vol. *5* (A. H. ROSE, ed.), pp. 523–573. Academic Press, London, New York.

VOSER, W. (1982). "Isolation of hydrophilic fermentation products by adsorption chromatography." J. Chem. Tech. Biotechnol. *32,* 109–118.

WADA, M., KATO, J., and CHIBATA, I. (1979). "A new immobilization of microbial cells. - Immobilized growing cells using carrageenan gel and their properties." Eur. J. Appl. Microbiol. Biotechnol. *8,* 241–247.

WANDREY, C., and FLASCHEL, E. (1979). "Process development and economic aspects of enzyme engineering. Acylase L-methionine system." Adv. Biochem. Eng. *12,* 147–218.

WANG, S. S., and KING, C. K. (1979). "The use of coenzymes in biochemical reactors." Adv. Biochem. Eng. *12,* 119–146.

WEIGAND, W. A. (1978). "Computer application to fermentation processes." Annu. Rep. Ferment. Proc. *2,* 43–72.

WHITESIDES, G. M., LAMOTTE, A., ADALSTEINSSON, O., and COLTON, C. K. (1976)." Covalent immobilization of adenylate kinase and acetate kinase in a polyacrylamide gel: Enzymes for ATP regeneration." Meth. Enzymol. *44,* 887–897.

WICHMANN, R., WANDREY, C., BÜCKMANN, A. F., and KULA, R. M. (1981)." Continuous enzymatic transformation in an enzyme reactor with simultaneous NAD(H) regeneration." Biotechnol. Bioeng. *23,* 2789–2802.

WINGARD, L. B., KATCHALSKI-KAZIR, E., and GOLDSTEIN, L. (Eds.) (1976). "Immobilized Enzyme Principles." Applied Biochemistry and Bioengineering, Vol. *1.* Academic Press, New York.

YAMAMOTO, K., SATO, T., TOSA, T., and CHIBATA, I. (1974 a). "Continuous production of L-citrulline by immobilized *Pseudomonas putida* cells." Biotechnol. Bioeng. *16,* 1589–1599.

YAMAMOTO, K., SATO, T., TOSA, T., and CHIBATA, I. (1974 b). "Continuous production of urocanic acid by immobilized *Achromobacter liquidum* cells." Biotechnol. Bioeng. *16,* 1601–1610.

YAMAMOTO, K., TOSA, T., YAMASHITA, K., and CHIBATA, I. (1976). "Continuous production of L-malic acid by immobilized *Brevibacterium ammoniagenes* cells." Eur. J. Appl. Microbiol. *3,* 169–183.

YAMAMOTO, K., TOSA, T., and CHIBATA, I. (1980). "Continuous production of L-alanine using *Pseudomonas dacunhae* immobilized with carrageenan." Biotechnol. Bioeng. *22,* 2045–2054.

YOKOZEKI, K., YAMANAKA, S., TAKINAMI, K., HIROSE, Y., TANAKA, A., SONOMOTO, K., and FUKUI, S. (1982). "Application of immobilized lipase to regio-specific interesterification of triglyceride in organic solvent." Eur. J. Appl. Microbiol. Biotechnol. *14,* 1–6.

Chapter 2

Steroids

Leland L. Smith

Division of Biochemistry
University of Texas Medical Branch
Galveston, Texas 77550, USA

I. Introduction
 A. Early Observations
 B. Early Biotechnology Applications
 C. Present Interests
 D. Steroid Raw Materials
II. Literature
III. Microbiological Reaction Types
 A. Oxidations
  1. Hydroxylations
  2. Alcohol oxidations
  3. Epoxidations
  4. Carbon-carbon bond scissions
  5. Double bond introductions
  6. Peroxidations
  7. Heteroatom oxidations
 B. Reductions
 C. Isomerizations
 D. Conjugations
 E. Hydrolyses
 F. Heteroatom Introductions
 G. Sequential Reactions
IV. General Process Features
 A. Fermentation Conditions
  1. Microorganisms used
  2. Operating parameters
 B. Analysis Methods
 C. Refining Methods
V. Hydroxylations
 A. Commercial Hydroxylation Processes

1. 11α-Hydroxylation
2. 11β-Hydroxylation
3. 16α-Hydroxylation

B. Alternative Hydroxylation Processes
1. Resting vegetative cells
2. Immobilized vegetative cells
3. Spore processes
4. Cell-free hydroxylases

C. By-Product Problems
1. 11α-Hydroxylation
2. 11β-Hydroxylation
3. 16α-Hydroxylation

VI. 1-Dehydrogenations
A. Commercial 1-Dehydrogenations
B. Alternative Processes
1. Acetone-dried vegetative cells
2. Spore processes
3. Cell-free enzymes
4. Immobilized cells and enzymes

C. By-Product Problems

VII. Controlled Multiple Transformations
VIII. Lactonization
IX. Stereospecific Reduction of Prochiral Steroids
X. Future Developments
XI. References

I. Introduction

The steroids are a diversified class of oxygenated tetracyclic isoprenoid derivatives bearing the ring system of Fig. 1 (or a modified system derived therefrom) that are vital in many ways to life of eukaryotic organisms. The sterols are essential for membrane stability and cell growth and proliferation and are precursors of the other steroids of importance. Bile salts are essential for lipid digestion and absorption. Vitamin D is required for calcium absorption and metabolism. Cardiotonic steroids are widely used in medical management of congestive heart failure. Additionally, a rich variety of hormonal activities is associated with steroids, and it is this interest that prevails in current applications of biotechnology to the steroids.

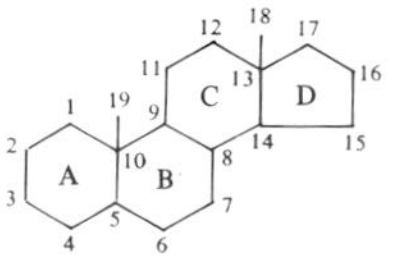

Figure 1. The steroid ring system.

Commercial biotechnology operations dealing with steroids center about the use of microbial agents for specific transformations of individual steroid substrates into useful intermediates or final products. This interest is thus distinct from biotechnology of, for example, sewage treatment, where fecal steroids are modified by microorganisms but not under direct control and where products of the process are not wanted. Other current interests in the microbial transformation of steroids by gastrointestinal microflora to cytotoxic, mutagenic, or carcinogenic derivatives implicated in human bowel cancer and by microbes of soil and water implicated in the disposition of lipids in the biosphere and in the origins of petroleum have less bearing on the biotechnology aspects discussed here.

There are currently two major biotechnology applications dealing with the steroids, that of use of microbial agents for processing raw materials into useful intermediates for general steroid production and that of specific transformations of steroid intermediates to finished products. Of the first matter, the microbial hydrolysis of plant saponins to sapogenins (dioscin to diosgenin) and the microbial degradation of the side-chain features of sterols are examples of commercially operated processes. The transformation of sterols to useful steroid intermediates and to the female sex hormone estrone is treated in a separate chapter (Chapter 3).

Still other aspects of biotechnology applied to steroid raw materials have loomed in the past and could become important given developments. The *de novo* biosynthesis of sterols, sapogenins, steroid acids, and related steroid alkaloids has been examined as means of providing controlled sources of steroid raw materials. For example, ergosterol production using *Saccharomyces* yeast in controlled fermentation may yield up to 750 g/m^3 in 30 h (KIESLICH, 1980d), but the process coupled with the necessary chemistry transforming ergosterol into useful manufacturing intermediates is uneconomical.

Production of eburicoic acid in *Polyporus sulfureus* fermentations has also been of interest (PAN and FRAZIER, 1962), but the chemical alteration of eburicoic acid to useful intermediates is not attractive. Finally, production of sapogenins such as diosgenin by callus tissue of higher plants grown in aerated fermentors, though successfully yielding up to 0.07 kg/m^3/d of diosgenin, is presently not economical (KIESLICH, 1980d). The costs of raw materials (media) for these *de novo* biosyntheses are approximately 40% of total costs, which additionally include 16% energy, 27% operations, and 17% general costs (KIESLICH, 1980c).

Biotechnology applications directed to the specific preparation of the manufacture of steroid hormones, hormone analogs, and related drugs as finished products will be emphasized further in this chapter.

A. Early Observations

The earliest interests in microbiological transformation of steroids developed out of 19th century interests in the metabolic origins of the fecal sterol coprosterol from cholesterol and from interest in the bacterial decomposition of bile acids in putrid bile (BONDI, 1908). At the same time, the capacity of *Mycobacterium* strains to utilize sterols as sole carbon source was recognized (SÖHNGEN, 1913). Nothing came of these early observations.

The earliest systematic investigation of the use of microbial agents for alteration of steroids was the work of MAMOLI and VERCELLONE dating from 1937. In their studies several stereospecific redox interconversions by yeast of C_{18}- and C_{19}-steroid 3- and/or 17-ketones and corresponding alcohols were demonstrated (MAMOLI and VERCELLONE, 1937a, 1937b, 1937c, 1938; MAMOLI, 1938; VERCELLONE and MAMOLI, 1937). Further related transformations of C_{21}-steroids and C_{24}-bile acids (KIM, 1937, 1939) established early the broad range of substrates subject to microbial transformation. These transformations had some utility for preparation of testosterone and of estradiol from 17-ketone precursors, but selective chemical methods soon supplanted microbial processes for these transformations.

Through inadvertant use of yeast cultures contaminated by bacteria (MAMOLI and VERCELLONE, 1938) it was demonstrated that bacteria were also capable of specifically transforming steroids. Thus, at the opening of World War II it was recognized that microbial transformations of steroids were possible, the early reports having established interconversions of ketones and alcohols, ester hydrolyses, and double bond isomerizations and reductions.

These early observations set the stage for post-war studies in which for the first time the two major microbial transformations of continuing biotechnology importance were recorded. With cholesterol as substrate both a 7-dehydrogenation by an *Azotobacter* sp. (HORVÁTH and KRÁMLI, 1947) and a 7-hydroxylation by *Proactinomyces roseus* (KRÁMLI and HORVÁTH, 1948, 1949) were described. Although the allylic hydroxylation of cholesterol in aqueous dispersions by non-enzymic means may well have accounted for the reported results (SMITH, 1981), the path was open for exploitation of microbial means for hydroxylations of steroids, a matter that was to become of great economic importance given that adrenal corticosteroids possess antiinflammatory activity.

A second independent line of investigation of the microbial metabolism of sterols to degraded products systematically initiated by TURFITT (1943, 1944) began at the same time and developed ultimately to commercial processes in which cholesterol and sitosterol are oxidized to useful C_{19}-steroid intermediates.

B. Early Biotechnology Applications

Although the male and female sex hormones testosterone, estrone, and estradiol, etc. had been identified before World War II, identification of cortisone and hydrocortisone as the important steroid hormones of the adrenal gland was a post-war event. Moreover, the discovery in 1949 of the powerful antiinflammatory activity of cortisone led to great interest in this group of steroids. Indeed, the first applications of biotechnology to the preparation of useful steroids were directed to making these adrenal steroids.

The first successful application of biotechnology to such effort did not utilize microorganisms but involved the perfusion of bovine adrenal glands. The report of HECHTER et al. (1949) that perfused adrenals transformed cortexone to corticosterone by specific enzymic 11β-hydroxylation (Fig. 2) clearly pointed directions taken up directly by several pharmaceutical companies. For preparation of adequate supplies of hydrocortisone for clinical evaluations adrenal perfusions were conducted by G. D. Searle & Co. (SEBEK and PERLMAN, 1979).

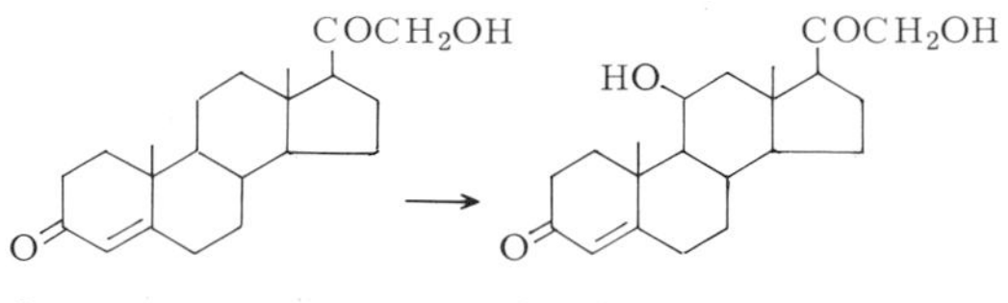

$COCH_2OH$ $COCH_2OH$
HO
Cortexone → Corticosterone

Figure 2. Enzymic 11β-hydroxylation.

These developments were rapidly superceded by those in which microbiological 11-hydroxylations were shown to be useful with C_{21}-steroids leading to the natural corticosteroid hormones. Three U.S. companies, Upjohn Co., E. R. Squibb, and Charles Pfizer Co. had entered the field, each to develop very important processes for the hydroxylation of steroids using microbial fermentation biotechnology. Additional application came from the later discovery by the Schering Corp. of microbial 1-dehydrogenation.

The first major systematic contributions were reports from the Upjohn Co. that strains of *Mucorales* molds, specifically *Rhizopus arrhizus,* grown in aerated culture were capable of the 11α-hydroxylation of progesterone and Reichstein's Substance S (Fig. 3). With the preferred *Rhizopus nigricans* ATCC 6227b yields as high as 90% of 11α-hydroxyprogesterone from progesterone were achieved (PETERSON et al., 1952; MURRAY and PETERSON, 1952).

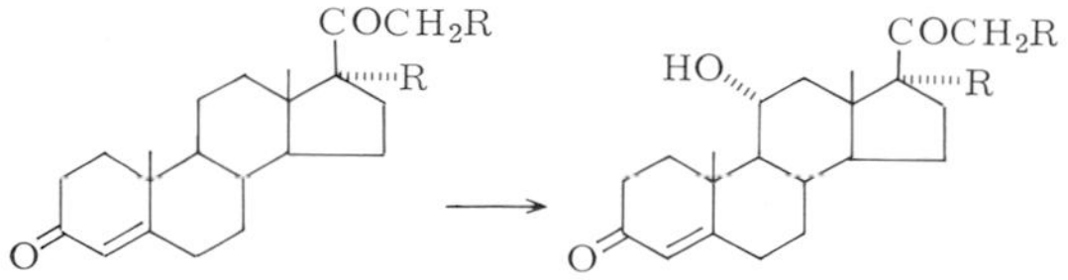

$COCH_2R$ $COCH_2R$
HO
Progesterone (R = H)
Substance S (R = OH)
→
11α-Hydroxyprogesterone (R = H)
11-*epi*-Hydrocortisone (R = OH)

Figure 3. Microbial 11α-hydroxylation.

Although the introduced 11α-hydroxyl group is not of the correct configuration required for hydrocortisone production or corticosteroid activity, the process offered access to the required 11β-hydroxyl function otherwise accessible chemically only with great difficulty. Of the three structural features of the hydrocortisone molecule, the Δ^4-3-ketone of the A-ring, the 11β-hydroxyl group of the C-ring, and the dihydroxyacetone side-chain of the D-ring, only the 11β-hydroxyl group presents difficulties for manufacture, and microbial 11-hydroxylation in either 11α- or 11β-configuration provides entry into the C-ring chemistry.

The inversion of the microbially introduced hydroxyl group is chemically conducted by chromic acid oxidation of the 11α-hydroxyl to 11-ketone group, with subsequent sodium borohydride reduction to the desired 11β-hydroxyl configuration, both chemical processes being conducted on suitably protected intermediates. The microbial 11α-hydroxylation of progesterone in high yield thus actually is not a barrier to its application to corticosteroid production. In fact, the process as elaborated is a very good one despite the early microbial step to give the 11α-hydroxylated intermediate.

The 11β-hydroxylation of Substance S directly to hydrocortisone using *Streptomyces fradiae* (COLINGSWORTH et al., 1952) and by *Cunninghamella blakesleeana* (HANSON et al., 1953) was also developed at the same time by the Upjohn Co. (Fig. 4).

Independent work at E. R. Squibb demonstrated that progesterone was 11α-hydroxylated by *Aspergillus niger* (FRIED et al., 1952) and 16α-hydroxylated by *Streptomyces argenteolus* (PERLMAN et al., 1952). At the time the observed microbial 16α-hydroxylation had no utility for steroid production, but the transformation was to become of major importance in the commercial production of the steroid hormone analog triamcinolone in 1957.

Entry into the field by the Charles Pfizer Co. came through their discovery of 11β-hydroxylation of Substance S by *Curvularia lunata* NRRL 2380, this transformation being the basis for commercial hydrocortisone

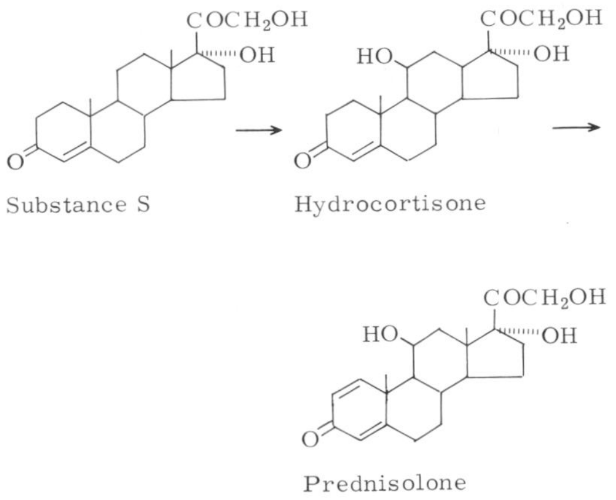

Substance S → Hydrocortisone →

Prednisolone

Figure 4. Microbial formation of hydrocortisone and prednisolone.

production (SHULL et al., 1953; SHULL and KITA, 1955).

The second great step in the application of microbial biotechnology to steroid production was the discovery that the organism *Arthrobacter (Corynebacterium) simplex* ATCC 6946 transformed hydrocortisone to its 1-dehydrogenated analog prednisolone (Fig. 4), the $\Delta^{1,4}$-3-ketone being a highly active antiinflammatory commercial product (NOBILE et al., 1955; NOBILE, 1958; HERZOG et al., 1962). The 1-dehydrogenation of other Δ^{4}-3-ketosteroids by *Streptomyces lavendulae* (with side-chain degradation) had been previously described (FRIED et al., 1953) but went unexploited at the time.

These hydroxylation and 1-dehydrogenation steps remain the most important in microbial transformation biotechnology of the steroids. At one time the manufacture of Substance S from 3β,21-diacetoxy-17α-hydroxypregn-5-en-20-one by Schering AG and Gist-Brocades involved a microbial ester hydrolysis, alcohol oxidation to 3-ketone, and Δ^{5}-double bond isomerization, although chemical methods are similarly suitable for these conversions (CHARNEY and HERZOG, 1967). Nevertheless this process is still used by Schering AG in modified form. The process using 21-acetoxy-3β,17α-dihydroxypregn-5-en-20-one as substrate, transformed by *Corynebacterium mediolanum* to Substance S is now also operated in the Soviet Union (KOLCHANOVA et al., 1979).

C. Present Interests

As the introduction of new steroids into commerce is limited of late, so new developments in microbial biotechnology of steroids are restricted. Manufacturing protocols for the several important steroid hormones and analogs have been set for some time, and although exact details of commercial manufacture are not publicly available, it appears that new developments with older processes are not receiving much attention.

There are continuing reports of special process developments, such as use of immobilized cells for transformations, and of searches for microbial steroid esterases, new sterol degradation systems, etc. but gradually large emphasis has shifted to less commercially oriented topics. Thus, fundamental studies of microbial ketosteroid reductase, steroid hydroxylases, and double bond isomerases are receiving attention, even to the purification of the enzymes implicated, with their characterization as proteins, including amino acid sequences and secondary and tertiary structures. Related attempts to develop submerged fermentation techniques for the production of steroids suitable for manufacture of hormones but using callus tissue from higher plants also is of continuing interest.

One application of microbial transformation that continues to be of general interest is the use of microbial agents for specific transformations and for specific syntheses. The use of microorganisms as chemical oxidizing agents in general organic synthesis is potentially of great importance (FONKEN and JOHNSON, 1972; SKRYABIN and GOLOVLEVA, 1976), and specific microbial transformation steps have been incorporated into numerous partial syntheses of new steroids wanted for evaluation as steroid hormone analogs and drugs (BEUKERS et al., 1972). A frequent application is the introduction of a wanted hydroxyl group at inaccessible sites and introduction of the Δ^1-double bond to give $\Delta^{1,4}$-3-ketones. Generally, the planned syntheses that employ microbial steps are devised so that the microbial transformation is conducted as a terminal or near-terminal event. Typical of these are the syntheses of 6α-fluocortolone (Fig. 5) involving 11β-hydroxylation with *Curvularia lunata* and 1-dehydrogenation with *Bacillus lentus* (KIESLICH et al., 1969b).

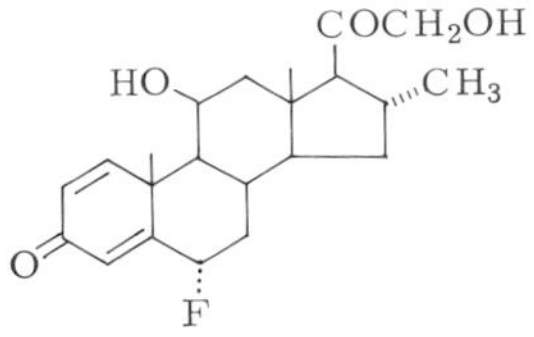

Figure 5. Fluocortolone.

In a few syntheses examples where Δ^5-3β-hydroxysteroid intermediates were used instead of the more common Δ^4-3-ketones, transformation of the Δ^5-3β-alcohol to the Δ^4-3-ketone can be conducted microbiologically, thus giving a synthesis sequence involving three microbiological steps: (1) Δ^5-3β-hydroxysteroid dehydrogenation to the corresponding Δ^4-3-ketone with *Flavobacterium dehydrogenans,* (2) 11β-hydroxylation with *Curvularia lunata,* and (3) 1-dehydrogenation with *B.lentus* (RASPÉ et al., 1964).

Although *C.lunata* has proven of very great versatility in such syntheses, there still may be advantages, even the need, to explore alternative 11β-hydroxylating microorganisms for specific substrates. This happens when unfavorable yields of desired 11β-hydroxylated product are encountered and also when the desired product is not obtained at all or with troublesome by-products. Thus, for example, *C.lunata* was much less useful in the introduction of the 11β-hydroxyl group into 6α-fluoro-21-hydroxy-16α,17α-isopropylidenedioxypregn-4-ene-3,20-dione, and *Tieghemella orchidis* was found to be better (RYZHKOVA et al., 1980). 11α-Hydroxylation by other organisms and hydroxyl inversion via the intermediate 11-ketone would also overcome an unfavorable situation with *C.lunata.*

Another special synthetic use of microorganisms lies in the preparation of isotopically labelled steroids for other metabolic studies. For example, *Streptomyces (Actinomyces) roseochromogenes* is particularly useful for introduction of the 16α-hydroxyl group into various [^{3}H]- and [^{14}C]-steroids of the C_{18}-, C_{19}-, and C_{21}-series (YOUNGLAI and SOLOMON, 1967). Other specialized synthesis involving microbial introduction of the 16α-hydroxyl group into 18-hydroxycortexone (BRAHAM et al., 1975) and the 18-hydroxyl group for preparation of aldosterone (KONDO et al., 1965) remain useful biotechnology applications though on a limited scale.

Other developments using diverse microbial transformations of selected substrates are of interest for specific synthetic purposes. Thus, 12α-hydroxylation of Substance S by *Cereospora kaki* provides access to 12α-halogenated derivatives of hydrocortisone (KIESLICH and KERB, 1980), and 12β,15α-dihydroxylation of 3β-acetoxy-5β-pregnan-20-one by *Calonectria decora* ATCC 14767 without concomitant ester hydrolysis is a useful entry into the synthesis of digoxigenin (VIDIC et al., 1974). 15α-Hydroxylation of 13β-ethylgon-4-ene-3,17-dione with *Penicillium* sp. or *Fusarium* sp. offers access to 15-dehydro derivatives from which improved contraceptives are synthesized (PETZOLDT and WIECHERT, 1976). Additionally, 1α-hydroxylations of 17α-ethinyl-19-nortestosterone with *Acremonium strictum* (AMBRUS et al., 1975), *Calonectria decora* ATCC 14767, or *Mucor spinosus* C.B.S. 29563 (PETZOLDT and ELGER, 1976, 1977) and 1β-hydroxylation with *Botryodiplodia malorum* (GREENSPAN et al., 1974) yield derivatives of altered estrogenic, antiestrogenic, and contraceptive properties.

Furthermore, preparation of new antiinflammatory agents for the market continues to utilize microbial steps. For instance, the D-homosteroids 17aα-acyloxy-21-deoxy-Dhomoprednisolone and 11β,17aα-dihydroxy-3-oxo-D-homoandrosta-1,4-diene-17aβ-carboxylic acid involve microbial 11β-hydroxylations and 1-dehydrogenations (ALIG et al., 1975, 1976).

Yet another aspect of microbial biotechnology of steroids involves production of enzymes that transform steroids. In this application, microbial alcohol dehydrogenases have been used for the quantitative estimation of bile acids in body fluids. Specific determinations of 3α-, 7α-, or 12α-hydroxylated bile acids can be made by use of the specific 3α-alcohol dehydrogenase of *Pseudomonas testosteroni,* the 7α-alcohol dehydrogenase of *Escherichia coli* ATCC 29532, or the 12α-hydroxysteroid dehydrogenase of a *Clostridium* sp. ATCC 29733 (MACDONALD et al., 1980).

A second more extensive application of microbial enzymes in analysis is that of microbial sterol oxidases for the conversion of the 3β-hydroxyl group of cholesterol to the 3-ketone of cholest-5-en-3-one (with subsequent isomerization to cholest-4-en-3-one), with the concomitant generation of hydrogen peroxide which can be measured and related to the initial cholesterol content. This assay procedure has become very popular for the clinical estimation of plasma cholesterol, and cholesterol oxidase preparations from *Nocardia erythropolis, Nocardia rhodochrous* NCIB 100554, *Brevibacterium sterolicum* ATCC 21387, *Schizophyllum commune* IFO 4928, and *Streptomyces violascens* have been described (SMITH and BROOKS, 1976, 1977). The cholesterol oxidase method has been adapted to automatic analysis equipment (RICHMOND, 1976) and to the use of a cholesterol electrode (SATOH et al., 1977).

D. Steroid Raw Materials

In order to appreciate the role that microbial biotechnology has played in steroid manufacture it is important to consider levels of world production of steroids and the source of raw materials. The world market for finished steroids appears to be ever increasing, both in amount of products and in dollar value. The latest year (1978) for which data are available saw production of over 1300 metric tons of steroid raw materials, calculated as diosgenin, considerably increased over 1973 (cf. Table 1).

From very low levels in 1951 the expansion of sales of finished steroids to the present $ 1 billion may be traced in reports of APPLEZWEIG (1959, 1969, 1974, 1977). Sales of antiinflammatory steroids in the U.S.A. of $ 120 million in 1959 jumped to $ 215 million by 1968, during which period sales of progestational (contraceptive) agents increased to $ 110 million. Markets for other steroids, including male and female sex hormones, anabolic agents, antialdosterone agents, anticancer agents, anesthetics, antilipemic agents, antiandrogenic agents, and tumor localizing agents, etc. though less in quantity and value, also continue to grow.

World steroid producers use three general processes: (1) direct isolation from natural sources, as the recovery of conjugated estrogens from horse urine and of cardiotonic steroids from *Digitalis* and related plants, (2) partial synthesis from relatively inexpensive steroid raw materials of animal and plant origins, and (3) total synthesis from non-steroidal materials. All three of these processes are presently commercially operated, but the first involving direct isolation of steroids does not use microbial biotechnology.

It is in the partial synthesis of steroid hormones and their analogs from naturally occurring steroid raw materials that the impact of microbial biotechnology has been greatest. World steroid production requiring microbial transformation steps has stabilized on but a few microbial processes and equally small numbers of steroids raw materials sources. Although exact manufacturing processes are not disclosed and details in most cases are secret or obscure, it is possible to provide a general outline of operations so as to see how the microbial biotechnology steps are integrated into the overall productions.

With the growth of sales of new steroid products the problem of supplies of basic raw materials developed. The early Schering AG process utilizing cholesterol and the Merck process using deoxycholic acid from slaughter house animals were supplanted in large part by processes using sterols from plant materials. Both diosgenin from Mexican (and other) *Dioscorea* plants and stigmasterol from soybeans afford useful C_{21}-intermediates from which the full complement of finished C_{18}-, C_{19}-, and C_{21}-steroids are manufactured today. The Merck deoxycholic acid process to cortisone in 37 steps was greatly simplified to 11 steps, including a microbial 11α-hydroxylation step by Upjohn.

Other raw materials for steroid manufacture are used, including retention of the older cholesterol and deoxycholic acid processes as well as newer processes from animal cholesterol and soybean sitosterol in which microbial degradations of the sterol side-chains yields useful C_{19}-steroid intermediates (cf. Chapter 3), together with limited use of other plant sapogenins such as hecogenin (Fig. 6). Microbial degradations of sitosterol and of cholesterol yield C_{19}-steroids used for production of other C_{19}-steroid sex hormones, anabolic agents, and for the conversion to the C_{18}-estrogens or for resynthesis of C_{21}-steroids as well.

However, it is generally conceded that C_{21}-steroid manufacture is operated from C_{21}-intermediates derived from diosgenin or stigmasterol. With the exceptions of cortexone and progesterone and its 17α-oxygenated derivatives most of the C_{21}-steroids intermediates go into corticosteroid manufacture, which phase requires at least the single microbial 11-hydroxylation step.

In Table 1 are listed the several steroid raw materials presently thought to be in use for finished steroid manufacture. Included in the list is the material of total synthesis, for although total synthesis does not start with a steroid raw material, the process leads to such an intermediate that must be resolved from unwanted optical enantiomer in order to produce the finished steroid hormone or hormone analog of the sought natural optical configuration. This resolution step is relevant in this context as a microbial fermentation is operated for this purpose.

In addition to these world production data, some production from other sources

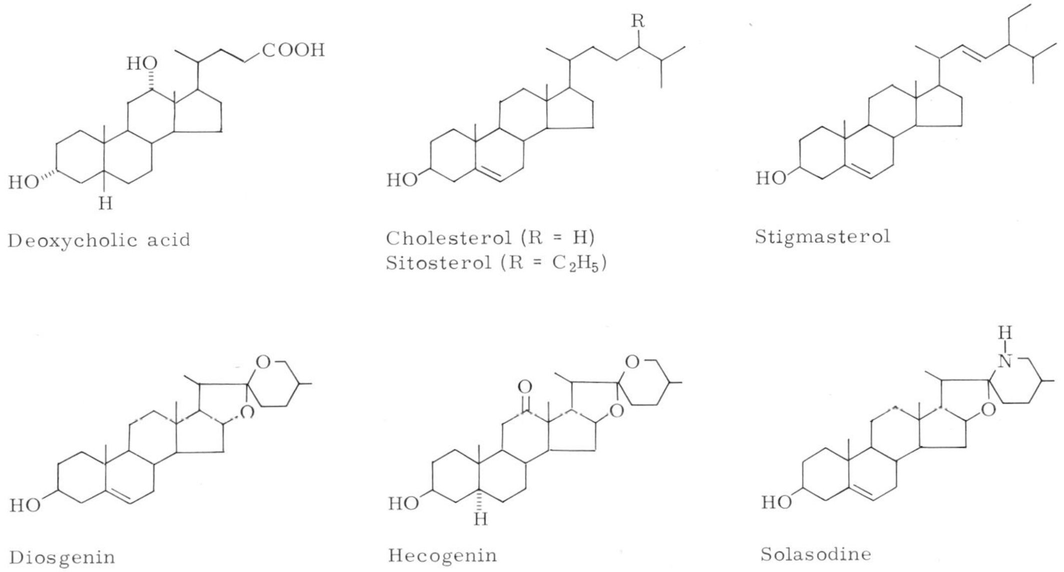

Deoxycholic acid

Cholesterol (R = H)
Sitosterol (R = C_2H_5)

Stigmasterol

Diosgenin

Hecogenin

Solasodine

Figure 6. Steroid raw materials.

Table 1. World Steroid Production[a)]

Production Process	Country	Production (metric tons) Diosgenin Equivalent 1963	1968	1973	1978
Recovery from natural sources					
Conjugated estrogens	Canada	40	60	100	—
Partial synthesis					
Diosgenin	Mexico	375	500	550	600
	Guatamala	10	30	—	20
	China	—	80	250	—
	Puerto Rico	—	20	—	5
	India	10	30	—	—
Smilagenin, sarsasapogenin	Mexico	—	10	—	15
Hecogenin	Africa	20	40	40	—
	Mexico	—	—	—	5
	USA	—	—	—	20
Cholesterol	Fed. Rep. Germany, Netherlands	5	10	—	—
Stigmasterol	USA	60	150	280	350
Sitosterol	USA	—	—	—	250
Bile acids	France	20	50	50	50
Total synthesis	Fed. Rep. Germany	—	—	70	100
	France	—	50	50	100
	USA	—	30	30	60
	Switzerland	—	—	—	40

[a] Data from APPLEZWEIG, 1969, 1977; SEBEK, 1977; ONKEN and ONKEN, 1980

such as solasodine in the Soviet Union is indicated (MULEVICH et al., 1977). Whatever the source of raw materials and problems in their supplies, alternative raw materials use will still require microbiological transformations of the sort presently in use (APPLEZWEIG, 1977).

These several steroid raw materials have been coupled with key microbial biotechnology steps to produce several effective manufacturing processes for finished steroids. Among these are those requiring microbial 11α-hydroxylations, such as the 11α-hydroxylation of progesterone by *Rhizopus nigricans* ATCC 6227b with subsequent chemical modifications yielding cortisone and those requiring 11β-hydroxylation of Substance S with *Curvularia lunata* NRRL 2380 to give hydrocortisone directly. Additional microbial 1-dehydrogenation of these active hormones then gives the more active analogs prednisone and prednisolone, respectively.

Yet other chemical modifications of initially 11-hydroxylated steroids allow elimination of the 11-hydroxyl function and introduction of the 9(11)-dehydro feature, from which the family of 9α-fluorosteroids are made. An alternative means of introduction of the 9α-fluoro-11β-hydroxyl feature derives from hecogenin and also from microbial 9α-hydroxylation of a C_{19}-intermediate, the 9α-hydroxyl being eliminated to give a 9(11)-dehydrosteroid. It appears that the bulk of 9α-fluorinated steroids may be manufactured from initially 11α-hydroxylated intermediates, although traces of byproducts are found based on the small amounts of 8(9)-dehydro-structures from the dehydration of 11α-hydroxyl groups, whereas the 9α- and the 11β-hydroxyl group yield exclusively 9(11)-dehydro-structures. Some dexamethasone may be derived in part from hecogenin.

II. Literature

The literature dealing with the microbial transformation of steroids is considerable and extends from individual journal articles through specialized reviews, major reviews, to substantial monographs, thus affording ready grasp of details of the field as it has developed since 1952. Additionally, a substantial patent literature exists. Indeed, in some cases the patent literature is more indicative of commercial operations than are technical papers.

Several reviews describe events in development, and PETERSON (1963) includes a list of early reviews to 1961 on the topic. The major monograph literature began to appear shortly thereafter with the contributions of AKHREM and TITOV (1965, 1970), ČAPEK et al. (1966), CHARNEY and HERZOG (1967), and IIZUKA and NAITO (1967). Those of AKHREM and TITOV (1965, 1970) and CHARNEY and HERZOG (1967) are encyclopedic in range and deserve the most careful scrutiny for entry into the field.

Progress in the field made after these publications and up to 1973 has also been reviewed by SMITH (1974) and to 1978 by KIESLICH (1980c) and KIESLICH and SEBEK (1979). A flurry of other reviews bring the field up to date (MARSHECK, 1971; MURRAY, 1976; SEBEK and PERLMAN, 1979; KIESLICH, 1980a, 1980b, 1980d; NOMINÉ, 1980). Additionally, reviews of related topics not addressed in the present chapter, including microbial transformations of bile acids (HAYAKAWA, 1973) and degradations of the sterol sidechain with and without A-ring aromatization (VÉZINA et al., 1971; MARTIN, 1977) are available. Those interested in conducting work in the field are well served with these many useful publications.

III. Microbiological Reaction Types

The diversity of enzymic reactions accomplished by microorganisms on steroids (as xenobiotic substances) rival that of animal and higher plant systems, but relatively little has been done to characterize the relevant microbial enzymes systematically. Accordingly, it is not possible to categorize the specific enzyme types implicated in most cases.

For instance, the oxidation of steroid alcohols to ketones may involve alcohol dehydrogenases (EC 1.1.1.-) or oxidases (EC 1.1.3.-); cleavage of a carboxylic acid ester may be via acyl transferase (EC 2.3.1.-) or ester hydrolase (EC 3.1.1.-) action. Necessarily, a general treatment of enzyme types cannot be given.

Some of the transformations (hydroxylations, 1-dehydrogenations, carbon-carbon bond scissions, and ketone reductions, cf. Table 2) have commercial importance. Others have limited synthetic utility, whereas yet others are of interest in studies of reaction mechanisms. Some reaction types, such as D-homoannulations, are undesired complications; some are simply oddities.

A. Oxidations

There are four oxidation reactions of importance commercially, hydroxylation, alcohol oxidation, 1-dehydrogenation, and carbon-carbon bond scission.

1. Hydroxylations

By far the most important microbial transformation is hydroxylation, and thorough reviews of the many reported examples involving large numbers of microor-

Table 2. Type Reactions with Steroids

Oxidations
1. Hydroxylation
   a) At all nuclear sites and angular methyl groups
   b) At some side-chain sites
2. Alcohol Oxidation
   a) Saturated alcohol to ketone
   b) Allylic or homoallylic alcohol to α,β-unsaturated ketone
3. Epoxidation
4. Carbon Carbon Bond Scission
   a) Oxygen insertions yielding esters, alcohols, ketones, lactones, and acid
   b) Retroaldol reactions
   c) Bond cleavages yielding unfunctionalized products
5. Double Bond Introduction
   a) 1-Dehydrogenation of saturated 3-ketones and Δ^4-3-ketones
   b) 4-Dehydrogenation of saturated 3-ketones and Δ^1-3-ketones
   c) 1-Dehydrogenation with A-ring aromatization
   d) Aromatization of $\Delta^{1(10),5}$-3β-alcohols
   e) Eliminations of hydroxyl, ester, or epidioxide
6. Peroxidation
   a) $5\alpha,8\alpha$-Epidioxides from 5,7-dienes
   b) Hydroperoxidation
7. Heteroatom Oxidation
   a) Stereospecific oxidation of thioether to sulfoxide
   b) Amine oxidation to ketone

Reductions
1. Ketone to alcohol
2. Aldehyde to alcohol
3. Hydroperoxide to alcohol
4. Enol to primary alcohol
5. Double bond saturation
   a) $\Delta^{1,4}$-3-Ketone to Δ^4-3-ketone
   b) Δ^4-3-Ketone to saturated 3-ketone
   c) $\Delta^{4,6}$-3-Ketone to Δ^4-3-ketone
   d) Δ^{16}-20-Ketone to saturated 20-ketone
   e) A-Ring to 19-nor-Δ^4-3-ketone
   f) Anaerobic dehydroxylation

Isomerizations
1. Δ^5-3-Ketone to Δ^4-3-ketone
2. D-Homoannulation
3. Retropinacolone rearrangement
4. Alcohol epimerization
5. $5\alpha,8\alpha$-Epidioxide rearrangement to epoxyalcohol

Conjugations
1. Acylation of hydroxyl groups
2. Glycosidation
3. Phenol ether formation
4. Amine acetylation
5. Steroid alkaloid conjugation with carboxylic acids

Hydrolyses
1. Ester Hydrolysis
   a) Carboxylic acid esters
   b) Lactones
   c) Sulfuric acid esters
2. Ether Cleavage
   a) Enol ether to ketone
   b) Phenol ether to phenol
   c) Glycoside to alcohol

Introduction of Heteroatoms
1. Nitrogen
   a) 21-Acetylaminosteroid from 21-alcohol
2. Halogen
   a) Haloperoxidases

ganisms and steroids are available (AKHREM and TITOV, 1965, 1970; CHARNEY and HERZOG, 1967; SMITH, 1974; KIESLICH, 1980c). Microbial hydroxylations proceed via stereospecific abstraction of the hydrogen atom present at the site to be oxidized. Furthermore, microbial hydroxylases appear to be mixed function oxidases utilizing molecular oxygen and requiring an electron transport system linked to NADPH-dependent dehydrogenases. Cytochrome P-450 serves as the terminal oxidase (BERG et al., 1975, 1976, 1979; BRESKVAR and HUDNIK-PLEVNIK, 1977a, 1977b). In these matters, microbial hydroxylations are exactly like those in mammalian systems.

Microbial hydroxylations at almost every possible position in the steroid nucleus are known, including 10β-hydroxylation of 19-norsteroids, 14β-hydroxylation of synthetic 14β-steroids (MUKAWA, 1971), and 3-hydroxylation of 3-deoxysteroids (CHERRY et al., 1966). Hydroxylations of both angular methyl groups and side-chain positions (C-21 in C_{21}-steroids, C-26 in C_{27}-sterols) are known. Additionally, retrosteroids (9β, 10α-configuration) and enantiomeric steroids serve as substrates for hydroxylations.

In addition to monohydroxylation, a given microorganism may di- or poly-hydroxylate a substrate, and several characteristic patterns are frequently encountered. For example, $6\beta,11\alpha$- and $12\beta,15\alpha$-dihydroxylations are common patterns. Dihydroxylation may occur as a result of two distinct, individual monohydroxylases acting independently or in a set sequence but also may occur in characteristic patterns that suggest possibly one enzyme with diminished regiospecificity is implicated. As an example of independent dihydroxylations, the $6\beta,11\alpha$-dihydroxylation of progesterone by *Aspergillus ochraceus* results from 6β-hydroxylation of initial product 11α-hydroxyprogesterone by a separate 6β-hydroxylase only induced by 11α-hydroxyprogesterone (SHIBAHARA et al., 1970). Perusal of any of the encyclopedic compendia (AKHREM and TITOV, 1965; CHARNEY and HERZOG, 1967) discloses many other dihydroxylation patterns commonly encountered.

Much work has been directed towards elucidation of the mechanisms of steroid hydroxylations. Studies both on isolated (cell-free) hydroxylases discussed in a later section of this chapter and on specially synthesized steroid substrates have been conducted (JONES, 1973), but a satisfactory understanding of details of hydroxylation is still elusive (HOLLAND, 1982).

2. Alcohol oxidations

The oxidation of secondary alcohols to ketones is a very common transformation, with examples of 3-, 7-, 11-, 17-, and 20-alcohol oxidations being plentiful. The transformation may also occur in connection with initial ester hydrolysis and with subsequent double bond isomerization, yielding α,β-unsaturated ketone products.

An important example of alcohol oxidation in company with ester hydrolysis and double bond isomerization is the transformation of $3\beta,17\alpha,21$-triacetoxypregn-5-en-20-one to Substance S 17α-acetate by *Flavobacterium dehydrogenans* (Fig. 7). Substance S 17α-acetate is the preferred substrate for subsequent 11β-hydroxylations by *Curvularia lunata* in the manufacture of hydrocortisone (KIESLICH, 1980a, 1980d).

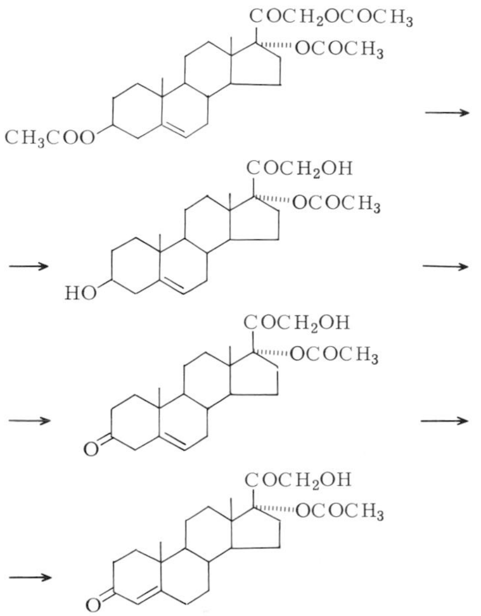

Figure 7. Synthesis of Substance S 17α-acetate.

3. Epoxidations

Olefinic bond epoxidations occur using microorganisms that are capable of hydroxylating the saturated steroid analog in the same positions. Thus, *Cunninghamella blakesleeana* and *Curvularia lunata* that 11β-hydroxylate saturated steroids transform corresponding $\Delta^{9(11)}$-substrates to $9\beta,11\beta$-epoxides, whereas an 11α-hydroxylating *Curvularia* sp. gave the corresponding $9\alpha,11\alpha$-epoxide (BLOOM and SHULL, 1955; KUROSAWA et. al., 1961). Likewise, the 9α-hydroxylating organism *Nocardia restrictus* ATCC 13 934 gave $9\alpha,11\alpha$-epoxides (SIH, 1962).

Notably, 9α-hydroxylase and $9\alpha,11\alpha$-epoxidase activities in cell-free systems of *N.restrictus* appear to be differentially induced but may involve the same active site on the enzyme (CHANG and SIH, 1964).

Other olefinic bonds have also been similarly epoxidized but not all (LIN and SMITH, 1970a). Furthermore, epoxidation by non-hydroxylating organisms is also known, as the 1-dehydrogenating organism *Arthrobacter simplex* ATCC 6946 epoxidized a $\Delta^{9(11)}$-steroid (CORONELLI et al., 1964).

Microbial epoxidations do not have synthetic utility and appear to be oddities rather than useful transformations.

4. Carbon-carbon bond scissions

Microbial scissions of carbon-carbon bonds are very important, as cleavage of sterol side-chains to useful C_{19}-intermediates and of the progesterone side-chain to testololactone derivatives are commercial processes. Bond scissions appear to be via oxygen insertion reactions of Bayer-Villiger type, the degradation of progesterone being a particularly clear case. Initial oxygen insertion into the C-17/C-20 bond yielding testosterone 17β-acetate is followed by ester hydrolysis and 17β-alcohol dehydrogenation to the corresponding 17-ketone. A second oxygen insertion into the C-13/C-17 bond gives testololactone (Fig. 8) (or 1-dehydrotestololactone if 1-dehydrogenation also occurs). Subsequent lactone hydrolysis may give the corresponding carboxylic acid. Thereby, both 17β-alcohols and their acetates, 17-ketones, D-secolactones, and D-secoacids result (CARLSTRÖM, 1973).

Other oxidative processes cleaving carbon-carbon bonds must also occur, as the microbial removal of the side-chain of certain sapogenins results in steroids unsubstituted at the C-17 site initially bearing the side-chain (KONDO and MITSUGI, 1966). Additionally, carbon-carbon cleavages may arise via retroaldol reactions following 9α-hydroxylation of $\Delta^{1,4}$-3-ketones. The unstable intermediate 9α-hydroxy-$\Delta^{1,4}$-3-ketone decomposes to the 9,10-secophenol product (Fig. 9) (DODSON and MUIR, 1958). This reaction sequence is implicated in the decomposition of steroids by soil microorganisms.

5. Double bond introductions

Although introduction of olefinic double bonds by microorganisms has been reported for several nuclear positions, only that of the Δ^1-double bond in Δ^4-3-ketosteroids is of importance commercially. Δ^4-3-Ketosteroids thereby yield $\Delta^{1,4}$-3-ketones, a structural feature of highly active corticosteroid hormone analogs. The 1-dehydrogenation occurs by stereospecific removal of

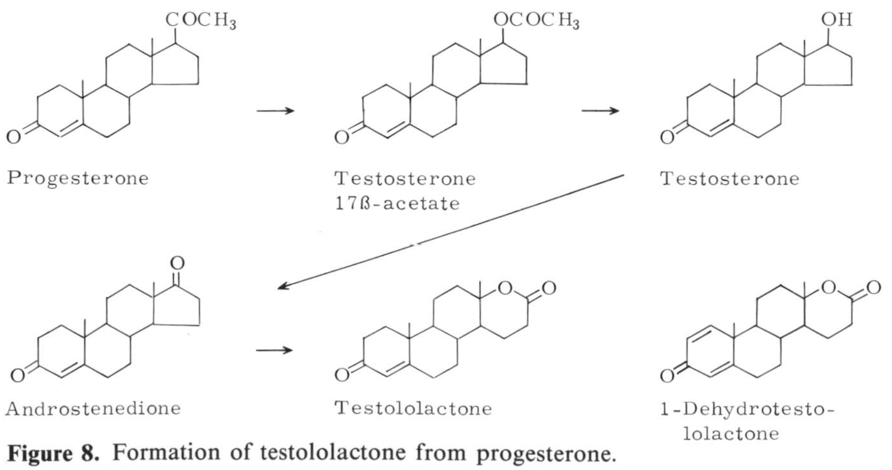

Figure 8. Formation of testololactone from progesterone.

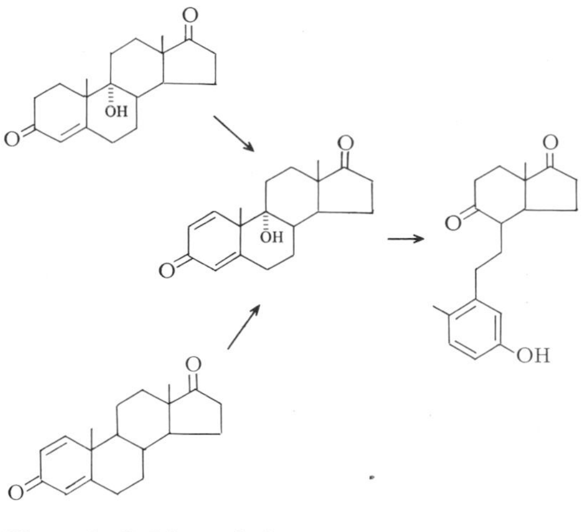

Figure 9. B-Ring scissions.

the 1α- and 2β-hydrogens and is reversible, that is, the Δ^1-double bond can be reduced by the same microorganism introducing it originally.

Microbial 1-dehydrogenation also occurs with saturated A-ring 3-ketones, and 4-dehydrogenations of 3-ketones can be effected microbiologically, both transformations ultimately giving $\Delta^{1,4}$-3-ketone products.

1-Dehydrogenation of 19-norsteroid Δ^4-3-ketones leads to A-ring aromatization and phenolic steroid products. Microbial 1-dehydrogenation of 19-hydroxy-Δ^4-3-ketones involves C-10/C-19 bond scission and A-ring aromatization, whereas 1-dehydrogenation of 9α-hydroxy-Δ^4-3-ketones is accompanied by C-9/C-10 bond scission and A-ring aromatization (with 9-ketone formation (DODSON and MUIR, 1958).

1-Dehydrogenation and resultant A-ring aromatization of suitable $\Delta^{4,7}$-3-ketone intermediates leads to the weak estrogen equilin, which has an unconjugated Δ^7-double bond feature not readily introduced into the molecule. 1-Dehydrogenation combined with 3β-alcohol dehydrogenation of substrates such as androsta-5,7-diene-$3\beta,17\beta$,19-triol also yields equilin (Fig. 10). A considerable amount of development work on these microbial syntheses of equilin, otherwise available only from horse urine, has been conducted (VÉZINA et al., 1971).

A-Ring aromatization of 19-norsteroid $\Delta^{9(10),5}$-3β-alcohols can also take place, the aromatization resulting from microbial 3β-alcohol dehydrogenation and subsequent double bond isomerization in this case (PROTIVA and SCHWARZ, 1974).

Double bonds can be introduced by elimination reactions conducted by microbial agents. Hydroxyl and esterified hydroxyl groups are subject to enzymic eliminations to isolated olefinic double bonds, and even steroid cyclic peroxides can be transformed via formal elimination reactions to yield double bonds (TOPHAM and GAYLOR, 1972).

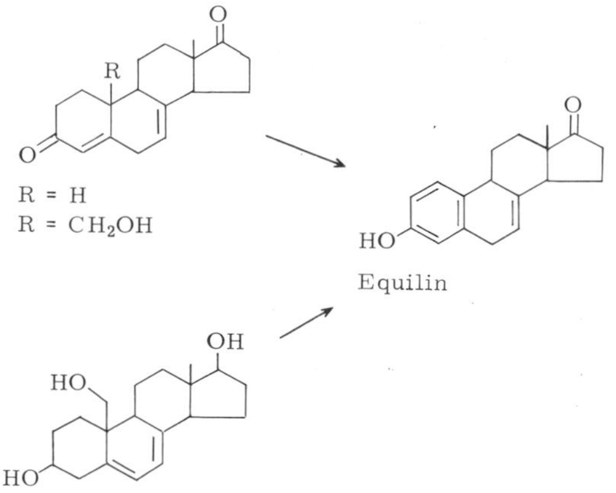

Figure 10. Microbial synthesis of equilin.

6. Peroxidations

Although unsaturated steroids are subject to autoxidation and enzyme-catalyzed generalized lipid peroxidations (SMITH, 1981), very few cases of peroxidation associated with microbial fermentations have emerged. The formation of a hydroperoxide 6β-hydroperoxy-17α-hydroxy-$6\alpha,16\beta$-dimethylprogesterone from 17α-hydroxy-$6,16\beta$-dimethylpregn-5-ene-3,20-dione during fermentation with *Flavobacterium dehydrogenans* (GARDNER et al., 1966) is surely such a case. However, evidence suggests that sterol peroxides such as ergosterol $5\alpha,8\alpha$-epidioxide are formed both by non-enzymic and enzymic peroxidations. Cultures of *Penicillium rubrum* and *Gibberella fujikuroi* appear to be capable of $5\alpha,8\alpha$-peroxidation of ergosterol (BATES et al., 1976).

7. Heteroatom oxidations

As there is but little work using steroid substrates bearing heteroatoms, so relatively few cases of oxidations involving the heteroatom are available. The stereospecific oxidation of steroidal thioethers to sulfoxides (HOLMLUND et al., 1962; MARSHEK and KARIM, 1973) and the oxidative deamination of the steroidal amine conessine (DE FLINES et al., 1962) are examples, but these reactions are clearly of very limited application.

B. Reductions

Reductions of carbonyl and olefinic features occur broadly. Primary alcohols are formed from aldehyde (SMITH et al., 1962a) and aldehyde enol (MANSON et al., 1965) reductions; secondary alcohols derive from ketone reductions. Alcohols are also products of microbial reductions of hydroperoxides (KULIG and SMITH, 1974).

Although regio- and stereospecificity are frequent in microbial reductions of ketones, chemical reductions have generally displaced microbial methods. Except for the technically used transformation of androst-4-ene-3,17-dione to testosterone by *Saccharomyces* species and the commercially useful reduction of prochiral steroid intermediates derived from total synthesis, microbial reductions are instead unwanted side-reactions often accompanying desired transformations such as hydroxylation or 1-dehydrogenation.

However, two microbial inducible oxidoreductases have been highly purified and used in fundamental investigations. A crystalline 20-ketone reductase (EC 1.1.1.53) from *Streptomyces hydrogenans* ATCC 19631 (HÜBENER and LEHMANN, 1958) giving 20β-hydroxysteroid products is a tetrameric enzyme with subunit M_r 27000 that has an associated 3-ketone reductase (giving 3α-alcohol products) possibly involving a single active center (GIBB and JEFFREY, 1971).

A second enzyme of fundamental interest is the 3β- and 17β-hydroxysteroid dehydrogenase (EC 1.1.1.50) from *Pseudomonas testosteroni* ATCC 11996 that is also a tetrameric enzyme but is composed of two different subunits (SCHULTZ et al., 1977).

Olefinic double bond reductions yield saturated steroids. Thus, $\Delta^{1,4}$-3- and $\Delta^{4,6}$-3-ketones are reduced to the corresponding Δ^4-3-ketones. Δ^4-3-Ketones are also reduced to saturated 3-ketones; Δ^{16}-20-ketones yield saturated 20-ketones. Moreover, aromatic systems can be reduced, A-ring phenols yielding 19-nor-10β-hydroxy-Δ^4-3-ketones (SCHUBERT et al., 1973). Also, allylic double bonds (BENN et al., 1964) can be reduced, and isolated double bonds are reduced in the anaerobic dehydroxylation of hydroxysteroids. In this latter reductive transformation, hydroxyl groups of bile acids are initially eliminated giving an olefinic steroid that is then reduced, thus formally providing an alcohol dehydroxylation (HAYAKAWA, 1973; KIESLICH, 1980c).

Still other potentially useful microbial transformations include reductions of olefinic double bonds. *Mycobacterium smegmatis* reduces the Δ^4-double bond only of progesterone and other Δ^4-3-ketosteroids (HÖRHOLD et al., 1977), whereas by selection of organisms, selective reduction to tetrahydro derivatives may be achieved. Thus, *Clostridium paraputrificum* yields $3\alpha,5\beta$-tetrahydro derivatives, and *Rhodotorula glutinis* yields $3\alpha,5\alpha$- and $3\beta,5\alpha$-tetrahydro products (SCHUBERT et al., 1972). Other similar reductive transformations include reductions of Δ^1-, Δ^4-, and Δ^{16}-double bonds by *Nocardia corallina* ATCC 13259, but the organism previously induced with progesterone is also capable of 1-dehydrogenations as well (LEFEBVRE et al., 1974).

C. Isomerizations

Several isomerization reactions have been described for steroids, the isomerization of Δ^5-3-ketones to Δ^4-3-ketones being one of importance. This transformation occurs in the frequently encountered sequence of dehydrogenation of Δ^5-3β-alcohols to Δ^5-3-ketones that are in turn isomerized to product Δ^4-3-ketones.

The isomerase (EC 5.3.3.1) from *Pseudomonas testosteroni* that has been extensively studied from a fundamental viewpoint is an aggregated enzyme composed of three subunits of 125 amino acid residues, the sequence of which is known (BENSON et al., 1971). The enzyme has an extraordinarily high molecular turnover number ($17 \cdot 10^6$ min^{-1}) and transfers a 4-hydrogen to the 6β-position in the isomerization (VIGER et al., 1981).

Other microbial isomerizations are relatively rare matters. Thus, epimerization of steroid secondary alcohols (BELIČ et al., 1977) and isomerization of cyclic peroxides to epoxyalcohols (PETZOLDT and KIESLICH, 1969) are very special cases. Moreover, as steroids are prone to a variety of rearrangements in chemical systems, it is not surprising that rearrangements also occur in microbial systems. Both enzymic and non-enzymic rearrangements have been recognized.

The D-homoannulation of 17α-hydroxyprogesterone (accompanied by 11α-hydroxylation) by *Aspergillus niger* (FRIED et al., 1952) must be enzymic in nature, but the analogous D-homoannulation of triamcinolone is non-enzymic and catalyzed by iron in media (Fig. 11). Sequestration of iron by added phosphates overcomes the otherwise disasterous D-homoannulation (SMITH et al., 1960b; GOODMAN and SMITH, 1960).

Angular methyl migrations occur in fermentations of sensitive substrates. The yeast reduction of 16α,17α-epoxyprogesterone was accompanied by migration of the C-18 methyl group (Fig. 12) (CAMERINO and VERCELLONE, 1956) whereas acidic medium caused migration of the C-19 methyl group during 11α-hydroxylations of C_{21}-steroids with *Aspergillus ochraceus*, giving 5β-methyl-19-norsteroid products (KIESLICH and SCHULZ, 1969).

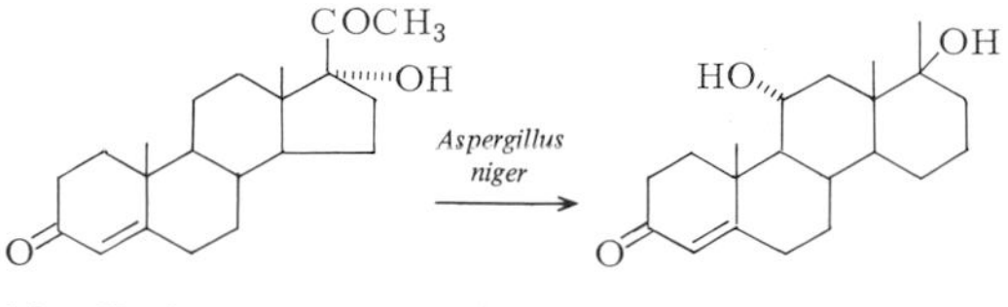

17α-Hydroxyprogesterone

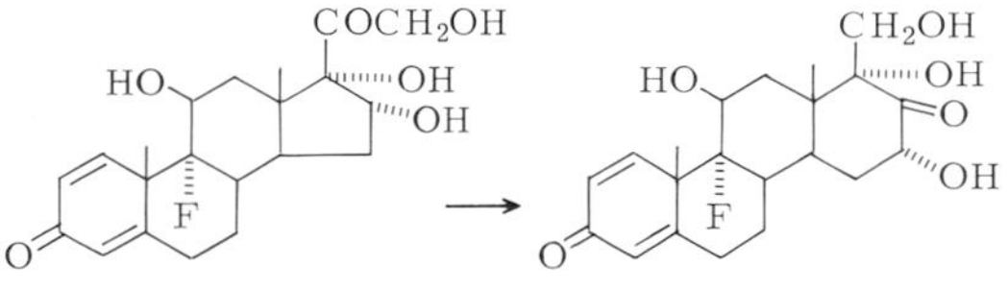

Triamcinolone

Figure 11. Microbial D-homoannulations.

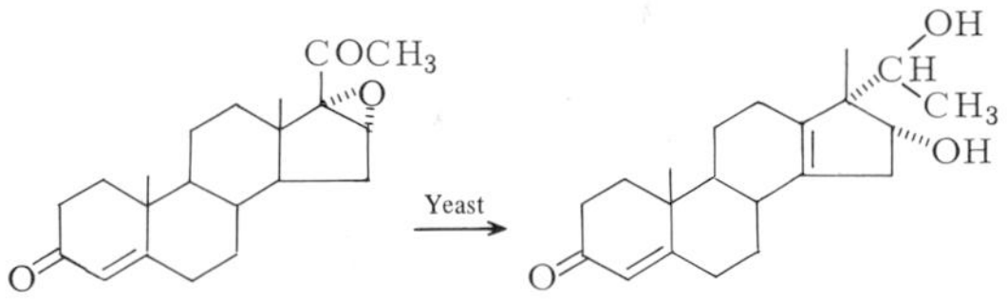

16α, 17α-Epoxy-
progesterone

Figure 12. Microbial induced methyl migration.

With the exception of the initially mentioned isomerization of steroid Δ^5-3-ketones, none of these transformations offers synthesis utility. Rather, they are undesired complications generally to be avoided.

D. Conjugations

The esterification of 3β-alcohols and of hydroxyl groups at other sites is a recognized microbial transformation, as is also the acetylation of steroid primary amines (SMITH et al., 1962b). Other conjugates between steroid alkaloids and carboxylic acids have also been suggested (BELIČ and KARLOVŠEK, 1981).

Microbial glucosidation of phenolic steroids (PETZOLDT et al., 1974) and 16α-hydroxysteroids (PAN and LERNER, 1971) by several common molds is also a recognized transformation, and phenolic steroid ether formation has been described (SCHUBERT et al., 1975). None of these conjugation reactions has been exploited in synthesis protocols.

E. Hydrolyses

The major hydrolytic microbial transformation is that of hydrolysis of steroid fatty acyl esters to steroid alcohol and fatty acid, and this transformation has been observed broadly for many microorganisms. Indeed, esterase actions accompany the important microbial hydroxylation reactions, thus allowing fermentations to be conducted with substrate esters. Strong esterase activity hydrolyzing steroid 21-acetates accompanies both *Curvularia lunata* 11β-hydroxylations and *Streptomyces roseochromogenes* 16α-hydroxylations. Furthermore, the use of microbial esterase action for the selective hydrolysis of steroid esters in the presence of other sensitive functional groups has been emphasized (KIESLICH, 1980b).

Steroid lactones and sulfate esters are also hydrolyzed microbially, and ether cleavages are observed. The transformation of steroid enol ethers to ketones (KIESLICH and KOCH, 1970), phenol ethers to phenols (SCHUBERT et al., 1975), and of sterol, alkaloid, and sapogenin 3β-glycosides to the corresponding steroid 3β-alcohols are typical examples.

F. Heteroatom Introductions

The microbial introduction of elements other than oxygen or hydrogen into steroid substrates is but rarely observed. Introduction of nitrogen has been observed in the transformation of 9α-fluorohydrocortisone to 21-acetylamino-9α-fluoro-11β,17α-dihydroxypregn-4-ene-3,20-dione by *S.roseochromogenes* (SMITH et al., 1962b). Additionally, the introduction of bromine and chlorine into steroid enolizable β-diketones, enol esters, and $\Delta^{9(11)}$-steroids by the haloperoxidase system of *Caldariomyces fumago* ATCC 16 373 (NEIDLEMAN et al., 1966; LEVINE et al., 1968; NEIDLEMAN and LEVINE, 1968; NEIDLEMAN and OBERC, 1968) has been described. The microbial introduction of either heteroatom has not proven useful in general synthesis reactions.

In addition to these enzymic reactions in which heteroatoms nitrogen or halogen are introduced, non-enzymic transformations have also been observed. For instance, the removal of the 16α-hydroxyl group of 16α-hydroxyprogesterone by *Eubacterium* sp. 144 gave 16-dehydroprogesterone, with which L-cysteine from the culture medium reacted to give a water soluble product that is probably a 16-*S*-cysteinyl progesterone conjugate (GLASS et al., 1982). As this transformation occurred anaerobically, the formation of such water soluble products probably is not likely to be encountered in the aerobic microbial transformations of current interest.

G. Sequential Reactions

Obviously, a given microbial transformation of a substrate may involve more than one individual reaction, and these reactions may be sequential and/or competing so that a complex spate of metabolites ultimately results. Thus, multiple hydroxylations such as the 2β- and 16α-hydroxylations of 9α-fluorohydrocortisone by *Streptomyces roseochromogenes* clearly involves both competing and sequential monohydroxylations (SMITH et al., 1961), whereas the transformation of Δ^5-3β-alcohol acetates to Δ^4-3-ketones is a typical example of ordered sequential transformations.

With respect to commercial operations, the desired transformation is often complicated by concomitant secondary reactions, such as 20-ketone reductions during 1-dehydrogenations with *Arthrobacter simplex* or 11β-hydroxylations with *Curvularia lunata*.

IV. General Process Features

It appears there are seven microbial transformations now in use for manufacture of finished steroids. As listed in Table 3, four of these (three hydroxylations and 1-dehydrogenation) are required for manufacture of corticosteroids and their analogs, some of which are listed in Table 4. A fifth process is a more complicated oxidation yielding 1-dehydrotestololactone. In contrast, the sixth process is a reduction of a prochiral 17-ketosteroid implicated in the total synthesis of estrogens and progestational agents. The same ketone reduction is applied for the production of testosterone from androst-4-ene-3,17-dione on a small scale. Yet a seventh process in used by Schering AG and in the Soviet Union, that of Δ^5-3β-alcohol dehydrogenation and Δ^5-isomerization.

By far the greatest commercial interest lies in the corticosteroids hydrocortisone

Table 3. Examples of Commercial Microbial Transformations[a]

Transformation	Substrate	Microorganism	Manufacturers
11α-Hydroxylation	Progesterone	*Rhizopus nigricans*	Upjohn
11β-Hydroxylation	Substance S	*Curvularia lunata*	Pfizer, Gist-Brocades
	Substance S 17α-derivatives	*Curvularia lunata*	Schering AG
	6α-Fluoro-16α-methyl-21-hydroxy-pregn-4-ene-3,20-dione	*Curvularia lunata*	Schering AG
	16-Methylene-Substance S	*Curvularia lunata*	Merck Darmstadt
16α-Hydroxylation	9α-Fluorohydrocortisone	*Streptomyces roseochromogenes*	Squibb
1-Dehydrogenation	Hydrocortisone	*Arthrobacter simplex*	Schering Corp.
	Hydrocortisone	*Bacillus lentus*	Schering AG
	6α-Fluoro-16α-methyl-corticosterone	*Bacillus lentus*	Schering AG
	Dienediol[b]	*Septomyxa affinis*	Upjohn
Lactonization	Progesterone	*Cylindrocarpon radicicola*	Squibb
17-Ketone reduction	Secosteroid[c]	*Saccharomyces uvarum*	Schering AG
	Androst-4-ene-3,17-dione	*Saccharomyces* sp.	Schering AG
Δ^5-3β-Alcohol dehydrogenation	21-Acetoxy-17α-hydroxypregnenolone	*Corynebacterium mediolanum*	Soviet Union
	3,17,21-Triacetoxy-pregn-5-ene-20-one	*Flavobacterium dehydrogenans*	Schering AG

[a] From SEBEK and PERLMAN (1979)
[b] Dienediol is 11β,21-dihydroxypregna-4,17(20)-dien-3-one
[c] Secosteroid is *rac*-3-methoxy-8,14-secoestra-1,3,5(10),9(11)-tetraene-14,17-dione

and cortisone and their synthetic analogs. Scrutiny of the structures of Table 4 reveals all to have 11β-hydroxyl groups and Δ^1-double bonds, and both features generally involve microbial steps. Additionally, triamcinolone manufacture involves a third microbial step, that of 16α-hydroxylation.

Table 4. Glucocorticoid Analogs

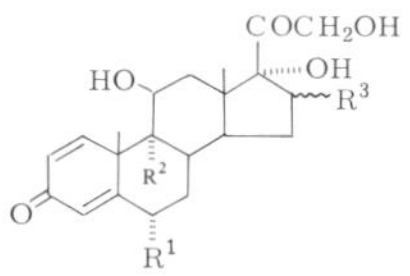

Steroid	R^1	R^2	R^3	
Prednisolone	H	H	H	
Medrol	α-CH_3	H	H	
Prednyliden	H	H	$=CH_2$	
Triamcinolone	H	F	α-OH	
Dexamethasone	H	F	α-CH_3	
Betamethasone	H	F	β-CH_3	
Paramethasone	F	H	α-CH_3	
Fluocinolone	F	F	α-OH	
Fluocortolone	F	H	α-CH_3	17-deoxy
Difluocortolone	F	F	α-CH_3	17-deoxy
Deoxymethasone	H	F	α-CH_3	17-deoxy

Clearly, optimum manufacturing processes must be developed with regard to substrate availability and cost as well as of product refining and any further chemical transformations necessary. Manufacturing appears to be centered about two general processes for 11-hydroxylations: (1) 11α-hydroxylation of progesterone or related C_{21}-derivative, leading to cortisone and hydrocortisone, and (2) 11β-hydroxylation of Substance S giving hydrocortisone directly. The required 1-dehydrogenation then follows as a second fermentation step generally at the end of the synthesis.

The requisite progesterone and Substance S substrates may be obtained from raw materials diosgenin or stigmasterol, as indicated in Fig. 13. Pregnenolone is also available from cholesterol in the older chemical processes. Steroid manufacture from bile acids and from hecogenin leads to 11-oxygenated intermediates outside this scheme, not to progesterone or Substance S. Use of sitosterol as raw material affords C_{19}-steroids useful in manufacture of sex hormones and progestational agents, but use of these C_{19}-intermediates would require resynthesis of the C_{21}-steroid side-chain, for which new chemical ways have been developed.

The manufacture of most finished C_{21}-steroid drugs, including the natural corticosteroids and their synthetic analogs, natural cortexone and 9α-fluorohydrocortisone (mineralocorticoids), and synthetic 17α-hydroxyprogesterone esters (progestational agents) proceeds from the raw materials diosgenin and stigmasterol and C_{21}-intermediates.

Alternatives for microbial introduction of the required 11-oxygen function include two means of synthesis of $\Delta^{9(11)}$-intermediates that can be used to advantage in manufacture of 9α-fluoro-11-oxygenated steroids. Hecogenin, containing a 12-ketone group, can be dehydrogenated chemically to a $\Delta^{9(11)}$-12-ketone, and thereafter transformed to $\Delta^{9(11)}$-intermediates. As supplies of hecogenin are limited, use appears most suited to synthesis of 9α-fluoro-11β-hydroxysteroids such as dexamethasone.

Another means of preparation of $\Delta^{9(11)}$-intermediates involves microbial 9α-hydroxylation of androstenedione (obtained from microbial degradation of sterols) using *Corynespora cassicola* ATCC 16718. The 9α-hydroxylated product is then transformed chemically via $\Delta^{9(11)}$-intermediates, including resynthesis of the side-chain, to 9α-fluorohydrocortisone (Fig. 14). 9α-Hydroxyandrostenedione is also available via microbial degradation of sitosterol by a *Mycobacterium* sp. NRRL B8119 (KIESLICH, 1980b). In this case the chemical resynthesis of the pregnane side-chain has an economical chance, as the introduction of the $\Delta^{9,11}$-double bond via 9α-hydroxylation coupled with the side-chain degradation replaces and eliminates the further fermentation step of the 11-hydroxylation.

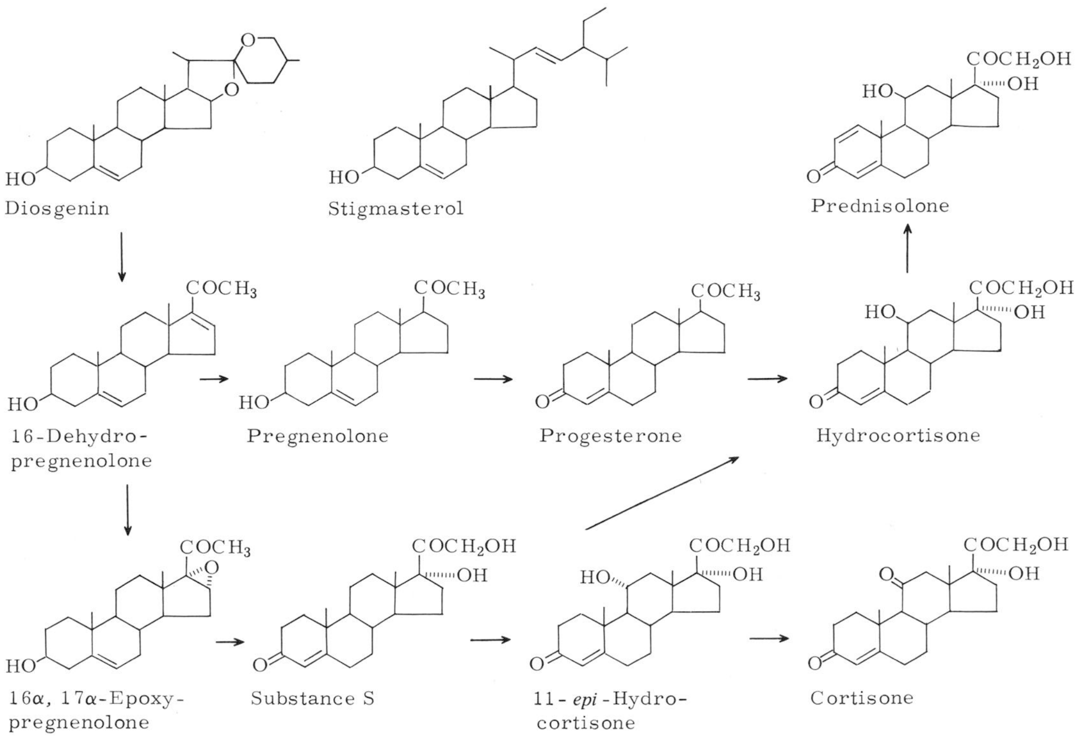

Diosgenin

Stigmasterol

Prednisolone

16-Dehydro-pregnenolone

Pregnenolone

Progesterone

Hydrocortisone

16α, 17α-Epoxy-pregnenolone

Substance S

11-*epi*-Hydro-cortisone

Cortisone

Figure 13. General pathways of manufacturing using microbial steps.

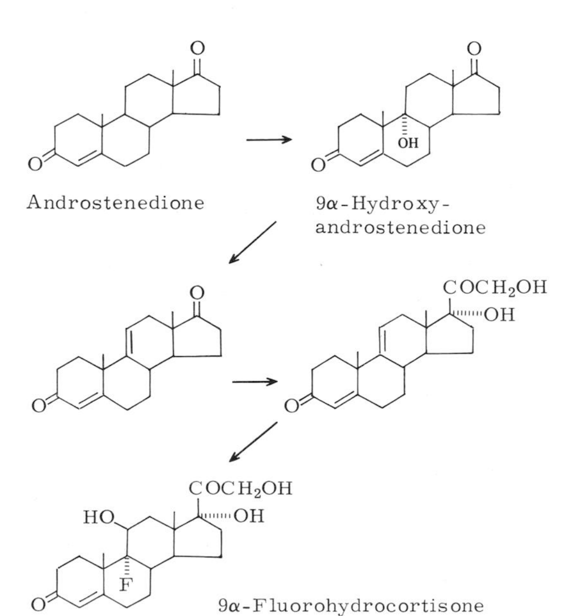

Figure 14. An alternative synthesis of 9α-fluoro-hydrocortisone.

A. Fermentation Conditions

There are two major items of steroid biotechnology requiring careful development work. The selection of microbial strains for production is obviously crucial, but perfection of operating parameters for the fermentation is also of great importance.

1. Microorganisms used

The capacity to transform in some fashion steroids introduced as xenobiotic substances has been discovered in almost every type of microorganism (bacteria, yeasts, fungi, algae, and protozoa). Over half of the 2000 or so organisms tested have some metabolic activity. Fungi are the most represented, with bacteria and actinomycetes next in number. There are but isolated cases for protozoa and algae. However, in terms of the number of known microorganisms, only a small fraction have been screened for these purposes. Of the many examined, less than a score are useful in current commercial biotechnology.

Microorganisms used in much of the world's investigations are from the important depositories such as the American Type Culture Collection (ATCC), Rockville, MD; Centraalbureau voor Schimmelcultures (CBS), Baaru (Netherlands); Institute for Fermentation (IFO), Osaka; Northern Regional Research Laboratories U.S. Department of Agriculture (NRRL), Peoria, IL; Quartermaster Culture Collection, U.S. Army (QM), Natick MA; Waksman Collection, Rutgers University (WC), New Brunswick, NJ; and the German Type Culture Collection, Göttingen. Yet other sources are known (KIESLICH, 1969).

Moreover, even though a culture is available from one of these depositories, production cultures are strains developed by the manufacturing company after very extensive strain selection and improvement. Both natural mutant organisms and those derived by treatment with mutagens, irradiation, etc. have been evaluated. As the steroid transformation process and its economics depend entirely on the effectiveness of the microorganism, much effort is expended in these studies.

2. Operating parameters

Although several different steroid substrates and transforming microorganisms are involved in steroid biotechnology work, the means of conducting such transformations are very much alike. Useful microorganisms are generally maintained as spores, lyophilized cultures, stored in liquid nitrogen or cultures on agar and must be inoculated into liquid nutrient medium for use in aerated fermentors.

Media must contain suitable sources of nitrogen, carbohydrate, mineral salts, and trace nutrients and may also incorporate pH adjustment or addition of buffering components such as calcium carbonate. The precise composition of media is adjusted along with other operating parameters for optimum mycelium growth (before substrate addition) and for maximum induction to the requisite enzyme necessary to the transformation (after substrate addition).

It is generally the policy that fermentation media must support maximum yield of desired product in the fermentor, it being assumed that refining can overcome problems generated by some media components (for example, lard oil as antifoam).

Steroid transformations are generally conducted commercially in stainless steel, glass-lined, or other stirred fermentors under sterile conditions. Inoculum prepared in shake flasks is transferred into seed tanks (250-5000 L) containing suitable medium and stirred with aeration under controlled conditions to obtain necessary mycelial growth. At the appropriate stage, seed tank contents are transferred to a second-stage fermentor of intermediate size, usually approximately 10% of the volume of the ultimate production unit. After mycelium growth is optimum, the contents are transferred into the production fermentor, possibly 50000–75000 L size.

After a suitable period for mycelium growth in the production unit (perhaps 24 h), as determined by cell mass, glucose and oxygen utilization, etc., the substrate is added under sterile conditions in a suitable solvent such as acetone, ethanol, dimethylformamide, propylene glycol, dimethylsulfoxide, etc. Final solvent level is generally kept at a minimum, of about 1% total volume, in order to limit toxicity of solvent to the organism.

Crucial operating parameters include temperature, stirring and aeration rates, and pH and foam control. Accordingly, controlled heating and cooling may be necessary, and a sterile source of air is required. Additionally, means for sterile additions for pH and foam control and of substrate addition and for sampling during the fermentation for assay work must be available (MÜLLER and KIESLICH, 1966).

B. Analysis Methods

The high costs of substrate and operations demand that microbial fermentations of steroids be under close monitoring by suitable analytical means. Product recovery for analysis is accomplished by solvent extractions, usually with the same solvents useful in isolation work. After removal of solvent under vacuum, the steroid sample is redissolved in medium suited to the intended analytical procedure.

Timely analyses over the course of a transformation are important to optimum operations, and rapid methods are needed. Direct instrumental analyses are attractive and can be used for total steroids content. For example, ultraviolet light absorption measurements and colorimetric methods (as for total reducing steroids) provide estimations of total steroids.

Where suitable chemical differences exist between substrate and product, direct instrumental analyses are also useful. The 1-dehydrogenation of Δ^4-3-ketone to $\Delta^{1,4}$-3-ketone product may be followed effectively by polarography (SMITH and HALWER, 1959), differential ultraviolet absorption spectrophotometry (IVASHKIV, 1971), or colorimetry in sulfuric acid (MIZSEI and SZABÓ, 1961). Where chemical differences are less pronounced, as in hydroxylations, resort to altered polarity between substrate and product allows for preliminary fractionation by extraction or partition methods, with direct instrumental analysis of partitioned components.

However, chromatographic methods are generally necessary for complete control, particularly where several products are formed. Paper chromatography originally of use was replaced early by more rapid partition column chromatography (SMITH et al., 1959), and thin-layer and high performance liquid column chromatography are now routine methods of choice. Gas chromatography of suitably protected (trimethylsilyl ethers) derivatives also may be useful (MCGREGOR et al., 1972).

Thin-layer chromatography of fermentation extracts provides perhaps the most versatile rapid assay method for qualitative purposes. Here a variety of differential means of detection of steroids *in situ*, applied sequentially, is important. For instance, steroids absorbing ultraviolet light are readily detected, and Δ^4- and $\Delta^{1,4}$-3-ketones can be distinguished from one another with isoniazid reagents (SMITH and FOELL, 1959, 1962). At the same time, reducing steroids (corticosteroids and analogs) are differentiated from their 20-ketone reduced by-products often encountered in fermentations by means of tetrazolium salts. Other general color tests (phosphomolybdic acid, sulfuric acid) then may detect all organic components present.

By eluting resolved components direct instrumental analyses can be applied for confirmation of identity and for quantitation (KOLCHANOVA et al., 1969). Direct photodensitometric evaluation of ultraviolet light absorbing components is also possible (CEJKA, 1976).

However, the broad applicability of high performance liquid column chromatography and its demonstrated capacity to resolve many steroid hormones and analogs (TYMES, 1977) makes this newer method the one of choice. Both adsorption and reversed phase systems have been successfully used for monitoring microbial hydroxylations and 1-dehydrogenations (LARSSON et al., 1979; SLOCUM and STUDEBAKER, 1975). Additionally, the method is of great importance in the analysis of finished steroids prior to release for sales.

Another aspect of analytical work needed with new organisms or substrates is that of identifying transformation products. Where products are already known, the matter becomes one of comparing metabolites with authentic samples, but unidentified products may require considerable study for identification. Spectral characterization of new products generally permits rapid progress, infrared spectra indicating oxygen functionalization and olefinic double bonds, confirmed by proton nuclear magnetic resonance spectra. Mass spectra indicate molecular weight, and elimination or fragment ions aid in the more detailed assignment of structure.

A common problem often encountered in hydroxylations is the question of location of the newly introduced hydroxyl group. The proximity of the hydroxyl group to other functionality may influence matters, allowing ultraviolet absorption spectra in alcohols, sulfuric acid, or alkaline ethanol to suggest structure (SMITH, 1963a, 1963b). Proton spectra are also very useful, as newly introduced hydroxylic substitution may affect chemical shifts of the readily distinguished angular methyl proton resonances. Moreover, both chemical shift and signal multiplicity of the proton geminal to hydroxyl newly introduced at methylene carbon are of importance in recognizing structure (SMITH, 1964; TORI and KONDO, 1964, 1966; BRIDGEMAN et al., 1970; JONES, 1973).

Another commonly used method in structure recognition of hydroxysteroids involves oxidation of the alcohol to carbonyl and location of the newly formed carbonyl group by infrared absorption spectra and related means (BOUL et al., 1971).

C. Refining Methods

The recovery of steroid products from microbial fermentations requires extraction of the fermentation medium with immiscible solvents, separation of the solvent extracts, and evaporation of solvent. Variations of this basic means may include such matters as pH adjustment and filtration of the whole broth, with solvent extraction of both mycelium and filtered broth or of either, depending on the location of wanted steroid products.

Only in the special case of microbial 1-dehydrogenations using very high substrate levels is it possible to isolate product by simple filtration of mycelium with crystalline steroid. Alternative recovery processes using activated charcoal to adsorb steroids from fermentation broths (EPPSTEIN and LEIGH, 1956) are not useful in comparison with solvent extraction.

Extractions with solvents such as chloroform, methylene chloride, and ethyl acetate have been widely used, but methyl isobutyl ketone appears to be a solvent of choice for commercial operations.

Crude steroid products are then recrystallized, filtered, and dried for analyses as finished materials. Just such simple and direct processes have been perfected for manufacture.

For usefulness, all refining processes must yield a product free from unaltered substrate, by-products, and other materials derived from the fermentation. Where proper strain and fermentation conditions have been selected, the single desired product is usually obtained from crude product by recrystallization.

However, in those cases where more than one product is obtained in approximately equal yield, for instance, with *Absidia orchidis* 310 where both 11α-and 11β-hydroxylated products are obtained from Substance S, special processes for separation of products are required. Only the 11β-dehydroxylated product is wanted for hydrocortisone production, but the 11α-hydroxylated product after separation is used for cortisone manufacture. Fractional crystallization from methanol affords hydrocortisone first, with 11-epihydrocortisone retained in mother liquors. Alternatively, acetylation of the product mixture gives hydrocortisone 21-acetate but 11-epidihydrocortisone 11α,21-diacetate, which are separated by recrystallization. Acetylation of the mixed products with only one equivalent of acetic anhydride yields the 21-monoacetates of both products, and these can also be separated by recrystallization (HANČ et al., 1962).

An additional matter besides product recovery requires attention, namely that of solvent recovery and disposal of fermentation wastes. Evaporated extraction solvent is recovered at that point, but as some solvent remains with the extracted fermentation broth, recovery from the aqueous medium may be necessary. This is particularly important where methyl isobutyl ketone is the extracting solvent.

If filtration precedes extraction, the mycelium cake may be handled separately from the filtered aqueous broth. Extracted aqueous broth stripped of solvent can then be disposed of via industrial waste-water treatment facilities. Extracted whole fermentation broths containing mycelium constitute special disposal problems requiring additional treatment.

V. Hydroxylations

Hydroxylation is that microbial reaction that has captured the most interest as it attacks unactivated sites and gives access to parts of the steroid molecule otherwise achieved only with great difficulty.

A. Commercial Hydroxylation Processes

There are three microbial hydroxylations of commercial importance, 11α-, 11β-, and 16α-hydroxylations. Each is presently operated in steroid manufacture; each has properties of unique interest.

1. 11α-Hydroxylation

The first microbial hydroxylation to be exploited in steroid manufacture was that of 11α-hydroxylation. Initially reported organisms *Rhizopus arrhizus* and *Aspergillus niger* were shortly replaced by the preferred *Rhizopus nigricans* ATCC 6227b initially classified as *R.stolonifer* (MURRAY, 1976), which is of immense practical importance. Additionally, *Aspergillus ochraceus* NRRL 405 is an important 11α-hydroxylator.

As in all commercial microbial transformations, many screening operations were conducted before these preferred species were recognized. For instance, 475 strains from 38 species of *Aspergillus* and 476 strains of *Penicillium* were screened by DULANEY et al. (1955a) for their 11α-hydroxylating capacities towards progesterone, the usual substrate of interest. Numerous other published reports tell of screening operations with hundreds of species and strains, and even more extensive strain selection by individual steroid manufacturers is the case.

As a result of much development work the 11α-hydroxylation of progesterone by strains of *Rhizopus nigricans* remains the most successful microbial hydroxylation in commercial operation, with yields over 90% of 11α-hydroxyprogesterone indicated (NOMINÉ, 1980).

Details of modern operations with *R.nigricans*ATCC 6227b are not available, but a general description of the process has been reported by MURRAY (1976). The strain is best maintained in the cold on agar slants containing 5% malt extract and 0.5% peptone. Slants less than two weeks old are used to seed cultures for steroid transformations. The organism grows optimally at 28 °C and does not grow above 32 °C. Fermentation flasks and tanks are then directly inoculated with these spore cultures to produce flourishing vegetative cell cultures for steroid tranformation. A heavy inoculum for larger fermentors is suggested, with fermentation for 11–18 h with moderate aeration, followed thereafter by addition of progesterone substrate.

A variety of media provide adequate nitrogen and carbohydrate sources for *R.nigricans*. Typical media are 5% dextrose – 3% corn steep liquor – 2% Edamine (enzymic digested lactalbumin), 1% dextrose – 2% corn steep liquor, 2% dextrose – 0.5% corn steep liquor – 0.5% soybean meal, and 2.5% glucose – 2% yeast extract, all adjusted to pH 4.3–4.5 to effect sterilization of the system. In production, medium composition depends on current economics of the supply of medium constituents (MURRAY and PETERSON, 1952; MURRAY, 1976; HANISCH et al., 1980).

Besides strain and medium selection, a number of other factors need control. Growth of the organism may be monitored by cell mass and glucose and oxygen consumption, but the crucial induction of the 11α-hydroxylase (requiring progesterone at >0.5 g/L) can be followed only by product assay. Factors such as age of the culture and time of addition of substrate are not critical. Contamination by other organisms is not a problem.

Progesterone is added as an acetone solution or as a micronized powder (no solvent) so as to give substrate levels of 0.5–5 g/L in the fermentation, which is then allowed to proceed. Powdered progesterone wetted with 0.01% Tween 80 is also useful to attain high substrate charges in *R.nigricans* fermentations (WEAVER et al., 1960). Progesterone added at 5 g/L as a powder is smoothly 11α-hydroxylated up to 50 h, but thereafter some 14% progesterone remains unaltered. Residual progesterone appears to be a mixed crystal with product 11α-hydroxyprogesterone and is unavailable for further transformation (MAXON et al., 1966).

The problem of high residual progesterone levels is overcome by continuous addition of substrate (SEBEK and PERLMAN, 1979). Indeed, the 11α-hydroxylation of progesterone by *R.nigricans* can be operated as a continuous process involving several separate stages. Mycelium growth in a first stage receives a continuous feed of progesterone at 0.5 g/L in the second stage, with subsequent conversion stages yielding 50–60% product recoveries by 5 h (REUSSER et al., 1961).

A factor of importance in the transformation is proper aeration. The mass transfer of oxygen from bulk liquid to mycelium surface and into the cell is crucial to good 11α-hydroxylation and is controlled by the rate of delivery of air to the fermentor and the rate of stirring. At high impeller tip speeds air is well dispersed in the medium but mycelium is also destroyed by shear. Moreover, at high aeration rates, foaming is a problem. Although foam can be controlled with antifoam agents, these interfere in subsequent refining operations.

At low impeller tip speeds aeration is inadequate, slowing the transformation and causing mycelium to clump. Accordingly, an optimum aeration rate is usually indicated (HANISCH et al., 1980).

Sampling of mycelium and relatively insoluble substrate and product over the course of the fermentation is a problem, but material balances of 96% have been attained. Recovery of products involves filtration of mycelium from broth, with separate extractions of each. Mycelium is washed with acetone and methylene chloride, and these washes are added to the filtered broth, which is then extracted twice with half-volumes of methylene chloride. The methylene chloride extracts are washed with 2% sodium bicarbonate and evaporated. Product 11α-hydroxyprogesterone is then crystallized from methylene chloride (MURRAY and PETERSON, 1952). 11α-Hydroxyprogesterone is the main product (90%) with but low levels of 11α-hydroxy-5α-pregnane-3,20-dione (increased by low aeration rates and high carbohydrate levels in media) and 6β,11α-dihydroxyprogesterone (formed at longer fermentation times).

Much development work has also been conducted with the efficient 11α-hydroxylator *Aspergillus ochraceus* NRRL 405, but it is uncertain whether the organism is currently used in manufacture.

Media composition and fermentation conditions are similar to those for *Rhizopus nigricans*. Thus, dextrose-lactalbumin-corn steep media at pH 6.3–6.5 are useful. Substrate progesterone can be added as a solution in propylene glycol at 4 g/L or as a powder wetted by Tween 80 at levels as high as 20–50 g/L(!) with yields of 11α-hydroxyprogesterone to 90%. Semi-continuous operations have been conducted, with incremental additions of substrate (KAROW and PETSIAVAS, 1956; WEAVER et al., 1960).

A new strain *A.ochraceus* TS is reported to 11α-hydroxylate progesterone at 0.2–20 g/L, with 100% transformations at 0.2–0.4 g/L (SAMANTA et al., 1978).

2. 11β-Hydroxylation

The requisite 11β-orientation of the 11-hydroxyl group can also be introduced microbially. As subsequent chemical steps to invert the 11α-hydroxyl group introduced by *R.nigricans* or *A.ochraceus* are then unnecessary, 11β-hydroxylation can be conducted on fully elaborated substrates such as Substance S. This transformation has been perfected with *Curvularia lunata* NRRL 2380, discovered via extensive screening operations. As with *R.nigricans* initially misidentified, so also *C.lunata* NRRL 2380 was previously designated *Curvularia falcata* QM120h (SHULL et al., 1953).

Other 11β-hydroxylating organisms, *Streptomyces fradiae* (COLINGSWORTH et al., 1952) and *Cunninghamella blakesleeana* (HANSON et al., 1953), are not competitive with *C.lunata* in production.

Curvularia lunata maintained on agar slants yields flourishing vegetative cell cultures in relatively simple media such as 5% malt extract – 1% sucrose or 1% tryptone – 1% sucrose, both with added salts and at pH 7.0 (SHULL et al., 1953). Also media containing 1% glucose – 1% corn steep solids and 5% glucose – 5% corn steep solids at pH 6.5 are useful (KIESLICH, 1969). Operations involve sterile media and additions.

Inoculated fermentors are stirred and aerated at 26–28 °C for mycelium growth, possibly through several stages, to production tanks before substrate addition. Substance S added as an alcohol solution to give 0.5 g/L levels is then transformed to products that are recovered by direct solvent extractions of the whole broth (SHULL et al., 1953). Alternatively, filtration of the broth, washing of the mycelium cake with methyl isobutyl ketone, and extracting the filtered broth with these washes and with added solvent is an effective refining process (KIESLICH, 1969).

Another alternative process involves filtration of the mycelium after suitable growth (without steroid) and resuspension of the mycelium in water. After substrate addition, the stirred aerated mycelium dispersion effectively 11β-hydroxylates the substrate, and product recovery is simplified (SHULL et al., 1953).

Although the 11β-hydroxylation of Substance S by *C.lunata* is immensely important, the process is blighted by unwanted side-products, particularly of other hydroxylated steroids. Consequently, much development work has been directed to the solution of this problem. However, neither strain selection (DULANEY and STAPLEY, 1959) nor media nor fermentation condition variation has been effective. Rather, in this case variation of substrate has proven successful, raising the yields of hydrocortisone from 40–65% of early operations (SHULL and KITA, 1955; DE FLINES, 1969) to about 90%.

Substance S 17β,21-diacetate and Substance S 17α-acetate have been shown to be good substrates. Although the 21-acetate was hydrolized during the transformation, the 17-acetate was not, as long as the medium remained slightly acid (<pH 6.5). The 11β-hydroxylation of Substance S 17α-acetate added as a water dispersion to give 0.5 g/L in flourishing *C.lunata* cultures in 0.8% glucose – 4% corn steep medium is described as giving hydrocortisone 17α-acetate, from which an overall yield of 88.6% of hydrocortisone was obtained upon saponification (KIESLICH, 1969).

Most of these same considerations also apply to fermentations by other organisms, for instance to *Tieghemella orchidis* иБфм F-398 that 11-hydroxylates Substance S and Substance S 21-acetate in both 11α- and 11β-positions (MOGIL'NITSKII et al., 1977).

3. 16α-Hydroxylation

Whereas 11-hydroxylation is essential to all highly active corticosteroid drugs, 16α-hydroxylation is limited to triamcinolone and the related fluocinolone (cf. Table 4). Introduction of the 16α-hydroxyl group by microbial means is indicated in both cases. For triamcinolone, 16α-hydroxylation of 9α-fluorohydrocortisone by *Streptomyces roseochromogenes* and subsequent 1-dehydrogenation constitutes the commercial manufacturing process (see Fig. 11) (THOMA et al., 1957). This two-step process replaced a fourteen-step (1% yield) chemical and microbiological process starting from Substance S (BERNSTEIN et al., 1959).

A parallel synthesis of fluocinolone involves *S.roseochromogenes* 16α-hydroxylation of 6α,9α-difluorohydrocortisone followed by 1-dehydrogenation (MILLS et al., 1960).

Triamcinolone is also derived by initial 1-dehydrogenation of 9α-fluorohydrocortisone, with subsequent 16α-hydroxylation by *S.roseochromogenes* of the intermediate product 9α-fluoroprednisolone. Although the original report of the process suggested lower yields for the 16α-hydroxylation of 9α-fluoroprednisolone (20% versus 50% for 9α-fluorohydrocortisone) (THOMA et al., 1957), development work on both processes shows that yields are nearly the same.

9α-Fluorohydrocortisone 21-acetate may also be used as substrate, as esterase rapidly hydrolyzes the 21-acetate (KITA, 1961; MARTÍNKOVÁ and DYR, 1965). As the water solubility of substrate is limited, attempts have also been made to increase substrate solubility by using 9α-fluorohydrocortisone 21-hemisuccinate (RYU et al., 1969) and 16α,17α-cycloborate (LEE et al., 1971) esters.

Several *S.roseochromogenes* strains are effective; chiefly *S.roseochromogenes* ATCC 3347 and Waksman No. 3689 are used. However, other species such as *Streptomyces halstedii* ATCC 13 499 and NRRL B-1238 are also useful for 16α-hydroxylation of 9α-fluorohydrocortisone (KITA, 1961). Furthermore, strain selection has been important in production, as unwanted by-products are a problem. The ATCC 3347 strain gives useful 16α-hydroxylation but also is capable of 2β-hydroxylation of substrate, whereas the Waksman No. 3689 has very low 2β-hydroxylation capacity. Strain selection from these parents gave derived strains with altered hydroxylation patterns. Thus, from strain ATCC 3347 was derived the Lederle C2 strain with strong 2β-hydroxylation but also the Lederle 409A strain with no 2β-hydroxylation. Moreover, the C402 strain with some 2β-hydroxylation was derived from the Waksman No. 3689 strain. Necessarily production strains devoid of unwanted 2β-hydroxylation are important (GOODMAN and SMITH, 1961).

Media for *S.roseochromogenes* fermentations include 1.5% soybean meal - 0.22% soybean oil - 1% glucose - 0.25% calcium carbonate, 3% glycerol - 2% corn steep - 0.7% calcium carbonate - 0.2% ammonium sulfate - 0.2% lard oil, and 4% starch - 2.5% corn steep - 0.5% calcium carbonate - 0.2% lard oil. Thus media contain calcium carbonate as buffer against acidification. Moreover, in order to prevent the unwanted isomerization of product, 0.3–0.5% dipotassium hydrogen phosphate must be added. A balance between calcium carbonate buffer and added phosphate needs to be achieved to control the unwanted side-chain reaction (GOODMAN and SMITH, 1960, 1961).

Fermentations are conducted at 26–28 °C (MARTÍNKOVÁ and DYR, 1965). Substrate is added in dimethylformamide. The solvent volume is limited to less than 2% final volume, and steroid levels are 0.2–0.5 g/L. The stirred aerated fermentation is continued until completion, as determined by assay, generally for about 80–100 h.

Product recovery involves filtration of acidified (pH 4) whole broth with the aid of diatomaceous earth, reslurry of the mycelium cake with water, and refiltration. The filtrate is then extracted twice with half-volumes of methyl isobutyl ketone, the solvent extracts reduced in volume under vacuum, and the crystallized product filtered. Yields of about 70% are attained.

Although 16α-hydroxylations of 9α-fluorohydrocortisone are usually conducted until residual substrate levels are quite low, there are cases where high levels of remaining substrate are present with product 16α-hydroxy-9α-fluorohydrocortisone. These can be handled by taking advantage of the water-solubilizing effects of boric acid on the 16α-hydroxylated product. Steroid 16α,17α-diols form water-soluble cycloborate esters that can be separated from residual substrate by contacting the mixture with alkaline aqueous sodium tetraborate solution and methyl isobutyl ketone. Substrate and steroids not forming 16α,17α-cycloborate esters are extracted into the solvent; desired product is retained in the water layer as the 16α,17α-cycloborate ester. Alternatively, methyl isobutyl ketone extracts of fermentation broth containing these steroids can be washed with alkaline sodium tetraborate solution to extract the desired products 16α-hydroxy-9α-fluorohydrocortisone. Water solutions of the cycloborate esters are then acidified to pH 1.5–2, thereby decomposing the cycloborate esters and causing the wanted steroids to precipitate (LEESON et al., 1961).

B. Alternative Hydroxylation Processes

Despite the commercial success of using flourishing vegetative cell cultures for steroid hydroxylations several alternative processes have been evaluated. Examinations of these alternatives have been with the view to increased yields, control of side-reactions, simplified processes, and improved economics. Among such alternatives are processes using resting vegetative cells, immobilized vegetative cells, spores and immobilized spores, and cell-free hydroxylases and immobilized enzymes. None of these alternative processes now appears to be commercially operated, but interest in these directions is continuing.

1. Resting vegetative cells

Resting vegetative cells freed from nutrient media but suitably aerated are also capable of hydroxylating steroid substrates. The very effective *Curvularia lunata* NRRL 2380 11β-hydroxylates the important substrates such as progesterone and Substance S very well by suspending mycelium in water alone (SHULL et al., 1953; SHULL and KITA, 1955).

Resting cells of *C.lunata* (previously grown in nutrient medium also containing 19-nortestosterone for induction of the requisite hydroxylase) also hydroxylate 19-nortestosterone when incubated in buffer at 4°C with substrate. All three of the recognized 10β-, 11β-, and 14α-hydroxylase activities are expressed. No hydroxylations occurred at 4°C for up to 8 h (or at 20°C for 1 h) unless the growing culture had been induced by addition of 19-nortestosterone or other suitable steroid inducer (LIN and SMITH, 1970b, 1970c).

The 11α-hydroxylation of steroids by *Aspergillus ochraceus* resting cells suspended in buffer is also an effective process. The 11α-hydroxylase must be induced in previous vegetative cell growth in nutrient medium (DULANEY et al., 1955a, 1955b; LEE et al., 1971). Furthermore, resting cells of *Rhizopus nigricans* ATCC 6227b induce progesterone 11α-hydroxylase when incubated at 28°C in pH buffer (BRESKVAR and HUDNIK-PLEVNIK, 1978). In several cases, 11α-hydroxylation by resting cells was more rapid than in flourishing vegetative cell cultures!

Furthermore, the 16α-hydroxylation of the retrosteroid 9β,10α-pregna-4,6-diene-3,20-dione by resting cells of *Sepedonium ampullosporum* ATCC 18217 or CBS 392.52 has been demonstrated in pilot plant equipment. The organism grown in nutrient media was recovered by filtration or centrifugation, resuspended in water or in buffer, and agitated and aerated in tanks up to 1400 L capacity, thereby giving a hydroxylating system stable up to 150 h at 28°C. The steroid substrate was best added as a micropulverized powder without solvent.

Of considerable advantage was the decrease in time required for the hydroxylation, conversion being complete by about 6 h, thus needing only 50–70% of the time required for comparable conversion by flourishing vegetative cell fermentations. Moreover, the resting cell suspensions could be charged with substrate several times, yielding product until endogenous reserves of nutrients and cofactors were exhausted (MCGREGOR et al., 1972).

The 16α-hydroxylase of *S.ampullosporum* did not appear to require induction and is thus distinct from the 11β-hydroxylase of *C.lunata* previously mentioned.

The 16α-hydroxylation of C_{18}-, C_{19}-, and C_{21}-steroids has also been demonstrated for resting cells of *Streptomyces roseochromogenes* NRRL B-1233. Mycelium recovered by centrifugation and resuspended in buffer in shake flasks transformed dehydroepiandrosterone, pregnenolone, androst-4-ene-3,17-dione, and estrone into their corresponding 16α-hydroxylated derivatives at 27°C. The system was pH-sensitive, with optimal transformation at about pH 7.5, and initial hydroxylation rates depended on cell concentration (IIDA et al., 1979).

Thus, all important 11- and 16α-hydroxylating microorganisms *Curvularia lunata, Aspergillus ochraceus, Rhizopus nigricans,* and *Steptomyces roseochromogenes* can be used satisfactorily for steroid hydroxylations as resting cells, without media.

2. Immobilized vegetative cells

Vegetative cells are capable of diverse steroid transformations even though the mycelium be entrapped in gels (KOSHCHEYENKO et al., 1983). Immobilization of vegetative cells for hydroxylations has been achieved in polyacrylamide, polyethyleneglycol diacrylate, calcium alginate, and agar gels with varying success.

The first demonstration of steroid hydroxylation using immobilized cells utilized *Curvularia lunata*. Vegetative cell growth of *C.lunata* induced with Substance S was harvested and incorporated into a polyacrylamide gel by mixing mycelium with monomer and catalyst, under nitrogen at 15°C. The gel formed in a sheet was fragmented and reduced to 30 mesh granules used as such.

The gel granules transformed Substance S to hydrocortisone and no other products should be detected. An important feature of the system was that the granules could be reactivated by incubation in nutrient media. It is uncertain whether reactivation involved reinduction of steroid hydroxylase or of renewed growth of the mold in the gel (MOSBACH and LARSSON, 1970).

The 11-hydroxylation of Substance S 21-acetate by *Tieghemella orchidis* immobilized in 20–30 mesh polyacrylamide gel granules has also been demonstrated, with 11β-hydroxylation to hydrocortisone in 22–25% yields, 11α-hydroxylation to epihydrocortisone in 40–55% yields, and with 6β-hydroxylation in 8% yield. These immobilized *T.orchidis* cells thus retained their full capacity to hydroxylate Substance S at the three recognized sites as well as to hydrolyze the 21-acetate ester moiety. Repeated use of the immobilized cells gave yields comparable to those of vegetative cells in nutrient media (SKRYABIN et al., 1974).

Polyacrylamide gels may not be suitable for steroid hydroxylations by all immobilized microbial agents. Thus, the 11α-hydroxylation of progesterone by *Rhizopus nigricans* ATCC 6227b was not successful. However, cells immobilized in calcium alginate or agar gels did successfully 11α-hydroxylate progesterone (MADDOX et al., 1981). Likewise, germinated spores of *C.lunata* ATCC 12017 immobilized in calcium alginate gels were considerably more active in the 11β-hydroxylation of Substance S than were those immobilized in polyacrylamide gels. The inactivation of microbial enzyme systems during exposure to the cytotoxic monomer, to temperature or pH changes during polymerization, or other factors may account for this failure with the otherwise very good 11α-hydroxylating organism *Rhizopus nigricans*.

Also, the 16α-hydroxylation of dehydroepiandrosterone by *Steptomyces roseochromogenes* NRRL B-1233 immobilized in photopolymerized polyethyleneglycol diacrylate gel has been reported. Mycelium was mixed with the hydrophilic prepolymer containing a polymerization initiator, and the preparation spread in layers was irradiated with ultraviolet light. The resultant polymeric sheet was cut into pieces and used in buffer for steroid hydroxylation.

In incubations of up to 60 h duration the immobilized mycelium gave improved yields of 16α-hydroxylated substrate in comparison with vegetative cells. Moreover, the immobilized cells could be used for three batch transformations before yields began to decrease (CHUN et al., 1981).

3. Spore processes

One of the most advanced alternative hydroxylation processes uses microbial spores instead of flourishing vegetative cell structures. The harvested spores are incubated with steroid substrates in the absence of vegetative cells and nutrient medium, so that the spores cannot germinate. The resultant hydroxylated products are thus products of spore transformations solely.

Spore processes are combinations of two separate processes, spore production and steroid transformation. As spores can be prepared and stored indefinitely frozen and the spore transformations are conducted in simple buffer containing glucose, under non-sterile conditions, the process has much to offer.

Such a spore process using *Aspergillus ochraceus* NRRL 405 spores for the 11α-hydroxylation of progesterone has proven quite successful (KNIGHT, 1962; SCHLEG and KNIGHT, 1962) and has received considerable attention at the Ayerst Laboratories, Montreal. Spores of *A.ochraceus* successfully 11α-hydroxylate a variety of other steroids as well and spores of other hydroxylating microorganisms are also capable of hydroxylating progesterone and other steroids (VÉZINA et al., 1963, 1968; SEHGAL et al., 1968; HAFEZ-ZEDAN and PLOURDE, 1971; SEDLACZEK et al., 1981).

Spores of *A.ochraceus* were produced from inocula grown on Sabouraud Dextrose Agar and suspended in 0.1% sodium lauryl sulfonate. Using these suspensions as inocula, sporulation on barley (28 °C, 5–7 days) yielded $2.5 \cdot 10^{11}$ spores, on wheat bran $3.5 \cdot 10^{11}$ spores, thus significantly more. These spores could be stored for over one year at −20 °C, or for over 3 months at 4 °C without deterioration in their hydroxylation capacity (SINGH et al., 1968).

The spore transformation process simply involves aerated incubations of steroid substrate with spores in buffer containing glucose under non-sterile conditions and recovering products as usual. Essentially the same enzymic transformations are observed as with flourishing vegetative cells. Progesterone is 11α-hydroxylated by *A.ochraceus* spores, but depending on substrate levels and incubation times, the 6β,11α-dihydroxylated product is also produced (VÉZINA et al., 1963). Spore transformations are but little affected by age or storage of conidia, aeration or agitation, pH, chelating agents, or metal ions but do require glucose or xylose. Spores may also be recovered from the transformation incubation and reused.

One potential advantage of the spore process is that it may be possible through species and strain selection to operate 11α-hydroxylation process with progesterone as substrate without the usual complication of unwanted 6β-hydroxylation or other byproduct formation (ZEDAN et al., 1976).

In distinction to the recognized induction of steroid hydroxylases in *A.ochraceus* and other microorganisms, an induction of hydroxylase in *A.ochraceus* spores was not observed, as spores derived from vegetative cell cultures exposed to steroid substrate or not yielded the same active spores.

Spores of *A.ochraceus* have been used for 11α-hydroxylations on a large scale. Up to 1 kg charges of progesterone or Substance S in 5–750 L stainless steel fermentors have been successful, and in the case of the latter substrate 400 g charges of steroid with spores in 400 L buffer (pH 5.9) in 750 L tanks aerated at 26–28 °C repeatedly gave yields of 85–98% of the desired 11α-hydroxylated product (SINGH et al., 1968)!

The Ayerst spore process strictly involves steroid 11α-hydroxylations achieved without germination of the spores during the transformation phase. Inclusion of glucose in the incubation medium was to meet energy requirements of the transformations but the medium did not support germination. However, there should be no bar to operating the process with some germination occurring or with immobilized spores and germinated spores.

Just such a process has been described for *Curvularia lunata* ATCC 12017. In this case, *C.lunata* spores incorporated under sterile conditions into calcium alginate gels did not possess 11β-hydroxylase activity towards Substance S. However, spores immobilized in calcium alginate provoked to germinate by addition of nutrients gave good 11β-hydroxylation rates. By-products were also detected, and the process in general resembled that using flourishing vegetative cells. An advantage of immobilized spores is that the preparations can be stored for months at 4°C and then germinated and used for 11β-hydroxylations in the same manner as with free spores (OHLSON et al., 1980).

4. Cell-free hydroxylases

The ultimate in processing would seem to be the use of isolated specific steroid hydroxylases for the transformation of substrates to desired products. Conceptually, such operations would avoid unwanted side-reactions and offer process flexibility and simplicity. However, it has not proven feasible to use cell-free hydroxylases for the preparation of steroid products, as the preparation of the requisite enzymes has been too great a problem.

Part of the difficulty in preparing effective steroid hydroxylase systems has been their instability on isolation. Hydroxylation systems from both prokaryote and eukaryote organisms have been examined, the soluble system from prokaryotes being much more amenable to study than the membrane-bound hydroxylases of eukaryotes. Furthermore, the hope that isolated hydroxylases would be free from other enzymic activities has not materialized, except for the case of the 15β-hydroxylase from the prokaryote *Bacillus megaterium* ATCC 13 368.

The seven microorganisms from which steroid hydroxylases have been prepared are listed in Table 5. In all cases the cell-free systems transformed steroid substrates to the same metabolites as did parent vegetative cell cultures. New or unexpected reactions were not observed.

In all cases except for that from *B.megaterium* prior induction of the steroid hydroxylase during vegetative cell growth was necessary.

Although the promise of productive use of cell-free hydroxylases has not been fulfilled, studies with cell-free systems have been important in elucidation of the microbial hydroxylation process. It is assumed that an electron transport system involving a NADPH-dependent flavoprotein, an iron-sulfur protein, and cytochrome P-450 is implicated, and in the case of the steroid 15β-hydroxylase system of *B.megaterium* these three components have been demonstrated. Additionally, the iron-sulfur protein and cytochrome P-450 have been purified to homogeneity (BERG et al., 1979; BERG, 1982). The participation of cytochrome P-450 in the 11α-hydroxylase system of *Rhizopus nigricans* has also been demonstrated (BRESKVAR and HUDNIK-PLEVNIK, 1977a, 1977b).

Studies with cell-free hydroxylases suggest the complexity of hydroxylation mechanisms involved with the steroids. The cell-free hydroxylases from *Aspergillus niger, A. ochraceus, Streptomyces roseochromogenes, Curvularia lunata,* and *Rhizopus nigricans* retained other non-hydroxylase enzymic activities, and those from *A.niger, Bacillus megaterium* KM, *C.lunata,* and *R.nigricans* retained multiple site hydroxylase activities. Thus, these studies have not answered the fundamental question whether multiple hydroxylations arise through the participation of multiple enzymes or of one enzyme of diminished regioselectivity.

In the cell-free 11α-hydroxylase of *A. ochraceus* clear evidence for the actions of two distinct enzymes, the 11α-hydroxylase induced by progesterone and a 6β-hydroxylase induced separately by 11α-hydroxyprogesterone, has been obtained (SHIBAHARA et al., 1970). Furthermore, the *B.megaterium* cytochrome P-450 enzyme purified to protein homogeneity also shows only 15β-hy-

Table 5. Microbial Cell-free Steroid Hydroxylases

Microorganism	Substrate	Site	References
Prokaryotes:			
Streptomyces roseochromogenes ATCC 3347	Progesterone	16α-[a)]	ELIN and KOGAN, 1966; ELIN et al., 1970
Bacillus megaterium KM	Cortexone	15β- 15α- 6β- 11α-	WILSON and VESTLING, 1965
Bacillus megaterium ATCC 13368	Progesterone	15β- 15α-	BERG et al., 1975, 1976, 1979; BERG and RAFTER, 1981
Nocardia restrictus	Progesterone	9α-[b)]	CHANG and SIH, 1964
Eukaryotes:			
Aspergillus niger	Progesterone	11α-	NGUYEN-DANG et al., 1971
Aspergillus niger 12Y	Progesterone	11α- 11β- 21-	ABDEL-FATTAH and BADAWI, 1975a, 1975b
Aspergillus ochraceus NRRL 405	Progesterone	11α-[c)]	SHIBAHARA et al., 1970; TAN and FALARDEAU, 1970; JAYANTHI et al., 1982
	19-Nortestosterone	11α-	SHIBAHARA et al., 1970
Aspergillus ochraceus TS	Progesterone	11α-	GOSH and SAMANTA, 1981
Curvularia lunata NRRL 2380	Substance S	11β-[a)] 14α-	ZUIDWEG et al., 1962; ZUIDWEG, 1968
	19-Nortestosterone	10β-[d)] 11β- 14α-	ZUIDWEG, 1968; LIN and SMITH, 1970b
Rhizopus nigricans REF 129	Progesterone	11α- 6β- 17α- 21-	SALLAM et al., 1971
Rhizopus nigricans ATCC 6227b	Progesterone	11α- 6β-	BRESKVAR and HUDNIK-PLEVNIK, 1977a, 1977b

[a] 20-Ketone reductase also observed
[b] 9α,11α-Epoxidase also present
[c] 17(20)-Lyase observed by TAN and FALARDEAU, 1970
[d] 17β-Alcohol dehydrogenase also observed

droxylase activity with molecular oxygen, but with sodium periodate as oxygen atom donor, 15α-hydroxylation also occurs (BERG et al., 1979). However, sodium periodate supported hydroxylations with cell-free systems from *R.nigricans* gave only 11α-hydroxylations (HANISCH and DUNNILL, 1980).

Commercial interest in cell-free hydroxylases has waned, but this state could change given improvement of active cell-free enzymes.

C. By-Product Problems

As the steroids are multifunctional and have many unsubstituted sites as well, it is not surprising that unwanted side-reactions may occur during microbial hydroxylations. Of the three important 11α-, 11β-, and 16α-hydroxylations, 11α-hydroxylations appear to be more free from the problem.

1. 11α-Hydroxylation

For the 11α-hydroxylation of progesterone the preferred *Rhizopus nigricans* ATCC 6227b also 6β-hydroxylates and reduces the Δ^4-double bond of the desired product 11α-hydroxyprogesterone, but these side-products appear to be minor and do not interfere with refining steps. Other C_{21}-substrates may be similarly 11α-hydroxylated with but little 6β-hydroxylation (PETERSON et al., 1952).

Aspergillus ochraceus NRRL 405 also 11α-hydroxylates progesterone with subsequent 6β-hydroxylation and Δ^4-double bond reduction of initial product. The accompanying 6β-hydroxylase is a distinctly separate enzyme, dependent upon Zn^{2+} in the medium (DULANEY et al., 1955a, 1955b) and apparently induced by product 11α-hydroxyprogesterone and not by progesterone (SHIBAHARA et al., 1970).

Satisfactory yields and control of by-products have been achieved with *R.nigricans* and *A.ochraceus* by strain selection, control of fermentation medium and conditions, etc., so that commercial use of these organisms is not compromised by side-reactions.

2. 11β-Hydroxylation

The most extensively used microorganism for 11β-hydroxylation is *Curvularia lunata* NRRL 2380, and the by-products of its 11β-hydroxylation of Substance S are a serious problem so that much work has been directed to the elimination of unwanted by-products. By-products derive from additional modes of hydroxylation, including 11α- and 14α-monohydroxylations and 7α,14α- and 11β,14α-dihydroxylations together with 20-ketone reduction in the substrate. Other strains of *C.lunata* additionally 6β- and 9α-monohydroxylate Substance S (KONDO and MITSUGI, 1961) and 11-alcohol dehydrogenations to cortisone have also been noted (GARCIA-RODRIGUES et al., 1978).

With other substrates, *C.lunata* gives desired 11β-hydroxylation but also unwanted 6β-, 7α-, 9α-, and 14α-hydroxylations as well as 6β- and 11β-alcohol dehydrogenations (RASPÉ et al., 1964; KIESLICH et al., 1971). As these results were all obtained on strains selected for their potential 11β-hydroxylation, the extent of the problem of unwanted by-products may be realized.

The co-occurrence of several monohydroxylations of Substance S and of other substrates suggests that perhaps only one enzyme of diminished regiospecificity is involved and not several independent monohydroxylases (RIEMANN et al., 1968). The case has been examined via preparation of cell-free hydroxylases (ZUIDWEG et al., 1962; ZUIDWEG, 1968; LIN and SMITH, 1970b) and by differential hydroxylase inductions and kinetics (LIN and SMITH, 1970b, 1970c) but without resolution.

Nonetheless, the unwanted α-face by-product hydroxylations accompanying desired 11β-hydroxylation are diminished substantially by employing alternative substrates that are substituted on the α-faces by more bulky substituents. Thus, Substance S derivatives acetylated at the 17α-position and analogs substituted by 16α-methyl-, 5α-bromo-, and 5α,6α-epoxy features have been variously helpful (KIESLICH and WIEGLEPP, 1971; KIESLICH et al., 1971). Indeed, the commercial process for hydrocortisone manufacture from Substance S is to involve the 17α-acetate derivative as substrate (DEFLINES and VAN DER WAARD, 1966).

By contrast, added substituents on the β-face suppress 11β-hydroxylation by *C.lunata* and favor α-face attack (KIESLICH et al., 1969a, LIN et al., 1969).

Other effective 11β-hydroxylating microorganisms are beset by the same problems. *Absidia orchidis* 310 acting on Substance S yields a mixture of 11α- and 11β-hydroxylated products as well as cortisone, their common 11-alcohol dehydrogenated derivative (HANČ et al., 1961) and some 1β-hydroxylation (SCHWARZ et al., 1964). *Corticium sasakii* acting on Substance S gives 6β-, 11α-, 11β-, and 19-monohydroxylated products, 11-alcohol dehydrogenated product cortisone, and the 1-dehydrogenated product prednisolone (HAGIWARA, 1960). *Cunninghamella blakesleena* ATCC 8688 11β-hydroxylates Substance S to hydrocortisone but also 6β- and 14α-monohydroxylates the substrate and dehydrogenates hydrocortisone to cortisone (SPALLA et al., 1962). Another *C.blakesleena* strain gives 11α- and 11β-hydroxylation, 11-alcohol dehydrogenation to cortisone, 20-ketone reduction of the cortisone, and 1-dehydrogenation of product hydrocortisone (GARCIA-RODRIGUES et al., 1978). *Tieghemella orchidis* 233 acting on Substance S 21-acetate gives 6β-, 11α-, and (wanted) 11β-hydroxylation but also 1-dehydrogenation of both substrate and primary product hydrocortisone (MOGIL'NITSKII et al., 1975).

3. 16α-Hydroxylation

A similar problem is associated with the 16α-hydroxylation of 9α-fluorohydrocortisone by *Streptomyces roseochromogenes* in the Squibb process for triamcinolone production. Because of unsatisfactory material losses early in the fermentation a thorough examination of the process has been made. Besides the desired 16α-hydroxylation, unwanted 2β-monohydroxylation and 2β,16α-dihydroxylation occurred, with D-homoannulation of the 16α-hydroxylated products taking place non-enzymically. Additionally, 20-ketone reduction of substrate and of product 9α-fluoro-16α-hydroxyhydrocortisone occurred, together with a unique transformation of substrate to 21-acetylamino-11β,17α-dihydroxypregn-4-ene-3,20-dione (Fig. 15) (GOODMAN and SMITH, 1960, 1961; SMITH et al., 1960b, 1961, 1962b). Traces of 1-dehydrogenation of the major product, yielding thereby triamcinolone, were also found. Additionally, where 9α-fluorohydrocortisone 21-acetate was used as substrate, an active 21-esterase action was observed.

These several transformation products from six processes (hydroxylation, 20-ketone reduction, 1-dehydrogenation, D-homoannulation, 21-amination, 21-ester hydrolysis) show *S.roseochromogenes* to be among the most metabolically varied microorganisms with respect to steroid transformations.

The alternative triamcinolone process employing 9α-fluoroprednisolone as substrate yields triamcinolone directly by 16α-hydroxylation with *S.roseochromogenes*. Although about the same yield of desired product could be recovered, the by-product spate was at least as complicated as that using 9α-fluorohydrocortisone. 2β-Hydroxylation did not occur, but reduction of the Δ^1-double bond was observed. 16α-Hydroxylations of other substrates, including hydrocortisone, 9α-fluoro-2β-hydroxyhydrocortisone, and 6α,9α-difluorohydrocortisone were of comparable complexity with respect to detected by-products.

Problems associated with these 16α-hydroxylations have been addressed by medium changes, strain selections, process variations, and by coupling the 16α-hydroxylation with the subsequent 1-dehydrogenation. The unwanted D-homoannulation of 16α-hydroxylated products catalyzed by iron was suppressed by inclusion of dipotassium hydrogen phosphate in media (GOODMAN and SMITH, 1960, 1961). Strain selection greatly reduced the unwanted 2β-hydroxylation, which in this case clearly involves a separate and distinct enzyme from the 16α-hydroxylase (GOODMAN and SMITH, 1961).

The accompanying 20-ketone reduction reaction observed in vegetative cell fermentations also occurred using cell-free preparations of the 16α-hydroxylase (ELIN and KOGAN, 1966; ELIN et al., 1970), but by harvesting products at maximum levels of 9α-fluoro-16α-hydroxyhydrocortisone, 20-ketone reduction was minimized.

By these means, yields of 9α-fluoro-16α-hydroxyhydrocortisone up to 70% were obtained, with finished product containing only low levels of by-products. Thus, using a paper chromatography assay, levels averaged 0.1% unaltered substrate, 3.0% 21-acetylamino-9α-fluoro-11β,17α-dihydroxypregn-4-ene-3,20-dione, 1.6% D-homoannulated product, and 0.2% 9α-fluoro-11β,17α,20,21-tetrahydroxypregn-4-en-3-one over several runs.

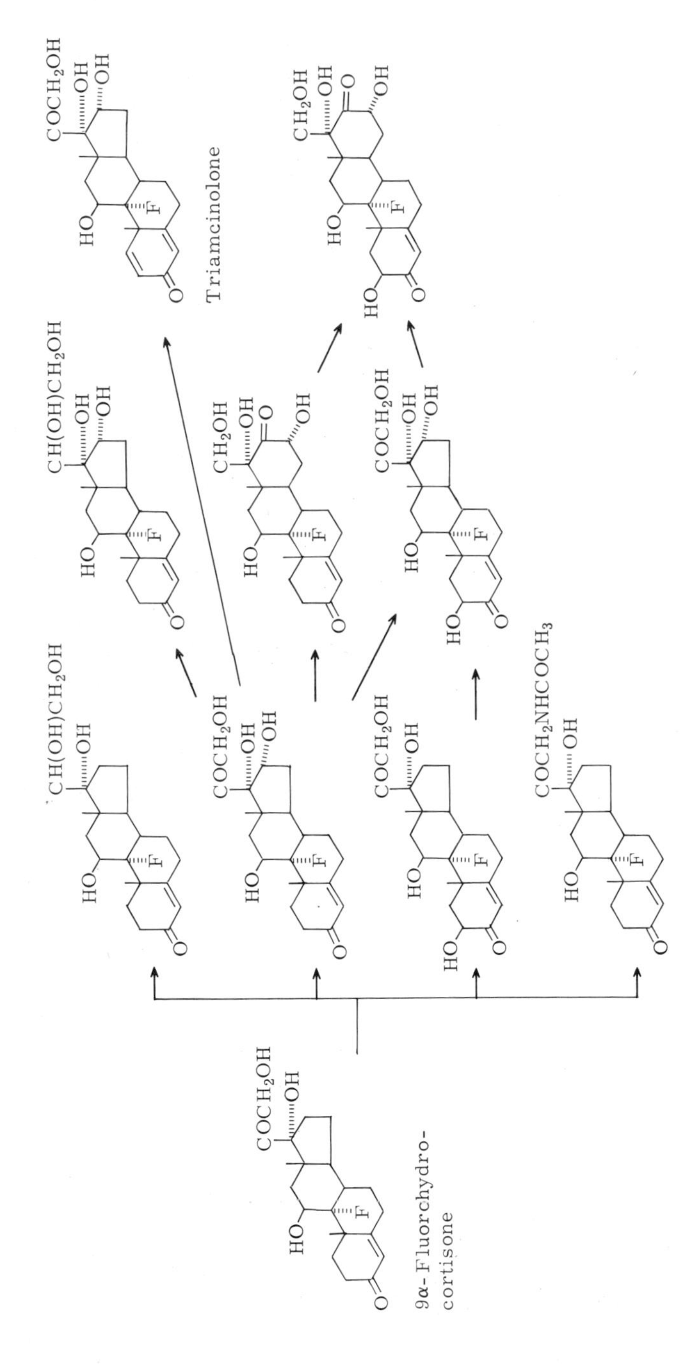

Figure 15. By-products of microbial 16α-hydroxylation.

VI. 1-Dehydrogenations

Microbial 1-dehydrogenations are essential to the manufacture of corticosteroid analogs, as chemical dehydrogenation methods do not compete effectively. The transformation is possible at several stages but in most cases is a terminal one, yielding finished drug products. Dehydrogenations are quite common among bacteria, with Δ^4-3-ketones, saturated 3-ketones, and Δ^5-3β-alcohols serving as substrates. Additionally, side-chain and B-ring degradations occur frequently, these unwanted transformations necessarily making the organism unsuitable for manufacture.

A. Commercial 1-Dehydrogenations

A large number of bacteria possess 1-dehydrogenating activities free from undesirable degradation activities, those with superior utility being *Arthrobacter (Corynebacterium) simplex* ATCC 6946, *Bacterium cyclooxydans* ATCC 12 673, *Bacillus lentus* ATCC 13 805, *Bacillus sphaericus* ATCC 7055, *Mycobacterium globiforme* 193, *Septomyxa affinis* ATCC 6737, and ATCC 13 425. Many other organisms, including other *Mycobacterium* and several *Nocardia* species, also appear to have superior 1-dehydrogenation capacities. In all cases, the 1-dehydrogenase enzymes are inducible.

In practice only three or four 1-dehydrogenators are used in manufacture. *Arthrobacter simplex* is very versatile and is used in prednisolone and triamcinolone manufacture. The use of *Septomyxa affinis* in the manufacture of medrol is indicated (SEBEK and PERLMAN, 1979); in this instance, the 1-dehydrogenation appears to be conducted at an early synthesis stage on 11β,21-dihydroxypregna-4,17(20)-dien-3-one ("dienediol") (REUSSER et al., 1961; CHEN et al., 1965). In the Soviet Union *Mycobacterium globiforme* appears to be the preferred organism (KRASIL'NIKOV et al., 1959).

Yet another organism in use is *Bacillus lentus*. Interestingly, the 1-dehydrogenation by *B.lentus* of hydrocortisone 17α-acetate yields prednisolone 17α-acetate (CEJKA, 1976).

Strain improvement of the key 1-dehydrogenators do not appear to be as important as in the case of hydroxylating organisms. High yields and minimal by-products are attained satisfactorily by control of fermentation conditions.

Media for 1-dehydrogenating organisms must contain nitrogen and carbohydrate sources. *Arthrobacter simplex* is maintained on 1% yeast extract-dextrose agar, and cultures may be inoculated into 3% yeast extract buffered at pH 6.9 for initial growth. Other media include 1% casein hydrolysate – 0.1% yeast extract, 1% fish solubles – 0.1% yeast extract, 0.1% yeast extract, and 0.5% peptone – 0.5% corn steep – 0.5% glucose, with various added salts and all adjusted to pH 6.8–7.0 (NOBILE, 1958; HERZOG et al., 1962). A 1% yeast extract – 1% glucose medium is useful for *Mycobacterium globiforme,* but the organism also grows well on synthetic medium (KRASIL'NIKOV et al., 1959).

1-Dehydrogenations are conducted with growing cultures of the selected organism, and all parameters important for microbial hydroxylations are important for 1-dehydrogenations. After suitable growth (12 h), substrate is added as a dimethylformamide solution to give 0.2–0.3 g/L substrate levels and fermentation is continued until conversions are above 90% assay (3–48 h). Fermentation temperature is usually higher than for hydroxylations, ranging from 25–37 °C (NOBILE, 1958).

Products are recovered by solvent extraction with methyl isobutyl ketone in the same manner as for hydroxylations.

Several operational innovations distinguish 1-dehydrogenations from hydroxylations. For instance, 1-dehydrogenations of 11β,21-dihydroxypregna-4,17(20)-dien-3-one by *Septomyxa affinis* is substantially accelerated by adding small amounts of another steroid such as testosterone or 3-oxo-bisnorchol-4-en-22-al (MURRAY and SEBEK, 1959; KOEPSELL, 1962). Multi-stage operation with *Septomyxa affinis,* with continuous feed of substrate at high levels, has been attained (REUSSER et al., 1961; CHEN et al., 1965). A two-stage continuous culture system for triamcinolone production using *Arthrobacter simplex* has also been described (RYU and LEE, 1975).

Yet another innovative means of microbial 1-dehydrogenation is available. "Pseudocrystallofermentation" is a term describing a variant process of 1-dehydrogenation with *A.simplex* in which very high levels, up to 500 g/L (!), of finely powdered substrate hydrocortisone are used. Only dissolved substrate appears to be transformed, and the product as it forms crystallizes from solution, thereby suggesting the equilibrium:

$$\underset{\text{(solid)}}{\text{Substrate}} \rightleftharpoons \underset{\text{(solution)}}{\text{Substrate}} \rightleftharpoons \underset{\text{(solution)}}{\text{Product}} \rightleftharpoons \underset{\text{(solid)}}{\text{Product}}$$

The transformation occurs slowly over 1-5 days but in high yield (to 97%) and is monitored by microscopy, the crystalline product prednisolone being readily distinguished from solid substrate (KONDO and MASUO, 1961). The technique also applies to 1-dehydrogenations of up to 400 g/L hydrocortisone using *M.globiforme* 193 (KOSHCHEYENKO et al., 1975).

Product recovery in these cases merely requires filtration of mycelium and crystalline prednisolone and recrystallization of the product.

B. Alternative Processes

As with hydroxylations, there are process variations that can be operated in which growing vegetative cell cultures are not directly used. Although some of these alternatives offer promise in process simplification, none is presently sufficiently developed to be competitive with vegetative cell cultures.

1. Acetone-dried vegetative cells

Cells from organisms capable of steroid 1-dehydrogenations are readily freeze-dried or dried with acetone and used as such for incubations with substrate. Acetone-dried cells of *Arthrobacter simplex* ATCC 6946 are particularly valuable in this manner (ERICKSON et al., 1967), but acetone-dried cells of *Bacillus lentus* are also useful (KIESLICH et al., 1971).

Incubations using dried cells are conducted in neutral buffered water containing steroid substrate and a synthetic electron acceptor such as menadione or phenazine methosulfate. The acetone-dried cells are quite useful for special preparations, but the process does not compete with vegetative cell fermentations for manufacture.

2. Spore processes

As with microbial hydroxylations, so also 1-dehydrogenations may be achieved using spores instead of vegetative cells. Demonstration of 1-dehydrogenations with spores of *Fusarium solani, Septomyxa affinis,* and *Streptomyces lavendulae* (VÉZINA et al., 1968; SINGH et al., 1968) has been made. Sporulation of *S.affinis* ATCC 6737 on barley or bran, with spore incubations at 25-28°C in aerated 5 L fermentors containing 2 g/L steroid substrate in non-sterile water led to about 85% transformation in 80 h (SINGH et al., 1968). Spore fermentations do not presently compare favorably with standard fermentations using vegetative cell cultures.

3. Cell-free enzymes

Cell-free 1-dehydrogenases from organisms such as *Pseudomonas testosteroni* and *Nocardia restrictus* have been investigated for fundamental interests, and similar cell-free enzymes have been described from the commercially useful *Bacillus sphaericus* (RINGOLD et al., 1963), *Bacterium cycloox-*

ydans (IIDA et al., 1965), *Mycobacterium globiforme* (LESTROVAYA et al., 1965), and *Arthrobacter simplex* (PENASSE and PEYRE, 1968; MOSBACH and LARSSON, 1970). The 1-dehydrogenases are inducible and competent in the 1-dehydrogenation of substrates, but their use for preparation of products is not indicated.

4. Immobilized cells and enzymes

The cell-free 1-dehydrogenase from *A.simplex* previously induced with hydrocortisone can be incorporated into polyacrylamide gels and used to transform hydrocortisone to prednisolone. Both batch and column operations were demonstrated, but it was necessary to incorporate an added electron acceptor such as phenazine methosulfate into the systems (MOSBACH and LARSSON, 1970).

However, it is not necessary to use isolated 1-dehydrogenase, as *A.simplex* vegetative, acetone-dried, or lyophilized cells can be incorporated into gels and used in similar operations. Polyacrylamide and calcium alginate gels transform hydrocortisone in simple incubations, but incubations in nutrient media allowing mycelium growth are more effective. Added synthetic electron acceptors are also stimulatory in these systems. The gels can be reused, in semi-continuous operations, and in the pseudocrystallofermentation mode, where 9 g/L hydrocortisone conversions of 80% were repeatedly obtained. Product recovery in this process simply involved screening out the gel and direct filtration of crystalline prednisolone (OHLSON et al., 1978, 1979; LARSSON et al., 1979).

Very good results have also been reported using *Mycobacterium globiforme (Arthrobacter globiformis)* 193 cells immobilized in several media for repeated conversions (95% transformation) of hydrocortisone to prednisolone (KOSHCHEYENKO et al., 1983).

Incorporation of dried *A.simplex* cells into mixed polyethyleneglycol and polypropyleneglycol polymers (SONOMOTO et al., 1979) and into collagen (CONSTANTINIDES, 1980) has also been successful. Although these approaches do not yet compete with established processes, their promise is obvious.

C. By-Product Problems

Microorganisms capable of steroid 1-dehydrogenations are often also capable of degradations of a side-chain and the steroid nucleus, but organisms of biotechnological interest have been selected that are free from these undesired complications. However, the selected organisms do have two accompanying activities leading to unwanted by-products, namely reduction of the 20-ketone of product or substrate and reduction of the newly introduced Δ^1-double bond. Thus, 1-dehydrogenation is reversible in practice.

Residual substrate in fermentations may be a steroid that was never transformed but may also be a reduction product of the original 1-dehydrogenation product. In unfavorable cases the ultimate product of such operations could be the 20-ketone reduction product of the original substrate (GOODMAN et al., 1960).

Control of these problems is achieved through control of fermentation conditions. The reversible 1-dehydrogenation is best maintained in the forward direction to product by adequate aeration. Low aeration rates or delay in refining (in unaerated condition) can result in substantial reversion of product to substrate. Sterilization of harvested broth also stops this reversion.

Adequate aeration is required for both 1-dehydrogenation and 20-ketone reduction, so control of unwanted 20-ketone reduction is best achieved via medium composition. Relatively lean media tend to suppress the reduction; rich media support the transformation (GOODMAN et al., 1960; SMITH et al., 1960a). It is also possible to inhibit selectively the 20-ketone reduction by added sodium iodoacetate (ROSS, 1962).

VII. Controlled Multiple Transformations

Inasmuch as microbial hydroxylations and 1-dehydrogenations are necessary for corticosteroid analog manufacture, various means of combining the two steps have been tried. Ordered sequential fermentations with vegetative cells have succeeded in yielding desired doubly transformed products (RYU et al., 1969), and modifications using resting cells or spores of hydroxylating organisms and acetone-dried cells of the 1-dehydrogenating organism *Arthrobacter simplex* have been described (LEE et al., 1971).

A preferred process involving sequential transformation of the 16α-hydroxylated derivative of Substance S solubilized to the extent of 200 mg/mL as the 16α,17α-cycloborate derivative has been described. The substrate was first 11α-hydroxylated using resting cells of *Aspergillus ochraceus* and then 1-dehydrogenated with acetone-dried cells of *Arthrobacter simplex* (LEE et al., 1971).

Yet better is the possibility of conducting the two microbial steps in one fermentation using mixed cultures of hydroxylating and 1-dehydrogenating organisms. In mixed fermentations it is essential that the two organisms be compatible and induce the desired enzymes in one another's presence. Satisfactory enzyme inductions occur with hydroxylators *Absidia coerulea, Curvularia lunata,* and *Streptomyces roseochromogenes* grown with the 1-dehydrogenator *A.simplex.* In these systems the unwanted 20-ketone reductase activity accompanying *A.simplex* 1-dehydrogenations and *S.roseochromogenes* and *C.lunata* hydroxylations was repressed, thereby providing an unexpected benefit. In this manner mixed cultures of *S.roseochromogenes* and *A.simplex* acting on 9α-fluorohydrocortisone gave triamcinolone directly without recovery of intermediates (LEE et al., 1969; RYU et al., 1969).

These double transformations may involve both ordered sequential steps and competing reactions. In the transformation of 9α-fluorohydrocortisone to triamcinolone, the process involves 9α-fluoroprednisolone as an intermediate, thus demonstrating that 1-dehydrogenation precedes 16α-hydroxylation. However, competing 11β-hydroxylations and 1-dehydrogenations occurred using *C.lunata* and *A.simplex* mixed cultures (RYU et al., 1969).

VIII. Lactonization

Microbial processes involving carbon-carbon bond scissions are important in the manufacture of steroid intermediates from sterols and in the relatively minor matter of synthesis of 1-dehydrotestololactone (see Fig. 8), a drug used in the adjunctive and palliative treatment of inoperable breast cancer in women. Although chemical processes are also available for the synthesis, the microbial degradation of progesterone by *Cylindrocarpon radicicola* ATCC 11011 is advantageous.

The degradation proceeds via a Baeyer-Villiger type oxidation as previously discussed to 1-dehydrotestololactone (FRIED et al., 1953; FRIED and THOMA, 1956). The oxidative degradation also occurs with Substance S and testosterone, but no advantage occurs over use of progesterone.

Other microorganisms also degrade progesterone in like manner to testololactone and/or 1-dehydrotestololactone. These include *Aspergillus, Cephalosporium, Fusarium, Gliocladium, Penicillium,* and *Streptomyces* species.

IX. Stereospecific Reduction of Prochiral Steroids

Total synthesis is now an established means of production of 19-norsteroid progestational agents, even though the products thereby obtained are racemic mixtures of the wanted active steroid of natural configuration and the unwanted optical enantiomer. Resolution of the racemic mixtures has been advantageously achieved by microbial fermentation technology.

Very early work directed to the total synthesis of naturally occurring steroid hormones demonstrated that microbial hydroxylation, 1-dehydrogenation, and 17-ketone reduction could resolve racemic mixtures, only the steroid eantiomer of natural configuration being metabolized (VISCHER et al., 1956; WETTSTEIN et al., 1958). The same transformations were also successful for resolutions of racemic 19-norsteroids, but racemic products were also encountered. Hydroxylations with *Aspergillus ochraceus* or *Curvularia lunata* (SMITH et al., 1966; LIN et al., 1969) and 17β-alcohol and 1-dehydrogenations with *Arthrobacter simplex* and other organisms (GREENSPAN et al., 1966) gave resolved but also unresolved products, vitiating such use commercially.

However, resolution of selected total synthesis intermediates has proven successful. Stereospecific reduction by *Rhizopus arrhizus* of the methyl 7-(2′,5′-dioxo-1′-methylcyclopentan-1′-yl)-5-oxoheptanoate intermediate in the VELLUZ total synthesis followed by cyclizations and other steps (BELLET et al., 1966) and reductions by *Saccharomyces* yeast of the intermediate 3-methoxy-8,14-secoestra-1,3,5(10),9(11)-tetraene-14,17-dione from the TORGOV total synthesis followed by cyclizations and other steps (KOSMOL et al., 1967; RUFER et al., 1967b) yield resolved estradiol 3-methyl ether and derivative 19-norsteroid progestational agents of natural configuration (Fig. 16).

Species and strain selection has been of crucial importance in improving the stereospecific reductions of total synthesis intermediates. The yeasts *Saccharomyces cerevisiae, Saccharomyces carlsbergensis, Saccharomyces pastorianus,* and *Saccharomyces uvarum* are useful organisms, with *S.uvarum* giving yields of 75% of product (RUFER et al., 1967a). Moreover, *S.uvarum* and *Saccharomyces chevalieri* which gives up to 80% yields appear to be useful in manufacture of the resolved contraceptive D-norgestrel

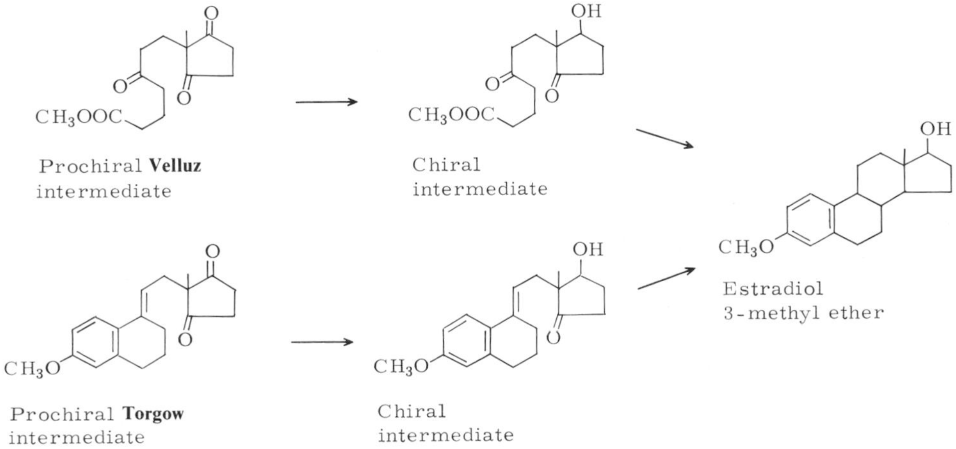

Figure 16. Total syntheses involving microbial introduction of chiralities into prochiral structures.

(Fig. 17) available only via total synthesis (KIESLICH, 1980a,d).

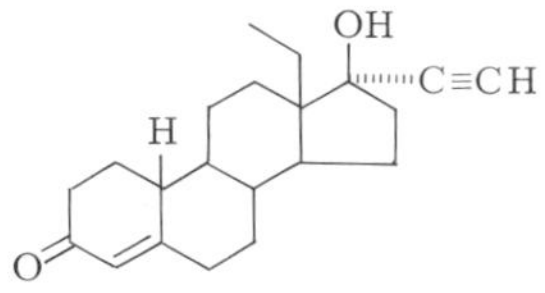

Figure 17. D-Norgestrel.

Other process improvements include use of *S.cerevisiae* VKMu-488 cells immobilized in polyacrylamide gels for resolutions of the TORGOV intermediate, where yields up to 90% are obtained (GULAYA et al., 1979), and of *Schizosaccharomyces pombe* ATCC 2476 enhanced by additions of allyl alcohol for resolution of the VELLUZ intermediate in 92% yield (LANZILOTTA et al., 1975).

X. Future Developments

The manufacture of steroids by present biotechnological means is achieved with established processes operating at high yield and controlled costs, with no apparent demands for changes. As these highly developed processes meet current medical needs, innovative new work and expansion of facilities is now limited. Moreover, in the absence of new steroid products requiring new biotechnology or of shortages of raw materials necessary to current production processes it appears that minimal developments in the mature field are warranted, as biotechnological resources are presently directed to processes and products of limited availability and greater cost.

An additional factor for change might be the cost of medium components or of electric power, these being important cost items in manufacture (KIESLICH, 1980d).

Applications of the newer concepts of genetic engineering of microorganisms for their improvement as steroid transforming agents or as agents for *de novo* steroid biosynthesis have not yet appeared in the literature (VALENTINE et al., 1982). However, several U.S. fermentation drug companies are entering the field of genetic engineering, and applications to steroid biotechnology may be anticipated.

Of obvious importance would be the genetic engineering of new microbial agents that would biosynthesize sterols, sapogenins, or steroidal alkaloids as raw materials or of new steroids not now recognized as raw materials and intermediates. Another major advance would be creation of an organism with single, specific hydroxylase that would offer further cost reductions in established hydroxylation fermentations or permit hydroxylations with immobilized cells to be more effective.

XI. References

ABDEL-FATTAH, A. F., and BADAWI, M. A. (1975a). J. Gen. Appl. Microbiol. *21,* 217.

ABDEL-FATTAH, A. F., and BADAWI, M. A. (1975b). J. Gen. Appl. Microbiol. *21,* 225.

AKHREM, A. A., and TITOV, Yu. A. (1965). "Mikrobiologicheskie Transformatsii Steroidov" (Microbiological Transformations of Steroids). U.S.S.R. Academy of Sciences, Moscow.

AKHREM, A. A., and TITOV, Yu. A. (1970). "Steroidy i Mikroorganizmy" (Steroids and Microorganisms). Nauka Publishing House, Moscow.

ALIG, L., FUERST, A., and MUELLER, M. (1975). U.S. Patent 3 939 193; German Patent 2 445 817.

ALIG, L., MUELLER, M., WIECHERT, R., NICKOLSON, R., FUERST, A., KERB, U., and KIESLICH, K. (1976). German Patent 2 614 079.

AMBRUS, G., SZARKA, E., BARTA. I., HORVATH, Gy., RADICS, L., and KAJTAR, M. (1975). Steroids *25,* 99.

APPLEZWEIG, N. (1959). Chem. Week (January 31), 38.

APPLEZWEIG, N. (1969). Chem. Week (May 17), 58.

APPLEZWEIG, N. (1974). Chem. Week (July 10), 31.

APPLEZWEIG, N. (1977). In "Crop Resources", pp. 149–163. Academic Press, New York.

BATES, M. L., REID, W. W., and WHITE, J. D. (1976). J. Chem. Soc. Chem. Commun., 44.

BELIČ, I., and KARLOVŠEK, M. (1981). J. Steroid Biochem. *14,* 229.

BELIČ, I., KOMEL, R., and SOČIČ, H. (1977). Steroids *29,* 271.

BELLET, P., NOMINÉ, G., and MATHIEU, J. (1966). C. R. Acad. Sci. Ser. C *263,* 88.

BENN, W. R., TIBERI, R., and NUSSBAUM, A. L. (1964). J. Org. Chem. *29,* 3712.

BENSON, A. M., JARABAK, R., and TALALAY, P. (1971). J. Biol. Chem. *246,* 7514.

BERG, A. (1982). Biochem. Biophys. Res. Commun. *105,* 303.

BERG, A., CARLSTRÖM, K., GUSTAFSSON, J.-Å., and INGELMAN-SUNDBERG, M. (1975). Biochem. Biophys. Res. Commun. *66,* 1414.

BERG, A., GUSTAFSSON, J.-Å., INGELMAN-SUNDBERG, M., and CARLSTRÖM, K. (1976). J. Biol. Chem. *251,* 2831.

BERG, A., INGELMAN-SUNDBERG, M., and GUSTAFSSON, J.-Å. (1979). J. Biol. Chem. *254,* 5264.

BERG, A., and RAFTER, J. J. (1981). Biochem. J. *196,* 781.

BERNSTEIN, S., LENHARD, R. H., ALLEN, W. S., HELLER, M., LITTELL, R., STOLAR, S. M., FELDMAN, L. I., and BLANK, R. H. (1959). J. Am. Chem. Soc. *81,* 1689.

BEUKERS, R., MARX, A. F., and ZUIDWEG, M. H. J. (1972). In "Drug Design" (E. J. ARIËNS, ed.), Vol. 3, pp. 1–131. Academic Press, New York.

BLOOM, B. M., and SHULL, G. M. (1955). J. Am. Chem. Soc. *77,* 5767.

BONDI, S. (1908). Wien. Klin. Wochenschr. *21,* 271.

BOUL, A. D., BLUNT, J. W., BROWNE, J. W., KUMAR, V., MEAKINS, G.D., PINHEY, J. T., and THOMAS, V. E. M. (1971). J. Chem. Soc. (C), 1130.

BRAHAM, R. L., DALE, S. L., and MELBY, J. C. (1975). Steroids *26,* 697.

BRESKVAR, K., and HUDNIK-PLEVNIK, T. (1977a). Croat. Chem. Acta *49,* 207.

BRESKVAR, K., and HUDNIK-PLEVNIK, T. (1977b). Biochem. Biophys. Res. Commun. *74,* 1192.

BRESKVAR, K., and HUDNIK-PLEVNIK, T. (1978). J. Steroid Biochem. *9,* 131.

BRIDGEMAN, J. E., CHERRY, P. C., CLEGG, A. S., EVANS, J. M., JONES, E. R. H., KASAL, A., KUMAR, V., MEAKINS, G. D., MORISAWA, Y., RICHARDS, E. E., and WOODGATE, P. D. (1970). J. Chem. Soc. (C), 250.

CAMERINO, B., and VERCELLONE, A. (1956). Gazz. Chim. Ital. *86,* 260.

ČAPEK, A., HANČ, O., and TADRA, M. (1966). "Microbial Transformations of Steroids" Publishing House of the Czechoslovak Academy of Sciences, Prague.

CARLSTRÖM, K. (1973). Acta Chem. Scand. *27,* 1622.

CEJKA, A. (1976). Eur. J. Appl. Microbiol. *3,* 145.

CHANG, F. N., and SIH, C. J. (1964). Biochemistry *3,* 1551.

CHARNEY, W., and HERZOG, H. L. (1967). "Microbial Transformations of Steroids. A Handbook". Academic Press, New York/London.

CHEN, J. W., HILLS, F. J., KOEPSELL, H. J., and MAXON, W. D. (1965). Ind. Eng. Chem. (Process Design Develop.) *4,* 421.

CHERRY, P. C., JONES, E. R. H., and MEAKINS, G. D. (1966). J. Chem. Soc. Chem. Commun., 587.

CHUN, Y. Y., IIDA, M., and IIZUKA, H. (1981). J. Gen. Appl. Microbiol. *27,* 505.

COLINGSWORTH, D. R., BRUNNER, M. P., and HAINES, W. J. (1952). J. Am. Chem. Soc. *74,* 2381.

CONSTANTINIDES, A. (1980). Biotech. Bioeng. *22,* 119.

CORONELLI, C., KLUEPFEL, D., and SENSI, P. (1964). Experientia *20,* 208.

DE FLINES, J. (1969). In "Fermentation Advances" (D. PERLMAN, ed.), pp. 385–390. Academic Press, New York/London.

DE FLINES, J., MARX, A. F., VAN DER WAARD, W. F., and VAN DER SIJDE, D. (1962). Tetrahedron Lett., 1257.

DE FLINES, J., and VAN DER WAARD, F. (1966). Dutch Patent 66 05 514.

DODSON, R. M., and MUIR, R. D. (1958). J. Am. Chem. Soc. *80,* 5004.

DULANEY, E. L., MCALEER, W. J., KOSLOWSKI, M., STAPLEY, E. O., and JAGLOM, J. (1955a). Appl. Microbiol. *3,* 336.

DULANEY, E. L., STAPLEY, E. O., and HLAVAC, C. (1955b). Mycologia *47,* 464.

DULANEY, E. L., and STAPLEY, E. O. (1959). Appl. Microbiol. *7,* 276.

ELIN, E. A., and KOGAN, L. M. (1966). Dokl. Akad. Nauk S. S. S. R. *167,* 1175.

ELIN, E. A., KOGAN, L. M., TARASOV, O. S., and TORGOV, I. V. (1970). Khim. Prirod. Soed., 47.

EPPSTEIN, S. H., and LEIGH, H. M. (1956). U.S. Patent 2 759 004.

ERICKSON, R. C., BROWN, W. E., and THOMA, R. W. (1967). U.S. Patent 3 360 439.

FONKEN, G. S., and JOHNSON, R. A. (1972). "Chemical Oxidations with Microorganisms". Marcel Dekker Inc., New York.

FRIED, J., and THOMA, R. W. (1956). U.S. Patent 2 774 120.

FRIED, J., THOMA, R. W., GERKE, J. R., HERZ, J. F., DONLIN, M. N., and PERLMAN, D. (1952). J. Am Chem. Soc. *74,* 3962.

FRIED, J., THOMA, R. W., and KLINGSBERG, A. (1953). J. Am. Chem. Soc. *75,* 5764.

GARCIA-RODRIGUES, L. K., KOROBAVA, Yu. N., MEDVEDEVA, I. V., SHNER, V. F., and MESSINOVA, O. V. (1978). Khim.-Farm. Zhur. *12*(9), 95.

GARDNER, J. N., CARLON, F. E., ROBINSON, C. H., and OLIVETO, E. P. (1966). Steroids *7,* 234.

GHOSH, D., and SAMANTA, T. B. (1981). J. Steroid Biochem. *14,* 1063.

GIBB, W., and JEFFREY, J. (1971). Eur. J. Biochem. *23,* 336.

GLASS, T. L., WINTER, J., BOKKENHEUSER, V. D., and HYLEMON, P. B. (1982). J. Lipid Res. *23,* 352.

GOODMAN, J. J., MAY, M., and SMITH, L. L. (1960). J. Biol. Chem. 235, 965.

GOODMAN, J. J., and SMITH L. L. (1960). Appl. Microbiol. *8,* 363.

GOODMAN, J. J., and SMITH, L. L. (1961). Appl. Microbiol. *9,* 372.

GREENSPAN, G., REES, R. W. A., LINK, G. D., BOYD, C. P., JONES, R. C., and ALBURN, H. E. (1974). Experientia *30,* 328.

GREENSPAN, G., SMITH, L. L., REES, R., FOELL, T., and ALBURN, H. E. (1966). J. Org. Chem. *31,* 2512.

GULAYA, V. E., ANANCHENKO, S. N., TORGOV, I. V., KOSHCHEYENKO, K. A., and BYCHKOVA, G. G. (1979). Bioorg. Khim. *5,* 768.

HAFEZ-ZEDAN, H., and PLOURDE, R. (1971). Appl. Microbiol. *21,* 815.

HAGIWARA, H. (1960). J. Pharm. Soc. Jpn. *80,* 1667.

HANČ, O., ČAPEK, A., and KAKÁČ, B. (1961). Folia Microbiol. *6,* 392.

HANČ, O., ČAPEK, A., and TADRA, M. (1962). Česk. Farm. *11,* 181.

HANISCH, W. H., and DUNNHILL, P. (1980). Biotechnol. Lett. *2,* 123.

HANISCH, W. H., DUNNHILL, P., and LILLY, M. D. (1980). Biotech. Bioeng. *22,* 555.

HANSON, F. R., MANN, K. M., NIELSON, E. D., ANDERSON, H. V., BRUNNER, M. P., KARNEMAAT, J. N., COLINGSWORTH, D. R., and HAINES, W. J. (1953). J. Am. Chem. Soc. *75,* 5369.

HAYAKAWA, S. (1973). Adv. Lipid Res. *11,* 143.

HECHTER, O., JACOBSON, R. P., JEANLOZ, R., LEVY, H., MARSHALL, C. W., PINCUS, G., and SCHENKER, V. (1949). J. Am. Chem. Soc. *71,* 3261.

HERZOG, H. L., PAYNE, C. C., HUGHES, M. T., GENTLES, M. J., HERSHBERG, E. B., NOBILE, A., CHARNEY, W., FEDERBUSH, C., SUTTER, D., and PERLMAN, P. L. (1962). Tetrahedron *18,* 581.

HOLLAND, H. L. (1982). Chem. Soc. Rev. *11,* 371.

HOLMLUND, C. E., SAX, K. J., NIELSEN, B. E., HARTMAN, R. E., EVANS, R. H., and BLANK, R. H. (1962). J. Org. Chem. *27,* 1468.

HÖRHOLD, C., GROH, H., DÄNHARDT, S., LESTROVAJA, N. N., and SCHUBERT, K. (1977). J. Steroid Biochem. *8,* 701.

HORVÁTH, J., and KRÁMLI, A. (1947). Nature (London) *160,* 639.

HÜBENER, H. J., and LEHMANN, C. O. (1958). Z. Physiol. Chem. *313.* 124.

IIDA, M., TOWNSLEY, J. D., HAYANO, M., and BRODIE, H. J. (1965). Steroids Suppl. I, 159.

IIDA, M., SHINOZUKA, T., and IIZUKA, H. (1979). Z. Allg. Mikrobiol. *19,* 557.

IIZUKA, H., and NAITO, A. (1967). "Microbial Transformation of Steroids and Alkaloids", pp. 3–250. University of Tokyo Press, Tokyo/ University Park Press, State College, PA.

IVASHKIV, E. (1971). Biotech. Bioeng. *13,* 561.

JAYANTHI, C. R., MADYASTHA, P., and MADYASTHA, K. M. (1982). Biochem. Biophys. Res. Commun. *106,* 1262.

JONES, E. R. H. (1973). Pure Appl. Chem. *33,* 39.

KAROW, E. O., and PETSIAVAS, D. N. (1956). Ind. Eng. Chem. *48,* 2213.

KIESLICH, K. (1969). Synthesis, 120.

KIESLICH, K. (1980a). Biotechnol. Lett. *2,* 211.

KIESLICH, K. (1980b). Bull. Soc. Chim. Fr. II, 9.

KIESLICH, K. (1980c). In "Microbial Enzymes and Bioconversions. Economic Microbiology" (A. H. ROSE, ed.), Vol. 5, pp. 369–465. Academic Press, London.

KIESLICH, K. (1980d). 13th Int. TNO-Conf., The Haague, p. 83.

KIESLICH, K., BERNDT, H.-D., WIECHERT, R., KERB, U., SCHULZ, G., and KOCH, H.-J. (1969a). Liebigs Ann. Chem. *726,* 161.

KIESLICH, K., PETZOLDT, K., KOSMOL, H., and KOCH, W. (1969b). Liebigs Ann. Chem. *726,* 168.

KIESLICH, K., WIEGLEPP, H., PETZOLDT, K., and HILL, P. (1971). Tetrahedron *27,* 445.

KIESLICH, K., and KERB, U. (1980). German Patent 2 919 984.

KIESLICH, K., and KOCH, J.-J. (1970). Chem. Ber. *103,* 610.
KIESLICH, K., and SCHULZ, G. (1969). Liebigs Ann. Chem. *726,* 152.
KIESLICH, K., and SEBEK, O. K. (1979). Annu. Rep. Ferment. Proc. *3,* 275.
KIESLICH, K., and WIEGLEPP, H. (1971). Chem. Ber. *104,* 205.
KIM, C. H. (1937). Enzymologia *4,* 119.
KIM, C. H. (1939). Enzymologia *6,* 105.
KITA, D. A. (1961). U.S. Patent 2 991 230.
KNIGHT, S. G. (1962). U. S. Patent 3 031 379.
KOEPSELL, H. J. (1962). Biotech. Bioeng. *4,* 57.
KOLCHANOVA, L. A., DOZOROVA, I. I., SAVROVA, O. D., and ISAVNINA, L. V. (1979). Khim.-Farm. Zhur. *13*(10), 101.
KONDO, E., and MASUO, E. (1961). J. Gen. Appl. Microbiol. *7,* 113.
KONDO, E., and MITSUGI, T. (1961). Nippon Nôgeikagaku Kaishi *35,* 521.
KONDO, E., and MITSUGI, T. (1966). J. Am. Chem. Soc. *88,* 4737.
KONDO, E., MITSUGI, T., and TORI, K. (1965). J. Am. Chem. Soc. *87,* 4655.
KOSHCHEYENKO, K. A., BORMAN, E. A., SOKOLOVA, L. V., SUVOROV, N. N., and SKRYABIN, G. K. (1975). Izv. Akad. Nauk Ser. Biol., 25.
KOSHCHEYENKO, K. A., TURKINA, M. V., and SKRYABIN, G. K. (1983). Enzyme Microb. Technol. *5,* 14.
KOSMOL, H., KIESLICH, K., VÖSSING, R., KOCH, H.-J., PETZOLDT, K., and GIBIAN, H. (1967). Liebigs Ann. Chem. *701,* 198.
KRÁMLI, A., and HORVÁTH, J. (1948). Nature (London) *162,* 619.
KRÁMLI, A., and HORVÁTH, J. (1949). Nature *163,* 219.
KRASIL'NIKOV, N. A., SKRYABIN, G. K., ASEEVA, I. V., and KORSUNSKAYA, L. O. (1959). Dokl. Akad. Nauk S.S.S.R. *128,* 836.
KULIG, M. J., and SMITH, L. L. (1974). J. Steroid Biochem. *5,* 485.
KUROSAWA, Y., HAYANO, M., and BLOOM, B. M. (1961). Agric. Biol. Chem. *25,* 838.
LANZILOTTA, R. P., BRADLEY, D. G., and BEARD, C. C. (1975). Appl. Microbiol. *29,* 427.
LARSSON, P.-O., OHLSON, S., and MOSBACH, K. (1979). In "Applied Biochemistry and Bioengineering" (L. B. WINGARD, E. KATCHALSKI-KATZIR, and L. GOLDSTEIN, eds.), Vol. 2, pp. 291–301. Academic Press, New York.
LEE, B. K., RYU, D. Y., THOMA, R. W., and BROWN, W. E. (1969). J. Gen. Microbiol. *55,* 145.
LEE, B. K., BROWN, W. E., RYU, D. Y., and THOMA, R. W. (1971). Biotech. Bioeng. *13,* 503.
LEESON, L. J., LOWREY, J. A., SIEGER, G. M., and MULLER, S. (1961). J. Pharm. Sci. *50,* 606.
LEFEBVRE, G., SCHNEIDER, F., GERMAN, P., and GAY, R. (1974). Tetrahedron Lett., 127.
LESTROVAYA, N. N., NASARUK, M. I., and SKRYABIN, G. K. (1965). Dokl. Akad. Nauk S.S.S.R. *163,* 768.
LEVINE, S. D., NEIDLEMAN, S. L., and OBERC, M. (1968). Tetrahedron *24,* 2979.
LIN, Y. Y., SHIBAHARA, M., and SMITH, L. L. (1969). J. Org. Chem. *34,* 3530.
LIN, Y. Y., and SMITH, L. L. (1970a). Biochim. Biophys. Acta *210,* 319.
LIN, Y. Y., and SMITH, L. L. (1970b). Biochim. Biophys. Acta *218,* 515.
LIN, Y. Y., and SMITH, L. L. (1970c). Biochim. Biophys. Acta *218,* 526.
MACDONALD, I. A., WILLIAMS, C. N., and MUSIAL, B. C. (1980). J. Lipid Res. *21,* 381.
MADDOX, I. S., DUNNILL, P., and LILLY, M. D. (1981). Biotech. Bioeng. *23,* 345.
MAMOLI, L. (1938). Ber. Dtsch. Chem. Ges. *71,* 2696.
MAMOLI, L., and VERCELLONE, A. (1937a). Ber. Dtsch. Chem. Ges. *70,* 470.
MAMOLI, L., and VERCELLONE, A. (1937b). Ber. Dtsch. Chem. Ges. *70,* 2079.
MAMOLI, L. and VERCELLONE, A. (1937c). Z. Physiol. Chem. *245,* 93.
MAMOLI, L., and VERCELLONE, A. (1938). Ber. Dtsch. Chem. Ges. *71B,* 1686.
MANSON, A. J., SJOGREN, R. E., and RIANO, M., (1965). J. Org. Chem. *30,* 307.
MARSHECK, W. J. (1971). Prog. Ind. Microbiol. *10,* 49.
MARSHECK, W. J., and KARIM, A. (1973). Appl. Microbiol. *25,* 647.
MARTIN, C. K. A. (1977). Adv. Appl. Microbiol. *22,* 29.
MARTÍNKOVÁ, J., and DYR, J. (1965). Coll. Czech. Chem. Commun. *30,* 2994.
MAXON, W. D., CHEN, J. W., and HANSON, F. R. (1966). Ind. Eng. Chem. (Process Design Develop.) *5,* 285.
MCGREGOR, W. C., TABENKIN, B., JENKINS, E., and EPPS, R. (1972). Biotech. Bioeng. *14,* 831.
MILLS, J. S., BOWERS, A., DJERASI, C., and RINGOLD, H. J. (1960). J. Am. Chem. Soc. *82,* 3399.
MIZSEI, A., and SZABÓ, A. (1961). J. Biochem. Microbiol. Technol. Eng. *3,* 21 and 119.
MOGIL'NITSKII, G. M., ANDREEV, L. V., and KOSHCHEYENKO, K. A. (1975). Mikrobiologiya *44,* 351.

MOGIL'NITSKII, G. M., KOSHCHEYENKO, K. A., PONOMAREVA, E. N., BUKHAR, M. I., and SKRYABIN, G. K. (1977). Khim.-Farm. Zh. *11* (No. 1), 94.

MOSBACH, K., and LARSSON, P.-O. (1970). Biotech. Bioeng. *12,* 19.

MUKAWA, F. (1971). J. Chem. Soc. Chem. Commun., 1060.

MULEVICH, V. M., BOGACHEVA, N. G., VOLOVEL'SKII, L. N., KISELEV, V. P., LEVANDOVSKII, G. S., and KOGAN, L. M. (1977). Khim.-Farm. Zhur. *11,* 138.

MÜLLER R., and KIESLICH, K. (1966). Angew. Chem. Int. Ed. (English) *5,* 653.

MURRAY, H. C. (1976). In "Industrial Microbiology" (B. M. MILLER and W. LITSKY, eds.), pp. 79–105. McGraw-Hill, New York.

MURRAY, H. C., and PETERSON, D. H. (1952). U.S. Patent 2 602 769.

MURRAY, H. C., and SEBEK, O. K. (1959). U.S. Patent 2 902 411.

NEIDLEMAN, S. L., DIASSI, P. A., JUNTA, B., PALMERE, R. M., and PAN, S. L. (1966). Tetrahedron Lett., 5337.

NEIDLEMAN, S. L., and LEVINE, S. D. (1968). Tetrahedron Lett., 4057.

NEIDLEMAN, S. L., and OBERC, M. A. (1968). J. Bacteriol. *95,* 2424.

NGUYEN-DANG, T., MAYER, M., and JANOT, M.-M. (1971). C. R. Acad. Sci. Ser. D *272,* 2032.

NOBILE, A. (1958). U.S. Patent 2 837 464.

NOBILE, A., CHARNEY, W., PERLMAN, P. L., HERZOG, H. L., PAYNE, C. C., TULLY, M. E., JEVNIK, M. A., and HERSHBERG, E. G. (1955). J. Am. Chem. Soc. *77,* 4184.

NOMINÉ, (1980). Bull. Soc. Chim. Fr. II, 18.

OHLSON, S., LARSSON, P. O., and MOSBACH, K. (1978). Biotech. Bioeng. *20,* 1267.

OHLSON, S., LARSSON, P.-O., and MOSBACH, K. (1979). Eur. J. Appl. Microbiol. Biotech. *7,* 103.

OHLSON, S., FLYGARE, S., LARSSON, P.-O., and MOSBACH, K. (1980). Eur. J. Appl. Microbiol. Biotech. *10,* 1.

ONKEN, D., and ONKEN, D. (1980). Pharmazie *35,* 193.

PAN, S. C., and FRAZIER, W. R. (1962). Biotech. Bioeng. *4,* 303.

PAN, S. C. and LERNER, L. J. (1971). U.S. Patent 3 616 227.

PENASSE, L., and PEYRE, M. (1968). Steroids *12,* 525.

PERLMAN, D., TITUS, E., and FRIED, J. (1952). J. Am. Chem. Soc. *74,* 2126.

PETERSON, D. H. (1963). In "Biochemistry of Industrial Micro-organisms" (C. RAINBOW and A. H. ROSE, eds.)., pp. 537–606. Academic Press, London/New York.

PETERSON, D. H., MURRAY, H. C., EPPSTEIN, S. H., REINEKE, L. M., WEINTAUB, A., MEISTER, P. D., and LEIGH, H. M. (1952). J. Am. Chem. Soc. *74,* 5933.

PETZOLDT, K., and ELGER, W. (1976). German Patent 2 450 106.

PETZOLDT, K., and ELGER, W. (1977). German Patent 2 539 261.

PETZOLDT, K., and KIESLICH, K. (1969). Liebigs Ann. Chem. *724,* 194.

PETZOLDT, K., and WIECHERT, R. (1976). German Patent 2 456 068.

PETZOLDT, K., KIESLICH, K., and STEINBECK, H. (1974). German Patent 2 326 084.

PROTIVA, J., and SCHWARZ, V. (1974). Folia Microbiol. *19,* 151.

RASPÉ, G., KIESLICH, K., and KERB, U. (1964). Arzneimittel Forsch. *14,* 450.

REUSSER, F., KOEPSELL, H. J., and SAVAGE, G. M. (1961). Appl. Microbiol. *9,* 346.

RICHMOND, W. (1976). Clin. Chem. *22,* 1579.

RIEMANN, J., RÖPKE, H., KIESLICH, K., KOCH, H.-J., and GIBIAN, H. (1968). Eur. J. Biochem. *6,* 60.

RINGOLD, H. J., HAYANO, M., and STEFANOVIC, V. (1963). J. Biol. Chem. *238,* 1960.

ROSS, J. W. (1962). U.S. Patent 3 022 226.

RUFER, C., KOSMOL, H., SCHRÖDER, E., KIESLICH, K., and GIBIAN, H. (1967a). Liebigs Ann. Chem. *702,* 141.

RUFER, C., SCHRÖDER, E., and GIBIAN, H. (1967b). Liebigs Ann. Chem. *701,* 206.

RYU, D. Y., and LEE, B. K. (1975). Process Biochem., 15.

RYU, D. Y., LEE, B. K., THOMA, R. W., and BROWN, W. E. (1969). Biotech. Bioeng. *11,* 1255.

RYZHKOVA, V. M., GUSAROVA, T. I., KLUBNICHKINA, G. A., GERASIMOVA, M. L., and GRINENKO, G. S. (1980). Khim.-Farm. Zhur. *14*(12), 69.

SALLAM, L. A. R., EL-REFAI, A. H., and EL-KADY, I. A. (1971). Z. Allg. Mikrobiol. *11,* 325.

SAMANTA, T., ROY, N., and CHATTOPADHYAY, S. (1978). Biochem. J. *176,* 593.

SATOH, I., KARUBE, I., and SUZUKI, S. (1977). Biotech. Bioeng. *19,* 1095.

SCHLEG, C., and KNIGHT, S. G. (1962). Mycologia *54,* 317.

SCHUBERT, K., SCHLEGEL, J., GROH, H., ROSE, G., and HÖRHOLD, C. (1972). Endokrinologie *59,* 99.

SCHUBERT, K., ROSE, G., and HÖRHOLD, C. (1973). J. Steroid Biochem. *4,* 283.

SCHUBERT, K., LÊ-DINH-PHÁI, KAUFMANN, G., and KNÖLL, R. (1975). Acta Biol. Med. Germ. *34,* 167.

SCHULTZ, R. M., GROMAN, E. V., and ENGEL, L. L. (1977). J. Biol. Chem. *252,* 3775.

SCHWARZ, V., ULRICH, M., and SYHORA, K. (1964). Steroids *4,* 645.

SEBEK, O. K. (1977). In "Biotechnological Applications of Proteins and Enzymes" (Z. BOHAK and N. SHARON, eds.), pp. 203–219. Academic Press, New York.

SEBEK, O. K., and PERLMAN, D. (1979). In "Microbial Technology" (H. J. PEPPLER and D. PERLMAN, eds.), 2nd Ed., Vol. 1, pp. 483–496. Academic Press, New York.

SEDLACZEK, L., JAWORSKI, A., and WILMAŃSKA, D. (1981). Eur. J. Appl. Microbiol. Biotech. *13,* 155.

SEHGAL, S. N., SINGH, K., and VÉZINA, C. (1968). Can. J. Microbiol. *14,* 529.

SHIBAHARA, M., MOODY, J. A., and SMITH, L. L. (1970). Biochim. Biophys. Acta, *202,* 172.

SHULL, G. M., and KITA, D. A. (1955). J. Am. Chem. Soc. *77,* 763.

SHULL, G. M., KITA, D. A., and DAVISSON, J. W. (1953). U.S. Patent 2 658 023.

SIH, C. J. (1962). J. Bacteriol. *84,* 382.

SINGH, K., SEHGAL, S. N., and VÉZINA, C. (1968). Appl. Microbiol. *16,* 393.

SKRYABIN, G. K., and GOLOVLEVA, L. A. (1976). "Isopol'zovanie Mikroorganizmov v Organicheskom Sinteze" (Utilization of Microorganisms in Organic Synthesis). Izdatel'stvo Nauka, Moscow, USSR.

SKRYABIN, G. K., KOSHCHEYENKO, K. A., MOGIL'NITSKII, G. M., SUROVTSEV, V. I., TYURIN, V. S., and FIKHTE, B. A. (1974). Izv. Akad. Nauk S.S.S.R., Ser. Biol., 857.

SLOCUM, S. A., and STUDEBAKER, J. F. (1975). Anal. Biochem. *68,* 242.

SMITH, A. G., and BROOKS, C. J. W. (1976). J. Steroid Biochem. *7,* 705.

SMITH, A. G., and BROOKS, C. J. W. (1977). J. Steroid Biochem. *8,* 111.

SMITH, L. L. (1963a). Steroids *1,* 544,

SMITH, L. L. (1963b). Steroids *1,* 570.

SMITH, L. L. (1964). Steroids *4,* 395.

SMITH, L. L. (1974). In "Specialist Periodical Reports, Terpenoids and Steroids" (K. H. OVERTON, ed.), Vol. 4, pp. 394–500. The Chemical Society, London.

SMITH, L. L. (1981). "Cholesterol Autoxidation", pp. 192–197, pp. 312–317. Plenum Press, New York.

SMITH, L. L., and FOELL, T. (1959). Anal. Chem. *31,* 102.

SMITH, L. L., and FOELL, T. (1962). J. Chromatogr. *9,* 339.

SMITH, L. L., and HALWER, M. (1959). J. Am. Pharm. Assoc., Sci. Ed. *48,* 348.

SMITH, L. L., FOELL, T., DE MAIO, R., and HALWER, R. (1959). J. Am. Pharm. Assoc., Sci. Ed. *48,* 528.

SMITH, L. L., GARBARINI, J. J., GOODMAN, J. J., MARX, M., and MENDELSOHN, H. (1960a). J. Am. Chem. Soc. *82,* 1437.

SMITH, L. L., MARX, M., GARBARINI, J. J., FOELL, T., ORIGONI, V. E., and GOODMAN, J. J. (1960b). J. Am. Chem. Soc. *82,* 4616.

SMITH, L. L., MENDELSOHN, H., FOELL, T., and GOODMAN, J. J. (1961). J. Org. Chem. *26,* 2859.

SMITH, L. L., FOELL, T., and GOODMAN, J. J. (1962a). Biochemistry *1,* 353.

SMITH, L. L., MARX, M., MENDELSOHN, H., FOELL, T., and GOODMAN, J. J. (1962b). J. Am. Chem. Soc. *84,* 1265.

SMITH, L. L., GREENSPAN, G., REES, R., and FOELL, T. (1966). J. Am. Chem. Soc. *88,* 3120.

SÖHNGEN, N. L. (1913). Zentrbl. Bakteriol. Parasitkd. (Abt. II) *37,* 595.

SONOMOTO, K., TANAKA, A., OMATA, T., YAMANE, T., and FUKUI, S. (1979). Eur. J. Appl. Microbiol. Biotech. *6,* 325.

SPALLA, C., AMICI, A. M., and BIANCHI, M. L. (1962). G. Microbiol. *9,* 255.

TAN, L., and FALARDEAU, P. (1970). J. Steroid Biochem. *1,* 221.

THOMA, R. W., FRIED, J., BONANNO, S., and GRABOWICH, P. (1957). J. Am. Chem. Soc. *79,* 4818.

TOPHAM, R. W., and GAYLOR, J. L. (1972). Biochem. Biophys. Res. Commun. *47,* 180.

TORI, K., and KONDO, E. (1964). Steroids *4,* 713.

TORI, K., and KONDO, E. (1966). Nippon Kagaku Zasshi *87,* 1117.

TURFITT, G. E. (1943). Biochem. J. *37,* 115.

TURFITT, G. E. (1944). J. Bacteriol. *47,* 487.

TYMES, N. W. (1977). J. Chromatogr. Sci. *15,* 151.

VALENTINE, R. C., RABSON, R., SEBEK, O., and HELINSKI, D. (1982). In "Genetic Engineering of Microorganisms for Chemicals" (A. HOLLAENDER, R. D. DE MOSS, S. KAPLAN, J. KONISKY, D. SAVAGE, and R. S. WOLFE, eds.), p. 445. Plenum Press, New York.

VERCELLONE, A., and MAMOLI, L. (1937). Z. Physiol. Chem. *248,* 277.

VÉZINA, C., SEHGAL, S. N., and SINGH, K. (1963). Appl. Microbiol. *11,* 50.

VÉZINA, C., SEHGAL, S. N., and SINGH, K. (1968). Adv. Appl. Microbiol. *10,* 221.

VÉZINA, C., SEHGAL, S. N., SINGH, K., and KLUEPFEL, D. (1971). Progr. Ind. Microbiol. *10,* 1.

VIDIC, H. J., KIESLICH, K., and LEHMANN, H. G. (1974). German Patent 2 306 529.

VIGER, A., COUSTAL, S., and MARQUET, A. (1981). J. Am. Chem. Soc. *103,* 451.

VISCHER, E., SCHMIDLIN, J., and WETTSTEIN, A. (1956). Experientia *12,* 50.

WEAVER, E. A., KENNY, H. E., and WALL, M. E. (1960). Appl. Microbiol. *8,* 345.

WETTSTEIN, A., VISCHER, E., and MEYSTRE, C. (1958). U.S. Patent 2 844 513.

WILSON, J. E., and VESTLING, C. S. (1965). Arch. Biochem. Biophys. *110,* 401.

YOUNGLAI, E., and SOLOMON, S. (1967). Endocrinology *80,* 141.

ZEDAN, H. H., EL-TAYEB, O. M., and ABDEL-AZIZ, M. (1976). Planta Med. *30,* 251.

ZUIDWEG, M. H. J. (1968). Biochim. Biophys. Acta *152,* 144.

ZUIDWEG, M. H. J., VAN DER WAARD, W. F., and DE FLINES, J. (1962). Biochim. Biophys. Acta *58,* 131.

Chapter 3

Sterols

Christoph K. A. Martin

BASF Aktiengesellschaft
Ludwigshafen, Federal Republic of Germany

I. Introduction
II. Microbial Sterol Metabolism
 A. Degradation Pathway of the Sterol Side Chain
 B. Degradation Pathway of the Steroid Ring System
III. Selective Side Chain Cleavage of Sterols
 A. Conversion of Sterols with Modified Structure
 B. Conversion in the Presence of Enzyme Inhibitors
  1. Enzyme inhibitors
  2. Processes employing enzyme inhibitors
 C. Conversion of Sterols by Mutants
  1. Production of mutants
  2. Processes employing mutants
IV. Degradation of Sterols to Hexahydroindan Derivatives
V. Conversion of Sterols in Organic Solvents or by Immobilized Cells
VI. Microbial Hydrogenation of Sterols
VII. Conclusion
VIII. References

I. Introduction

Since the discovery that the side chain of various abundant, naturally occurring sterols such as cholesterol, *β*-sitosterol or campesterol can be degraded selectively by microorganisms, this conversion method has attracted much attention and several useful processes have been developed. In addition to the microbial step, chemical methods for the conversion of the fermentation products, e. g., 17-ketosteroids or 20-carboxy-pregnane derivatives, to steroid hormones have been worked out. Therefore, an increasing amount of pharmacologically active steroids annually produced in industry is prepared by the initial microbial transformation of sterols.

Numerous reviews on the transformation of steroids by microorganisms, covering also the conversion of sterols, have been published (e.g., CHARNEY and HERZOG, 1967; MARSHECK, 1971; BEUKERS et al., 1972; VOISHVILLO and KAMERNITSKII, 1976; MARTIN, 1977; SATO, 1979; NOMINÉ, 1980; KIESLICH, 1980; SCHOEMER and MARTIN, 1980; IMADA et al., 1981). It is the purpose of this chapter to summarize the literature on, firstly, the selective side chain cleavage of sterols for the synthesis of C_{19}- or C_{22}-steroids and, secondly, the partial degradation of the sterol ring system to hexahydroindan propionic acid derivatives. The latter can be used for the preparation of steroids with unnatural configuration.

II. Microbial Sterol Metabolism

Since 1913 it has been known that numerous organisms of the genera *Nocardia, Pseudomonas, Mycobacterium, Corynebacterium* and *Arthrobacter* are capable of utilizing sterols such as cholesterol or *β*-sitosterol as sole source of carbon. In the ensuing years the metabolic pathway of sterol degradation has been elucidated completely.

The steroid ring structure and the side chain are metabolized by different mechanisms. These enzymatic reactions do not follow a given order but occur simultaneously and independently. Thus, if the structure of the side chain is modified so that the enzymes normally involved in the degradation are unable to catalyze the fission, the ring system will be attacked resulting in accumulation of metabolites with partly oxidized ring structures. On the other hand, in substrates with a modified ring structure blocking the first enzymic reactions in the ring oxidation, the side chain will be degraded resulting in the formation of 17-ketosteroids.

A. Degradation Pathway of the Sterol Side Chain

The pathway of the C-17 sterol side chain degradation during the microbial conversion of cholesterol to 17-ketosteroids has been elucidated by SIH and coworkers (SIH and WHITLOCK, 1968). In contrast to mammalian systems where 17-ketosteroids are formed via cleavage of the C-20 – C-22 bond followed by cleavage of the C-17 – C-20 bond, microorganisms shorten the side chain of cholesterol by a mechanism similar to the *β*-oxidation of fatty acids (Fig. 1).

Following C-27-hydroxylation and oxidation of the resulting alcohol to a C-27 carboxylic acid, propionic acid, acetic acid, and propionic acid again are removed, resulting in the formation of C-24 and C-22 carboxylic acids as intermediates and finally in the formation of the C-17 keto compounds. The conversion of the dinorcholanic acid derivative IV also takes place under anaerobic conditions. Therefore, this last step involves dehydrogenation and ad-

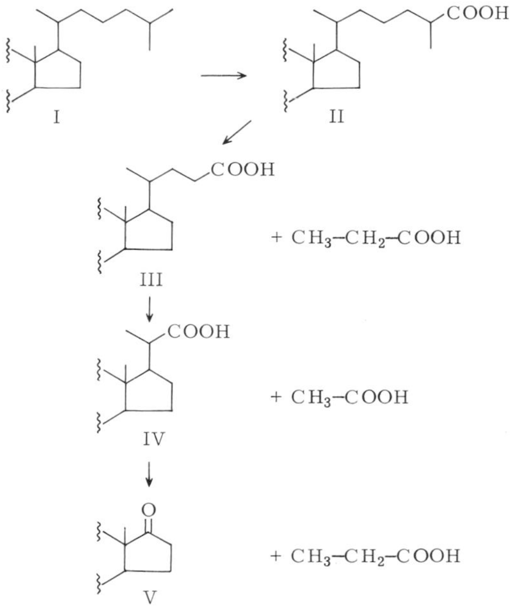

Figure 1. Microbial side chain cleavage of cholesterol.

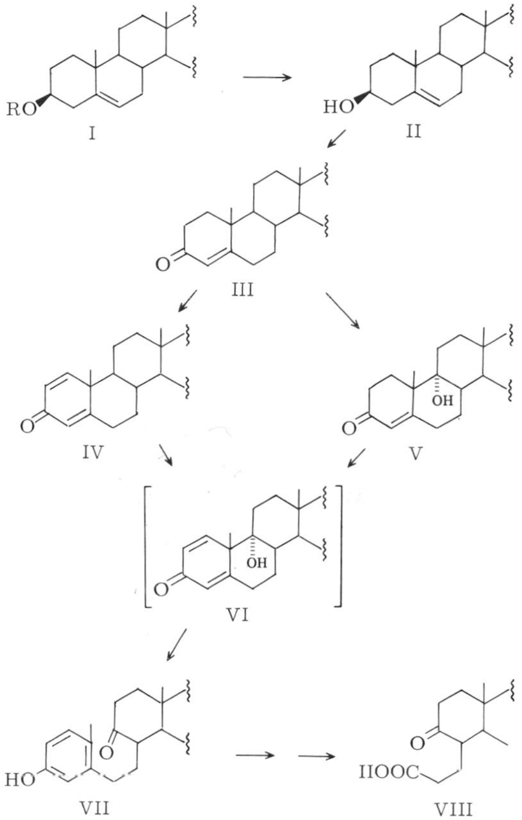

Figure 2. Degradation pathway of the steroid ring system.

dition of water, followed by aldolytic cleavage to yield the 17-keto function and propionic acid. Phytosterols branched at C-24 are degraded via 24-oxo intermediates (Knight and Wovcha, 1980).

B. Degradation Pathway of the Steroid Ring System

Sterols esterified at the C-3 position have to be saponified before degradation can occur. This reaction is of interest for the analytical determination of cholesterol esters by cholesterol oxidase. Several processes for the production of cholesterol esterase have been patented (Kieslich, 1980). As is illustrated in Fig. 2, sterols containing the 3β-hydroxy-5-ene configuration II are transformed first to the corresponding 3-keto-4-ene compounds III. The enzymes involved in this step are either NAD-dependent dehydrogenases or oxidases, requiring only molecular oxygen for their action (Smith and Brooks, 1976). Various sterols are substrates for these enzymes; however, the length of the C-17 side chain determines the reaction rate. 4,4-Dimethylsterols are not oxidized by these enzymes; therefore, lanosterol has to be decomposed by a different mechanism. Cholesterol oxidase preparations are produced commercially and are used for the determination of cholesterol in body fluids.

Oxidation of the 3β-hydroxy function is followed by the isomerization of the Δ^5-double bond, a reaction which has been shown to be catalyzed by an isomerase as well as various cholesterol oxidases. This isomerization step can also occur nonenzymatically.

The subsequent enzymatic steps are similar in the degradation of sterols and steroids. References to these reactions are cited in the comprehensive review by SMITH (1974). Depending on the organism studied, the further metabolism of the 3-keto-4-ene compounds involves 9α-hydroxylation followed by C-1(2)-dehydrogenation or *vice versa*. The resulting metabolite VI undergoes simultaneous aromatization and cleavage of the B-ring, via a nonenzymic reverse aldol type reaction, to produce a 9,10-secophenolic derivative VII. It has been shown that the mechanism of the C-1(2)-dehydrogenase proceeds via a transdiaxial loss of the 1α,2β-hydrogens. The second enzyme acting at the early stage of ring degradation, the 9α-hydroxylase, can be inhibited by complexing agents for ferrous ions. It has been shown to be a monoxygenase requiring molecular oxygen for action. In subsequent enzymic reactions, ring A is degraded to yield the hexahydroindan propionic acid derivative VIII.

III. Selective Side Chain Cleavage of Sterols

As mentioned in Section II, numerous species of bacteria are capable of utilizing cholesterol and phytosterols as carbon and energy source. This complete degradation reaction is of no commercial interest.

However, methods have been developed to selectively cleave the side chain of sterols by microorganisms. The resulting C_{19}- or C_{22}-steroids are important sources for the manufacture of steroid drugs. These processes are based on the inhibition of the key enzymes for steroid ring degradation, the C-1(2)-dehydrogenase and 9α-hydroxylase.

Three different methods have been employed to inhibit one or both of these enzymes:

- structural modification of the substrates, thus preventing enzymic attack on the ring system;
- inhibition of the 9α-hydroxylase by chemical means such as complexing agents for Fe^{2+};
- mutation of the microorganisms.

In the following sections, these various methods will be discussed.

A. Conversion of Sterols with Modified Structure

The microbial transformations of sterols with chemically modified structure are summarized in Table 1. The fermentation products – 17- or 16-ketosteroids, respectively – are modified likewise and cannot serve as substrates for the enzymes attacking the ring structure.

Conversion of 19-hydroxysterols (I, II in Fig. 3) by *Nocardia restrictus* ATCC 14 887 or *Nocardia* sp. ATCC 19 170 yields up to 30% estrone (V). Incomplete degradation of the substrates takes place under these conditions, since the estrone formed cannot be further metabolized by these organisms. 3β-Acetoxy-19-hydroxy-5-cholestene (III), which can be prepared from cholesterol in three chemical steps, is an even better substrate and can be converted to estrone by *Nocardia* sp. ATCC 19 170 in about 70% yield (96 h). 19-Norcholesta-1,3,5(10)-triene-3-ol (IV) is not oxidized by this strain; however, prolonged incubation (240 h) of this compound with *N.restrictus* ATCC 14 887 yields estrone in small amounts. In addition to *Nocardia* spp., different mycobacteria have also been found to convert 19-nor- or 19-hydroxysterols to estrone.

Estrone can also be formed from 19-hydroxy-4,7-cholestadiene-3-one by microbial degradation with *Nocardia* sp. ATCC 19 170. Evidence has been found that reduction of the C-7(8)-double bond must have taken place before the complete removal of the cholestane side chain and aromatization of ring A. However, using *Mycobacterium* sp. the same substrate is con-

Table 1. Selective Side Chain Cleavage of Sterols with Modified Structure[a)]

Substrate	Product	Microorganism
19-Hydroxysterols, 19-norsterols, 3-hydroxy-19-nor-$\Delta^{1,3,5}$-sterols	Estrone	*Nocardia restrictus* ATCC 14887 *Nocardia* sp. ATCC 19170 *Arthrobacter simplex* IAM 1660 *Corynebacterium* sp. Mycobacteria
19-Hydroxy-$\Delta^{4,7}$-sterols, 3-hydroxy-19-nor-$\Delta^{1,3,5,7}$-sterols	Equilin, equilenin, estrone	*Mycobacterium* sp. *Corynebacterium simplex* *Nocardia rubra*
6β,19-Oxidostenones, 3β-acetoxy-5α-chloro(fluoro)-6β,19-oxido-sterols	6β,19-Oxido-4-androstene-3,17-dione	*Nocardia* sp. ATCC 19 170 Mycobacteria
3β-Acetoxy-5α-bromo-6β,19-oxidosterols	5α-Bromo-6β,19-oxidoandrostane-3,17-dione	*Nocardia* sp. ATCC 19170
5α,5α-Cyclosterols	3α,5α-Cycloandrostane-17-one	*Mycobacterium phlei*
3α,5α-Cyclo-6β,19-oxido-sterols	3α,5α-Cyclo-6β,19-oxido-androstane-17-one	*Arthrobacter* spp. Corynebacteria
Sterol-3-oximes	4-Androstene-3,17-dione (after hydrolysis)	*Mycobacterium* sp.
4-Hydroxycholestenone	3β-Hydroxy-5α-androstane-4,17-dione, 3α-hydroxy-5α-androstane-4,17-dione, 3β,4α-dihydroxy-5α-androstane-17-one	*Mycobacterium phlei*
25D-Spirost-4-ene-3-one	1,4-Androstadiene-3,16-dione, 20α-hydroxy-4-pregnene-3,16-dione, 3α,11β,20α-trihydroxy-5α-pregnane-16-one	*Fusarium solani* *Verticillium theobromae* *Stachylidium bicolor*
Crustecdysone, makisterone	Poststerone, rubrosterone	*Rhizopus arrhizus* *R.nigricans* *Curvularia lunata*
Ponasterone A	Rubrosterone	*Fusarium lini* ATCC 9593

[a] For references, see MARTIN (1977)

verted to 16% equilin, 20% estrone and traces of equilenin. Similarly, 19-norcholesta-1,3,5(10),7(8)-tetraene-3-ol is transformed to equilenin and traces of equilin by *Corynebacterium simplex,* and to equilenin by *Nocardia rubra.*

3β-Acetoxy-5α-chloro-6β,19-oxidocholestane (I in Fig. 4) is an intermediate in the chemical preparation of 3β-acetoxy-19-hydroxy-5-cholestene and may be prepared from cholesterol in only two steps. Therefore, this compound has also been used as substrate for the microbial transformation.

Nocardia sp. ATCC 19 170 is capable of converting I into II in 36% yield (80 hours). The conversion of the corresponding bromo-compound V results in accumulation of 5α-bromo-6β,19-oxidoandrostane-3,17-dione (VI).

In addition to these halogen derivatives of sterols, 6β,19-oxido-4-cholestene-3-one III and the corresponding sitosterol derivative can also be transformed to 6β,19-oxido-4-androstene-3,17-dione II by *Nocardia* sp. and various *Mycobacterium* spp. in yields of up to 57% (70 hours). Since 6β,19-oxido-

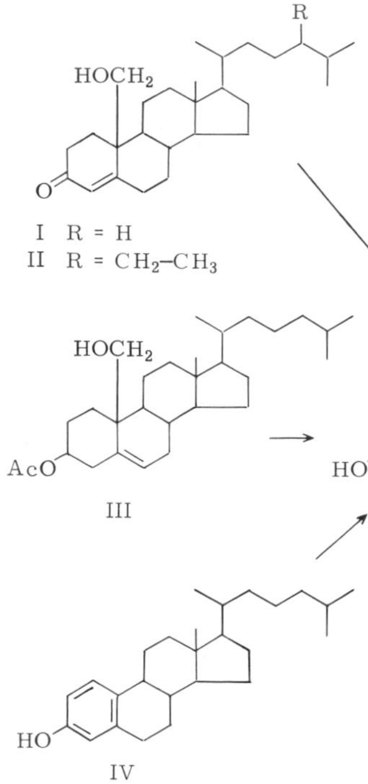

(I: 19-Hydroxycholesterol,
II: 19-Hydroxysitosterol,
III: 3β-Acetoxy-19-hydroxy-5-cholestene,
IV: 19-Norcholesta-1,3,5 (10)-triene-3-ol,
V: estrone)

Figure 3. Conversion of 19-nor- or 19-hydroxysterols by microorganisms.

9α-hydroxy-4-androstene-3,17-dione IV is found in smaller amounts as by-product, it can be concluded that presence of the 6β,19-oxido-bridge blocks the introduction of the C-1(2)-double bond, whereas the C-19-hydroxy function prevents 9α-hydroxylation.

Completely different substrate modifications also prevent the breakdown of the steroid ring system when organisms capable of utilizing cholesterol are used. Exposure of 3α,5α-cyclocholestane-6β-ol (I in Fig. 5) or the corresponding stigmastane derivatives to cultures of *Mycobacterium phlei* results in accumulation of 3α,5α-cycloandrostane-6,17-dione (II, 22.5% in 96 hours). 17-Ketosteroids may also be prepared by combining different methods of substrate structure modification. 3α,5α-Cyclo-6β,19-oxido-5α-cholestane (IV) and the corresponding substrates of the ergostane and stigmastane series can be converted into 3α,5α-cyclo-6β,19-oxido-5α-androstane-17-one (V) by *Corynebacterium equi* IAM 1 038, *C.xerosis, Arthrobacter simplex* IAM 1 660 and *A.ureafaciens* IAM 1 658.

Incubation of 4-hydroxy-4-cholestene-3-one with *Mycobacterium phlei* yields 3β- and 3α-hydroxy-androstane-4,17-dione and the further reduced compounds 3β,4α-dihydroxy-5α-androstane-17-one (TÖMÖRKENI et al., 1975).

Sterol-3-carboxymethyloxime derivatives have been described to be ideal substrates with respect to solubility and transforma-

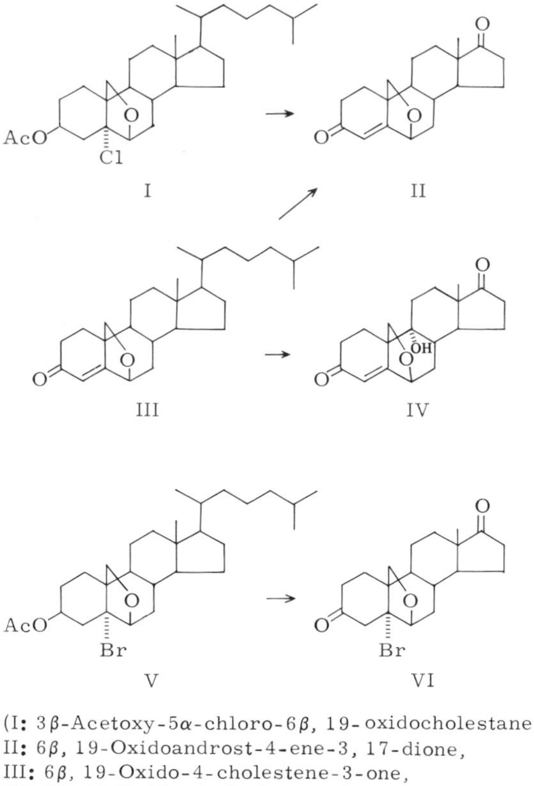

(I: 3β-Acetoxy-5α-chloro-6β, 19-oxidocholestane,
II: 6β, 19-Oxidoandrost-4-ene-3, 17-dione,
III: 6β, 19-Oxido-4-cholestene-3-one,
IV: 6β, 19-Oxido-9α-hydroxyandrost-4-ene-3,17-dione,
V: 3β-Acetoxy-5α-bromo-6β, 19-oxidocholestane,
VI: 5α-Bromo-6β, 19-oxidoandrostane-3,17-dione)

Figure 4. Microbial conversion of 6β,19-oxidosterols.

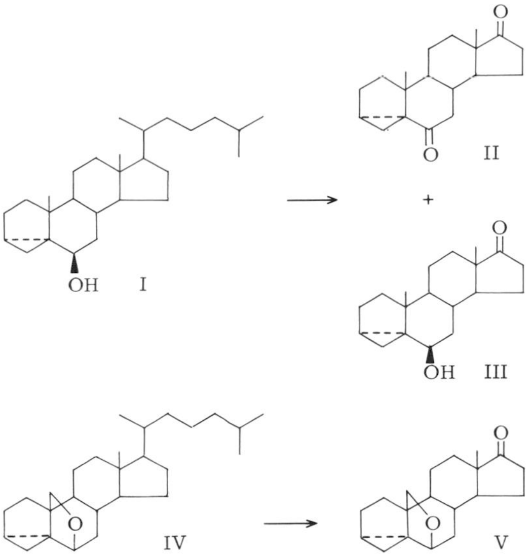

(I: 3α,5α-Cyclocholestane-6β-ol,
II: 3α,5α-Cycloandrostane-6,17-dione,
III: 3α,5α-Cyclo-6β-hydroxyandrostane-17-one,
IV: 3α,5α-Cyclo-6β, 19-oxido-5α-cholestane,
V: 3α,5α-Cyclo-6β, 19-oxido-5α-androstane-17-one)

Figure 5. Microbial conversion of 3α,5α-cyclosterols.

tion rate by strains of *Mycobacterium* (BOEHME and HOERHOLD, 1980). The resulting androstane-3-oxime can easily be saponified to the corresponding ketone under acidic conditions.

Sapogenins and insect molting hormones are naturally occurring steroid precursors which may be partly degraded without adding enzyme inhibitors or utilizing mutants. Incubation of 25D-spirost-4-ene-3-one (diosgenone, I in Fig. 6) its derivatives with cultures of *Fusarium solani* results in the accumulation of up to 65% 1,4-androstadiene-3,16-dione (II). Similar substrates may be converted to 20α-hydroxy-4-pregnene-3,16-dione (III) and 3α,11β,20α-trihydroxy-5α-pregnane-16-one (IV) by *Verticillium theobromae* and *Stachylidium bicolor*.

Mycelia of *Rhizopus arrhizus, R.nigricans,* and *Curvularia lunata* are capable of degrading crustecdysone (I in Fig. 7) and makisterone (II) to give poststerone (IV) and subsequently rubrosterone (V). Similarly, *Fusarium lini* ATCC 9 593 converts pon-

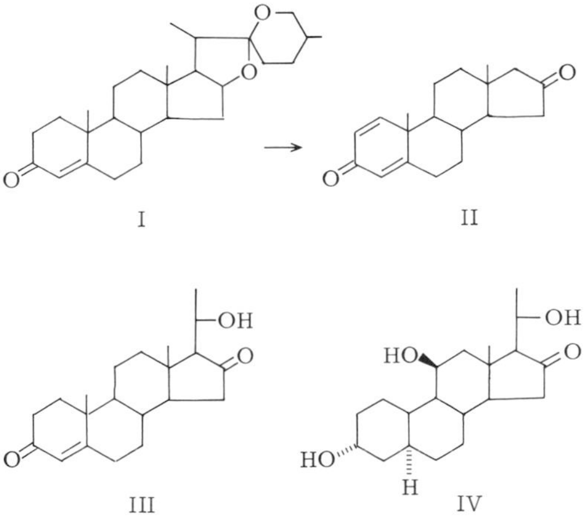

(I: Diosgenone, II: 1,4-Androstadiene-3,16-dione,
III: 20α-Hydroxy-4-pregnene-3,16-dione,
IV: 3α,11β,20α-Trihydroxy-5α-pregnan-16-one)

Figure 6. Microbial side chain cleavage of sapogenins.

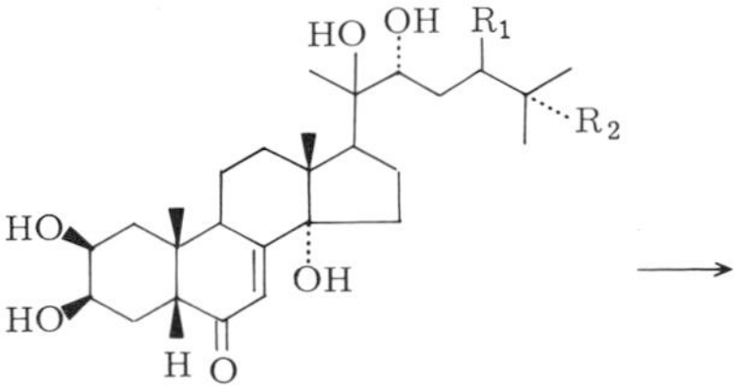

I R_1 = H, R_2 = OH
II R_1 = CH_3, R_2 = OH
III R_1 = H, R_2 = H

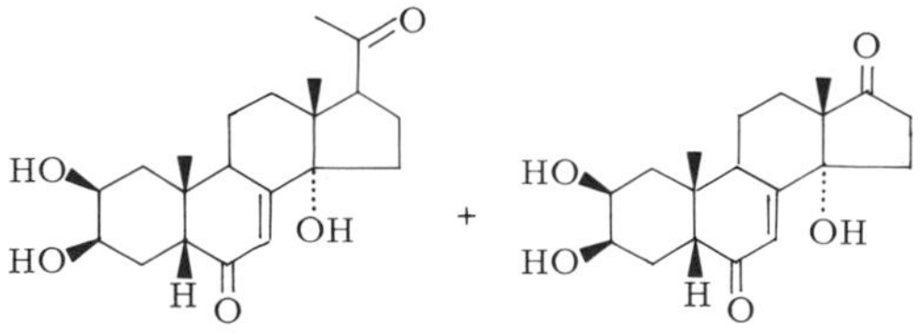

(I: Crustecdysone, II: Makisterone,
III: Ponasterone A, IV: Poststerone,
V: Rubrosterone)

Figure 7. Microbial transformation of insect molting hormones.

asterone A (III) into rubrosterone (V) in 15% yield.

In order to change the pharmacological activity of cardenolids and bufanolids attempts have been made to transform these compounds by microorganisms. *Arthrobacter simplex* ATCC 6946, *Nocardia restrictus* ATCC 14 887, and *Nocardia asteroides* ATCC 3 308 are capable of dehydrogenating scillarenin (I in Fig. 8) and scillarenone at C-1(2) in yields of almost 70% (GÖRLICH, 1973a). Similarly, Δ^6-dehydroscillarenone and Δ^6-dehydrocanaringenone (II) are dehydrogenated effectively with *Arthrobacter simplex* at the 1,2-position (GÖRLICH, 1973b).

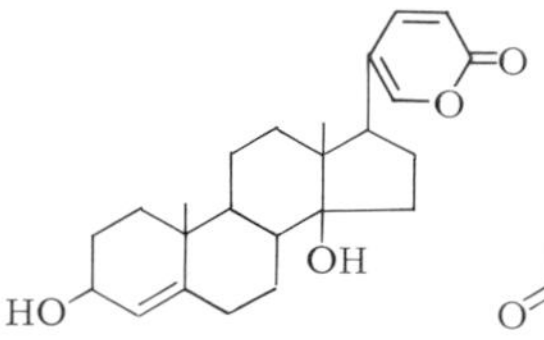

I Scillarenin

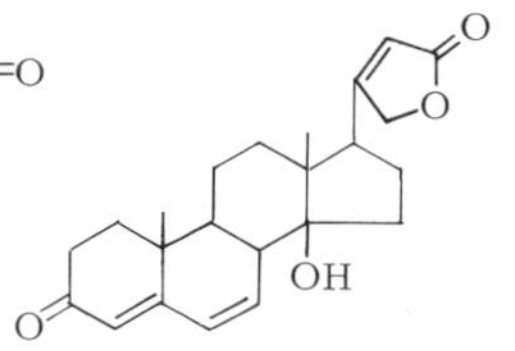

II Δ^6-Dehydrocanaringenone

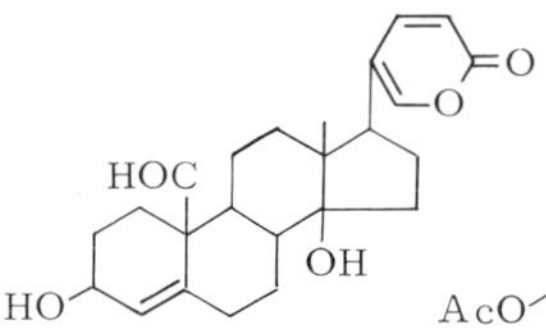

III Scilloglaucosidone

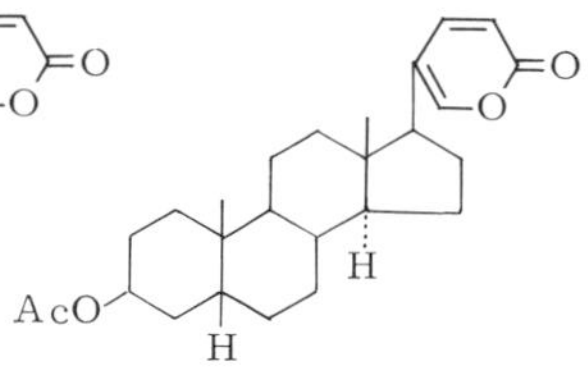

IV 3β-Acetoxy-5β, 14α-bufa-20,22-dienolid

Figure 8. Microbial transformation of cardenolids and bufanolids.

9α-Hydroxyscillarenone can be prepared from scillarenin (48% yield) or scillarenone (66% yield) by *Nocardia corallina* (GÖRLICH, 1973c). As in the case of C-19 oxygenated sterols, the microbial transformation of scilloglaucosidone III with its 19-aldehyde group or 19-hydroxyscillarenone leads to ring A aromatic compounds when *Arthrobacter simplex, Nocardia restrictus* ATCC 14 887, *Nocardia asteroides* ATCC 3 308, or *Nocardia corallina* ATCC 4 137 are used (GÖRLICH, 1973d).

Microbial hydroxylation steps have been utilized in the chemical synthesis of bufanolids and cardenolids. For instance, 14α-hydroxylation with *Helminthosporium buchloes* of 3β-acetoxy-5β,14α-bufa-20,22-dienolid IV can be used to introduce the 14β,15β-epoxide function in order to produce resibufogenin (DIAS and PETIT, 1972).

B. Conversion in the Presence of Enzyme Inhibitors

The key enzyme in steroidal ring fission, the 9α-hydroxylase, has been found to be a monoxygenase consisting of several proteins forming an electron transfer chain. Some of these proteins contain Fe^{2+} as essential metal ions. Removal or replacement of these ions results in complete inactivation of the enzymatic activity. Numerous processes for the selective side chain cleavage of sterols employing enzyme inhibitors have been developed.

1. Enzyme inhibitors

Accumulation of 17-ketosteroids has been observed with chemicals classified as:

- lipophilic chelating agents,
- metal ions with similar ion radii to Fe^{2+},
- inorganic SH-reagents,
- autoxidizable redox dyes.

The most effective inhibitors are listed in Table 2. α,α'-Dipyridyl, 1,10-phenanthroline and 8-hydroxyquinoline have been used most frequently. Since the compounds are toxic or at least decrease the growth rate of microorganisms they are added after the cultures have grown for some time.

Table 2. Compounds Effective for the Inhibition of Steroid Ring Degradation

Compound	Mechanism of Action
α,α'-Dipyridyl	Chelating agents for Fe^{2+}
1,10-Phenanthroline	
8-Hydroxyquinoline	
5-Nitro-1,10-phenanthroline	
Cupferron	
Diphenylthiocarbazone	
Diethyldithiocarbamate	
Isonicotinic acid hydrazide	
Xanthogenic acid	
o-Phenylenediamine	
4-Isopropyltropolone	
Tetraethylthiuram-disulfide	
Ni^{2+}, Co^{2+}, Pb^{2+}	Metal ions replacing iron or blocking SH-functions
SeO_3^{2-}, AsO_2^-	
Methylene blue	Redox dyes
Resazurine	

2. Processes employing enzyme inhibitors

In the presence of enzyme inhibitors cholesterol, as well as campesterol, β-sitosterol, stigmasterol or their mixtures, can be converted by microorganisms to the products shown in Fig. 9.

1,4-Androstadiene-3,17-dione (ADD) is usually isolated as main product, accompanied by smaller amounts of 4-androstene-3,17-dione (AD) and other androstane derivatives such as 1-dehydrotestosterone, testosterone etc. 23,24-Dinorcholanic acid derivatives are only found in significant amounts under special conditions, e.g., when organic adsorbents or selected strains are used. The ratio of 4-ene-3-oxosteroids to 1,4-diene-3-oxosteroids is enhanced if the aerobic degradation phase is followed by an anaerobic incubation phase (Komel et al., 1980).

Processes in which enzyme inhibitors are employed are summarized in Table 3. In addition to enzyme inhibitors, many different chemicals such as oils or adsorbents have been found to increase the yields of 17-ketosteroids.

The microorganisms used are usually isolated in screening programs, in which they are selected for their capability to decompose sterols rapidly and completely. After isolation of a suitable strain, this breakdown process is optimized with regard to culture medium and incubation conditions. Finally, the selective inhibition of steroid ring fission is optimized with respect to the nature of the enzyme inhibitor, its concentration and time of addition to the culture.

The preferred organisms used in such processes are *Arthrobacter simplex, Nocardia* spp., *Brevibacterium lipolyticum* and *Corynebacterium* spp., but also strains belonging to the genera *Bacillus, Microbacterium, Serratia, Streptomyces* and *Pseudomonas* have been used occasionally. Mixed cultures of *A.simplex* and *Nocardia corallina* or *Pseudomonas aeruginosa* have been described to show enhanced productivity.

α,α'-Dipyridyl, 1,10-phenanthroline and 8-hydroxyquinoline are the preferred chelating agents. They are used in the concentration range 0.1–1 mM. Inorganic ions or redox dyes have been employed only very rarely. Adjustment of the optimal concentration of the chelating agent in the fermentation broth is essential for high yields but very difficult to achieve. Iron or other metal ions in the culture medium react with these compounds and neutralize their inhibitory effect on the enzymatic activity, as they are no longer available for complexing of intracellular protein-bound Fe^{2+}. Therefore, the components of the culture broth have to be controlled rather exactly, especially if complex ingredients such as cornsteep liquor or molasses are used.

If the concentration of the inhibitor in the fermentation broth is too low, complete degradation of the substrate occurs to a large extent, and only traces of 17-ketosteroids are accumulated, since the 9α-hydroxylase is inhibited incompletely. On the other hand, if the inhibitor concentration is higher than optimal, the C-26-hydroxylase (initiating the side chain oxidation) is also inhibited, partly resulting in incomplete substrate conversion as well as low yields of 17-ketosteroids. On further development of the culture chelating agents occasionally display adverse effects, even when they are

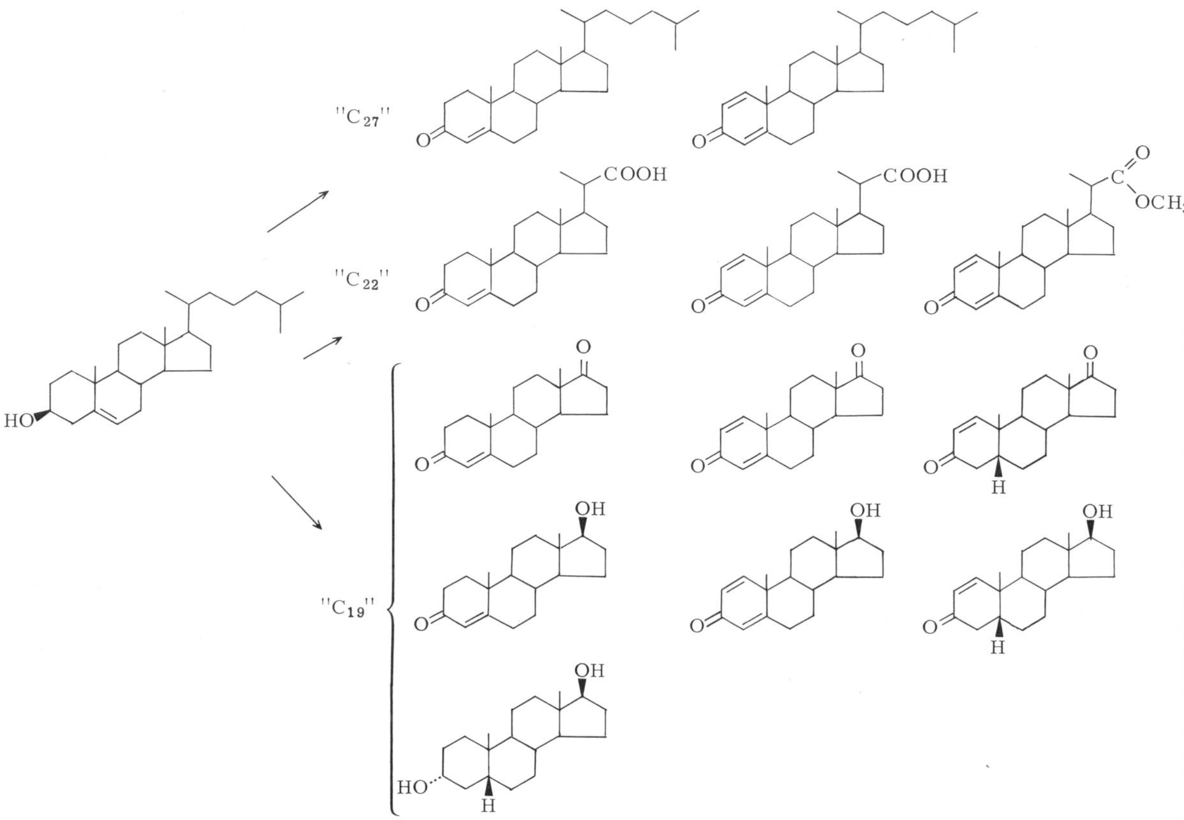

Figure 9. Microbial conversion of cholesterol in the presence of enzyme inhibitors.

Table 3. Microbial Transformation of Sterols in the Presence of Enzyme Inhibitors[a)]

Inhibitor	Products
Chelating agents	4-Androstene-3,17-dione (AD)
α,α-Dipyridyl	1,4-androstadiene-3,17-dione (ADD)
1,10-Phenanthroline	other androstane derivatives
8-Hydroxyquinoline	23,24-dinorcholanic acids
Chelating agents and active carbon	ADD
Chelating agents and styrene-divinyl-benzene-copolymers	AD, ADD, 23,24-dinorcholanic acids
Chelating agents and fats, oils, glycerides	ADD, 1-dehydrotestosterone
Chelating agents and antibiotics	ADD
Inorganic ions	AD, ADD, other androstane derivatives
Redox dyes	ADD
Rhamnolipids	AD, ADD

[a] References summarized by MARTIN (1977) and SCHOEMER and MARTIN (1980)

added to fully grown cultures. To overcome these problems, methods have been developed either to trap the accumulating 17-ketosteroids, thus preventing their further degradation, or to add the chelating agent in a form in which its concentration can be maintained in the optimal range.

The toxic effect is diminished by partial adsorption of the inhibitor on styrene-divinylbenzene-copolymers such as Amberlite XAD-2, although the ability to trap Fe^{2+} ions is not hindered. Furthermore, these resins are capable of adsorbing the C_{19}- and C_{22}-steroids selectively, since sterols form micelles in aqueous suspension and for this reason are not adsorbed. The addition of such polymers results in 2- to 3-fold enhancement of ADD yields. Presumably, the stimulation of the transformation by addition of active carbon is based on a similar mechanism.

A substantial increase in the ADD yield has also been reported for addition of oils or fats to the fermentation broth together with chelating agents. Of the compounds tested, linseed or soya oil were among the most effective, although experiments with rape oil and olive oil also worked well. Using substrate concentrations of 3 to 4 g/L and incubation times of approximately 100 hours, yields of ADD higher than 50% could be obtained. The strains used in these experiments were *Arthrobacter simplex* IAM 1 660, *Brevibacterium lipolyticum* IAM 1 398 or *Mycobacterium phlei* IFO 3 158. The ability of *A.simplex* to degrade the side chain of various sterols decreases in the following sequence: lithocholic acid (63% ADD) > cholesterol (58% ADD) > β-sitosterol (39% ADD) > campesterol (38% ADD) > cholestanol (33% ADD) > stigmasterol (29% ADD) > 7-dehydrocholesterol (16% ADD) > ergosterol (5% ADD).

C. Conversion of Sterols by Mutants

Mutagenic treatment has been employed for the production of organisms capable of selectively degrading the sterol side chain. Such mutants are biochemically blocked from degrading the nucleus and can be used to efficiently produce steroids from sterols without the necessity of modifying the substrate or of adding chemical inhibitors. Thus, sterol conversions become much more reproducible.

1. Production of mutants

The generation and isolation of mutants are well established processes in microbial genetics. Mutagens of choice include ultraviolet light, *N*-methyl-*N'*-nitro-*N*-nitrosoguanidine, etc. CARGILE and MCCHESNEY (1974), WOVCHA et al. (1978) and HILL et al. (1982) have described methodologies that allow for the production and selection of mutant organisms capable of a specific side chain cleavage of sterols. These methodologies are based upon the mutation of wild type strains that are capable of completely degrading cholesterol, and upon selection of mutants blocked at the desired conversion step.

The production of mutants may be carried out as follows: cultures of a suitable strain are grown in a nutrient broth, then treated with *N*-methyl-*N'*-nitro-*N*-nitrosoguanidine. Clones from the mutagenized cell population growing on nutrient agar are replicated on selective plates containing minimal medium plus an appropriate carbon source, such as 4-androstene-3,17-dione, and then replicated on control plates containing minimal medium plus glycerol. After a sufficient incubation time isolates showing a phenotype different from the parent culture on the selective plates are then tested for accumulation of steroids in shake flasks containing sterols as carbon source. For the enrichment of mutants the penicillin technique can also be used.

Working with *Mycobacterium fortuitum* ATCC 6 842, WOVCHA et al. (1978) were able to isolate mutants blocked at different sites and therefore to convert β-sitosterol to various steroids, as shown in Fig. 10.

2. Processes employing mutants

Each of the mutants described in Fig. 10 (with the exception of *M.fortuitum* SCM-1, producing 24-oxo-sterols) has potential industrial use, since all have been found to efficiently convert *β*-sitosterol and other sterols to products, some of which are useful as intermediates in the manufacture of medically important steroids. For instance, AD can be converted to testosterone, ADD to estrone. 9*α*-Hydroxy-1,4-androstadiene-3,17-dione can be dehydrated to give a $\Delta^{9,11}$-double bond, and, since several chemical procedures are known for the construction of the 17*α*-hydroxyprogesterone side chain, this compound represents a cheap starting material for the synthesis of 9*α*-halogen-substituted corticoids without a microbiological 11-hydroxylation. Ring A-degraded tricyclic compounds have been used as starting material for the total synthesis of 19-nor-steroids.

Further mutation of *M.fortuitum* NRRL B-8119 yields strain SCM-1, which cannot degrade sterol side chains that are branched at C-24. Conversion of *β*-sitosterol by this mutant results in the accumulation of 9*α*-hydroxy-27-nor-4-cholestene-3,24-dione,

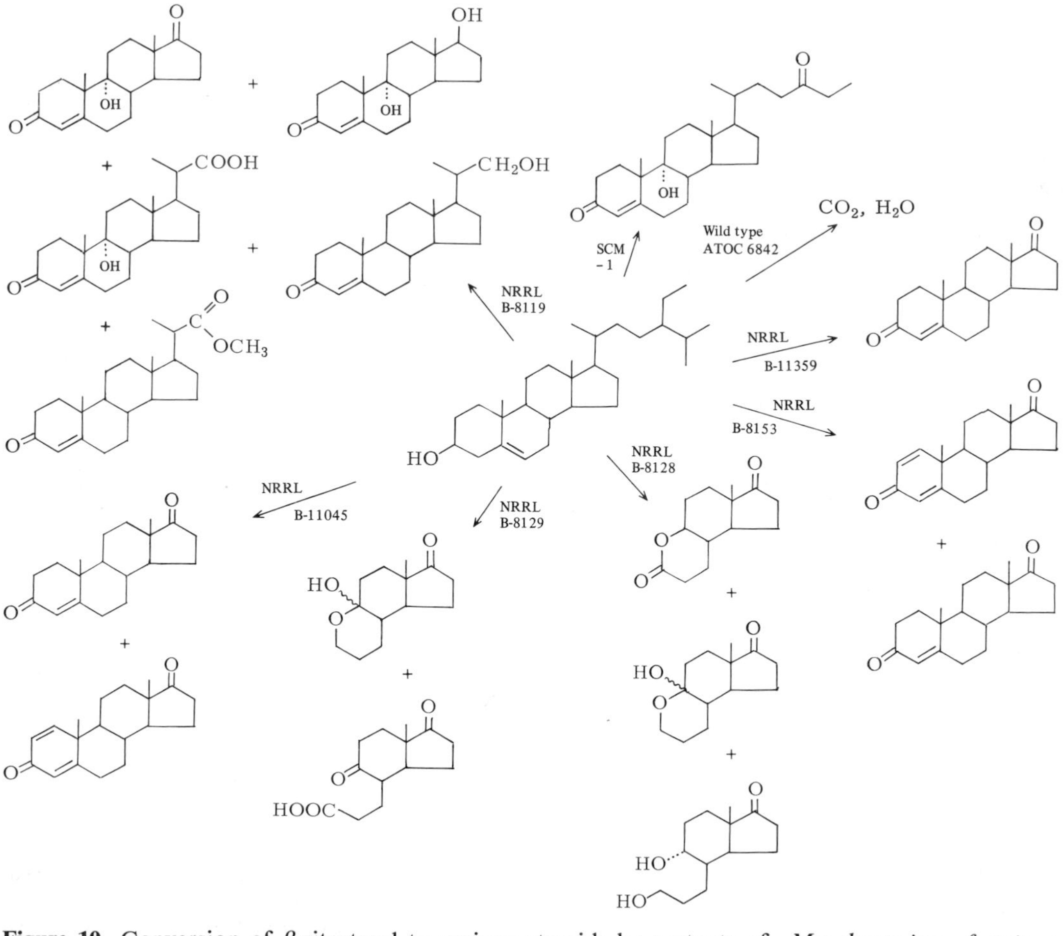

Figure 10. Conversion of *β*-sitosterol to various steroids by mutants of *Mycobacterium fortuitum* ATCC 6842.

whereas cholesterol is degraded mainly to 9α-hydroxy-4-androstene-3,17-dione (KNIGHT and WOVCHA, 1980).

All processes for the conversion of sterols to C_{19}- or C_{22}-steroids by mutants of *M.fortuitum* have been patented by the Upjohn Company. Several other groups have prepared mutants of different sterol-degrading strains which are blocked from degrading the steroid ring system. Likewise, most of these processes have been patented. They are summarized in Table 4.

Since 23,24-dinorcholanic acids are readily converted to compounds with the pregnane side chain by chemical methods, they are potentially useful intermediates for the production of corticosteroids from a cheap sterol source. Furthermore, these compounds can also be used as starting material for a multistep synthesis of chenodeoxycholic acid (HOFFMANN-LA ROCHE, 1981). Dinorcholanic acids are frequently found in small amounts when sterols are transformed by microorganisms. However, HILL et al. (1982) reported the isolation of mutants (*Corynebacterium* sp. Chol 73 DSM 1 444 and ATCC 31 385) capable of converting cholesterol nearly quantitatively into 3-oxo-23,24-dinorchola-1,4-diene-22-oic acid. In order to find a wild-type organism with a highly active side chain degradation system they used (4-^{14}C)- and (26-^{14}C)-cholesterol as substrates. The preferred organism showed a higher $^{14}CO_2$ evolution rate from (26-^{14}C)-cholesterol than from (4-^{14}C)-cholesterol.

Table 4. Mutants Used for the Selective Side Chain Cleavage of Sterols

Microorganism	Main Product	Company
Mycobacterium fortuitum NRRL B-8119	9α-Hydroxyandrost-4-ene-3,17-dione 9α-Hydroxytestosterone	Upjohn
M.parafortuitum FERM P-4926	9α-Hydroxyandrost-4-ene-3,17-dione	Mitsubishi
M.vaccae FERM P-4741+4754	9α-Hydroxyandrost-4-ene-3,17-dione	Mitsubishi
Mycobacterium sp. NRRL B-3683	1,4-Androstadiene-3,17-dione (ADD)	Searle
M.fortuitum NRRL B-8153	ADD	Upjohn
M.vaccae FERM P-4739, 4740, 4753	ADD	Mitsubishi
M.diernhoferi FERM P-4927, 4928	ADD	Mitsubishi
Arthrobacter simplex FERM P-4261	ADD	Mitsubishi
Brevibacterium lipolyticum FERM P-4474	ADD	Mitsubishi
Mycobacterium sp. NRRL B-3805	4-Androstene-3,17-dione (AD)	Searle
M.fortuitum NRRL B-11359, 11045	AD	Upjohn
M.parafortuitum MCI 0807	AD	Mitsubishi
M.fortuitum NRRL B-8119	9α-Hydroxy-3-oxo-23,24-dinorchol-4-ene-22-ol	Upjohn
M.parafortuitum MCI 0617	3-Oxo-23,24-dinorchola-1,4-diene-22-ol	Mitsubishi
Rhodococcus corallinus FERM P-4812	3β-Hydroxy-23,24-dinorchol-5-ene-22-oic acid	Mitsubishi
M.parafortuitum FERM P-5261+5262	24-Norchola-1,4-diene-3,22-dione	Mitsubishi
Corynebacterium equi	9α-Hydroxy-3-oxo-23,24-dinorchola-4,17-diene-22-oic acid	Mitsubishi
DSM 1435, 1437, 1439, 1442, 1443, 1444, 1445, ATCC 31385	3-Oxo-23,24-dinorchol-4-ene-22-oic acid, 3-oxo-23,24-dinorchola-1,4-diene-22-oic acid	Henkel
ATCC 31636+1990	3-Oxo-23,24-dinorchola-1,4,17-triene-22-oic acid	Henkel
A.simplex FERM P-4477+0868	3-Hydroxy-9-oxo-9,10-seco-pregna-1,3,5-triene-20-carboxylic	Mitsubishi
M.fortuitum FERM P-4809	3,9-Dihydroxy-9,10-secoandrosta-1,3,5-triene-17-one	Mitsubishi

9,10-Seco-derivatives such as 3-hydroxy-9-oxo-9,10-seco-pregna-1,3,5(10)-triene-20-carboxylic acid formed by *A.simplex* mutants are useful intermediates in the production of vitamin D_3.

As in processes in which enzyme inhibitors have been employed, certain additives to the fermentation broth stimulate steroid formation by mutants. For instance, admixture of small amounts of vegetable oils, such as soybean oil, or egg yolk enhances the androstane yield from sterols by *Nocardia, Mycobacterium* and *Arthrobacter* mutants. The formation of C_{22}-steroids seems to be stimulated by boron compounds. Although adsorbents such as styrene-divinylbenzene resins increase androstane yields in fermentations with enzyme inhibitors partly because of adsorption of the chelating agent, these compounds also stimulate the product formation in fermentations with mutants. Therefore, their main effect seems to be selective adsorption of 17-ketosteroids owing to micelle formation of the sterol substrate. Using *Mycobacterium* sp. NRRL B-3 683, AD and ADD yields of up to 80% can be obtained in the presence of 5 g/L Amberlite XAD-2 and 7 g/L Tween 20 (SAUERBAUM et al., 1978).

In addition to sterols in their naturally occurring form, various sterol derivatives can be transformed by microorganisms in a similar way to 17-ketosteroids. *Mycobacterium* sp. NRRL B-3 805 is capable of cleaving the side chains of 6α-fluoro, 6α-methyl-, 7α-methyl-cholest-4-ene-3-one, 1α-methyl-, 1α,2α-methylene-5α-cholestane-3-one, and 1α-methyl-cholest-4-ene-3-one or 1α,2α-methylene-4,6-cholestadiene-3-one to the corresponding 17-keto-steroids. Since the substituents can generally be introduced more successfully in the 17-ketostructures, these processes are of only limited interest. However, since retention of the 3β-hydroxy-5-ene structure during degradation is desirable, the C-3 position can be protected. Incubation of 3-methoxy-, 3-ethoxy-, or 3,3-ethylenedioxy-sterols and 3β-methoxymethoxy-cholest-4-ene with *Mycobacterium* sp. NRRL B-3 805 yields the corresponding 5-androstene-17-one derivatives.

Instead of pure uniform sterols, naturally occurring sterol mixtures have also been used as substrates. For instance, tall oil sterols (about 17 different compounds, of which sitosterol and campesterol comprise about 85%) are potentially available in large amounts (20 000 tons per year in the U.S.). CONNER et al. (1976) studied their microbial conversion to C_{19}-steroids. Compared to soybean sterols, tall oil sterols were transformed with similar efficiency. Similarly, residues of soybean and sun flower oil can be used.

IV. Degradation of Sterols to Hexahydroindan Derivatives

Besides their use as substrates for microbial C_{19}- or C_{22}-steroid production, sterols are also starting material for the microbial synthesis of hexahydroindan derivatives, which contain intact steroid rings C and D. These compounds are also available by microbial degradation of androstanes and pregnanes. However, these substrates are more expensive than sterols.

Hexahydroindan derivatives have gained some interest because they are useful precursors for chemical synthesis of steroids which are not accessible by microbial transformation, for instance, retrosteroids with reversed stereochemistry between rings A and B.

As described in Sect. II. B., the steroid ring system is degraded by microorganisms via these ring C/D-compounds. However, with wild type strains they are only accumulated when androstane or pregnane derivatives such as androst-4-ene-3,17-dione or progesterone are used. No formation from sterols, even in the presence of various enzyme inhibitors, has been reported. Therefore, strain improvement, by mutation, of organisms capable of metabolizing sterols completely is the only approach leading to production of hexahydroindan derivatives from sterols. Table 5 lists those processes in which mutants have been employed for the preparation of these compounds. A changed growth pattern or colony morphology on steroids as substrates after mutation served as criterion for the isolation of useful mutants.

V. Conversion of Sterols in Organic Solvents or by Immobilized Cells

In the processes described above, conventional techniques well known in the microbial transformation of steroids are usually employed as methods for substrate addition and conversion. However, one of the factors operating against facile microbiological degradation of sterols is their aggregation in aqueous solution. Cholesterol has a maximum solubility in water of 1.8 mg/L (4.7 μM) and undergoes a thermodynamically reversible monomer micelle equilibrium with a critical micelle concentration of 25–40 nM at 25 °C (HABERLAND and REYNOLDS, 1973). To avoid the problems of aggregation, nonionic surface active agents such as the Tweens or Spans are added to the cultures. Furthermore, in order to obtain finely distributed substrate suspensions, the sterols are frequently first dissolved in an organic solvent such as ethanol, acetone, DMF or DMSO and then mixed with the culture broth.

Attempts have been made to overcome these solubility problems by converting the sterols in nonaqueous solvents. BUCKLAND

Table 5. Microbial Preparation of Hexahydroindan Compounds from Sterols[a)]

Microorganism	Main Products
Mycobacterium fortuitum NRRL B-8128	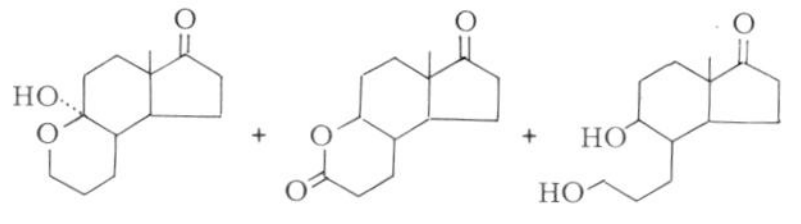
Mycobacterium fortuitum NRRL B-8129	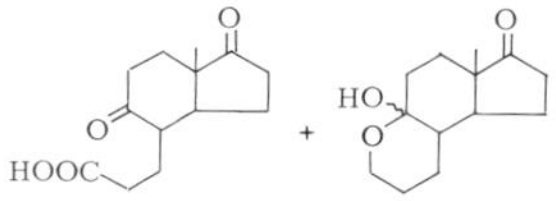
Nocardia sp. M 29–40	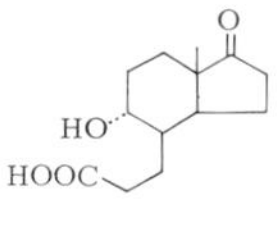
Nocardia corallina mutant NG-511	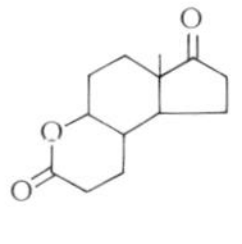
Nocardia corallina mutant NG-34	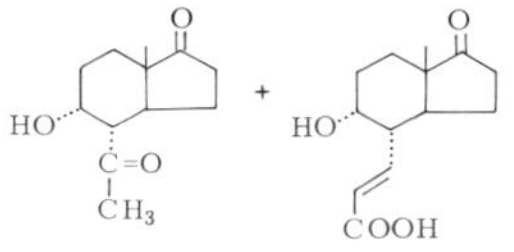

[a] References in SCHOEMER and MARTIN (1980)

et al. (1975) demonstrated that cholesterol could be transformed to 4-cholestene-3-one very efficiently by *Nocardia* sp. NCIB 10 554 in the presence of high concentrations of water-immiscible solvents such as carbon tetrachloride. The enzymes involved (cholesterol oxidase and catalase) continued to function under these conditions. Unfortunately, no further breakdown of the substrate occurred indicating that several other enzymes essential for degradative reactions had been inactivated.

Immobilization by entrapment in polymers represents a method for stabilization of enzymes and cells. Therefore, this method was employed for sterol transformation in organic solvents. However, cholesterol was only converted to 4-cholestene-3-one by *Nocardia rhodochrous* (FUKUI and TANAKA, 1980) or *N. erythropolis* (ATRAT et al., 1980) when immobilized in hydrophobic photocrosslinkable polymers or by adsorption. As in the case of free suspended cells, further breakdown of the sterol did not occur in the presence of organic solvents. On the other hand, side chain degradation of sterols by immobilized cells in aqueous buffers seems to occur readily. For instance, viable cells of *Mycobacterium phlei*, immobilized in polyacrylamide gel, converted 4-cholestene-3-(*O*-carboxymethyl)-oxime to 4-androstene-17-one-3-(*O*-carboxymethyl)-oxime in a continuous process over more than 40 days (ATRAT et al., 1981).

VI. Microbial Hydrogenation of Sterols

The discussion of oxidative degradation reactions of sterols by microorganisms has been emphasized in previous sections since the resulting fermentation products are of industrial value. However, microorganisms are also capable of reducing double bonds or keto functions in various sterol compounds.

Depending on the organism used, 5β- or 5α-H-sterol derivatives can be obtained from cholesterol or analogs. *Eubacterium* sp. ATCC 21 048 hydrogenates cholesterol, campesterol, β-sitosterol and 7-dehydrocholesterol to the corresponding 5β-H-structures in good yield (EYSSEN and PARMENTIER, 1974). 5α- and 5β-cholestan-3-one are reduced to the 3β-alcohols. Furthermore, since Δ^4-double bonds are also reduced by this organism, 4-cholestene-3β-ol and 4-cholestene-3-one are converted to coprostanol, similarly (EYSSEN et al., 1973, 1974). Using double labelled substrates the authors postulated the following reaction sequence: 3β-OH-Δ^5 → 3-oxo-Δ^4 → 3-oxo-5β-H → 3β-OH-5β-H (PARMENTIER and EYSSEN, 1974). Although these hydrogenated sterols are of no commercial value these reactions may have other applications because of their stereoselectivity.

Besides 5β-H-compounds, the corresponding 5α-H-structures can be prepared by microbial hydrogenation with *Nocardia corallina* ATCC 13 259 (LEFEBVRE et al., 1974a) or *Nocardia opaca* ATCC 4 176 (LEFEBVRE et al., 1974b).

VII. Conclusion

As a result of a shortage of diosgenin and an increased demand for steroid drugs, the conversion of plant sterols to intermediates useful for the synthesis of medically important steroid hormones has been investigated intensively.

Based upon studies on the pathway of sterol metabolism by certain bacteria, methods for selective cleavage of the 17-alkyl side chain and for partial degradation of the ring structure have been developed on a laboratory scale. Since these processes have to compete with chemical methods, high conversion efficiencies in the presence of large substrate concentrations have to be obtained.

Some of the processes described for the synthesis of 17-ketosteroids, using mutants blocked in their ability to degrade the steroid ring system, are now utilized on an industrial scale.

The fermentation products are very useful starting compounds for the synthesis of pharmacologically active steroid drugs: 4-androstene-3,17-dione can be converted to testosterone and other steroids with androgenic activity; estrogens like estrone are accessible by pyrolytic aromatization of ring A of 1,4-androstadiene-3,17-dione, and 9α-hydroxy-4-androstene-3,17-dione, after chemical addition of the pregnane side chain, is important for the preparation of 9α-halogen-hydrocortison.

Furthermore, in addition to 17-ketosteroids, other steroids prepared by oxidative microbial degradation of sterols are of industrial importance. For instance, bisnorcholanic acid derivatives are considered useful starting materials for the synthesis of new corticoids.

As mentioned earlier, hexahydroindan derivatives are precursors for chemical synthesis of retrosteroids or steroids containing hetero atoms.

VIII. References

Atrat, P., Hueller, E., and Hoerhold, C. (1980). Z. Allg. Mikrobiol. *20,* 79.

Atrat, P., Hueller, E., and Hoerhold, C. (1981). Eur. J. Appl. Microbiol. Biotechnol. *12,* 157.

Beukers, R., Marx, A. F., and Zuidweg, M. H. J. (1972). Med. Chem., Ser. Monogr. *11,* 1.

Boehme, K. H., and Hoerhold, C. (1980). Z. Allg. Mikrobiol. *20,* 85.

Buckland, B. C., Dunnill, P., and Lilly, M. D. (1975). Biotechnol. Bioeng. *17,* 815.

Cargile, N. L., and McChesney, J. D. (1974). Appl. Microbiol. *27,* 991.

Charney, W., and Herzog, H. (1967). "Microbial Transformation of Steroids." Academic Press, New York.

Conner, A. H., Nagaoka, M., Rowe, J. W., and Perlman, D. (1976). Appl. Environ. Microbiol. *32,* 310.

Dias, J. R., and Petit, G. R. (1972). US Patent 3 661 941.

Eyssen, H. J., and Parmentier, H. (1974). Am. J. Clin. Nutr. *27,* 1329.

Eyssen, H. J., Parmentier, H., Compernolle, F. C., de Pauw, G., and Piessens-Denef, M. (1973). Eur. J. Biochem. *36,* 411.

Eyssen, H. J., de Pauw, G., and Parmentier, H. (1974). J. Nutr. *104,* 605.

Fukui, S., and Tanaka, A. (1980). Advances in Biotechnology, Proc. 6th Int. Ferm. Symp., London (Ontario), p. 343.

Görlich, B. (1973a). Ger. Patent 2 226 846.

Görlich, B. (1973b). Ger. Patent 2 226 930.

Görlich, B. (1973c). Ger. Patent 2 226 997.

Görlich, B. (1973d). Planta Medica *23,* 39.

Haberland, M. E., and Reynolds, J. A. (1973). Proc. Natl. Acad. Sci. USA *70,* 2 313.

Hill, F. F., Schindler, J., Schmid, R. D., Wagner, R., and Voelter, W. (1982). Eur. J. Appl. Microbiol. Biotechnol. *15,* 25.

Hoffmann-La Roche (1981), U.S. Patent 4 301 246.

Imada, Y., Ishikawa, H., and Nishikawa, D. (1981). Nippon Nogei Kagaku Kaishi *55,* 713.

Kieslich, D. (1980). In "Economic Microbiology" (A. H. Rose, ed.), Vol. 5, p. 370, Academic Press, New York.

Knight, J. C., and Wovcha, M. G. (1980). Steroids *36,* 2711.

Komel R., Groh, H., and Hoerhold, C. (1980). Z. Allgem. Mikrobiol. *20,* 637.

Lefebvre, G., Germain, P., and Gray, P. (1974a). Tetrahedron Lett., 127.

Lefebvre, G., Germain, P., and Gray, P. (1974b). C. R. Seances Soc. Biol. Fil. *274,* 449.

Marsheck, W. J. (1971). Progr. Ind. Microbiol. *10,* 49.

Martin, C. K. A. (1977). Adv. Appl. Microbiol. *22,* 29.

Nominé, G. (1980). Bull. Soc. Chim. Fr. II, 18.

Parmentier, G., and Eyssen, H. J. (1974). Biochim. Biophys. Acta *348,* 279.

Sato, A., (1979). Kagaku Kojo *23,* 31.

Sauerbaum, J., Martin, C. K. A., and Wagner, F. (1978). 1st Eur. Congr. Biotechnol., Interlaken 1978, p. 138.

Schoemer, U., and Martin, C. K. A. (1980). Biotechnol. Bioeng. *22,* Suppl. *1,* 11.

Sih, C. J., and Whitlock, H. W. (1968). Annu. Rev. Biochem. *37,* 661.

Smith, A. G., and Brooks, C. J. W. (1976). J. Steroid Biochem. *7,* 705.

Smith, L. L. (1974). Terpenoids Steroids *4,* 394.

Tömörkeni, E., Toth, G., Horvath, G., and Büki, K. G. (1975). Acta Chim. Acad. Sci. Hung. *87,* 409.

Voishvillo, N. E., and Kamernitskii, A. V. (1976). Biol. Nauki (Moscow) *19,* 7.

Wovcha, M. G., Antosz, F. J., Knight, J. C., Kominek, L. A., and Pyke, T. R. (1978). Biochim. Biophys. Acta *531,* 308.

Chapter 4

Terpenoids

Victor Krasnobajew

Givaudan Research Company Ltd.
CH-8600 Dübendorf, Switzerland

I. Introduction
II. Methods
  A. Transformation with Pure Cultures from Collections
  B. Substrate Addition
  C. Enrichment Cultures
  D. Various Techniques
  E. Isolation of Transformation Products
III. Prediction of Transformation Site
IV. Microbial Transformations of Terpenoid Compounds
  A. Monoterpenoids
    1. Acyclic monoterpenoids
      a) Citronellal
      b) Citral
      c) Linalool
    2. Monocyclic monoterpenoids
      a) Limonene
      b) (−)-Menthol
    3. Bicyclic monoterpenoids
      a) α-Pinene
      b) β-Pinene
      c) 1,8-Cineole
  B. Ionones and Related Compounds
    1. β-Ionone
    2. (±)-3-Acetoxy-4-oxo-β-ionone
    3. α-Ionone
    4. Oxo-isophorone
    5. β-Damascone
  C. Sesquiterpenoids
    1. Patchouli alcohol
  D. Diterpenoids
  E. Triterpenoids
    1. Squalene-2,3-epoxide
V. References

I. Introduction

In recent years microbial transformations of terpenoid compounds have collectively been discussed by CIEGLER (1969) and VOISHVILLO et al. (1970); certain chapters within reviews on microbial transformations of organic compounds (ABBOTT and GLENDHILL, 1971; KIESLICH, 1976; SEBEK and KIESLICH, 1977; JOHNSON, 1978) have also dealt with the topic.

This review focuses on microbial transformations of terpenoid compounds potentially interesting for practical application especially in the flavor and fragrance industry, and furthermore, on compounds with a potential for further biotechnological development.

Since antiquity, flavorists and perfumers have developed sophisticated formulas of naturally occurring terpenes and their oxygenated derivatives. Steam distillation of plant materials yields essential oils with terpenoids representing the major components. Some are present in low concentrations whereas others constitute the principal components of these oils. Terpenoid compounds are also found in the animal kingdom, especially in insects (HEROUT, 1970), microorganisms (COLLINS, 1976), and marine organisms (GOAD, 1978; LIAAEN-JENSEN, 1978; TURSCH et al., 1978).

After it was recognized that terpenoid compounds are basically formed from biological C_5-isoprene units, RUZICKA (1959) formulated the so-called isoprene rule. The biosynthesis of monoterpenoids has been recently reviewed by CHARLWOOD and BANTHORPE (1978), and of terpenoids in general by CROTEAU (1980).

Together with newly acquired knowledge of chemical reaction mechanisms it became possible to explain the vast variety of naturally occurring terpenoid structures. With the rapid progress in chemical technology, scientists devised simple procedures to synthesize many of the important natural terpenoids (MEULY, 1970; HOFFMANN, 1975).

Microbial transformations of terpenoids were carried out by MAYER and NEUBERG as early as 1915. However, most of the more recent studies are confined to microbial breakdown of acyclic as well as cyclic compounds isolated from essential oils (WOOD, 1969). The results were more of theoretical than of practical value.

Difficulties encountered in terpenoid transformation are due to the fact that many microorganisms selected through screening and adaptation often transform or degrade the added terpenoid into a variety of useless or difficultly separable metabolites. Furthermore, the volatile terpenoid compounds may get lost through the required strong aeration, especially in deep-tank fermentation.

On the other hand, various reports indicate a high potential for the industrial use of microbial transformations of terpenoids (CIEGLER, 1969).

Due to the concepts of earlier studies, industry has widely neglected the possibility of applying fermentative technologies on terpenoid compounds much in contrast to the current employment of these techniques in steroid chemistry. Over the last years, however, attempts for industrial use of terpenoid fermentations were described in patent literature of various research groups; these will be discussed in later sections of this chapter. Although microbiological transformations of organic compounds including terpenoids may appear economically discouraging on lab-scale, compared to synthesis it should be pointed out that biological as well as petrochemical procedures follow the cost/volume scale laws (KING, 1982).

II. Methods

Valuable practice oriented techniques and procedures in studying microbial transformations of organic compounds in general, were described by PERLMAN (1976) and in Chapter 1 of this volume by LEUENBERGER. In the following a brief outline of the methods regarding terpenoid compounds has been compiled.

A. Transformation with Pure Cultures from Collections

In most cases, a transformation of terpenoids and related compounds can be achieved by choosing pure cultures of microorganisms from public collections.

The chances for success are significantly higher by selecting organisms reportedly known to perform the desired type of reaction on other compounds; similarly, strains may be chosen which through personal experience have shown to be efficient "allround transformers".

The selected microorganisms are cultivated by standard microbiological techniques in appropriate growth media. For small-scale transformation studies, conical flasks on rotation shakers will be sufficient. For larger runs, fermenters with efficient stirring and aeration are necessary.

B. Substrate Addition

The terpenoid compound, in most cases, is added to the culture when the maximum growth rate has been obtained. This is especially the case when transformations are carried out with cultures which were randomly selected from culture collections. In many cases, the addition of the terpenoid will inhibit culture growth and lead to cell lysis. Continuous or repeated addition of non-toxic quantities of the terpenoid during incubation can extend the transformation capacity of the transforming organism. Substrates may be added in pure form but are mostly administered as concentrates in solvents non-toxic to the organisms (PERLMAN, 1976).

If the terpenoid simultaneously being utilized by the microorganism as a carbon source the compound should be added to the culture medium from the early beginning of growth. In order not to overload the organisms, especially during the initial growth phase concentrations lower than 1% are usually employed.

As lower terpenoids are fairly volatile, care must be taken to maintain the desired concentration (e. g. repeated addition). The course of transformation can be monitored by gas chromatography or thin layer chromatography of culture samples. A more advanced method is the monitoring of volatile terpenoids by an attached mass spectrometer (HEINZLE and DUNN, 1983).

C. Enrichment Cultures

A method often employed to select microorganisms with a desired enzymatic activity is the enrichment of cultures obtained from soils, decomposing materials, sewage plants, river waters, factory grounds, or from the surroundings of plants which produce terpenoids. In this case, the sample is incubated in a growth medium containing the terpenoid compound to be transformed as the sole carbon source.

Descriptions of the media used for culture enrichment are summarized by BECK (1971), practice-oriented methods are summarized in Volume 1 of "Biotechnology", p. 355, by NAKAYAMA (1981).

In nature terpenoid compounds are easily decomposed by microorganisms, and, therefore, there is a fair chance to quickly obtain a selection of one or several organisms which will grow on a particular terpenoid substrate.

An often observed drawback of such isolates is the uncontrolled course of transformation resulting in numerous metabolites or in a total breakdown of the terpenoid, followed by their incorporation into biomass and CO_2. Only persistent metabolites, if any, will accumulate. In such a situation variation of the growth media, e. g., by addition of competing carbon sources, can be helpful. A separation of the microorganism and resuspension in buffer solution containing the terpenoid may prevent uncontrolled degradation.

D. Various Techniques

A further technique is the immobilization of the separated cells, thus allowing repetitious use or continuous operations in bioreactors.

In certain cases a transformation has to be performed with isolated enzymes. Moreover, biotransformations by spores are feasible (VEZINA et al., 1968).

Treatment with mutational agents may produce mutants of the selected microorganisms which lack the ability to grow on a certain terpenoid. The mutation obviously leads to an enzyme defect; however, it may be possible to maintain the essential first steps of metabolism which allow the accumulation of the desired transformation product.

On the other hand, it can not be excluded that mutants which still grow on the terpenoid may also accumulate the desired transformation product. Mutagenic techniques are summarized by JACOBSON (1981) in "Biotechnology," Volume 1, p. 279.

A more recent method for selective enrichment of mutants defective in the dissimilatory pathway was described by WIGMORE and RIBBONS (1980) with respect to the transformation of *p*-cymene. An enrichment of mutants unable to grow on the added compound as sole carbon source can be effected by halogenated substrate analogs.

E. Isolation of the Transformation Products

Even on lab-scale the isolation of terpenoid transformation products from culture broth may pose considerable problems. Due to cell-lysis effected by the substrate, but also due to the presence of emulgators and other additives, attempts of organic solvent extraction of the metabolites have led to the formation of stable emulsions.

On small scale, such emulsions can be broken by simple centrifugation in beakers or by separation through filters supported with filter aids such as diatomaceous earth. Kutscher-Steudel perforators have proved efficient for the purpose. On a production scale, where continuous extraction of the fermentation broth is envisaged, suspended solids may be removed by centrifugal separators connected to counter-current extractors (EDWARDS, 1969). A recent review on work-up procedures in biotechnology has been given by SCHMIDT-KASTNER and GÖLKER (1982).

An excellent but expensive method for extracting sensitive terpenoid metabolites is by supercritical CO_2, a method which has been established in the coffee industry (GRIMMETT, 1981), as well as in the flavor and perfumery industry (CALAME and STEINER, 1982).

III. Prediction of Transformation Site

The site of transformation in the terpenoid molecule is in many cases a matter of speculation as is generally the case for organic compounds. Generalizations or analogies are mostly impossible, except when molecular structures are very similar. Specific terpenoid transformations must usually be established empirically by screening, mutational techniques, and variation of fermentation conditions.

Microbial transformations of terpenoids mostly result in an enzymatic attack at several sites of the molecule leading to mixtures of metabolites. Accumulation of single transformation products is rare. The initial metabolites may in turn very well fit into the active site of the transforming enzymes and this may eventually lead to a cascade of transformations and to undesired degradation.

In many cases different terpenoid metabolites have been observed during the first stages of transformation than those found more towards the end, e. g., after several days. Such findings may indicate an induction of new enzymes by metabolites, or an alteration of the cellular membrane by the substrate.

Reaction sites of terpenoid substrate molecules, which are highly susceptible to chemical reagents are also easily attacked by corresponding enzyme systems. This is summarizes various aspects of the functionalization of organic compounds including terpenoids. Much work has already been done on steroids. Only little is known on lower terpenoids, however.

IV. Microbial Transformation of Terpenoid Compounds

A. Monoterpenoids

The majority of microbial transformations of terpenoids has been performed on monoterpenoids. Most of these studies are of theoretical value the main accomplishment being the description of various new

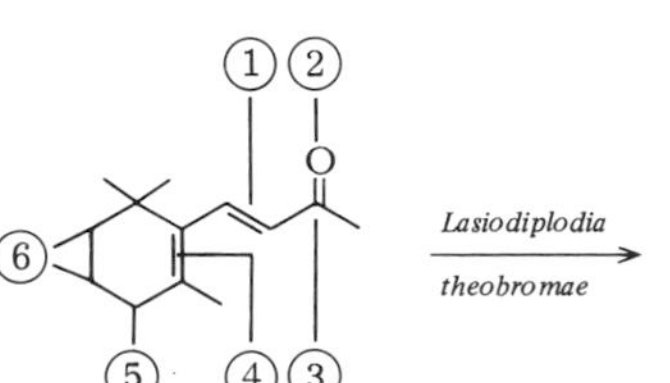

① Hydrogenation
② Reduction
③ Baeyer-Villiger oxidation
④ Epoxidation
⑤ Allylic oxidation
⑥ Hydroxylation (non-activated)

Scheme 1

illustrated in Scheme 1 for the example of *β*-ionone (**64**), a constituent of many essential oils. *β*-Ionone has at least five reaction sites. Indeed, five different chemical reactions can be performed by the fungus *Lasiodiplodia theobromae* ATCC 28 570 (KRASNOBAJEW and HELMLINGER, 1982). In addition, non-activated, chemically inert carbon atoms are unpredictably functionalized. By changing the fermentation conditions the type of reaction can be influenced.

In his comprehensive review on oxygenations with microorganisms JOHNSON (1978)

transformation products rather than the development of practical applications for industrial purposes. Nevertheless, microbial transformations of various monoterpenoids have challenging industrial possibilities.

Monoterpenoids and their derivatives are widely occurring in nature and are the main constituents of many essential oils. Their strong and pleasant odors make them important components in the manufacturing of flavors and fragrances. On the other hand, various monoterpenes, such as *α*-pinene (**47**), *β*-pinene (**56**), citral (**9**), citronellal (**1**), and limonene (**23**) are used in large quantities in the chemical industry for the production of more valuable terpenoids (NEWMAN, 1972; ERICSON, 1976).

1. Acyclic monoterpenoids

The earliest studies of microbial transformations were on acyclic terpenoids.

a) Citronellal (**1**)
MAYER and NEUBERG (1915) described the reduction of (+)-citronellal (**1**) to (+)-citronellol (**2**) by yeast (see Scheme 2).

Citronellol is of great importance in the fragrance industry. The optical antipodes β(+)-citronellol and β(−)-citronellol exhibit different odors (RIENÄCKER and OHLOFF, 1961). β(−)-Citronellol is preferred by perfumers.

A selective microbiological reduction of β(−)-citronellal in racemic citronellal to β(−)-citronellol as it occurs in Bulgarian rose oil (FENAROLI, 1975) was described in a Japanese patent application (TAKASAGO, 1970). Yeast strains such as *Candida reukaufii* AHU 3032 were the preferred microorganisms and an optical purity of more than 80% has been obtained after relatively short incubation times of several hours.

In a later research paper on a broad spectrum of (±)-citronellal reducing microorganisms YAMAGUCHI et al. (1976) showed that optical selectivity was higher in yeasts than in filamentous fungi. The authors point out that the key to industrial success of such a process is to find ways for an efficient regeneration of stoichiometrically necessary cofactors for the redox enzyme systems of the cells.

The microbial transformation of citronellal has attracted various investigators and, depending on the type of microorganisms employed, different compounds were produced. In most cases, however, the transformations resulted in mixtures of compounds.

Using *Pseudomonas aeruginosa* JOGLEKAR and DHAVLIKAR (1969) obtained citronellic acid (**3**) in 65% yield from concentrations of 0.35% (w/w) citronellal. Minor compounds were citronellol (**2**) (0.6%); dihydrocitronellol (**4**) (0.6%), 3,7-dimethyl-1,7-octane-diol (**5**) (1.7%), and interestingly, menthol (**6**) (0.75%). After incubation times

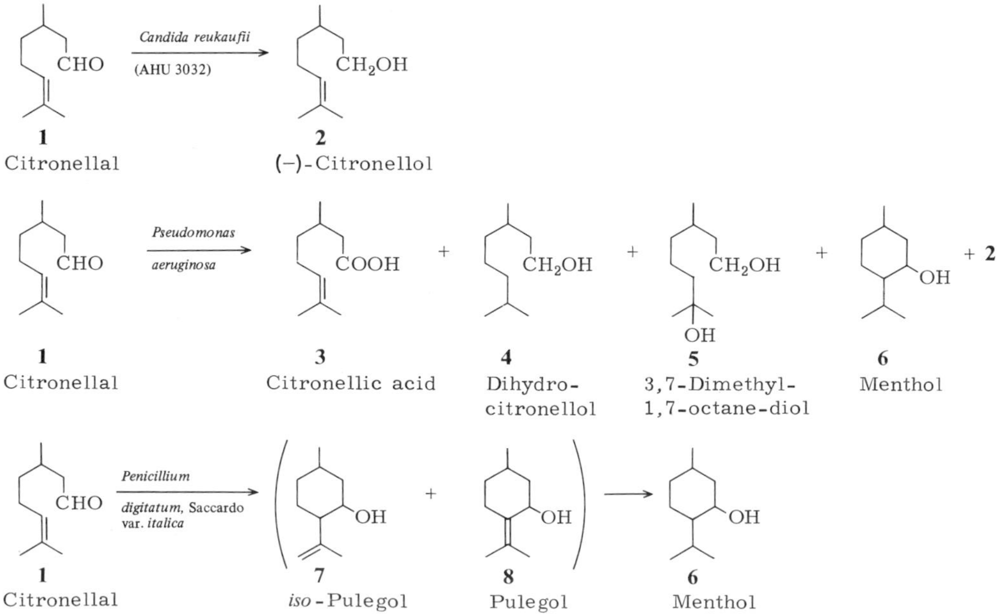

CHO —*Candida reukaufii* (AHU 3032)→ CH_2OH
1 Citronellal **2** (−)-Citronellol

CHO —*Pseudomonas aeruginosa*→ COOH + CH_2OH + CH_2OH / OH + OH + **2**
1 Citronellal **3** Citronellic acid **4** Dihydrocitronellol **5** 3,7-Dimethyl-1,7-octane-diol **6** Menthol

CHO —*Penicillium digitatum*, Saccardo var. *italica*→ (OH + OH) → OH
1 Citronellal **7** *iso*-Pulegol **8** Pulegol **6** Menthol

Scheme 2

of 96 h there were still 20% of unreacted citronellal present.

Microbial cyclization of citronellal by *Penicillium digitatum* Saccardo var. *italica* to menthol (**6**) in yields of 93% was claimed in a Czechoslovakian patent, by BABIČKA and VOLF (1955), where the unreacted citronellal is recirculated. Pulegol (**8**) and isopulegol (**7**) may be formed as intermediates. A mixture of these two latter compounds has yielded 95% menthol (**6**) when incubated with *Penicillium digitatum*. Optical purity of (**6**) has not been determined in this case. On the other hand, transformation times were extremely long. Citronellal (**1**) was left in contact with a surface culture for 28 days, the pulegol (**8**)/isopulegol (**7**) mixture for 21 days with a submerged culture.

b) Citral (**9**)

The conversion of citral (**9**), the main constituent of lemon grass, to geranic acid (**10**) was described by HAYASHI et al. (1967) using *Pseudomonas convexa* isolated from soil.

Geranic acid (**10**) was obtained in yields of 62% from 0.35% (w/w) citral (**9**) by *Pseudomonas aeruginosa* as well as minor quantities of 2,6-dimethyl-8-hydroxy-7-oxo-2-octene (**11**) (0.8%), 6-methyl-5-heptenoic acid (**12**) (0.5%), and 3-methyl-2-butenoic acid (**13**) (1%) as reported by JOGLEKAR and DHAVLIKAR (1969).

Citral (**9**) (Scheme 3) is less readily transformed by microorganisms than citronellal (**1**), probably due to the double bond in conjugation to the aldehyde function.

To date, no industrially important compound other than geranic acid (**10**) could be obtained via microbial transformation using shake cultures. The situation may look different by employing fermenters with efficient aeration, and higher yields of geranic acid (**10**) from citral (**9**), or citronellic acid (**3**) from citronellal (**1**) could be expected.

c) Linalool (**14**)

Linalool, a widespread constituent of essential oils used extensively in the fragrance industry (ARCTANDER, 1969) and in synthetic chemistry, has so far only yielded complex mixtures of metabolites in microbial transformations. MIZUTANI et al. (1971) described the conversion of linalool by a newly isolated *Pseudomonas pseudomallei* (strain A). Camphor (**15**) may result as a cyclization product. Other products were 2,6-dimethyl-6-hydroxy-*trans*-2,7-octadienoic acid (**16**) as well as the degradation products 5-methyl-5-vinyl-tetrahydro-2-furanone (**17**) and 4-methyl-*trans*-3-hexenoic acid (**18**) (see Scheme 4).

DEVI and BHATTACHARYYA (1977) obtained an even more complex metabolite mixture in their degradation studies of linalool by *Pseudomonas incognita*. The latter strain was isolated by enrichment culture techniques using linalool as the sole carbon source. Formation of camphor (**15**) was not observed. However, a cyclic compound (**17**) and a hydroxy acid (**16**) were identified. New metabolites were 8-hydroxy-linalool (**19**), 2-methyl-2-vinyl-5-hydroxyisopropyl-tetrahydrofuran (**22**), linalool-10-carboxylic acid (**21**), and oleuropeic acid (**20**).

Though these metabolite mixtures contain various interesting compounds such as linalool oxide and the lavender lactone (**17**) which are used in perfumery, their isolation remains laborious. Extensive care in the selection of strains as well as the use of muta-

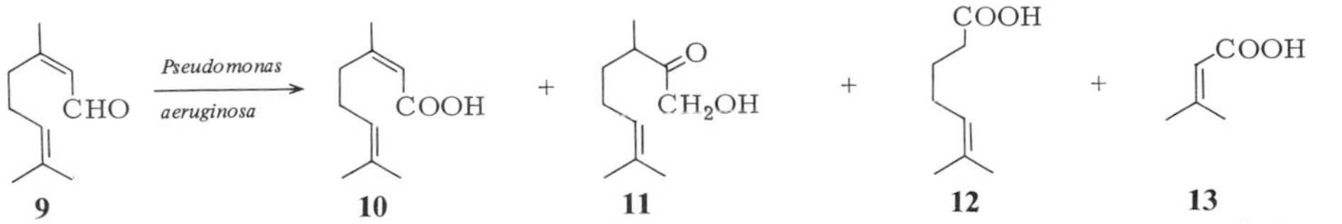

CHO —*Pseudomonas aeruginosa*→ COOH + CH_2OH + COOH + COOH

9 Citral

10 Geranic acid

11 2,6-Dimethyl-8-hydroxy-7-oxo-2-octene

12 6-Methyl-5-heptanoic acid

13 3-Methyl-2-butenoic acid

Scheme 3

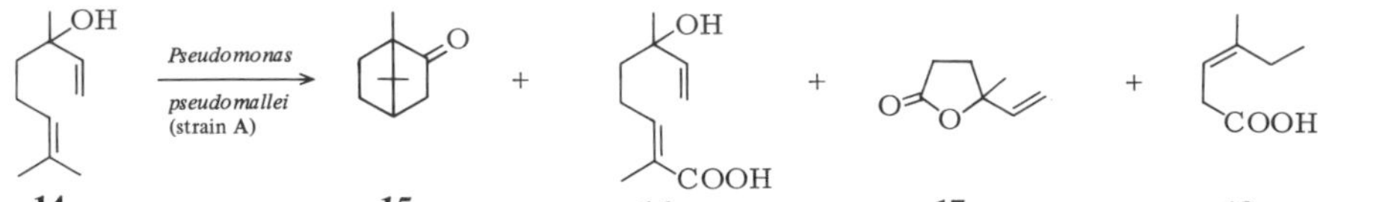

14 Linalool → *Pseudomonas pseudomallei* (strain A) → 15 Camphor + 16 2,6-Dimethyl-6-hydroxy-*trans*-2,7-octadienoic acid + 17 Lavenderlactone + 18 4-Methyl-*trans*-3-hexenoic acid

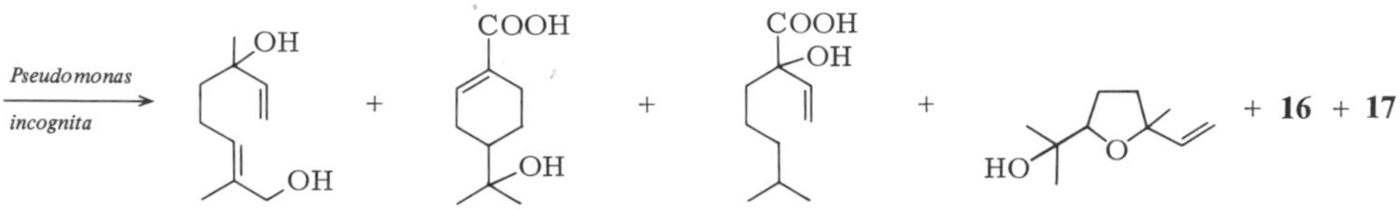

→ *Pseudomonas incognita* → 19 8-Hydroxy-linalool + 20 Oleuropeic acid + 21 Linalool-10-carboxylic acid + 22 2-Methyl-2-vinyl-5-hydroxyisopropyl-tetrahydrofurane + 16 + 17

Scheme 4

tional techniques may be helpful to produce single metabolites. On the other hand, such complex metabolite mixtures could possibly be used in compositions of essential oils.

2. Monocyclic monoterpenoids

a) Limonene (**23**)

Much work has been done on the microbial transformation of the inexpensive limonene (Bowen, 1975), which is, besides α-pinene (**47**), the most widely distributed terpene in nature (Gildemeister and Hoffmann, 1960a).

Limonene (**23**) is easily attacked by the complex enzyme systems of all kinds of microorganisms, and strains growing on limonene as the sole carbon source can be found within a short time by enrichment culture techniques (refer to the previously cited reviews on terpene transformations). However, little success has been achieved in the attempt to convert the numerous research data into industrial practice. Limonene undoubtedly has great potential in fermentation technology, and there is a good possibility of producing numerous flavor and fragrance chemicals.

The microbial transformation of limonene has been studied in detail by Dhavalikar and Bhattacharyya (1966), and a large number of neutral and acidic metabolites have been isolated and identified. Several of these are of industrial importance such as carveol (**24**), carvone (**25**), dihydro carvone (**28**), perillyl alcohol (**31**), and perillaldehyde (**32**). As can be seen from Scheme 5, Dhavalikar et al. (1966) propose three pathways in the conversion of limonene (**23**). Transformation products from pathway 1 are (+)-*cis*-carveol (**24**) and (+)-carvone (**25**), both important constituents of caraway seed and dill-seed oils (Henry, 1982), besides 1-*p*-menthene-6,9-diol (**26**). Interestingly, the industrially more important (−)-carvone which contributes to spearmint flavor (Fenaroli, 1975) has not been described in microbiological transformations.

Pathway 2 yields (+)-dihydrocarvone (**28**) via the intermediate limonene epoxide (**27**), as well as 8-*p*-menthene-1-ol-2-one (**30**), an oxidation product of limonene-diol-1,2 (**29**).

Interesting transformation products of pathway 3 are perillyl alcohol (**31**), perillaldehyde (**32**), and perillic acid (**33**). They are constituents of various essential oils and

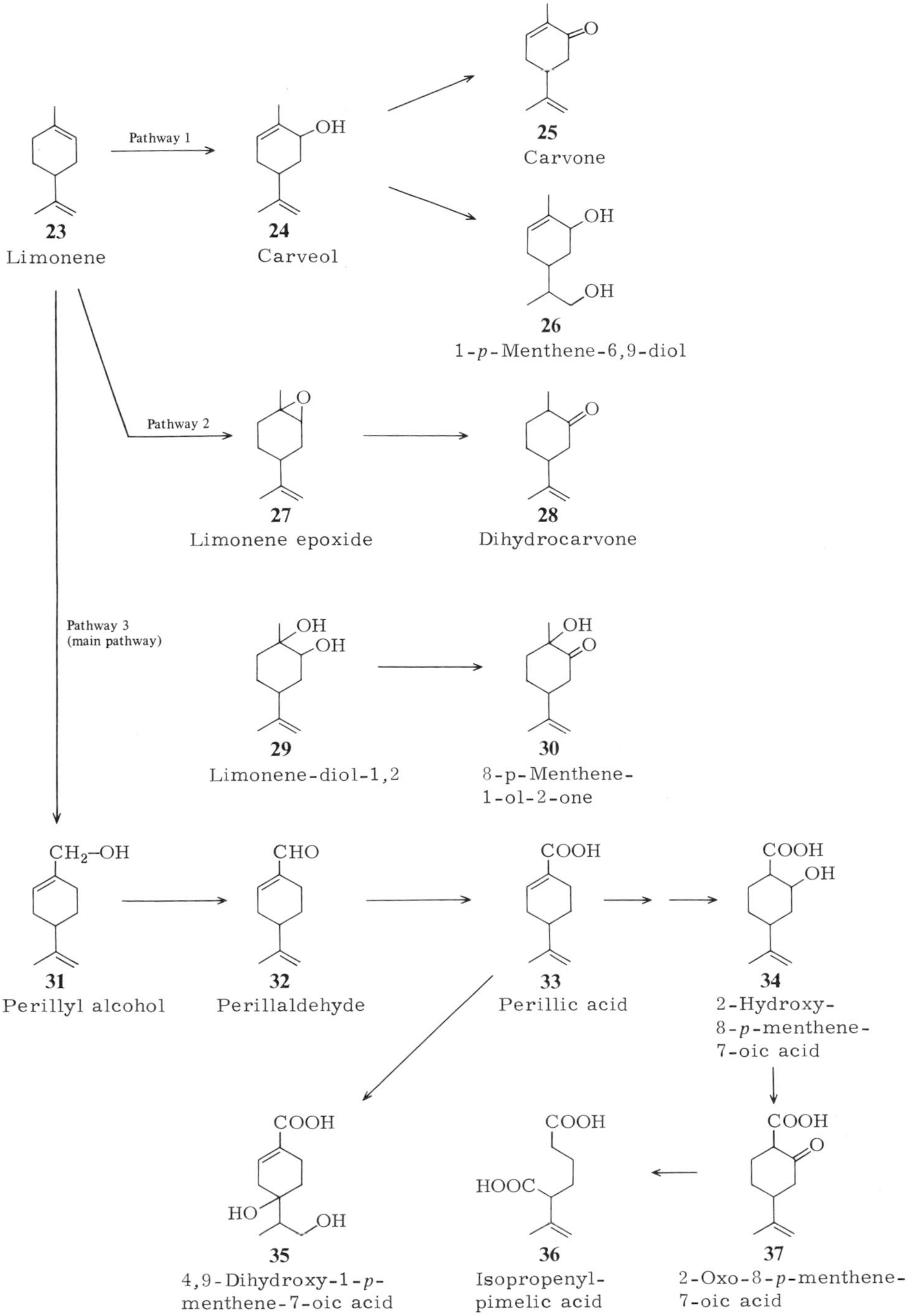

Pathway 1
OH
25
Carvone
23
Limonene
24
Carveol
OH
OH
26
1-p-Menthene-6,9-diol
Pathway 2
O
O
27
Limonene epoxide
28
Dihydrocarvone
Pathway 3
(main pathway)
OH
OH
OH
O
29
Limonene-diol-1,2
30
8-p-Menthene-
1-ol-2-one
CH2-OH
CHO
COOH
COOH
OH
31
Perillyl alcohol
32
Perillaldehyde
33
Perillic acid
34
2-Hydroxy-
8-p-menthene-
7-oic acid
COOH
COOH
COOH
O
HOOC
HO
OH
35
4,9-Dihydroxy-1-p-
menthene-7-oic acid
36
Isopropenyl-
pimelic acid
37
2-Oxo-8-p-menthene-
7-oic acid

Scheme 5

used in the flavor and fragrance industry (FENAROLI, 1975).

Other products of pathway 3 degradation are 2-hydroxy-8-*p*-menthene-7-oic acid (**34**), its oxidation product (**37**), isopropenyl-pimelic acid (**36**), and 4,9-dihydroxy-1-*p*-menthene-7-oic acid (**35**). The aim of these studies was not the production of industrially important compounds but the desire to differentiate between enzymic pathways and identify the metabolites. For industrial application mutation techniques as well as genetic engineering could possibly provide modified strains performing only one single pathway leading to desired products, without degradation.

An effort to increase the value of cheap (+)-limonene (**23**) isolated from citrus was described by BOWEN (1975), who used molds responsible for the common post-harvest diseases of citrus fruits (SMOOT, 1971). Pure cultures of *Penicillium digitatum* and *P.italicum* were isolated from overripe oranges by enrichment culture techniques on Czapek-Dox broth (DIFCO MANUAL, 1977) containing 1% limonene as the sole carbon source.

Transformation experiments with both fungi were conducted in Erlenmeyer flasks. Deep-tank fermentations were not reported. *P.italicum* converted limonene concentrations of 0.5% (v/v) with an efficiency of up to 80%. Higher concentrations of limonene were found to be inhibitory.

The composition of metabolites at the end of fermentation after 9 days was similar in both cases and six main products were identified. Transformation by *P.italicum* yielded the following products: *cis*- and *trans*-carveol (**24**) as main products at 26%, carvone (**25**) at 6%, *cis*- and *trans-p*-mentha-2,8-dien-1-ol (**38**) at 18%, *p*-mentha-1,8-dien-4-ol (**39**) at 4%, perillyl alcohol (**31**) at 3%, and *p*-menth-8-ene-1,2-diol (**29**) at 3% (cf. Scheme 6).

Most of these transformation products are of commercial value, especially carvone and carveol. Interestingly, the two alcohols

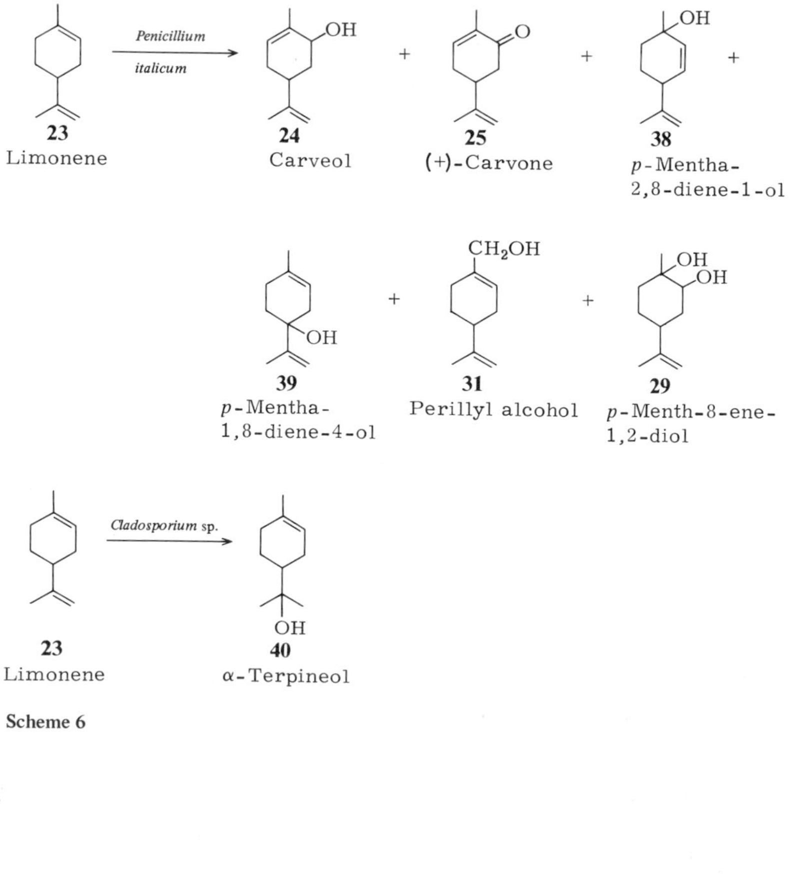

Penicillium italicum

23 Limonene → **24** Carveol + **25** (+)-Carvone + **38** *p*-Mentha-2,8-diene-1-ol + **39** *p*-Mentha-1,8-diene-4-ol + **31** Perillyl alcohol + **29** *p*-Menth-8-ene-1,2-diol

Cladosporium sp.

23 Limonene → **40** α-Terpineol

Scheme 6

38 and **39** were not described in the transformation studies of DHAVALIKAR and BHATTACHARYYA (1966).

A process for the microbiological production of carvone (**25**) from limonene (**23**) was patented by HASEGAWA Co. (1972). The inventors use the technique of cooxidation which is frequently applied to hydrocarbon oxidations (RAYMOND and JAMISON, 1971). A strain of *Corynebacterium hydrocarboclastum* (FERM-P 401) which grows on aliphatic hydrocarbons has been found to produce and accumulate carvone by cooxidation of limonene.

The results indicate, however, that only low amounts of limonene may undergo cooxidation in the described system. From 10 liters of culture medium only 32 mg of purified carvone were obtained; for an industrial application, this undoubtedly interesting process still requires optimization.

From their investigations on microbial transformations of monocyclic terpenes MUKHERJEE et al. (1973) reported the conversion of (+)-limonene (**23**) to limonene-1,2-diol (**29**). The latter diol appears to be a common metabolite in microbial transformations of limonene and has also been isolated from the transformation product mixtures mentioned above (DHAVALIKAR et al., 1966; BOWEN, 1975).

Through transformation of (+)-limonene by an isolated fungal strain, tentatively identified as *Cladosporium* sp. T_7, MUKHERJEE et al. (1973) obtained 1.5 g of *trans*-limonene-1,2-diol (**29**) and 0.2 g of the corresponding *cis*-diol from one liter of growth medium.

This conversion represents the rare case where practically only one single transformation product is obtained. Limonene-1,2-diol has so far not attained industrial importance, even though the authors have shown the possibility of chemically converting the diol (**29**) into carvone (**25**).

By following the enrichment technique described by the above authors, microorganisms growing on limonene or other lower terpenoids can be obtained.

The same authors (KRAIDMAN et al., 1969) reported the production of (+)-α-terpineol (**40**) from (+)-limonene by fungal and bacterial microorganisms, especially by a *Cladosporium* species designated T_{12}, isolated as described above from terpene-soaked soil.

α-Terpineol is widely distributed in nature and is one of the most commonly used perfume chemicals (ARCTANDER, 1969; FERNAROLE, 1975). As α-terpineol is a rather inexpensively synthesized compound from limonene, a microbial process for its production would have to achieve high yields.

KRAIDMAN et al. reported a modest yield of about 1 g/L of a growth medium composed of minerals and (+)-limonene as the sole carbon source. The optimization of α-terpineol production on an inexpensive growth medium on a continuous production scale could be of interest to the chemical industry.

b) (−)-Menthol

(−)-Menthol (**42**) is one of the most important terpene alcohols and is used extensively in the pharmaceutical, perfumery, and flavoring industry. Natural (−)-menthol is obtained by crystallization from peppermint oils of mint plants of the genus *Mentha* (GUENTHER, 1966).

Chemists have been trying for a long time to produce (−)-menthol from inexpensive terpenoid sources which unavoidably have also yielded the (±)-isomers: isomenthol (**44**), neomenthol (**45**), and neoisomenthol (**46**). The (−)-isomers of the latter compounds are depicted in Scheme 7 (for a summary on menthol refer to CROTEAU, 1980).

The optical resolution of (±)-menthol from such mixtures has been realized on the industrial scale (HAARMANN and REIMER GmbH, 1971) by esterification with benzoic acid derivatives and fractional crystallization.

Biochemists have conducted elaborate investigations to make use of the discovery that microbial esterases preferably hydrolyze (−)-menthyl esters while only barely affecting hydrolysis of contaminating stereoisomers.

The separation of liberated (−)-menthol is subsequently achieved by chromatography, fractional distillation, or crystallization. Patents have been filed by several companies covering a wide variety of bacteria as well as fungi and their esterases.

MOROE et al. (1971) from the Takasago Perfumery Co. Ltd. claim that certain selected species of *Absidia, Penicillium, Rhizopus, Trichoderma, Bacillus, Pseudomonas* and many others asymmetrically hydrolyze esters of (±)-menthol isomers such as formiates, acetates, propanoates, caproates, and esters of higher fatty acids.

The complex isomer composition of racemic menthol esters used as substrates for industrial production, causes difficulties in separation of (−)-menthol even after enzymatic hydrolysis, as shown by the following example.

Example: Asymmetric hydrolysis of (±)-menthyl acetate (**41**) by *Trichoderma reesei* AHU 9 485.

A 30 L jar-fermenter containing 20 kg of growth medium composed of 2% glucose, 0.5% KH_2PO_4 and 0.7% yeast extract was inoculated after heat sterilization with 500 g of a *Trichoderma reesei* AHU 9 485 preculture, grown on the same medium. The mixture was cultured for 24 h at 27 °C with an airflow of 5 L/min and a stirring rate of 300 r.p.m. Then 200 g of the acetates of a (±)-menthol isomer mixture (63.8% (±)-menthol, 5.8% (±)-neomenthol and 30.4% (±)-isomenthol) were added and the mixture was incubated at 28 °C by stirring at 300 r.p.m. for 24 hours.

Product isolation

Subsequently, the fermenter content was steam-distilled and the watery distillate extracted with toluene. After evaporation of the solvent, 173 g of oil was obtained. Analysis by gas chromatography showed a content of 5% (±)-neomenthol acetate, 48.5% (+)-menthyl acetate and (+)-isomenthyl acetate, 28.8% (−)-menthol, and 17.4% (−)-isomenthol.

100 g of the isolated oil was chromatographed on 1 kg alumina, and developed with hexane. Elution with 2 L of hexane yielded 50 g of a mixture consisting of (±)-neomenthol acetate, (+)-menthyl acetate, and (+)-isomenthyl acetate. Further elution with 2 L ethyl acetate yielded 45 g of a mixture of 62.3% (−)-menthol and 37.7% (−)-isomenthol. The latter mixture was esterified with monochloroacetic acid and recrystallized from methanol to yield pure (−)-menthyl monochloroacetic acid ester. Hydrolysis of this ester yielded ca. 23 g of (−)-menthol, $[\alpha]_D^{18} = -50$ ° (in ethanol).

A later publication by YAMAGUCHI et al. (1977) describes the asymmetric hydrolysis of (±)-menthyl acetate (**41**) by *Rhodotorula mucilaginosa* AHU 3 243 on an industrial scale; a detailed study of parameters affecting this hydrolysis was also included.

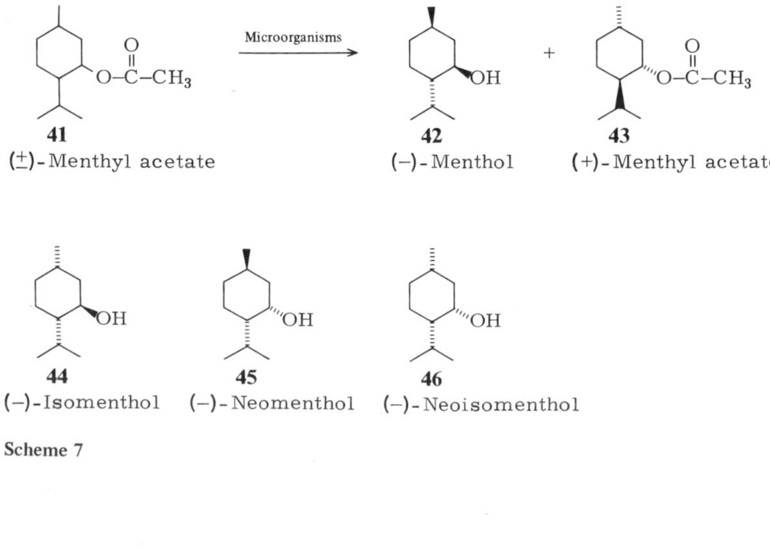

Microorganisms

41 (±)-Menthyl acetate → **42** (−)-Menthol + **43** (+)-Menthyl acetate

44 (−)-Isomenthol **45** (−)-Neomenthol **46** (−)-Neoisomenthol

Scheme 7

The latter yeast strain was found to have the highest optical selectivity and stereospecificity of all previously tested microorganisms.

From UV irradiation a mutant (UV-144-strain) was selected which showed a ca. 70% higher esterase activity than the original strain. An impressive yield of 44.4 g of (−)-menthol was obtained in 24 h from a 30% (±)-menthyl acetate mixture per liter of culture medium as the best result.

The acid group of the menthyl esters plays an important role for the reaction velocity of hydrolysis, and various modifications are listed in the patent literature. WATANABE and INAGAKI (1977) have patented an asymmetric hydrolysis of the monochloroacetates of (±)-menthol by *Arginomonas non-fermentans* FERM-P-1 924. The monochloroacetate is described to increase the reaction velocity and, furthermore, to decrease the amount of waste water. A large scale biochemical production of (−)-menthol is described in the aforementioned patent. The use of α-halo fatty acid esters of (±)-terpene alcohols including (±)-menthol has been claimed in 1974 by TAGASAGO PERFUMERY KK in combination with a large number of microorganisms.

NELBOECK-HOCHSTETTER et al. (1977) have patented a procedure using lactic acid esters of (±)-menthol which are asymmetrically hydrolized by an esterase from *Bacillus subtilis* CCM-110.

A reverse approach to separate (−)-menthol from racemic menthol has been described by the MEITO SANGYO Co. Ltd., (1975). Here, separation is achieved by making use of the ability of certain esterases, such as lipases, which preferably esterify the (−)-menthol isomer in combination with fatty acids.

The asymmetric hydrolysis of (±)-menthylsuccinate and of various other organic compounds, including terpenoids, has been reported by ORITANI and YAMASHITA (1973), and FUKUI (1981).

3. Bicyclic monoterpenoids

a) α-Pinene (**47**)

The most abundant terpene in nature is α-pinene which is industrially obtained by fractional distillation of turpentine (GILDEMEISTER and HOFFMANN, 1960b). Several microbiological transformation studies have been reported.

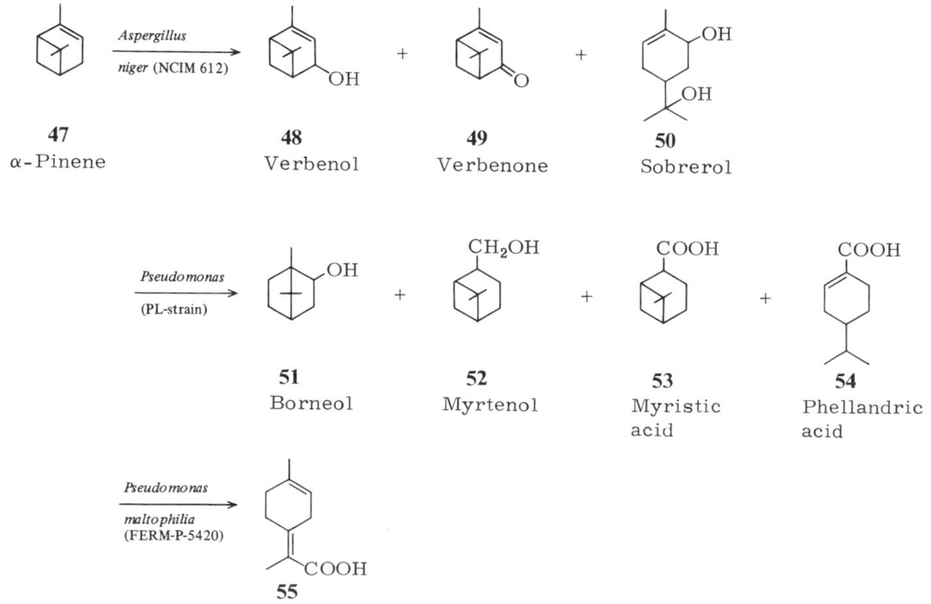

Aspergillus niger (NCIM 612)

47 α-Pinene → **48** Verbenol + **49** Verbenone + **50** Sobrerol

Pseudomonas (PL-strain)

→ **51** Borneol + **52** Myrtenol + **53** Myristic acid + **54** Phellandric acid

Pseudomonas maltophilia (FERM-P-5420)

→ **55** 2-(4-Methyl-3-cyclohexenylidene)-propionic acid

Scheme 8

The chemical industry produces numerous important substances for perfumes from the inexpensive α-pinene (TRAAS, 1982), and microbiological approaches for the conversion of α-pinene are challenging. On the other hand, the high volatility of this bicyclic terpene and its low conversion rates pose serious problems for an economic production of transformation products by fermentation.

Microbial degradation of α-pinene has been reported by HUNGUND et al. (1970) and by GIBBON and PIRT (1971) describing complex pathways which lead to the formation of organic acids through ring cleavage. PREMA and BHATTACHARYYA (1962a) transformed α-pinene by several fungi with the intention of obtaining oxygenated compounds valuable to the perfume industry (Scheme 8).

Aspergillus niger NCIM 612 from their culture collection exhibited maximum transformation capacity in shake cultures. Transformation products of importance, such as (+)-verbenone (**49**) and mainly (+)-*cis*-verbenol (**48**) were identified besides some (+)-*trans*-sobrerol (**50**). The authors found that relatively short fermentation times of 4 to 8 hours resulted in considerably better yields of transformation products than long contact periods. From ten pooled shake flasks containing 100 mL of culture and 0.5 mL of α-pinene each, an oily residue was extracted which consisted mainly of unreacted α-pinene (3.15 g) together with 0.88 g of oxygenated products. Encountered difficulties were the mentioned volatility of α-pinene and its inhibition when concentrations exceeded 0.6%. Deep-tank fermentations have not been successful.

A later publication by SHUKLA et al. (1968) described the fermentation of α-pinene (**47**) and β-pinene (**56**) in shake cultures by a soil *Pseudomonas* sp. (PL-strain) growing on α-pinene as the sole carbon source. A complex metabolite mixture was obtained composed of neutral as well as acidic compounds. Here, the valuable compounds verbenon (**49**) or verbenol (**48**) were not found in contrast to the case of *Aspergillus niger* mentioned above. Identified transformation products were (+)-borneol (**51**), myrtenol (**52**), myristic acid (**53**), phellandric acid (**54**), together with various compounds which had also been obtained from limonene (**23**) transformation product mixtures.

For industrial purposes, *Pseudomonas* strains would be of interest in which the degradation pathways are blocked, and which show higher resistance against the added terpene.

A deep-tank fermentation of α-pinene has been described in a patent application of the NISSAN Chem. Ind. (1980) by using *Pseudomonas maltophilia* FERM-P-5 420. The transformation product obtained was 2-(4-methyl-3-cyclohexenylidene)-propionic acid (**55**) which is formed by oxidative ring cleavage of α-pinene. The yield of 2 g of the acid **55** in the described process starting from 75 mL of α-pinene reflects a modest conversion rate.

b) β-Pinene (**56**)

Only very little is known about microbial transformations of β-pinene, which also is an abundantly occurring natural terpene.

From β-pinene SHUKLA et al. (1968) obtained a similarly complex mixture of transformation products as from α-pinene (**47**) through degradation by a *Pseudomonas* sp. (PL-strain). This indicates that the same intermediate products may enzymatically be formed from α- as well as from β-pinene and that these intermediates subsequently are converted to a large variety of products.

On the other hand, BHATTACHARYYA and GANAPATHY (1965) indicated that fungi such as *Aspergillus niger* NCIM 612, act differently and more specifically on the pinenes by preferably oxidizing β-pinene in the allylic position to form the interesting products pinocarveol (**57**) and pinocarvone (**58**), besides myrtenol (**52**) (see Scheme 9). Myrtenol is also formed by the above mentioned *Pseudomonas* sp.

c) 1,8-Cineole (**59**)

An interesting type of microbial transformation has been reported by MACRAE et al. (1979) for 1,8-cineole (**59**) or eucalyptol by

Aspergillus niger (NCIM 612)

56 β-Pinene → **57** Pinocarveol + **58** Pinocarvone + **52** Myrtenol

Scheme 9

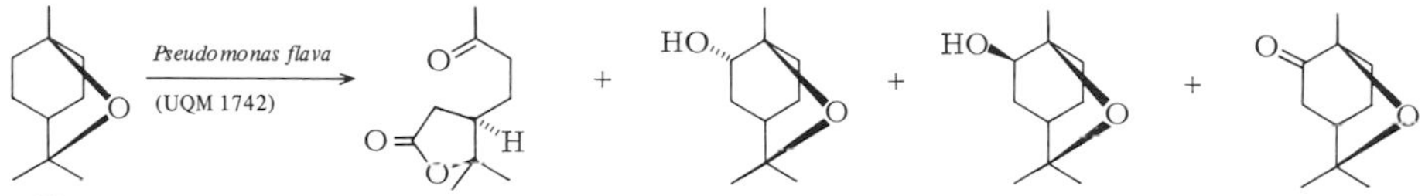

Pseudomonas flava (UQM 1742)

59 1,8-Cineole → **60** (*R*)-5,5-Dimethyl-4-(3'-oxobutyl)-4,5-dihydrofuran-2-(3 H)-one + **61** *exo*-2-Hydroxy-cineole + **62** *endo*-2-Hydroxy-cineole + **63** 2-Oxo-cineole

Scheme 10

a bacterium closely related to *Pseudomonas flava,* isolated from eucalyptus leaves, which utilized 1,8-cineole as the only carbon source. 1,8-Cineole, which is the main constituent of *Eucalyptus globulus,* is widely distributed in the plant kingdom and is one of the least expensive natural terpenoids (ARCTANDER, 1969). As indicated by MADYASTHA et al. (1977) the compound had, until that time, resisted microbial transformations.

Pseudomonas flava UQM-1 742 grew well on 1,8-cineole up to a concentration of 0.5 g/L; higher concentrations had a toxic effect on the bacterium.

In shake cultures the bacterium produced at least four compounds while transformations in fermenters were not reported. Analysis of the extracted transformation products from shake cultures showed mainly unreacted 1,8-cineole (**59**) at 68% and 14% of 2-oxo-cineole (**63**) as well as two isomeric alcohols, 2-hydroxy-cineole, 2% of an *endo*-alcohol (**62**), 5% of an *exo*-alcohol (**61**), and 10% of an interesting ketolactone (**60**), which obviously results from enzymatic ring cleavage and subsequent lactonization. The structure of the ketolactone (**60**) had been proposed as (*R*)-5,5-dimethyl-4-(3′-oxobutyl)-4,5-dihydrofuran-2(3H)-one (**60**). These aforementioned four terpenoids have not been previously reported from a natural source.

Example: Transformation of 1,8-cineole (**59**) by *Pseudomonas flava* UQM-1 742.

1. Isolation of *Pseudomonas flava* UQM-1 742. *Pseudomonas flava* was isolated from the surface of eucalyptus leaves by an enrichment culture on a medium composed of 0.2 g/L K_2HPO_4, 1.0 g/L $(NH_4)_2HPO_4$, 0.4 g/L $MgSO_4 \cdot 7\,H_2O$, 0.002 g/L $FeSO_4 \cdot 7\,H_2O$, and 0.02 g/L $Ca(NO_3)_2$ in distilled water. Leaf samples were added to 50 mL of medium in 250 mL Erlenmeyer flasks. 1,8-Cineole, as the carbon source, was sterilized by filtration through a Millipore Solvinert membrane (pore size 0.22 μm) and added to the heat-sterilized mineral salt medium just prior to inoculation yielding a final concentration of 0.1 g/L.

Incubation was performed at 30 °C under shaking. Following a growth phase 1 mL of the culture was transferred to a fresh flask containing the same medium. After a number of these transfers, pure cultures were obtained by inoculating an agar medium of the same composition.

2. Transformation. *Pseudomonas flava* UQM-1 742 was grown in batch and continuous cultures in the above mineral salt medium containing 0.5 g/L of 1,8-cineole (**59**). The cultures were extracted with dichloromethane during the early stationary phase of growth after acidification with H_2SO_4 to pH 2. The extract was washed with water, dried over Na_2SO_4, and evaporated to yield a yellow oil. The separation of the formed metabolites was achieved by liquid column chromatography on aluminia.

B. Ionones and Related Compounds

1. β-Ionone (64)

Ionones and their derivatives are widely distributed in nature. They represent important constituents of many essential oils (NAVES, 1976) and are presumably generated from carotenoids by complex enzymatic activities (ISOE, 1970).

Over the last years, much attention has been paid to fragrant nor-isoprenoid degradation products possessing a trimethyl cyclohexane ring (OHLOFF, 1978).

The need for pure chiral ionone derivatives as precursors for synthesis of naturally occurring carotenoids also has evoked an interest to investigate biotechnological approaches for their access.

MIKAMI et al. (1978) reported the hydroxylation of β-ionone (**64**) by *Aspergillus niger* JTS 191. Two major metabolites were isolated and the structure proposed as (2*S*)-hydroxy-β-ionone (**65**) and (4*R*)-hydroxy-β-ionone (**66**) (Scheme 11). The complex was found to be very effective for tobacco flavoring at ppm level.

The fermentative production of (2*S*)-hydroxy-ionone (**65**) had also been reported by ITO et al. (1977) for the synthesis of (2*S*)-hydroxy-β-carotin through the reduction of 2-oxo-β-ionone (**65a**) by yeast.

A considerably different type of β-ionone transformation has been recently reported by KRASNOBAJEW and HELMLINGER (1982).

By employing the closely related genera of *Bothryodiplodia, Bothryosphaeria,* and *Lasiodiplodia,* an essential-oil-type product with tobacco flavor could be obtained from β-ionone.

Various enzyme systems in these fungi such as hydrogenases and hydroxylases, but most of all an oxygenase-type enzyme system are responsible for the conversion of β-ionone into a complex mixture of metabolites. In analogy to the BAEYER-VILLIGER (1899) oxidation the latter enzyme system efficiently removes a C_2-unit from the β-ionone molecule giving rise to the main product β-cyclo-homogeraniol (**67**). Hydroxylation products of this alcohol (**67**) were found to be 2-hydroxy- (**68**), 3-hydroxy- (**69**), and 4-hydroxy-β-cyclo-homogeraniol (**70**) as well as 4-oxo-β-cyclo-homogeraniol (**71**) (Fig. 1).

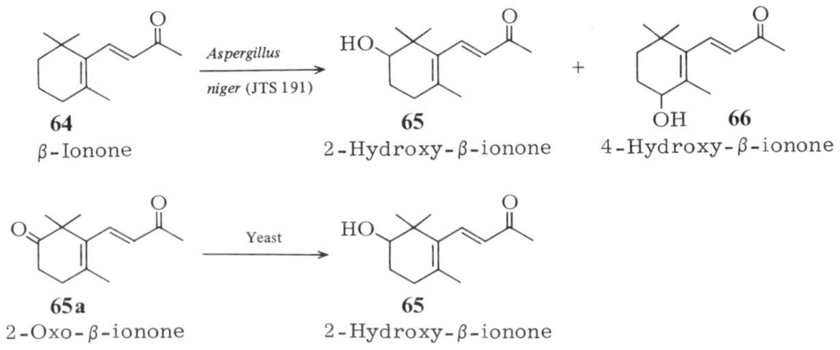

64 β-Ionone → *Aspergillus niger* (JTS 191) → 65 2-Hydroxy-β-ionone + 66 4-Hydroxy-β-ionone

65a 2-Oxo-β-ionone → Yeast → 65 2-Hydroxy-β-ionone

Scheme 11

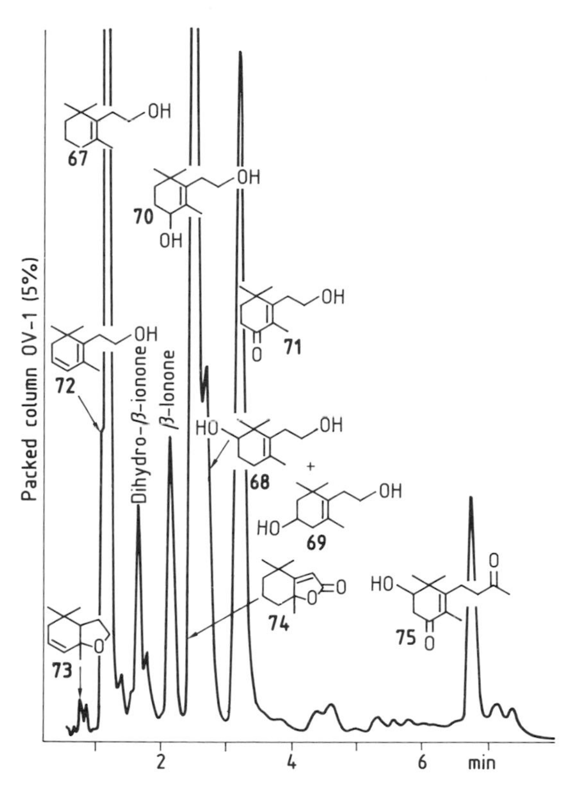

Figure 1. Gas chromatogram of *β*-ionone metabolites at the end of fermentation with *Lasiodiplodia theobromae*. – Conditions: Packed glass column (3 mm × 2 m) with OV-1, 5% on Gas-chrom Y (60–80 mesh); temperature 150 °C.

Other interesting metabolites, such as 3,4-dehydro-*β*-cyclo-homogeraniol (**72**), a bicyclic ether (**73**), dihydro-actinidiolide (**74**), and 2-hydroxy-4-oxo-7,8-dihydro-*β*-ionone (**75**) were identified. It may be of interest that the metabolite mixture found during early fermentation differs widely from the composition at the end of fermentation. Initial, major metabolites result from hydroxylation of *β*-ionone. The degradation product *β*-cyclo-homogeraniol (**67**) appears several hours later, indicating that the responsible oxygenase enzyme system must be induced.

Lasiodiplodia theobromae ATCC 28 570 was found most suitable for *β*-ionone transformation, due to a remarkable resistance towards the substrate. Fermentations of *β*-ionone must be carried out with pre-grown mycelia as *β*-ionone inhibits growth and leads to cell lysis. Usually mycelia are used which are at the end of the logarithmic growth phase.

The fermentation process is depicted in Fig. 2 showing the characteristic growth of the fungus in a 75 L fermenter. After near glucose depletion where the biomass had reached more than 20 g of dry matter per liter, *β*-ionone was added.

In order not to overload the mycelia with the toxic substrate, the substrate was added continuously over several days, so that *β*-ionone concentrations in the metabolite mixture never exceeded 10%. The formation of metabolites was pursued by gas chromatography. Under optimal conditions, about 10 g of *β*-ionone per liter of culture medium were metabolized.

Example: Transformation of *β*-ionone (**64**) by *Lasiodiplodia theobromae* ATCC 28 570.

A 75 L fermenter (Bioengineering, Rüti), equipped with axial flow-impeller with draft-tube, containing 45 L of growth medium consisting of: 5% glucose, 1% corn-steep liquor, 0.05% KCl, 0.1% KH_2PO_4, and 0.1% $NaNO_3$, tap water, at pH 6.3 (before sterilization), with 10 mL of polypropylene glycol (P2000) as antifoaming agent was inoculated with 1.6 L of a 24 h old pre-culture of *Lasiodiplodia theobromae,* grown on the same medium.

Fermentation conditions: Growth temperature was 28 °C at pH 6.3 with a stirring rate of 800 r.p.m. and an aeration of 20 L/min. After about 48 h of growth all glucose was consumed and the biomass reached more than 20 g dry matter per liter. At this point the pH, which had dropped to a value of about 4.5, was readjusted to about pH 6, and the continuous addition of *β*-ionone was initiated. Conversion of *β*-ionone was monitored by withdrawing 50 mL samples of fermentation broth, which were extracted with ethyl acetate. The concentrated extract was used for to gas chromatography. Over a fermentation period of 4 days, about 420 g of *β*-ionone were converted.

Product separation: Final extraction of metabolites from the fermented broth was

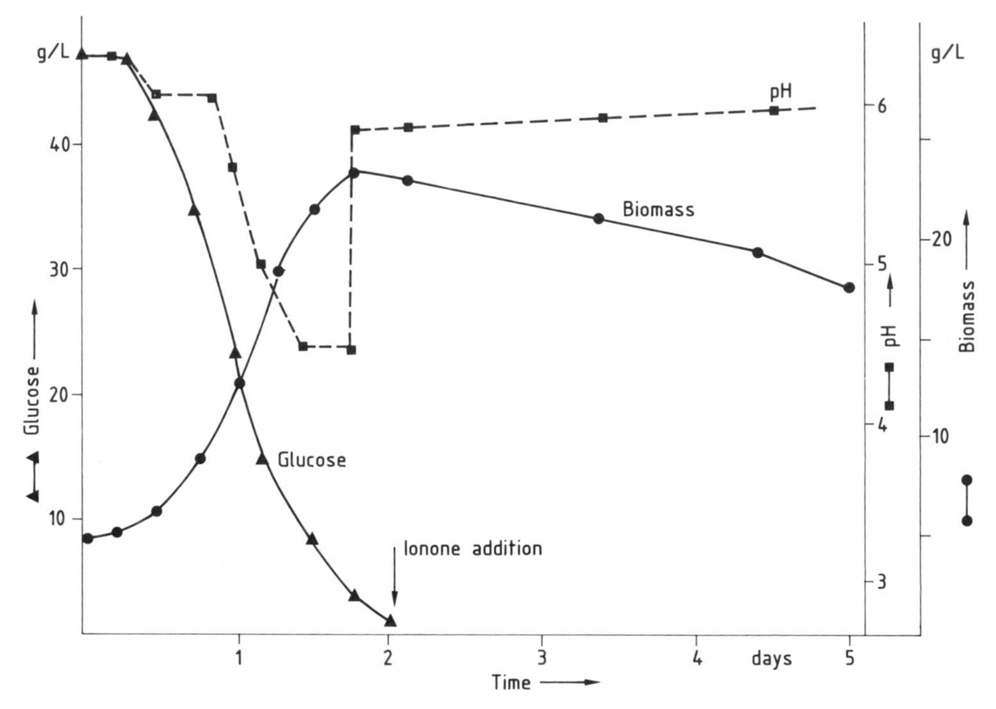

Figure 2. Growth of *Lasiodiplodia theobromae* and transformation of *β*-ionone (**64**).

achieved by means of a centrifugal separator (Gyrotester from Alpha-Laval); using methylene chloride or ethyl acetate as extraction solvents. The broth (about 43 liters) was stirred with 30 L methylene chloride and the phases separated by passing the emulsion through the Gyrotester. This procedure was repeated with the water phase. After combining the extracts, drying them over Na_2SO_4, and evaporating the solvent under vacuum, about 400 g of the crude *β*-ionone transformation product was obtained having a pleasant odor resembling tobacco flavor.

2. (±)-3-Acetoxy-4-oxo-*β*-ionone (76)

For the synthesis of astaxanthin (**79**), a carotenoid pigment made from *β*-ionone (**64**), BECHER et al. (1981a) described a microbiological resolution of (±)-3-acetoxy-4-oxo-*β*-ionone (**76**) in (*S*)-3-acetoxy-4-oxo-ionone (**77**) and (*R*)-3-hydroxy-4-oxo-*β*-ionone (**78**) (Scheme 12).

In reference to asymmetric hydrolysis of corresponding acetic acid esters of various terpenoids described by ORITANI and YAMASHITA (1973), several microorganisms were selected and examined. In a laboratory fermenter the fungus *Absidia hyalospora* Lender IFO 8 022 hydrolyzed an acetate (**76**) with an optical purity of 60%, in high yields.

A bacterium isolated from soil and identified as a *Bacillus* species showed an exceptionally high enantioselectivity. The latter strain selectively hydrolyzed the (*R*)-ester to yield the optically active 3-hydroxy-4-oxo-*β*-ionone (**78**), leaving the (*S*)-ester of the racemate (**77**) practically unchanged. By an optimization of the growth medium and fermentation conditions, as well as by selection of mutants (UV irradiation), the concentration of the racemate (**76**) could be raised to 30 g/L.

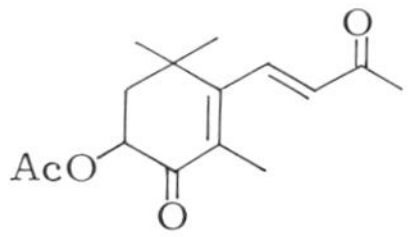

Bacillus sp.

76
(±)-3-Acetoxy-4-oxo-ß-ionone

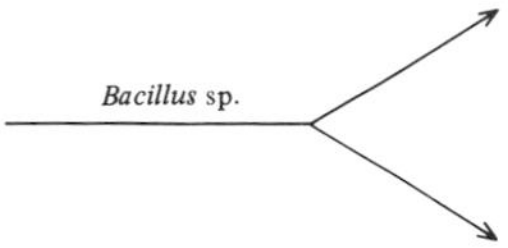

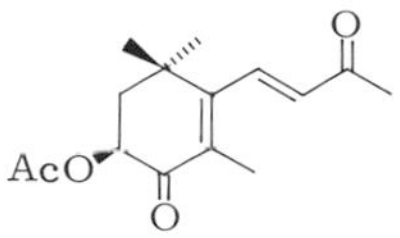

77
(S)-3-Acetoxy-4-oxo-ß-ionone

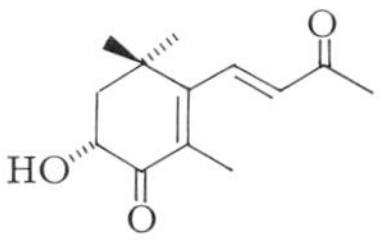

78
(R)-3-Hydroxy-4-oxo-ß-ionone

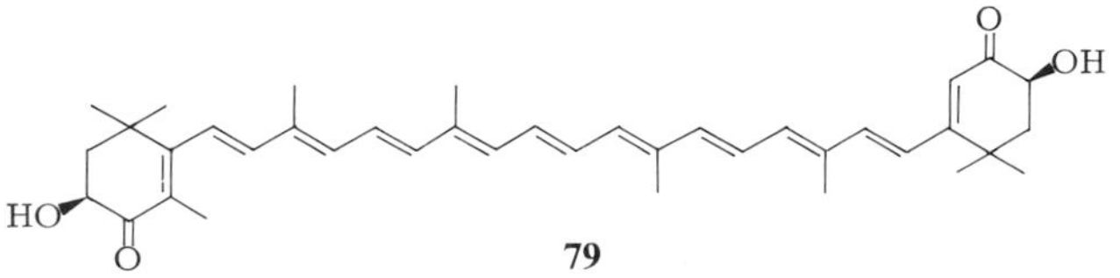

79
Astaxanthin

Scheme 12

3. α-Ionone (80)

The same fungi which had been found efficient in transforming β-ionone (**64**) also proved to transform α-ionone (**80**) (KRASNOBAJEW and RYTKOENEN, 1981; KRASNOBAJEW, 1982). α-Ionone, a constituent of various essential oils is used in large quantities in the fragrance industry (ARCTANDER 1969).

The same fermentation conditions as elaborated for β-ionone transformations were also optimal for α-ionone fermentation, although α-ionone was somewhat less readily transformed. The displacement of the endocyclic double-bond out of conjugation obviously renders α-ionone more resistant towards microbial attack.

Fermentation of α-ionone yielded an essential oil-type product with tobacco odor containing various unexpected bicyclic compounds which had been previously reported as flavoring components of Burley tobacco (ROBERTS and ROHDE, 1972). The latter flavoring components seem to be naturally derived from degradation of carotenoids.

As observed for β-ionone, α-ionone transformations with *Lasiodiplodia theobromae* ATCC 28 570 also suggested an oxygenase-type enzyme system to be responsible for the degradation of the molecule by loss of one C_2-unit. Under optimal aeration an oxygenase, besides a hydroxylase, was found to be the most active enzyme system. The resulting main transformation product α-cyclo-homogeraniol (**81**) does not accumulate in large quantities but is further hydroxylated in the allylic position to yield 3-hydroxy-α-cyclo-homogeraniol (**82**) together with its oxidation product 3-oxo-α-cyclo-homogeraniol (**83**). All three compounds are fairly unstable and cyclize either by enzyme catalysis or spontaneously to benzofuran derivatives.

The following benzofurans were found: 4,4,7a-trimethyl-2,3,3a,4,5,7a-hexahydrobenzofuran (**84**), 4,4,7a-trimethyl-6-oxo-

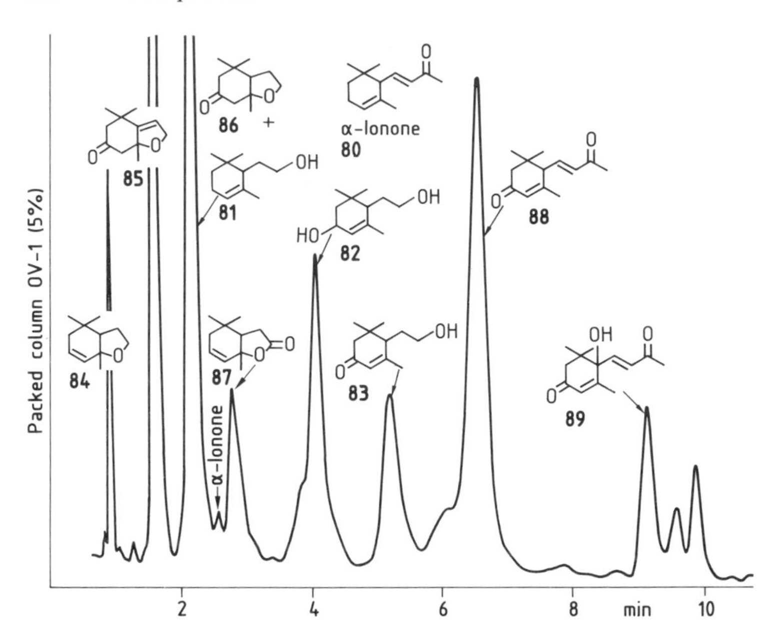

Figure 3. Gas chromatogram of α-ionone metabolites at the end of fermentation with *Lasiodiplodia theobromae*. – Conditions: Packed glass column (3 mm × 2 m) with OV-1, 5% on Gaschrom Q (60–80 mesh); temperature 145 °C.

2,4,5,6,7,7a-hexahydro-benzofuran (**85**), and 4,4,7a-trimethyl-6-oxo-2,3,3a,4,5,6,7,7a-octahydro-benzofuran (**86**). The 4,4,7a-trimethyl-2-oxo-2,3,3a,4,5,7a-hexahydro-benzofuran (**87**) may be formed through oxidation of 3-hydroxy-α-cyclo-homogeraniol (**82**) followed by cyclization. One of the major compounds in the metabolite mixture was identified as 3-oxo-α-ionone (**88**); under non-optimal aeration this compound appeared as the main metabolite in shake culture experiments. In this case the oxygenase activity responsible for degradation of α-ionone remains fairly low.

Interestingly, dehydro-vomifoliol (**89**), a hydroxylation product of 3-oxo-α-ionone (**88**), was also found in the metabolite mixture. In contrast to β-ionone transformation there was no hydrogenation of the exocyclic double-bond.

The α-ionone fermentation follows the same procedures described for β-ionone.

4. Oxo-isophorone (90)

A stereoselective fermentative reduction of oxoisophorone (**90**), one of the smallest monocyclic nor-isoprenoid aroma compounds resulting in nature through carotenoid degradation (Ohloff, 1978), was reported by Leuenberger et al. (1976).

In their synthetic studies of optically active, naturally occurring carotenoids such as zeaxanthin (**94**) and structurally related compounds, the authors elaborated a technical synthesis of the required precursor (4*R*,6*R*)-4-hydroxy-2,2,6-trimethylcyclohexanone (**93**) by reducing the readily available oxo-isophorone (**90**) by baker's yeast. The yeast introduces chirality at C-6 by a stereoselective hydrogenation of the double bond to yield (6*R*)-2,2,6-trimethyl-1,4-cyclohexane dione (**91**). The carbonyl function at C-4 can subsequently be reduced selectively and stereospecifically to the corresponding alcohol (**93**) (Scheme 13).

The optimized process performed in a 200 L deep tank fermenter requires gradual addition of oxo-isophorone over a period of 17 days up to a total concentration of 65 g/L. The reduction is carried out with 5% baker's yeast and a total of 75 g of sucrose per liter. Fermentation is initiated by adding 10 g/L of oxo-isophorone to the yeast which is suspended in a 2.5% sucrose solution. To avoid toxicity at a higher concentration, the substrate (**90**) is successively added in portions of 5 g/L as soon as the

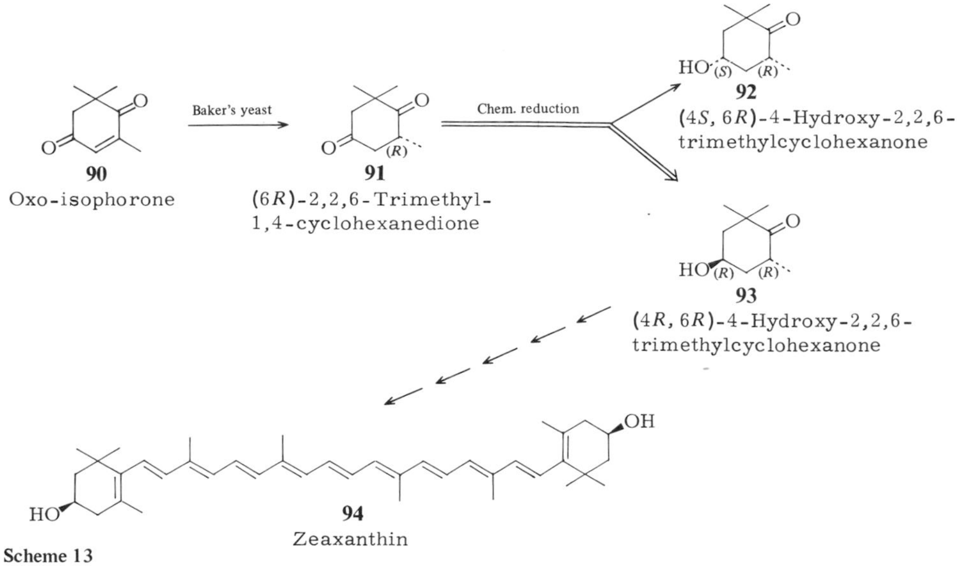

Baker's yeast
Chem. reduction

90
Oxo-isophorone

91
(6R)-2,2,6-Trimethyl-1,4-cyclohexanedione

92
(4S, 6R)-4-Hydroxy-2,2,6-trimethylcyclohexanone

93
(4R, 6R)-4-Hydroxy-2,2,6-trimethylcyclohexanone

94
Zeaxanthin

Scheme 13

substrate concentration becomes less than 5 g/L. Sucrose is fed at intervals of 3 days in amounts of 10 g/L.

The reaction velocity remains practically constant during the first 8 days of fermentation. The yield of crude product was 87%, of which 80% was optically pure diketone (**91**).

The precursor (**93**) has been also used in the synthesis of optically active plant growth regulators, as well as for various flavoring compounds (KIENZLE et al., 1978).

Example: Reduction of oxo-isophorone (**90**) to (6*R*)-2,2,6-trimethyl-1,4-cyclohexane dione (**91**) by baker's yeast

A fermenter equipped with a circulation stirrer (Biologische Verfahrenstechnik, Basel) with a working volume of 200 liters, containing a solution of 5 kg sucrose in 200 L deionized water at 20 °C is supplied with 10 kg of baker's yeast. As antifoam agent, 20 mL of propylene glycol monobutyl ether were added. Under stirring (800 r.p.m.) and aeration (3 200 L/h) the yeast suspension was allowed to ferment for 30 minutes. Thereafter, 2 kg of oxo-isophorone (**90**) were added. The conversion of the substrate (**90**) was monitored analytically. For each converted kilogram of oxo-isophorone (**90**) a further kilogram was added.

After a fermentation period of 17 days, a total amount of 13 kg of oxo-isophorone was converted. Furthermore, 2 kg of sucrose was added after 3, 6, 12, and 15 days.

Isolation of product

The fermenter broth consisting of mycelia and crystalline product was stirred with 5 kg of Dicalite-Speedex filter aid and subsequently filtered. The filtrate as well as the solid were worked-up separately but in the same way. Exhaustive extraction of the solids with methylene chloride and evaporation of the extract yielded 9.43 kg of crystalline material; the filtrate contained 3.63 kg of a semi-crystalline product. Both crude products were separately dissolved in diisopropyl ether, treated with active carbon, and crystallized from diisopropyl ether. A total of 10.57 kg (80.3%) of pure (6*R*)-2,2,6-trimethyl-1,4-cyclohexane dione (**91**) was isolated.

5. β-Damascone (95)

The damascones represent an important discovery in the field of carotenoid-derived flavor compounds (OHLOFF, 1978). They were first isolated from Bulgarian rose oil

(DEMOLE et al., 1970). The microbiological transformation of β-damascone (**95**), also found in Burley tobacco and various other plants, has been reported by HELMLINGER and KRASNOBAJEW (1980) using fungi of genera such as *Aspergillus, Botryosphaeria* and *Lasiodiplodia*.

Mainly, these fungi convert β-damascone (**95**), by hydroxylation of the allylic ring position, to 4-hydroxy-β-damascone (**96**) together with some 2-hydroxy-β-damascone (**97**). The metabolite mixtures were found to be excellent tobacco flavorings. *Lasiodiplodia theobromae* ATCC 28 570 produced two other transformation products in minor quantity which were determined as 10-hydroxy-β-damascone (**98**) and 4,10-dihydroxy-β-damascone (**99**) (HELMLINGER et al., 1981). In contrast to β-ionone (**64**), no degradation of β-damascone was observed. Depending on the fungus employed, the 4-hydroxy-damascone (**96**) showed different optical rotations (Scheme 14).

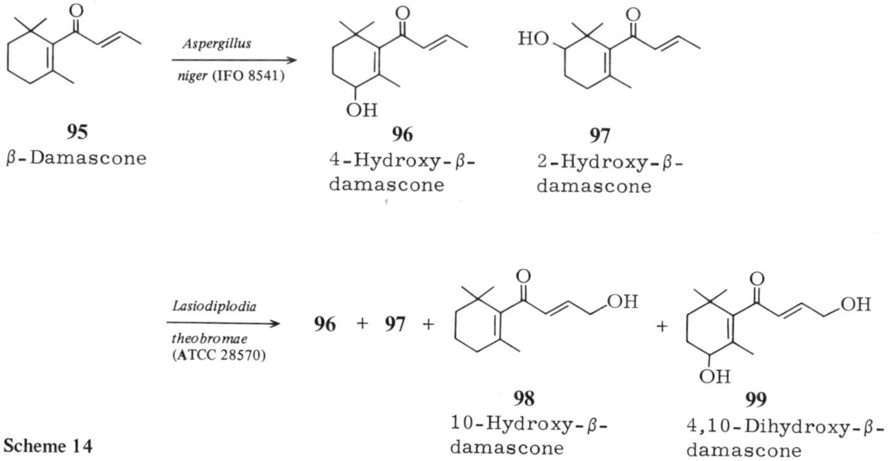

C. Sesquiterpenoids

Sesquiterpenes and their derivatives are found together with monoterpenoids in essential oils. As sesquiterpenoids contain one more isoprene unit than monoterpenoids, a greater variety of structures is possible which is manifested in nature by a tremendous diversity of this group of compounds (CROTEAU, 1980). By the end of 1976 their number was estimated to amount to at least 2 000 naturally occurring known representatives (OHLOFF, 1978). Many sesquiterpenoids are important flavor and perfume compounds. As their derivatives are often not accessible through chemical synthesis, microbiological transformations pose powerful means for their production. Reports on microbial transformation of sesquiterpenoids are rare compared with monoterpenoids. One reason may be the difficult access to the biologist of pure compounds which mostly first have to be separated from essential oils.

Aspergillus niger (IFO 8541)

95 β-Damascone → **96** 4-Hydroxy-β-damascone, **97** 2-Hydroxy-β-damascone

Lasiodiplodia theobromae (ATCC 28570)

→ **96** + **97** + **98** 10-Hydroxy-β-damascone + **99** 4,10-Dihydroxy-β-damascone

Scheme 14

Aspergillus niger IFO-8 541 produced 4-hydroxy-damascone (**96**) with a positive optical rotation of $[\alpha]_D^{22} = +24.3°$, whereas *Botryosphaeria rhodina* CBS 175,25 and *Lasiodiplodia theobromae* ATCC 28 570 yielded the same compound with a negative optical rotation of $[\alpha]_D^{22} = -30.8°$ and $-42.2°$, respectively.

In general, microbial transformations of sesquiterpenoids yield less degradation products and fewer metabolites. Microorganisms have also been found to grow on sesquiterpenoids as sole carbon source. Accumulation of metabolites has been regis-

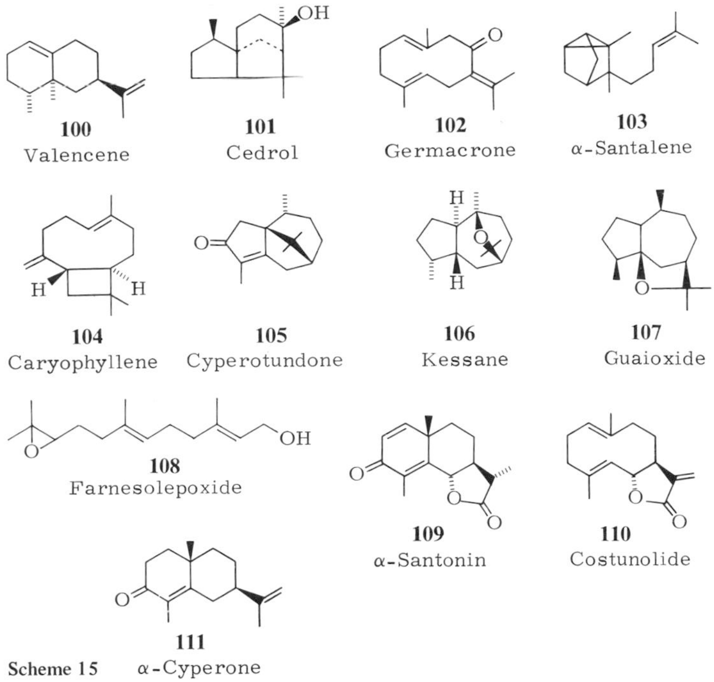

100 Valencene
101 Cedrol
102 Germacrone
103 α-Santalene
104 Caryophyllene
105 Cyperotundone
106 Kessane
107 Guaioxide
108 Farnesolepoxide
109 α-Santonin
110 Costunolide
111 α-Cyperone

Scheme 15

tered for microorganisms grown on caryophyllene (**104**) (DEVI, 1979) and valencene (**100**) (PAKNIKAR and DHAVLIKAR, 1975) (Scheme 15).

Microbial transformations have furthermore been reported on cedrol (**101**) (WANG et al., 1972), costunolide (**110**) (CLARK and HUFFORD, 1979), α-cyperone (**111**) (HIKONO et al., 1975), cyperotundone (**105**) (HIKINO et al., 1968), farnesol epoxide (**108**) (SUSUKI and MARUMO, 1972) germacrone (**102**) (HIKINO et al., 1971) guaioxide (**107**) (ISHII et al., 1970), kessane (**106**) (HIKINO et al., 1969), α-santalene (**103**) (PREMA and BHATTACHARYYA, 1962b), and α-santonin (**109**) (HIKINO et al., 1970). Elucidation of the metabolite structures has been emphasized in several of these studies.

1. Patchouli alcohol (112)

An instructive example of the use of biotechnology in the synthesis of precious fragrant trace compounds found in patchouli oil has been described in the patent literature by BECHER et al. (1981b).

Patchouli alcohol or patchoulol (**112**) (Scheme 16) is a major constituent (30–45%) of the patchouli essential oil which is extensively being used in perfumery. The essential oil is obtained by steam distillation of the dried leaves of *Pogostenom cablin* Benth. (GILDEMEISTER and HOFFMANN, 1961). Patchoulol itself has a characteristic faintly earthy, camphorous odor (NÄF et al., 1981). A nor-sesquiterpenoid (0.4–0.5%) of the oil, named norpatchoulenol (**115**), however, is thought to be one of the powerful main carriers of the odor (TEISSEIRE et al., 1974). The problematic chemical synthesis of the compound **115** has been achieved only recently (GRAS, 1977; OPPOLZER, 1981).

Preceding metabolic studies by BANG and OURISSON (1975) revealed that patchoulol (**112**) is hydroxylated in the liver of rabbits or dogs at the C-10 position, and that the isolated 10-hydroxy-patchoulol (**113**) could be chemically converted to the desired norpatchoulenol (**115**). Intense

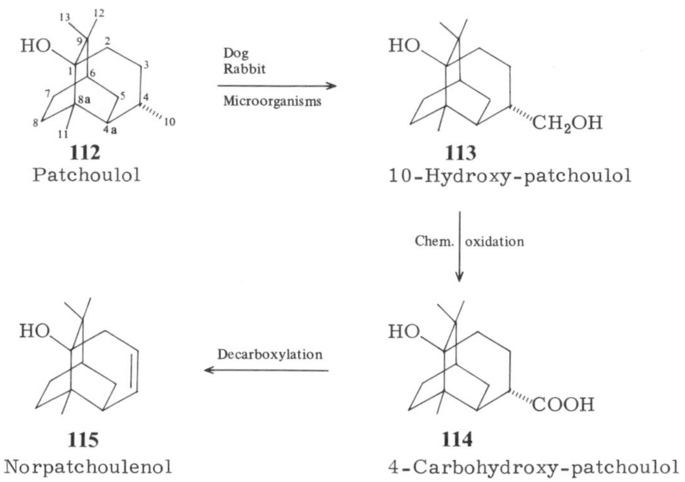

112
Patchoulol

Dog
Rabbit
Microorganisms

113
10-Hydroxy-patchoulol

Chem. oxidation

114
4-Carbohydroxy-patchoulol

Decarboxylation

115
Norpatchoulenol

Scheme 16

screening efforts for microorganisms which would perform a C-10 hydroxylation were subsequently initiated (BECHER et al., 1981b). In fact, a surprising number of fungi were found which perform the hydroxylation of the C-10 methyl group. On the other hand, various interesting transformation products were found, hydroxylated not only at the C-10 position, but preferably at other secondary and/or tertiary carbon atoms.

Depending on the employed fungi, the latter metabolites can become the major transformation products; the identified compounds were 5-hydroxy-patchoulol (**117**), 6-hydroxy-patchoulol (**116**), and 6,8-dihydroxy-patchoulol (**118**) (Scheme 17).

The metabolite 8-hydroxy-patchoulol (**119**) preferably produced by *Curvularia lunata* NRRL-2 380 is a precursor of another important trace component of patchouli oil, namely patchoulione (**120**).

The desired 10-hydroxy-patchoulol (**113**) was obtained only in combination with other metabolites and had to be separated by chromatography.

Among the large number of fungi yielding 10-hydroxy-patchoulol rank strains such as *Calonectaria decora* ATCC 14 767, *Cephalosporium coremioides* NRRL 11 003, *Paecilomyces carneus* FERM P-3 797, and *Penicillium rubrum* FERM P-3 796. The reported conversion rates of patchoulol are fairly low in view of industrial production. However, the yields of valuable metabolites could certainly be increased by optimization of fermentation conditions, as well as by improvement of strains.

Example: Hydroxylation of patchoulol (**112**) by *Paecilomyces carneus* FERM P-3 797.

Ten 500 mL conical flasks with baffles were filled with 100 mL of medium each, composed of: 5% Dextrin, 1% Pharma media (Traders Oil Mill Co., Texas), 0.05% KCl, 0.1% KH_2PO_4, and 0.1% $NaNO_3$. The flasks were inoculated with a spore suspension of *Paecilomyces carneus* FERM P-3 797 and shaken at 180 r.p.m. at 26.5 °C for 3 days. Thereafter, a solution of 1 gram patchoulol (**112**) dissolved in 10 mL of dimethylsulfoxide was equally distributed between the flasks. The bioconversion proceeded for 7 days under shaking (180 r.p.m.) at 26.5 °C. Then, the culture broth was filtered and the mycelia extracted with 80% aqueous acetone. The acetone was removed and the remaining aqueous extract combined with the culture filtrate which was then extracted twice with equal volumes of chloroform. The chloroform extracts were concentrated under vacuum to a small volume and chromatographed on silicagel. Using a solvent mixture of chloro-

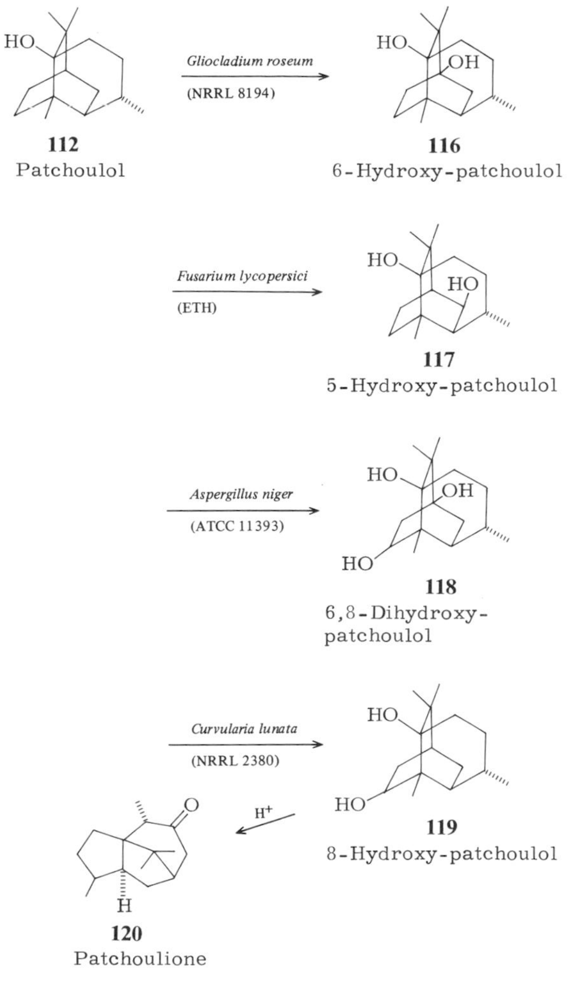

112 Patchoulol

Gliocladium roseum (NRRL 8194) → 116 6-Hydroxy-patchoulol

Fusarium lycopersici (ETH) → 117 5-Hydroxy-patchoulol

Aspergillus niger (ATCC 11393) → 118 6,8-Dihydroxy-patchoulol

Curvularia lunata (NRRL 2380) → 119 8-Hydroxy-patchoulol

H^+ → 120 Patchoulione

Scheme 17

form/hexane (1:1), 10-hydroxy-patchoulol (**113**) was eluted to yield 454 mg of a solid which, after two crystallizations, gave 260 mg of 10-hydroxy-patchoulol (**113**) as white needles.

D. Diterpenoids

Sources of non-volatile diterpenoids are the oleoresins of various conifers which mainly contain diterpenic acids (KARRER, 1976a; ZINKEL, 1981). Their microbial degradation has been studied by using bacteria isolated from soil of pine forests (RAYNAUD et al. 1966). Additionally, fungal hydroxylations have been reported (BRANNON et al., 1968).

In connection with the biosynthesis of gibberellins, which affect cell growth and division, various studies were reported in which mainly kaurane derivatives were microbiologically transformed (GHISALBERTI et al., 1977). A detailed review on microbial transformations of diterpenoids was presented by KIESLICH (1976), and SEBEK and KIESLICH (1977). Other than resin acids, only few diterpenoids have so far attained industrial importance.

E. Triterpenoids

Triterpenoids found in various plant sources are mainly polycyclic (tetra- and penta-cyclic) C_{30} compounds, with the exception of squalene (**121**) which is linear (KARRER, 1976b).

Microbial transformations of a few triterpenoids were reviewed by KIESLICH (1976).

A more recent study in this field has been reported by YAMADA et al. (1977) on linear terpenoids, including squalene (**121**) and squalene-2,3-epoxide (**122**).

1. Squalene-2,3-epoxide (122)

Squalene (**121**), a naturally abundant triterpene, has been degraded by an *Arthrobacter* species isolated from soil on a medium containing squalene (**121**) as the sole carbon source (YAMADA, 1975) leading to an accumulation of *trans*-geranylacetone (**123**) (Scheme 18).

Subsequently, the authors found that squalene-2,3-epoxide (**122**) was degraded by the same *Arthrobacter* sp. to *trans*-geranylacetone (**123**) and to a further com-

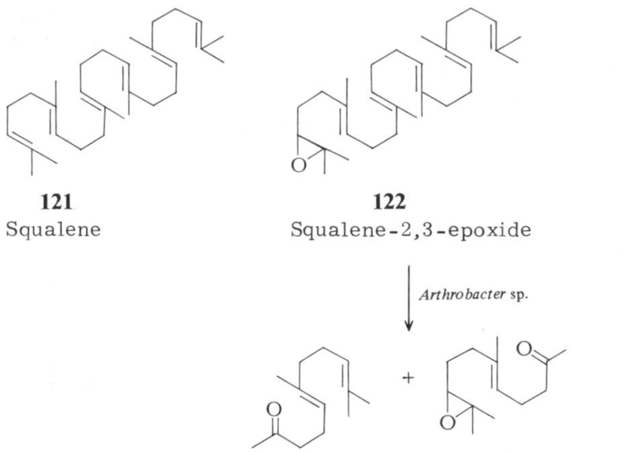

121
Squalene

122
Squalene-2,3-epoxide

Arthrobacter sp.

123
trans-Geranylacetone

124
trans-9,10-Epoxy-geranylacetone

Scheme 18

pound proposedly *trans*-9,10-epoxy-geranylacetone (**124**). As the enzyme systems of *Arthrobacter* partially recognize the chirality at the terminal ends of the substrates, the authors propose the use of such oxidative degradation processes for preparing optically active starting materials for the synthesis of physiologically active compounds, such as juvenile hormones.

V. References

ABBOT, B. J., and GLEDHILL, W. E. (1971). Adv. Appl. Microbiol. *14,* 249.

ARCTANDER, S. (1969). "Perfume and Flavor Chemicals." S. Arctander Publ., PO Box 114, Elisabeth, N. J.

BABIČKA, J., and VOLF, J. (1955). Czech. Patent 84 320.

BAEYER, A., and VILLIGER, V. (1899). Ber. Dtsch. Chem. Gesellsch. *32,* 3 625.

BANG, L., and OURISSON, G. (1975). Tetrahedron Lett. *26,* 2 211.

BECHER, E., ALBRECHT, R., BERNHARD, K., LEUENBERGER, H. G. W., MAYER, H., MÜLLER, R. K., SCHÜEP, W., and WAGNER, H. P. (1981a). Helv. Chim. Acta *64,* 2 419.

BECHER, E., SCHÜEP, W., MATZINGER, P. K., TEISSEIRE, P. J., EHRET, CH., MARUYAMA, H., SUHARA, Y., ITO, S., OGAWA, M., YOKOSE, K., SAWADA, T., FUJIWARA, A., FUJIWARA, M., TAZOE, M., and SHLOMI, Y. (1981b). (Fa. Hoffmann-La Roche & Co.) Br. Patent 1 586 759. (Prior. date: Sept. 2. 1976)

BECK, J. V. (1971). Methods Enzymol. *22,* pp. 57–64.

BHATTACHARYYA, P. K., and GANAPATHY, K. (1965). Indian J. Biochem. *2,* 137.

BOWEN, E. R. (1975). Proc. Fla. State Hortic. Soc. *88,* 304.

BRANNON, D. R., BOAZ, H., MABE, J., and HORTON, D. R. (1968). Chem. Comm., 681.

CALAME, J. P., and STEINER, R. (1982). Chem. Ind. 19 June, 399.

CHARLWOOD, B. V., and BANTHORPE, D. V. (1978). Prog. Phytochem. *5,* 65–121.

CIEGLER, A. (1969). In "Fermentation Advances" (D. PERLMAN, ed.), pp. 689–714. Academic Press, New York.

CLARK, A. M., and HUFFORD, C. D. (1979). J. C. S. Perkin I, 3 022.

COLLINS, R. P. (1976). Lloydia *39,* 20.

CROTEAU, R. (1980). In "Fragrance and Flavor Substances." Proc. Second Int. Haarmann and Reimer Symposium on Fragrance and Flavor Substances (R. CROTEAU, ed.) pp. 13–36. D & PS. Verlag, Pattensen, W.-Germany.

DEMOLE, E., ENGGIST, P., SÄUBERLI, U., STOLL, M., and KOVÁTS, E. (1970). Helv. Chim. Acta *53,* 541.

DEVI, J. R. (1979). Indian J. Biochem. Biophys. *16,* 76.

DEVI, J. R., and BHATTACHARYYA, P. K. (1977). Indian J. Biochem. Biophys. *14,* 359.

DHAVALIKAR, R. S., and BHATTACHARYYA, P. K. (1966). Indian J. Biochem. *3,* 144.

DHAVALIKAR, R. S., RANGACHARI, P. N., and BHATTACHARYYA P. K. (1966). Indian J. Biochem. *3,* 158.

EDWARDS, V. H. (1969). In "Fermentation Advances" (D. PERLMAN ed.), pp. 273–298. Academic Press, New York.

ERICKSON, R. E. (1976). Lloydia *39,* 8.

FENAROLI, G. (1975). "Handbook of Flavor Ingredients". 2nd Ed. CRC Press, Cleveland, Ohio.

FONKEN, G. S., and JOHNSON, R. A. (1972). "Chemical Oxidations with Microorganisms" (J. S. BELEV, ed.). Marcel Dekker, New York.

FUKUI, S. (1981). Appl. Microbiol. Biotechnol. *11,* 199.

GIBBON, G. H., and PIRT S. J. FEBS Lett. *18,* 103.

GILDEMEISTER, E., and HOFFMANN, Fr. (1960a). In "Die ätherischen Oele", 4th Ed., Vol. 3a (W. TREIBS and D. MERKEL, eds.), p. 45. Akademie-Verlag, Berlin.

GILDEMEISTER, E., and HOFFMANN, Fr. (1960b). In "Die ätherischen Oele", 4th Ed., Vol. 3a (W. TREIBS and D. MERKEL, eds.), p. 127. Akademie-Verlag, Berlin.

GILDEMEISTER, E., and HOFFMANN, Fr. (1961). In "Die ätherischen Oele", 4th Ed., Vol. 7 (W. TREIBS and D. MERKEL, eds.), p. 449. Akademie-Verlag, Berlin.

GHISALBERTI, E. L., JEFFERIES, P. R., SEFTON, M. A., and SHEPPARD, P. N. (1977). Tetrahedron *33,* 2 451.

GOAD, L. J. (1978). In "Marine Natural Products", Vol. 2 (P. J. SCHEUER, ed.), pp. 75–159. Academic Press, New York.

GRAS J. L. (1977). Tetrahedron Lett. *47,* 4 117.

GRIMMET, CH. (1981). Chem. Ind. May, 359.

GUENTHER, E. (1966). "The Essential Oils", Vol. 2, Fifth Printing, p. 216. D. Van Nostrand, Princeton.

HAARMANN and REIMER GmbH (1971). W. Ger. Patent DT-109 456. (Prior. date February 27, 1971)

HASEGAWA Co. (1972). Jap. Patent 7 238 998

HAYASHI, T., TAKASHIBA, H., UEDA, H., and TATSUMI, Ch. (1967). J. Agric. Chem. Soc. (Japan) *41,* 254.

HEINZLE, E., and DUNN, I. J. (1983). "Application of on-line mass spectrometry for measurement of gases and volatiles in fermentation." Submitted to Biotechnol. Bioeng.

HELMLINGER, D., KRASNOBAJEW, V., RYTKOENEN, S., and STAUCH, W. (1981). Abstr. 2nd Eur. Symposium on Organic Chemistry, Stresa; p. 333.

HELMLINGER, D., and KRASNOBAJEW, V. (1980). (L. Givaudan & Cie S. A.) Swiss Patent 3966.80.

HENRY, B. S. (1982). Perfumer Flavorist *7,* 39.

HEROUT, V. (1970). Prog. Phytochem. *2,* 143–171.

HIKINO, H., AOTA, K., TOKUOKA, Y., and TAKEMOTO, T. (1968). Chem. Pharm. Bull. (Tokyo) *16,* 1 088.

HIKINO, H., KOHAMA, T., and TAKEMOTO, T. (1969). Chem. Pharm. Bull. (Tokyo *17,* 1 659.

HIKINO, H., KONNO, C., IKEDA, Y., IZUMI, N., and TAKEMOTO, T. (1975). Chem. Pharm. Bull. (Tokyo) *23,* 1 231.

HIKINO, H., KONNO, C., NAGASHIMA, T., KOHAMA, T., and TAKEMOTO, T. (1971). Tetrahedron Lett. *4,* 337.

HIKINO, H., TOKUOKA, Y., and TAKEMOTO, T. (1970). Chem. Pharm. Bull. (Tokyo) *18,* 2 127.

HOFFMANN, W. (1975). Seifen Öle Fette Wachse *101,* 89.

HUNGUND, B. L., BHATTACHARYYA, P. K. (1970). Arch. Mikrobiol. *71,* 258.

ISHII, H., TOZYO, T., NAKAMURA, M., and MINATO, H. (1970). Tetrahedron *26,* 2 751.

ISOE, S. (1970). Kagakuto Seibutsu *8,* 575.

ITO, M., MASAHARA, R., and TSUKIDA, K. (1977). Tetrahedron Lett., 2 767.

JACOBSON, G. K. (1981). In "Biotechnology", Vol. 1 (H.-J. REHM and G REED, eds.), pp. 280-304. Verlag Chemie, Weinheim - Deerfield Beach/Florida - Basel.

JOGLEKAR, S. S., and DHAVLIKAR, R. S. (1969). Appl. Microbiol. *18,* 1 084.

JOHNSON, R. A. (1978). In "Oxidation in Organic Chemistry", Part C. (W. S. TRAHANOVSKY, ed.), pp. 131–210. Academic Press, New York.

KARRER, W. (1976a). "Konstitution und Vorkommen der organischen Pflanzenstoffe", 2nd Ed., pp. 781–795. Birkhäuser Verlag, Basel - Stuttgart.

KARRER, W. (1976b). "Konstitution und Vorkommen der organischen Pflanzenstoffe", 2nd Ed., pp. 796–843. Birkhäuser Verlag, Basel - Stuttgart.

KIENZLE, F., MAYER, H., MINDER, R. E., and THOMMEN, H. (1978). Helv. Chim. Acta *61,* 2 616.

KIESLICH, K. (1976). "Microbial Transformation of Non-Steroid Cyclic Compounds". Georg Thieme Verlag, Stuttgart.

KING, P. P. (1982). J. Chem. Tech. Biotechnol. *32,* 2.

KRAIDMAN, G., MUKHERJEE, B. B., and HILL, I. D. (1969). Bacteriol. Proc., 63.

KRASNOBAJEW, V. (1982). (L. Givaudan & Cie. S. A.) U.S.-Patent 4.311.860. (Prior. date Nov. 24, 1978)

KRASNOBAJEW, V., and HELMLINGER, D. (1982). Helv. Chim. Acta *65,* 1 590.

KRASNOBAJEW, V., and RYTKOENEN, S. (1981). Experientia *37,* 1 218.

LEUENBERGER, H. G., BOGUTH, W., WIDMER, E., and ZELL, R. (1976). Helv. Chim. Acta *59,* 1 832.

LIAAEN-JENSEN, S. (1978). In "Marine Natural Products", Vol. 2 (P. J. SCHERRER, ed.), pp. 1–64. Academic Press, New York.

MACRAE, J. C., ALBERTS, V., CARMAN, R. M., and SHAW, J. M. (1979). Aust. J. Chem. *32,* 917.

MADYASTHA, K. M., BHATTACHARYYA, P. K., and VAIDYANATHAN, C. J. (1977). Can. J. Microbiol. *23,* 230.

MAYER, P., and NEUBERG, C. (1915). Biochem. Z. *71,* 174.

MEITO SANGYO Co. Ltd. (1975). Japan. Patent 126 375 and J5-2 051-095.

MEULY, W. C. (1970). Am. Perfumer Cosmet., *85,* 123.

MIKAMI, Y., WATANABE, E., FUKUNAGA, Y., and KISAKI, T. (1978). Agric. Biol. Chem. *42,* 1 075.

MIZUTANI, S., HAYASHI, T., UEDA, H., and TATSUMI, Ch. (1971). J. Agric. Chem. Soc. (Japan) *45,* 368.

MOROE, T., HATTORI, S., KOMATSU, A., and YAMAGUCHI, Y. (1971). (Takasago Perfumery Co.). W. Ger. Patent DT 2 036 875. Jap. prior. 1969, also U.S. Patent 3 607 651 (1971).

MUKHERJEE, B. B., KRAIDMAN, G., and HILL, I. D. (1973). Appl. Microbiol. *25,* 447.

NÄF, F., DECORZANT, R., GIERSCH, W., and OHLOFF, G. (1981). Helv. Chim. Acta *64,* 1 387.

NAKAYAMA, K. (1981). In "Biotechnology", Vol. 1 (H.-J. REHM and G. REED, eds.), pp. 355–410. Verlag Chemie, Weinheim – Deerfield Beach – Basel.

NAVES, Y. R. (1976). Riv. Ital. E.P.P.O.S. *10,* 505.

NELBOECK-HOCHSTETTER, M., SEIDEL, H., and GAUHL, H. (1977). (Boehringer Mannheim GmbH) W.-Ger. Patent DT 25 37 339.

NEWMAN, A. (1972). "Chemistry of Terpenes and Terpenoids" (A. NEWMAN, ed.). Academic Press, London – New York.

NISSAN Chemical Ind. KK (1981). Japan. Patent J5-6 124-388. (Prior. date March 3, 1980)

OHLOFF, G. (1978). In "Progress in the Chemistry of Organic Natural Products", Vol. 35 (W. HERZ, H. GRISEBACH, and G. W. KIRBY, eds.), pp. 431–527. Springer Verlag, Wien – New York.

OPPOLZER, W. (1981). (L. Givaudan & Cie. S. A.). U.S. Patent 4 277-631. (Prior. date Aug. 10 1978).

ORITANI, T., and YAMASHITA, K. (1973). Agric. Biol. Chem. *37,* 1 697, 1 923.

ORITANI, T., and YAMASHITA, K. (1974). Agric. Biol. Chem. *38,* 1961, 1965.

ORITANI, T., and YAMASHITA, K. (1975). Agric. Biol. Chem. *39,* 89.

PAKNIKAR, S. K., and DHAVLIKAR, R. S. (1975). Chem. Ind. May 17, 432.

PERLMAN, D. (1976). In "Applications of Biochemical Systems in Organic Chemistry", Part I (J. B. JONES, J. J. SIH, D. PERLMAN, and J. WILEY, eds.), pp. 47–68. John Wiley, New York.

PREMA, B. R., and BHATTACHARYYA, P. K. (1962a). Appl. Microbiol. *10,* 524.

PREMA, B. R., and BHATTACHARYYA, P. K. (1962b). Appl. Microbiol. *10,* 529.

RAYMOND, R. L., and JAMISON, V. W. (1971). Adv. Appl. Microbiol. *14,* 93.

RAYNAUD, M., BIELLMANN, J.-F., and DASTE, P. (1966). C. R. Soc. Biol. *160,* 371.

RIENÄCKER, R., and OHLOFF, G. (1961). Angew. Chem. *73,* 240.

ROBERTS, D. L., and ROHDE, W. A. (1972). Tobacco Sci. *16,* 107.

RUZICKA, L. (1959). Proc. Chem. Soc. (London), p. 341.

SCHMIDT-KASTNER, G., and GÖLKER, Ch. (1982). In "Handbuch der Biotechnologie" (P. PRÄVE, U. FAUST, W. SITTIG, and D. A. SUKATSCH, eds.), pp. 215–245. Akademische Verlagsgesellschaft, Wiesbaden.

SEBEK, O. K., and KIESLICH, K. (1977). In "Annual Reports on Fermentation Processes", Vol. 1 (D. PERLMAN, ed.), pp. 267–297. Academic Press, New York.

SHUKLA, O. P., MOHOLAY, M. N., and BHATTACHARYYA, P. K. (1968). Indian J. Biochem. *5,* 79.

SMOOT, J. J., HOUCK, L. G., and JOHNSON, H. B. (1971). U.S. Dept. of Agric., Handbook No. 398, pp. 9–12.

SUSUKI, Y., and MARUMO, S. (1972). Tetrahedron Lett. *19,* 1 887.

TAKASAGO Perfumery Co. Ltd. (1970). Japan. Patent 73 161 91.

TAKASAGO Perfumery KK (1974). Japan. Patent 106 767 and J5-1 035-491.

TEISSEIRE, P., MAUPETIT, P., CORBIER, B., and ROUILLER, P. (1974). Recherches (BRD) *19,* 8.

TRAAS, P. C. (1982). In "Fragrance Chemistry" (E. T. THEIMER, ed.), pp. 165–219. Academic Press, New York.

TURSCH, B., BRAEKMAN, J. C., DALOZE, D., and KAISIN, M. (1978). In "Marine Natural Products", Vol. 2 (P. J. SCHEUER, ed.), pp. 247–291. Academic Press, New York.

VEZINA, C., SEHGAL, S. N., and SING, K. (1968). Adv. Appl. Microbiol. *10,* 211.

VOISHVILLO, N. E., AKHREM, A. A., and TITOV, Y. A. (1970). Prikl. Biokhim. Mikrobiol. *6,* 491.

WANG, K. C., HO, L. Y., and CHENG, Y. S. (1972). J. Chin. Biochem. Soc. *1,* 53.

WATANABE, Y., and INAGAKI, T. (1977). (Nippon Terpene Chemical Co., Ltd.), Japan. Patent 77 122 690.

WIGMORE, G. J., and RIBBONS, D. W. (1980). J. Bacteriol. *143,* 816.

WOOD, B. J. B. (1969). Process Biochem., 50.

YAMADA, Y., KUSUHARA, N., and OKADA, H. (1977). Appl. Environ. Microbiol. *33,* 771.

YAMADA, Y., MOTOI, H., KINOSHITA, S., TAKADA, N., and OKADA, H. (1975). Appl. Microbiol. *29,* 400.

YAMAGUCHI, Y., KOMATSU, A., and MOROE, T. (1976). J. Agric. Chem. Soc. (Japan) *50,* 443.

YAMAGUCHI, Y., KOMATSU, A., and MOROE, T. (1977). J. Agric. Chem. Soc. (Japan) *51,* 411.

ZINKEL, D. F. (1981). In "Organic Chemicals from Biomass" (I. S. GOLDSTEIN, ed.), pp. 163–187. CRC Press Inc., Boca Raton, Florida.

Chapter 5

Alicyclic and Heteroalicyclic Compounds

Alain Kergomard

Université de Clermont-Ferrand II
Aubière, France

I. Introduction
II. Simple Conversions
  A. Reduction of Ketones
    1. Introduction
    2. Compounds with a bicyclo[4.4.0]decane skeleton and analogs
    3. Reduction of substrates with steroid-related skeletons
    4. Stereochemistry of reduction
  B. Reduction of Carbon-Carbon Double Bonds
  C. Oxidation of Alcohols
  D. Hydrolysis Reactions
  E. Hydroxylation Reactions
  F. Degradation Reactions
  G. Other Reactions
III. Miscellaneous Applications
  A. Synthesis of Estradiol and Analogs
  B. Hexahydroindene Derivatives
  C. Intermediates in the Biosynthesis of Aromatic Amino Acids
  D. Biosynthesis and Degradation of Biotin
  E. Biodegradation and Metabolization
    1. Biodegradation of pesticides
    2. Degradation of diazepines
  F. Applications in Prostaglandin Synthesis
    1. Production of prostaglandin synthons
    2. Production of optically active prostaglandins
    3. Degradation and metabolism of prostaglandins
  G. Biotransformation of Cannabinoids
IV. References

I. Introduction

Extensive accounts of chemical conversions using microorganisms have been published by, e.g., KIESLICH (1976), FONKEN and JOHNSON (1972), SKRYABIN and GOLOVLEVA (1976), and a review by PERUZZOTTI (1982). Here, we shall deal first of all with conversions involving a single reaction, namely, reduction of ketones and double bonds, oxidation of alcohols, hydrolysis of esters, and hydroxylation of various organic compounds. We shall then look at conversions involving more than one reaction, namely, degradation and aromatization. Finally, some results will be described which involve series of compounds of theoretical importance (e. g., in the study of biosynthesis) or of practical interest (e. g., sterol and prostaglandin synthons).

In the course of this review, particular emphasis will be laid on the stereochemistry of these reactions. As already stressed by SIH (1976), "the application of microorganisms to resolution of synthetic problems of organic chemistry will continue to grow in importance, for microorganisms are prodigious sources of organic chiral reagents".

II. Simple Conversions

A. Reduction of Ketones

1. Introduction

The microbiological or enzymatic reduction of ketones can create a new asymmetric carbon. Accordingly, this type of reaction has received much attention. The simpler cyclic ketones were the first to be studied. Thus, cyclopentanone was reduced to cyclopentanol by yeasts with a yield of 42% (NEUBERG, 1950). 2-Methylcyclohexanone gave (+)-2-methylcyclohexanol (AKAMATSU, 1923; NEUBERG, 1949), while (±)-*cis*- and *trans*-2,5-dimethylcyclohexanone, 2-methylcyclohexanone, and cyclopentanone were reduced by *Pseudomonas ovalis* (NOMA, 1977).

Other racemic substituted cyclohexanones were reduced with the aim of obtaining optically active compounds (Scheme 1). (±)-2-Chlorocyclohexanone **1** was reduced by *Saccharomyces cerevisiae* to (−)-chlorocyclohexanol **2** and 2*S*-starting ketone (CRUMBIE et al., 1977). Less interestingly, (±)-2-acetoxycyclohexanone **3** gave a mixture of (+)- and (−)-2-acetoxycyclohexanol **4** along with 15% (−)-*R*-starting ketone. Finally, the same yeast reduced (±)-*cis*-2-acetoxy-4-*tert*-butylcyclohexanone **5** to the (−)-(1*R*,2*S*,4*S*)-alcohol **6** and (+)-(2*R*,4*S*)-starting ketone.

A series of ketones derived from spiro[4.4]nonane were reduced by *Curvularia lunata* IFO 6288 (NAKAZAKI et al., 1981a).

The monoketone **7** gave the optically pure *S*-alcohol **8**. Racemic 1,6-diketospiro[4.4]nonane **9** yielded the *cis*-alcohol **10** with low optical purity and the *trans*-alcohol **11.** The recovered starting ketone **9** was (−) and of high optical purity. Finally, the racemic *trans*-monoalcohol **11** was reduced to the diols **12** and **13** with high optical purities. The recovered starting ketol was (−).

The racemic cyclopentenyl acetone **14** (Scheme 2) was reduced by *Rhodotorula mucilaginosa* to the diastereoisomeric alcohols **15** (2*S*,1′*R*) and **16** (2*S*,1′*S*) (SIEWINSKI et al., 1979). Under the same conditions, racemic cyclopentenyl-2 butanone **17** gave the alcohol **18** (2*S*,1′*R*) and left the (−)-ketone **17** (1′*S*). The *β*-oxo esters **19** and **21** were reduced by baker's yeast to the 2*S*-alcohols **20** and **22** (DEOL et al., 1976). Finally, the ketopantoyl lactone **23** was reduced to the (2*R*)-pantoyl lactone **24** by *Byssochlamys fulva* (LANZILOTTA et al., 1974).

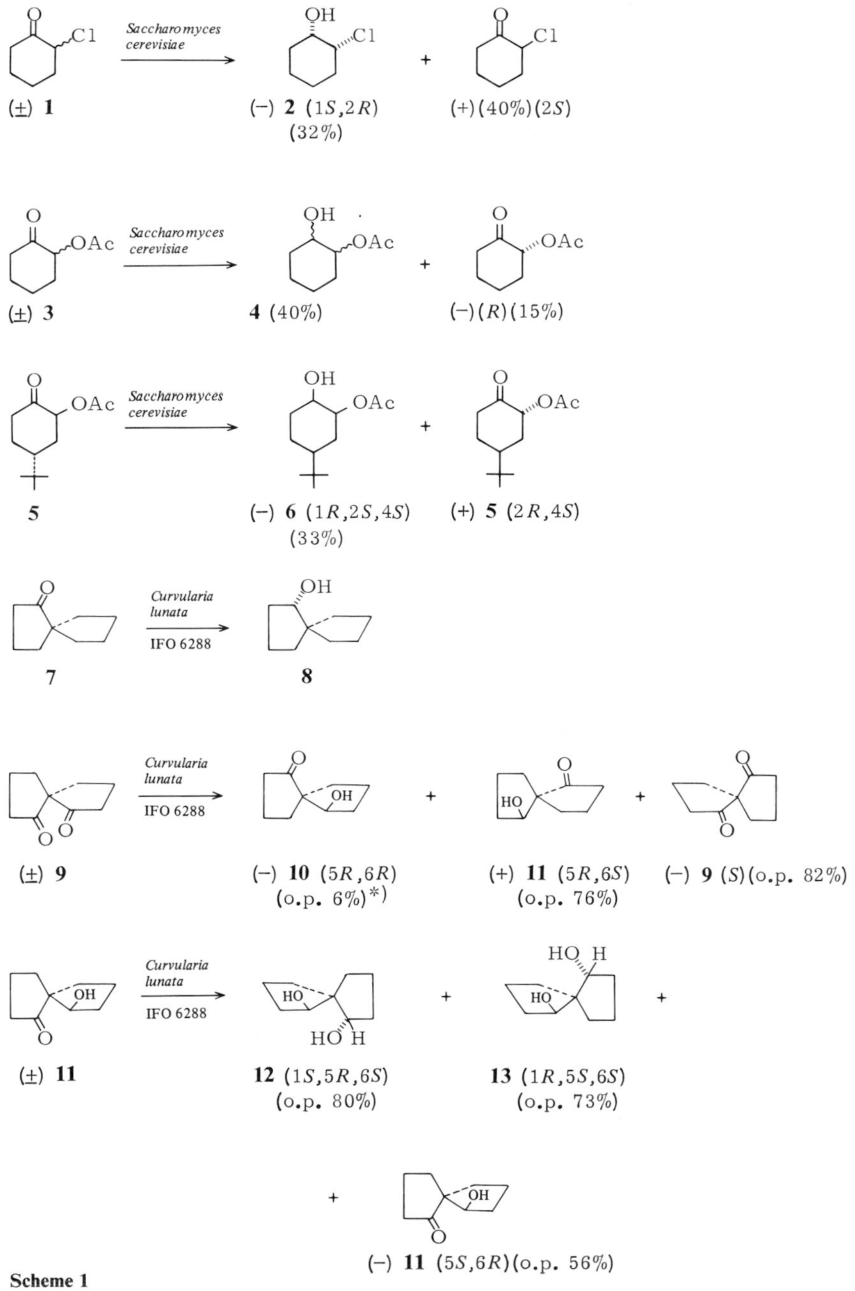

Saccharomyces cerevisiae

(±) **1** → (−) **2** (1*S*,2*R*) (32%) + (+)(40%)(2*S*)

Saccharomyces cerevisiae

(±) **3** → **4** (40%) + (−)(*R*)(15%)

Saccharomyces cerevisiae

5 → (−) **6** (1*R*,2*S*,4*S*) (33%) + (+) **5** (2*R*,4*S*)

Curvularia lunata IFO 6288

7 → **8**

Curvularia lunata IFO 6288

(±) **9** → (−) **10** (5*R*,6*R*) (o.p. 6%)*) + (+) **11** (5*R*,6*S*) (o.p. 76%) + (−) **9** (*S*)(o.p. 82%)

Curvularia lunata IFO 6288

(±) **11** → **12** (1*S*,5*R*,6*S*) (o.p. 80%) + **13** (1*R*,5*S*,6*S*) (o.p. 73%) + (−) **11** (5*S*,6*R*)(o.p. 56%)

Scheme 1

* o.p. = optical purity

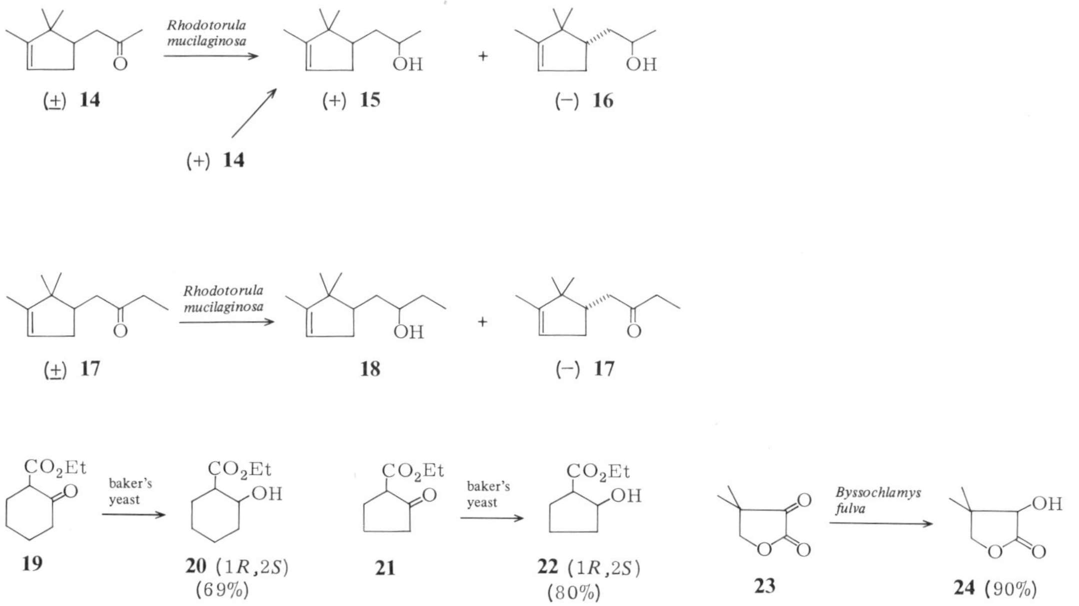

Rhodotorula mucilaginosa

(±) 14 → (+) 15 + (−) 16

(+) 14 ↗ (+) 15

Rhodotorula mucilaginosa

(±) 17 → 18 + (−) 17

19 →(baker's yeast) 20 (1*R*,2*S*) (69%)

21 →(baker's yeast) 22 (1*R*,2*S*) (80%)

23 →(*Byssochlamys fulva*) 24 (90%)

Scheme 2

2. Compounds with a bicyclo[4.4.0]decane skeleton and analogs

Compounds of this series have been extensively studied by PRELOG and his school.

The reduction of carbonyl functions by *Curvularia falcata* were found to follow a general rule summarized in Fig. 1. Few exceptions have been found. The direction of attack on the carbonyl can be deduced from the substitution pattern (PRELOG, 1962; 1964).

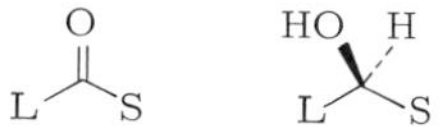

Figure 1. Steric course of the reduction of ketones by *Curvularia falcata.*

It appears that cyclohexanones are reduced preferentially from the equatorial side.

The first rule applies to 1-decalones. In all cases, *C.falcata* formed the *S*-alcohol. Thus, (±)-*trans*-decalone **1** (Scheme 3) gave the diastereoisomeric *trans*-decanols **2** (1*S*,9*R*) and **3** (1*S*,9*S*); (±)-*cis*-decalone **4** gave the *cis*-1-decanols **5** (1*S*,9*R*) and **6** (1*S*,9*R*) (AKLIN et al., 1965).

Reduction of (±)-*trans*-2-decalone **7** was less stereospecific, probably because the L- and S-sides are less different than in the previous examples (PRELOG et al., 1959a).

The two 9*R*-alcohols **8** and **9** were obtained from the corresponding decalone and the two 9*S*-alcohols **8′** and **9′** from the other decalone isomer. Likewise, the two *cis*-decalones **10** each give the two isomers **11** and **12** from **10** (9*R*,10*S*) and **11′** and **12′** from **10′** (9*S*,10*R*).

The same strain of *Curvularia falcata* was also used to reduce decalindiones (BAUMANN and PRELOG, 1958a; FELDMAN and PRELOG, 1958). The reaction was stereospecific for the 1,4-diones, but mixtures of hydroxydecalones and dihydroxydecalins were obtained. The racemic *trans*-dione **13** (Scheme 4) gave the hydroxydecalones **14** and **15** in which the alcohol function has

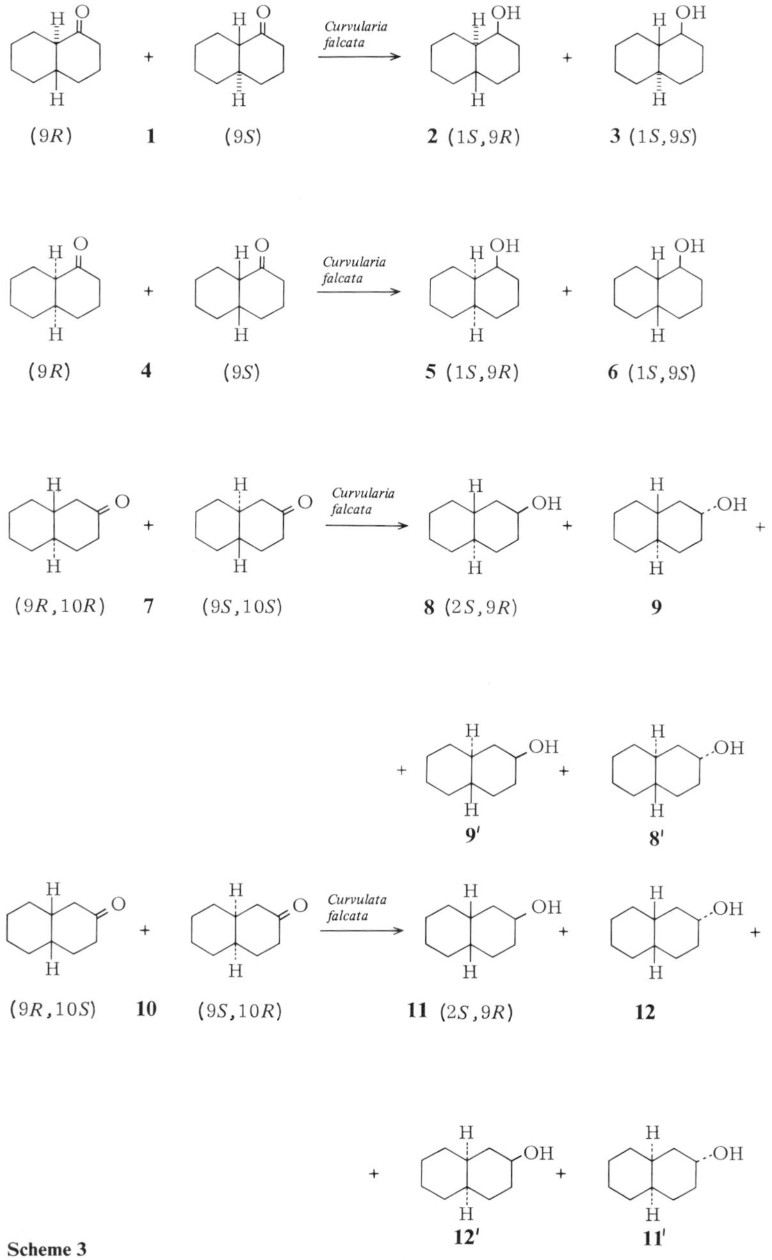

(9*R*) **1** (9*S*) *Curvularia falcata* → **2** (1*S*,9*R*) + **3** (1*S*,9*S*)

(9*R*) **4** (9*S*) *Curvularia falcata* → **5** (1*S*,9*R*) + **6** (1*S*,9*S*)

(9*R*,10*R*) **7** (9*S*,10*S*) *Curvularia falcata* → **8** (2*S*,9*R*) + **9** + **9′** + **8′**

(9*R*,10*S*) **10** (9*S*,10*R*) *Curvulata falcata* → **11** (2*S*,9*R*) + **12** + **12′** + **11′**

Scheme 3

the *S*-configuration as predicted by the Prelog rule, and the dihydroxydecalone **17** in which the two hydroxyl groups have the same configuration. However, reduction of **14** led to anomalous results, the alcohol obtained had *R*-configuration.

Rhizopus nigricans showed different results. A single diol **17** was obtained with two *S*-alcohols, hydroxy ketones **14** and **15** (4*S*,9*S*) (these last three compounds formed by *C.falcata*) and also the enantiomer of **15**, **15′** of 4*R*,9*R*-configuration.

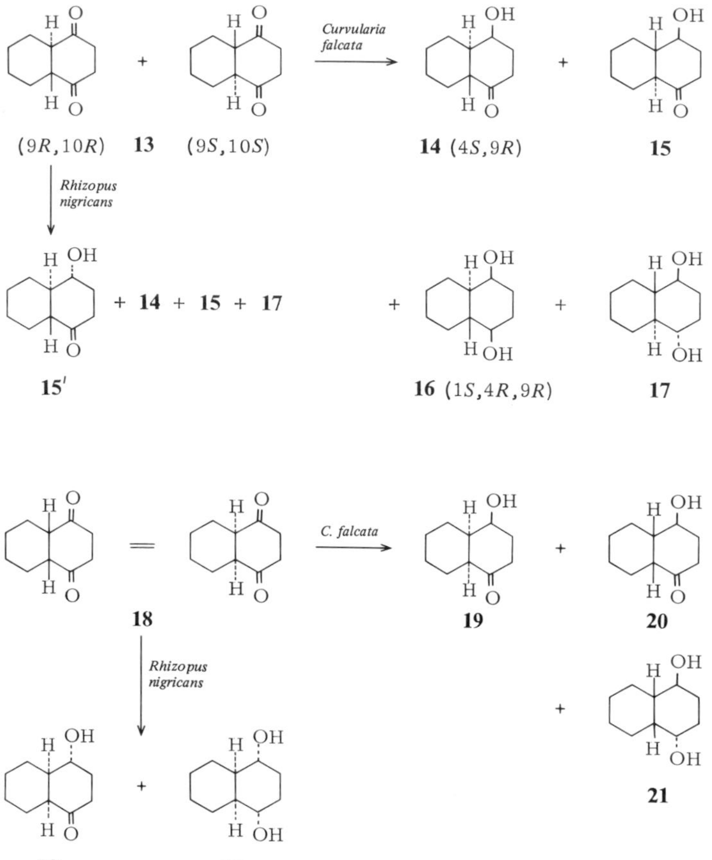

(9*R*,10*R*) **13** (9*S*,10*S*) — *Curvularia falcata* → **14** (4*S*,9*R*) + **15**

Rhizopus nigricans → **15′** + **14** + **15** + **17** + **16** (1*S*,4*R*,9*R*) + **17**

18 — *C. falcata* → **19** + **20** + **21**

18 — *Rhizopus nigricans* → **22** + **23**

Scheme 4

Curvularia falcata reduced *meso-cis*-1,4-dioxodecalin **18** to both hydroxyketones **19** and **20** with the 4*S*-configuration and a diol **21** having the 1*S*,4*S*-configuration. *Rhizopus nigricans* behaved differently, yielding diol **22** with the 4*R*-configuration and diol **23** with the configuration 1*R*,4*S*.

The 1,5-dioxodecalins have also been studied (BAUMANN and PRELOG, 1958b). *Curvularia falcata* transformed *meso*-1,5-dioxo-*trans*-decalin **25** (group S_2) (Scheme 5) into the two possible monoalcohols with 5*S*-configurations, **26** and **27,** along with the diol **28** with a 1*S*,5*R*-configuration. *Rhizopus nigricans* yields the monoalcohol **27,** the diol **28,** as well as the *meso*-diol **29** which thus has the configuration 5*R*. *Streptomyces* ETH A 7 747 gave in addition to diol **28** the racemic monoalcohol **26** and **26′**. *C.falcata* reduced racemic 1,5-dioxo-*cis*-decalin **30** to the two monoalcohols **31** and **32** and the two diols **33** and **34.** All resulting alcohols had the *S*-configuration. On the other hand, *R.nigricans* made diols **35** and **35′** (racemate with a function possessing the *R*-configuration), the diol **33** (1*S*,5*S*), and the monoalcohol **31.**

The reduction of octalindiones also has been studied. Δ^6-1,4-Octalindiones were reduced by *Curvularia falcata* (AKLIN et al., 1965). The racemic *trans*-compound **36** (Scheme 6) gave the two monoalcohols **37** and **38** with 4*S*-configuration and the three alcohols **39, 40,** and **41,** the first two having the 1*S*,4*S*-configuration but the third with an *R*-configuration hydroxyl unpredicted by the Prelog rule though this had been pre-

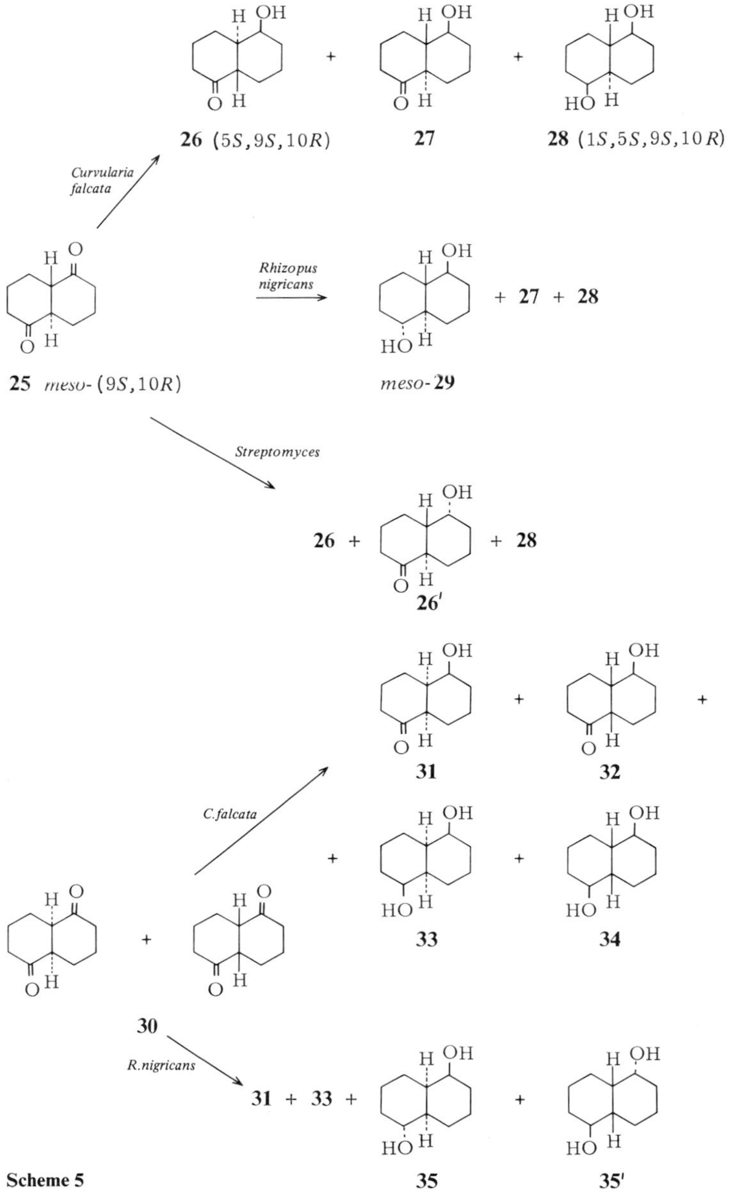

26 (5*S*,9*S*,10*R*) + 27 + 28 (1*S*,5*S*,9*S*,10*R*)

Curvularia falcata

25 *meso*-(9*S*,10*R*)

Rhizopus nigricans → *meso*-29 + 27 + 28

Streptomyces → 26 + 26′ + 28

30

C.falcata → 31 + 32 + 33 + 34

R.nigricans → 31 + 33 + 35 + 35′

Scheme 5

viously observed in the corresponding decalindione **13** (Scheme 4).

C.falcata transformed Δ^6-*cis*-dioxo-octalin **42** (*meso*) into the two monoalcohols **43** and **44** with 4*S*-configurations and the 1*S*,4*S*-configurational diol **45.**

Finally, Δ^9-1,5-dioxooctalin **46** (Scheme 7) was reduced by *C.falcata* and *R.nigricans* (BAUMANN and PRELOG, 1959) (resting cells). Only one ketone function was reduced stereospecifically giving an *S*-configuration. The high yield of the reaction is of interest since a single chiral product is obtained.

Other authors have studied variously substituted tetralones. 1-Tetralone **47** was

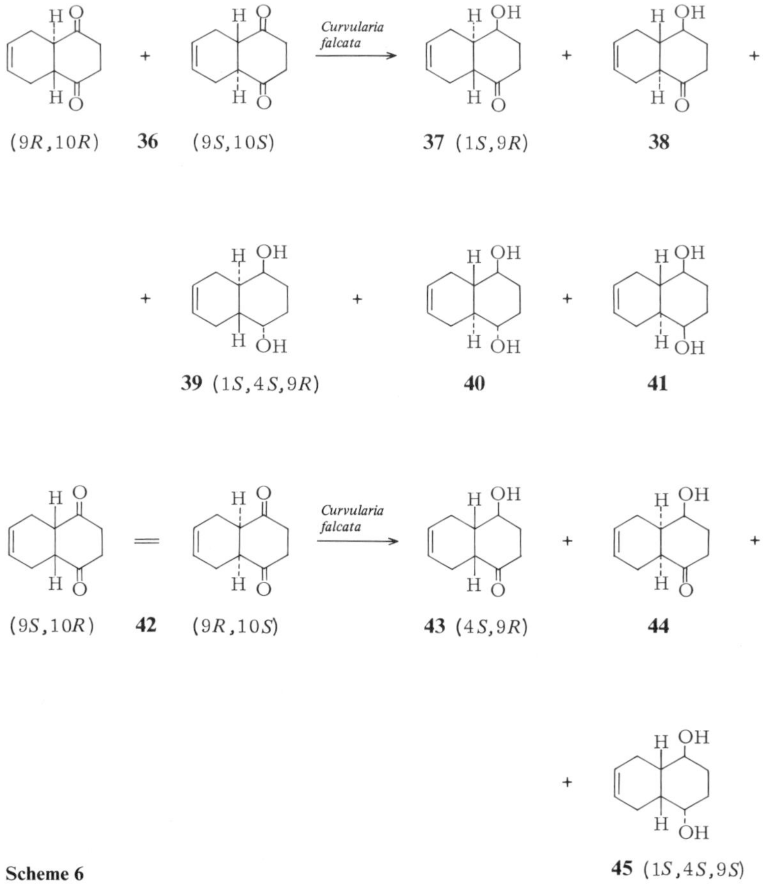

(9*R*,10*R*) **36** (9*S*,10*S*) *Curvularia falcata* → **37** (1*S*,9*R*) + **38** +

+ **39** (1*S*,4*S*,9*R*) + **40** + **41**

(9*S*,10*R*) **42** (9*R*,10*S*) *Curvularia falcata* → **43** (4*S*,9*R*) + **44** +

+ **45** (1*S*,4*S*,9*S*)

Scheme 6

reduced by *Cryptococcus macerans* to the alcohol **48** (1*S*) with a yield of 15% only; 70% of the starting ketone was recovered (KABUTO et al., 1978). 2-Methyl-tetralone **49** was reduced with a low yield by *Sporobolomyces pararoseus* ATCC 11 386 into the alcohols **50** (1*S*,2*S*) and **51** (1*S*,2*R*). However, reduction of racemic 2-acetoxy-1-tetralone **52** by *S.pararoseus* gave the alcohols **53** (1*S*,2*R*) and **54** (1*S*,2*S*) with higher yields (23%). The same authors studied the reduction of 2,3-dihydroindanone **55** to the 1*S*-alcohol **56** by *S.pararoseus* (90% yield) and that of benzosuberone **57** to the 1*S*-alcohol **58** (90% yield) by *Cryptococcus macerans.*

Heterocyclic analogs of decalones were studied by USKOKOVIC et al. (1973) with the aim of obtaining optically active ketones. The (4a*S*,8a*R*)-enantiomer of **59** was reduced by *Sporotrichum exile* QM 1 250 faster than the (4a*R*,8a*S*)-enantiomer, to the alcohol **60.** Hence, the compound with the *cis*-configuration can be resolved, unlike that with the *trans*-configuration.

3. Reduction of substrates with steroid-related skeletons

The preparation of chiral steroid synthons through ketone reduction reactions will be dealt with in Sect. III. A., while Sect. III. B. will focus on reduction reactions involving indane derivatives from steroid degradation.

Here, we shall be concerned with bicyclic and tricyclic substrates related in structure to parts of the steroid skeleton.

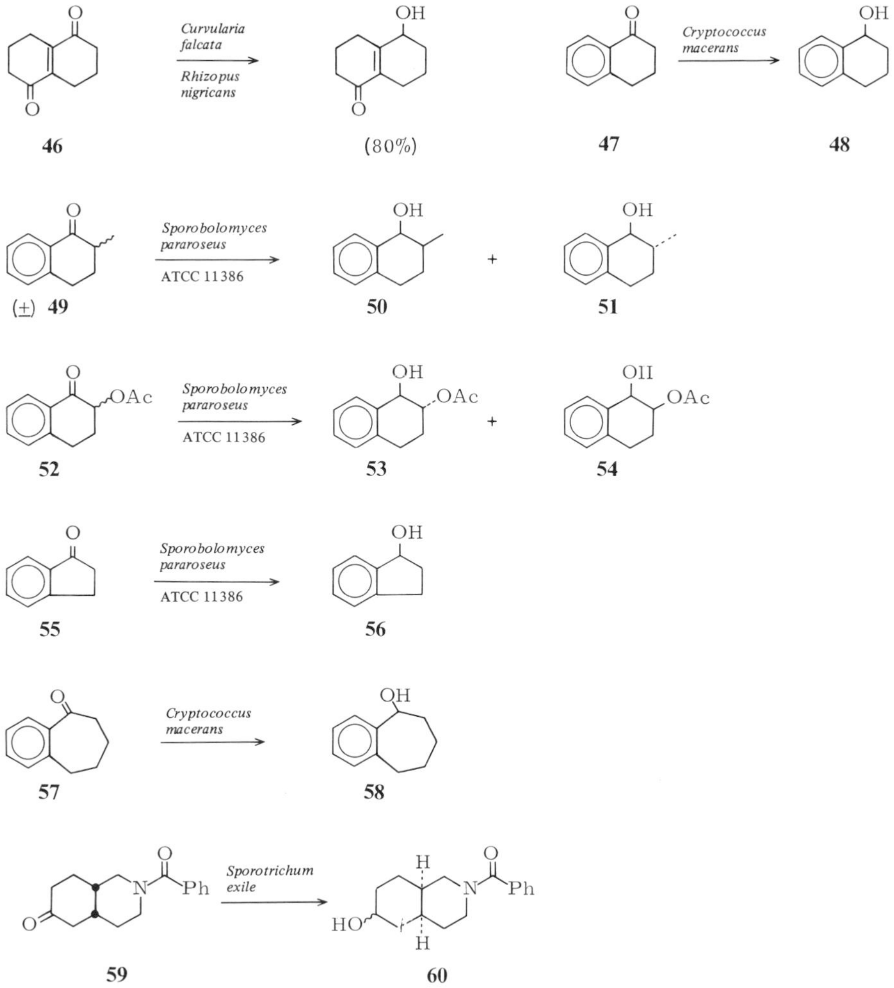

46 → *Curvularia falcata* / *Rhizopus nigricans* → (80%)

47 → *Cryptococcus macerans* → 48

(±) 49 → *Sporobolomyces pararoseus* ATCC 11386 → 50 + 51

52 → *Sporobolomyces pararoseus* ATCC 11386 → 53 + 54

55 → *Sporobolomyces pararoseus* ATCC 11386 → 56

57 → *Cryptococcus macerans* → 58

59 → *Sporotrichum exile* → 60

Scheme 7

The first of these have the 9-methyl-decalin skeleton and are thus analogous to the A + B part of the steroid structure. Racemic 1,6-dioxomethyl-9-*trans*-decalin **1** (Scheme 8) was converted by *Curvularia falcata* into a complex mixture containing a single 1-monoalcohol **2** (1*S*,9*S*) and four 6-monoalcohols, the two enantiomers of **3**, and the two enantiomers of **4**, with a preference for the 6*S*-compounds (PRELOG and ZÄCH, 1959). The diols **5** and **6** were also obtained, both with 1*S*,6*S*-configurations. Thus, the 1*S*- and 6*S*-derivatives are always favored in accordance with the Prelog rule. This preference is particularly marked in position 1 as already observed with the corresponding unmethylated diketones. Reduction by *Rhizopus nigricans* was studied in the case of the analogous *cis*-configuration compounds (AKLIN et al., 1958a). This fungus preferentially attacked the 6-ketone to yield a monoalcohol **8** and a diol **10,** though it also produces a monoalcohol **9** with the 1*S*-configuration.

Finally, reduction by *C.falcata* was studied on 1,5-dioxo derivatives which are less close to a steroid structure (AKLIN et al., 1965). The racemic diketone **11** gave three

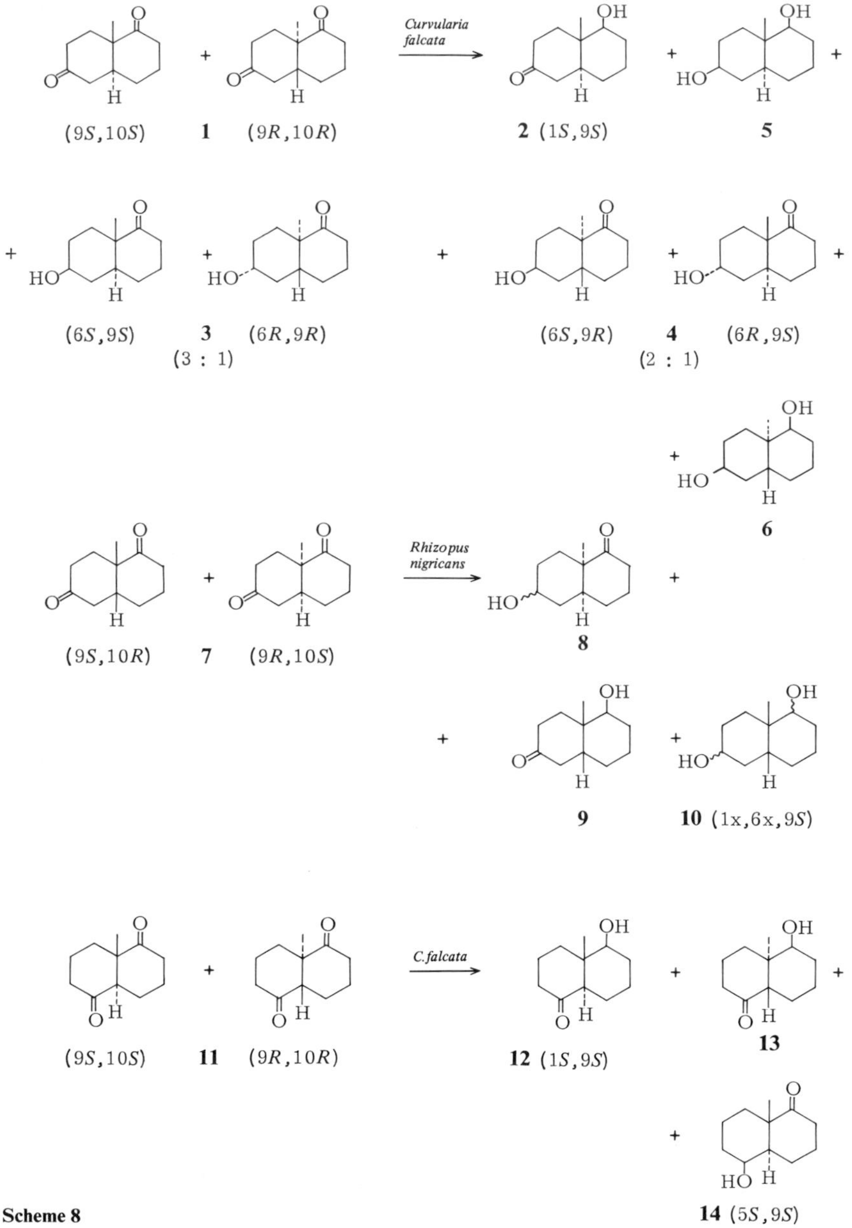

Scheme 8

monoalcohols **12, 13,** and **14,** all having the *S*-configuration.

Among those compounds analogous to the A + B part of the steroid skeleton, 4-unsaturated 1,6-diketones were studied. With *Curvularia falcata,* a first rapid reaction reduced the 1-ketone (PRELOG and AKLIN, 1956). The racemic diketone **15** (Scheme 9) initially gave the two diastereoisomeric monoalcohols **16** and **17** with 1*S*-configurations, which could thus be separated. **16** could then be used as a steroid synthon (PRELOG, 1956). Next, a slower reaction led to small amounts of two *cis*-decalin com-

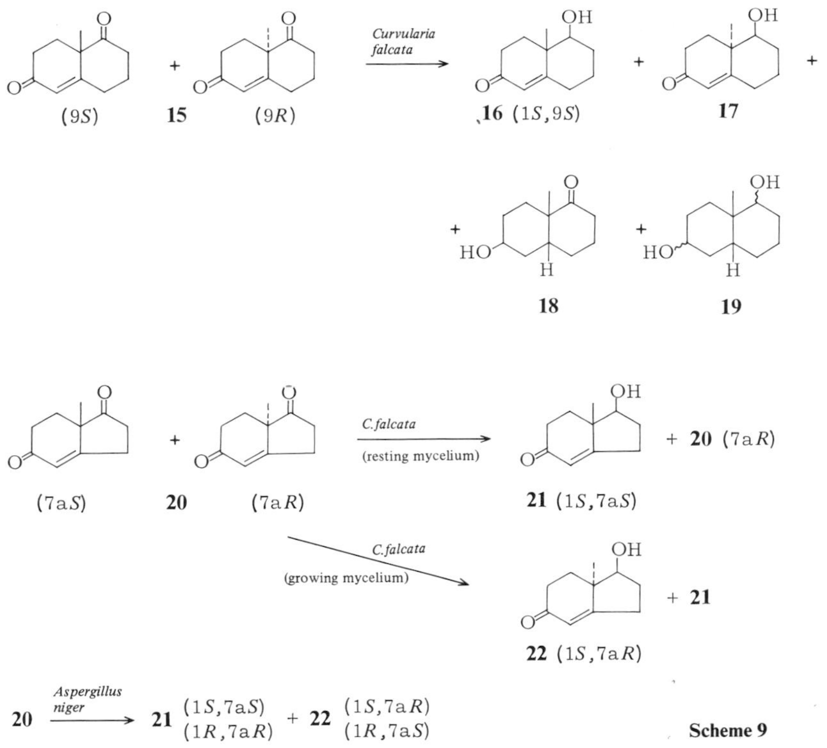

(9*S*) **15** (9*R*) — *Curvularia falcata* → **16** (1*S*,9*S*) + **17** + **18** + **19**

(7a*S*) **20** (7a*R*) — *C. falcata* (resting mycelium) → **21** (1*S*,7a*S*) + **20** (7a*R*)

— *C. falcata* (growing mycelium) → **22** (1*S*,7a*R*) + **21**

20 — *Aspergillus niger* → **21** (1*S*,7a*S*) (1*R*,7a*R*) + **22** (1*S*,7a*R*) (1*R*,7a*S*)

Scheme 9

pounds (9*S*,10*R*), the monoalcohol **18** (6*S*) and the diol **19** of unknown structure (AKLIN et al., 1963). *Aspergillus niger* behaved differently. The main product was **17** obtained as a racemate along with racemic **16** (AKLIN et al., 1958b).

Studies have been carried out on substrates analogous to the C+D part of the steroid ring system.

Racemic 1,5-dioxo-7a-methyl-Δ^4-hexahydroindene **20** gave different products depending on the strain used and on the experimental conditions.

Curvularia falcata (resting cultures) only reduced the 7a*S*-enantiomer to the 1*S*-hydroxy derivative **21** (AKLIN and PRELOG, 1959). This procedure thus can be used to separate racemates (PRELOG, 1956).

Procedure for 1*S*,8*S*(+)-$\Delta^{4,9}$-1-hydroxy-8-methyl-hexahydro-5-indenone **21.**

Fermentation was conducted by a two-stage procedure with a medium of the following composition: saccharose, 10 g; Difco tryptone, 10 g; $NaNO_3$, 2 g; K_2HPO_4, 1 g; $MgSO_4 \cdot 7\,H_2O$, 0.5 g; KCl, 0.5 g; $FeSO_4 \cdot 7H_2O$, 0.01 g; and tap water to a total volume of one liter. The culture was incubated at 25 °C on a rotary shaker with 280 rpm.

Stage II culture was seeded with a 24-hour old stage I culture of *Curvularia falcata* (50 mL in 450 mL) and incubated for 16 hours at 25 °C in a 2-liter Erlenmeyer flask. The broth was then centrifuged at 0 °C and 17800 × *g*, the residue washed and centrifuged twice. Twenty grams of this wet mycelium were suspended in 500 mL of phosphate buffer (pH 7) containing 10 g of saccharose and 200 mg of (±)-$\Delta^{4,5}$-8-methyl-hexahydro-1,5-indenedione **20.** The mixture was placed in a 2-liter flask and agitated for 30 hours at 25 °C on a rotary shaker. The mycelium was then filtered and washed three times with 100 mL of water added to the filtrate; the whole mixture was extracted with ethyl acetate. After drying over anhydrous sodium sulfate and concentrating under vacuum, a crude yellow oil was obtained (270 mg) and purified by chromatography on alumina (AKLIN and PRELOG, 1959).

On the other hand, in growing cultures, *Curvularia falcata* reduced both isomers thereby yielding **21** with the 7a*S*-configuration as well as **22** with the 7a*R*-configuration and a 1*S*-hydroxyl group (AKLIN et al., 1958c). This stereospecificity in the 1-position was not observed with *Aspergillus niger,* which produced racemates of both **21** and **22.**

It should be added that fungi such as *C.falcata, Rhizopus nigricans,* and *A.niger* are known to hydroxylate steroid substrates. The decalin and indane derivatives described here were merely reduced, while tricyclic substrates, which come closer in structure to steroids would generally be hydroxylated. Reduction still mainly occurs when washed mycelia are employed.

Further on, the classical nomenclature for steroids will be used. Compound **23** (Scheme 10) contains the A, B, and C rings, a conjugated ketone in position 3, a 14-oxo group, and an 11-hydroxy group. The racemate was reduced by *Curvularia falcata* to the two separable diastereoisomers **24** and **25.** In both cases, reduction gave a hydroxylated carbon with an *S*-configuration (AKLIN et al., 1965). The same authors reduced the Sarett ketone **26** by using the same microorganism. The results were perfectly comparable. Compound **29** may be regarded as an analog of the BCD part of the steroid ring system, with a 6-membered D-ring. *C.falcata* reduced the racemate to the separable diastereoisomers **30** and **31** both having a 1*S*-hydroxylated carbon.

Further steroid analogs ought to be cited as well. Compound **32** (Scheme 11) has an aromatic B-ring and a 17-oxo group on the D-ring. This analog was reduced by *Saccharomyces cerevisiae* var. *ellipsoides* only to the

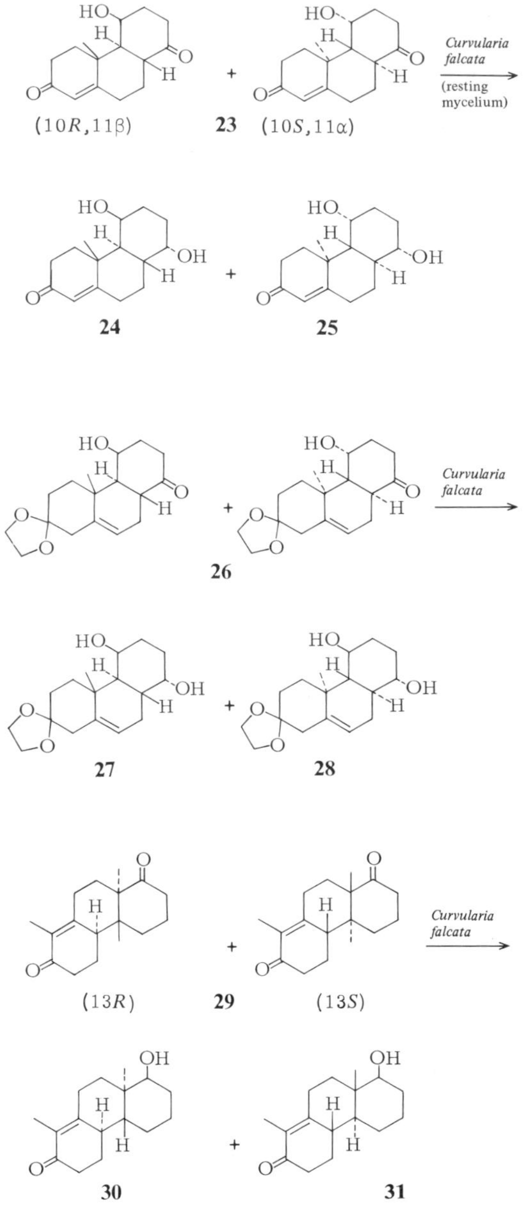

(10*R*,11β) **23** (10*S*,11α) — *Curvularia falcata* (resting mycelium) →

24 + **25**

26 — *Curvularia falcata* →

27 + **28**

(13*R*) **29** (13*S*) — *Curvularia falcata* →

30 + **31**

Scheme 10

alcohol **33** in its natural configuration. Here again, the alcohol had the *S*-configuration (KUROSAWA et al., 1965).

Derivatives with a phenanthrene skeleton have been studied by various authors. For instance, the ketone **34** was reduced to the

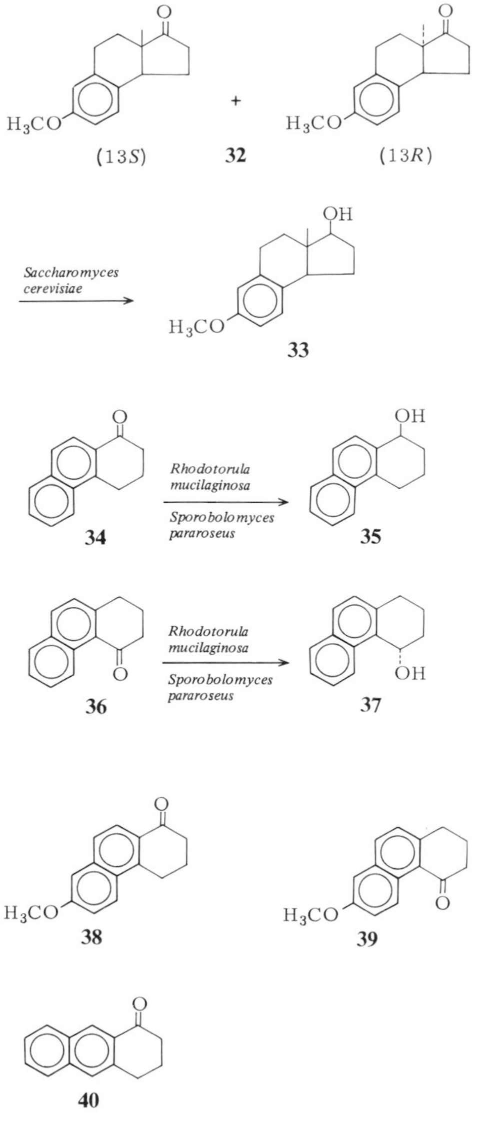

(13S) 32 (13R)

Saccharomyces cerevisiae

33

Rhodotorula mucilaginosa / *Sporobolomyces pararoseus*

34 35

Rhodotorula mucilaginosa / *Sporobolomyces pararoseus*

36 37

38 39

40

Scheme 11

(−)-9*S*-alcohol **35** by *Rhodotorula mucilaginosa* (SIEWINSKI, 1969) and by *Sporobolomyces pararoseus* ATCC 11 386 (KABUTO et al., 1978). Ketone **36** was transformed into the (−)-*S*-alcohol **37** by the same microorganisms, of which the former gave the better yield. Finally, ketones **38** and **39** bearing a 3-methoxy group were reduced by *R.mucilaginosa* to the (−)-alcohol (40% yield) and the (+)-alcohol (45% yield), respectively.

One result with an anthracene analog should also be mentioned, in which ketone **40** underwent reduction to the 1*S*-alcohol by *S.pararoseus*.

4. Stereochemistry of reduction

The rule proposed by Prelog on the stereochemistry of ketone reduction was presented in Fig. 1. The listed examples, most of which were reported by the Zurich re-

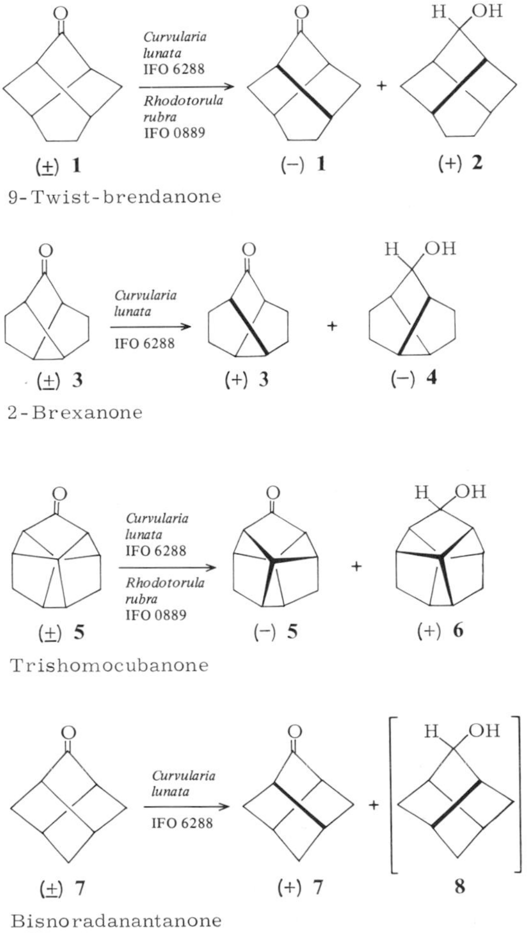

Curvularia lunata IFO 6288 / *Rhodotorula rubra* IFO 0889

(±) 1 (−) 1 (+) 2

9-Twist-brendanone

Curvularia lunata IFO 6288

(±) 3 (+) 3 (−) 4

2-Brexanone

Curvularia lunata IFO 6288 / *Rhodotorula rubra* IFO 0889

(±) 5 (−) 5 (+) 6

Trishomocubanone

Curvularia lunata IFO 6288

(±) 7 (+) 7 8

Bisnoradanantanone

Scheme 12

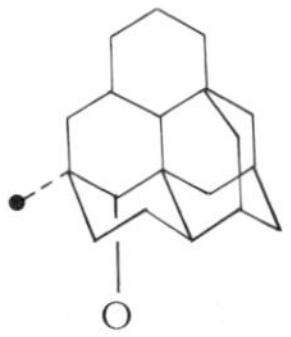

Figure 2. Diamond lattice section for a oxidoreductase from *Curvularia falcata*. – • Forbidden position

search group, include several exceptions to this rule, mainly in compounds with polar substituents.

Furthermore, PRELOG and coworkers have proposed a rule regarding the rate of reduction of carbonyls and of the steric hindrance in these reactions.

In a diamond-lattice model (Fig. 2) the respective positions of the carbonyl and forbidden substituents can be seen (PRELOG, 1963, 1964).

Japanese authors have reported on systems which do not fit into a diamond-lattice. These complex molecules have C_2-symmetries and two enantiomeric faces (NAKAZAKI et al., 1979). 9-Twist-brendanone **1** (Scheme 12) was reduced by *Curvularia lunata* IFO 6 288 to the (+)-alcohol **2** (optical purity 85%, yield 24%) and the (−)-ketone **1** (optical purity 64%, yield 30%). 2-Brexanone **3** gave the (−)-alcohol **4** (o.p. 100%, yield 12%) and unreacted (+)-ketone (o.p. 21%, yield 35%). D_3-Trishomocubanone **5** gave the (+)-alcohol **6** (o.p. 61%, yield 75%) together with unaltered (−)-ke-

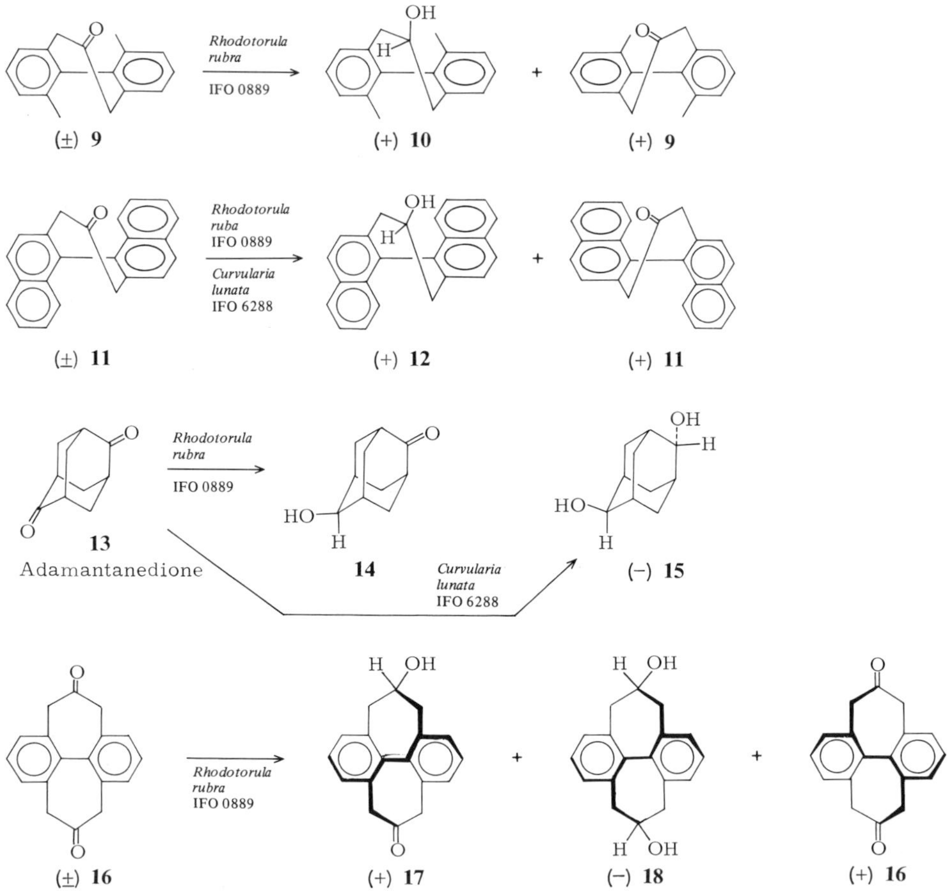

(±) 9 → *Rhodotorula rubra* IFO 0889 → (+) 10 + (+) 9

(±) 11 → *Rhodotorula ruba* IFO 0889 / *Curvularia lunata* IFO 6288 → (+) 12 + (+) 11

13 Adamantanedione → *Rhodotorula rubra* IFO 0889 → 14; → *Curvularia lunata* IFO 6288 → (−) 15

(±) 16 → *Rhodotorula rubra* IFO 0889 → (+) 17 + (−) 18 + (+) 16

Scheme 13

tone (o.p. 40%, yield 40%). Finally, with the same microorganism, bisnoradamantanone **7** was reduced leaving the (+)-ketone (o.p. 30%, yield 20%); the expected alcohol **8** was not isolated, however. The same workers also studied the reduction of **1** and **5** by *Rhodotorula rubra* IFO 0 889, which gave results inferior to those obtained with *C.lunata*. Finally, they studied the reduction of bridged biphenyl ketones **9** and binaphthyl ketones **11** (Scheme 13). **9** showed best results with *R.rubra* at 34% (+)-alcohol **10** (o.p. 94%) and 31% (+)-ketone **9** (o.p. 100%). *C.lunata* converted **11** into the (+)-alcohol **12** (o.p. 65%, yield 12%) and the (+)-ketone **11** (o.p. 33%, yield 46%). The authors proposed a "C_2-ketone rule" stating that these microorganisms preferentially reduce the enantiomers with P-helicity.

To test the applicability of this rule, the same investigators (NAKAZAKI, 1981b) studied an adamantanedione **13** and a bridged diketone **16.** Adamantanedione **13** converted by *C.lunata* gave 18% of the (−)-diol **15** (o.p. 75%), while *R.rubra* yielded the monoalcohol **14.** The latter strain also reduced the diketone **16** to (−)-*cis*-diol **18** and (+)-ketol **17;** the (+)-diketone was recovered. The ketones described above (**1, 3, 5, 9, 11, 13,** and **16**) all have a C_2-axis passing through the carbonyl. They are said to be gyrochiral (NAKAZAKI et al., 1978).

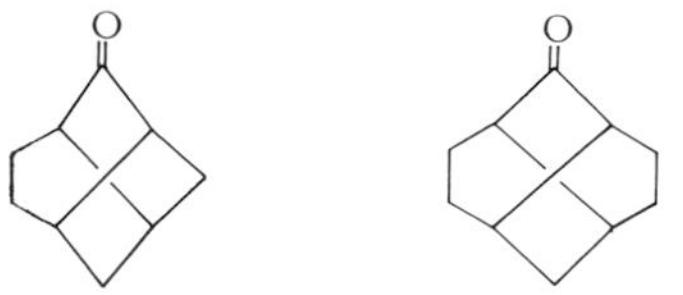

19
2-Twist-brendanone

20
2-Twistanone

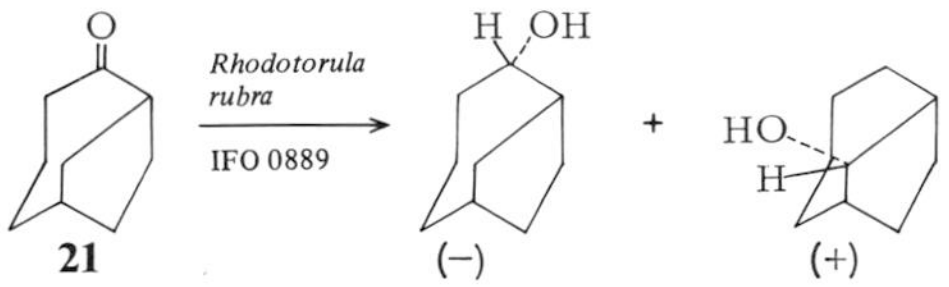

21 Protoadamantanone → *Rhodotorula rubra* IFO 0889 → (−) + (+)

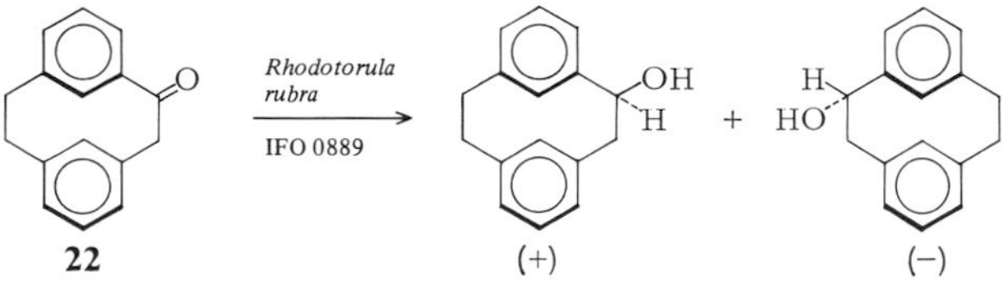

22 1-Oxo[2.2]metacyclophane → *Rhodotorula rubra* IFO 0889 → (+) + (−)

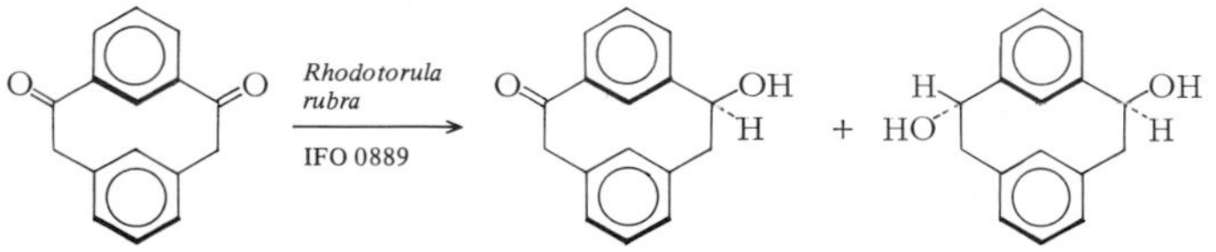

23 1,10-Dioxo[2.2]metacyclophane → *Rhodotorula rubra* IFO 0889 → (+) **24** + (−) **25**

Scheme 14

However, 2-twist-brendanone **19** (Scheme 14) and 2-twistanone **20** were not reduced by the microorganisms cited above (NAKAZAKI et al., 1979).

The same authors proposed their C_2-ketone rule as a particular case of the more general quadrant rule, which is applicable to C_s and C_1 group molecules. However, in these two cases, the Prelog rule seems to be applicable.

This applies for (±)-protoadamantanone **21** which was reduced by *Rhodotorula rubra* to both *S*-configuration alcohols. The same microorganism reduced (±)-1-oxo[2.2]metacyclophane **22** also to the two *S*-configuration alcohols, while reduction of 1,10-dioxo[2.2]metacyclophane **23** led to the ketol **24** and diol **25.** All the hydroxyl groups are attached to *S*-configuration carbons (NAKAZAKI et al., 1978).

The same team studied the reduction of bicyclic ketones. Bicyclo[2.2.2]octane-2-one **26** (Scheme 15) (C_s group) was reduced by several microorganisms. *R.rubra* showed the best result. After 24 hours of incubation, **26** gave 37% of the *S*-configuration alcohol (o.p. 98%). The symmetry is lost in norbornanone **27** and norbornenone **28.** Best results were obtained with *Curvularia lunata,* where (±)-**27** mainly afforded the (−)-*endo*-product **29** (o.p. 91%) leaving the (+)-ketone **27** (o.p. 67%). The *exo*-compound was probably also obtained, but quantities were too low for characterization.

(±)-Norbornenone **28** gave similar results with *C.lunata,* now with slightly higher amounts of the *exo*-product, but selectivity was rather low: 54 and 15% optical purity, respectively, for the (2*S*)-(−)-*endo*-alcohol **30** and the recovered (+)-ketone **28.**

Bicyclo[2.2.2]oct-5-ene-2-one **31** lacks the C_s-symmetry of **26.** Here, *Rhodotorula rubra* gave the highest optical purities, of 26, 74, and 90%, respectively, for the

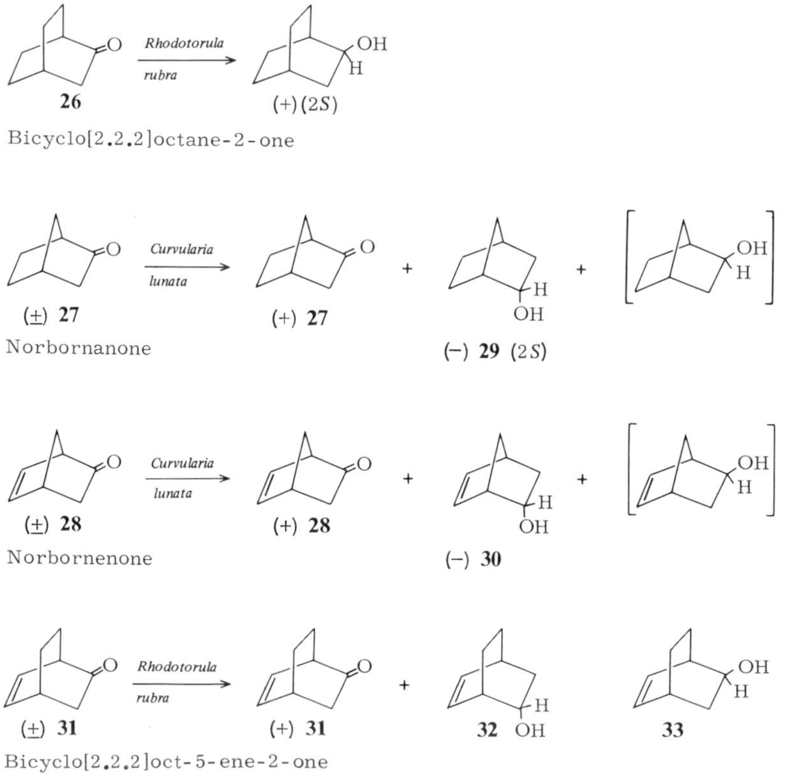

26 —*Rhodotorula rubra*→ (+)(2*S*)

Bicyclo[2.2.2]octane-2-one

(±) **27** —*Curvularia lunata*→ (+) **27** + (−) **29** (2*S*) + [exo-alcohol]

Norbornanone

(±) **28** —*Curvularia lunata*→ (+) **28** + (−) **30** + [exo-alcohol]

Norbornenone

(±) **31** —*Rhodotorula rubra*→ (+) **31** + **32** + **33**

Bicyclo[2.2.2]oct-5-ene-2-one

Scheme 15

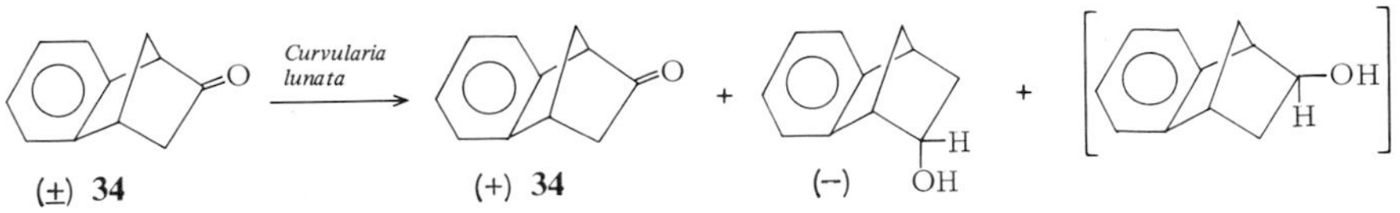

Curvularia lunata

(±) 34 — Benzonorbornanone → (+) 34 (yield 9%, o.p; 53%) + (−) (yield 37,5%, o.p 82%) + [ ]

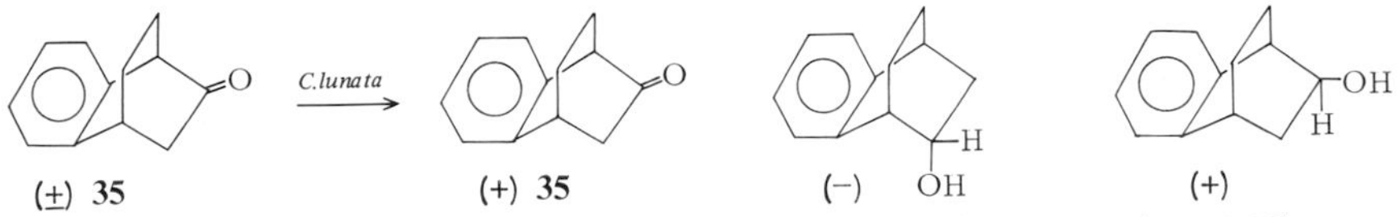

C.lunata

(±) 35 — Benzobicyclo[2.2.2] octane-2-one → (+) 35 (yield 33%, o.p. 95%) + (−) (yield 27%, o.p; 85%) + (+) (yield 3%, o.p. 77%)

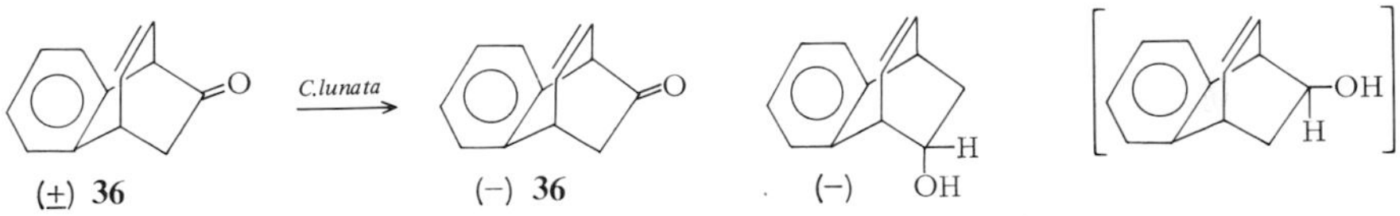

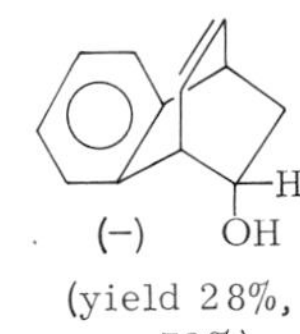

C.lunata

(±) 36 — Benzobicyclo[2.2.2] oct-5-ene-2-one → (−) 36 (yield 22%, o.p. 93%) + (−) (yield 28%, o.p. 72%) + [ ]

Scheme 16

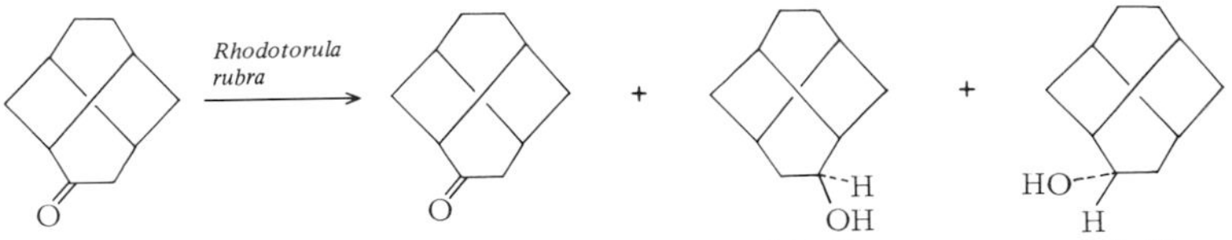

Rhodotorula rubra

(±) 37 — 4-Twistanone → (+) 37 + (−) 41 + (+) 40

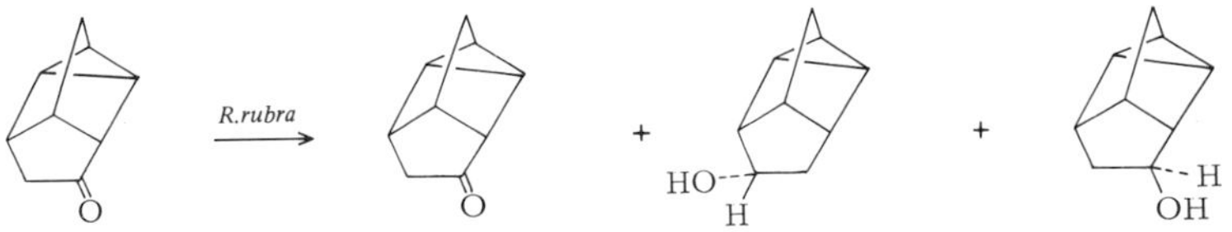

R.rubra

(±) 38 — 8-Deltacyclanone → (+) 38 + (−) 43 + (−) 42

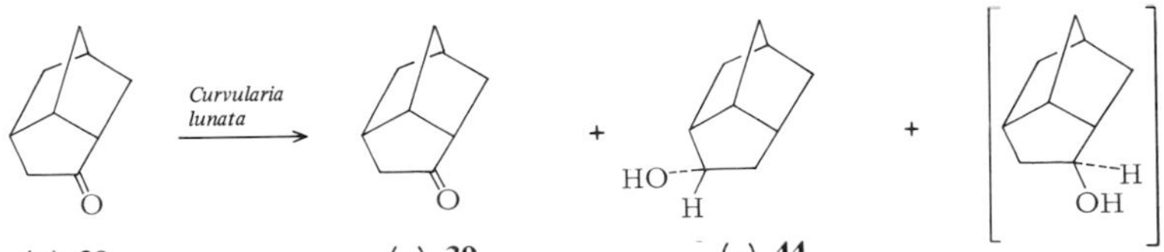

Curvularia lunata

(±) 39 — 4-Brendanone → (−) 39 + (−) 44 + [ ]

Scheme 17

recovered (+)-ketone, the (−)-*endo*-alcohol **32,** and the (+)-*exo*-alcohol **33.** Preference for the *S*-configuration is evident here.

Benzoanalogs of the above mentioned ketones have also been studied. Benzonorbornanone **34** (Scheme 16), benzobicyclo[2.2.2]octane-2-one **35,** and benzobicyclo[2.2.2]oct-5-ene-2-one **36** were all reduced to products with high optical purities by *C.lunata,* **35** being converted into the *endo*- and *exo*-alcohols (NAKAZAKI et al., 1980).

Finally, the same authors studied bicyclic and tetracyclic compounds with C_1-symmetry: 4-twistanone **37** (Scheme 17) (NAKAZAKI et al., 1980), 8-deltacyclanone **38,** and 4-brendanone **39** (NAKAZAKI et al., 1981c).

Rhodotorula rubra showed best results on **37** in forming the (−)-(4*S*)-*exo*-alcohol **41** (yield 24%, optical purity 94%), and the (+)-(4*S*)-*endo*-alcohol **40** (yield 4%, optical purity 81%), leaving the (+)-ketone **37** (yield 27%, optical purity 92%).

8-Deltacyclanone **38** was reduced by *Curvularia lunata* and *R.rubra.* With the latter, 12% of the (−)-*endo*-alcohol **43** could be obtained (yield 12%, o.p. 83%) along with 6% of the (−)-*exo*-alcohol **42** (yield 6%, o.p. 53%); 32% of the (+)-ketone **38** was recovered (o.p. 28%). 4-Brendanone **39** was reduced by *C.lunata* to the (−)-*endo*-alcohol **44** (yield 20%, o.p. 88%), leaving 48% of the (−)-ketone **39** (o.p. 35%). The alcohols derived from **39** and **38** had the *S*-configuration.

B. Reduction of Carbon-Carbon Double Bonds

In all following cases the carbon-carbon double bonds are α,β to a carbonyl, or sometimes to a hydroxyl group. Apparently, isolated double bonds are only reduced with great difficulty. The oldest example is the reduction of 3-methyl-2-cyclohexenone to the corresponding saturated alcohol (FISCHER and WIEDEMANN, 1935) by yeasts. Since then, the reduction of a number of α,β-unsaturated cyclenones has been studied, particularly with *Beauveria sulfurescens* ATCC 7 159. Cyclopentenone **1** (Scheme 18) and cyclohexenone **2** were reduced by mycelia of the aforementioned strain under very low aeration to the corre-

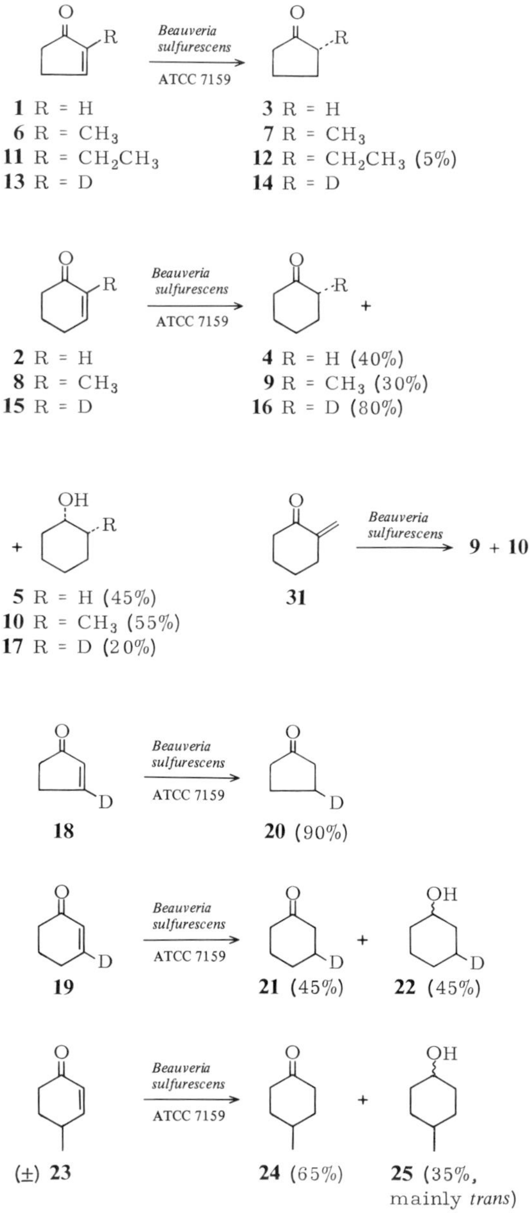

Beauveria sulfurescens ATCC 7159

1 R = H; **6** R = CH_3; **11** R = CH_2CH_3; **13** R = D → **3** R = H; **7** R = CH_3; **12** R = CH_2CH_3 (5%); **14** R = D

2 R = H; **8** R = CH_3; **15** R = D → **4** R = H (40%); **9** R = CH_3 (30%); **16** R = D (80%) + **5** R = H (45%); **10** R = CH_3 (55%); **17** R = D (20%)

31 → *Beauveria sulfurescens* → **9** + **10**

18 → **20** (90%)

19 → **21** (45%) + **22** (45%)

(±) **23** → **24** (65%) + **25** (35%, mainly *trans*)

Scheme 18

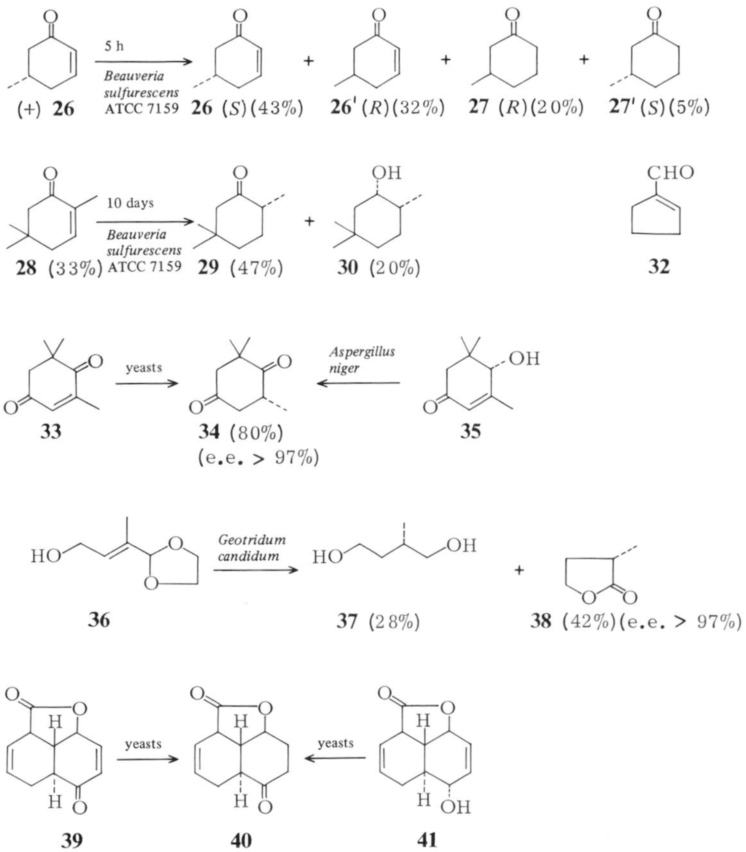

(+) **26** —5 h, *Beauveria sulfurescens* ATCC 7159→ **26** (*S*)(43%) + **26'** (*R*)(32%) + **27** (*R*)(20%) + **27'** (*S*)(5%)

28 (33%) —10 days, *Beauveria sulfurescens* ATCC 7159→ **29** (47%) + **30** (20%)

CHO

32

33 —yeasts→ **34** (80%) (e.e. > 97%) ←*Aspergillus niger*— **35**

36 —*Geotridum candidum*→ **37** (28%) + **38** (42%)(e.e. > 97%)

39 —yeasts→ **40** ←yeasts— **41**

Scheme 19

sponding saturated ketones **3** and **4.** For cyclohexenone, the reduction proceeds to the saturated alcohol **5** (Kergomard et al., 1978). The analogous ketones bearing 2-methyl groups (**6** and **8**) gave saturated ketones **7** and **9** with an *R*-configuration. The saturated alcohol *cis*-(1*S*,2*R*) was also obtained which is in accordance with the rules given in Fig. 1. The ethyl group in the cyclopentenone **11** considerably slowed the reaction. *Beauveria sulfurescens* also reduced the 2-deuterated ketones **13** and **15** (Dauphin et al., 1980a; Schwab, 1981), as well as the 3-deuterated ketones **18** and **19** (Dauphin et al., 1980b). The stereochemistry of this reaction shows that hydrogen is added in *trans*-position. The position of substituents on these substrates influences the reduction. A 3-methyl group prevented reduction in both cyclopentenone and cyclohexenone (Kergomard et al., 1978). Other substituents were studied in cyclohexenone. For instance, 4-methylcyclohexenone **23** was reduced in an expected way (Kergomard et al., 1982a). After 5 h (±)-5-methylcyclohexenone **26** (Scheme 19) was reduced to a mixture of saturated ketones, the proportions of which are given in Scheme 19. (±)-6-Methylcyclohexanone gave an analogous result. On the other hand, 2,5,5-trimethylcyclohexenone **28** reacted much slower and 2,6- and 6,6-dimethylcyclohexenone failed to react. Finally, 2-methylenecyclohexanone **31** showed the same results as 2-methylcyclohexenone **8.**

These results, along with others obtained in different series of substrates, led the authors to propose a stereochemical rule analogous to that of Prelog for the reduction of carbonyl groups.

The carbonyl is contained either in group L or in group S. Another cyclic compound, the α,β-unsaturated aldehyde **32** gave the corresponding saturated alcohol (DESRUT et al., 1981).

The reduction of cyclic unsaturated ketones can be performed with microorganisms other than *Beauveria sulfurescens*. Thus, **1** was reduced by *Curvularia lunata* and *Cunninghamella blakesleeana* (KERGOMARD et al., 1982b). In addition, various types of bacteria reduced **2** and **8** (DESRUT et al., 1983). A strain of *Clostridium* DSM 1 460 anaerobically reduced these ketones as well as **15** and **19** giving the same stereochemistry as *B.sulfurescens*. Further, three actinomycetes, two *Streptomyces* and one *Nocardia* species reduced **2** and **8**, with the latter ketone leading to the same stereochemistry as also observed through *B.sulfurescens*. The reduction of **8** by *Pseudomonas ovalis* has been reported (NOMA, 1977).

A number of other examples should be mentioned here.

The diketone **33** was reduced by yeasts to the saturated diketone **34** (LEUENBERGER et al., 1976). The ketol **35** was converted to the same diketone **34** by *Aspergillus niger* (MIKAMI, 1981), presumably via the diketone **33.**

The carbon-carbon double bond of the unsaturated dioxolane **36** was reduced by *Geotrichum candidum* to the diol **37** along with the lactone **38** through oxidation of the initial aldehyde function (LEUENBERGER et al., 1979). Finally, "Woodward's lactone" **39** was reduced by yeasts and by other microorganisms (PROTIVA, 1961; MORI et al., 1965; CAPEK et al., 1962).

C. Oxidation of Alcohols

The oxidation of cyclanols to cyclanones is not of very great interest except where polyols are to be selectively oxidized, since chemical methods are generally adequate. Cyclopentanol and cycloheptanol can be oxidized by *Acetobacter suboxydans* (POSTERNAK and REYMOND, 1953) and cyclohexanol by soil bacterium JOB5 (OOYAMA and FOSTER, 1965); alcohol **1** (Scheme 20) gave the corresponding aldehyde **2** with a strain of *Pseudomonas* PL (BALLAL et al., 1968) and the sterol analog **3** was oxidized by *Pseudomonas testosteroni* ATCC 11 996 (SIH and WANG, 1963).

Several strains opened the lactone ring of lactonic acid **4** and oxidized the alcohol function to ketone **5** (IGUCHI et al., 1964). Other strains oxidized the substrate to glutamic acid.

The selective oxidation of cyclane polyols has been extensively studied. Among these, we shall deliberately omit cyclohexane hexols and pentols which belong to the

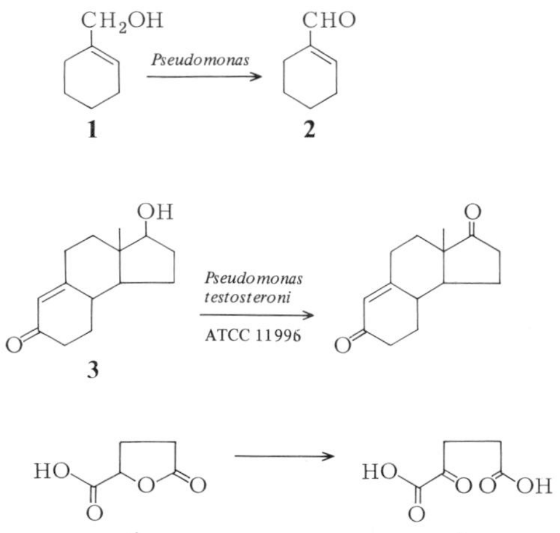

Microbacterium ammoniaphilum ATCC 15354
Micrococcus glutamicus ATCC 13032 and 13068
Brevibacterium sibercutum NRRL B2311 and B2312
Corynebacterium lilium NRRL B2243

Scheme 20

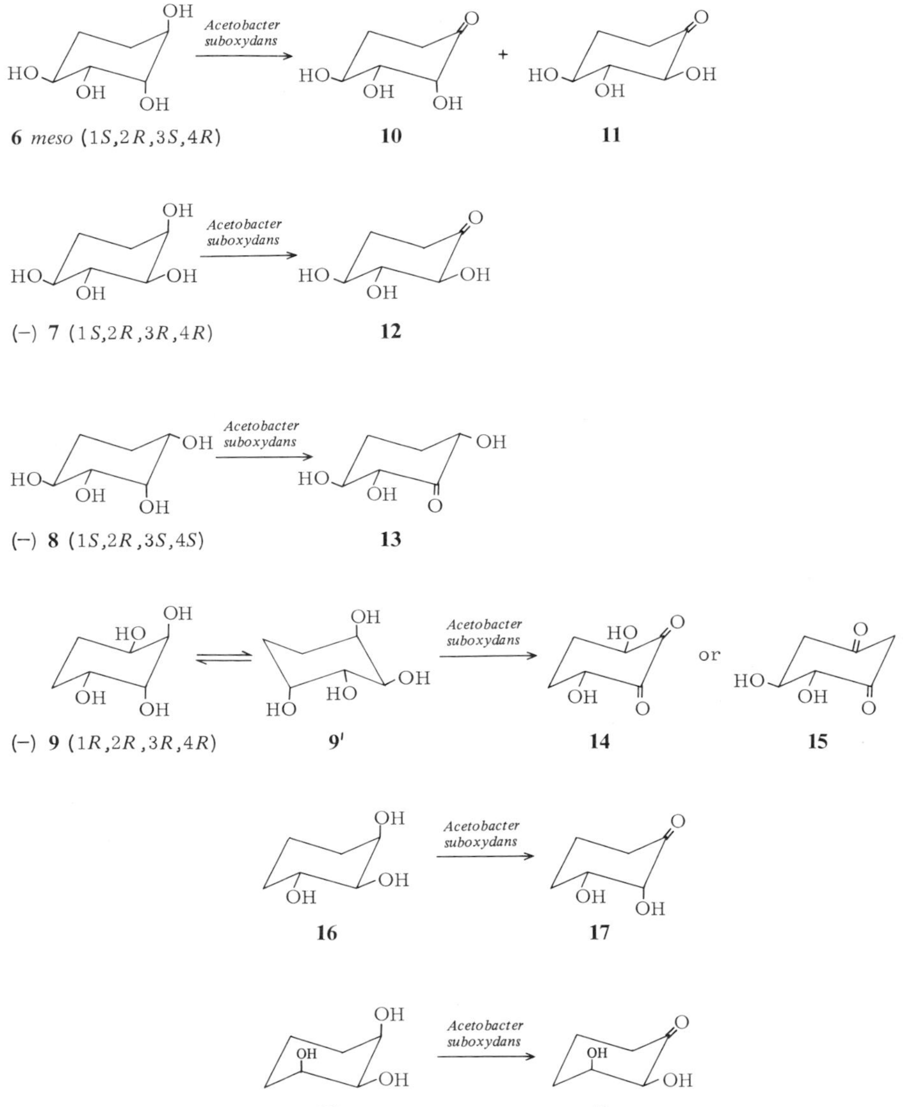

6 *meso* (1*S*,2*R*,3*S*,4*R*) → *Acetobacter suboxydans* → **10** + **11**

(−) **7** (1*S*,2*R*,3*R*,4*R*) → *Acetobacter suboxydans* → **12**

(−) **8** (1*S*,2*R*,3*S*,4*S*) → *Acetobacter suboxydans* → **13**

(−) **9** (1*R*,2*R*,3*R*,4*R*) ⇌ **9′** → *Acetobacter suboxydans* → **14** or **15**

16 → *Acetobacter suboxydans* → **17**

18 → *Acetobacter suboxydans* → **19**

Scheme 21

sugars and deoxysugars. Strains of *Acetobacter suboxydans* were used for the cyclane polyol studies.

A fairly general rule states that oxidation of hydroxyl groups takes place axially (POSTERNAK et al., 1957). This may be the counter part of reduction by the equatorial side. Scheme 21 shows the results of the oxidation of the four tetrahydroxycyclohexanes **6, 7, 8,** and **9**. The rule was closely obeyed by **7** and **8**. From **6**, the two products **10** and **11** were obtained. Product **11** was unexpected, and arises from epimerization of the 3-hydroxyl.

The oxidation product of **9** is the undetermined diketone **14** or **15** (POSTERNAK and REYMOND, 1955). Either one may be formed from **9** in conformational equilibrium with **9′**. Oxidation of **6** was also performed with resting cells of *Acetobacter sub-*

oxydans. Trihydroxycyclohexanes have also been studied.

Acetobacter suboxydans oxidized **16** (1,*cis*2,*trans*3) to the *trans*-dihydroxy ketone **17**, and the triol **18** (*cis*1,2,3) to the *cis*-dihydroxy ketone **19.** The authors explain this result in terms of Hudson's rule (HANN, 1938; POSTERNAK and RAVENNA, 1947). The same rule would explain why the racemic *trans*-diols **20, 22, 24,** and **26** (Scheme 22) gave the corresponding ketols **21, 23, 25,** and **27** with the L-configuration while the *cis*-diols **28, 30,** and **32** gave the corresponding D-ketols (POSTERNAK et al., 1955). The oxidation of *trans*-1,2-cyclohexanediol to the ketol and further to the diketone has been considered as an initial step in a more extensive degradation by *Acinetobacter* TD 63 (DAVEY and TRUDGILL, 1977).

The oxidation of ketol **35** (Scheme 19) by *Aspergillus niger* to the saturated diketone **34** via the unsaturated ketone **33** was mentioned earlier (Sect. II. B.). Similarly, alcohol **41** (Scheme 19) was oxidized to ketone **40** by yeasts and many other microorganisms, most likely via **39** (PROTIVA et al., 1961; MORI et al., 1965; CAPEK et al., 1962).

D. Hydrolysis Reactions

The hydrolysis of a racemic ester RCO_2R' yields an acid RCO_2H and an alcohol R′OH one of which possesses an asymmetric carbon. Since the hydrolysis rates will be different for *R*- and *S*-configurations, it would be expected that an optically active acid or alcohol might be recovered. For instance, the hydrolysis of ethylcyclohepta-2,5-diene carboxylate **1** (Scheme 23) by a suspension of lyophilized *Rhodotorula minuta* var. *texensis* at 14% completion showed an enantiomeric excess of 89% for acid **2** and 14% for the recovered ester. At 55% completion these figures were 11% and 78%, respectively. Hence, the *S*-ester was hydrolyzed faster (KAJIWARA et al., 1981).

A systematic study was carried out by KAWAI et al. (1981) using *Rhizopus nigricans*. Some of their results are given in Scheme 23. The substrates were acetic esters of benzocyclopentanol **3**, benzocyclohexanol **4**, benzocycloheptanol **5**, and alcohols with a hydroanthracene skeleton **6**, and a hydrophenanthrene ring system **7** and

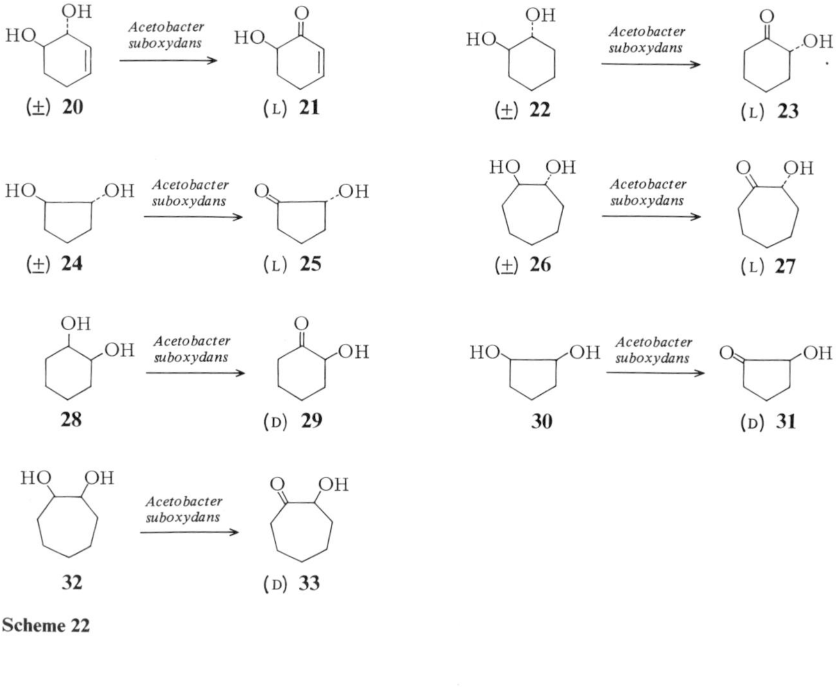

(±) **20** → *Acetobacter suboxydans* → (L) **21**

(±) **22** → *Acetobacter suboxydans* → (L) **23**

(±) **24** → *Acetobacter suboxydans* → (L) **25**

(±) **26** → *Acetobacter suboxydans* → (L) **27**

28 → *Acetobacter suboxydans* → (D) **29**

30 → *Acetobacter suboxydans* → (D) **31**

32 → *Acetobacter suboxydans* → (D) **33**

Scheme 22

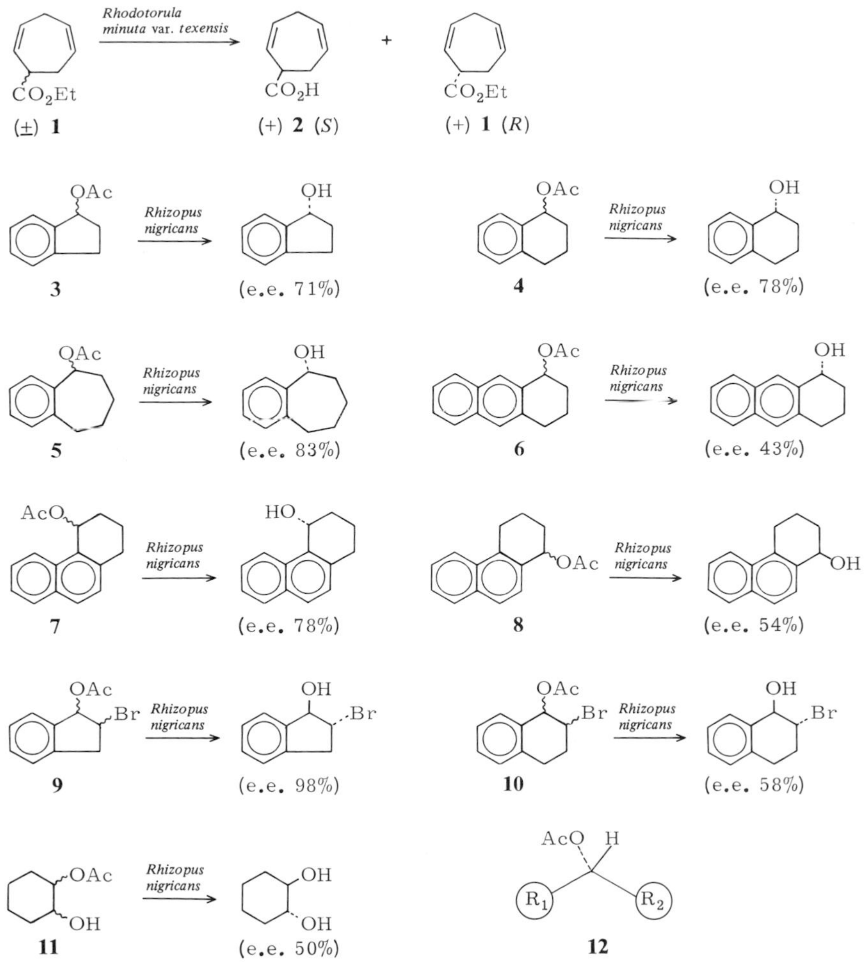

Rhodotorula minuta var. texensis

CO_2Et (±) **1** → CO_2H (+) **2** (*S*) + CO_2Et (+) **1** (*R*)

OAc **3** —*Rhizopus nigricans*→ OH (e.e. 71%)

OAc **4** —*Rhizopus nigricans*→ OH (e.e. 78%)

OAc **5** —*Rhizopus nigricans*→ OH (e.e. 83%)

OAc **6** —*Rhizopus nigricans*→ OH (e.e. 43%)

AcO **7** —*Rhizopus nigricans*→ HO (e.e. 78%)

OAc **8** —*Rhizopus nigricans*→ OH (e.e. 54%)

OAc, Br **9** —*Rhizopus nigricans*→ OH, Br (e.e. 98%)

OAc, Br **10** —*Rhizopus nigricans*→ OH, Br (e.e. 58%)

OAc, OH **11** —*Rhizopus nigricans*→ OH, OH (e.e. 50%)

AcO H, R_1, R_2 **12**

Scheme 23

8. Bromhydrins **9** and **10** as well as *trans*-cyclohexanediol monoacetate **11** were also studied.

A rule was proposed for acetates of secondary alcohols bearing two substituents R_1 and R_2 (**12**): The acetate hydrolyzed fastest has $R_1 > R_2$. This rule applies to the hydrolysis of (Scheme 24) the racemic *trans*-2,*trans*-4-dimethylcyclohexanol **13** (Scheme 24) by *Bacillus subtilis* var. *niger* IFO 3 108 (ORITANI et al., 1982). When hydrolysis was 46% complete the (−)-alcohol **14** was obtained (94% e. e.)*) and the (+)-acetate **13** (92% e. e.) was recovered. The production of these alcohols is of interest for the synthesis of cycloheximide. The resolution of racemic allethrolone is important in the synthesis of pyrethrinoids. (±)-Acetyl-allethrolone **15** was hydrolyzed by *Trichoderma reesei* AHU 9 484 (ORITANI and YAMASHITA, 1976). (−)-Allethrolone was thus obtained, while (+)-allethrolone could be formed by chemical hydrolysis of the recovered starting material.

A biochemical study of the hydrolysis of cyclohexanol acetate by a strain of *Streptomyces hygroscopicus* has been carried out (REUTER and HUETTNER, 1977).

* e. e. enantiomeric excess

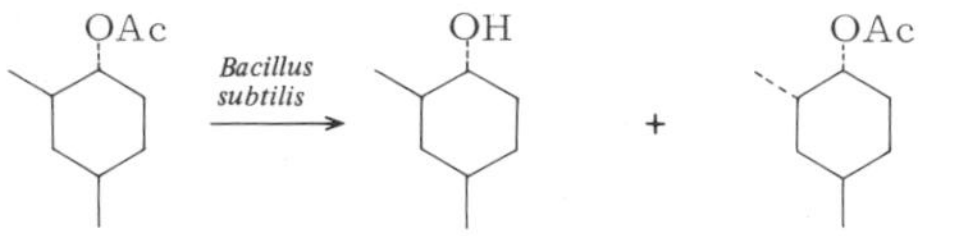

The hydrolysis of a number of nitrogen-containing compounds has also been reported. The sulfuric amide of cyclohexylamine **16** was hydrolyzed by *Pseudomonas* and *Corynebacterium* to the corresponding amine and some cyclohexanone (CAPEK et al., 1963). Pyrrolidone **17** was hydrolyzed to γ-aminobutyric acid and reduced to butyric acid by ruminal bacteria (LAROCHE, 1960). Allantoin **18** was hydrolyzed by *Streptococcus allantoicus* to oxamic, lactic, acetic, and formic acids (BARKER, 1943). Finally, the cyclic nitriles 2-cyanopiperidine **19** (Scheme 25) and 2,6-dicyanopiperidine **20** were hydrolyzed by strains of *Micrococcus, Brevibacterium imperiale,* and *Brevibacterium* species (ARNAUD et al., 1976; JALLAGEAS et al., 1979).

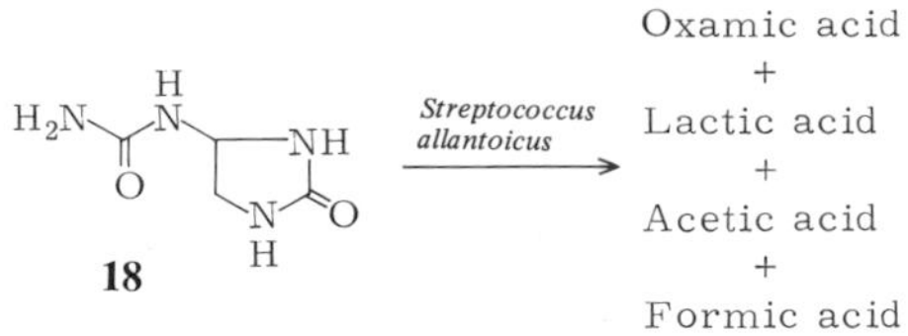

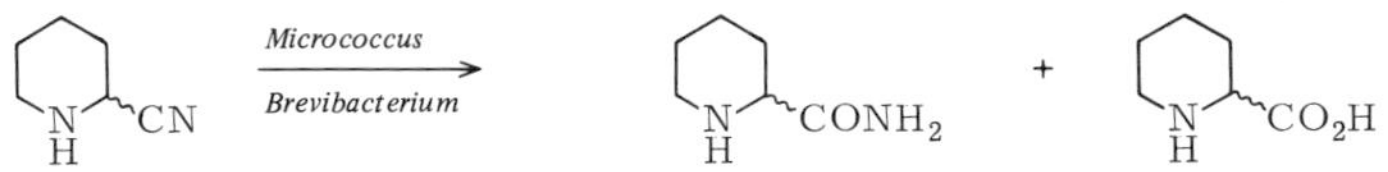

(+) **13** —*Bacillus subtilis*→ (−) **14** + (+) **13**

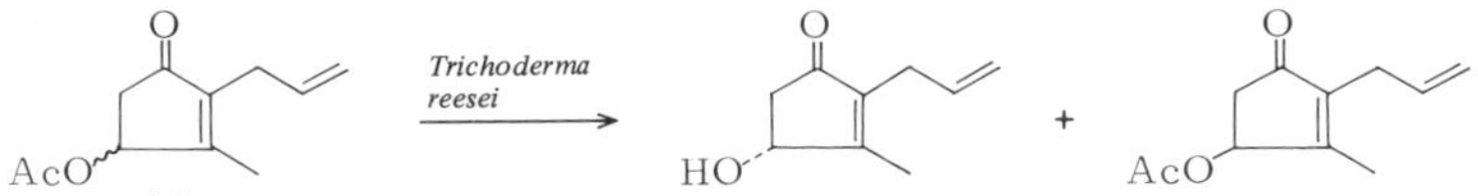

15 (±) *O*-Acetyl-allethrolone —*Trichoderma reesei*→

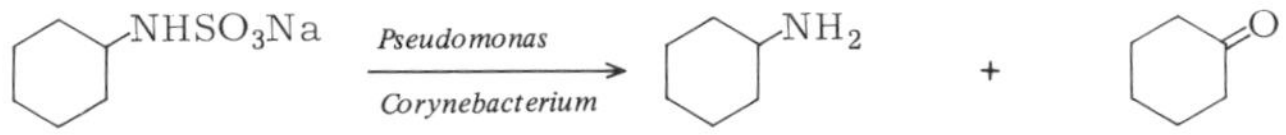

16 —*Pseudomonas*, *Corynebacterium*→

17 —rumen bacteria→ $CH_2NH_2(CH_2)_2CO_2H$ + $CH_3(CH_2)_2CO_2H$

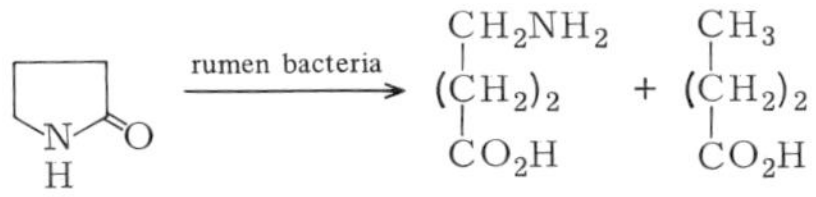

18 —*Streptococcus allantoicus*→ Oxamic acid + Lactic acid + Acetic acid + Formic acid

Scheme 24

Another reaction may be included among these hydrolysis reactions, albeit somewhat arbitrarily, namely, the hydration of methylcyclohexene **21** by *Aspergillus niger* NCIM 612 (GANAPATHY et al., 1966).

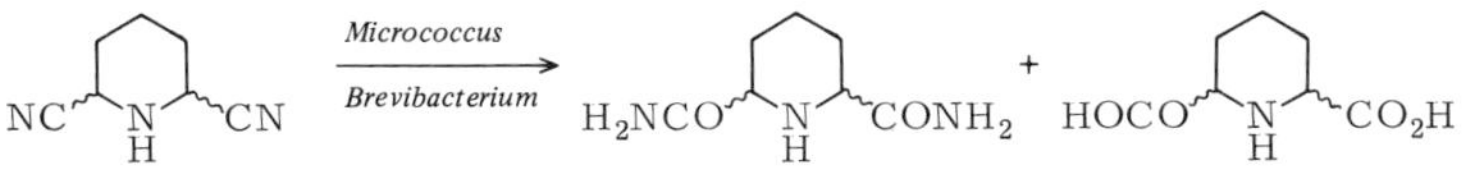

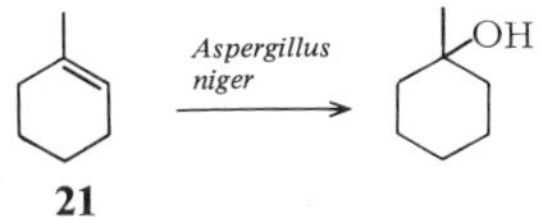

19 —*Micrococcus*, *Brevibacterium*→

20 —*Micrococcus*, *Brevibacterium*→

21 —*Aspergillus niger*→

Scheme 25

E. Hydroxylation Reactions

Hydroxylation by microorganisms is one of the most extensively studied areas in the field of bioconversion. The important work conducted by FONKEN and JOHNSON (1972) (Upjohn Comp.) using mainly fungi must be acknowledged.

The simplest hydrocarbons can be hydroxylated. Cyclopropane **1** (Scheme 26) and methylcyclopropane **2** were oxidized by *Methylococcus capsulatus* (crude enzyme preparation) to cyclopropanol **3** and methylolcyclopropane **4**, respectively (DALTON et al., 1981). Cyclopentane **5** was oxidized to cyclopentanol **6** and cyclopentanone **7** by soil bacterium JOB 5 and *Nocardia* R9 (OOYAMA et al., 1965). The same bacterium JOB 5 converted methylcyclopentane **8** into methyl-3-cyclopentanone **9**.

The latter strain can also hydroxylate cyclohexane and some of its derivatives. Cyclohexane **10** was oxidized to cyclohexa-

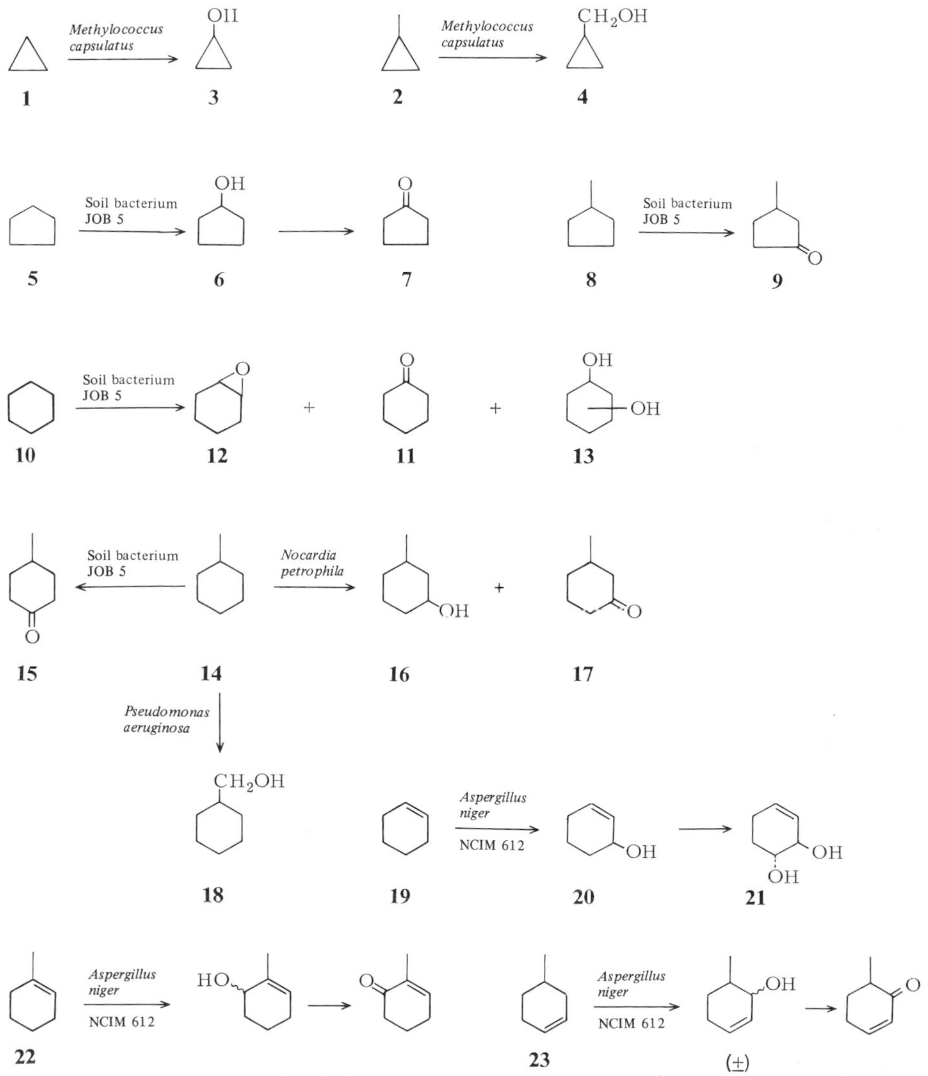

Methylococcus capsulatus
Methylococcus capsulatus
Soil bacterium JOB 5
Soil bacterium JOB 5
Soil bacterium JOB 5
Soil bacterium JOB 5
Nocardia petrophila
Pseudomonas aeruginosa
Aspergillus niger NCIM 612
Aspergillus niger NCIM 612
Aspergillus niger NCIM 612

Scheme 26

none **11** and epoxycyclohexane **12.** Cyclohexanone gave a cyclohexanedione of undetermined structure, and the epoxide gave cyclohexenone and cyclohexanediol **13.**

A strain of *Pseudomonas* converted cyclohexane into cyclohexanol (DE KLERK and VAN DER LINDEN, 1974). The bacterium JOB 5 has turned methylcyclohexane **14** into 4-methylcyclohexanone **15** (OOYAMA et al., 1965). *Nocardia petrophila* NCIB 9 438 converted **14** into 3-methylcyclohexanol **16** and the corresponding ketone **17** (TONGE and HIGGINS, 1974). A strain of *Pseudomonas aeruginosa* transformed **14** into high yields of methylolcyclohexane **18** (BOLSMAN and BAILEY, 1982). A strain of *Alcaligenes faecalis* oxidized ethylcyclohexane to 4-ethylhydroxycyclohexane (mixture of *cis* and *trans*) (ARAI and YAMADA, 1969). The bacterium JOB 5 converted cycloheptane and cyclooctane into the corresponding monoketones (OOYAMA et al., 1965).

Cyclohexenes can also be hydroxylated. Cyclohexene itself **19** was hydroxylated by *Aspergillus niger* NCIM 612 to cyclohexenol **20** and *trans*-cyclohexenediol **21** (BHATTACHARYYA and GANAPATHY, 1965). The same fungus hydroxylated 1-methylcyclohexene **22** and 4-methylcyclohexene **23** to the alcohols and ketones (GANAPATHY et al., 1966).

Extensive work has been done on a series of variously substituted cyclohexanes. Table 1 shows results concerning phenylcyclohexanes bearing an ether function on the aromatic ring (FONKEN et al., 1964). Table 2 shows results for phenylcyclohexanes bearing a carbamate function on the aromatic ring (FONKEN et al., 1964). The results given in these two tables all correspond to hydroxylation at position 4. The same authors showed that phenylcyclohexane **24** (Scheme 27) was hydroxylated at position 4 by *Beauveria sulfurescens* ATCC 7 159, *Trichothecium roseum* ATCC 8 685, and *Streptomyces mediocidicus* ATCC 13 279, but at position 3 by *Rhizopus arrhizus* ATCC 11 145.

Penicillium concavorugulosum IFO 6 226 also hydroxylated the substituted cyclohexane **27** at the 4-position with a yield of 71% (TAKEDA, 1974; KISHIMOTO et al., 1976). *Beauveria sulfurescens* hydroxylated **28** at both the 3- and 4-positions, and also hydroxylated the cyclohexane group of the carbamate **29.** *Rhizopus arrhizus* hydroxylated **29** (Scheme 28) at both the 3- and 4-carbons.

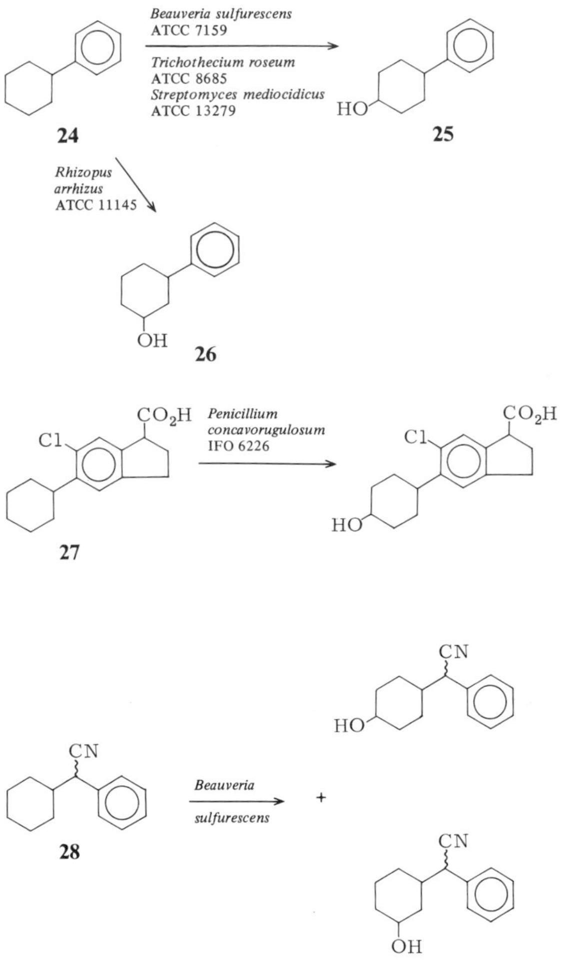

Scheme 27

Variously substituted bicyclohexyls have been thoroughly studied. Results are shown in Table 3. Large ring hydrocarbons have been hydroxylated by various strains. Results are shown in Table 4. Table 5 shows results for sulfones bearing two hydrocarbon rings (FONKEN et al., 1964).

Table 1. Hydroxylation of Substituted Phenylcyclohexanes

(Reaction scheme: phenylcyclohexane–OR → HO-cyclohexyl-phenyl–OR)

R	o, m, p	Microorganism
Me	*o*	*Gongronella butleri* CBS
Me	*m*	*Gloniopsis brevisaccata* CBS
Me	*p*	*Beauveria sulfurescens* ATCC 7159
Me	*p*	*Cladosporium resinae* NRRL 2778
Et	*o*	*Absidia glauca* ATCC 7852a
Et	*m*	*Glonium clavisporium* CBS
Et	*m*	*Absidia glauca*
Et	*p*	*Hysterium angustatum* CBS
Et	*p*	*Curvularia lunata* ATCC 12017
Pr	*o*	*Calonectria decora* CBS
Pr	*p*	*Mytilidion tortile* CBS
*i*Pr	*m*	*Hypomyces haematococcus* CBS
*i*Pr	*p*	*Corticium sasakii*
Bu	*m*	*Brachysporium oryzae* ATCC 11571
*i*Bu	*m*	*Adelopus nudus* CBS
*t*Bu	*o*	*Boletus* sp. Peck. 168
Pe	*p*	*Penicillium atrovenetum* CBS
$(CH_2)_2CHMe_2$	*m*	*Alnicola escharoides* CBS
$C(CH_3)_2CH_2CH_3$	*p*	*Septomyxa affinis* ATCC 6737
$CH_2C(CH_3)_3$	*o*	*Diplodia natalensis*
$(CH_2)_5CH_3$	*m*	*Mycobacterium rhodochrous* ATCC 4277
$(CH_2)_3CH(CH_3)_2$	*p*	*Keratinomyces ajelloi* CBS

(FONKEN et al., 1964)

Table 2. Hydroxylation of Substituted Phenylcyclohexanes

(Reaction scheme: phenylcyclohexane–OCONRR' → HO-cyclohexyl-phenyl–OCONRR')

R	R'	o, m, p	Microorganism
Me	H	*o*	*Circinella spinosa* ATCC 9025
Me	H	*p*	*Beauveria sulfurescens* ATCC 7159
Et	H	*m*	*Chaetomium globosum* ATCC 6205
Et	H	*p*	*Guignardia bidwelli* ATCC 9560
Me	Me	*p*	*Gibberella saubinetti* CBS
Pr	H	*m*	*Cylindrocarpon radicicola* ATCC 11011
Bu	H	*o*	*Deconia coprophila* CBS
*i*Bu	H	*m*	*Micrococcus rubens* ATCC 186
*t*Bu	H	*p*	*Cyathus poeppigii* CBS
Et	Et	*p*	*Endothia parasiticus* ATCC 9414
Ph	H	*o*	*Cenangium abietis* CBS
Ph	H	*p*	*Dermea balsama* CBS
pNO_2Ph	H	*p*	*Aspergillus niger* ATCC 9027
*o*MePh	H	*o*	*Pseudomonas aeruginosa* ATCC 8689
*p*MePh	H	*p*	*Nocardia corallina* CBS
*p*Ph-Ph	H	*o*	*Wojnowicia graminis* CBS
mm'Me_2Ph	H	*p*	*Rhizoctonia solani* ATCC 6221

(FONKEN et al., 1964)

There are numerous examples of hydroxylations of hydrocarbons bearing nitrogen-containing substituents, e. g., substituted sulfonylureas attached to rings of various sizes (Table 6), amides *N*-substituted with various rings (Table 7), and amides with two cycloalkane substituents (Table 8).

The benzamides **32** of cyclohexylamine and **33** of a spirane cyclohexylamine were both 4-hydroxylated by *Beauveria sulfurescens* (JOHNSON et al., 1970). Similarly, amides of perhydroquinolines have been hydroxylated by *B.sulfurescens*. The 4a*S*,8a*R*-isomer **34** was also hydroxylated by *B.sulfurescens*. **34** was hydroxylated at the 5-carbon (59%) and the 6-carbon, while the 4a*R*,6a*S*-enantiomer **35** was hydroxylated at the 6- and 7-carbons (JOHNSON et al., 1968b).

The hydroxylation of adamantamine derivatives has mainly been studied using *B.sulfurescens* (Table 9). *Rhizopus nigricans* hydroxylated diamantan-1-ol **36** and diamantan-4-ol **38** (Scheme 29). The isomer **36** gave the 1,7-diol **37** and the isomer **38** gave the 4,9- and 1,9-diols **39** and **40** with high optical purities (BLANEY, 1974).

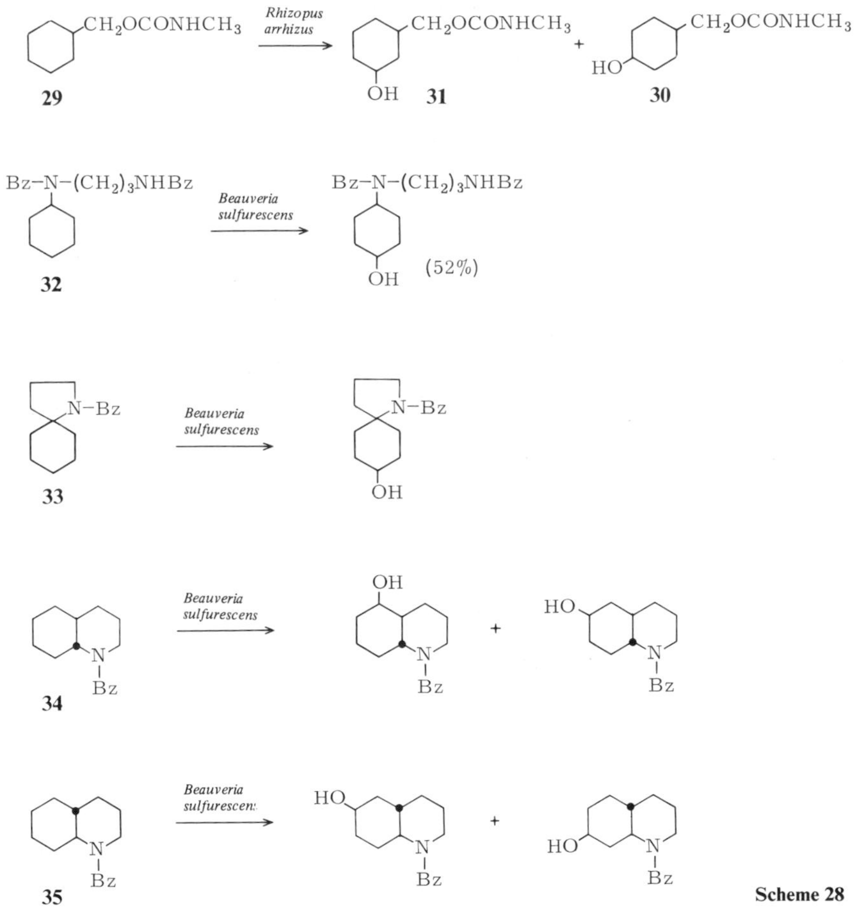

In addition to the systematic investigations referred to above, a number of isolated examples may be cited. Hydroxylation can occur in saturated rings. Thus, *Corynebacterium cyclohexenicum* oxidized cy-

Scheme 28

Table 3. Hydroxylation of Some Bicyclohexyl Derivatives

X	OH	Microorganism	Reference
H	4,1′; 4,2′; 4,3′; 4,4′; 3,1′; 3,2′; 2,1′; 2,2′; 1,1′	Many strains	
1′ OH	4	*Beauveria sulfurescens* ATCC 7159	a), b), c), d)
2′ OH	3,4	*B.sulfurescens*	a), b), c), d)
2′ OH	3,4	*Cunninghamella blakesleeana*	a), b), c), d)
3′ OH	3,4	*Beauveria sulfurescens*	a), b), c), d)
3′ OH	3,4	*Alnicola escharoides* CBS	a), b), c), d)
4′ OH	1, 2, 4, 4 (C=O)	*Beauveria sulfurescens*	a), b), c), d)
4′ OH	1, 2, 4, 4 (C=O)	+3 other strains	a), b), c), d)
2′ C=O	3,4	*B.sulfurescens*	a), b), c), d)
2′ C=O	3,4	*Cyathus poeppigii* CBS	a), b), c), d)
2′ C=NOH	4	*Adelopus nudus* CBS	a), b), c), d)
C(O)(O) (cyclic ketal)	4	*Mytilidion tortile* CBS	a)
3′ C=NOH	3 (9%)	*Nocardia corallina* CBS	a), b), c), d)
3′ C(O)(O) (cyclic ketal)	4	*Hysterium angustatum* CBS	a), b), c), d)
id.	4	*Glonium stellatum* CBS	a)
id.	4	*Rhizoctonia solani* ATCC 6221	a)
4′ C=O	4	*Beauveria sulfurescens*	a), b), c), d)
4′ C=N–OH	3,4	*B.sulfurescens*	a), b), c), d)
4′ C(O)(O) (cyclic ketal)	4	*B.sulfurescens*	a), b), c), d)
3′-N (ring)	4	*Permea balsama* CBS	a), b), c), d)
4′-N (ring)	4	*Boletus* sp. Peck 168	a), b), c), d)
4′-N (ring) O	4	*Beauveria sulfurescens*	a), b), c), d)
4′-N (ring) S	4	*Cylindrocarpon radicicola* ATCC 11811	a), b), c), d)
4′ O (ring) N	4	*Cenangium abietis* CBS	a), b), c), d)
3′-N (ring) N–Ac	4	*Gloniopsis brevisaccata* CBS	a), b), c), d)

[a] (FONKEN et al., 1964), [b] (FONKEN et al., 1968a), [c] (FONKEN et al., 1970), [d] (FONKEN et al., 1968b)

clohexane carboxylic acid **41** (Scheme 30) to the *para*-ketone **42.** *Aspergillus niger* ATCC 11 394 hydroxylated the tetramethylated spiranedione **43** at one methyl group forming **44** (80%). Furthermore one of the ketone groups was reduced in **45** (2%) (GONZALEZ et al., 1976). *Nocardia salmonicolor* and *N.corallina* oxidized 1,3-dipropylcyclohexane **46** to ketol **47** and diketone **48** (HESLER and RAYMOND, 1973).

Finally, *Curvularia falcata,* which reduces many decalones, hydroxylated the

Table 4. Hydroxylation of Some Cyclic Compounds

$(CH_2)_{n-1}$ CH–X ⟶ $(CH_2)_{n-1}$ (OH) CH–X

n	X	OH (C=O)	Microorganism	Ref.
11	$CONH_2$	6 (C=O) (13%)	*Beauveria sulfurescens* ATCC 7159	a), b)
11	OH	5,5 (C=O), 1 and 5 (C=O) 6,6 (C=O), 1 and 6 (C=O)	*B.sulfurescens*	a)
12	OH	5,5 (C=O), 1 and 5 (C=O) 6,6 (C=O), 1 and 6 (C=O) 7,7 (C=O), 1 and 7 (C=O)	*B.sulfurescens* +6 other strains	a), b)
13	OH	5,5 (C=O), 1 and 5 (C=O) 6,6 (C=O), 1 and 6 (C=O) 7,7 (C=O), 1 and 7 (C=O)	*B.sulfurescens*	a), b)
14	OH	5,5 (C=O), 1 and 5 (C=O) 6,6 (C=O), 1 and 6 (C=O) 7,7 (C=O), 1 and 7 (C=O)	*B.sulfurescens*	a), b)
15	C=O	9 (26%) 2,9 (11%)	*Calonectria decora*	a)

[a] (FONKEN et al., 1964), [b] (FONKEN et al., 1967)

Table 5. Hydroxylation of Some Sulfones

$(CH_2)_{n-1}$ CH–SO_2–CH $(CH_2)_{m-1}$ ⟶ $(CH_2)_{n-1}$ (OH) CH–SO_2–R

n	m	OH (or C=O)	Microorganism
5	5	3	*Mycobacterium rhodochrous* ATCC 4276
5	8	2 and 3	*Curvularia lunata* ATCC 12017
6	5	3 and 4	*Deconia coprophila* CBS
6	6	3 (11%) 4 (5%)	*Beauveria sulfurescens* ATCC 7159
6	8	3 and 4	*Aspergillus niger* ATCC 8740
8	6	5	*Aspergillus niger* ATCC 8740
6	12	3,4	*Cunninghamella blakesleeana* ATCC 8688a
12	6	6	*Cunninghamella blakesleeana* ATCC 8688a
6	13	3 and 4	*Gibberella saubinetti* CBS
7	6	4 (17%) + ketone (18%)	*Beauveria sulfurescens* ATCC 7159
7	7	4	*Calonectria decora* CBS
10	9	6	*Cyathus poeppigii* CBS
14	15	7	*Ascochyta linicola* NRRL 2923

(FONKEN et al., 1964)

trans- and *cis*-5-decalones **49** and **50** (PRELOG and SMITH, 1959) concurrently with reduction.

Variously substituted cyclohexenes and cyclopentenes have been hydroxylated. Thus, methylcyclohexenes **51** and **52** (Scheme 31) 2-substituted with a phenylpropanone or thiophenepropanone group were oxidized to the ketones **53** by *Mycobacterium smegmatis* (PLEMENITAS and KOMEL, 1982).

The isophorone **54** was oxidized by *Aspergillus niger* JTS 191 (MIKAMI et al., 1981). Hydroxylation occurred at the 3-methyl- (**55**) or the 4-carbon giving ketol **56** and subsequently dione **57.** The oxidoreduction of **56** to **58** has already been mentioned.

Table 6. Hydroxylation of Some Substituted Sulfonylureas

$(CH_2)_n$ CH–NH–CO–NH–SO_2–C_6H_4–X ⟶ $(CH_2)_n$(OH) CH–NH–R

n	X	OH	Microorganism	Ref.
5	*p*CH_3	3 (15% *cis*, 7% *trans*)	*Beauveria sulfurescens* ATCC 7159	a), b), c)
5	*m*OC_2H_5	3	*Cladosporium resinae* NRRL 2728	a)
5	*o*Bu	3	*Circinella angarensis* NRRL 2628	a)
5	*p*Cl	3	*Absidia glauca* ATCC 7852a	a)
6	none	3 and 4	*Guignardia bidwelli* ATCC 9559	a)
6	*p*$COCH_3$	3 and 4	*Beauveria sulfurescens* ATCC 7159	a), b), c)
6	*o*CH_3	3 and 4	*Nocardia corallina* ATCC 4273	a), b), c)
6	*m*CH_3	3 and 4	*Calonectria decora* CBS	a), b), c)
6	*p*CH_3	3 and 4	*Rhizopus arrhizus* ATCC 11145	a), b), c)
6	*i*Pr	3 and 4	*Corticium sasakii* NRRL 2705	c)
6	di*o*,*p*Me	3 and 4	*Chaetomiun globosum* ATCC 6205	a)
6	(S)*)	2 and 4	*Streptomyces griseolus*	d)
7	*p*CH_3	3,3 (C=O), 4	*Beauveria sulfurescens* ATCC 7159	a), b), c)
7	di*m*,*p*Me	3 and 4	*Endothia parasiticus* ATCC 9414	a), b), c)
7	*n*Bu	3 and 4	*Gongronella butleri* CBS	a), b), c)
7	*m*OCH_3	3, 4, and 5	*Cyathus poeppigii* CBS	a), b), c)
7	*p*Cl	3 and 5	*Cunninghamella blakesleeana* ATCC 8688	a), b), c)
7	*p*Br	3, 4, and 5	*Deconia coprophila* CBS	a), b), c)
7	di*m*,*p*Cl	3 and 4	Many strains	a), b), c)
7	di*o*,*m*$COCH_3$	3 and 4	*Pseudomonas aeruginosa* ATCC 8689	a), b), c)
8	di*m*,*p*Et	3, 4, and 5	*Aspergillus niger* ATCC 9027	a), b), c)
9	*p*Et	3, 4, and 5	*Curvularia lunata* 12017	a), b), c)
9	*n*Bu	3, 4, and 5	*Septomyxa affinis* ATCC 6737	a), b), c)
9	*p*OC_6H_{13}	3, 4, and 5	*Cylindrocarpon radicicola* ATCC 1101	a), b), c)
10	di*o*,*p*Cl	3, 4, and 5	*Wojnowicia graminis* CBS	a), b), c)
12	*p*CH_3	5, 6, and 7	*Beauveria sulfurescens* ATCC 7159	a), b), c)
12	*m*Cl	5, 6, and 7	*Ascochyta linicola* NRRL 2923	a), b), c)

[a] (FONKEN et al., 1964), [b] (FONKEN et al., 1968), [c] (FONKEN et al., 1979), [d] (SIEWERT et al., 1973)

* (S) = *p*-CH_2CH_2NH CO–C_6H_3(H_3CO)(Cl)

Table 7. Hydroxylation of Some Cycloalkanes Amides

$(CH_2)_{n-1}$ CH–NH–X (ring, substituent R) ⟶ $(CH_2)_{n-1}$ CH–NH–X (ring, substituents R, OH)

n	R	X	n(OH)	Microorganism	Ref.
5	H	Bz	3	*Cyathus poeppigii* CBS	c)
6	H	Cyclo C_6H_{11}—CO	4 (25%)	*Beauveria sulfurescens* ATCC 7159	a)
6	H	Cyclo C_6H_{11}—CO	4	*Mycobacterium rhodochrous* ATCC 4273	a)
6	H	Cyclo C_6H_{11}—CO	4	17 other strains	a)
6	H	$C_6H_5CH_2OCO$	4 (25%)	*Beauveria sulfurescens*	a)
6	H	Ts	4 (31%)	*Guignardia bidwelli* ATCC 9559	a), b)
6	H	CH_3CO	3,3 (C=O), 4,4 (C=O)	*Wojnowicia graminis* CBS	a)
6	H	CH_3CH_2CO	3,3 (C=O), 4,4 (C=O)	*Gibberella saubinetti* CBS	a)
6	H	$CH_3CH_2CH_2CO$	3,3 (C=O), 4,4 (C=O)	*Boletus* sp. Peck 168	a)
6	H	Bz	3,3 (C=O), 4	*Rhizopus arrhizus* ATCC 11145	a), b), c)
6	H	Bz	4 (35%)	*Beauveria sulfurescens*	a), b), c)
6	4 CH_3 *(cis)*	Bz	*trans* 4 (7%)	*B.sulfurescens*	e)
6	4 CH_3 *(cis + trans)*	Bz	*cis* 4 (30%), *trans* 4 (0,8%)	*B.sulfurescens*	e)
6	2 CH_3 *(trans)*	Bz	*trans* 1-4 (13%) + 4 (C=O) (5%) + *cis* 1-4 (5%)	*B.sulfurescens*	d)
6	2 CH_3 *(cis)*	Bz	*trans* 1-4	*B.sulfurescens*	d)
7	H	$C_6H_5CH_2OCO$	4,4 (C=O)	*B.sulfurescens*	a), b)
7	H	TsNHCO	3,3 (C=O), 4	*B.sulfurescens*	a), b), c)
7	H	Ts	3,4	*B.sulfurescens*	a), b)
7	H	Bz	4,4 (C=O) (48%)	*B.sulfurescens*	a), b), c)
8	H	Bz	4 (C=O) (15%), 15 (C=O) (45%)	*B.sulfurescens*	b)
8	H	CH_3CO	4,4 (C=O), 5,5 (C=O)	*B.sulfurescens*	a)
8	H	Ts	4 (C=O) (30%), 5 (C=O) (11%)	*Wojnowicia graminis* CBS	b)
9	H	Bz	4,4 (C=O), 5,5 (C=O)	*Beauveria sulfurescens*	c)
10	H	Bz	5 (C=O), 6 (C=O)	*B.sulfurescens*	c)
10	H	CH_3CO	5 (C=O), 6 (C=O)	*B.sulfurescens*	a)
10	H	CH_3CO	5 (C=O), 6 (C=O)	Other strains	a)
12	H	CH_3CO	6 (C=O) (20%), 7 (C=O) (6%)	*B.sulfurescens*	a), b)
12	H	CH_3CO	6 (C=O), 7 (C=O)	Many other strains	a), b)
12	H	CH_3CH_2CO	6 (C=O)	*B.sulfurescens*	a), b)
12	H	H	6,6 (C=O) (10%), 7,7 (C=O) (4,5%)	*B.sulfurescens*	a), b)
12	H	Bz	6 (C=O), 7 (C=O)	*Rhizopus arrhizus*	c)
15	H	CH_3CO	6 (C=O)	*Deconia coprophila* CBS	a)
15	H	CH_3CO	6 (C=O)	*Cyathus poeppigii* CBS	a)
15	H	CH_3CO	6 (C=O)	*Rhizopus arrhizus*	a)
15	H	CH_3CO	6 (C=O)	*Beauveria sulfurescens*	a)

[a] (FONKEN et al., 1964), [b] (FONKEN et al., 1968a), [c] (FONKEN et al., 1970), [d] (JOHNSON et al., 1970), [e] (JOHNSON et al., 1971)

Table 8. Hydroxylation of Amides with two Cycloalkane Substituents

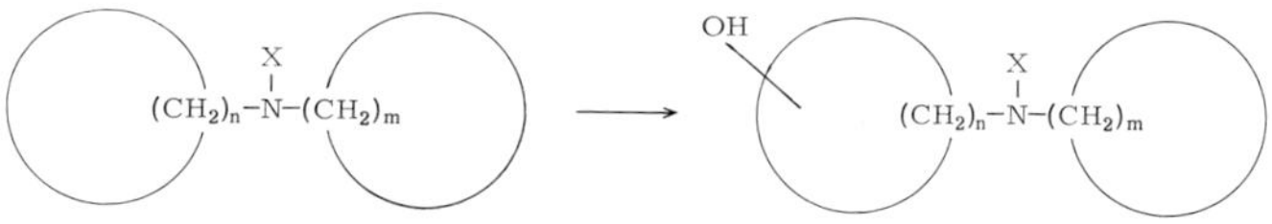

n	m	X	OH (or C=O)	Microorganism	Ref.
6	5	Ac	4 (40%)	*Beauveria sulfurescens* ATCC 7159	a), b)
6	6	Bz	4	*B.sulfurescens*	a)
6	6	Ac	3	*Rhizopus arrhizus* ATCC 11145	a)
6	6	Ac	4	*Beauveria sulfurescens*	a), b), c)
6	6	Ac	4	*Calonectria decora*	a), b), c)
7	6	Ac	4	*Beauveria sulfurescens*	a), b)
7	7	Ac	4 (35%), 4 (C=O) (13%)	*B.sulfurescens*	a), b)
8	6	Ac	4,5	*B.sulfurescens*	a)

[a] (FONKEN et al., 1964), [b] (FONKEN et al., 1968a), [c] (FONKEN et al., 1970)

Table 9. Hydroxylation of Amides of Adamantamines

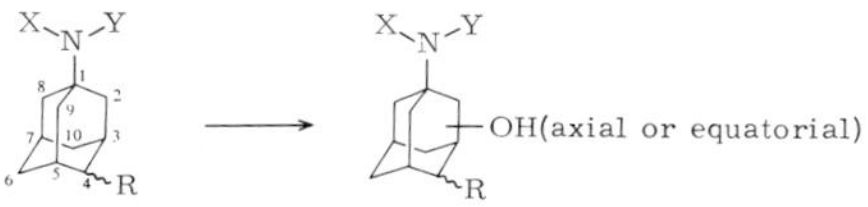

X	Y	R	OH	Microorganism	Ref.
Ac	H	H	4 (a) (48%)	*Beauveria sulfurescens* ATCC 7159	a), b)
Ac	H	H	3 (12%)	*Rhizopus arrhizus* ATCC 11145	a), b), c)
Ac	H	H	4 (e) (1%)	*Curcularia lunata* ATCC 12017	a), b)
Ac	H	H	4 (a) (8%)	*C.lunata*	c)
Bz	H	H	4 (a) (25%)	*Beauveria sulfurescens*	a), b)
$C_6H_5CH_2CO$	H	H	4 (a), 4 (a)-6 (a) (70%)	*B.sulfurescens*	a), b)
Ac	CH_3	H	4 (a) (36%), 3 (7%)	*B.sulfurescens*	a), b)
Ac	H	H	4 (a)	*B.sulfurescens*	c)
Bz	CH_3	H	4 (a) (15%), 4 (a)-6 (a) (58%)	*B.sulfurescens*	a), b)
$C_6H_{11}CO$	CH_3	H	4 (a)-6 (a) (9%)	*B.sulfurescens*	a)
Bz	CH_3	CH_3 (e)	4 (a)-7 (18,5%), 4 (e)-6 (a) (22%)	*B.sulfurescens*	d)
Bz	CH_3	CH_3 (a)	6 (a) (6%)	*B.sulfurescens*	d)
p		H	4 (a), 4 (a)-6 (a)	*B.sulfurescens*	a), b)
p'		H	4 (a)-6 (a)	*B.sulfurescens*	b)

p = ; p' =

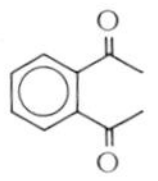

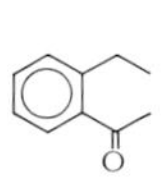

[a] (HERR et al., 1968), [b] (HERR et al., 1969), [c] (HERR et al., 1972a, b), [d] (HERR et al., 1970)

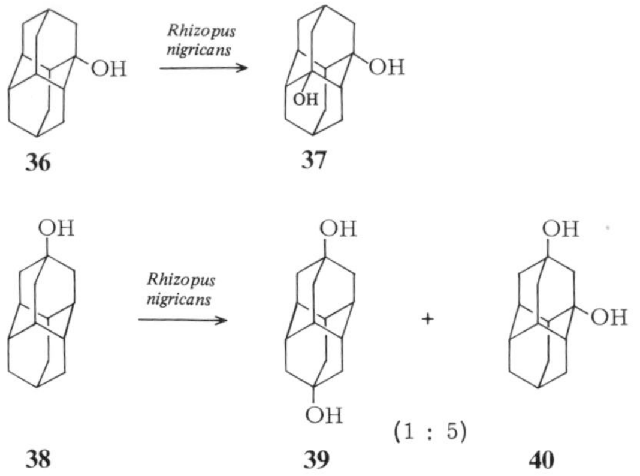

Rhizopus nigricans

36 37

Rhizopus nigricans

38 39 (1 : 5) 40

Scheme 29

Allethrone **59** was oxidized by *Aspergillus niger* NRRL 3 228 mainly to allethrolone **60** along with small amounts of **61** and **62** oxidized at the allyl or methyl groups (TABENKIN et al., 1969a). Cinerone **63** was hydroxylated by a large number of strains to cinerolone **64,** the reduction product of the butenyl group of cinerolone, **65,** and compound **66** (TABENKIN et al., 1969a, b). Octalone **67** (Scheme 32) was 6-hydroxylated by *Calonectria decora* (SCHUBERT et al., 1958). *Absidia coerulea* IFO 4 420 converted **69** (R = H) to **70** (R = OH) which is an inhibitor of cholesterol biosynthesis (TERAHARA and TANAKA, 1981).

Racemic perhydrodiethylstilbene **71** (R=R′=H) was hydroxylated by *Curvu-*

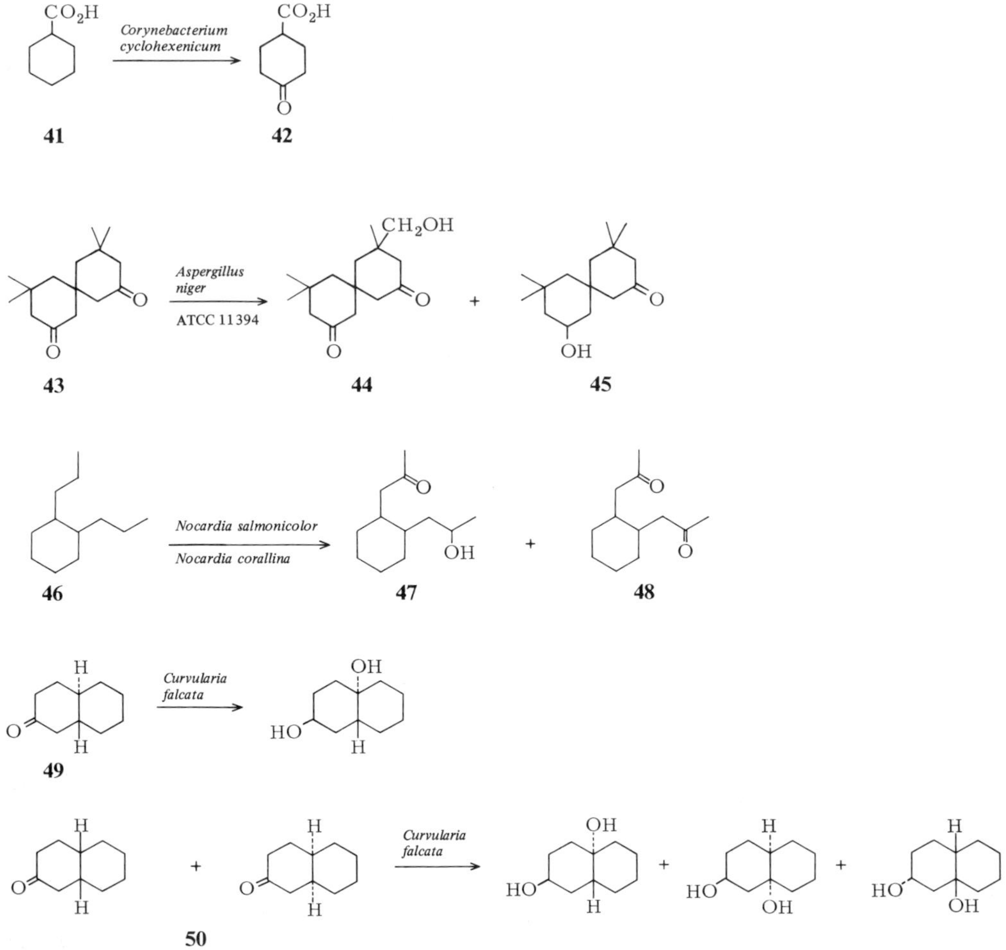

Corynebacterium cyclohexenicum

41 42

Aspergillus niger ATCC 11394

43 44 45

Nocardia salmonicolor / Nocardia corallina

46 47 48

Curvularia falcata

49

Curvularia falcata

50

Scheme 30

laria lunata; the 2*S*,3*S*-isomer was degraded leaving the monohydroxylated 2*R*,3*R*-products **72** and **73** and the dihydroxylated product **74** (KREISER and LANG, 1976).

tial distances between the nitrogen atom and the hydroxylation sites.

A number of bicyclic heterocycles have been studied. In the azabicyclo[2.2.2] series,

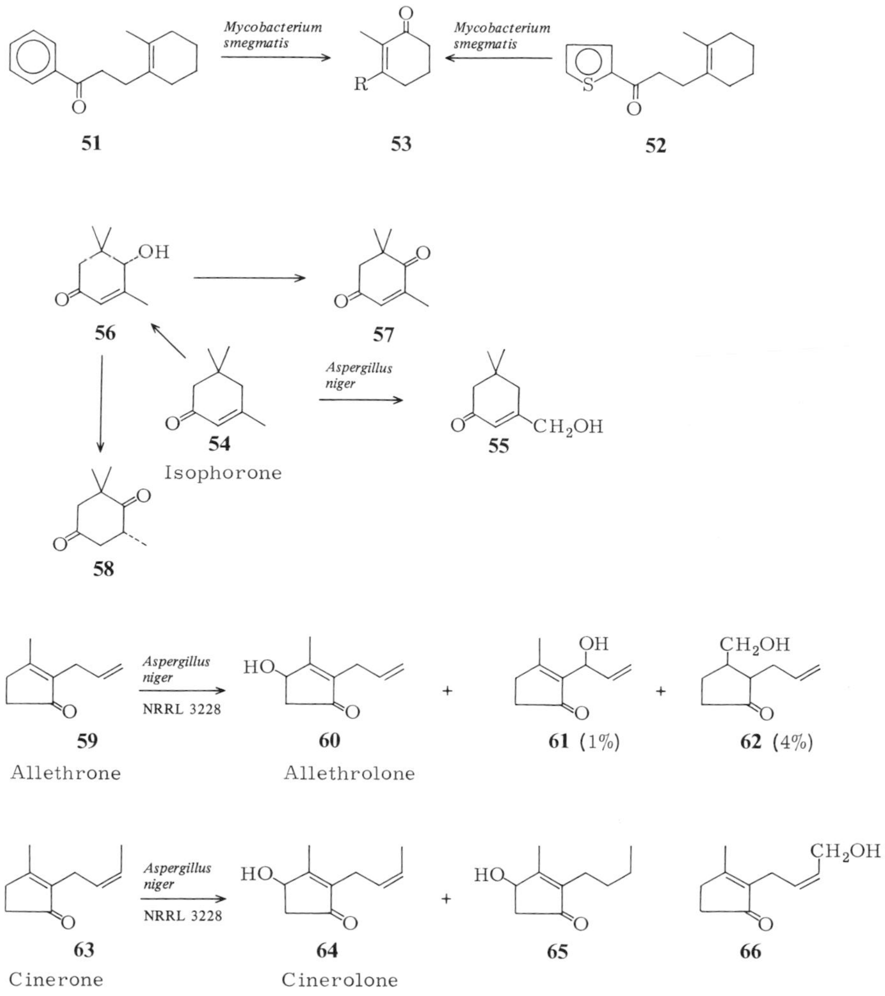

Mycobacterium smegmatis *Mycobacterium smegmatis*

R

S

51 **53** **52**

OH

56 **57**

Aspergillus niger

CH_2OH

54 **55**

Isophorone

58

OH CH_2OH

Aspergillus niger NRRL 3228

HO

59 **60** **61** (1%) **62** (4%)

Allethrone Allethrolone

CH_2OH

Aspergillus niger NRRL 3228

HO HO

63 **64** **65** **66**

Scheme 31 Cinerone Cinerolone

Finally, phenanthrene-type substrate **75** (R=R′=H) was 7-hydroxylated to **76** (R=H, R′=OH) and 9-hydroxylated to **77** (R=OH, R′=H) by *Cunninghamella bainieri* ATCC 9 244 (DAUM et al., 1967).

Variously substituted heterocycles of different size have been hydroxylated by *Beauveria sulfurescens*. Results are shown in Table 10. The authors determined preferen-

amide **77** (Scheme 33) was oxidized by *B.sulfurescens* at position 5 or 6, forming the *endo*-product in both cases. In the azabicyclo[3.2.2] series, amide **80** was *endo*-hydroxylated at position 6 and the resulting alcohol **81** was oxidized to ketone **82** (JOHNSON et al., 1968c).

In the azabicyclo[3.3.1] series, amide **83** gave different results according to the mi-

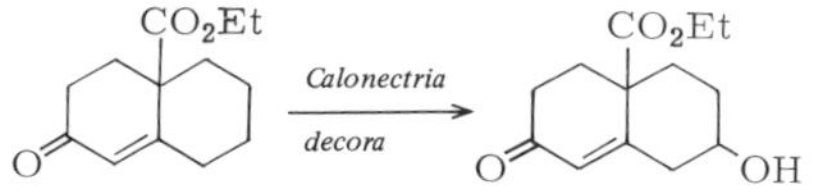

Calonectria decora

67

68

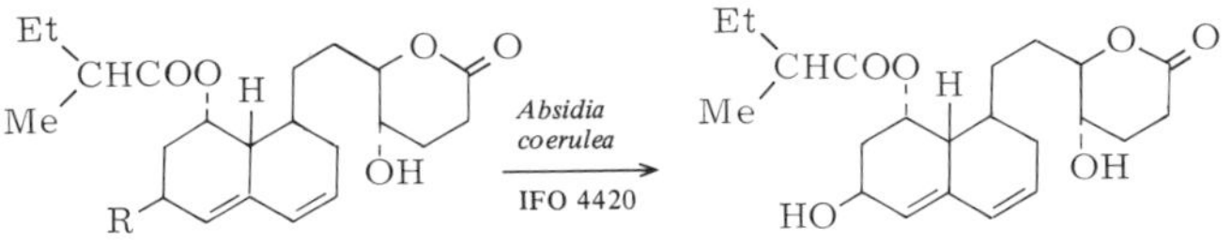

Absidia coerulea IFO 4420

69 R = H

70

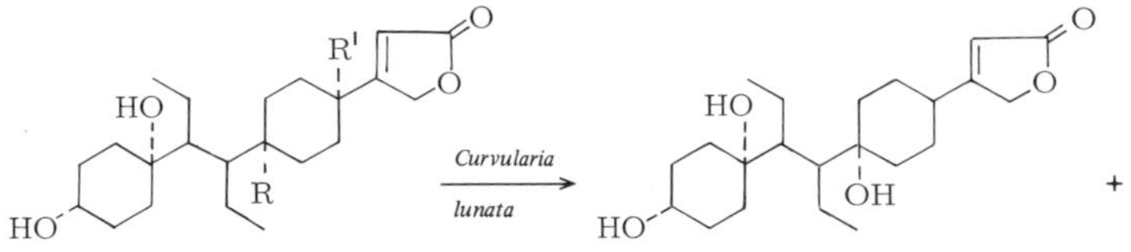

Curvularia lunata

71 R = R' = H

72

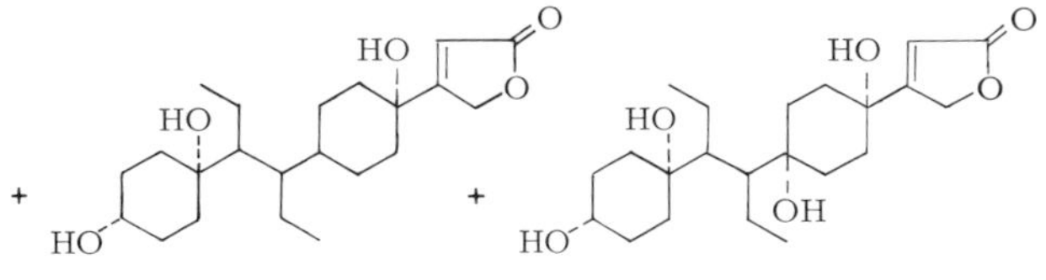

73

74

Cunninghamella bainieri ATCC 9244

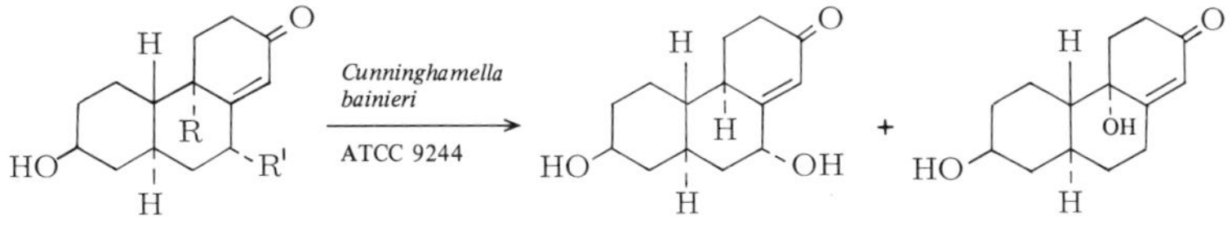

75 R = R' = H

76

77

Scheme 32

croorganism used. Thus, *B.sulfurescens* 6-*endo*-hydroxylated **83** to **84** (JOHNSON et al., 1968c) while *Rhizopus arrhizus* 1-hydroxylated **83** to **85** (JOHNSON et al., 1969b).

Another team has studied bicyclic systems (FURSTOSS et al., 1980), with the amide function either on or in the ring. In all cases, hydroxylation was carried out with *B.sulfurescens*.

The amide **86** with the carbonyl carbon in the ring is the azabicyclo[2.2.2] equivalent of **77** where the carbonyl carbon is outside the ring. **86** gave **87** analogous to **78.** In the azabicyclo[3.3.1] series, the internal amide **88** was hydroxylated like **83** (FURSTOSS

Table 10. Hydroxylation of Some Derivatives of Piperidine and Higher Ring Homologs by *Beauveria sulfurescens* ATCC 7159

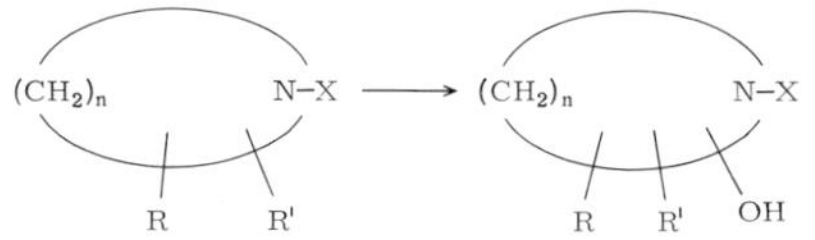

$(CH_2)_n$ N–X ⟶ $(CH_2)_n$ N–X (R, R′ → R, R′, OH)

n	R	R′	X	OH (or C=O)	Ref.
5	H	H	Bz	4 (20%)	a)
5 (±)	2-Me	H	Bz	3 (2*S*, 3*S*) (10%), 4 (2*R*, 4*S*) (21%), 4 (2*R*, C=O) (1%)	b)
5 (±)	3-Me	H	Bz	3 (±) (7%), 4 (−) (6%)	b)
5	4-Me	H	Bz	4 (13%), 4′ (23%)	b)
5 (±)	2-Et	H	Bz	3 (+) (20%), 4 (8%)	b)
5	2 Me	6 Me	Bz	3 *cis* (−) (49%)	b)
5	2 Pr	H	Bz	2′ (2%), 3 (+) (15%), 4 (8%)	b)
5	4 Pr	H	Bz	2′ (C=O) (30%)	b)
6	H	H	Bz	4 (55%), 4 (C=O), 3 (C=O)	a)
6	H	H	Ts	4 (C=O) (15%)	a)
6	4 CH_3	H	Bz	4 (29%), 5 (11%)	c)
7	H	H	Bz	4 (C=O) (20%), 5, 5 (C=O) (63%)	b)
8	H	H	Bz	4 (C=O) (19%), 5 (C=O) (46%)	b)

[a] (Johnson et al., 1968a), [b] (Johnson et al., 1969a), [c] (Johnson et al., 1971)

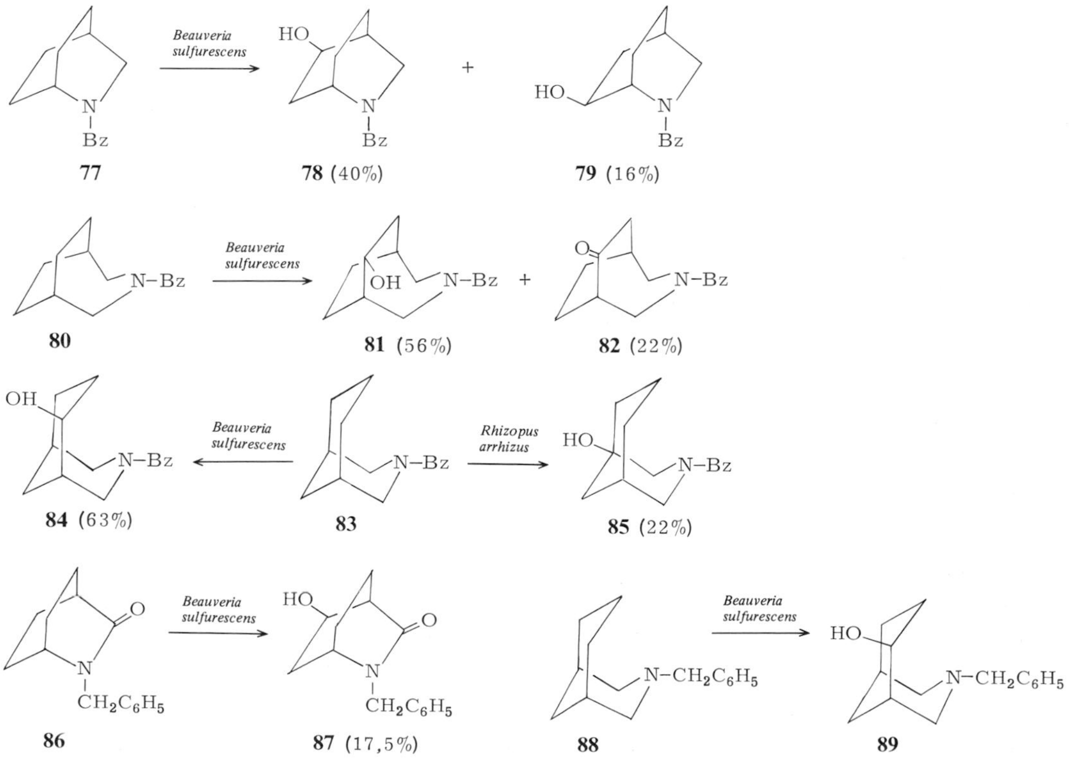

77 —*Beauveria sulfurescens*→ 78 (40%) + 79 (16%)

80 —*Beauveria sulfurescens*→ 81 (56%) + 82 (22%)

84 (63%) ←*Beauveria sulfurescens*— 83 —*Rhizopus arrhizus*→ 85 (22%)

86 —*Beauveria sulfurescens*→ 87 (17,5%)

88 —*Beauveria sulfurescens*→ 89

Scheme 33

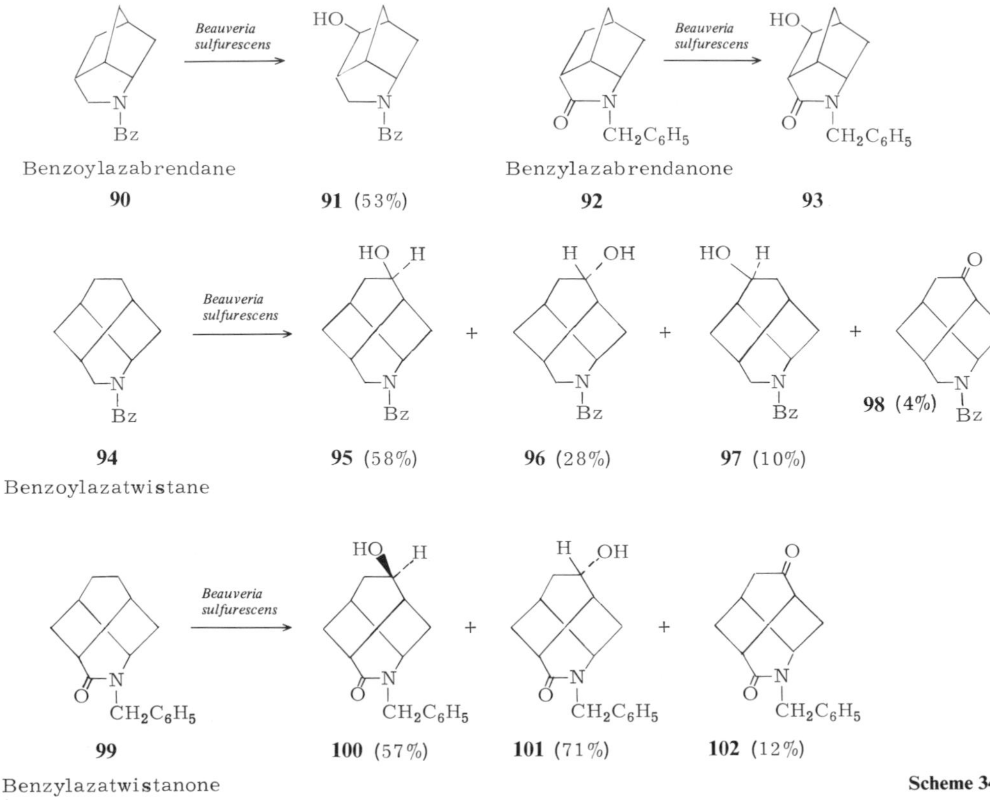

Beauveria sulfurescens

Benzoylazabrendane **90** → **91** (53%)

Benzylazabrendanone **92** → **93**

Benzoylazatwistane **94** → **95** (58%) + **96** (28%) + **97** (10%) + **98** (4%)

Benzylazatwistanone **99** → **100** (57%) + **101** (71%) + **102** (12%)

Scheme 34

et al., 1980). In the azabrendane series, the same results were obtained with the amide on (**90**) (Scheme 34) or in (**92**) the ring; similar results were obtained in the azatwistane series from **94** and **99** (FURSTOSS et al., 1981).

F. Degradation Reactions

A number of degradation reactions have been described. The mechanisms most often encountered are β-oxidation and the microbiological Baeyer-Villiger reaction. However, other mechanisms do occur. In addition, degradation often begins by hydroxylation of a type described in Sect. II. E.

Cyclopropane was transformed into propanal by the bacterium JOB 5 (OOYAMA et al., 1965). Cyclopentanol **1** (Scheme 35) was oxidized to glutaric acid **5** by *Pseudomonas* NCIB 9 872 (GRIFFIN and TRUDGILL, 1972). Valerolactone **3** is an intermediary resulting from Baeyer-Villiger oxidation of cyclopentanone. *Pseudomonas fluorescens* 8 027 reacted similarly with cyclopentanone, though no intermediate stage was reported (SHAW, 1966). The Baeyer-Villiger reaction evidently occurs with alkylcyclopentanones **6** (n = 4 and n = 6) by *Pseudomonas oleovorans* yielding the corresponding hydroxy acids **7**.

Adipic and other acids are formed during the degradation of cyclohexane, cyclohexanol, and cyclohexanone. *Pseudomonas aeruginosa* degraded cyclohexene to cyclohexanol, adipic, formic, and valeric acids (IMELIK, 1948). Lactone **8** appears to be an intermediate in degradation reactions due to some bacteria, including *Nocardia globerula* CI1 and *Acinetobacter* NCIB 9 871

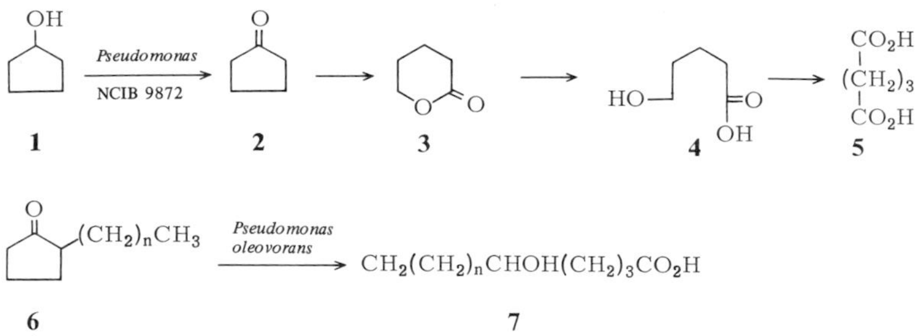

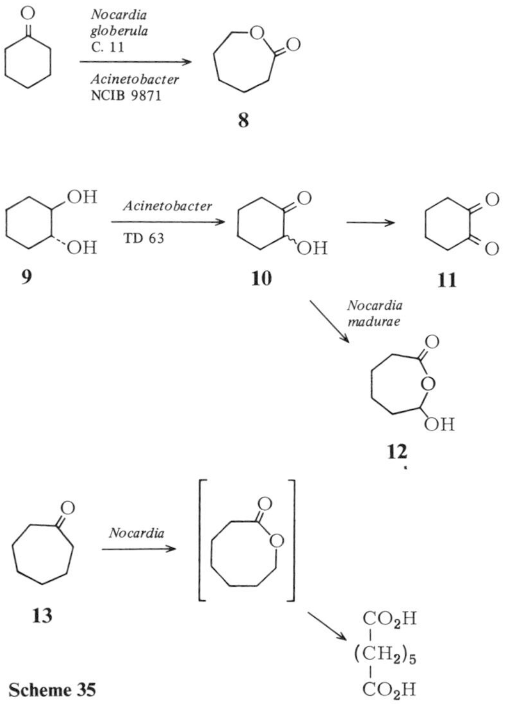

Pseudomonas NCIB 9872

1 → 2 → 3 → 4 → 5

Pseudomonas oleovorans

6 → $CH_2(CH_2)_nCHOH(CH_2)_3CO_2H$ 7

Nocardia globerula C. 11; *Acinetobacter* NCIB 9871 → 8

9 → (*Acinetobacter* TD 63) → 10 → 11; 10 → (*Nocardia madurae*) → 12

13 → (*Nocardia*) → [lactone] → $CO_2H(CH_2)_5CO_2H$

Scheme 35

(Donoghue et al., 1976). With the latter, in addition to adipic acid, 6-hydroxy hexanoic acid and adipic semialdehyde were also isolated (Donoghue and Trudgill, 1975).

Analogous results were obtained with growing cells of *Pseudomonas* (Anderson et al., 1980) as well as with resting cells of the same bacteria (Tanaka et al., 1977a, b), and with *Nocardia* (Stirling et al., 1977).

A somewhat different mechanism is apparently involved with cyclohexanediol **9** and ketol **10.** The latter can either form dione **11** or semiacetal lactone **12** which directly leads to adipic semialdehyde. The microorganisms used were *Acinetobacter* TD 63 (Davey and Trudgill, 1977) and *Nocardia madurae* (Murray et al., 1974). The mechanism via a lactone intermediate like **8** occurs in the degradation of cycloheptanone **13** which yields pimelic acid when converted by *Nocardia* (Hasegawa et al., 1982).

Several studies have dealt with the degradation of acids attached to cycloalkanes. A strain of *Corynebacterium* attacked cyclopentane carboxylic acid **14** (Scheme 36) leading to the 2-ketone and further to adipic acid via β-oxidation (Hasegawa et al., 1980).

Bacterium PRLW-19 reacted analogously with cyclohexane carboxylic acid **15;** here, further intermediates were detected (Blakley, 1978). *Alcaligenes faecalis* attacked cyclohexene 3-carboxylic acid **16** to form 3-heptenedioic acid **17** (Blakley and Papish, 1982). Other β-oxidations have been observed, e. g., with cyclohexane butanoic acid **18,** which was degraded to cyclohexane acetic acid **19** by *Nocardia* 107-332 (Davis and Raymond, 1961) and *Arthrobacter* (Ougham and Trudgill, 1982). The latter acid further reacted by β-oxidation to cyclohexanone via **20.**

In addition, butylcyclohexane was oxidized by *Nocardia* 107-332 to cyclohexane acetic acid (41% yield) by chain-end oxidation and subsequent β-oxidation (Davis and Raymond, 1959, 1961). The same cyclohexane acetic acid was obtained in high yield by degradation of dodecylcyclohexane and hexylcyclohexane with a strain of *Mycobacterium rhodochrous* (Feinberg et al., 1980).

Finally, adipic acid was produced by degradation of a decalin by *Flavobacterium* (Colla and Treccani, 1960).

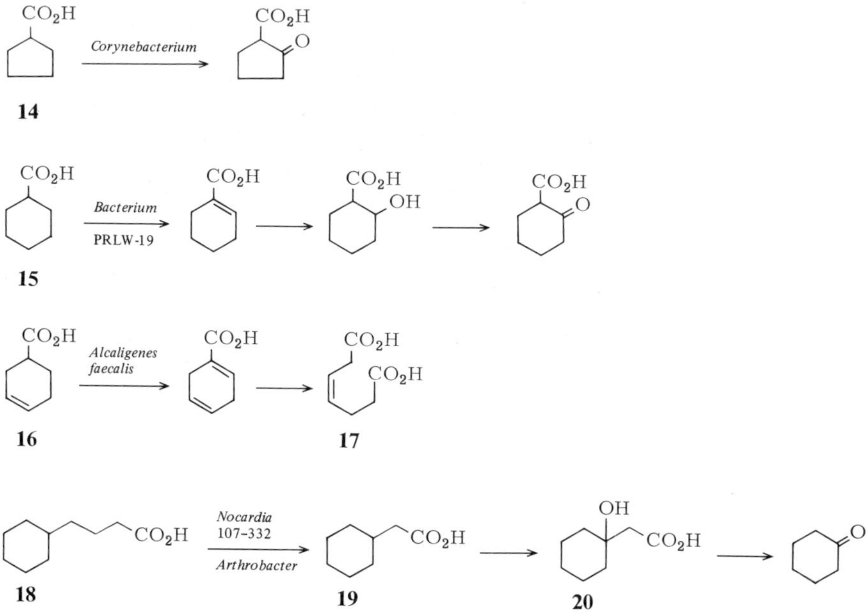

CO_2H *Corynebacterium* CO_2H O

14

CO_2H *Bacterium* PRLW-19 CO_2H CO_2H OH CO_2H O

15

CO_2H *Alcaligenes faecalis* CO_2H CO_2H CO_2H

16 **17**

CO_2H *Nocardia* 107-332 *Arthrobacter* CO_2H OH CO_2H O

18 **19** **20**

Scheme 36

CO_2H *Alcaligenes* sp. W1 CO_2H OH CO_2H O CO_2H O CO_2H OH

1 **3**

CO_2H *Pseudomonas putida* CO_2H OH CO_2H CO_2H OH OH CO_2H OH

5 **4** **2**

Cl HO HO *Pseudomonas putida* Cl HO HO

6

Pseudomonas putida OH OH + OH OH

(±) **7** (+) (−)

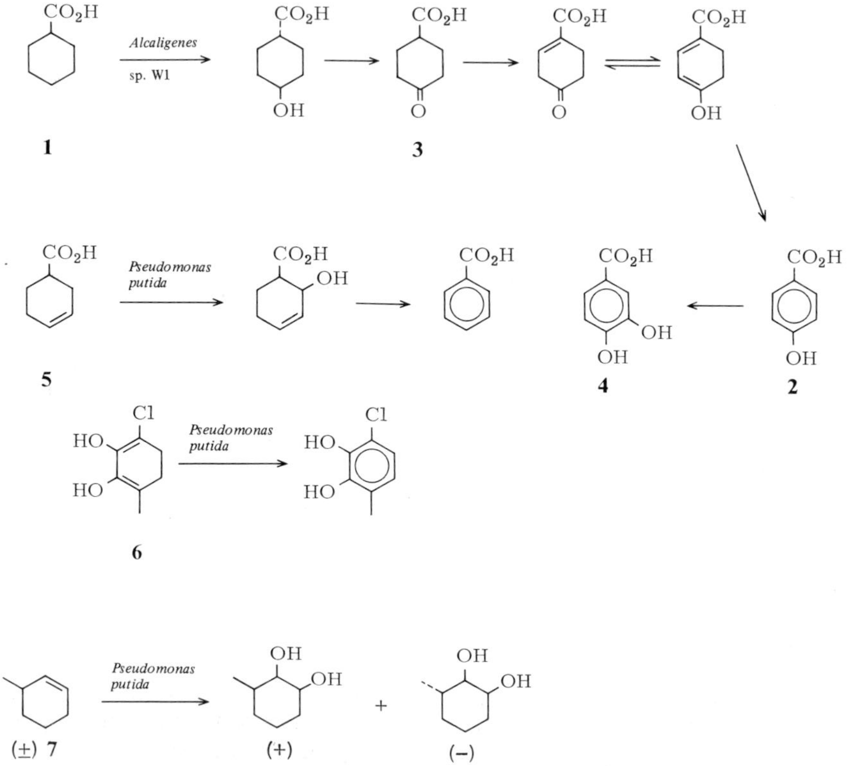

Scheme 37

G. Other Reactions

A number of other reactions which do not fall readily into any of the above categories will be described in the following.

Aromatization of cyclic compounds has been reported.

Cyclohexane carboxylic acid **1** (Scheme 37) was converted into *p*-hydroxybenzoic acid **2** by *Alcaligenes* sp. W1 (TAYLOR and TRUDGILL, 1978). The intermediates are shown in the scheme.

Cyclohexanone carboxylic acid **3**, which is an intermediate in the above reaction, also gave **2** with *Corynebacterium cyclohexanicum* (KANEDA, 1974). An *Arthrobacter* converted **1** into **2** and protocatechuic acid **4** (SMITH and CALLELY, 1975). *Pseudomonas putida* transformed cyclohex-3-ene carboxylic acid **5** into benzoic acid (BLAKLEY and PAPISH, 1982).

Finally, *P.putida* dehydrogenated **6** to the corresponding aromatic derivative (GIBSON et al., 1968). The oxidation of 3-methylcyclohexene **7** by *P.putida* (ZIFFER and GIBSON, 1975) led to the two *cis*-diols. Two strains of a microorganism able to develop on morpholine have been isolated. The morpholine is probably degraded to ethanolamine (KNAPP et al., 1982).

Two other heterocycles **8** and **9** (Scheme 38) were degraded by *Rhizopus japonicus* by a reduction process (WALLNÖFER, 1968, 1969). Nitroxide radicals used as spin labels, **10, 11,** and **12,** were reduced to saturated heterocycles by *Staphylococcus aureus* ATCC 25 923 (GOLDBERG et al., 1977).

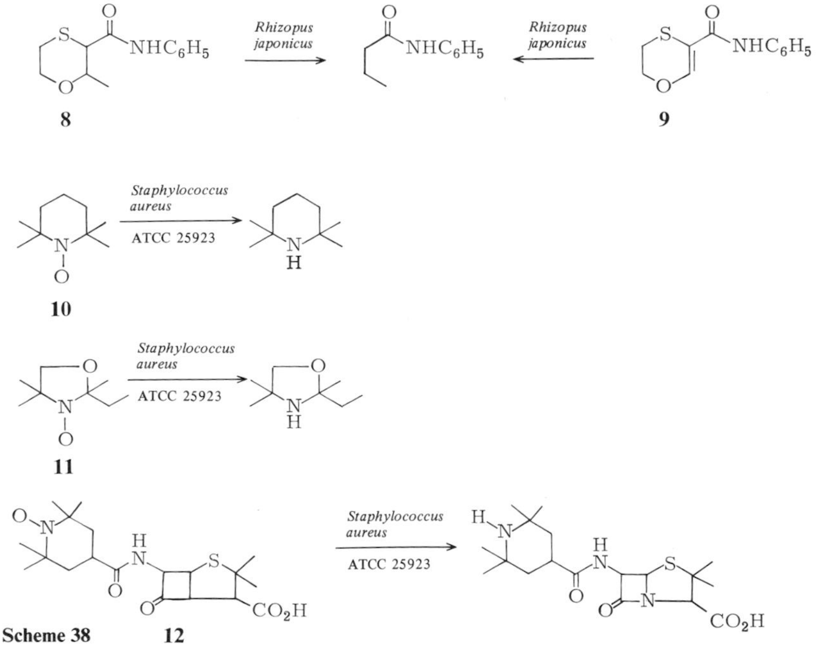

Scheme 38

III. Miscellaneous Applications

This section deals with a number of special bioconversions which are of particular theoretical or practical interest.

A. Synthesis of Estradiol and Analogs

In one of the classical syntheses of estradiol, the starting material is compound **1** (Scheme 39) which can be transformed into **2** and then to estrone by hydrogenation (see Chapter 2). **1** is synthesized chemically and is non-chiral, while **2** is racemic. If we consider the conformation of **1** with the 18-methyl pointing forward (β-position) with two ketone groups at C-14 and C-17, the microbiological reduction of C-17 creates an asymmetric C-13 with the proper configuration, and the 14-ketone remains free for ring closure. Hence, the single useful enantiomer of **3** will be obtained. With the aim of performing this step in the synthesis, a very large number of strains have been tested. Either α- or β-reduction of the 17-ketone has been observed. The β-OH(17*S*)-configuration is predicted by the Prelog rule and was obtained with *Curvularia falcata*. Reduction of the 14-ketone gave a 14α-OH(14*S*) configuration (BELLET and VAN THUONG, 1965; GIBIAN et al., 1965, 1966; KOSMOL et al., 1967; ISONO et al., 1967; KRAYCHY et al., 1967; WANG, 1969; GREENSPAN et al., 1968). The disemicarbazide of **1** acted upon by *Streptococcus lactis* or *S.cremonis* gave a monosemicarbazide (GIBIAN et al., 1965).

Results with microorganisms are given in Table 11. Scheme 39 depicts a possible synthesis pathway of estradiol from **1**.

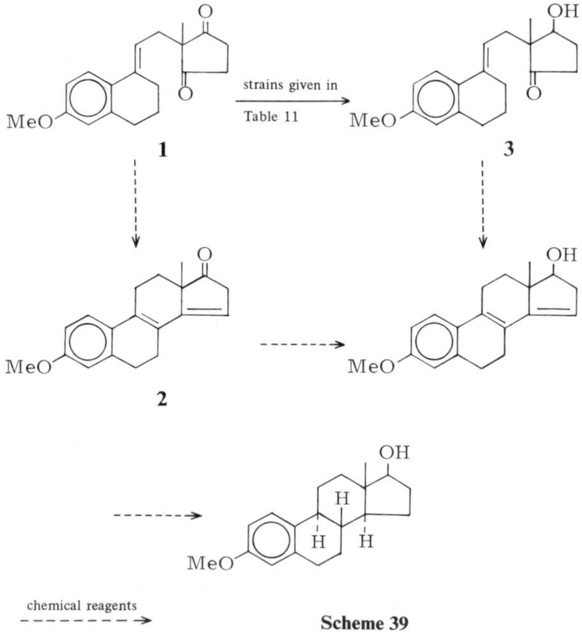

strains given in Table 11

chemical reagents

Scheme 39

Table 11. Reduction of Synthon **1** of Estradiol

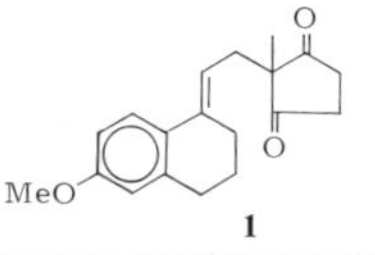

a) Reduction to the 17βOH-derivative

Microorganism		Ref.
Candida utilis		c), d)
Cryptococcus species		h)
Kloeckera africana	ATCC 20111	e)
K.javanica	ATCC 20112	e)
K.jensenii	ATCC 20110	e)
K.magna	ATCC 20109	e)
Rhizopus arrhizus	ATCC 11145	a)
Rhodotorula torulopsis		h)
R. species	ATCC 18101	f)
Saccharomyces carlbergensis	CBS 2354	c), d)
S.cerevisiae		c), d)
S.ellipsoidus		c), d)
S.exiguus	ATCC 20113	e)
S.humicola sp.	ATCC 18100	f)
S.uvarum	CBS 1508 (74%)	b), c)
Schizosaccharomyces octosporus ATCC 2479		g)

b) Reduction to the 14αOH-derivative

Microorganism	Ref.
Aspergillus ochraceus CBS	c), d)
Bacillus thuringiensis (87%)	b), c)
Curvularia lunata NRRL 2434	c), d)
Cylindrocarpon radicicola	c), d)
Fusarium solani	c), d)
Penicillium albidum	c), d)
P.nigricans	c), d)
P.notatum NRRL 2284	c), d)
Saccharomyces uvarum CBS 1508 (resting cells)	d)

[a] BELLET et al., 1965, [b] GIBIAN et al., 1966, [c] KOSMOL et al., 1967, [d] GIBIAN et al., 1965, [e] ISONO et al., 1967, [f] KRAYCHY et al., 1967, [g] WANG, 1969, [h] GREENSPAN et al., 1968

Procedure for preparing 3-methoxy-8,14-*seco*-1,3,5(10),9-estratetraene-17β-ol-14-one **3**.

Microbial transformation was conducted with *Saccharomyces uvarum* (CBS 1 508) grown in a 50-liter fermenter containing 30 liters of medium.

Cells were grown superficially on agar for three days and washed from the surface with physiological saline solution. The cell suspension was used to inoculate a 2-liter Erlenmeyer flask containing 500 mL of a medium of the following composition: 50 g of glucose, 20 g of cornsteep liquor, and water up to a total volume of 1 liter. After 24 hours of incubation at 30 °C on a rotary shaker, the culture was used to seed the fermenter containing the same medium. Substrate **1** dissolved in ethanol (30 g in 800 mL) was added immediately together with an antifoam agent (silicon SH). Fermentation was run at 29 °C with an agitation of 220 rpm and an aeration of about 2 m^3/h.

After 2, 4, and 6 hours, additional 30 g of **1** were added. After 40 hours the fermentation was terminated, the contents harvested and filtered; 95% of the product were thus obtained for further purification (TLC analytic: Kieselgel GF-254 Merck, eluant $CHCl_3$-acetone, 9:1).

The liquor was then shaken twice with 15 liters of methylethylketone and the extract evaporated in vacuo at 45 °C to yield 350–400 mL. After 16 hours, the crystallized product was filtered, washed with 15 mL of methylethylketone, and dried in vacuo at 60 °C.

Yield: ~70%.

F: 110–112 °C $(\alpha)_D^{20}$ = 35.5 °

(C = 1, dioxane).

After concentration of the mother liquor and recrystallization in 4 parts of ethanol, it was possible to obtain an extra 5–10% of **3** (KOSMOL et al., 1967).

Table 12. Reduction of 3-Methoxy-8,14-*seco*-1,3,5(10),9,15-estrapentaene-14,17-dione

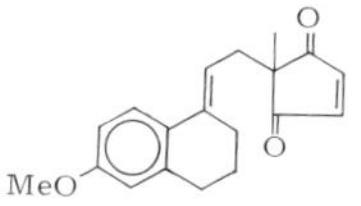

Microorganism		Ref.
**Candida utilis* IFO 1086		a)
**Debaryomyces nicotianae* IFO 855		
D.vanriji IFO 1285		
Hansenula capsulata IFO 984		
H.holstii IFO 980 (78%)		
**H.saturnus* IFO 993		
Kloeckera magna IFO 868	17α	a)
**Pichia etchellsii* IFO 1283		a)
P.piperi IFO 1280		
P. (Debaryomyces) vini IFO 1214		
P.wickerhamii IFO 1278		
**Rhodotorula rubra* IFO 8891		a)
**Schizosaccharomyces octosporus* ATCC 2479		b)
S.pombe IFO 362		a)

[a] WANG (1969), [b] TAKEDA (1967)

* Asterisk indicates reduction of the 15(16) double bond

Table 12 presents an analogous study of the compound with an unsaturated D-ring.

A number of analogs of **1** have been studied. *Saccharomyces uvarum* CBS 1 508 reduced the 14-ketone of 1- or 2-methylated **1**, and the 17-ketone of 4-methylated **1** (to 17β-OH) (GIBIAN et al., 1965). *Schizosaccharomyces pombe* ATCC 2 478 reduced the 17-ketone of 3-ethoxyl **1** (to 17β-OH), and *Saccharomyces uvarum* reduced the 17-ketone of 1,3-dimethoxyl **1** (to 17β-OH) (GIBIAN et al., 1965) (Table 13).

Another modification concerns the 13-substituent. When the methyl was replaced by an ethyl, the 17-ketone could be reduced either to a 17α-OH with *Kloeckera magna* ATCC 20 109 (ISONO et al., 1967) or to a 17β-OH with other strains (GREENSPAN et al., 1968; PLONKA et al., 1974; RUFER et al., 1967). When the methyl was replaced by a propyl, the 17-ketone gave a 17β-OH with *S.uvarum* (GIBIAN et al., 1965).

In **4** (Table 13), where the 6-CH_2 was replaced by sulfur, *S.uvarum* reduced the 14-ketone (GIBIAN et al., 1965). In **5**, having a five-membered B-ring, *Saccharomyces chevalieri* NCYC 91 reduced the 17-ketone to 17β-OH (GIBIAN et al., 1965; HEIDEPRIEM et al., 1968) (Table 13).

The method used to obtain a chiral compound from cyclopentane-1,3-dione has been applied to other molecules bearing two various substituents. One of these may be, e. g., the 18-methyl (or ethyl), the other a chain subsequently used to form the three remaining rings. **6** (Scheme 40) has a 3-carbon chain, **7** a 4-carbon chain, and **8** a 7-carbon chain.

with the reduction product of the 17-ketone (ZHOU et al., 1982a, b).

The 9-position of **7** can be protected by an ethyleneketal function. The obtained derivative **12** (Scheme 41) was reduced at C-17 by *S.cerevisiae* (ZHOU et al., 1982a, b). Finally, **13,** which can also be considered as a protected compound **7**, was reduced to a 17α-product by *R.arrhizus* (BELLET et al., 1965). **14,** the cyclohexane homolog of **7,** was reduced at C-14 by the same strain (BELLET et al., 1965). Compound **8** (Scheme 42) was reduced to a 17β-product **15** by

Table 13. Reduction of Synthons of Estradiol Analogs (references in the text)

Compound **1**

R	X	Y	Microorganism	Attacked Ketone
CH_3	OCH_3	1-CH_3	*Saccharomyces uvarum*	14
CH_3	OCH_3	2-CH_3	*S.uvarum*	14
CH_3	OCH_3	4-CH_3	*S.uvarum*	17(β)
CH_3	OCH_2CH_3	H	*Schizosaccharomyces pombe*	17(β)
CH_3	OCH_3	1-OCH_3	*S.uvarum*	17(β)
CH_2CH_3	OCH_3	H	*Kloeckera magna*	17(α)
CH_2CH_3	OCH_3	H	Other strains	17(β)
$CH_2CH_2CH_3$	OCH_3	H	*S.uvarum*	17(β)
4			*S.uvarum*	14
5			*Saccharomyces chevalieri* NCYC 90	17(β)

6 was reduced by *Schizosaccharomyces pombe* at position 17. The same strain reduced **9** to a 17β-product (WANG, 1969). It also reduced the 17-ketone of **7** (to 17β-OH) as also achieved by *Rhizopus arrhizus* ATCC 11 145 (BELLET et al., 1966). *Saccharomyces cerevisiae* can convert **10** to the cyclic hemiacetal **11** (CHOU et al., 1980) along

R.arrhizus ATCC 11 145 (yield 70%) (BELLET et al., 1966) and by various other strains (BELLET et al., 1965; WANG, 1969). On the other hand, it was reduced at C-14 by *Saccharomyces* and *Schizosaccharomyces* (SYNTEX, 1974; LANZILOTTA et al., 1975). *Schizosaccharomyces pombe* reduced the 13-ethyl homolog and the cyclohexane homo-

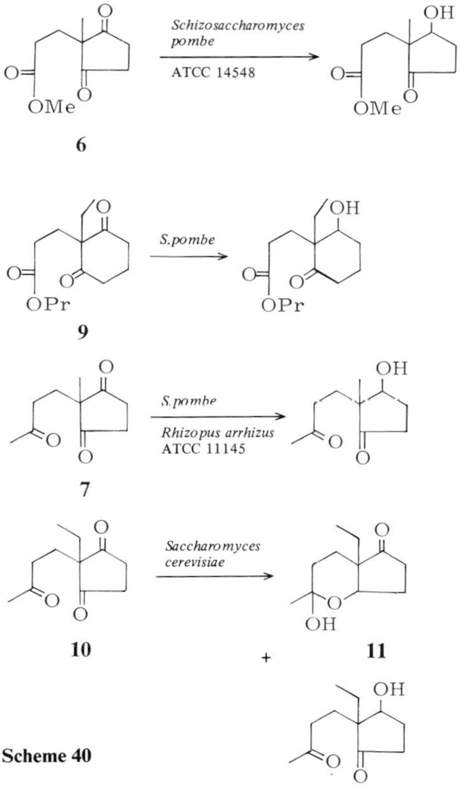

Schizosaccharomyces pombe ATCC 14548

6

S. pombe

9

S. pombe
Rhizopus arrhizus ATCC 11145

7

Saccharomyces cerevisiae

10 + 11

Scheme 40

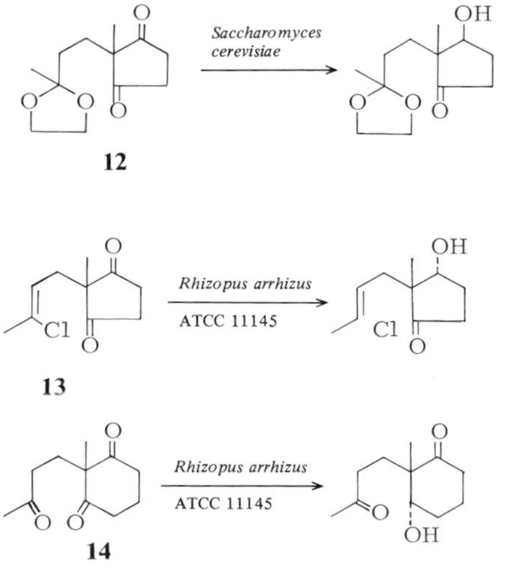

Saccharomyces cerevisiae

12

Rhizopus arrhizus ATCC 11145

13

Rhizopus arrhizus ATCC 11145

14

Scheme 41

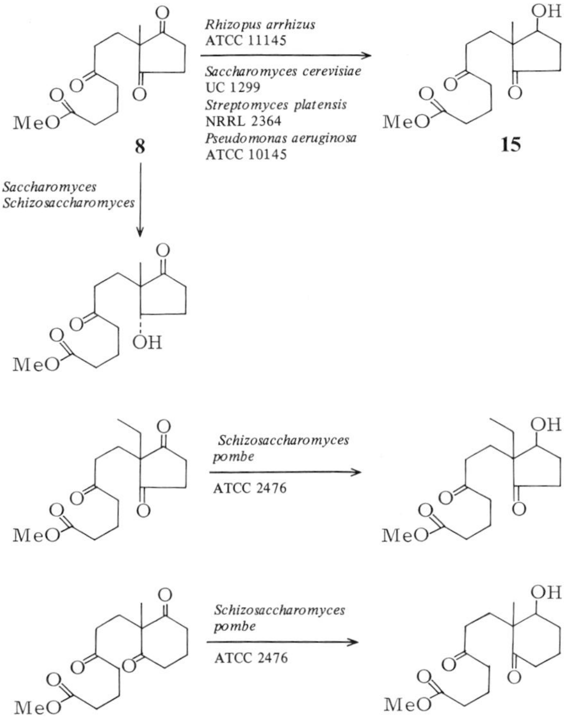

Rhizopus arrhizus ATCC 11145
Saccharomyces cerevisiae UC 1299
Streptomyces platensis NRRL 2364
Pseudomonas aeruginosa ATCC 10145

8 15

Saccharomyces
Schizosaccharomyces

Schizosaccharomyces pombe ATCC 2476

Schizosaccharomyces pombe ATCC 2476

Scheme 42

log of **8** at C-17 (WANG, 1969). Scheme 43 presents a possible synthesis of estradiol from **8**.

Preparation of 2α-(6-Carbomethoxy-3-oxohexyl)-3β-hydroxy-2β-methylcyclopentanon **15**.

A vegetative inoculum of *Rhizopus arrhizus* ATCC 11 145 was grown in a 250 mL-flask containing 100 mL of a medium consisting of: glucose, 10 g; cornsteep liquor, 10 g; soybean meal, 10 g; malt extract, 5 g; $CaCO_3$, 1 g; NaCl, 5 g; and water filled up to 1 liter. After sterilization for 30 min at 120 °C, the medium was inoculated with $5 \cdot 10^5$ spores of *R.arrhizus* and incubated for 24 hours at 27 °C on a rotary shaker. The resulting culture was used to provide a 10% inoculum for the production flasks (100 mL of the same medium in 250 mL-flasks). After 24 hours on a rotary shaker, 0.2

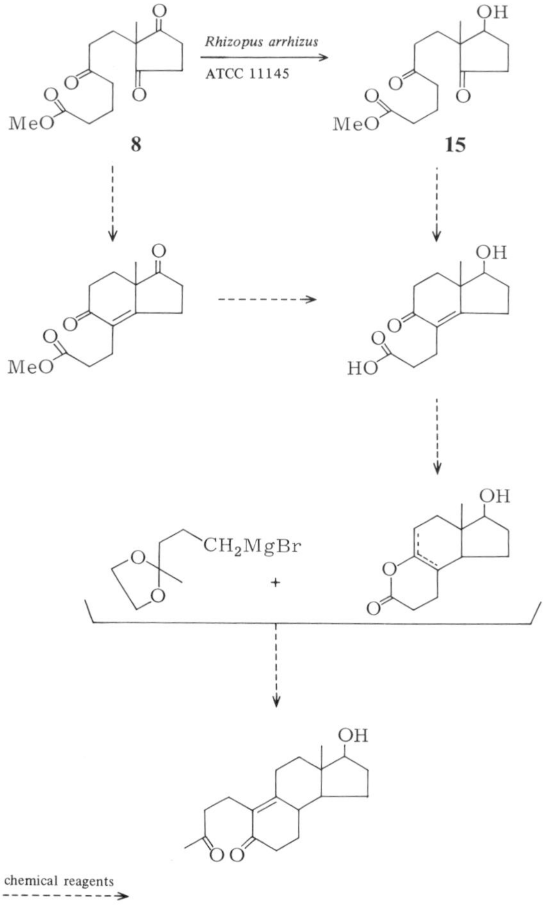

Rhizopus arrhizus ATCC 11145

chemical reagents

Scheme 43

g of **8** dissolved in 1 mL of ethanol were added. Additional incubation was carried out for 24 hours at 34 °C, then for 45 hours at 27 °C. The broths from fifteen flasks were filtered and the mycelium washed with water added to the filtrate. The whole broth was extracted with chloroform, the chloroform extract was washed with water, dried, and concentrated: about 3 g of an oil was obtained. Separation of components was accomplished by adsorption chromatography on a Florisil column; the pure product **15** was extracted as an oil: $(\alpha)_D^{20}$ −35 °C (C = 0.02, $CHCl_3$) (BELLET et al., 1965).

B. Hexahydroindene Derivatives

The degradation of steroids by various microorganisms gives products in which only the C- and D-rings remain intact (the conventional steroid nomenclature will be used). Generally, C-5, C-6, and C-7 remain attached at position 8 as a chain, C-17 may or may not be functionalized and may bear the C-20 and C-21 carbons. These molecules have been extensively studied.

Some of these molecules can also be produced by chemical synthesis (see Sect. III. A.).

Compound **1** (Scheme 44) (non-functionalized at C-17) was degraded by various strains of *Nocardia* and *Mycobacterium.* It underwent 8,14-dehydrogenation followed by *β*-oxidation of the chain at C-8. Compound **2** (non-functionalized at C-9) also underwent *β*-oxidation, and was additionally oxidized at C-17 by *Nocardia restrictus.* Compound **3**, bearing two functions on each ring was reduced at C-9 by *Pseudomonas testosteroni* to lactone **4** (LEE and SIH, 1967).

N.restrictus converted the acetic ester **5** of **3** to a 2-molecule condensation product; the acid function of **6** (resulting from hydrolysis of **5** followed by oxidation) esterified the free hydroxyl after simple hydrolysis of **5** (HÖRHOLD et al., 1969). *N.corallina* converted diketone **6** to ketolactone **7** and ketol **3**. Diketone **8** differs from **6** by the presence of an 8(13) double bond. It can be reduced by the same strain to **6** (23%) and then to lactone **7** (31%) (LEE and SIH, 1967).

Differently, *N.opaca* hydrated the double bond of **8** to yield **9** (HÖRHOLD et al., 1969). Compounds **10** and **8** are isomers, **10** having a 1(2) double bond. It was also hydrated by *N.opaca* to a 1-hydroxy derivative (KONDO et al., 1969).

Derivatives bearing a 17-acetyl group have also been studied. *Mycobacterium smegmatis* reduced the acid function of **11. 12** then lost the acetyl group by a Baeyer-Villiger reaction (SCHUBERT et al., 1964, 1965).

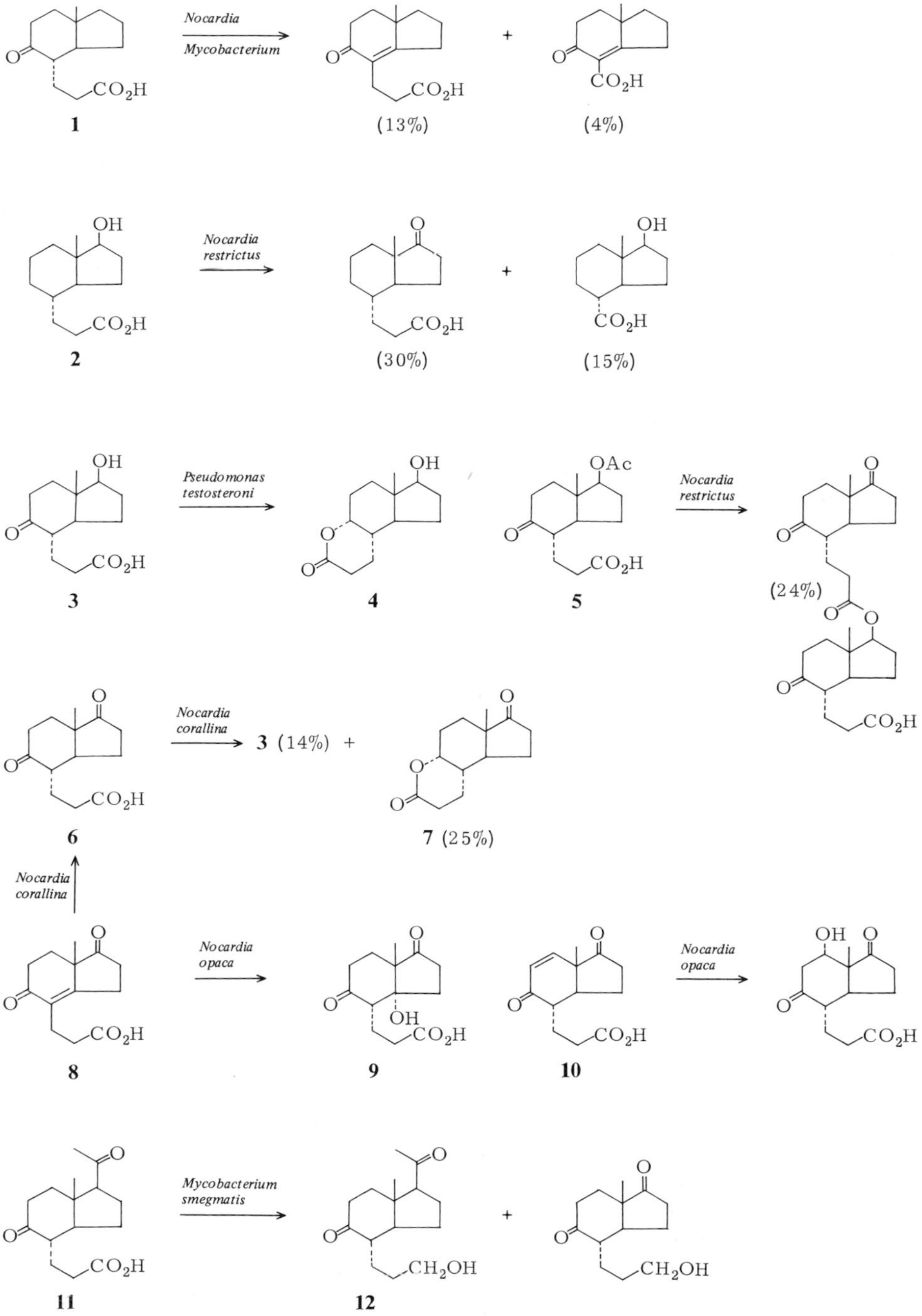

Nocardia / Mycobacterium: 1 → (13%) + (4%)

Nocardia restrictus: 2 → (30%) + (15%)

Pseudomonas testosteroni: 3 → 4

Nocardia restrictus: 5 → (24%)

Nocardia corallina: 6 → 3 (14%) + 7 (25%)

Nocardia corallina: 8 → 6

Nocardia opaca: 8 → 9

Nocardia opaca: 10 → product

Mycobacterium smegmatis: 11 → 12 + product

CO_2H, CH_2OH, OH, OAc

Scheme 44

Finally, *N.opaca* can degrade **11** to α-ketoglutaric and succinic acids (SCHUBERT et al., 1967) by classical degradation pathways.

C. Intermediates in the Biosynthesis of Aromatic Amino Acids

The biosynthesis of aromatic amino acids involves four alicyclic precursors: 5-dehydroquinic acid **1**, 5-dehydroshikimic acid **2**, shikimic acid **3**, and chorismic acid **4** (Scheme 45).

These acids, and quinic acid **5**, undergo reactions with numerous different microorganisms.

Quinic acid can be converted to a large number of different aromatic compounds. Two derivatives have been detected which may be intermediates in the aromatization process. These are the 3- and 5-dehydroquinic acids **6** and **1**. The first was obtained with *Pseudomonas ovalis* (EMMERLING and ABDERHALDEN, 1903), the second with *Acetobacter suboxydans* at a yield of 48% (WHITING and COGGINO, 1967).

The aromatic derivative most often encountered is protocatechuic acid **7** obtained with numerous strains (BEIJERINK, 1911; LOEW, 1981; EMMERLING and ABDERHALDEN, 1903; BUTKEWITSCH, 1924, 1925; PERVOZWANSKY, 1930; BERNHAUER and WAELSCH, 1932; RATLEDGE, 1964, 1967; RATLEDGE and WINDER, 1966). Others include *p*-hydroxybenzoic acid **8**, anthranilic acid **9** produced from quinic acid by *Aerobacter* (RATLEDGE, 1964, 1967) and *Mycobacterium smegmatis,* which also gave salicylic acid **10** (RATLEDGE and WINDER, 1966) and hydroxy-4-phenylacetic acid **8'**.

Aerobacter also gave 2,3-dihydroxy benzoic acid **11.** Pyrocatechol **12** was also obtained with *Aerobacter* (RATLEDGE, 1964, 1967), with *Aspergillus niger,* and an undefined bacterium (BUTKEWITSCH, 1925). Finally, simpler degradation products of quinic acid have been isolated, such as oxalic acid with *A.niger* (BERNHAUER and WAELSCH, 1932) and propionic and formic acid with *Schizosaccharomyces* species (LOEW, 1981).

Quinic acid gave dihydroshikimic acid **13** with *Lactobacillus pastorianus* var. *quinicus* (WHITING and COGGINO, 1967; CARR et al., 1957). This acid was oxidized to **14** by *Acetobacter suboxydans* (WHITING and COGGINO, 1967).

Shikimic acid **3** (Scheme 46) reacted analogously to quinic acid by aromatization of the ring. Prior oxidation to 5-dehydroshikimic acid **2** can be achieved, e. g., with *Acetobacter suboxydans* (55% yield) (WHITING and COGGINO, 1967) or with *Pseudomonas ovalis* (YOSHIDA, 1964). This bacterium also gave protocatechuic acid **7**. *Mycobacterium smegmatis* gave, in addition to **7**, salicylic acid **10,** catechol **12,** and anthranilic acid **9** (RATLEDGE and WINDER, 1966). Acid **9**, either free or as an amide was produced from shikimic acid by *Aerobacter aerogenes* (RATLEDGE, 1964, 1967). Finally, labeled shikimic acid was converted to 2,5-dihydroxy-phenylalanine by a strain of *Streptomyces* (SCANNEL et al., 1970).

Chorismic acid **4** is a late step in the biosynthesis of aromatic amino acids. A cell-free system from *Saccharomyces cerevisiae* converted **4** into prephenic acid **16** (SPRÖSSLER et al., 1970), while an analogous system from *Aerobacter* (*Enterobacter*) *aerogenes* formed the acids **17** and **18** (YOUNG and GIBSON, 1969).

D. Biosynthesis and Degradation of Biotin

Use of microorganisms in both the synthesis and the degradation of biotin **1** (Scheme 47) has been reported. One use is for the resolution of racemates. Thus, (±)-*N*-benzylbiotin **2** was selectively hydrolyzed by *Pseudomonas* and *Corynebacterium primorioxydans* FERM-P-1 427 (OGINO et al., 1975a). The analog **3**, devoid of the acid function, was also debenzylated selectively by the same microorganisms (OGINO et al.,

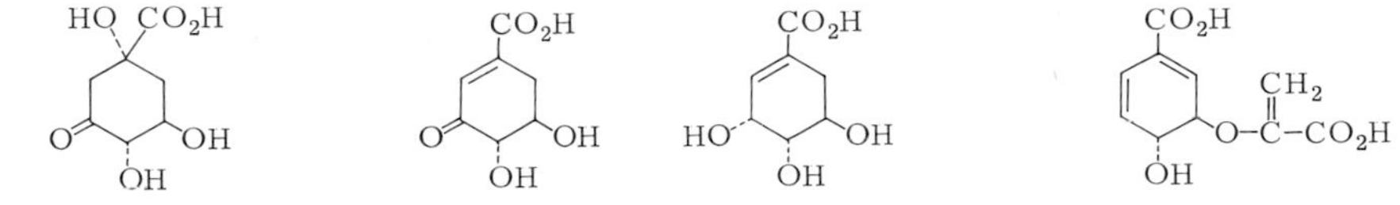

HO CO$_2$H / O, OH, OH — **1** 5-Dehydroquinic acid

CO$_2$H / O, OH, OH — **2** 5-Dehydroshikimic acid

CO$_2$H / HO, OH, OH — **3** Shikimic acid

CO$_2$H / CH$_2$ / O–C–CO$_2$H / OH — **4** Chorismic acid

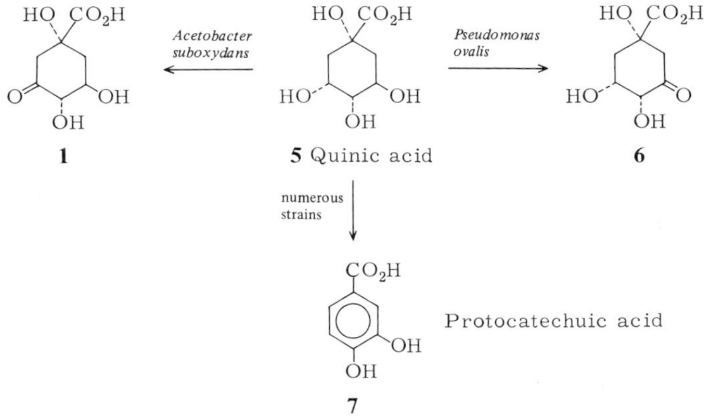

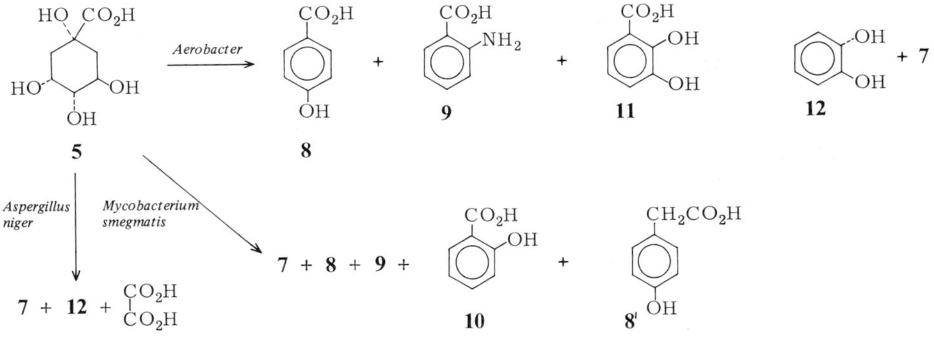

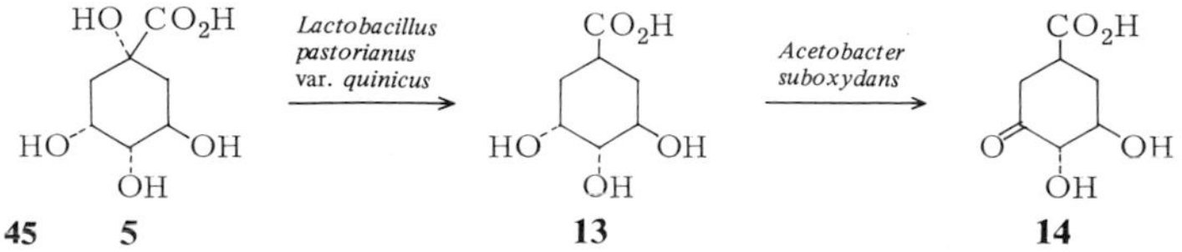

5 Quinic acid —*Acetobacter suboxydans*→ **1**

5 Quinic acid —*Pseudomonas ovalis*→ **6**

5 Quinic acid —numerous strains→ **7** Protocatechuic acid

5 —*Aerobacter*→ **8** + **9** + **11** + **12** + 7

5 —*Aspergillus niger*→ 7 + **12** + CO$_2$H–CO$_2$H

5 —*Mycobacterium smegmatis*→ 7 + **8** + **9** + **10** + **8'**

Scheme 45 **5** —*Lactobacillus pastorianus* var. *quinicus*→ **13** —*Acetobacter suboxydans*→ **14**

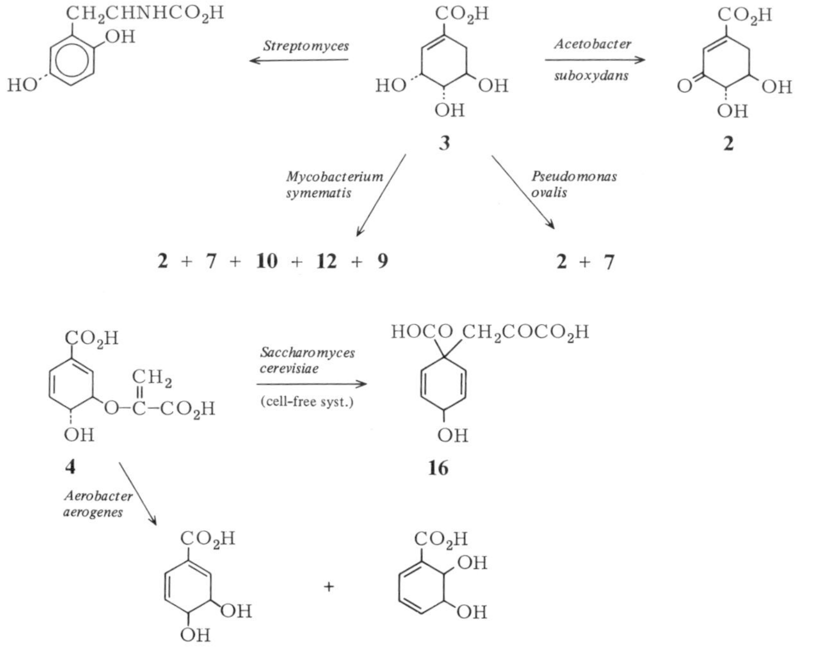

$CH_2CHNHCO_2H$ (OH, HO) ← *Streptomyces* — **3** (CO_2H; HO, OH, OH) — *Acetobacter suboxydans* → **2** (CO_2H; O, OH, OH)

3 — *Mycobacterium symematis* → **2** + **7** + **10** + **12** + **9**

3 — *Pseudomonas ovalis* → **2** + **7**

4 (CO_2H; CH_2 ‖ O–C–CO_2H; OH) — *Saccharomyces cerevisiae* (cell-free syst.) → **16** (HOCO CH_2COCO_2H; OH)

4 — *Aerobacter aerogenes* → **17** (CO_2H; OH, OH) + **18** (CO_2H; OH, OH)

Scheme 46

1975b). **4**, a debenzylated isomer of **3**, was converted to (±)-biotinol (80% yield) and (±)-biotin by *Corynebacterium* (OGINO et al., 1974a,b).

This reaction is interesting for the production of biotin.

The industrial production of biotin is mainly carried out by chemical synthesis, but the process is very complicated. The cooxidation by microorganisms gives a new and simplified method in conjunction with the chemical method. A mutant strain of *Corynebacterium* B-321, strain M-6 318 produces large amounts of (±)-biotin from 5-[2-oxo-hexahydro-1*H*-thieno-(3,4)-(+)-imidazolyl-4*H*]-pentane **4**. A conversion rate of ca. 80% is obtained by growing cultures of strain M-6 318 in the constant presence of *n*-paraffin, since the inability of the microbe to assimilate *n*-alkane resulted in repression of biotin degradation.

Procedure for biotin production. 100 mL of a bacterial suspension grown on a medium containing Na_2HPO_4, 3.5 g; KH_2PO_4, 2.5 g; $MgSO_4 \cdot 7\ H_2O$, 1.0 g; $CaCl_2$, 0.1 g; NaCl, 0.1 g; $ZnSO_4 \cdot 7\ H_2O$, 8 mg; $MnSO_4 \cdot n\ H_2O$, 0.8 mg; $CuSO_4 \cdot 5\ H_2O$, 40 µg; urea, 20 g; $FeSO_4 \cdot 7\ H_2O$, 20 mg; sodium acetate, 10 g; 0.5% *n*-paraffin, and 0.1% corn-steep liquor in 1 000 mL distilled water is transferred to 1 900 mL of the same medium in a 3 000 mL-glass jar (aeration 2 vvm at 28 °C). After 15 h of cultivation 1.6 g of 5-[2-oxo-hexahydro-1*H*-thieno-(3,4)-(+)-imidazolyl-4*H*]-pentane **4** in 5 mL of dimethylformamide is added. The formation of the intermediate biotinol and the reaction product biotin is followed up by gas chromatography (OGINO et al., 1974a, b).

Several single-ring compounds such as dethiobiotin **5** can be converted to biotin using a wide variety of strains (IWAHARA et al., 1966a). It was shown with *Aspergillus niger* ATCC 1 004 that the first carbon (C-5) was not involved in the stereochemistry of the chain on the thienyl ring (PARRY and

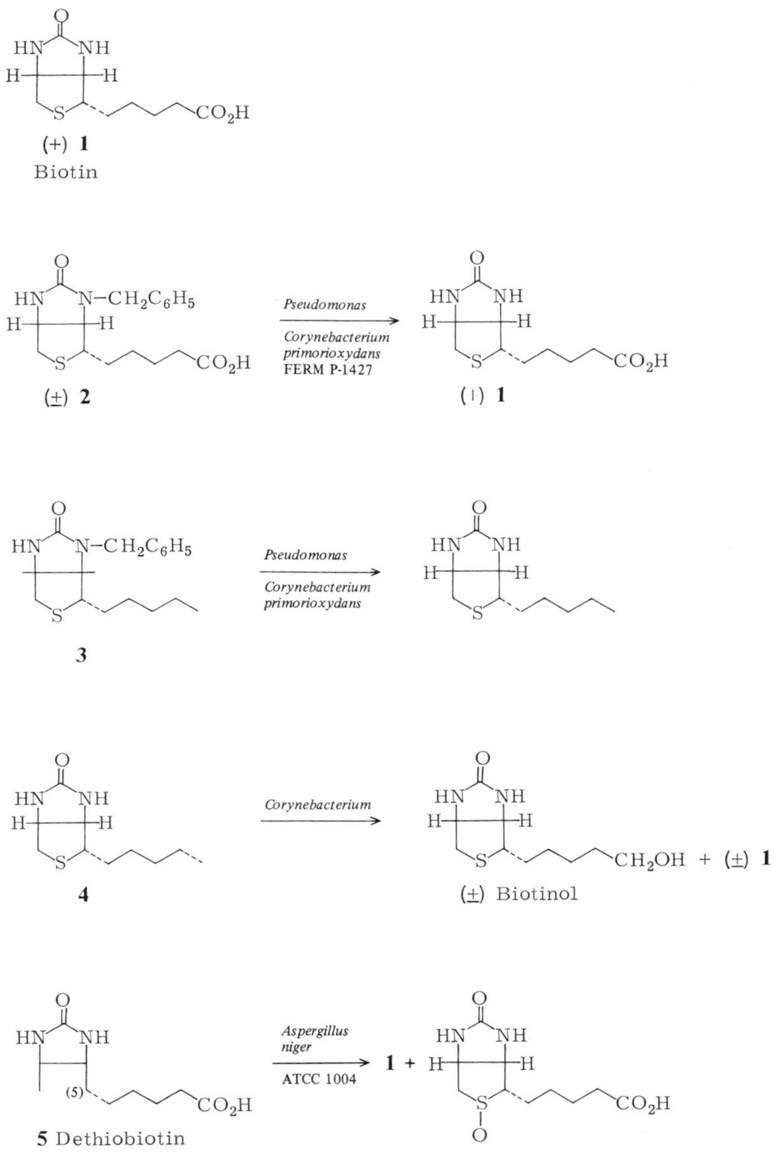

(+) **1**
Biotin

(±) **2** — *Pseudomonas* / *Corynebacterium primorioxydans* FERM P-1427 → (+) **1**

3 — *Pseudomonas* / *Corynebacterium primorioxydans* →

4 — *Corynebacterium* → (±) Biotinol + (±) **1**

5 Dethiobiotin — *Aspergillus niger* ATCC 1004 → **1** +

Scheme 47

NAIDU, 1980). *Aspergillus niger* also converted **5** to biotin, along with biotin sulfoxide (TEPPER et al., 1966).

Certain bacteria also converted bisnorbiotin **6** (Scheme 48) to biotin (IZUMI et al., 1973). Similarly, bisnordethiobiotin was transformed into dethiobiotin **5** by the same bacteria. *Rhodotorula flava* formed the amide of dethiobiotin (SEKIJO et al., 1969).

Degradation of biotin generally involves β-oxidation, though occasionally the formation of a sulfoxide has been observed.

A cell-free system of *Pseudomonas* converted biotin to bisnorbiotin **6**, the unsaturated derivative **7**, and tetranorbiotin **8** (IWAHARA et al., 1969). A soil *Pseudomonas* oxidized the sulfur to sulfoxide and β-hydroxylated the acid function (the first step

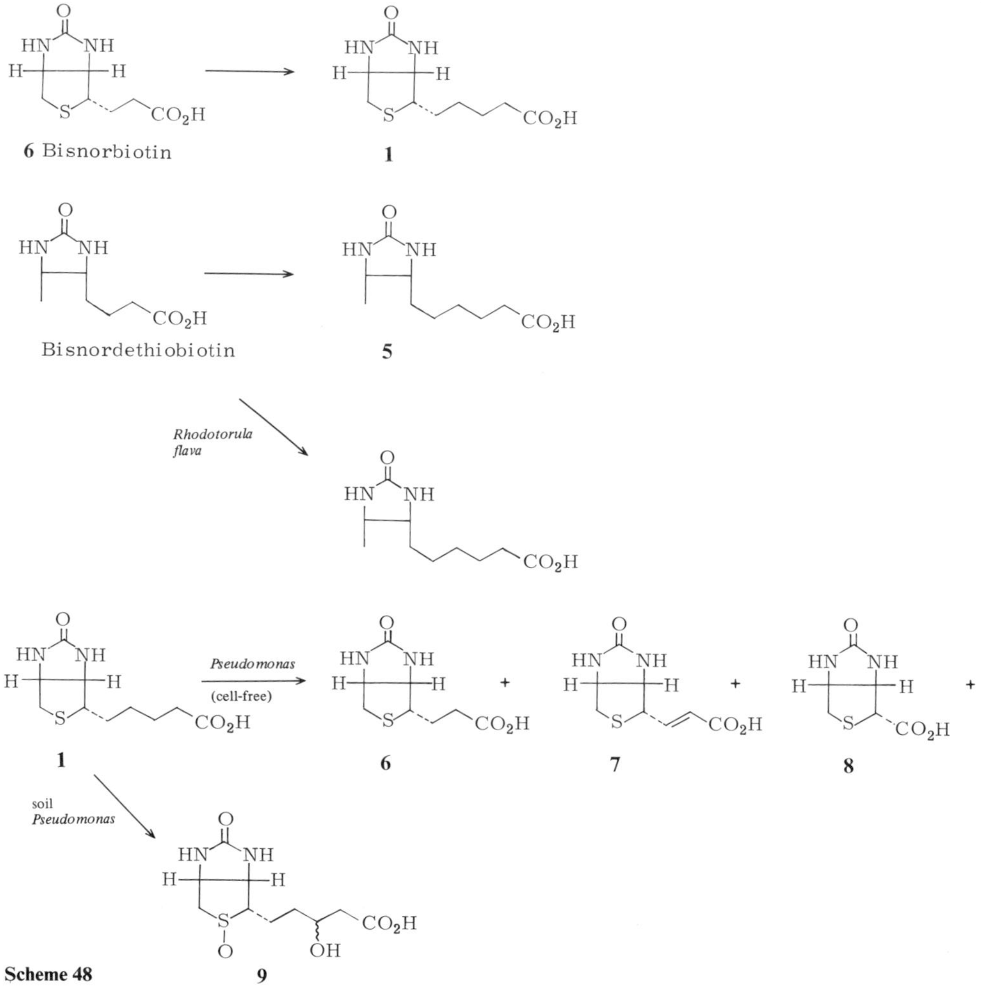

6 Bisnorbiotin → **1**

Bisnordethiobiotin → **5**

Rhodotorula flava

1 → *Pseudomonas* (cell-free) → **6** + **7** + **8** +

1 → soil *Pseudomonas* → **9**

Scheme 48

of β-oxidation) forming **9** (YANG et al., 1969; (IWAHARA et al., 1969; WHA et al., 1970).

Endomycopsis converted biotin to bisnorbiotin **6** (Scheme 49) and its sulfoxide (YANG et al., 1968, 1970). Resting or growing cells of *Pseudomonas* converted biotin to its sulfone, to bisnorbiotin **6**, and tetranorbiotin **8** (WHA et al., 1970). β-Oxidation again occurred during the degradation of homobiotin by *Pseudomonas* which gave norbiotin and trisnorbiotin (RUIS et al., 1968). Finally, *Pseudomonas* converted biotin to bisnorbiotin and (+)-*allo*-bisnorbiotin **13** (WHA et al., 1973).

Aspergillus oryzae degraded dethiobiotin **5** (Scheme 50) to bisnordethiobiotin (IWAHARA et al., 1966b).

Certain microorganisms, such as *Aspergillus niger* (WRIGHT et al., 1954), *Rhodotorula, Penicillium,* and *Endomycopsis* (YANG et al., 1969) formed the sulfoxide of biotin.

AURET et al. (1968) studied a reaction related to the oxidation of the sulfur atom of biotin. Compound **10** was oxidized by *A.niger* NRRL 337 to (+)-sulfoxide **11** (yield 38%) and (+)-sulfone **12** (yield 28%, optical purity 29%). The (±)-sulfoxide was oxidized by the same fungus to the (+)-sulfone (yield 29%, optical purity 16%).

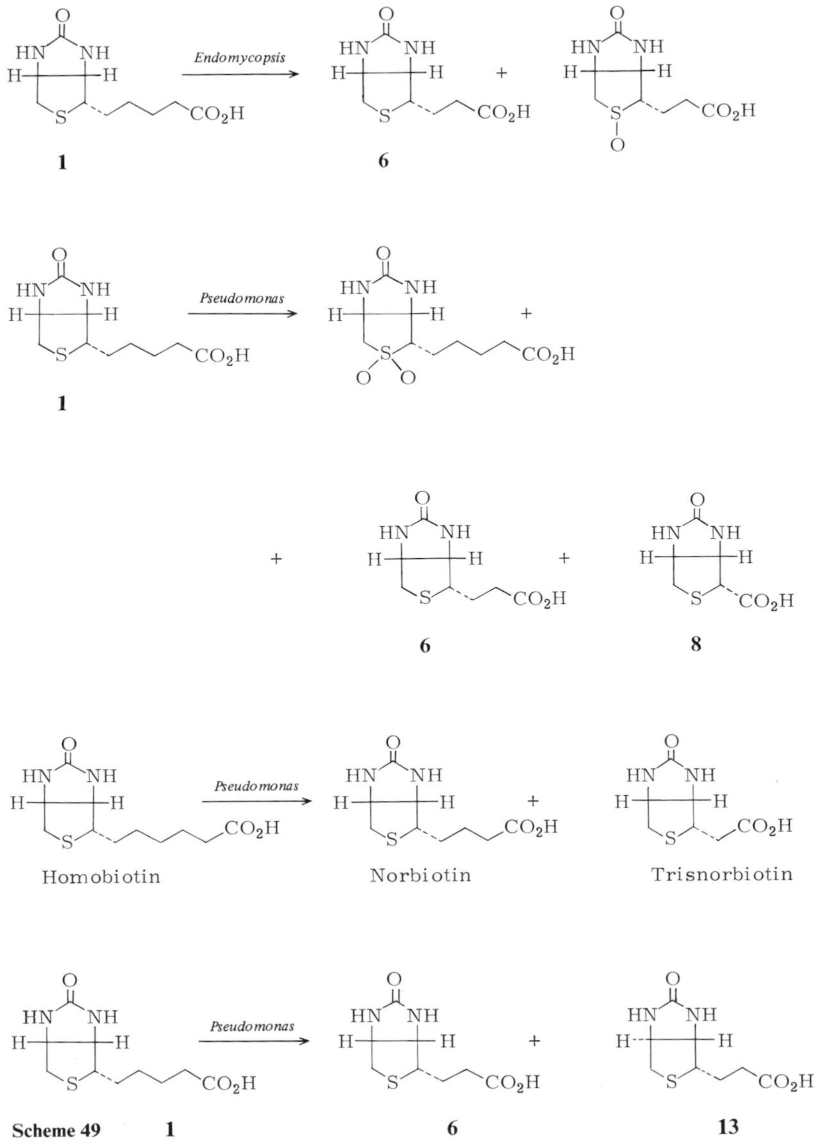

Endomycopsis

1 → 6 + [sulfoxide]

Pseudomonas

1 → [sulfone] + 6 + 8

Pseudomonas

Homobiotin → Norbiotin + Trisnorbiotin

Pseudomonas

Scheme 49 1 → 6 + 13

E. Biodegradation and Metabolization

The biodegradation of several particular classes of substrates will be dealt within the following, namely, chlorine-containing insecticides, herbicides, and diazepines, and their analogs. The biodegradation of pesticides is clearly an important area of research given its usefulness in environmental protection (see Chapter 8). The biodegradation of diazepines is of interest insofar as microbiological degradation can be regarded as a model of drug metabolization by higher organisms (ROSAZZA and SMITH, 1979).

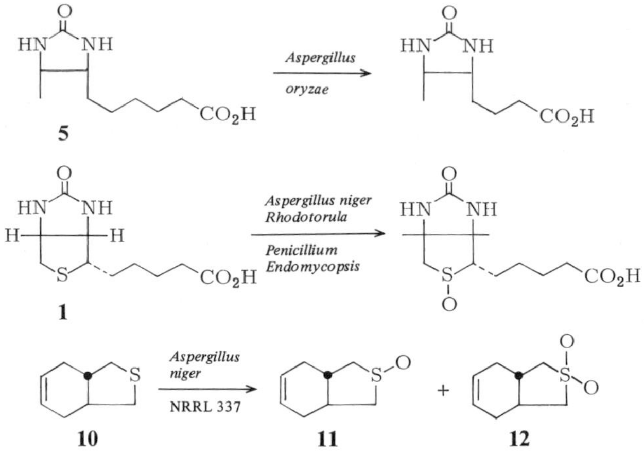

5 → (*Aspergillus oryzae*)

1 → (*Aspergillus niger*, *Rhodotorula*, *Penicillium*, *Endomycopsis*)

10 → (*Aspergillus niger* NRRL 337) 11 + 12

Scheme 50

---

2 ← (*Pseudomonas putida*) 1 γHCH → (*Pseudomonas putida*, *Clostridium sphenoides*) 3

4 αHCH → (*Clostridium sphenoides*) 5

6 → (*Clostridium butyricum*) 7 + 8 + 9

10 → (*Clostridium butyricum*) 7 + 8

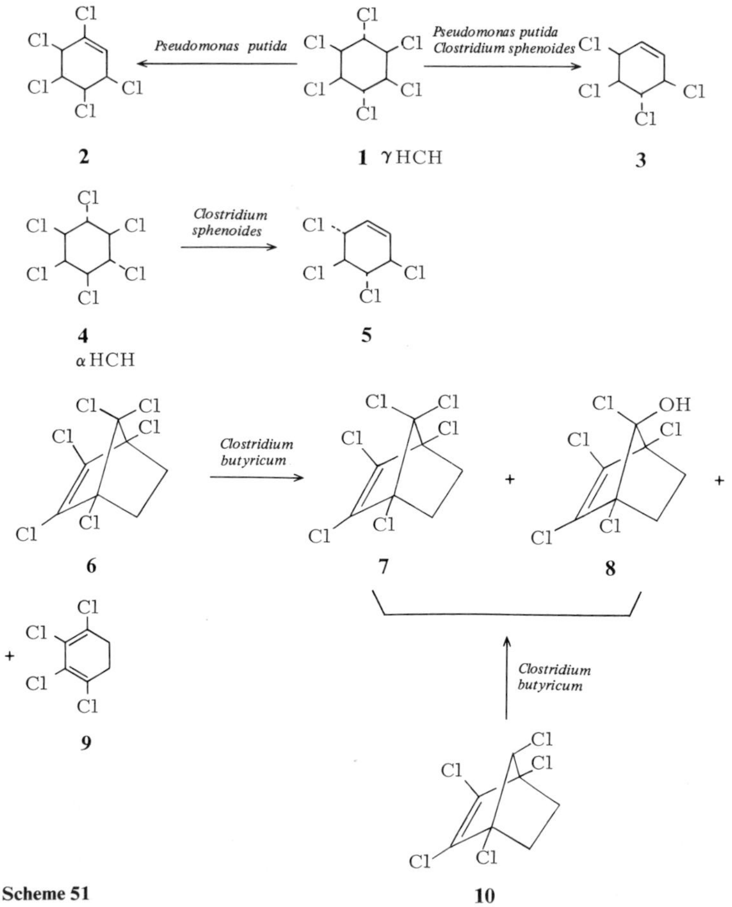

Scheme 51

1. Biodegradation of pesticides

The biodegradation of hexachlorocyclohexanes has been the subject of a good deal of work. Lindane **1** (Scheme 51) (γHCH, γBCH) was converted by *Pseudomonas putida* via two pathways, one forming γ-pentachlorocyclohexene **2** (γPCCH), the other tetrachlorocyclohexene **3** which can subsequently be degraded to CO_2. The α-isomer **4** was also found in small amounts among the degradation products (MATSUMURA et al., 1976). *Clostridium sphenoides* converted γHCH **1** to the tetrachloro derivative **3** and the isomer αHCH **4** to another tetrachloro derivative **5** (HERITAGE and MAC RAE, 1977).

The degradation of hexachloronorbornene **6** by *Clostridium butyricum* was studied under anaerobic conditions. A tetrachloro derivative **7** was obtained along with

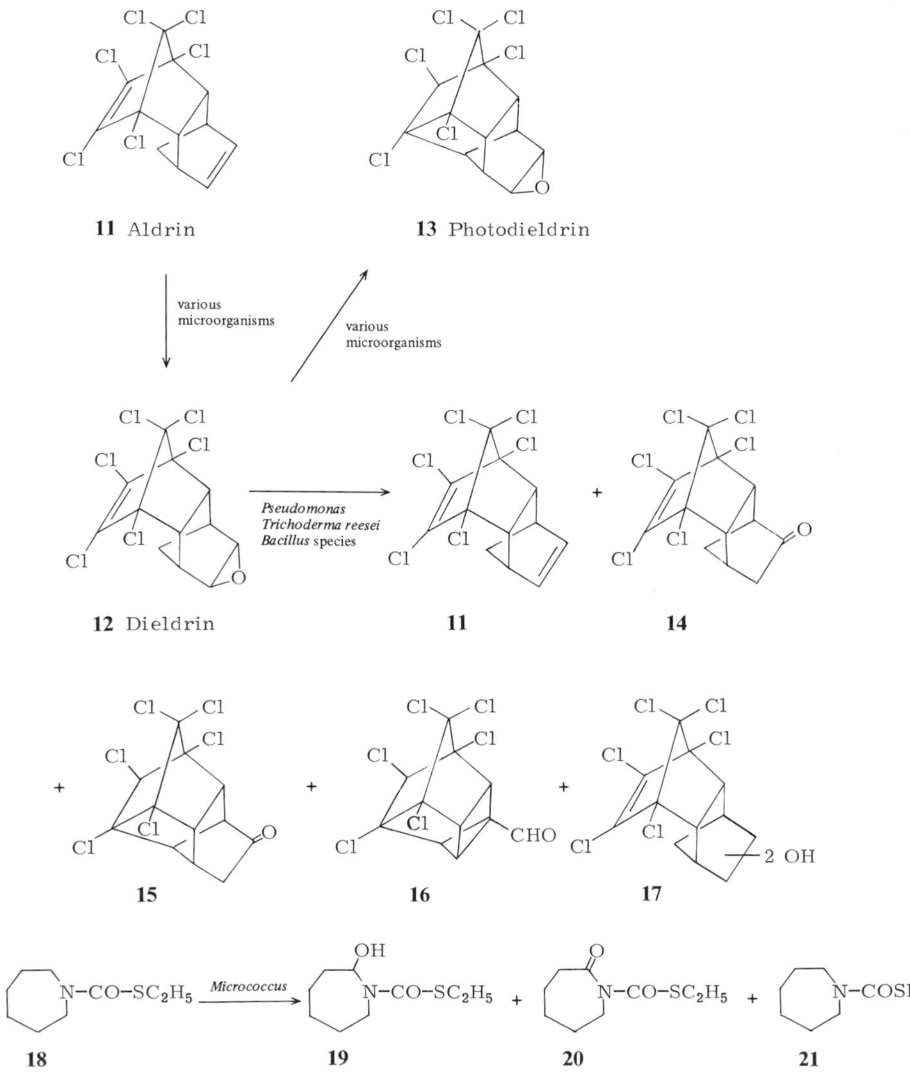

11 Aldrin

13 Photodieldrin

various microorganisms

various microorganisms

12 Dieldrin

Pseudomonas *Trichoderma reesei* *Bacillus* species

11

14

15

16

17

18

Micrococcus

19

20

21

Scheme 52 Ordram

the hydroxylated derivative of **7**, **8**, and tetrachlorocyclohexadiene **9**. The hydroxylation of **7** was confirmed separately. Finally, pentachloronorbornene **10** also gave **7** and **8** (SCHUPHAN and BALLSCHMITER, 1972).

The biodegradations of aldrin **11** and dieldrin **12** (Scheme 52) have also been investigated. A large number of strains have been found to convert aldrin to dieldrin (KORTE et al., 1962; KORTE and STIASHI, 1964; FERGUSON and KORTE, 1977); the *exo*-configuration was always obtained. Various microorganisms convert dieldrin to photodieldrin **13** (MATSUMURA et al., 1970). *Pseudomonas, Trichoderma reesei,* and a *Bacillus* species gave a large number of degradation products with dieldrin. Conversion to aldrin, the inverse of the conversion cited above, occurred. In addition, isomerization of the epoxide to **14** giving **15** by additional ring closure (analogous to the conversion of **12** to **13**) occurred, followed by an additional isomerization to aldehyde **16** which subsequently oxidized to an acid. Finally, a dihydroxylated product **17** of undetermined stereochemistry was isolated (MATSUMURA and BOUSCH, 1967; MATSUMURA et al., 1968).

The herbicide ordram **18** was degraded to an sulfoxide by *Nocardia globerula,* a *Bacillus,* and an *Enterobacter* species, with subsequent ring cleavage giving $HCOO(CH_2)_4NHCOSC_2H_5$ and $HOCO(CH_2)_3NHCOSC_2H_5$ (ZYAKUN et al., 1983). A *Micrococcus* species hydrolyzed the thioester to **21,** and then hydroxylated the carbon in α-position to the nitrogen yielding the alcohol **19** and finally the amide **20** (GOLOVLEVA et al., 1978).

2. Degradation of diazepines

Benzodiazepines devoid of an *N*-alkyl substituent of type **1** (Scheme 53) are hydrolyzed to **2**. This reaction was carried out for X = Cl, Y = H with *Streptomyces* (SUMITOMO 1971) and *Pellicularia filamentosa* spp. *sasakii* (GREENSPAN et al., 1967), for X = Cl, Y = H, Cl, F with *Streptomyces* (KISHIMOTO et al., 1975), and for X = NO_2, Y = H (SUMITOMO, 1971).

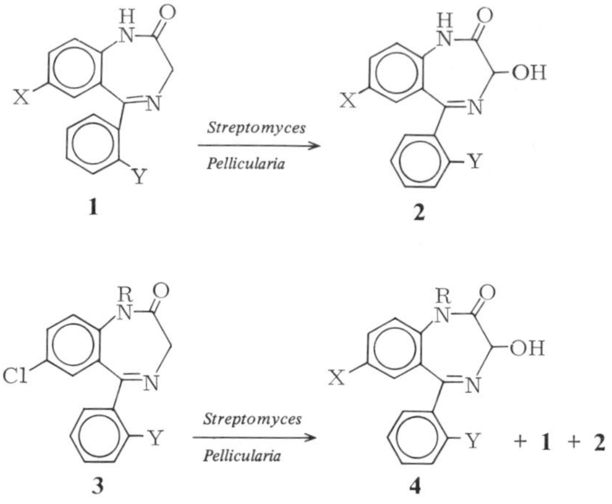

Streptomyces / Pellicularia: 1 → 2

Streptomyces / Pellicularia: 3 → 4 + 1 + 2

Scheme 53

With *N*-alkylbenzodiazepines of type **3**, two kinds of reaction were observed; in almost all cases dealkylation occurred, occasionally with additional hydroxylation as above yielding **2**. Thus, *Streptomyces* converted **3** (R = CH_3, Y = H) (R = cyclopropyl methyl, Y = H) (R = CH_3, Y = F) to **1**, **2**, or **4** (KISHIMOTO et al., 1975). A strain of *Pellicularia filamentosa* spp. *microsclerotia* CBS gave an analogous result with **3** R = CH_3, Y = H) (GREENSPAN et al., 1967).

F. Applications in Prostaglandin Synthesis

Microbiological conversions have been used extensively in prostaglandin synthesis. Chiral synthons have been obtained, and racemic molecules formed at later stages in synthesis have been resolved. Much work has also been done on the biodegradation of prostaglandins.

A general review has been published recently by PERUZZOTTI (1982). Scheme 54 gives the nomenclature of various prostaglandins along with the numbering of the carbon skeleton.

1. Production of prostaglandin synthons

The simplest synthons are derived from the selective hydrolysis of 3,5-diacetoxycyclopentenes. The *cis-(meso)*isomer **1** was hydrolyzed by *Bacillus subtilis* var. *niger* (TAKANO et al., 1976) to **3** with a yield of 56%. A mixture of *cis* and *trans* was hydrolyzed by baker's yeasts; after 48 h a small amount of *trans*-monoester **4** (3.2%) was obtained with an optical purity of 90%. The recovered diol **5** (*cis* and *trans*) was weakly active (MIURA et al., 1976).

Selective hydrolysis gave another interesting alcohol **8** from the racemic acetate **7**. This ester was hydrolyzed by the esterase of *Trichoderma* giving alcohol **8** (*R*) (yield 38%, e. e. 82%). The *S*-ester **7** was recovered (yield 42%, e. e. 96%) (ORITANI and YAMASHITA, 1975).

The racemic bicyclic ketone **9** was reduced by baker's yeast into the two diastereoisomeric *S*-alcohols **10** (yield 38,5%, e. e. 90%) and **11** (yield 18%, e. c. 90%). They were separated by column chromatography and subsequently submitted to further conversions by ordinary chemical means (NEWTON et al., 1979) yielding the two enantiomeric ketones **9**. Ketone **12** (Scheme 55) gave a prostaglandin synthon by ozonization of the double bond. This racemic ketone was reduced by *Saccharomyces droso-*

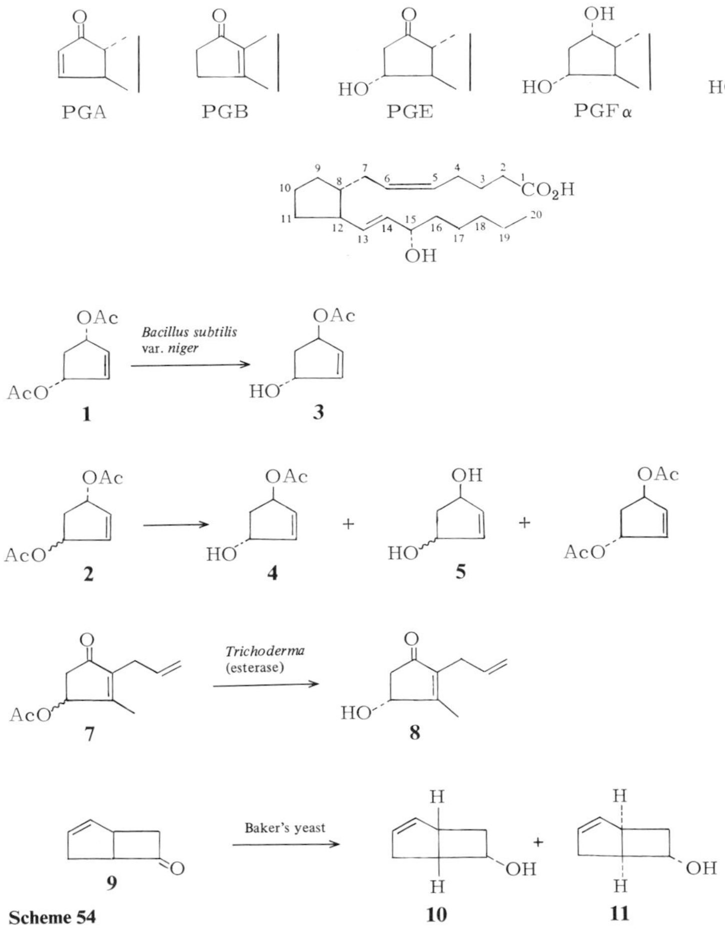

PGA PGB PGE PGFα PGFβ

CO_2H OH

OAc *Bacillus subtilis* var. *niger* OAc AcO HO **1** **3**

OAc OAc OH OAc AcO HO HO AcO **2** **4** + **5** +

Trichoderma (esterase) AcO HO **7** **8**

Baker's yeast H H OH OH H H **9** **10** + **11**

Scheme 54

phylarum to the (−)-*endo*- and the (+)-*exo*-alcohols (of unreported configuration) which were used as starting materials for the synthesis of active derivatives (AMBRUS et al., 1974). The microbiological reduction of substituted cyclopentanediones (**13**) has provided useful optically active ketols. *Dipodascus albidus* reduced acids **13** (R = CH_3, R′ = CH_2CO_2H, and R = C_2H_5, R′ = CH_2CO_2H) and their methyl esters to lactones **14** (TRUCKENBRODT et al., 1978; SCHWARZ et al., 1981). A yield of 65% was obtained from **13** (R = CH_3, R′ = CH_2CO_2H). *D.albidus* and *Absidia coerulea* also reduced **13** (R = CH_3, R′ = $CH{=}CH_2$) to **15**.

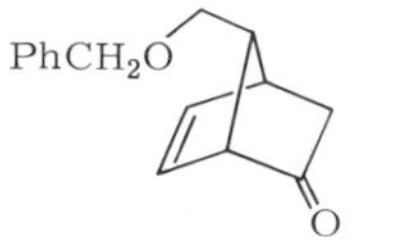

$PhCH_2O$

12

13 —*Dipodascus albidus*→ **14**

R = CH_3 or C_2H_5

R′ = CH_2CO_2H or CH_2CO_2CH

13 → **15**

R′	Microorganism
$CH{=}CH_2$	*D.albidus*
	Absidia coerulea
*n*Pr	Baker's yeast
$CH_2{-}C{\equiv}CH$	*Absidia coerulea*

13 —Baker's yeast→ **16** + **15**

R′	16 : 15
$CH_2{-}CH{=}CH_2$	1 : 10
$CH_2{-}C{\equiv}CH$	1 : 2

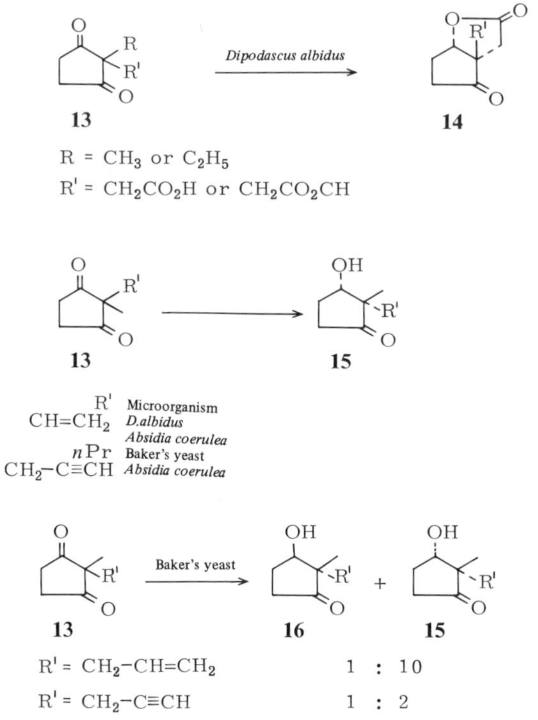

Scheme 55

Baker's yeast reduced the methyl ketones **13** (R = CH_3) with R′ = *n*Pr, R′ = $CH_2{-}CH{=}CH_2$, and R′ = $CH_2{-}C{\equiv}CH$. In the first case, only **15** was obtained; in the other two, a mixture of **15** and **16** was obtained, 10:1 for R′ = $CH_2{-}CH{=}CH_2$ (yield 75%) and 2:1 for R′ = $CH_2{-}C{\equiv}CH$ (BROOKS et al., 1982). The diketone **13** (R = CH_3, R′ = $CH_2{-}C{\equiv}H$) was reduced by *Absidia coerulea* to **15** (SCHWARZ et al., 1981).

Some synthons have a 7-membered carboxylic acid side-chain, e. g., the triketones **17** and **18** (Scheme 56), which were reduced by *Dipodascus uninucleatus* to the diketols **19** with a 4*R*-configuration, with 75% yield of **17.** Reduction of **17** by *Mucor rammanianus* gave the 4*S*-isomer of **19** (yield 43%) (SIH et al., 1975; SIH, 1977).

Production of (4*R*)-2-(6-Carboxy-2-hexenyl)-4-hydroxy-1,3-cyclopentanedione **19.**

The surface growth from a one week old agar of *Dipodascus uninucleatus* was suspended in 5 mL of saline (0.85%) solution. Portions (2 mL) of this suspension were used to inoculate 50 mL of the soybean-dextrose medium (soybean meal, 5 g; glucose, 20 g; yeast extract, 5 g; K_2HPO_4, 5 g; NaCl, 5 g; with distilled water filled up to 1 L; pH adjusted to 6.5 with HCl) held in 250 mL-Erlenmeyer flasks. The flasks were incubated at 25 °C on a rotary shaker (250 rpm/2-in. radius) for 24 h, after which a 10 vol-% transfer was made to each of four 2 L-Erlenmeyer flasks, containing 500 mL of the soybean-dextrose medium. After 24 h of incubation on a rotary shaker, 250 mg of **18,** dissolved in 2 mL of dimethylformamide, was added to each 2 L-flask. The flasks were then incubated for an additional 24 h under the conditions used in the incubation of 250 mL-flasks. 24 h after the addition of **18,** the cells were removed by centrifugation. The supernatant was adjusted to pH 2.5 with 6N HCl and was extensively extracted three times

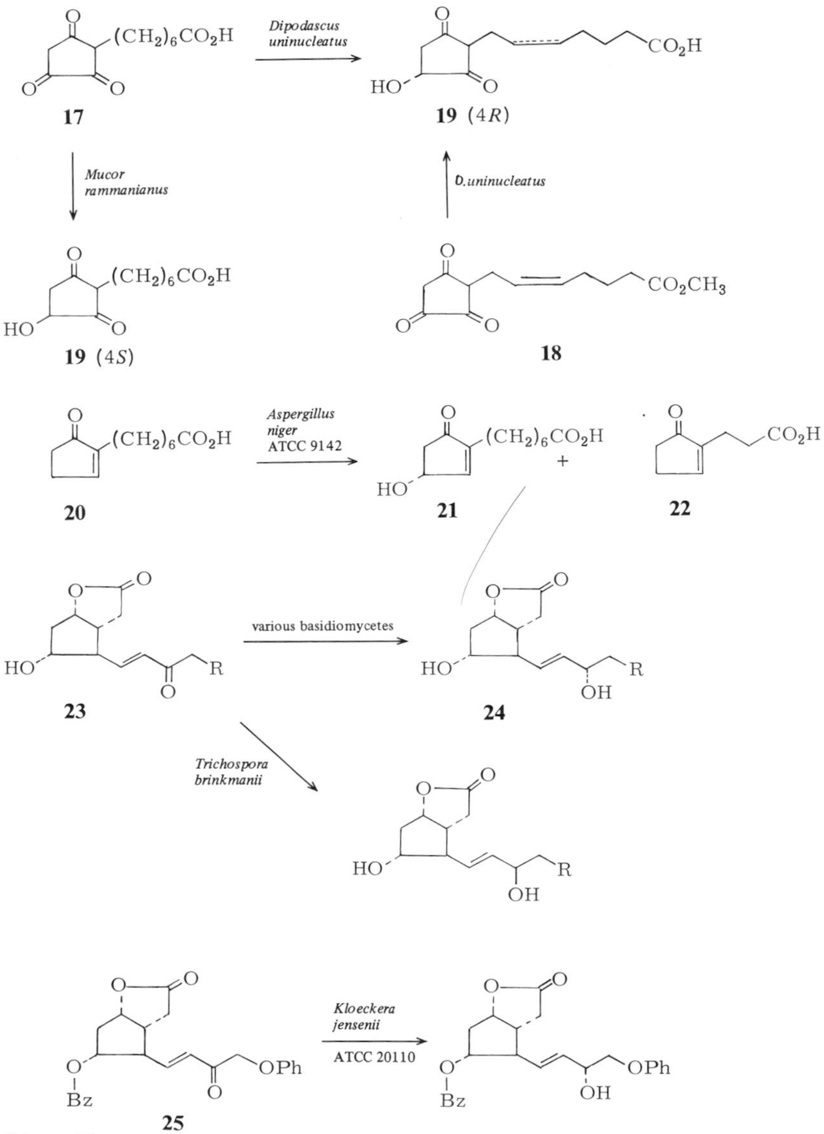

Scheme 56

with 1.5 L of ethyl acetate. The ethyl acetate was dried over Na_2SO_4 and evaporated to yield an oily residue. This residue was chromatographed over a silicic acid-celite (85/15) column (35 × 2.5 cm). The column was eluted with a gradient system consisting of 500 mL of ethyl acetate-benzene (1:1) in the mixing chamber and 500 mL of ethyl acetate in the reservoir, and 7 mL fractions were collected. Fractions 42–65 were pooled to yield 0.48 g of **19:** mp 57.5–59 °C; $(\alpha)_D^{23}$ +19.0 ° (SIH et al., 1975).

The achiral compound **20** was 4α-hydroxylated by *Aspergillus niger* ATCC 9 142 to **21.** β-Oxydation gave the side product **22** (KUROZUMI et al., 1973).

Lactones of the general structure of **23** and **25** can serve as starting materials for the synthesis of PGFα. Reduction to the 15α-alcohol of the 15-ketone **23** (R = Bu, *m*-chlorophenoxy, *m*-trifluoromethylphenoxy) was carried out by a considerable number of basidiomycetes (MOORE, 1975). Similarly, lactone **25** was reduced to the corresponding 15α-alcohol by *Kloeckera jensenii* ATCC 20 110 (KIESLICH et al., 1980).

Trichospora brinkmanii and four other microorganisms reduced the 15-ketone of the lactone **23** (R = Bu) to the 15β-alcohol (MOORE, 1974).

2. Production of optically active prostaglandins

Important procedures for this purpose involve hydrolysis reactions and reduction of carbonyls as well as carbon-carbon double bonds.

Hydrolysis reactions have been used to obtain derivatives from methyl or ethyl esters under gentler conditions than those of the usual chemical methods. For instance, *Rhizopus oryzae* has been shown to hydrolyze the methyl esters of (−)-PGE$_1$ **1** and (−)-PGE$_2$ **2** (Scheme 57) (SIH et al., 1975).

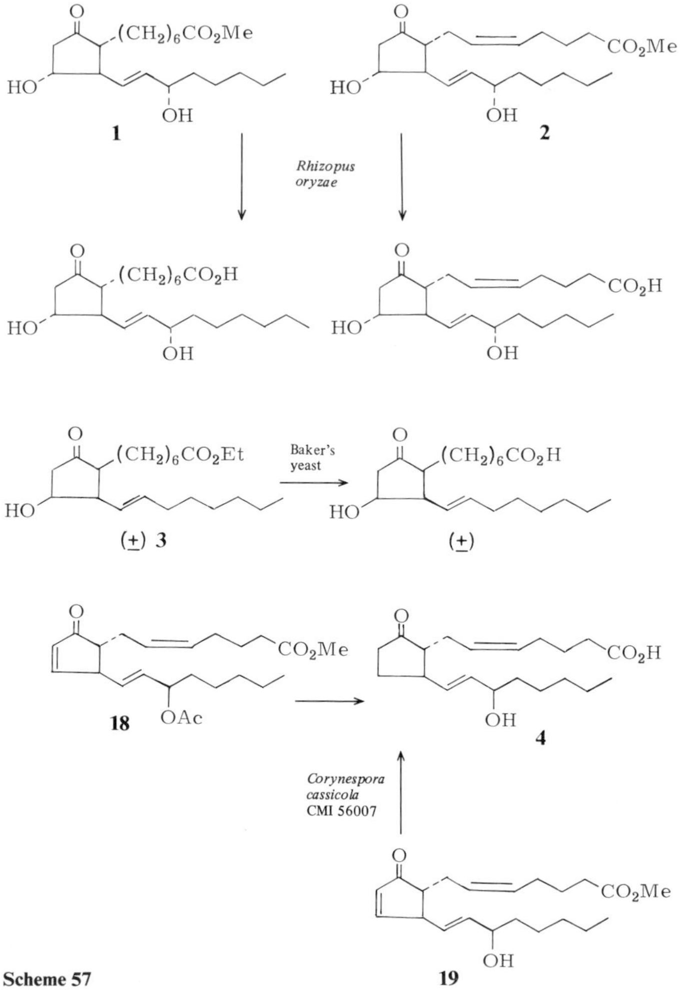

Scheme 57

Hydrolysis of (−)-PGE_1 methyl ester by *Rhizopus oryzae*.

The mold was grown in a soybean-dextrose medium for 24 h. The cells were then harvested by filtering through cheesecloth; approximately 56 mg of wet mycelia were obtained from 500 mL of medium in a 2 L-Erlenmeyer flask. After the cells were washed twice with 0.1 M phosphate buffer (pH 7.0) the mycelia were suspended in 500 mL of the same phosphate buffer. (−)-PGE_1 methyl ester **1** (500 mg) dissolved in 5 mL of ethanol was added to the flask which was incubated on a rotary shaker (280 rpm) at 26 °C. After 24 h the mycelia were removed by filtration and the filtrate was extracted twice with ether and then acidified to pH 2.5 with 6 N HCl. The acidified aqueous layer was then extracted three times with ethyl acetate-hexane which yielded 298 mg of (−)-PGE_1, mp 115–116 °C, $(\alpha)_D^{24} - 54$ ° (C = 0.7, THF). A second crop of crystals (120 mg), mp 110–112 °C, was obtained from the mother liquor (Sih et al., 1975).

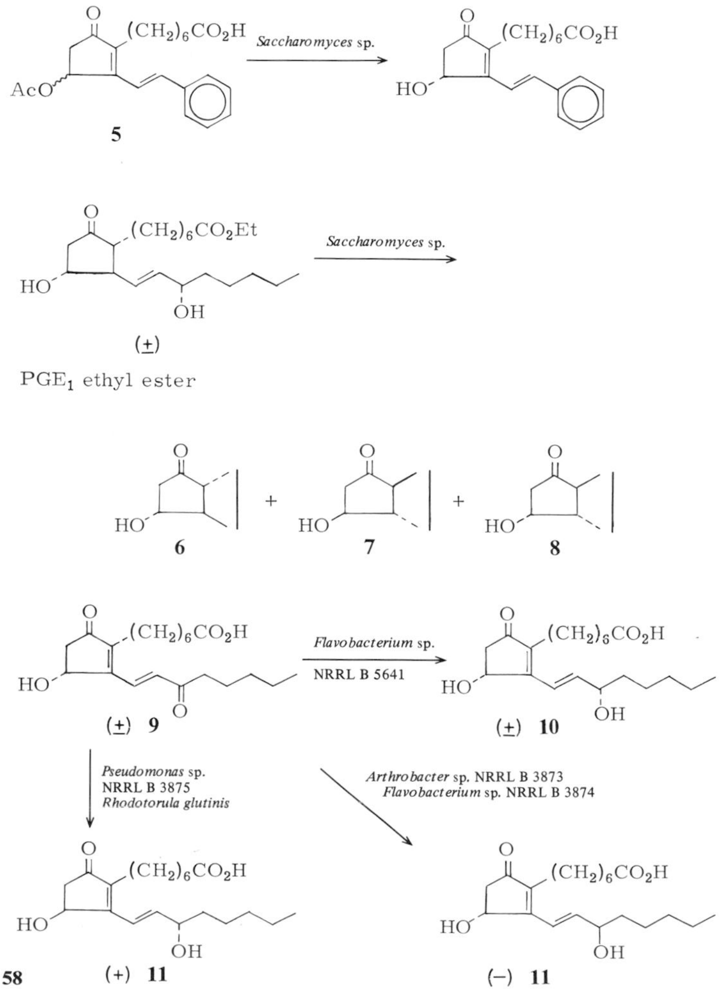

O, $(CH_2)_6CO_2H$, AcO, **5** — *Saccharomyces* sp. → O, $(CH_2)_6CO_2H$, HO

O, $(CH_2)_6CO_2Et$, HO, OH, (±) PGE_1 ethyl ester — *Saccharomyces* sp. →

HO **6** + HO **7** + HO **8**

O, $(CH_2)_6CO_2H$, HO, O, (±) **9** — *Flavobacterium* sp. NRRL B 5641 → O, $(CH_2)_6CO_2H$, HO, OH, (±) **10**

Pseudomonas sp. NRRL B 3875 *Rhodotorula glutinis* ↓ O, $(CH_2)_6CO_2H$, HO, OH, (+) **11**

Arthrobacter sp. NRRL B 3873 *Flavobacterium* sp. NRRL B 3874 → O, $(CH_2)_6CO_2H$, HO, OH, (−) **11**

Scheme 58

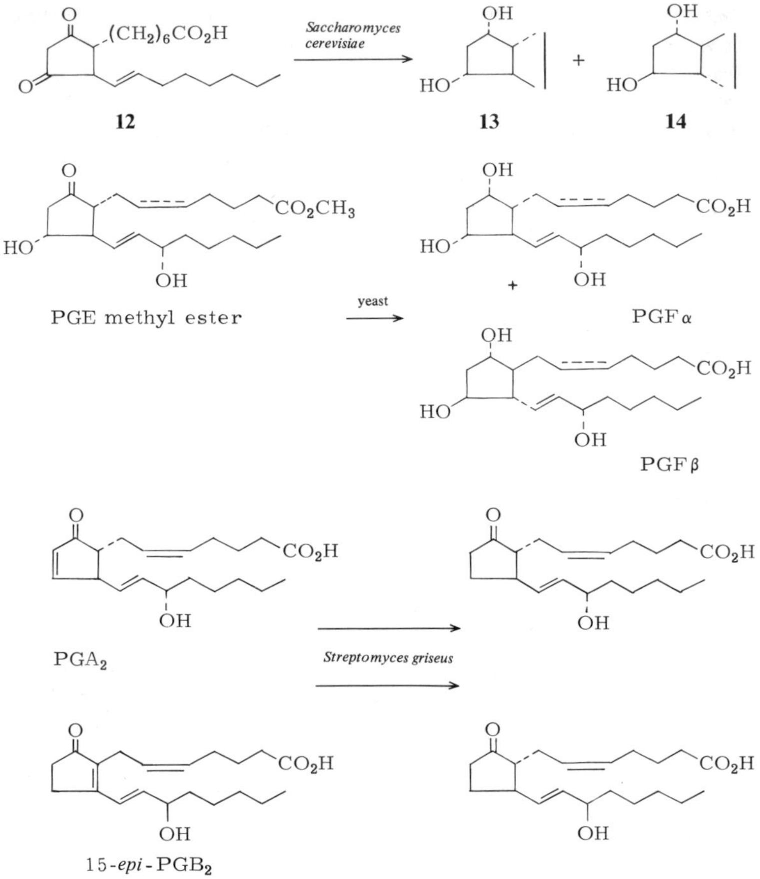

O
$(CH_2)_6CO_2H$
Saccharomyces cerevisiae
OH
HO
+
OH
HO
12
13
14

O
CO_2CH_3
HO
OH
PGE methyl ester
yeast
OH
CO_2H
HO
+
OH
PGF α
OH
CO_2H
HO
OH
PGF β

O
CO_2H
OH
PGA_2
Streptomyces griseus
O
CO_2H
OH
O
CO_2H
OH
15-*epi*-PGB_2
O
CO_2H
OH

Scheme 59

O
CO_2H
OH
PGA_2
Dactylium dendroides
O
CO_2H
OH
15
+
+
O
CO_2H
O
16
+
O
CO_2H
O
17

OH
CO_2H
HO
O
20
Saccharomyces cerevisiae
OH
CO_2H
HO
O

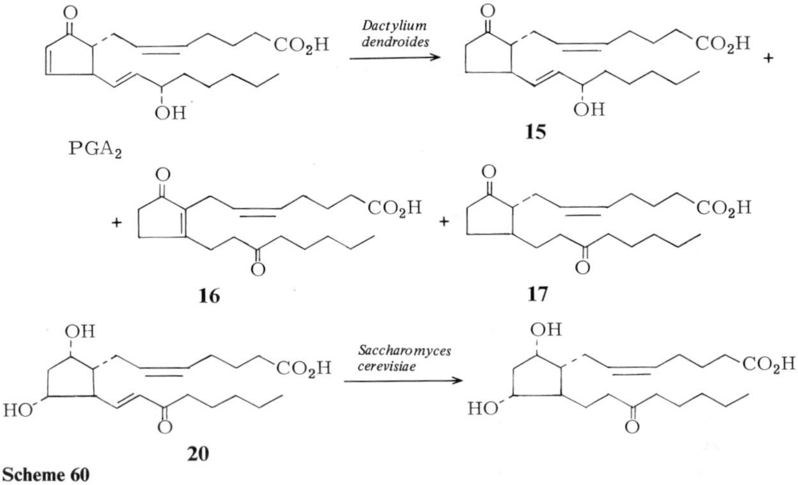

Scheme 60

Baker's yeast hydrolyzed (±)-deoxy-PGE_1 ethyl ester **3** and also gave the racemic free acid (SIH et al., 1972). Likewise, *Corynespora cassicola* CMI 56 007 hydrolyzed **18,** 15-*epi*-PGA_2, the acid function being methylated and the 15-alcohol acetylated (LEEMING and GREENSPAN, 1973).

The same reaction occurred with 15-*epi*-PGA_2 with only a methylated acid function **19** (GREENSPAN and LEEMING, 1975). The same most likely occurred with *Cladosporium resinae* ATCC 11 274 (cell-free system) which hydrolyzed the methyl ester of PGE_1 (AXEN and MURRAY, 1968). Selective hydrolysis has also been used to separate racemates. Thus, the prostaglandin analog **5** (Scheme 58) was hydrolyzed by *Saccharomyces* sp. NRRL Y 7 342 to the 11α-alcohol (COLTON et al., 1973). Similarly, the ethyl ester of (±)-PGE_1 formed three optically active compounds **6** (yield 22%), **7** (yield 23%), and **8** (yield 6%). Only **6** is of interest for subsequent synthetic steps.

Selective hydrogenation of ketones has been used to create chiral centers. Didehydro PGE_1 **9** was reduced by various microorganisms, such as *Flavobacterium* sp. NRRL B 5 641 which reduced racemic **9** to racemic **10,** as well as *cis*-(11α,15α)- and (11β,15β)-derivatives of **10** (MARSHECK and MIYANO, 1974a, b), while *Pseudomonas* sp. NRRL B 3 875 gave the D-*trans*-(11β,15α)-derivative (+) **11.** L-*trans*-(11α,15β) (−) **11** was obtained with *Arthrobacter* sp. NRRL B 3 873 and *Flavobacterium* NRRL B 3 874 (MIYANO et al., 1971) (MARSHECK and MIYANO, 1974b; COLTON et al., 1972). *Rhodotorula glutinis* NRRL Y 842 also formed a *trans*-derivative (COLTON et al., 1972).

The reduction of 9- and 11-keto groups can also lead to optically active derivatives.

Thus, 15-deoxydiketone **12** (Scheme 59) gave deoxy-$PGF_{1\alpha}$ **13** and the derivative **14** (SIH and ROSAZZA, 1976).

Finally, the racemic methyl ester of PGE_2 underwent a double conversion with yeasts. The ester function was hydrolyzed yielding (+)-$PGF_{2\alpha}$ and (+)-$PGF_{2\beta}$. The same reaction occurs with PGE_1 methyl ester (SCHNEIDER and MURRAY, 1973).

The second reaction type involves reduction of carbon-carbon double bonds. These reactions mainly occur with compounds of the A-series, but double bonds other than those at positions 10(11) also can be reduced.

A large number of strains reduced PGA_2 (JIU et al., 1973) forming compounds of the 11-deoxy E-series. *Streptomyces griseus* performed this conversion and an analogous reduction of 15β-PGB_2 (GREENSPAN and LEEMING, 1976). In addition, *Dactylium dendroides* NRRL 2 575 gave **15, 16,** and **17** with PGA_2 through reduction of the 10(11) double bond (Scheme 60). **16** is a B-series compound saturated at 13(14) and with the hydroxyl oxidized to a ketone. In **17** the 10(11) and 13(14) double bonds have been reduced and the 15-alcohol dehydrogenated (HSU et al., 1975; JIU et al., 1973).

It was shown above (Scheme 57) that the methyl ester of 15-*epi*-PGA_2 **19** was converted to **4** by *Corynespora cassicola;* the ester was hydrolyzed and the 10(11) double bond reduced (GREENSPAN and LEEMING, 1975). *C.cassicola* gave an analogous reaction with **18** (R = Ac) (LEEMING and GREENSPAN, 1973). The 13(14) double bond of **20** (15-keto-$PGF_{2\alpha}$) was reduced by *Saccharomyces cerevisiae* (yield 46%) (SEBEK et al., 1976).

3. Degradation and metabolism of prostaglandins

Two types of reactions are encountered: 18-, 19-, or 20-hydroxylation and β-oxidation on the 7-membered chain. However, other reactions occur simultaneously, e. g., reduction of 10(11) or 13(14) double bonds, hydrolysis of esters, and oxidation of hydroxyls to ketones.

Cunninghamella blakesleeana ATCC 9 245 converted PGA_2 to the 10(11)-hydrogenated derivative **1** (Scheme 61) (X = H) and **1′** (X = OH), which is additionally 18-hydroxylated (HSU et al., 1974). *Streptomyces ruber* converted the 15-acetic ester of PGA_2 to **2** and **2′**, hydroxylation occurring at position 19 (MARSHECK and MIYANO, 1975).

Penicillium M. 8 904 also reduced PGA_2 to **1** (X = H) but also degraded the 7-membered side chain to form **3**. The same *Penicillium* also converted PGB_2 **4** (Scheme 62) (X = Y = H) to the 18-hydroxy derivative **4** (X = OH, Y = H) and the 19-hydroxy derivative **4** (X = H, Y = OH). A third product resulted from the shortening of the acid-bearing chain to 3 carbon atoms (HSU et al., 1977). *Aspergillus niger* ATCC 9 142 20-hydroxylated PGB_2 **4** (MARX and DOODEWARD, 1978). The racemic prostaglandin analog **5** (n = 6) was degraded by *Penicillium* sp. M. 8 904 to racemic **5** (n = 4). *Mortierella* sp. M. 8 872 gave an active product (11*S*) **5** (n = 2) (HSU et al., 1977).

E-series prostaglandins have also been hydroxylated. *Streptomyces* sp. CBS 18 874 hydroxylated PGE_1 at positions 18 and 19 (MARX and DOODEWARD, 1978) (Scheme 63). *Streptomyces* sp. UC 5 761, on the other hand, converted PGE_2 **6** (X = Y = H) to a more complex mixture. All the compounds were hydroxylated at position 18 (X = OH, Y = H) or 19 (X = H, Y = OH), but they belong either to series E or A (**7**), or **8** (SEBEK et al., 1976).

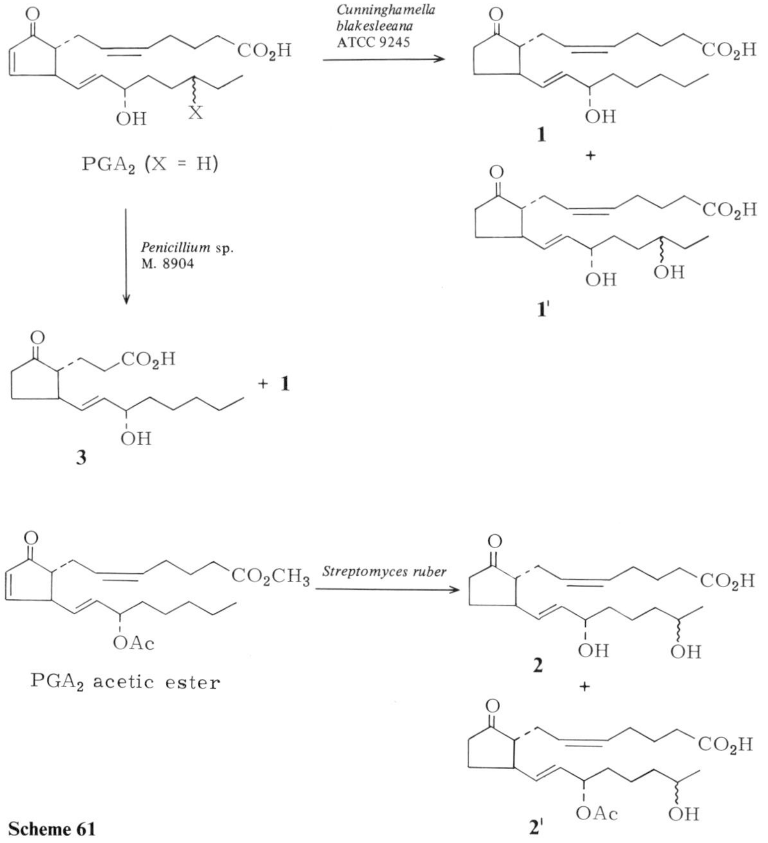

PGA_2 (X = H) —*Cunninghamella blakesleeana* ATCC 9245→ **1** + **1'**

PGA_2 (X = H) —*Penicillium* sp. M. 8904→ **3** + **1**

PGA_2 acetic ester —*Streptomyces ruber*→ **2** + **2'**

Scheme 61

With $PGF_{2\alpha}$ results were simpler where 18- or 19-hydroxylation occurred either with *Streptomyces* CBS 18 871 (MARX and DOODEWARD, 1978), or with *Streptomyces* spp. UC 5 762 (SEBEK et al., 1976) (Scheme 64). *Mycobacterium rhodochrous* degraded a 6-keto analog of $PGF_{1\alpha}$ **9** (n = 4) to **9** (n =

2) via β-oxidation. **10** was also obtained as a result of reduction of the 13(14) double bond and dehydrogenation of the 15-alcohol group (SUN et al., 1980). Finally, we should mention the bioconversion of racemic derivatives of the 11-deoxy E-series **11** and **12** (Scheme 65) by *Microascus trigonosporus* NRRL 1199. The *cis*-derivative **11** gave 35 and 19% of 18- and 19-hydroxylated derivatives, respectively. The *trans*-isomer **12** gave 33 and 20% of the corresponding analogs. In the latter case, the compounds obtained were active with the non-natural configuration (8*S*,12*R*) (LANZILOTTA et al., 1976).

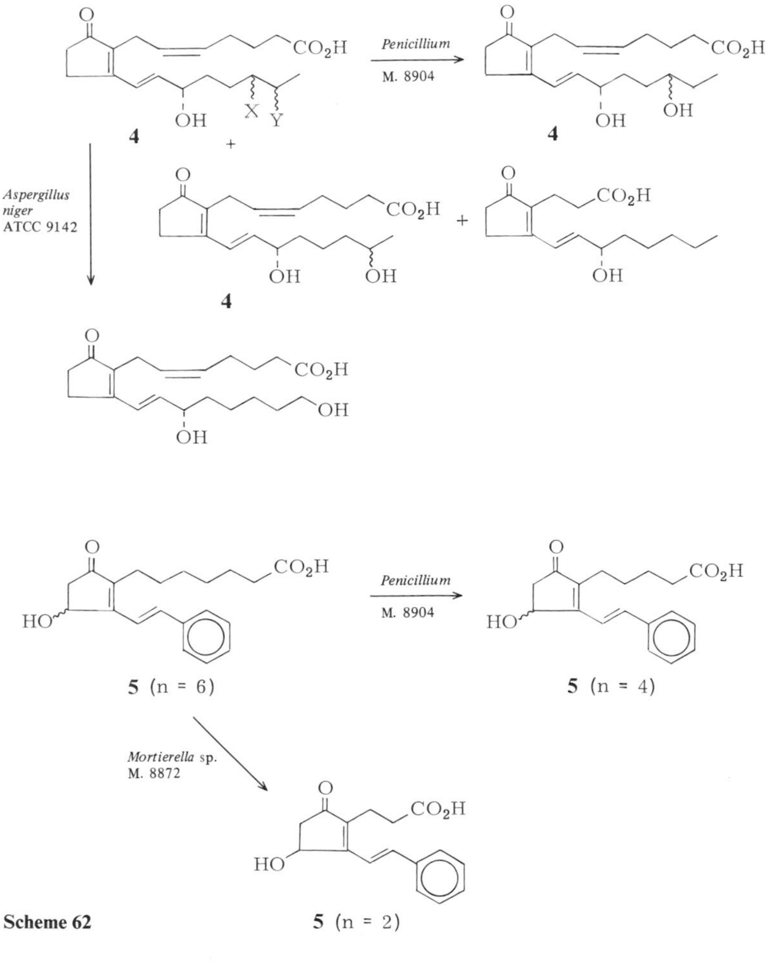

Scheme 62

G. Biotransformations of Cannabinoids

The naturally occurring cannabinoids possess a number of interesting pharmacological activities and their metabolism is of considerable importance in view of its possible role in pharmacology. As mentioned before, microbial systems show a considerable potential to serve as models for studying drug metabolism, and in particular, for the preparation of specific metabolites for further pharmacological studies. Biotransformation studies have been performed on

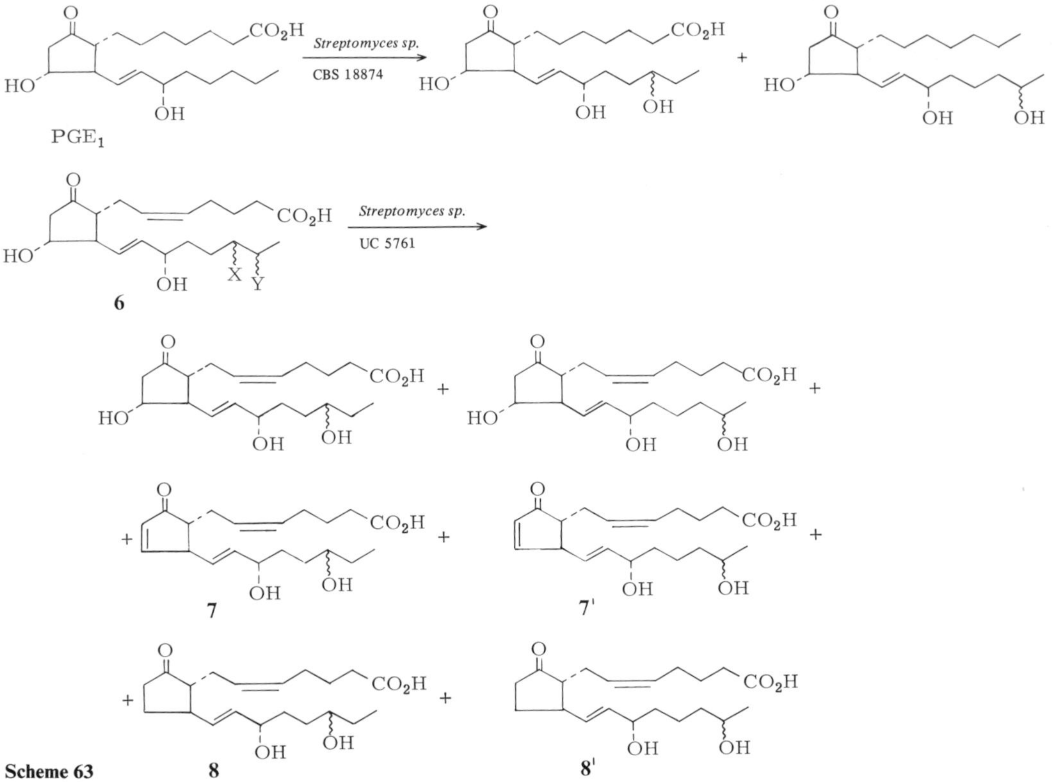

PGE_1 → *Streptomyces sp.* CBS 18874

6 → *Streptomyces sp.* UC 5761 → 7, 7', 8, 8'

Scheme 63

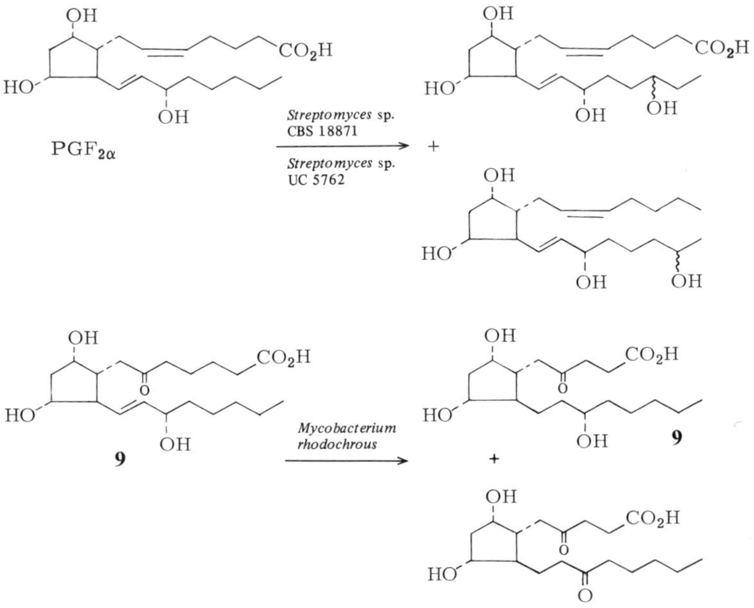

$PGF_{2\alpha}$ → *Streptomyces* sp. CBS 18871 / *Streptomyces* sp. UC 5762

9 → *Mycobacterium rhodochrous* → 9 + 10

Scheme 64

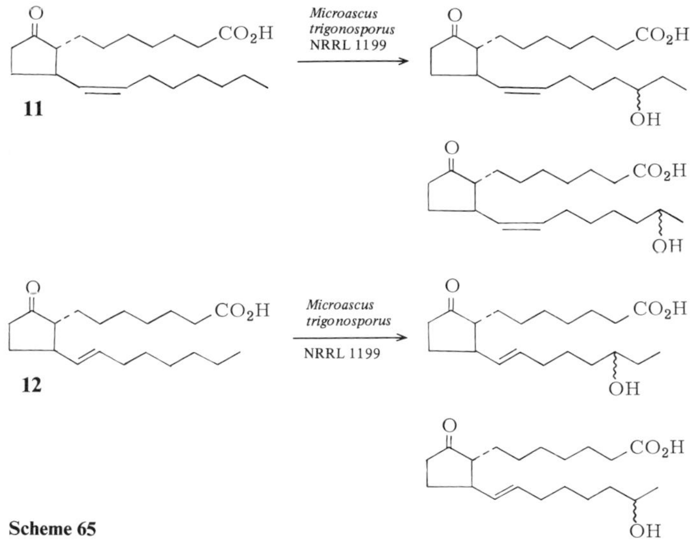

O CO$_2$H

11

Microascus trigonosporus NRRL 1199

O CO$_2$H OH

O CO$_2$H OH

O CO$_2$H

12

Microascus trigonosporus NRRL 1199

O CO$_2$H OH

O CO$_2$H OH

Scheme 65

OH NO$_2$ Pe

39

OH NO$_2$ Pe NO$_2$

40

Pseudomonas putida

9 8 7 10 OH 1 2 3 4 O 2' 4' 1' 3' 5'

1 Cannabinol CBN

Mycobacterium rhodochrous ATCC 19067

OH O CO$_2$H

6

\+

OH CO$_2$H OH

7

\+

\+ OH CO$_2$H

8

\+

OH CO$_2$H

9

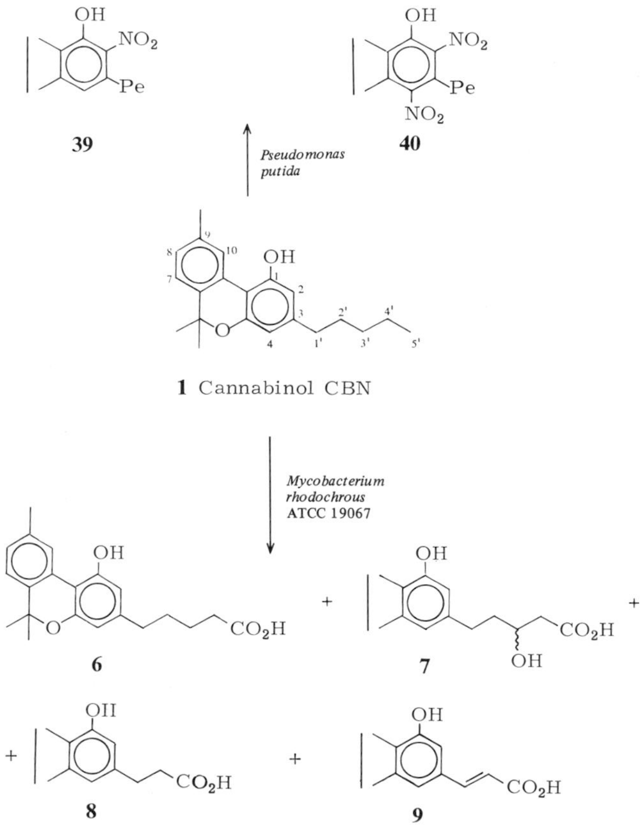

Scheme 66

cannabinoids with several strains of microorganisms. A general review was published recently by ROBERTSON (1982).

The substrates used in those studies were cannabinol (CBN) **1** (Scheme 66), cannabidiol (CBD) **2**, $\Delta^{6a\text{-}10a}$-tetrahydrocannabinol (Δ^{6a}-THC) **3**, as well as Δ^{8}- and Δ^{9}-tetrahydrocannabinol (Δ^{8}-THC, Δ^{9}-THC), **4** and **5**. The numbering for the substrates is given for **1** and **2**. The numbering of **3**, **4**, and **5** is the same as for **1**.

The other identifiable products are 3′-hydroxy-CBN-5′-oic acid **7**, CBN-3′-oic acid **8**, and the 1′,2′-unsaturated CBN-3′-oic acid **9** (major metabolites) (ROBERTSON et al., 1978).

The degradation of cannabidiol CBD **2** (Scheme 67) was carried out with *Syncephalastrum racemosum* ATCC 18 192, a fungal organism used industrially in steroid hydroxylations. The most abundant product was 4″,5″-bis-nor-3″-hydroxy-CBD **10**.

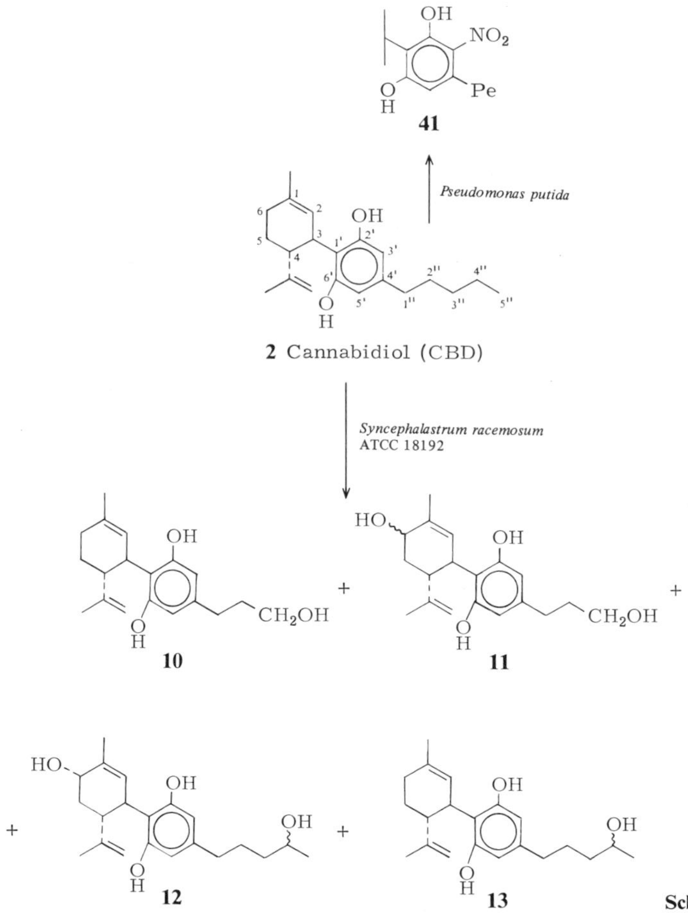

Scheme 67

Mycobacterium rhodochrous ATCC 19 067 is known to oxidize hydrocarbons. It degrades the side chain of CBN **1** by terminal oxidation to the 5′-carboxylic acid **6**, followed by 1 or 2 cycles of β-oxidation.

Other products were compound **11** which is analogous to **10** hydroxylated at C-6, **12** hydroxylated at positions 6 and 4″, and **13** hydroxylated at 4″ only (ROBERTSON et al., 1978).

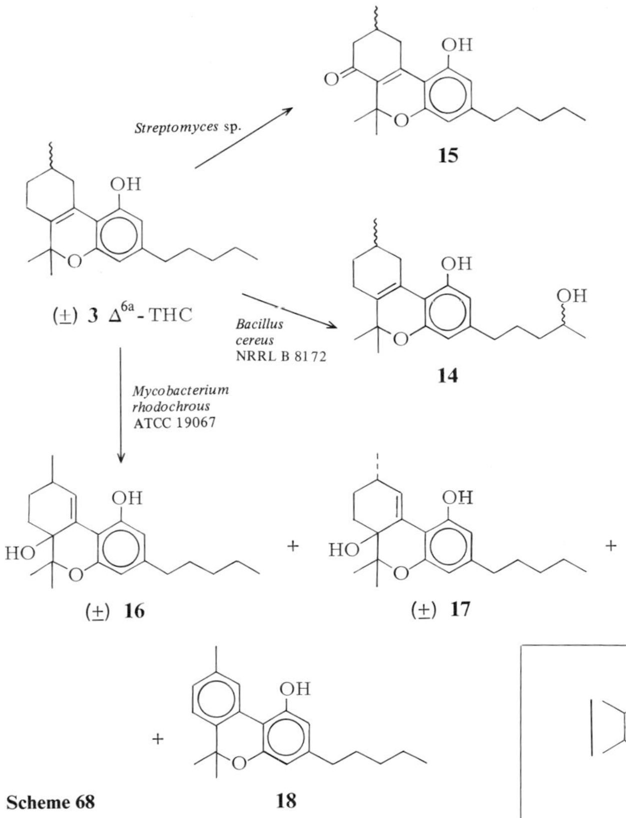

Streptomyces sp.

15

(±) 3 Δ^{6a}-THC

Bacillus cereus NRRL B 8172

14

Mycobacterium rhodochrous ATCC 19067

(±) 16 + (±) 17 +

+ 18

Scheme 68

42

43

Pseudomonas putida

4 Δ^{8}-THC

Pellicularia filamentosa ATCC 13289

19 + 20 +

+ 21 + 22

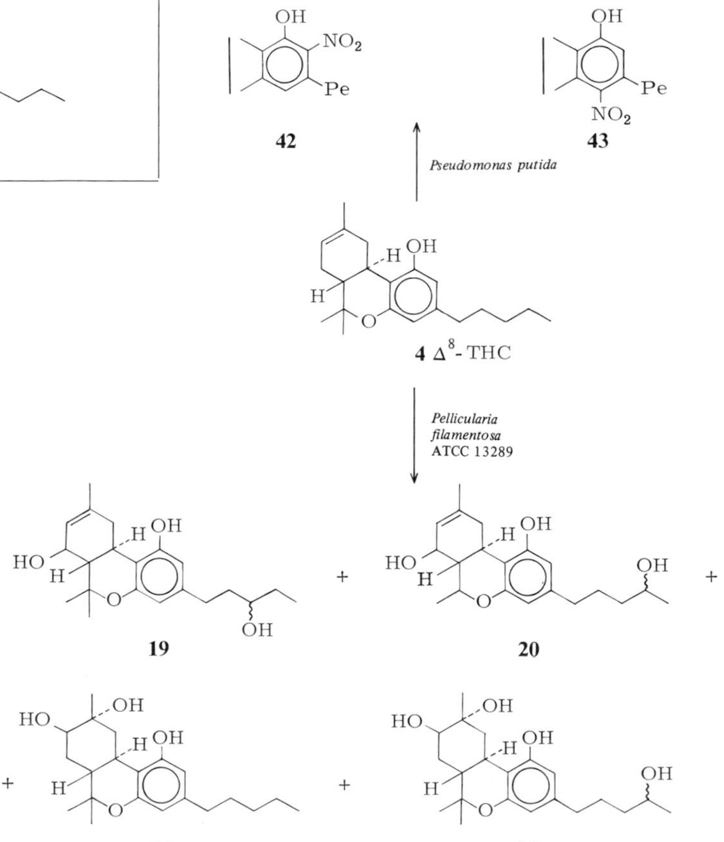

Scheme 69

Biotransformation of Δ^{6a-10a}-tetrahydrocannabinol (Δ^{6a}-THC **3**) (Scheme 68), an easily available synthetic compound, occurred with different strains. *Bacillus cereus* NRRL B 8 172 transformed **3** to **14** by primary oxidation of the side chain at C-4′. Ketone formation (**15**) was observed with *Streptomyces* sp. A 41 596 by attack on the cyclohexane ring at C-7. *Mycobacterium rhodochrous* ATCC 19 067 appeared to oxidize the cyclohexane ring of **3**. Two optically inactive allylic compounds, the **16** (6aβ,9β) and **17** (6aβ,9β) diastereoisomers were obtained. In addition, the same strain led to an aromatization of the cyclohexane ring yielding **18** (Abbott et al., 1977; Fukuda et al., 1977).

Hydroxylation of Δ^8-tetrahydrocannabinol (Δ^8-THC), (Scheme 69), **4**, was studied with some strains. *Pellicularia filamentosa*

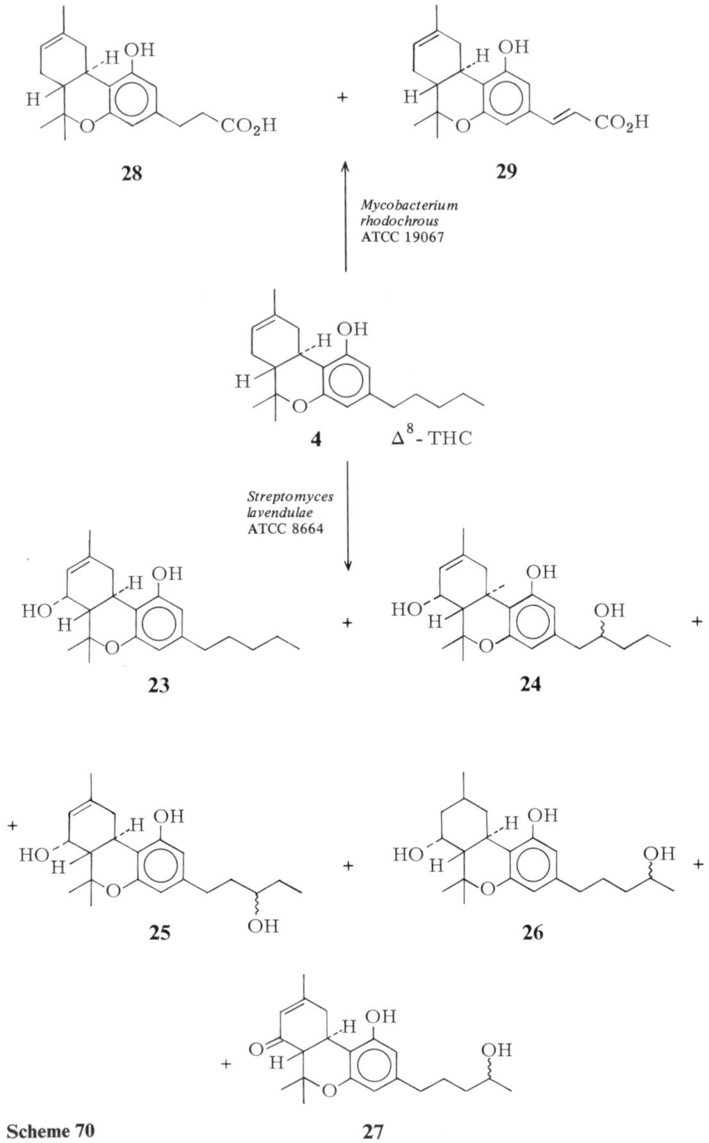

28 + 29

Mycobacterium rhodochrous ATCC 19067

4 Δ^8-THC

Streptomyces lavendulae ATCC 8664

23 + 24 + 25 + 26 + 27

Scheme 70

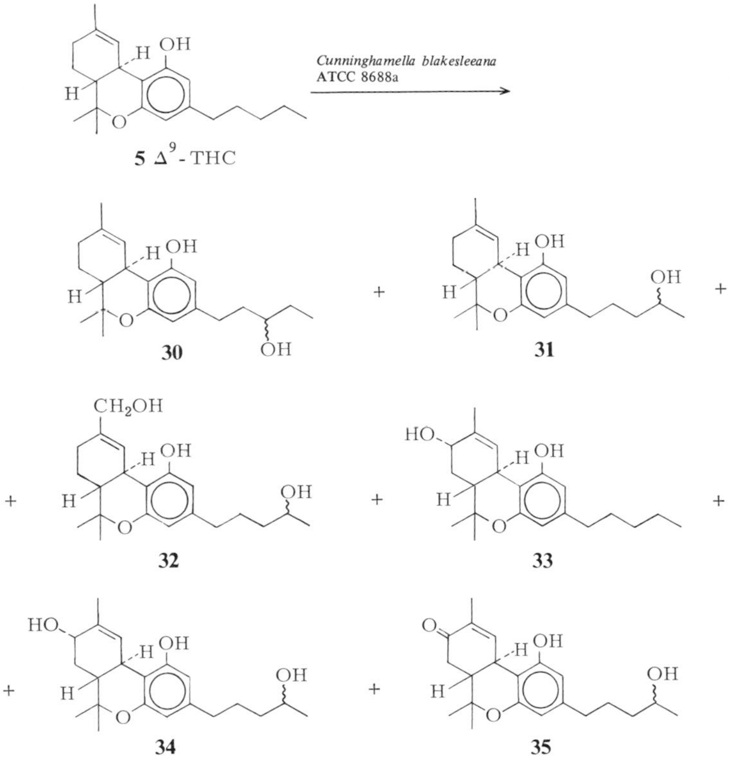

Scheme 71

spp. *sasakii* ATCC 13 289 hydroxylated both the cyclohexane ring and the side chain. **19** and **20** were hydroxylized at C-7 and C-3′, and at C-7 and C-4′, respectively. In **21** and **22** the double bond at C-8(9) of **4** was transformed to a diol and additionally, C-4′ of **22** was hydroxylated.

Streptomyces lavendulae ATCC 8 664 gave five metabolites (Scheme 70): **23, 24, 25,** and **26** hydroxylated at 7α-position, **27,** bearing a ketone function at C-7. Furthermore, hydroxylation of the side chain was observed at 2′-, 3′-, and 4′-positions for **24, 25,** and **26,** respectively. Hydroxylation at C-4′ was observed for **27** (VIDIC et al., 1976).

The transformation of Δ^8-THC **4** was studied with *Mycobacterium rhodochrous* (ROBERTSON et al., 1978). The major product from incubation of **4** was Δ^8-THC-3′-oic acid **28,** and small amounts of 1′,2′-unsaturated Δ^8-THC-3′-oic acid **29.**

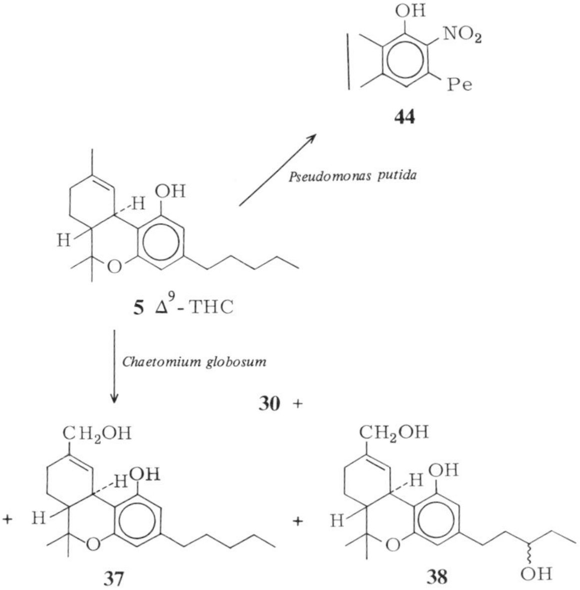

Scheme 72

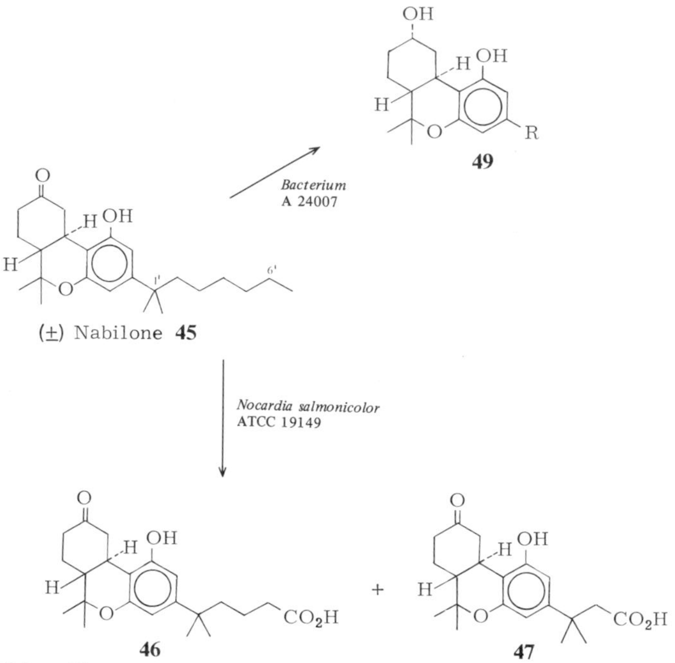

Scheme 73

Finally, Δ^9-THC **5** (Scheme 71) was studied. Incubation of (−)-Δ^9-THC **5** with stationary cultures of *Cunninghamella blakesleeana* ATCC 8 688a yielded a number of metabolic conversion products (BINDER, 1976). Hydroxylations leading to products **30** and **31** occurred only on the side chain at C-3′ and C-4′, respectively. In metabolite **32** hydroxylation occurred at C-11 and on the side chain at C-4′. Metabolites **33, 34,** and **35** were hydroxylated at C-8 on the cyclohexane ring (ketone function in **35**) and, the C-4′ located on the side chain was hydroxylated in **34** and **35.**

CHRISTIE et al. (1978) investigated the metabolism of (−)Δ^9-THC **5** by a strain of *Chaetomium globosum* (Scheme 72). **5** can be hydroxylated at C-11 to **37,** at C-3′ to **30,** and at both C-11 and C-3′ to **38.** The identification of **37** is the first demonstration of the production of strongly psychoactive mammalian metabolites of Δ^9-THC by a microorganism.

A nitration reaction of aromatic rings of some cannabinoids was described with *Pseudomonas putida* (TSAI, 1980). In all cases, nitration was observed at C-2 (eventually identical with 3′) and C-4.

With cannabinol, **1**, nitration occurs at C-2 leading to **39** and at C-2 and C-4 to form **40.** With cannabidiol, **2**, nitration occurs at C-3 (equivalent to C-2) yielding only **41.** Δ^8-THC 4 is nitrated at C-2 and C-4 into **42** and **43,** and Δ^9-THC **5** at C-2 forming **44.** Finally, some microorganisms were screened for their ability to modify the synthetic cannabinoid nabilone **45** (ABBOTT et al., 1977; FUKUDA et al., 1977) (Scheme 73). From a culture of *Nocardia salmonicolor* ATCC 19 149, the carboxylic acids **46** and **47** were isolated: the alkyl side chain was either two or four carbon atoms shorter than the side chain of the starting substrate. These acids may arise from β-oxidation.

Diastereoisomeric 9*S*-alcohols were formed by stereospecific reduction of the 9-ketone of racemic nabilone **45** (Scheme 74). Compound **48** arises from the reduction of the 6a*S*,10a*S*-isomer by *Bacillus cereus* NRRL B 8 172. Compound **49** results from the reduction of the 6a*R*,10a*R*-isomer by *Bacterium* A 24 009. Other products of the

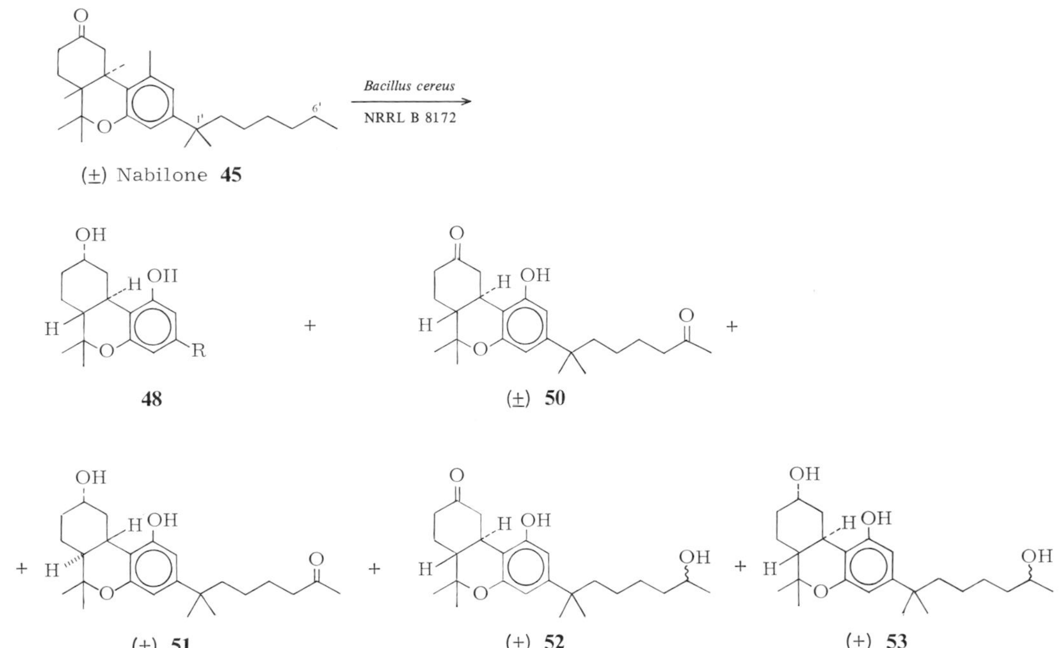

Bacillus cereus
NRRL B 8172

(±) Nabilone **45**

48 (±) **50**

(±) **51** (±) **52** (±) **53**

Scheme 74

attack by *Bacillus cereus* NRRL B 8 172 correspond to the oxidation of the side chain at C-6′. Compound **50** is a diketone; **51** is 9*S*-hydroxy-6-oxo; **52** is the reverse 9-oxo-6′-hydroxy; **53** is the 9*S*,6′-dihydroxy compound.

IV. References

ABBOTT, B. J., FUKUDA, D. S., and ARCHER, R. A. (1977). Experienta *33,* 718.

AKAMATSU, S. (1923). Biochem. Z. *142,* 188.

AKLIN, W., and PRELOG, V. (1959). Helv. Chim. Acta *42,* 1239.

AKLIN, W., PRELOG, V., and ZÄCH, D. (1958a). Helv. Chim. Acta *41,* 1428.

AKLIN, W., DÜTTING, D., and PRELOG, V. (1958b). Helv. Chim. Acta *41,* 1424.

AKLIN, W., PRELOG, V., and PRIETO, A. P. (1958c). Helv. Chim. Acta *41,* 1416.

AKLIN, W., PRELOG, V., and SERDAREVIC, B. (1963). Helv. Chim. Acta *46,* 2440.

AKLIN, W., PRELOG, V., SCHENKER, F., SERDAREVIC, B., and WALTER, P. (1965). Helv. Chim. Acta *48,* 1725.

AMBRUS, G., SZENTIRMAI, A., MEHESFALVI, C., KOVACS G., SZANTAI, C., and NOVAK, L. (1974). Hung. Teljes 8066; Chem. Abstr. *81,* 49 337 (1974).

ANDERSON, M. S., HALL, R. A., and GRIFFIN, M. (1980). J. Gen. Microbiol. *120,* 89.

ARAI, Y., and YAMADA, K. (1969). Agric. Biol. Chem. *33,* 63.

ARNAUD, A., GALZY, P., and JALLAGEAS, J. C. (1976). C. R. Acad. Sci. Ser. D. *283,* 571.

AURET, B. J., BOYD, D. R., and HENBEST, H. B. (1968). J. Chem. Soc., 2 374.

AXEN, U. F., and MURRAY, H. C. (1968). Ger. Offen. 1937678 (1970), US Prior. (1968); Chem. Abstr. *72,* 109 830 (1970).

BALLAL, N. R., BHATTACHARYYA, P. K., and RANGACHARI, P. N. (1968). Ind. J. Biochem. *5,* 1.

BARKER, H. A. (1943). J. Bacteriol. *45,* 251.

BAUMANN, P., and PRELOG, V. (1958a). Helv. Chim. Acta *41,* 2 362.

BAUMANN, P., and PRELOG, V. (1958b) Helv. Chim. Acta *41,* 2 379.

BAUMANN, P., and PRELOG, V. (1959). Helv. Chim. Acta *42,* 736.

BEIJERINK, M. W. (1911). Proc. Acad. Wetenschapen *13,* 1066.

BELLET, P., and VAN THUONG, T. (1965). US Patent 3 432 293 (1969). Fr. Prior. (1965). Neth. Appl. 6 605 744 (1966). Chem. Abstr. *67,* 2 097 (1967).

BELLET, P., NOMINÉ, G., and MATHIEU, J. (1966). C. R. Acad. Sci., *263,* 88.

BERNHAUER, K., and WAELSCH, H. H. (1932). Biochem. Z. *249,* 223.

BHATTACHARYYA, P. K., and GANAPATHY, K. (1965). Ind. J. Biochem. *2,* 137.

BINDER, M. (1976). Helv. Chim. Acta *59,* 1674.

BLAKLEY, E. R. (1978). Can. J. Microbiol, *24,* 847.

BLAKLEY, E. R., and PAPISH, B. (1982). Can. J. Microbiol. *28,* 1 037.

BLANEY, F., JOHNSTON, D. E., MC KERVEY, M. A., JONES E. R. H., and PRAGNELL, J. (1974). J. Chem. Soc. Chem. Comm., 297.

BOLSMAN, T. A. B., and BAILEY, M. L. (1982). Eur. Patent 54 987.

BROOKS, D. W., GROTHAUS, P. G., and IRWIN, W. L. (1982). J. Org. Chem. *47,* 2 820.

BUTKEWITSCH, W. (1924). Biochem. Z. *145,* 442.

BUTKEWITSCH, W. (1925). Biochem. Z. *159,* 408.

CAPEK, A., TADRA, M., KAKAC, B., ERNEST, I., and PROTIVA, M. (1962). Folia Microbiol. (Prague) *7,* 253.

CAPEK, A., SVATEK, E., and FADRA, M. (1963). Folia Microbiol. (Prague) *8,* 304.

CARR, J. G., POLLARD, A., WHITING, C. G., and WILLIAMS, A. H. (1957). Biochem. J. *66,* 283.

CHOU, W. S., HUANG, T. C., and WANG, C. C. (1980) K'O Hsueh T'urg Pao *25,* 561; Chem. Abstr. *93,* 200 713 (1980).

CHRISTIE, R. M., RICKARDS, R. W., and WATSON, W. P. (1978). Aust. J. Chem. *31,* 1799.

COLLA, C., and TRECCANI, V. (1960). Ann. Microbiol. Enzimol. *10,* 77.

COLTON, F. B., MARSHECK, W. J., and MIYANO, M. (1972). US Patent 3 687 811; Chem. Abstr. *77,* 163 040 (1972).

COLTON, F. B., MARSHECK, W. J., and MIYANO, M. (1973). Ger. Offen. 2 318 594; Chem. Abstr. *80,* 25 943 (1974).

CRUMBIE, R. L., RIDLEY, D. D., and SIMPSON, G. W. (1977). J. Chem. Soc. Chem. Comm., 315.

DALTON, H., GOLDING, B. T., WATERS, B. W., HIGGINS, R., and TAYLOR, J. A. (1981). J. Chem. Soc. Chem. Comm., 482.

DAUM, S. J., RIANO, M. H., SHAW, PH. E., and CLARKE, R. L. (1967). J. Org. Chem. *32,* 1 435.

DAUPHIN, G., GRAMAIN, J. C., KERGOMARD, A., RENARD, M. F., and VESCHAMBRE, H. (1980a). Tetrahedron Lett., 4 275.

DAUPHIN, G., GRAMAIN, J. C., KERGOMARD, A., RENARD, M. F., and VESCHAMBRE, H. (1980b). J. Chem. Soc. Chem. Comm., 318.

DAVEY, J. F., and TRUDGILL, P. W. (1977). Eur. J. Biochem. *74,* 115.

DAVIS, J. B., and RAYMOND, R. L. (1959). US Patent 3 057 784; Chem. Abstr. *59,* 13 319 (1963).

DAVIS, J. B., and RAYMOND, R. L. (1961). Appl. Microbiol. *9,* 383.

DE KLERK, H., and VAN DER LINDEN, A. C. (1974). Antonie van Leeuwenhoek J. Microbiol. Serol. *40,* 7.

DEOL, B. S., RIDLEY, D. D., and SIMPSON, G. W. (1976). Aust. J. Chem. *29,* 2 459.

DESRUT, M., KERGOMARD, A., RENARD, M. F., and VESCHAMBRE, H. (1981). Tetrahedron *37,* 3 825.

DESRUT, M., KERGOMARD, A., RENARD, M. F., and VESCHAMBRE, H. (1983). Biochem. Biophys. Res. Comm. *110,* 908.

DONOGHUE, N. A., and TRUDGILL, P. W. (1975). Eur. J. Biochem. *60,* 1.

DONOGHUE, N. A., NORRIS, D. B., and TRUDGILL, P. W. (1976). Eur. J. Biochem. *63,* 175.

EMMERLING, O., and ABDERHALDEN, E. (1903). Zentralbl. Bakteriol. Parasitenkd. Infektionskr. Hyg. Abt. 2 *10,* 337.

FEINBERG, E. L., RAMAGE, P. I. N., and TRUDGILL, P. W. (1980). J. Gen. Microbiol. *121,* 507.

FERGUSON, J. A., and KORTE, F. (1977). Appl. Environ. Microbiol. *34,* 7.

FELDMAN, W. R., and PRELOG, V. (1958). Helv. Chim. Acta *41,* 2 396.

FISCHER, and WIEDEMANN (1935). Liebigs Ann. Chem. *520,* 52.

FONKEN, G. S., and JOHNSON, R. A. (1972). "Chemical Oxidations with Microorganisms". Marcel Dekker Inc. New York.

FONKEN, G. S., HERR, M. E., and MURRAY, H. C. (1964). US Patent 3 281 330; Chem. Abstr. *66,* 9 974 (1967).

FONKEN, G. S., HERR, M. E., MURRAY, H. C., and REINEKE, L. M. (1967). J. Am. Chem. Soc. *89,* 672.

FONKEN, G. S., HERR, M. E., MURRAY, H. C., and REINEKE, L. M. (1968a). J. Org. Chem. *33,* 3 182.

FONKEN, G. S., HERR, M. E., and MURRAY, H. C. (1968b). US Patent 3 392 171; Chem. Abstr. *69,* 58 586 (1968).

FONKEN, G. S., HERR, M. E., and MURRAY, H. C. (1970). Fr. Patent 1 585 856; Chem. Abstr. *66,* 9 974 (1967).

FUKUDA, D. S., ARCHER, R. A., and ABBOTT, B. J. (1977). Appl. Environ. Microbiol. *33*, 1134.

FURSTOSS, R., ARCHELAS, A., WAEGELL, B., LE PETIT, J., and DEVEZE, L. (1980). Tetrahedron Lett., 451.

FURSTOSS, R., ARCHELAS, A., WAEGELL, B., LE PETIT, J., and DEVEZE, L. (1981). Tetrahedron Lett., 445.

GANAPATHY, K., KHANCHANDANI, K. S., and BATTACHARYYA, P. K. (1966). Ind. J. Biochem. *3*, 66.

GIBIAN, H., KIESLICH, K., KOSMOL, H., RUFER, C., and SCHRÖDER, E. (1965). Ger. Offen. 1 493 171. Neth. Appl. 6 612 878 (1967); Chem. Abstr. *67*, 72 459 (1967).

GIBIAN, H., KIESLICH, K., KOCH, H. J., KOSMOL, H., RUFER, C., SCHRÖDER, E., and VOSSING, R. (1966). Tetrahedron Lett., 2 321.

GIBSON, D. T., KOCH, J. R., SCHULD, C. L., and KALLIO, R. E. (1968). Biochemistry *7*, 3 795.

GOLDBERG, J. S., RAUCKMAN, E. J., and ROSEN, G. M. (1977). Biochem. Biophys. Res. Comm. *79*, 198.

GOLOVLEVA, L. A., GOLOVLEV, E. L., ZYAKUN, A. M., SHURUKIN, YU. V., and FINKELSTEIN, Z. I. (1978). Izv. Akad. Nauk SSSR Ser. Biol., 44.

GONZALEZ, A. G., FAGUNDO, C. R., PADRON, M. P., and TRUJILLO, J. M. (1976). An. Quim. *72*, 565.

GREENSPAN, G., and LEEMING, M. R. G. (1975). US Patent 3 928 134; Chem. Abstr. *84*, 103 848 (1976).

GREENSPAN, G., and LEEMING, M. R. G. (1976). US Patent 3 930 952; Chem. Abstr. *85*, 31 622 (1976).

GREENSPAN, G., RUELIUS, H. W., and ALBURN, H. E. (1967). US Patent 3 453 179; Chem. Abstr. *71*, 79 747 (1969).

GREENSPAN, G. N., BUZBY, G. C., and BETON, J. L. (1968). Ger. Offen. 1 922 035 (1969). US Prior. (1968). Chem. Abstr. *72*, 65 395 (1970).

GRIFFIN, M., and TRUDGILL, P. W. (1972). Biochem. J. *129*, 595.

HANN, R. M., TILDEN, E. B., and HUDSON, C. S. (1938). J. Am. Chem. Soc. *60*, 1 201.

HASEGAWA, Y., OBATA, H., and TOKUYAMA, T. (1980). J. Ferment. Technol. *58*, 215; Chem. Abstr. *93*, 65 700 (1980).

HASEGAWA, Y., HAMANO, H., OBATA, H., and TOKUYAMA, T. (1982). Agric. Biol. Chem. *46*, 1 139.

HEIDEPRIEM, H., RUFER, C., KOSMOL, H., SCHRÖDER, E., and KIESLICH, K. (1968). Liebigs Ann. Chem. *712* 155.

HERITAGE, A. D., and MAC RAE, I. C. (1977). Appl. Environ. Microbiol. *33*, 1 295.

HERR, M. E., JOHNSON, R. A., MURRAY, H. C., REINEKE, L. M., and FONKEN, G. S. (1968). J. Org. Chem. *33*, 3 201.

HERR, M. E., JOHNSON, R. A., KRUEGER, W. C., MURRAY, H. C., and PSCHIGODA, L. M. (1970). J. Org. Chem. *35*, 3 607.

HERR, M. E., MURRAY, H. C., and FONKEN, G. S. (1972a). US Patent 3 649 453; Fr. Patent 1 511 936 (1969); Chem. Abstr. *70*, 87 149 (1969).

HERR, M. E., MURRAY, H. C., and FONKEN, G. S. (1972b). US Patent 3 706 801.

HESLER, P., and RAYMOND, R. L. (1973). US Patent 3 748 231.

HÖRHOLD, C., BÖHME, K. H., and SCHUBERT, K. (1969). Z. Allg. Mikrobiol *9*, 235.

HSU, C. F., JIU, J., and MIZUBA, S. S. (1974). US Patent. 3 856 852; Chem. Abstr. *82*, 139 489 (1975).

HSU, C., JIU, J., and MIZUBA, S. S. (1975). US Patent 3 868 412; Chem. Abstr. *83*, 129 994 (1975).

HSU, C. F., JIU, J., and MIZUBA, S. S. (1977). Dev. Ind. Microbiol. *18*, 487.

IGUCHI, T., KAWAMURA, T., and TAKESHI, T. (1964). (Asahi. Chem. Ind.) Japan Patent 7 226 312; Chem. Abstr. *73*, 108 243 (1970).

IMELIK, B. (1948). C. R. Acad. Sci. *226*, 2 082.

ISONO, M., TAKAHASHI, T., YAMASAKI, Y., and MIKI, T. (1967). Brit. Patent 1 230 455 (1968). Jap. Prior. (1967). Fr. Patent 1 565 986 (1968). Chem. Abstr. *72*, 90 738 (1970).

IWAHARA, S., TAKASAWA, S., and OGATA, K. (1966a). Agric. Biol. Chem., *30*, 385.

IWAHARA, S., TATSUROKURO, T., and OGATA, K. (1966b). Agric. Biol. Chem. *30*, 1 069.

IWAHARA, S., MC CORMICK, D. B., WRIGHT, L. D., and LI, H. C. (1969). J. Biol. Chem. *244*, 1 393.

IZUMI, Y., TANI, Y., and OGATA, K. (1973). Biochim. Biophys. Acta *326*, 485.

JALLAGEAS, J. C., ARNAUD, A., and GALZY, P. (1979). C. R. Acad. Sci. Ser. D *288*, 655.

JIU, J., HSU, C. F., and MIZUBA, S. (1973). Dev. Ind. Microbiol. *15*, 345.

JOHNSON, R. A., HERR, M. E., MURRAY, H. C., and FONKEN, G. S. (1968a). J. Org. Chem. *33*, 3 187.

JOHNSON, R. A., MURRAY, H. C., REINEKE, L. M., and FONKEN, G. S., (1968b). J. Org. *33*, 3 207.

JOHNSON, R. A., HERR, M. E., MURRAY, H. C., REINEKE, L. M., and FONKEN, G. S. (1968c). J. Org. Chem. *33*, 3 195.

JOHNSON, R. A., MURRAY, H. C., REINEKE,

L. M., and FONKEN, G. S. (1969a). J. Org. Chem. *34,* 2 279.
JOHNSON, R. A., MURRAY, H. C., and REINEKE, L. M. (1969b). J. Org. Chem. *34,* 3 207.
JOHNSON, R. A., HERR, M. E., MURRAY, H. C., and FONKEN, G. S. (1970). J. Org. Chem. *35,* 622.
JOHNSON, R. A., HERR, M. E., MURRAY, H. C., and FONKEN, G. S. (1971). J. Am. Chem. Soc. *93,* 4 880.
KABUTO, K., IMUTA, M., KEMPNER, E. S., and ZIFFER, H. (1978). J. Org. Chem. *43,* 2 357.
KAJIWARA, T., SASAKI, Y., KIMURA, F., and HATANAKA, A. (1981). Agric Biol. Chem. *45,* 1 461.
KANEDA, T. (1974). Biochem. Biophys. Res. Comm. *58,* 140.
KAWAI, K., IMUTA, M., and ZIFFER, H. (1981). Tetrahedron Lett., 2 527.
KERGOMARD, A., RENARD, M. F., and VESCHAMBRE, H. (1978). Tetrahedron Lett., 5 197.
KERGOMARD, A., RENARD, M. F., and VESCHAMBRE, H. (1982a). J. Org. Chem. *47,* 792.
KERGOMARD, A., RENARD, M. F., and VESCHAMBRE, H. (1982b). Agric. Biol. Chem. *46,* 97.
KIESLICH, K. (1976). "Microbial Transformation of Non-steroid Cyclic Compounds". G. Thieme Verlag, Stuttgart.
KIESLICH, K., RADUECHEL, B., SKUBALLA, W., VORBRUEGGEN, H., and DAHL, H. (1980). Ger. Offen. 2 853 637; Chem. Abstr. *93,* 130 606 (1980).
KISHIMOTO, F., SUZUKI, K., OGINO, S., and YAMAMOTO, H. (1975). Agric. Biol. Chem. *39,* 953.
KISHIMOTO, S., SUGINO, H., TANAKA, K., KAKINUMA, A., and NOGUCHI, S. (1976). Chem. Pharm. Bull. *26,* 584.
KNAPP, J. S., CALLELY, A. G., and MAINPRIZE, J. (1982). J. Appl. Bacteriol. *52,* 5.
KONDO, E., STEIN, B., and VOGEL, J. (1969). Biochim. Biophys. Acta *176,* 135.
KORTE, F., and STIASHI, M. (1964). Liebigs Ann. Chem. *673,* 146.
KORTE, F., LUDWIG, G., and VOGEL, J. (1962). Liebigs Ann. Chem. *656,* 135.
KOSMOL, H., KIESLICH, K., VÖSSING, R., KOCH, H. J., PETZOLDT, K., and GIBIAN, H. (1967). Liebigs Ann. Chem. *701,* 198.
KRAYCHY, S., GARLAND, R. B., MIZUBA, S. S., and SCOTT, W. M. (1967). US Patent 3 481 974; Chem. Abstr. *72,* 55 761 (1970).
KREISER, W., and LANG, S. (1976). Chem. Ber. *109,* 3 318.
KUROSAWA, Y., SHIMOJIMA, H., and OSAWA, Y. (1965). Steroids Suppl. *1,* 185.
KUROZUMI, S., TORU, T., and ISHIMOTO, S. (1973). Tetrahedron Lett., 4 959.
LANZILOTTA, R. P., BRADLEY, D. G., and MC DONALD, K. M. (1974). Appl. Microbiol. *27,* 130.
LANZILOTTA, R. P., BRADLEY, D. G., and BEARD, C. C. (1975). Appl. Microbiol. *29,* 427.
LANZILOTTA, R. P., BRADLEY, D. G., MC DONALD, K. M., and TÖKES, L. (1976). Appl. Environ. Microbiol. *32,* 726.
LAROCHE, A. M. (1960). C. R. Acad. Sci. *250,* 2 773.
LEE, S. S., and SIH, C. J. (1967). Biochemistry *6,* 1 395.
LEEMING, M. R. G., and GREENSPAN, G. (1973). US Patent 3 726 765; Chem. Abstr. *78,* 157 863 (1973).
LEUENBERGER, H. G. W., BOGUTH, W., WIDMER, E., and ZELL, R. (1976). Helv. Chim. Acta *59,* 1832.
LEUENBERGER, H. G. W., BOGUTH, W., BARNER, R., SCHMID, M., and ZELL, R. (1979). Helv. Chim. Acta *62,* 455.
LOEW, O. (1981). Chem. Ber *14,* 450.
MARSHECK, W., and MIYANO, M. (1974a). US Patent 3 799 841; Chem. Abstr. *81,* 48 568 (1974).
MARSHECK, W., and MIYANO, M. (1974b). Biochim. Biophys. Acta *348,* 283.
MARSHECK, W., and MIYANO, M. (1975). US Patent 3 878 046; Chem. Abstr. *83,* 76 968 (1975).
MARX, A. F., and DOODEWARD, J. (1978). Brit. Patent 1 501 864; Chem. Abstr. *89,* 127 756 (1978).
MATSUMURA, F., and BOUSCH, G. M. (1967). Science *156,* 959.
MATSUMURA, F., BOUSCH, G. M., and TAI, A. (1968). Nature *219,* 965.
MATSUMURA, F., PATIL, K. C., and BOUSCH, G. M. (1970). Science *170,* 1 206.
MATSUMURA, F., BENEZET, H. J., and PATIL, K. C. (1976). Nippon Noyaku Gakkaishi 1 3; Chem. Abstr. *85,* 106 533 (1976).
MIKAMI, Y., FUKUNAGA, Y., ARIK, M., OBI, Y., and KISAKI, T. (1981). Agric. Biol. Chem. *45,* 791.
MIURA, S., KUROZUMI, S., TORU, T., TANAKA, T., KOBAYASHI, M., MATSUBARA, S., and ISHIMOTO, S. (1976). Tetrahedron *32,* 1893.
MIYANO, M., DORN, C. R., COLTON, F. B., and MARSHECK, W. J. (1971). J. Chem. Soc. Chem. Comm., 425.
MOORE, R. H. (1974). Ger. Offen. 2 401 761.
MOORE, R. H. (1975). US Patent 3 880 712.
MORI, K., WHELER, D. M. S., JILEK, J. O., KA-

KAC, B., and PROTIVA, M. (1965). Coll. Czech. Chem. Comm. *30,* 2 236.

MURRAY, J. R., SCHEIKOWSKI, T. A., and MACRAE, L. C. (1974). Antonie van Leeuvenhoek J. Microbiol. Serol. *40,* 17.

NAKAZAKI, M., CHIKAMATSU, H., NAEMURA, K., HIROSE, Y., SHIMIZU, T., and ASAO, M. (1978). J. Chem. Soc. Chem. Comm., 668.

NAKAZAKI, M., CHIKAMATSU, H., NAEMURA, K., NISHINO M., MURAKAMI, H., and ASAO, M. (1979). J. Org. Chem. *44,* 4588.

NAKAZAKI, M., CHIKAMATSU, H., NAEMURA, K., and ASAO M. (1980). J. Org. Chem. *45,* 4 432.

NAKAZAKI, M., CHIKAMATSU, H., and ASAO, M. (1981a). J. Org. Chem. *46,* 1 147.

NAKAZAKI, M., CHIKAMATSU, H., NISHINO, M., and MURAKAMI, H. (1981b). J. Org. Chem. *46,* 1 151.

NAKAZAKI, M., CHIKAMATSU, H., FUMII, T., and NAKATSUKII, T. (1981c). J. Org. Chem. *46,* 585.

NEUBERG, C. (1949). Adv. Carbohydrate Chem. *4,* 74.

NEUBERG, C. (1950). Biochim. Biophys. Acta *4,* 170.

NEWTON, R. F., PATON, J., REYNOLDS, D. P., and YOUNG, S. (1979). J. Chem. Soc. Chem. Comm., 908.

NOMA, Y. (1977). Nippon Nogei Kasaku Kaishi *51,* 57; Chem. Abstr. *88,* 168 467 (1978).

OGINO, S., FUJIMOTO, S., and AOKI, Y. (1974a). Agric. Biol. Chem. *38,* 275.

OGINO, S., FUJIMOTO, S., and AOKI, Y. (1974b). Agric. Biol. Chem. *38,* 707.

OGINO, S., FUJIMOTO, S., WADA, H., and AOKI, Y. (1975a). Japan. Kokai 75 77 591; Chem. Abstr. *84,* 3 289 (1976).

OGINO, S., FUJIMOTO, S., WADA, H., and AOKI, Y. (1975b). Japan Kokai 75 77 592; Chem. Abstr. *84,* 3 306 (1976).

OOYAMA, J., and FOSTER, J. W. (1965). Antonie van Leeuwenhoek J. Microbiol. Serol. *31,* 45.

ORITANI, T., and YAMASHITA, K. (1975). Agric. Biol. Chem. *39,* 89.

ORITANI, T., and YAMASHITA, K. (1976). Japan. Kokai 76 35 490; Chem. Abstr. *85,* 61 429 (1976).

ORITANI, T., KUDO, S., and YAMASHITA, K. (1982). Agric. Biol. Chem. *46,* 757.

ORNSTON, L. N., and STANIER, R. Y. (1966), J. Biol. Chem. *241,* 3 776.

OUGHAM, H. J., and TRUDGILL, P. W. (1982). J. Bacteriol. *150,* 1 172.

PARRY, J. P., and NAIDU, M. V. (1980). Tetrahedron Lett., 4 783.

PERUZZOTTI, G. P. (1982). "Microbial transformation of prostaglandins" in "Microbial Transformations of Bioactive Compounds" (J. P. ROSAZZA, ed.), Chap. 5, p. 109. CRC Press, Boca Raton, Florida.

PERWOZWANSKY, W. W. (1930). Zentralbl. Bakteriol. Parasitenkd. Infektionskr. Hyg. Abt. 2 *81,* 372.

PLEMENITAS, A., and KOMEL, R. (1982). Vestn. Slov. Kem. Drus. *29,* 35; Chem. Abstr. *97,* 22 034 (1982).

PLONKA, G., MEIER, A., TRUCKENBRODT, G., and TEICHMUELLER, G. (1974) Ger. (East) Patent 114 809; Chem. Abstr. *84,* 178 226 (1976).

POSTERNAK, T., and RAVENNA, F. (1947). Helv. Chim. Acta *30,* 441.

POSTERNAK, T., and REYMOND, D. (1953). Helv. Chim. Acta *36,* 260.

POSTERNAK, T., and REYMOND, D. (1955). Helv. Chim. Acta *38,* 195.

POSTERNAK, T., REYMOND, D., and FRIEDLI, H. B. (1955). Helv. Chim. Acta *38,* 205.

POSTERNAK, T., RAPIN, A., and HAENNI, A. L. (1957). Helv. Chim. Acta *40,* 1594.

PRELOG, V. (1956). Can. Patent 596 063, Prior. Switzerland.

PRELOG, V. (1962). Ind. Chim. Belge *27,* 1 309.

PRELOG, V. (1963). 14. Colloquium der Gesellschaft für Physiologische Chemie in Mosbach/Baden, p. 288.

PRELOG, V. (1964). Pure Appl. Chem. *9,* 119.

PRELOG, V., and AKLIN, W. (1956). Helv. Chim. Acta *39,* 748.

PRELOG, V., and SMITH, H. E. (1959). Helv. Chim. Acta *42,* 2 624.

PRELOG, V., and ZÄCH, D. (1959). Helv. Chim. Acta *42,* 1862.

PROTIVA, M., CAJEK, A., TILEK, T. O., KAKAC, B., and TADRA, M. (1961). Collect. Czech. Chem. Comm. *30,* 2 236.

RATLEDGE, C. (1964). Nature *203,* 428.

RATLEDGE, C. (1967). Biochim. Biophys. Acta *14,* 55.

RATLEDGE, C., and WINDER, F. G. (1966). Biochem. J. *101,* 274.

REUTER, G., and HUETTNER, R. (1977). Z. Allg. Mikrobiol. *17,* 149.

ROBERTSON, L. W. (1982). "Microbial transformations of cannabinoids". In "Microbial Transformations of Bioactive Compounds" (J. P. ROSAZZA, ed.), p. 91. CRC Press, Boca Raton, Florida.

ROBERTSON, L. W., KOH, S. W., HUFF, S. R., MALHOTRA, R. K., and GHOSH, A. (1978). Experientia *34,* 1 020.

ROSAZZA, J. P., and SMITH, R. V. (1979). "Microbial models for drug metabolism" in "Advances in Applied Microbiology", Vol. 25 (D. PERLMAN, ed.). Academic Press, New York.

RUFER, C., KOSMOL, H., SCHROEDER, E., KIESLICH, K., and GIBIAN, H. (1967). Liebigs Ann. Chem. *702,* 141-148.

RUIS, H., BRADY, R. N., MC CORMICK, D. B., and WRIGHT, L. D. (1968). J. Biol. Chem. *243,* 547.

SCANNEL, J. P., PRUESS, D. L., DEMMY, T. C., and WILLIAMS, T. (1970). J. Antibiot. *23,* 618.

SCHNEIDER, W. P., and MURRAY, H. C. (1973). J. Org. Chem. *38,* 397.

SCHUBERT, A., RIECKE, A., HILGETAG, G., SIEBERT, R., and SCHWARZ, S. (1958). Naturwissenschaften *45,* 623.

SCHUBERT, K., BÖHME, K. H., and HÖRHOLD, C. (1964). Steroids *4,* 581.

SCHUBERT, K., BÖHME, K. H., and HÖRHOLD, C. (1965). Biochim. Biophys. Acta *111,* 529.

SCHUBERT, K., BÖHME, K. H., and HÖRHOLD, C. (1967). Acta Biol. Med. Germ. *18,* 295; Chem. Abstr. *67,* 41 185 (1967).

SCHUPHAN, J., and BALLSCHMITER, K. (1972). Nature *237,* 100.

SCHWAB, J. M. (1981). J. Amer. Chem. Soc., 103, 1 876.

SCHWARZ, S., TRUCKENBRODT, G., MEYER, M., ZEPTER, R., WEBER, G., CARL, C., WENTZKE, M., SCHICK, H., and WELZEL, H. P. (1981). J. Prakt. Chem. *323,* 729.

SEBEK, O. K., LINCOLN, F. H., and SCHNEIDER, W. P. (1976). 5th Int. Ferment. Congr., West Berlin, (Abstr.), 17.05.

SEKIJO, C., TSOBOI, T., and YOSHIMURA, Y. (1969). Agric. Biol. Chem. *33,* 683.

SHAW, R. (1966). Nature *209,* 1 369.

SIEWERT, G., KIESLICH, K., HOYER, G. A., and ROSENBERG, D. (1973). Chem. Ber. *106,* 1 290.

SIEWINSKI, A. (1969). Bull. Acad. Pol. Sci. *17,* 475.

SIEWINSKI, A., DMOCHOWSKA-GLADYSZ, J., KOLEK, T., ZABZA, A., and DERDZINSKI, K. (1979). Tetrahedron *35,* 1 409.

SIH, C. H. (1976). In "Microbial Transformations of Non-steroid Cyclic Compounds" (K. KIESLICH, ed.). G. Thieme Verlag, Stuttgart.

SIH, C. J. (1977). Am. J. Pharm. Educ. *41,* 432.

SIH, C. J., and ROSAZZA, J. P. (1976). "Microbial transformations in organic synthesis", Chap. 3, p. 74 in "Application of Biochemical Systems in Organic Chemistry", Part I (J. B. JONES, C. J. SIH, and D. PERLMAN, eds.). John Wiley, New York.

SIH, C. H., and WANG, K. C., (1963). J. Am. Chem. Soc. *85,* 2 135.

SIH, C. J., SALOMON, R. G., PRICE, P., PERUZZOTTI, G., and SOOD, R. (1972). J. Chem. Soc. Chem. Comm., 240.

SIH, C. J., HEATHER, J. B., SOOD, R., PRICE, P., PERUZZOTTI, G., HSU LEE, L. F., and LEE, S. S. (1975). J. Am. Chem. Soc. *97,* 865.

SKRYABIN, G. K., and GOLOVLEVA, L. A. (1976). "Ispolgovanie Microorganisms v Organicheskom Sinteze." Nauka Publ., Moscow.

SMITH, D. I., and CALLELY, A. G. (1975). J. Gen. Microbiol. *91,* 210.

SPRÖSSLER, B., LENSSEN, V., and LINGENS, F. (1970). Z. Physiol. Chem. *351,* 1 178.

STIRLING, L. A., WATKINSON, R. J., and HIGGINS, I. J., (1977). J. Gen. Microbiol, *99,* 119.

SUMITOMO CHEM. CO LTD. (1971). Belg. Patent 785 008.

SUN, F. F., TAYLOR, B. M., LINCOLN, F. H., and SEBEK, O. K. (1980). Prostaglandins *20,* 729.

SYNTEX CORP. (1974). Neth. Patent 7 310 071.

TABENKIN, B., LE MAHIEU, R. A., BERGER, J. and KIERSTEAD, R. W. (1969a). Appl. Microbiol. *17,* 714.

TABENKIN, B., LE MATIEU, R. A., BERGER, J., and KIERSTEAD, R. W. (1969b). Bacteriol. Proc. *3,* A16.

TAKANO, S., TANIGAWA, K., and OGASAWARA, K. (1976). J. Chem. Soc. Chem. Comm., 189.

TAKEDA COMP. (1967). Fr. Patent 158 108 (1969). Japan. Prior. (1967); Chem. Abstr. *73,* 75 654 (1970).

TAKEDA COMP. (1974). Japan. Patent 4 9001 552.

TANAKA, H., SHIKATA, K., OBATA, H., TOKUYAMA, T., and UENO, T. (1977a). Hakko Kogaku Kaishi, *55,* 57; Chem. Abstr. *86,* 167 644 (1977).

TANAKA, H., OBATA, H., TOKUYAMA, T., UENO, T., YOSHIGAKO, F., and NISHIMURA, A. (1977b). Hakko Kogaku Kaishi, *55,* 62; Chem. Abstr. *86,* 167 645 (1977).

TAYLOR, D. G., and TRUDGILL, P. W. (1978). J. Bacteriol. *134,* 401.

TEPPER, J. P., MC CORMICK, D. B., and WRIGHT, L. D. (1966). J. Biol. Chem. *241,* 5 734.

TERAHARA, A., and TANAKA, M. (1981) Ger. Offen. D. E. 3 122 499; Chem. Abstr. *97,* 4 664 (1982).

TONGE, G. M., and HIGGINS, I. J. (1974). J. Gen. Microbiol. *81,* 521.

TRUCKENBRODT, G., SCHWARZ, S., SCHICK, H., HESS, L., MEYER, M., WEBER, G., PORWOL, K., EBERHARDT, U., and HENKEL, H. (1978).

Ger. (East) Patent 134 953; Chem. Abstr. *91,* 173 388 (1979).

TSAI, M. M. (1980). Diss. Abstr. Int. B. *40,* 3 691.

USKOKOVIC, M. R., PRUESS, D. L., DESPREAUX, C. W., SHIUEY, S., PIZZOLATO, G., and GUTZWILLER, J. (1973). Helv. Chim. Acta *56,* 2 834.

VIDIC, H. J., HOYER, G. A., KIESLICH, K., and ROSENBERG, D. (1976). Chem. Ber. *109,* 3 606.

WALLNÖFER, P. (1968). Naturwissenschaften *55,* 351.

WALLNÖFER, P. (1969). Arch. Mikrobiol. *64,* 319.

WANG, K. (1969). Ger. Offen. 2 038 926 (1971). US Prior. (1969). Chem. Abstr. *74,* 139 563 (1971).

WHA, Bin Im., ROTH, J. A., MC CORMICK, D. B., and WRIGHT, L. D. (1970). J. Biol. Chem. *245,* 6 269.

WHA, Bin Im., MC CORMICK D. B., and WRIGHT, L. D. (1973). J. Biol. Chem. *248,* 7 798.

WHITING, G. C., and COGGINO, R. A. (1967). Biochem. J., *102,* 283.

WRIGHT, L. D., CRESSON, E. L., VALIANT, J., WOLF, D. E., and FOLKERS, K. (1954). J. Am. Chem. Soc. *76,* 4 163.

YANG, H. C., KUSUMOTO, M., IWAHARA, S., and TOCHIKURA, T. (1968). Agric. Biol. Chem. *32,* 399.

YANG, H. C., KUSUMOTO, M., IWAHARA, S., TOCHIKURA, T., and OGATA, K. (1969). Agric. Biol. Chem. *33,* 1 730.

YANG, H. C., KUSUMOTO, M., TOCHIKURA, T., and OGATA, K. (1970). Agric. Biol. Chem. *34,* 370.

YOSHIDA, S. (1964). Bot. Mag. Tokyo *77,* 10; Chem. Abstr. *61,* 12 352h (1964).

YOUNG, I. G., and GIBSON, F. (1969). Biochim. Biophys. Acta *177,* 182.

ZHOU, W., HUANG, D., DONG, Q., and WANG, Z. (1982a). Huaxue Xuebao *40,* 648; Chem. Abstr. *98,* 31 096 (1983).

ZHOU, W., ZHUANG, Z., and WANG, Z., (1982b). Huaxue Xuebao *40,* 666, Chem. Abstr. *98,* 31 097 (1983).

ZIFFER, H., and GIBSON, D. T. (1975). Tetrahedron Lett; 2 137.

ZYAKUN, A. M., NEFEDOVA, M. Yu, BASKUNOV, B. P., and FINKELSTEIN, Z. I. (1983). Izv. Akad. Nauk SSSR Ser. Biol., 126.

Chapter 6

Natural and Semi-synthetic Alkaloids

Patrick J. Davis

Division of Medicinal and Natural Products Chemistry
College of Pharmacy, University of Texas
Austin, Texas 78712, USA

I. Introduction
  A. Biotechnological Applications of Alkaloid Transformations
  B. Organic Synthetic Applications
  C. Microbial Models of Mammalian Metabolism
II. Transformations According to Alkaloid Class
  A. Indole Alkaloids
    1. Vinca *(Catharanthus)* alkaloids
    2. Lysergic acid derivatives
    3. Clavine and ergoline alkaloids
    4. Rauwolfia (yohimbine) alkaloids
    5. Pyridocarbazole alkaloids (ellipticine)
    6. β-Carboline derivatives
    7. Miscellaneous indole alkaloids
  B. Isoquinoline Alkaloids
    1. Simple tetrahydroisoquinolines
    2. Sanguinarine
    3. Aporphines
    4. Benzylisoquinolines
    5. "Dimeric" isoquinoline alkaloids
  C. Quinoline Alkaloids
  D. Morphine Alkaloids
  E. Colchicine and Related Alkaloids
  F. Steroidal Alkaloids
  G. Miscellaneous Alkaloids
    1. Nicotine and related pyridine alkaloids
    2. Pyrrolizidine alkaloids
    3. Quinolizidine alkaloids
    4. Tropane alkaloids
    5. Ephedrine
III. References

I. Introduction

A. Biotechnological Applications of Alkaloid Transformations

In any discussion of microbial transformation of "alkaloids", one must first acknowledge the heterogeneity of members of the class. Unlike discrete chemical classes (steroids, prostaglandins) or pharmacological classes (antibiotics, antitumor agents), members covered by the broad umbrella "alkaloids" show little structural similarity other than the presence of a nitrogen atom (usually in a heterocycle), and little pharmacological similarity other than initial observations of profound pharmacological (or toxicological) activity of the plant source (often the impetus for its further investigation). It is precisely this diversity, however, that lends this class of compounds to a wealth of microbial transformations of interest to diverse disciplines, as indicated by several recent reviews in this area (IIZUKA and NAITO, 1967; VINING, 1969; VINING, 1980; KIESLICH, 1976; HOLLAND, 1981; DAVIS, 1982). These often complex chemical structures yield a multitude of functional groups amenable to selective modification by microbial systems which complement chemical conversion methods (see Sect. I.B. "Organic Synthetic Applications"). Further, the profound pharmacological activities of these agents have stimulated their examination as medicinals, toxicological agents, and biochemical tools, which further necessitates a determination of metabolic routes of bioactivation and bioinactivation. Microbial transformations have application here as well in facilitating the elucidation of mammalian routes of metabolism, as described below (Sect. I.C. "Microbial Models of Mammalian Metabolism").

B. Organic Synthetic Applications

Microbial transformations have proven to be useful adjuncts to organic synthetic techniques, since they often exhibit selected advantages over classical methods, as summarized previously (SIH and ROSAZZA, 1976). First, these reactions take place under mild, physiological reaction conditions. Consequently, decomposition due to extremes of pH and heat usually does not affect yield. Secondly, reactions may be effected at chemically "non-activated" positions to yield metabolites difficult to obtain by chemical routes. A classic example is the microbial 11α-hydroxylation of progesterone, a reaction difficult to conduct chemically with the efficiency needed for commerical demands (PETERSON and MURRAY, 1952). Thirdly, cultures may be exploited in a predictable manner for their type-reaction specificity to obtain a desired transformation. While "random screens" continue to be used in some types of studies (e. g., if the investigator desires all possible metabolites), a more refined approach, the "type-reaction" screen is more useful to the chemist, and involves the selection of organisms for screening based on their reported ability to exhibit a specific reaction (e. g., ketone reductions, ester hydrolysis). Compilations such as that by KIESLICH (1976) have rapidly advanced this approach by allowing the chemist a more judicious choice of cultures. As illustrated in Table 1, a variety of useful type-reactions have been observed in the microbial transformations of alkaloidal substrates. Another excellent approach in this regard is "active site mapping" in which sufficient substrate analogs are examined to define the structural requirements (and restrictions) allowed for transformations within that substrate series. This has been accomplished with several microbial strains exhibiting reactions of synthetic utility, as recently summarized in an excellent review by ABBOTT (1979).

An additional advantage is that microbial metabolism can exhibit a high degree of regioselectivity in which multiple similar functional groups (a common occurrence in alkaloids) can be enzymatically distinguished, a process often difficult to execute chemically. Finally, and perhaps most importantly, one often observes a high degree of stereoselectivity in microbial transformations based on the chiral specificity of the enzymes involved. This may be expressed as substrate and/or product stereoselectivities, and thus involves existing chiral centers, or the creation of new chiral centers by metabolism at enzymatically distinguishable prochiral ligands or faces (TESTA, 1979; FLOSS, 1970). Due to the complex chiral nature of many alkaloids, stereochemically controlled syntheses and modifications remain a major challenge (BROSSI, 1979). With microbial systems, a predominance (or exclusive production) of one stereoisomeric alkaloidal metabolite may occur, or a destructive resolution may be effected by the selective metabolism of one enantiomer in a racemic substrate.

Table 1. Type-Reactions Observed in the Microbiological Transformations of Alkaloidal Substrates

Oxidations
- *O*-, *N*-, and *S*-Dealkylations
- Hydroxylations (aliphatic and aromatic)
- *N*-Oxide formation
- Dehydrogenation (to olefins, ketones, enamines)
- Deaminations
- Decarboxylations
- Oxidative phenolic coupling
- Oxidative benzylic cleavage
- Intramolecular ether formation

Reductions
- Ketones
- *N*-Oxides
- Imines
- Reductive cyclization

Hydrolysis
- Esters
- Glycosides
- Nitriles
- Hydration of Olefins

These chemical principles will be exemplified in the various sub-classes of alkaloids listed below. Since "semi-synthetic" alkaloids (i. e., those derived by chemical modification of the parent alkaloid) often maintain the structural and stereochemical complexities of the parent, the same principles apply. Consequently, microbial transformations of semi-synthetic alkaloids are included in the discussion. Because of the limited biotechnological potential of alkaloidal biodegradation, this subject will receive only limited coverage.

C. Microbial Models of Mammalian Metabolism

The pharmacological and toxicological aspects of alkaloid activity, as well as their potential development into useful therapeutic agents demand the elucidation of metabolic patterns responsible for bioactivation and detoxification in mammals. The concept that microbial systems could be used to mimic those routes of metabolism observed in higher organisms (including man) was first put forth as "microbial models of mammalian metabolism" by ROSAZZA and SMITH (1974), and has been the subject of several recent reviews (SMITH and ROSAZZA, 1975; ROSAZZA and SMITH, 1979, 1982). Rather than simply representing the fortuitous production of similar metabolites in diverse genera, this concept has a firm foundation in comparative enzymology (ROSAZZA and SMITH, 1982; SMITH and DAVIS, 1980). The discipline has tremendous potential in facilitating mammalian metabolic studies for several reasons. First, experimental parameters such as aeration, pH, and temperature are easily manipulated in culture, as opposed to whole animal systems. Also, these reactions are easily scaled upwards to produce milligram or even gram quantities of metabolites for full biological evaluation, for use as analytical standards in mammalian metabolism studies, and for full structure elucidation (particularly useful when mammalian results are equivocal). Thirdly, cultures may be examined which exhibit the full complement of metabolic pathways observed with a particular animal system, or a selected culture may serve as a model of one route in lieu of others. An example of the former case is that of COUTTS et al. (1979b) in which *Cunninghamella echinulata* ATCC 9244 catalyzed *C*- and *N*-oxidation and *N*-acetylation of *N*-(*n*-propyl) amphetamine to produce nine known oxidative mammalian metabolites, and a tenth (the *N*-acetyl derivative) which has not been observed in mammals. Conversely, metabolic reduction of the anticoagulant warfarin, a relatively minor mammalian route overshadowed by aromatic hydroxylation, was the sole transformation observed with *Nocardia corallina* ATCC 19 070 and *Arthrobacter* sp. (ATCC 19 140) (DAVIS and RIZZO, 1982).

Finally, once a microbial model is fully developed, it should serve in a predictable manner to aid further mammalian studies. While most studies to date have been retrospective modeling (i. e., the mammalian metabolism is elucidated first), investigators are now crossing the threshold of using such systems to predict routes of bioactivation, bioinactivation, and detoxification, as well as to predict the occurrence of minor undetected mammalian metabolites. It should be mentioned that such model systems will probably never totally replace animal studies, any more than animal models will replace human investigations. Instead, microbial models have excellent potential, especially in the early stages of metabolism studies, for facilitating the elucidation of and confirming complex mammalian pathways using relatively simple techniques. Numerous comparisons of microbial and mammalian metabolism of alkaloidal subclasses are discussed below to illustrate these principles. The excellent review by HOLLAND (1981) also compares alkaloidal transformations by mammalian and microbial enzymatic systems.

II. Transformations According to Alkaloid Class

A. Indole Alkaloids

1. Vinca (*Catharanthus*) alkaloids

The Vinca alkaloids derived from the periwinkle *Catharanthus roseus* (*Vinca rosea*) have been extensively investigated for metabolism in microbiological systems. Major interest in these compounds is based on the profound antitumor activity of the dimeric alkaloids vinblastine (**1**) and vincristine (**2**) (CREASEY, 1975; 1977). The goals of such studies include an understanding of general routes of metabolism of these dimeric and certain monomeric alkaloids in terms of bioactivation and detoxification, to obtain active metabolites, and to determine biogenetic mechanisms of dimeric alkaloid formation. Results of these studies have yielded some of the most diverse and interesting routes of metabolism in microbial transformation.

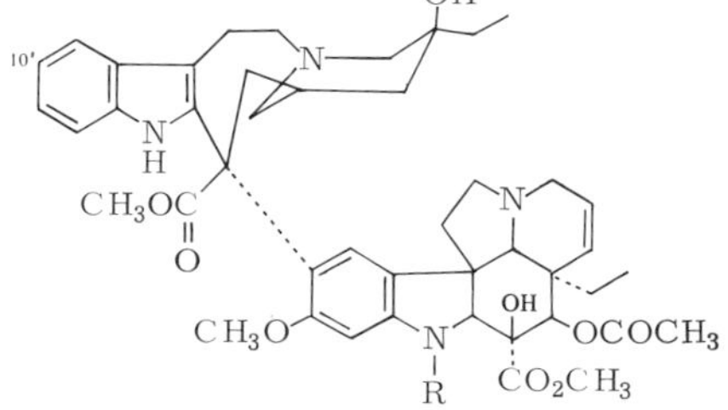

1 R = CH_3
2 R = CHO

Early studies by MALLETT and co-workers at Eli Lilly (MALLET et al., 1964) involved the screening of 437 cultures (mostly soil Streptomycetes) with the monomeric alkaloid vindoline (**3**), with the result that approximately 25% of these cultures exhibited bioconversions. An unidentified bacterium (A16 237) produced desacetylvindoline (**4**) quantitatively, while *Streptomyces cinnamonensis* A15 167 further metabolized this same product (**4**) to desacetyldihydrovindoline ether (**9**). The *O*-acetyl-derivativc of vindoline (**5**) was first deacetylated to vindoline (**3**), followed by the expected conversion to **4** and **9**. Desacetyl dihydrovindoline (**6**), however, was not converted, indicating the importance of the 14,15-olefin in intramolecular ether formation by this organism. NEUSS et al. (1973a) described the utility of *Streptomyces* sp. (A17 000) in the conversion of vindoline (**3**) to dihydrovindoline ether (**10**), as well as a unique ring contraction lactam (**12**). *Streptomyces albogriseolus* A17 178 also produced a unique metabolite in which an acetonyl-moiety was added at position three of **10** to yield **11**. It is interesting to note that in these studies, the substrate was added in acetone. Labeling studies later demonstrated that acetone itself could be added to this position during incubation (NEUSS et al., 1973b), indicating that position 3 is metabolically activated for addition as proposed by other authors in related transformations (*vide infra*). An additional reaction observed was the *N*-demethylation of vindoline to norvindoline (**7**) in 10% yield by *S.albogriseolus* A17 178, NRRL 5748.

ROSAZZA and co-workers have recently extended these studies on the metabolism of vindoline, by demonstrating that this substrate is subject to *O*-demethylation (position 11 on **3**) by growing and resting cell

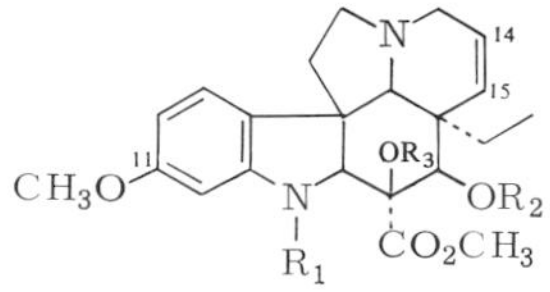

3 $R_1 = CH_3$, $R_2 = COCH_3$, $R_3 = H$
4 $R_1 = CH_3$, $R_2 = R_3 = H$
5 $R_1 = CH_3$, $R_2 = R_3 = COCH_3$
6 14,15-dihydro-**4**
7 $R_1 = R_3 = H$, $R_2 = COCH_3$
8 14,15-dihydro-**3**

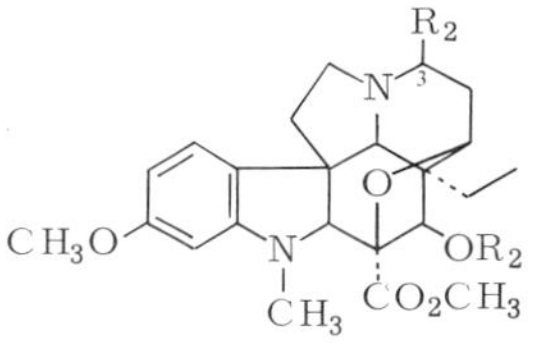

9 $R_1 = R_2 = H$
10 $R_1 = COCH_3$, $R_2 = H$
11 $R_1 = COCH_3$, $R_2 = CH_2COCH_3$

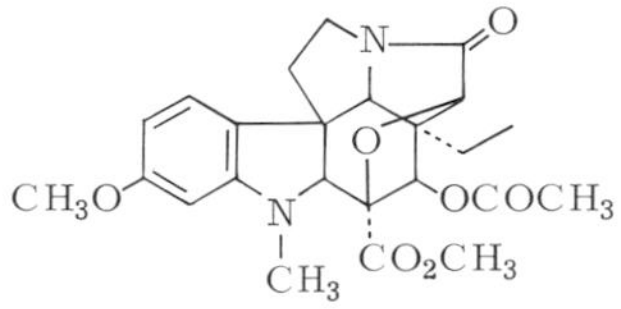

12

cultures of *Sepedonium chrysospermum* ATCC 13 378 in 33% isolated yield, a process which is difficult to accomplish by chemical methods (WU et al., 1978). Previous studies by others (LIN et al., 1975) demonstrated that a related alkaloid, aspidospermine (**13**) was similarly *O*-dealkylated (position 12) by *Streptomyces griseus* 14 823, while the corresponding *N*-desacetyl- (**14**) and *N*-desacetyl-*N*-ethyl- (**15**) analogs were

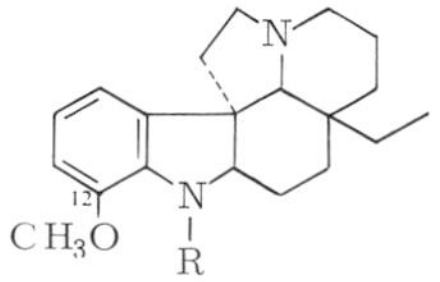

13 $R = COCH_3$
14 $R = H$
15 $R = CH_2CH_3$

not attacked. ECKENRODE and ROSAZZA (1982) demonstrated that *Streptomyces griseus* UI 1158w, NRRL B8090 attacks dihydrovindoline (**8**) to yield the corresponding 11-*O*-desmethyl-metabolite. This is a particularly interesting observation in view of the fact that this organism does not *O*-demethylate vindoline (**3**), but instead extensively attacks the 6-membered nitrogen heterocycle. As mentioned above, MALLETT et al. (1964) observed that *S.cinnamonensis* (which forms **9**) was incapable of metabolizing the dihydro-derivative **6**. In their studies with *S.griseus* using vindoline as the substrate, ROSAZZA and co-workers observed the production of dihydrovindoline ether (**10**, 20%–40%), as well as a dimer of this compound, **16** (NABIH et al., 1978;

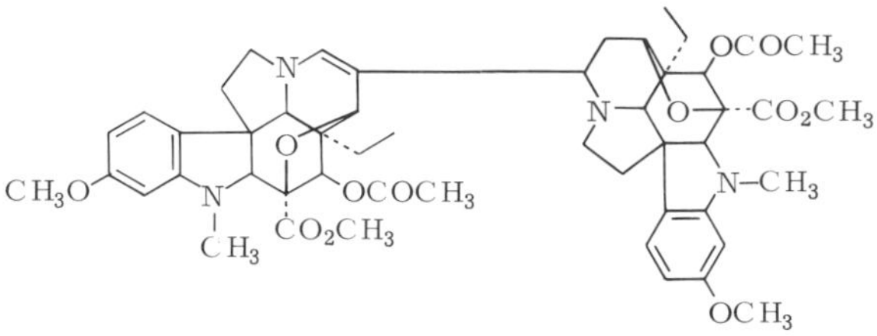

16

GUSTAFSON and ROSAZZA, 1979). As shown in Fig. 1, the authors suggested that vindoline is first oxidized to a reactive *α,β*-unsaturated iminium species (**17**), allowing for the intramolecular nucleophilic attack by the C-16 hydroxyl group to yield the enamine-ether, **18**. Reduction of **18** (or the tautomer **19**) would yield dihydrovindoline ether (**10**), or the enamine **18** could be dimerized by condensation with the tautomeric iminium species, **19**, as shown. This proposal was confirmed by using medium modification experiments to limit dimerization which allowed for the isolation of the enamine **18**, and determination of the site of the double bond by isotopic reductive labeling (GUSTAFSON and ROSAZZA, 1979). Further, the use of *O*-acetyl vindoline (**5**) as substrate prevented the intramolecular ether formation shown in **17**, allowing for isolation of **17** as the *O*-acetyl derivative.

The formation of the dimer **16** was also demonstrated using an extracellular laccase (a copper oxidase) from the basidiomycete *Polyporus anceps* (ECKENRODE et al., 1979; ROSAZZA and SMITH, 1982). The presence of the reactive dihydrovindoline ether enamine (**18**) was again demonstrated. A laccase from a plant source (*Rhus vernicifra*) and a mammalian copper oxidase, ceruloplasmin, also catalyzed the formation of these products, demonstrating a firm link between microbial, plant, and mammalian xenobiotic metabolism.

Reports on the microbial metabolism of the "dimeric" Vinca alkaloids (**1** and **2**) are few. NEUSS and co-workers (NEUSS et al., 1974b) described the formation of an indolenine-indoline product (**21**) from vinblas-

21

tine (**1**) metabolism by *Streptomyces albogriseolus* A17 178, involving the formation of an intramolecular ether on the catharantine portion of the dimeric alkaloid. In contrast, *S.punipalus* A36 120 achieved aromatic hydroxylation to yield the 10′-phenol of **1** (i. e., again on the catharanthine portion), with no other changes. The elucidation of the position of hydroxylation was based primarily on PMR studies showing the expected *meta*-coupling between H-9′ and H-11′, and *ortho*-coupling between H-11′ and H-12′. In a similar study, SCHUM-

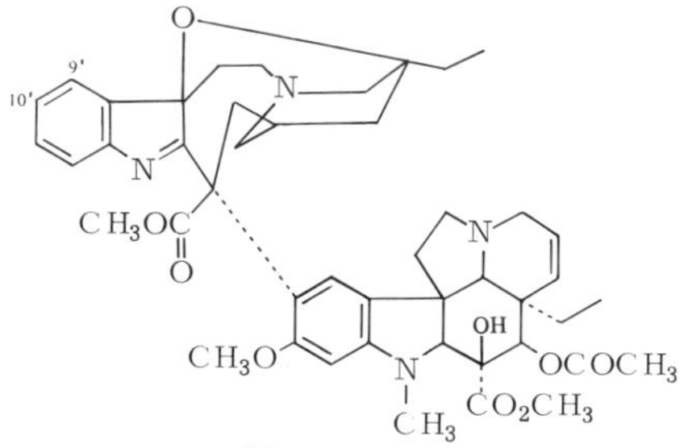

17 18 19

20

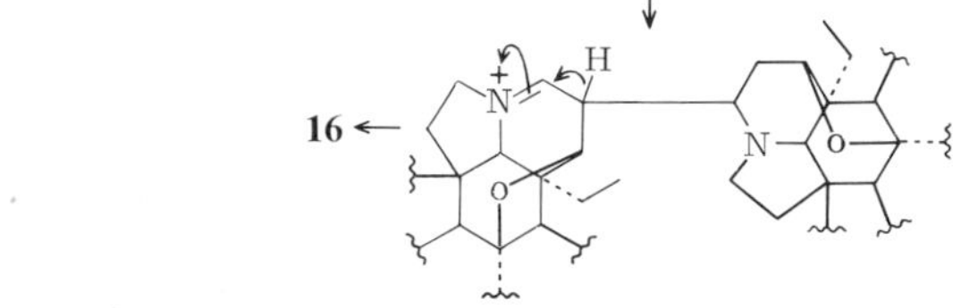

Figure 1. Proposed mechanism for the formation of dihydrovindoline ether dimer by *Streptomyces griseus* (from NABIH et al., 1978) (reproduced with permission of the authors).

BERG and ROSAZZA (1981) observed that several *Streptomyces* species, including resting cell suspensions of *S.griseus* UI 1158 hydroxylated leurosine (a 15′,20′-epoxide analog of vinblastine) to 10′-hydroxyleurosine, **22**. Similar PMR arguments were used to define the site of hydroxylation. These authors also indicated that this position of attack (in lieu of 9′ or 11′) would be expected on the basis of electrophilic aromatic substitution. It is indeed the most common position of metabolic hydroxylation with indolic compounds (KIESLICH, 1976).

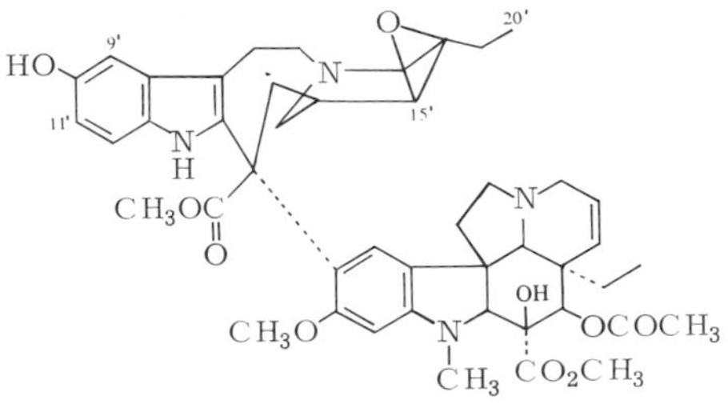

22

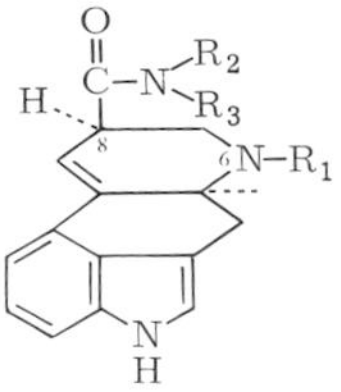

23 $R_1 = CH_3$, $R_2 = R_3 = CH_2CH_3$
24 $R_1 = H$, $R_2 = R_3 = CH_2CH_3$
25 $R_1 = CH_3$, $R_2 = CH_2CH_3$, $R_3 = H$
26 $R_1 = CH_3$, $R_2 = CH_2CH_3$, $R_3 = CH{=}CH_2$
27 $R_1 = CH_3$, $R_2 = CH_2CH_3$, $R_3 = CH_2CH_2OH$

2. Lysergic acid derivatives

The lysergic acid amides, most notably lyseric acid diethylamide (LSD, **23**) are of interest because of their profound hallucinogenic activity, use as pharmacological tools, and abuse potential. The elucidation of mammalian metabolic pathways is important in attempting to rationalize detoxification, and to evaluate potentially active metabolites, but such studies have been hampered by a lack of availability of minor metabolites. For this reason, ISHII and coworkers examined a series of microorganisms and mammals for parallel routes of metabolism of LSD and related compounds. Initial studies with LSD (**23**) indicated that numerous cultures were capable of attacking the N-6 and amide *N*-alkyl substituents (NIWAGUCHI et al., 1975; ISHII et al., 1979a; 1980). *Streptomyces lavendulae* IFM 1031 attacked only the N-6-position to yield nor-LSD (**24**). Conversely, *Streptomyces roseochromogenes* IFM 1081 attacked only the *N*-amide alkyl group to yield lysergic acid ethylamide (**25**), lysergic acid ethyl vinylamide (**26**), and lysergic acid ethyl 2-hydroxyethylamide (**27**). Other *Streptomyces* and *Cunninghamella* strains produce all four metabolites (**24–27**). Both **24** and **25** are known metabolites of LSD in mammals, and the authors were able to use the vinyl analog **26** generated in the microbial studies to determine its presence in mammals. Contrary to initial expectation, this metabolite gave rise to **27** and **25**, rather than the expected production of **26** by dehydration of the hydroxy metabolite **27**. The high degree of substrate stereoselectivity in *Streptomyces roseochromogenes* was demonstrated by the fact that this organism could not metabolize *iso*-LSD (epimeric at C-8), while, in contrast, *S.lavendulae* yielded *iso*-nor-LSD.

The C-8-amide dealkylations were examined in greater detail using a series of lower and higher alkyl homologs of LSD (ISHII et al., 1979b). Results observed were as follows (see Fig. 2): lysergic acid dimethylamide (**28**) was only dealkylated to the monomethylamide, **29**; lysergic acid diethylamide was also dealkylated to yield **25**, as well as the other metabolites mentioned above (**26, 27**); neither lysergic acid di-*n*-propylamide (**28**) nor lysergic acid di-*n*-butylamide (**32**) were *N*-dealkylated, but instead yielded the two epimeric alcohols resulting from ω-1 hydroxylation (that is, adjacent to the terminal carbon) (i. e., **29** and **30** from **28**, and **33** and **34** from **32**), as well as the further oxidation products, ketones **31** and **35**, respectively. Based on these results, the

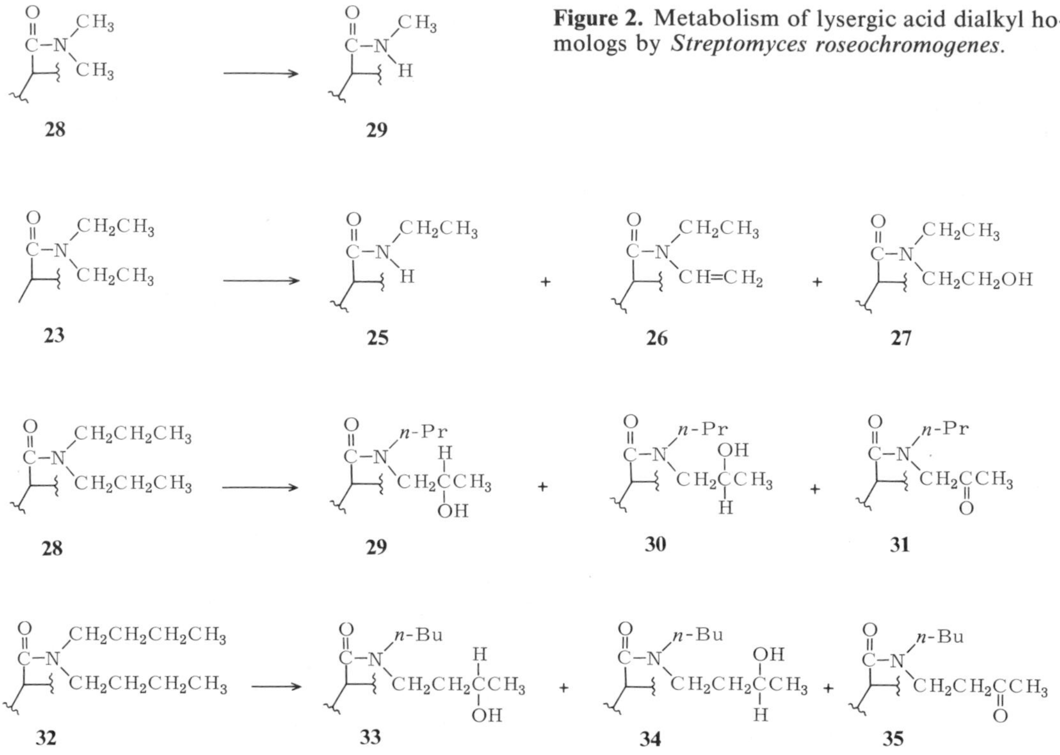

Figure 2. Metabolism of lysergic acid dialkyl homologs by *Streptomyces roseochromogenes*.

authors proposed that the chain length regulates the site of oxygenation, with ω-1 hydroxylation occurring if possible. The dimethyl-derivative **28** simply does not have an ω-1 position, consequently $C\alpha$-(methyl)hydroxylation yields the carbinolamine with resultant *N*-demethylation.

These studies culminated in a proposed active site map for the hydroxylase (*N*-dealkylase) of *S.roseochromogenes* which accounts for the general mode of metabolism of the homologs (see Fig. 3). The authors also use the diagram to explain the stereochemical control of the enzyme system, based on the observation that one epimeric alcohol predominates in the hydroxylation of **28** or **32**. This argument is based on a favored binding of one alkyl group over the other, implying, as the authors explain, that the two alkyl groups are not equivalent. An alternative explanation would be based on the prochirality of the ω-1 position (HANSON, 1966; FLOSS, 1970; TESTA, 1979)

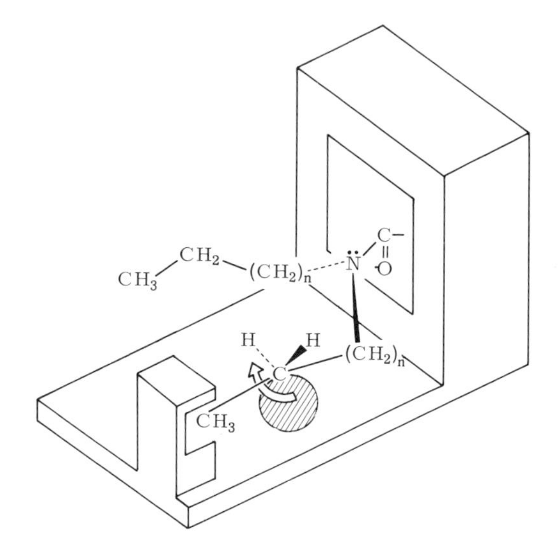

Figure 3. Proposed active site for the ω-1-hydroxylase of *Streptomyces roseochromogenes* (from ISHII et al., 1979b; reprinted with permission of the authors).

where the enzyme could favor oxygen insertion in the (e. g.) C-$H_{pro\text{-}S}$ bond (as indicated) on either (equivalent) alkyl group, with only minor C-$H_{pro\text{-}R}$ insertion. Such an "active-site map" should facilitate predicting acceptable substrates and stereochemical outcome on related compounds as has been accomplished with other classes of substrates in microbial transformations (ABBOTT, 1979).

3. Clavine and ergoline alkaloids

The major microbiological studies associated with clavine-type alkaloids have been directed towards an elucidation of alkaloidal biosynthetic interrelationships in *Claviceps purpurea* (BEREDE and SCHILD, 1978; REHACEK, 1980). Nevertheless, various organisms have been shown to catalyze specific type-reactions observed with other indole alkaloids. For example, a variety of *Streptomyces* species including *S.roseochromogenes, S.rimosus,* and *S.purpurescens* catalyze the *N*-dealkylation of agroclavine (**36**) to noragroclavine (**37**). *Corticium sasakii* converts **36** to 2-hydroxyagroclavine (**38**) (YAMANO et al., 1962).

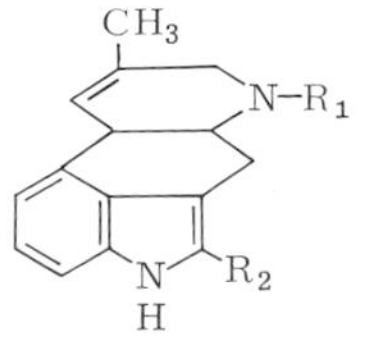

36 $R_1 = CH_3$, $R_2 = H$
37 $R_1 = R_2 = H$
38 $R_1 = CH_3$, $R_2 = OH$

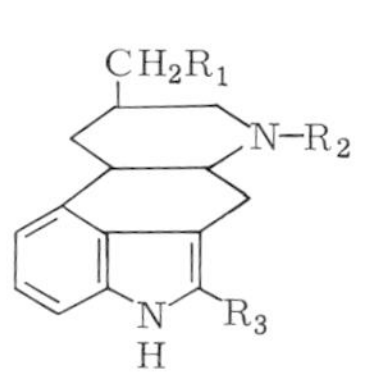

39 $R_1 = CN$, $R_2 = CH_3$, $R_3 = Cl$
40 $R_1 = CN$, $R_2 = H$, $R_3 = Cl$
41 $R_1 = S{-}CH_3$, $R_2 = n\text{-}Pr$, $R_3 = H$
42 $R_1 = SO{-}CH_3$, $R_2 = n\text{-}Pr$, $R_3 = H$
43 $R_1 = SO_2{-}CH_3$, $R_2 = n\text{-}Pr$, $R_3 = H$

The semi-synthetic ergoline lergotrile (**39**) is of interest because of its potent dopaminergic activity, more specifically, its ability to inhibit prolactin secretion. Toxicity in clinical trials prompted the exploration of microbial transformations as a method of producing less toxic metabolites, and to draw parallels to mammalian metabolism. Of thirty-eight cultures examined, *Cunninghamella echinulata* UI 3655, *Streptomyces rimosus* ATCC 23 955, *S.platensis* NRRL 2364, *S.spectabilus* UI-C632, and *S.flocculus* ATCC 25 453 produced norlergotrile (**40**), a route of metabolism also observed in mammals. *S.platensis* exhibited complete conversion, and preparative-scale incubations allowed for a 50% isolation of **40** (DAVIS et al., 1979).

Microbial systems also parallel mammalian metabolism with another dopaminergic ergoline, pergolide (**41**). Metabolism in mammals centers on the methylsulfide moiety, which is sequentially oxidized to the sulfoxide **42** and the sulfone **43**. Similarly, *Aspergillus alliaceus* UI-315 catalyzed the same sequential oxidative transformations to yield **42** and **43**. In contrast, *Helminthosporium* sp. (NRRL 4761) stops at the sulfoxide stage, and also catalyzes reduction of the sulfoxide back to pergolide (SMITH et al., 1983). No stereoselectivity in sulfoxide formation was observed, in contrast to the high degree of product stereoselectivity often observed in this microbial type-reaction (AURET et al., 1981).

4. Rauwolfia (yohimbine) alkaloids

Alkaloids from *Rauwolfia serpentina* have been used for centuries in numerous folklore medications, and have been used clinically in the treatment of hypertension (SANNERSTEDT and CONWAY, 1970) and certain psychiatric disorders (GOODMAN et al., 1980). They are also of interest as pharmacological tools for differentiating α-

adrenergic receptor sub-types in mammalian systems. Several systematic investigations of the ability of microorganisms to modify these complex alkaloids were conducted over the last three decades. The initial paper by PAN and WEISENBORN (1958) described the use of *Streptomyces aureofaciens* ATCC 11 834 and *S.rimosus* NRRL 2234 to catalyze the 18α-hydroxylation of yohimbine (**44**) and α-yohimbine (**45**) to form the corresponding glycols **46** and **47**, respectively. The related alkaloids β-yohimbine methyl ether (**48**), apoyohimbine (**49**), and 3-*epi*-apoyohimbine (**50**) were metabolized by *Cunninghamella blakesleeana* (Lendner) by aromatic hydroxylation in the 10-position (GODTFREDSEN, 1960; GODTFREDSEN et al., 1958). In addition, apoyohimbine appeared to yield the 18-hydroxy derivative. LOO and REIDENBERG (1959) also oberved 10-hydroxylation with yohimbine (**44**) using *Streptomyces platensis* NRRL 2364. Later studies by PATTERSON and co-workers (PATTERSON et al., 1963) described the same reaction for β-yohimbine (**51**). In this case, the addition of ascorbic acid as an antioxidant facilitated isolation of the labile hydroxyindole product. This reaction was later reported to be catalyzed by several Streptomycetes and fungi (MEYERS and PAN, 1961).

The first report of 11-hydroxylation of yohimbine alkaloids was that of PATTERSON et al. (1963). A species-dependent regioselectivity was observed in that *Cunninghamella bertholletiae* NRRL A-11 497 hydroxylated both yohimbine (**44**) and α-yohimbine (**45**) to the 11-phenols, while *C.echinulata* NRRL A-11 498 produced the 10-phenols. *Calonectria decora* (CBS) hydroxylated yohimbine only, and in the 18α-position to give **46**. These three hydroxylations were later extended to other *Cunninghamella* and *Streptomyces,* and were also shown to be strain specific (HARTMAN et al., 1964). For example, *C.bainieri* ATCC 9244 catalyzed the 10-hydroxylation of yohimbine (**44**), while the Campbell X-48 strain hydroxylated the 11-position. The regiospecificity of hydroxylation with all strains was upheld on extension to the related alkaloid, corynanthine (**52**).

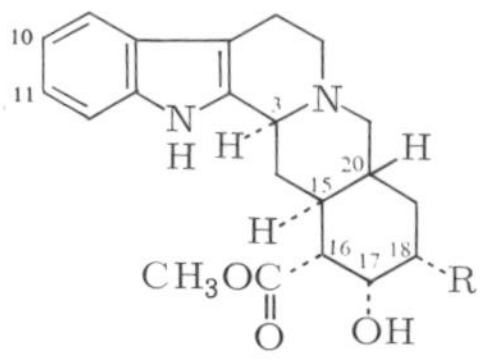

44 R = H
46 R = OH

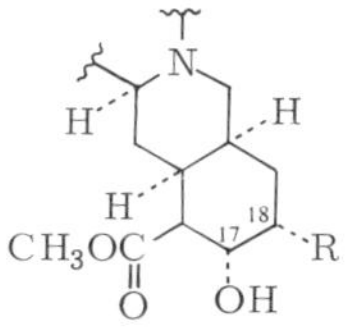

45 R = H
47 R = OH

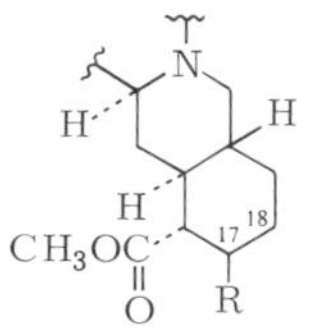

48 R = OCH_3
51 R = OH

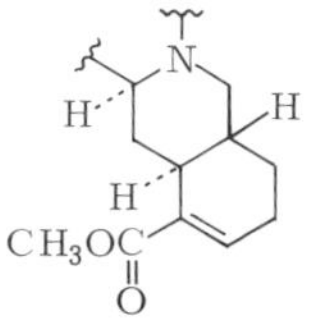

49

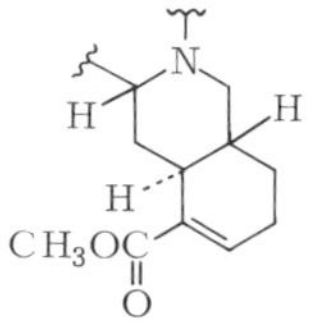

50

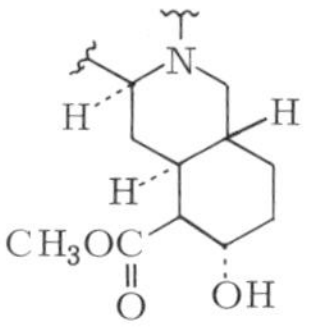

52

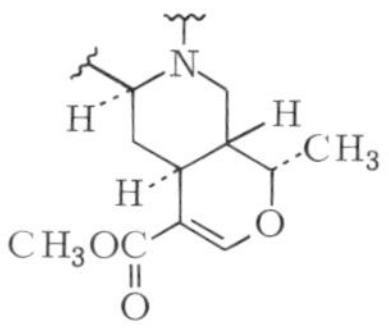

53

While the majority of the foregoing examples involved low (<15%) conversions to hydroxylated yohimbine alkaloid metabolites, the study by BELLET and VAN-THUONG (1970a; 1972) demonstrated the clear preparative utility of these reactions in their synthesis of anticonvulsant derivatives. Specifically, *Gongronella urceolifera* (CBS) catalyzed a 90% conversion of corynanthine (**52**) to the 10-hydroxy metabolite, and also converted ajmalicine (**53**) to its 10-hydroxy metabolite (40%). A similar chemical transformation in this series has recently been reported (HANNART, 1971).

A systematic study of the influence of substitution at the 16-, 17-, and 18-positions on C-10 hydroxylation by *Cunninghamella blakesleeana* allowed ADAM et al. (1973) to establish minimum structural requirements, substituent effects, and stereochemical features allowing for hydroxylation. Generally, the presence of electron-rich (i. e., carbonyl or sulfate) groups on the α-face of the molecule at C-16 or C-17 was required for transformation. By contrast, substitution with hydroxyl groups in these positions yield compounds which are not significantly metabolized. This study represents a most significant approach to understanding structural features allowing for metabolism in an alkaloidal series.

5. Pyridocarbazole alkaloids (ellipticine)

Ellipticine (**54**), a pyridocarbazole alkaloid from *Ochrosia* and *Aspidosperma* sp., is noted for its activity as an antitumor agent, and as a potent inhibitor of microsomal monooxygenases (LESCA et al., 1979). CHIEN and ROSAZZA (1979a; 1980) explored the microbial metabolism of ellipticine and its 9-methoxy analog **55** (CHIEN and ROSAZZA, 1979b) and observed that *Aspergillus alliaceus* NRRL 315 oxidized ellipticine to a mixture of 8- and 9-hydroxyellipticines (**56** and **57**, respectively) in an approx. 1:4 ratio. The latter compound is a known mammalian metabolite, and is also the most potent ellipticine derivative in antitumor activity. 8-Hydroxyellipticine (**56**) is a new compound (CHIEN and ROSAZZA, 1979a). When medium and incubation alteration studies failed to enhance yields, attempts were made to enhance solubility using non-toxic solubilizing agents including polyvinylpyrrolidone and polyoxyethylated vegetable oils (Emulphor 620). These agents doubled the extent of conversion, while the ratio of products was unaffected, indicating the possibility of a common intermediate, perhaps an 8,9-arene oxide (CHIEN and ROSAZZA, 1980).

The related alkaloid 9-methoxyellipticine (**55**) is *O*-dealkylated to 9-hydroxyellipticine (**57**) by *Cunninghamella echinulata* NRRL 1386, *Botrytis allii* NRRL 2502, and

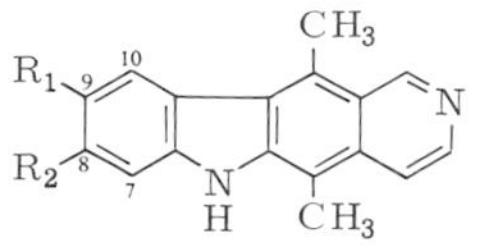

54 $R_1 = R_2 = H$
55 $R_1 = OCH_3$, $R_2 = H$
56 $R_1 = H$, $R_2 = OH$
57 $R_1 = OH$, $R_2 = H$

Penicillium brevi-compactum ATCC 10 418, the latter used to prepare the product in 38% yield (CHIEN and ROSAZZA, 1979b). Numerous other cultures were also shown to catalyze the same reaction on an analytical scale. Structure proof included chemical *O*-dealkylation of **55** to the same product.

6. β-Carbolines

Interest in β-carbolines stems from their reported CNS activities and the possibility that compounds such as β-carboline-3-carboxylic acids may serve a role in neuromodulation *in vivo* (BRAESTRUP et al., 1980; TENEN and HIRSCH, 1980). The earliest report of microbial transformations in this group of alkaloids was the observation by SKRYABIN et al. (1974) that the alkaloid brevicarine (**58**) from *Carex brevicollis* is converted to brevicolline (**59**) by the action of several *Penicillium* and *Rhizopus* species. The reaction presumably involves benzylic oxidation followed by cyclization.

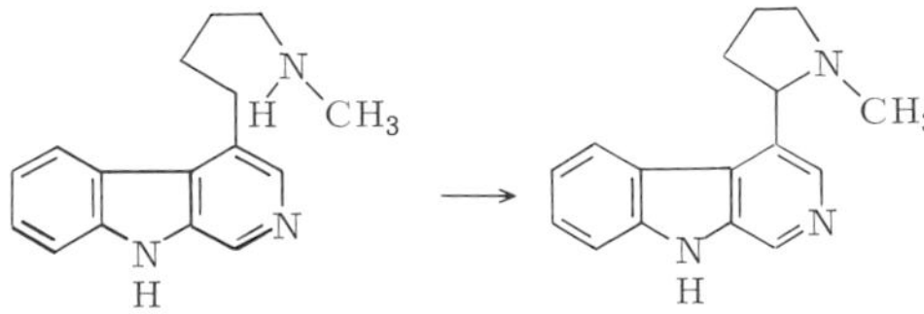

58 → **59**

β-Carbolines of the homologous series **60** to **63** were examined by NEEF et al. (1982) who observed that *Sporotrichum* (*Beauveria*) *sulfurescens* ATCC 7159 catalyzed hydroxylation of **60** in the electronically predictable 6- and 8-positions, followed by conjugation, resulting in the isolation of the

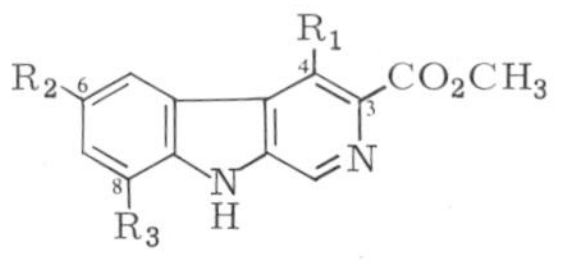

60 $R_1 = R_2 = R_3 = H$
61 $R_1 = CH_3$, $R_2 = R_3 = H$
62 $R_1 = CH_2CH_3$, $R_2 = R_3 = H$
63 $R_1 = CH_2CH_2CH_3$, $R_2 = R_3 = H$
64 $R_1 = R_3 = H$, $R_2 = OH$
65 $R_1 = R_3 = H$, $R_2 =$ *O*-methylglucoside
66 $R_1 = R_2 = H$, $R_3 =$ *O*-methylglucoside
67 $R_1 = CH_3$, $R_2 =$ *O*-methylglucoside, $R_3 = H$
68 $R_1 = CH_3$, $R_2 = H$, $R_3 =$ *O*-methylglucoside
69 $R_1 = CH_2CH_3$, $R_2 = H$, $R_3 =$ *O*-methylglucoside
70 $R_1 = CH_2CH_2CH_3$, $R_2 = H$, $R_3 =$ *O*-methylglucoside

6-hydroxy derivative **64** as the major metabolite, and minor amounts of the 6- and 8-*O*-4′-*O*-methyl-*β*-glucosides **65** and **66.** Further examination in the homologous series indicated a shift in regioselectivity, with the methyl analog **61** producing a 50:50 mixture of the 6- and 8-hydroxy conjugates **67** and **68**, while the ethyl (**62**) and propyl (**63**) homologs produced only the corresponding 8-hydroxy conjugates, **69** and **70**, respectively. Only aliphatic hydroxylation was observed using *Streptomyces lavendulae* ATCC 8664 and *S.griseus* ATCC 10 137, yielding (presumably by intramolecular transesterification) the lactones **71** and **72.** The production of an optically active lactone **72** indicated that a high degree of product enantioselectivity was operative (absolute configuration not determined).

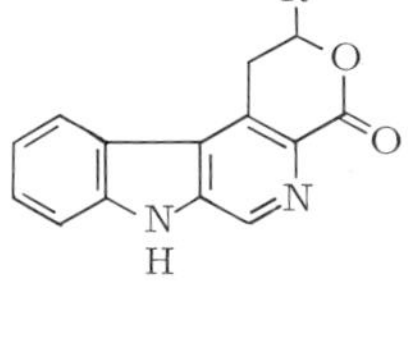

71 R = H
72 $R = CH_3$

7. Miscellaneous indole alkaloids

Several other microbial conversions representing principally oxidative reactions have been described for several misc. indole alkaloids. For example, coronaridine (**73**)

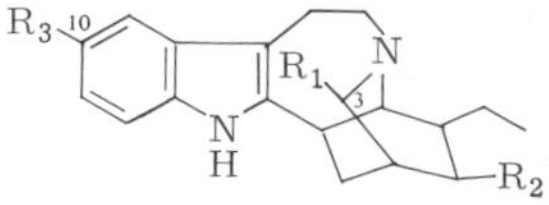

73 $R_1 = R_2 = R_3 = H$
74 $R_1 = R_2 = H$, $R_3 = OH$
75 $R_1 = {=}O$, $R_2 = R_3 = H$
76 $R_1 = {=}O$, $R_2 = OH$, $R_3 = H$
77 $R_1 = R_3 = H$, $R_2 = OH$

undergoes aromatic hydroxylation with *Sporotrichum sulfurescens* ATCC 7159 to yield the 10-phenol **74**, oxidation to the lactam **75**, and further hydroxylation to the 15-hydroxy lactam **76**. *Cunninghamella blakesleeana* metabolizes **73** to 15-hydroxycoronaridine (**77**) (GARNIER, 1975). Ajmaline (**78**) is first *N*-dealkylated to **79**, and then hydroxylated in the 10-position to yield **80** in 20% yield by *Streptomyces platensis* NRRL 2364 (BELLET and VANTHUONG, 1970a). Several other Streptomycetes simply *N*-dealkylate **78** to **79** (BELLET and VANTHUONG, 1969).

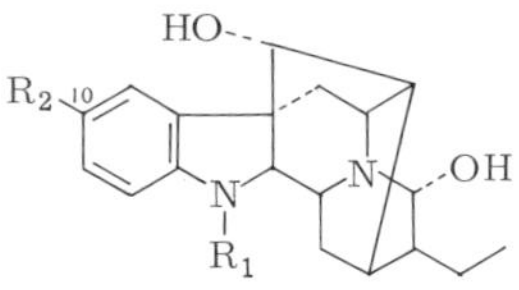

78 $R_1 = CH_3$, $R_2 = H$
79 $R_1 = R_2 = H$
80 $R_1 = H$, $R_2 = OH$

Finally, strychnine (**81**) and brucine (**82**) are oxidized to the corresponding *N*-oxides by *Bacillus thuringiensis* ATCC 10 791 (BELLET and GERARD, 1962; 1965; PHILIP-

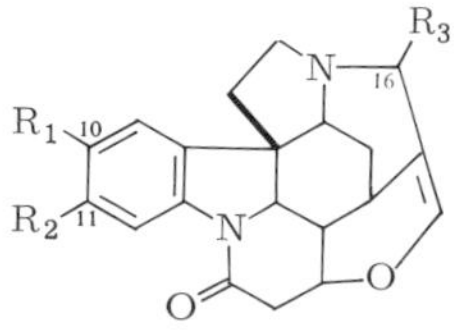

81 $R_1 = R_2 = R_3 = H$
82 $R_1 = R_2 = OCH_3$, $R_3 = H$
83 $R_1 = R_2 = H$, $R_3 = OH$

SON, 1971) and by *Helicostylum piriforme* in preparative (50%) yield (NGUYEN-DANG and DISSET, 1968). *H.piriforme* also catalyzed the 16α-hydroxylation of strychnine to yield **83**. The biodegradation of these compounds has also been reported with *Arthrobacter* sp. (BUCHERER, 1965; BUCHERER and NIEMER, 1962; NIEMER and BUCHERER, 1965).

B. Isoquinoline Alkaloids

1. Simple tetrahydroisoquinolines

Recently, COUTTS, HAMBLIN, et al. (1979a) described the use of a laccase from the basidiomycete *Polyporus versicolor* induced with 2,5-dimethylaniline, to catalyze the oxidative decarboxylation of several tetrahydroisoquinoline-1-carboxylic acids (e. g., **84**) to yield the corresponding 3,4-dihydroisoquinolines (e. g., **85**). Non-phenolic substrates were not oxidized. The enzymatic system could also oxidize the corresponding tetrahydro-1-benzylisoquinoline-1-carboxylic acids (e. g., **86**) to yield **87**. These results paralleled those obtained using horseradish peroxidase/hydrogen peroxide, and were designed to expand upon current theories that the isoquinoline alkaloids are derived by enzymatic decarboxylation of analogous precursors.

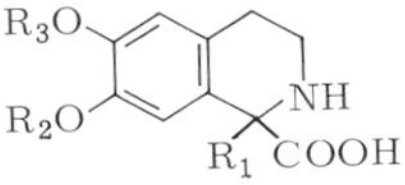

84 $R_1 = R_2 = CH_3$, $R_3 = H$
86 $R_1 = CH_2Ph$, $R_2 = H$, $R_3 = CH_3$

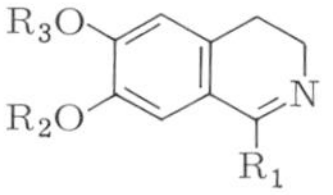

85 $R_1 = R_2 = CH_3$, $R_3 = H$
87 $R_1 = CH_2Ph$, $R_2 = H$, $R_3 = CH_3$

2. Sanguinarine

The cytotoxic quaternary isoquinoline alkaloid sanguinarine (**88**) derived from *Sanguinaria canadensis* undergoes reduction at the quaternary iminium site to yield dihydrosanguinarine (**89**) using *Verticillium dahliae* (HOWELL et al., 1972). The strain examined was initially observed to be resistant to the toxic effects of this agent, now explainable based on the fact that this route of metabolism represents a mode of detoxification for the organism. The reduction of imines and iminium species by microorganisms represents a largely unexplored reaction with excellent potential in natural products synthesis related to alkaloids. The utility of microbial systems in ketone reductions is well documented (SIH and ROSAZZA, 1976), and is especially useful where a high degree of stereoselectivity is observed. In direct electronic analogy, imines should be as readily reduced with the potential of enzymatically introducing a chiral center adjacent to the resultant amine.

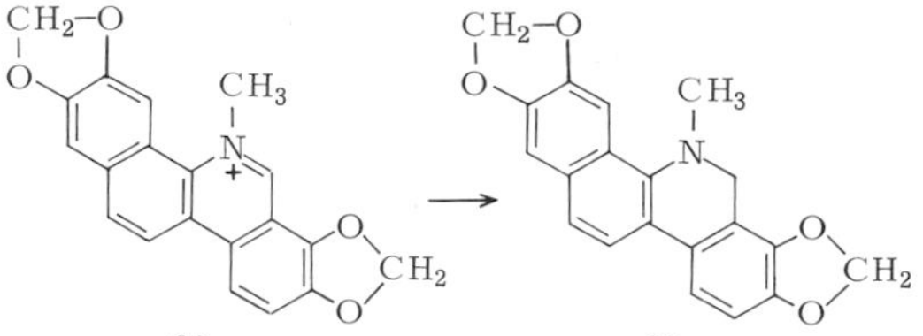

88 → **89**

3. Aporphines

Aporphines represent a pharmacologically interesting class of isoquinoline alkaloids based on profound CNS and peripheral effects (SHAMMA, 1972). The first report of microbial transformations in this class was that of ROSAZZA et al. (1975) involving derivatives of the dopaminergic aporphine, apomorphine (**90**), and described the regiospecific *O*-demethylation of apomorphine dimethyl ether (**91**). Specifically, *Streptomyces griseus* SP-WI-1158, NRRL B-8090 catalyzed the *O*-demethyla-

90 $R_1 = R_2 = H$
91 $R_1 = R_2 = CH_3$
92 $R_1 = H$, $R_2 = CH_3$
93 $R_1 = CH_3$, $R_2 = H$

tion of **91** to yield a mixture of apocodeine (**92**) and isoapocodeine (**93**), *Streptomyces rimosus* ATCC 23 955 catalyzed a highly regioselective reaction yielding **92** with only traces of **93**, and *Cunninghamella elegans* ATCC 9245 produces **93** exclusively (all on analytical scale). The latter reaction is of particular interest, since isoapocodeine represents only a minor product in the enzymatic (catechol *O*-methyl transferase) and chemical methylation of apomorphine (CANNON et al., 1972) as well as in chemical methods of demethylating **91** by nucleophiles (AHMAD et al., 1977). This is explainable based on a less sterically hindered C-10 position for alkylation or dealkylation. Hence, the availability of **93** for pharmacological testing has been limited. The preparative-scale utility of the production of **93** using *C.elegans* was demonstrated (59% preparative yield) utilizing growing cultures in the presence of ascorbic acid (2 mg/mL) as an antioxidant, and was also shown to be quantitative in cell-suspension cultures over a pH range of 3–6 (SMITH and DAVIS, 1978). These studies clearly show the high degree of regioselectivity exhibited by microbial systems, and their adjunct role in organic synthesis.

The aporphine *S*-(+)-glaucine (**94**) exhibits potent antitussive action (DIERCKX, 1981; CHACHAJ et al., 1972; ALESHINSKAYA, 1976) and serves as a model aporphine for certain antitumor alkaloids (*vide infra*). DAVIS et al. (1977b) described the *N*-dealkylation and regiospecific *O*-dealkylation of glaucine using *Streptomyces griseus* UI 1158 to yield norglaucine (**95**, 11%) and 2-*O*-demethylglaucine (predicentrine, **96**, 14%), respectively. Neither dealkylation was stereoselective. The *N*-dealkylation of aporphines represents a novel reaction of excellent synthetic potential, since classical chemical methods yield ring cleavage products (SMISSMAN et al., 1979; BORGMAN et al., 1975). The position of *O*-dealkylation was determined in part by the base (NaOD) induced shift of the aromatic protons in PMR spectra, with H-3 (*ortho* to the phenol) exhibiting the expected upfield shift. A third major reaction observed with **94** was oxidation to 6a,7-dehydroglaucine (**97**) catalyzed by *Fusarium solani* ATCC 12 823. Preliminary studies indicated that a stereoselective oxidation was operative with this organism, since optically enriched R-(−)-glaucine (**98**) was isolated after a 7 day incubation when racemic glaucine (**94**/**98**) was used as substrate.

94 $R_1 = R_2 = CH_3$ (6a*S*, as shown)
95 $R_1 = H$, $R_2 = CH_3$
96 $R_1 = CH_3$, $R_2 = H$
98 $R_1 = R_2 = CH_3$ (6a*R*-enantiomer)

97 R = H
99 $R = CH_3$

This initial observation was confirmed by examining the pure enantiomers *R*-glaucine (**98**) and *S*-glaucine (**94**) as well as racemic glaucine as substrate (DAVIS, 1980; DAVIS and ROSAZZA, 1981). *S*-(+)-Glaucine was quantitatively oxidized to dehydroglaucine while *R*-glaucine was not metabolized. Racemic glaucine was oxidized to the expected extent of 50%. Thus, this reaction proceeds with complete substrate stereospecificity, and is quantitative. The results with *RS*-glaucine suggested the utility of this method in the "destructive resolution" of racemic glaucine to prepare the unnatural *R*-enantiomer (**98**). This was accomplished at the gram level by metabolism of *RS*-glaucine while monitoring residual substrate concentration (which levels at 50% consumption) and optical enrichment in **98**, leading to an isolation of **98** in 90% theoretical yield and 96% enantiomeric excess (DAVIS and TALAAT, 1981). Since most natural aporphines are of the *S*-chirality (SHAMMA, 1972), and are easily chemically racemized, or are chemically synthesized as the racemate (SHAMMA, 1972), a convenient tool is available for preparing (by resolution) the unnatural *R*-enantiomers for pharmacological testing and use in natural products chemistry.

The use of *Fusarium solani* ATCC 12823 in the destructive resolution of *RS*-glaucine can be used to illustrate the general "two-stage" fermentation procedure used by many investigators for either analytical-scale or preparative-scale studies (DAVIS and ROSAZZA, 1981). Readers interested in an extensive discussion of the various techniques utilized in microbial transformations are referred to the excellent review by GOODHUE (1982). *F.solani* stock cultures were maintained on refrigerated (2 °C) slants on Sabouraud-maltose agar, and transferred to fresh slants at 4–6 month intervals to maintain viability. It is generally desirable with most sporulating cultures to insure heavy sporulation on slants before cultures are subjected to refrigerator temperatures, and this is particularly critical with this organism. The medium for submerged culture experiments consisted of the following: dextrose, 20 g; Acidicase peptone, 5 g; dibasic potassium phosphate, 5 g; sodium chloride, 5 g; yeast extract, 5 g; distilled water, 1000 mL (pH adjusted to 7.0 with HCl prior to sterilization in individual flasks at 121 °C for 15 minutes). "First-stage" cultures were generated in 125 mL-Bellco DeLong culture flasks (which use a metal closure to prevent contamination and allow reproducible aeration), with each flask containing 25 mL of medium. Growth was initiated by suspending spores from slants into sterile medium by agitation with an inoculating needle, and then transferring the spore suspension to the 125 mL-flasks under aseptic conditions. Incubations were conducted on a gyrorotary (circular stroke) shaker at 250 rpm and 27 °C. After 72 hours, 2 mL of the fluid, homogeneous growth could be used to inoculate each "second-stage" of the same size (i.e., 125 mL) for analytical-scale experiments. Alternatively, 10 mL of the first-stage growth could be used to inoculate each 1 liter-flask containing 200 mL of the medium for preparative-scale experiments. After the incubation was allowed to continue for 24 hours, the substrate (in this case, glaucine) was added to a concentration of 500 micrograms per mL using dimethylformamide as the solvent (50 μL or less DMF per 12.5 mg glaucine). In the destructive resolution of *RS*-glaucine using fifteen 1 liter-culture flasks, 1.5 g of *RS*-glaucine in 5 mL DMF was aseptically distributed equally among the flasks. Samples of 5.0 mL could be harvested at convenient (24 hour) time intervals for analysis by high performance liquid chromatography and optical rotation determination. As expected, the conversion proceeded to an extent of 50% (only *S*-glaucine **94** was metabolized) and optical rotation indicated enrichment in (unmetabolized) *R*-(−)-glaucine (**98**), with the reaction completed after 96 hours. As described elsewhere (DAVIS and TALAAT, 1981) the product of the destructive resolution (**98**) was isolated and purified to give a 90% theoretical yield in 96% enantiomeric excess.

R-(−)-Glaucine (**98**) is also a substrate for oxidation by certain microorganisms. Indeed, studies with *Aspergillus flavipes* ATCC 1030 indicated that this culture catalyzed the quantitative and stereospecific oxidation of this enantiomer to dehydroglaucine (**97**) (DAVIS and TALAAT, 1981). Presumably this system could be used in a destructive resolution analogous to that described for *Fusarium solani* in the production of certain unnatural *S*-aporphines. This observation illustrates that organisms catalyzing the same reaction with reverse stereoselectivity ("*R*-" and "*S*-organisms") exist with respect to aporphine oxidations, in analogy to other synthetically useful micro-

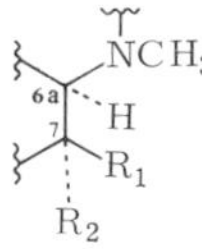

100 $R_1 = H_S$, $R_2 = H_R$
102 $R_1 = CH_3$, $R_2 = H$ (6a*S*, 7*S*)
103 $R_1 = H$, $R_2 = CH_3$ (6a*S*, 7*R*)

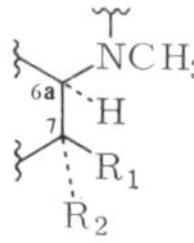

101 $R_1 = H_S$, $R_2 = H_R$
104 $R_1 = H$, $R_2 = CH_3$ (6a*S*, 7*R*)
105 $R_1 = CH_3$, $R_2 = H$ (6a*R*, 7*S*)

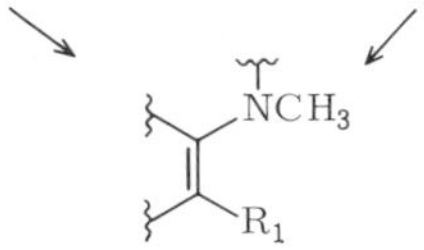

97 R = H
99 $R = CH_3$

Figure 4. Stereochemical course of glaucine oxidation by *Fusarium solani* and *Aspergillus flavipes.*

bial reactions such as ketone reductions (SIH and ROSAZZA, 1976) and ester hydrolyses (SIH et al., 1975). Oxidations with both organisms also compliment chemical methods of aporphine oxidations which are non-stereoselective (CAVA et al., 1972; 1975).

As illustrated in Fig. 4, oxidation of *R*- or *S*-glaucine to **97** involves not only the chiral position at C-6a, but also the prochiral position 7, since the protons at this position are not equivalent, and can be considered diastereotopic (TESTA, 1979). Oxidation could thus involve either a *cis*- or *trans*-elimination of hydrogen (or no stereoselectivity at C-7 might be observed). For example, oxidation of *S*-(+)-glaucine by *F.solani* could proceed by the loss of the 6a-*S* and 7-*proS*-hydrogens in a *trans*-elimination, or the 6a*S* and 7-*proR*-hydrogens in a *cis*-elimination. The approach used to determine the overall stereochemistry of these reactions involved the metabolism of 7-methylglaucine analogs **102–105** which would restrict the stereochemical course of elimination (KERR and DAVIS, 1983). These four stereoisomers represent two diastereomeric sets of enantiomers, *cis*-7-methylglaucine (**102/104**) and *trans*-7-methylglaucine (**103/105**). In rationalizing expected routes of metabolism by *F.solani,* one would expect the oxidation of the 6a*S*,7*S*-enantiomer (**102**) to 7-methyldehydroglaucine (**99**) if a *cis*-elimination occurs, or the 6a*S*,7*R*-enantiomer (**103**) if a *trans*-elimination is operative. Similar arguments apply to **104** and **105** metabolism by *Aspergillus flavipes*. When racemic *trans*-7-methylglaucine was used as substrate (**103/105**), neither organism allowed oxidation, whereas racemic *cis*-7-methylglaucine (**102/104**) was metabolized to the expected extent of 50% by each organism. The use of chiral shift reagents and ORD confirmed the stereochemistry of the residual substrate in each case, confirming that *Fusarium solani* metabolized preferentially **102** and *Aspergillus flavipes* only **104**. Thus both organisms catalyze an overall "*cis*"-elimination of hydrogen, and maintain stereochemical preference at the 6a-position (KERR and DAVIS, 1983).

Recently ROSAZZA and co-workers have developed methods using a commercially available microbial enzyme in a crucial step for the synthesis of analogs of the antitumor alkaloids thalicarpine and hernandaline (*vide infra*) (ROSAZZA and YANG, 1982). Boldine diacetate (**106**) (synthesized from boldine, **107**, using acetyl chloride and trifluoroacetic acid) was regioselectively hydrolyzed by a lipase from *Candida cylindracea* to yield boldine-9-acetate (**108**). Methylation and acetate cleavage yielded the monophenol synthon **109** for coupling with various aryl synthons for the preparation of the analogs. This study is an excellent ex-

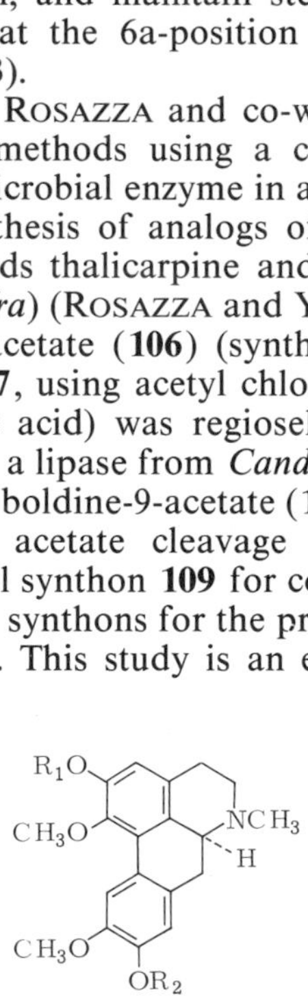

106 $R_1 = R_2 = COCH_3$
107 $R_1 = R_2 = H$
108 $R_1 = H$, $R_2 = COCH_3$
109 $R_1 = CH_3$, $R_2 = H$

ample of the high degree of regioselectivity available using microbial systems and exploitation in synthesis.

The aporphine *R*-(−)-nuciferine (**110**) also serves as a substrate for *Streptomyces griseus* UI 1158 with resultant *N*- and *O*-dealkylation, as well as aromatic hydroxylation to yield principally nornuciferine (**111**), 2-*O*-desmethylnuciferine (**112**), *N*(2)-bis-nornuciferine (**113**), and 9-hydroxynuciferine (**114**) (GUSTAFSON and ROSAZZA, 1979). Product **113** was shown to form by sequential *N*-, then *O*-dealkylation. The authors proposed a model for the active site of the oxygenase (dealkylase, hydroxylase) based on intramolecular distances of accepted substrates and positions of oxygenation, including apomorphine dimethyl ether (**91**), glaucine (**94**), papaverine (**115**, *vide infra*), vindoline (**3**, *vide supra*), and simpler structural analogs. This model should be of excellent utility for predicting substrate specificity with this versatile organism. Similar models have been proposed for other classes of microbial substrates (ABBOTT, 1979).

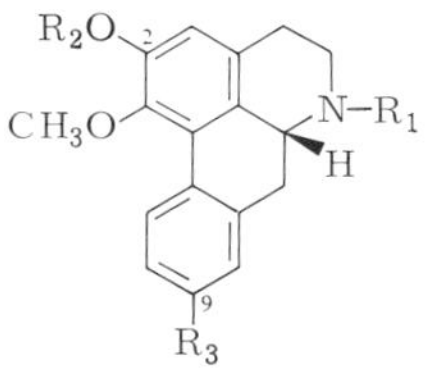

110 $R_1 = R_2 = CH_3$, $R_3 = H$
111 $R_1 = R_3 = H$, $R_2 = CH_3$
112 $R_1 = CH_3$, $R_2 = R_3 = H$
113 $R_1 = R_2 = R_3 = H$
114 $R_1 = R_2 = CH_3$, $R_3 = OH$

4. Benzylisoquinolines

The alkaloid papaverine (**115**), derived from the opium poppy *Papaver somniferum*, has no narcotic activity, but is widely prescribed as a smooth muscle relaxant, especially for cerebral vascular disorders. ROSAZZA et al. (1977) conducted parallel microbial and *in vitro* mammalian metabolism studies on this substrate, resulting in the demonstration of clear parallels in patterns

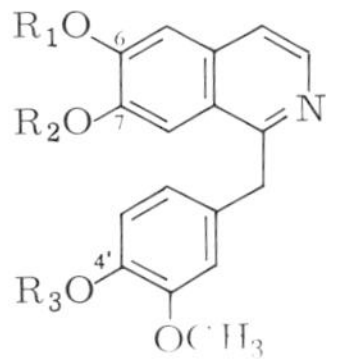

115 $R_1 = R_2 = R_3 = CH_3$
116 $R_1 = H$, $R_2 = R_3 = CH_3$
117 $R_1 = R_3 = CH_3$, $R_2 = H$
118 $R_1 = R_2 = CH_3$, $R_3 = H$

of *O*-dealkylation. In hepatic microsomal preparations from phenobarbital-induced rats, *O*-dealkylation in the 6-, 7-, and 4′-positions yielded the corresponding monophenols **116–118**, respectively, with **117** predominating. In comparison, hepatic microsomes from guinea pigs yielded the 6- and 4′-*O*-dealkylated products, **116** and **118** in equal amounts. Similar patterns of *O*-dealkylation were reported previously by other authors. Ten of sixty cultures initially screened yielded the same metabolites obtained in mammals, in varying proportions. In preparative-scale experiments, *Aspergillus alliaceus* UI 315 produced 6-*O*-desmethylpapaverine (**116**) exclusively (40%), while *Cunninghamella echinulata* ATCC 9244 produced 4′-*O*-desmethylpapaverine (**118**, 27%) with minor amounts of other phenols. Comparisons of microbial and mammalian metabolism are particularly interesting in this case, since the phenolic metabolites, like the parent agent, are active in inhibiting cyclic-AMP-phosphodiesterase.

Other studies concerning the metabolism of papaverine include the isolation by enrichment culture of a *Nocardia* strain capable of the biodegradation of the alkaloid by complex pathways involving oxidative ring and benzylic cleavages (HAASE-ASCHOFF and LINGENS, 1979). Studies using mutant strains of this organism blocked at various stops in the biodegradation allowed a refined and more extensive proposal for the pathways of biodegradation (HAUER et al., 1982).

SCHOENEWALDT et al. (1974) described the use of several growing or resting cell cultures of Schizomycetes and Eumycetes for the intramolecular oxidative (*ortho*-

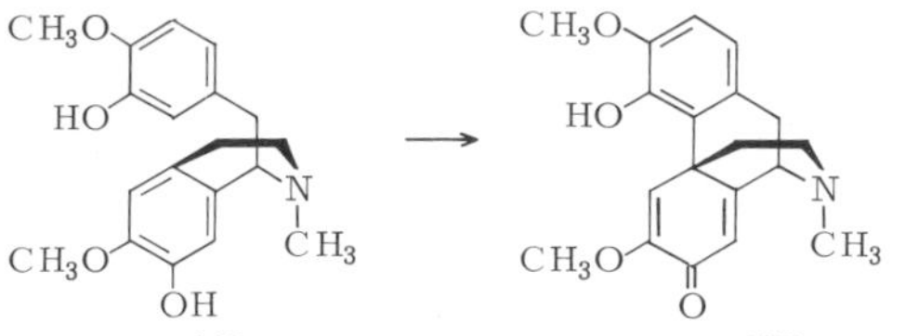

119 → 120

para) coupling of (−)-reticuline (**119**) to (+)-salutaridine (**120**). Racemic reticuline also served as a substrate. Pseudomonads and Streptomycetes exhibited the highest activity, although yields were typically low (1–2%). It is of interest to note that the oxidative phenolic coupling of reticuline has also been reported in mammals, although an alternative mode of coupling (*para-para*) yields a different alkaloid, pallidine (KAMETANI et al., 1977). This reaction could be of particular interest for the production of narcotic analgesics of the morphine type. Two major limitations to this approach have existed in the past: first, a lack of a ready source of reticuline, has now been solved (RICE and BROSSI, 1980; RICE et al., 1980); the second problem is the lack of an efficient method for the conversion illustrated above (BROSSI, 1982). If yields could be enhanced, this could develop into a commercially useful process.

The benzyltetrahydroisoquinoline *RS*-laudanosine (**121**), serving as a model for more complex dimeric alkaloids, was metabolized by *O*-dealkylation by five of sixty cultures screened (DAVIS and ROSAZZA, 1976). *Cunninghamella blakesleeana* ATCC 9244 exhibited regiospecific dealkylation of **121** in the 4′-position to yield pseudocodamine (**122**). The position of *O*-demethylation was deduced by mass spectral correlations and chemical degradation. Since neither unreacted laudanosine nor the product were optically active, no substrate enantioselectivity was exhibited. *O*-Dealkylation of the less common 7,8,4′-trioxygenated tetrahydrobenzylisoquinolines was also demonstrated with a related strain, *C.blakesleeana* ATCC 8688a. REIGHARD et al. (1981) demonstrated the complete oxidation of **123**, and in preparative-scale experiments, isolated the 4′-phenolic metabolite **124** (5%), the 7-phenol **125** (2%), and trace amounts of the 8-phenol **126**. The latter compound was produced in highest yield by chemical demethylation, while the former two products are formed in only minor amounts by this method.

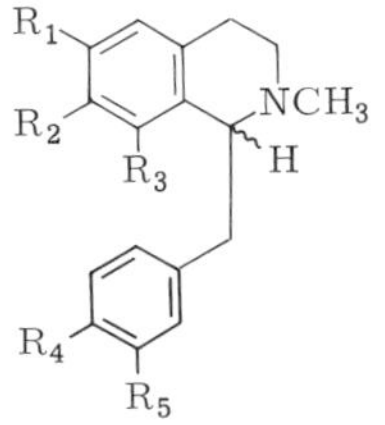

121 $R_1 = R_2 = R_4 = R_5 = OCH_3$, $R_3 = H$
122 $R_1 = R_2 = R_5 = OCH_3$, $R_3 = H$, $R_4 = OH$
123 $R_1 = R_5 = H$, $R_2 = R_3 = R_4 = OCH_3$
124 $R_1 = R_5 = H$, $R_2 = R_3 = OCH_3$, $R_4 = OH$
125 $R_1 = R_5 = H$, $R_2 = OH$, $R_3 = R_4 = OCH_3$
126 $R_1 = R_5 = H$, $R_2 = R_4 = OCH_3$, $R_3 = OH$

5. "Dimeric" isoquinoline alkaloids

With the antitumor bis-benzyltetrahydroisoquinoline (+)-tetrandrine (**127**) the type-reaction observed in microbial transformations shifted from *O*-dealkylation (as seen in simple benzylisoquinolines) to *N*-dealkylation. Specifically, *Streptomyces griseus* UI 1158 regiospecifically *N*-dealkylated **127** to only *N*(2′)-nortetrandrine (**128**) in 50% preparative yield (DAVIS and ROSAZZA, 1976). Seven other cultures yielded the same metabolite with lower yield. The preparative-scale reaction was uncomplicated by side-products, and unreacted (+)-tetrandrine was quantitatively recovered. In contrast to these results, *Cunninghamella blakesleeana* ATCC 8688a gave *N*(2)-nortetrandrine (**129**) in 20% yield, again, uncom-

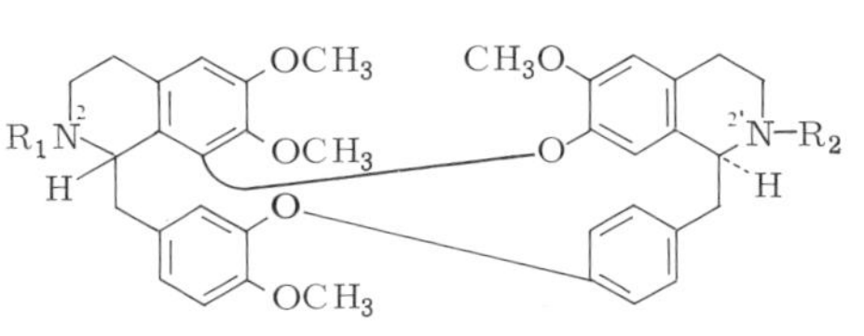

127 $R_1 = R_2 = CH_3$
128 $R_1 = CH_3$, $R_2 = H$
129 $R_1 = H$, $R_2 = CH_3$

plicated by side-products (Davis et al., 1977a). The structures of the metabolites were determined principally on the basis of the presence or absence of the *N*(2′)-methyl (2.62 ppm) or *N*(2)-methyl (2.33 ppm) in the PMR spectra, as well as comparison to related bis-benzylisoquinoline alkaloids. The authors discussed a "models approach" whereby simpler aporphine and benzylisoquinoline alkaloids were first examined by microbial transformation to narrow the field from sixty to twenty-two cultures which were then examined with tetrandrine. Of these twenty-two, twelve demonstrated the ability to metabolize **127**, all by *N*-dealkylation. Ten produced the *N*(2′)-nor-metabolite **128**, one produced the *N*(2)-nor-metabolite (**129**) (*vide supra*), and *C.echinulata* NRRL 3655 produced both **129** and **128**. Thus a high degree of regioselectivity could be obtained by proper choice of microorganisms, in contrast to chemical methods of *N*-dealkylation, which gives a mixture of both nor-compounds as well as bis-*N*(2),*N*(2′)-nortetrandrine (Davis et al., 1977).

Thalicarpine (**130**) is a benzyltetrahydroisoquinoline-aporphine dimer from *Thalictrum* and *Hernania* species, and is of interest because of its potential antitumor activity. In an attempt to propose routes of mammalian metabolism and to prepare metabolites for further biological testing, Nabih et al. (1977) reported that five of twenty-two microbial cultures were capable of metabolizing the alkaloid. In preparative-scale experiments, *Streptomyces punipalus* NRRL 3524 catalyzed an oxidative cleavage at the benzylic position of the benzylisoquinoline portion to yield hernandalinol (**131**), so-named because of its structural similarity to a known alkaloid, hernandaline (**132**). The metabolic intermediacy of hernandaline (**132**) was suggested based on the fact that this organism efficiently reduced the aldehyde to the corresponding benzyl alcohol (**131**) under the fermentation conditions employed. The structure was confirmed by spectral interpretation, comparison to hernandaline itself, and oxidative cleavage of thalicarpine using potassium permanganate (to yield **132**) followed by reduction with sodium borohydride to yield **131**. A similar type of benzylic cleavage was proposed by Haase-Aschoff in the biodegradation of papaverine, involving a retro-benzoin condensation (Haase-Aschoff and Lingens, 1979). The optical purity of the alcohol metabolite was determined by the use of a chiral shift reagent $Eu(facam)_3$, employing *RS*-glaucine as a model analyte. Under the conditions employed, the C-11 proton of glaucine appeared at 0.255 ppm further downfield in the *S*-(+)-enantiomer than in *R*-(−)-glaucine. With **131**, no separation of enantiomeric signals was observed, indicating that the metabolite was optically pure.

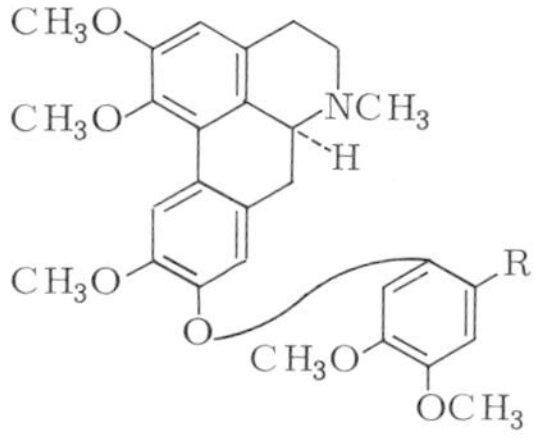

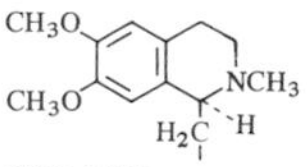

130 R = (benzyltetrahydroisoquinoline moiety)

131 R = CH_2OH

132 R = CHO

C. Quinoline Alkaloids

The acridone alkaloid acronycine (**133**) which was isolated from *Acronychia baueri,* has exhibited potent antitumor activity. Microbial transformations of this alkaloid were examined by several groups in an attempt to prepare metabolites with more promising clinical utility. Betts et al. (1974) screened forty-seven cultures with the result that ten (mostly *Cunninghamella*) produced more polar metabolites. *C.echinulata* NRRL 3655 produced a 30% yield of 9-hydroxyacronycine (**134**), which is also the principle metabolite in mammals. Traces of other oxygenated mammalian metabolites from this and other *Cunninghamella* were tentatively identified based on chromatographic characteristics, and included 11-hydroxyacronycine (**135**), the 9,11-dihydroxy-derivative **136**, and 3-hydroxymethyl-11-hydroxyacronycine (**137**). The authors

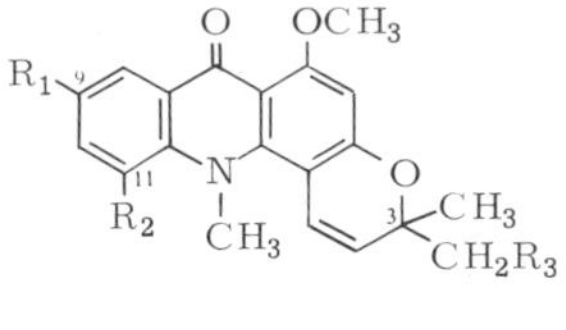

133 $R_1 = R_2 = R_3 = H$
134 $R_1 = OH, R_2 = R_3 = H$
135 $R_1 = R_3 = H, R_2 = OH$
136 $R_1 = R_2 = OH, R_3 = H$
137 $R_1 = H, R_2 = R_3 = OH$
138 $R_1 = R_2 = H, R_3 = OH$

pointed out that the site aromatic hydroxylation follows predictable chemical rules of electrophilic aromatic substitution, in that they occur *ortho* or *para* to the amine and *meta* to the ketone. BRANNON et al. (1974) simultaneously explored the microbial metabolism of acronycine, and observed that *Aspergillus alliaceus* QM 1915 produced 9-hydroxyacronycine (**134**), while *Streptomyces spectabilis* NRRL 2492 produced 3-hydroxymethylacronycine (**138**). Neither of these metabolites were active as an antitumor or antiviral agent, in contrast to the parent alkaloid, indicating that these oxidative routes of metabolism represent bioinactivation. These studies clearly indicate the utility of microbial transformations in the production of sufficient quantities of otherwise insufficient mammalian metabolites for full structure elucidation and biological evaluation.

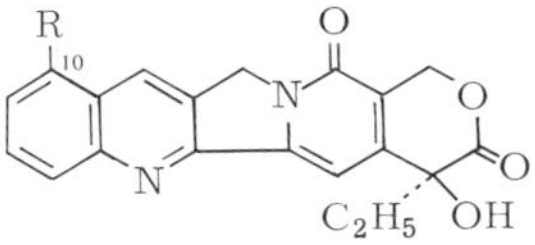

139 R = H
140 R = OH

The *Camptotheca* alkaloid camptothecin (**139**) is also of interest as a potent antitumor agent, although toxicity and insolubility limit its utility (BROSSI, 1980). Indications that more water soluble phenolic derivatives may prove more efficaceous are hampered by a lack of availability from natural sources (WANII and WALL, 1967). Thus, it is of considerable interest that *Aspergillus* sp. T-36 (a soil isolate) converts camptothecin to 10-hydroxycamptothecin (**140**) in low yield (CHU, 1978). It would be interesting to determine if enhanced solubility using solubilizing agents would facilitate bioconversion as was observed with the insoluble indolic alkaloid ellipticine (*vide supra*).

D. Morphine Alkaloids

Morphine and its derivatives, derived from the opium poppy *Papaver somniferum*, are pharmacologically important because of the profound analgesic action of the parent agent, now known to be based on the presence of a family of "opiate receptors" in the brain (GOODMAN et al., 1980). Considera-

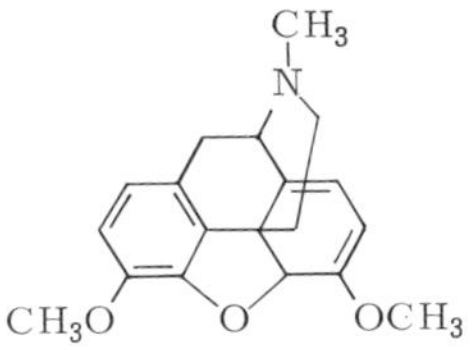

141

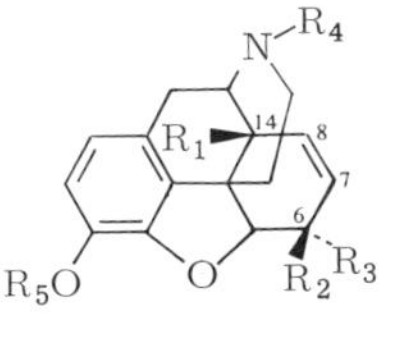

142 $R_1 = OH, R_2/R_3 = {=}O, R_4 = R_5 = CH_3$
143 $R_1 = R_3 = OH, R_2 = H, R_4 = R_5 = CH_3$
144 $R_1 = OCOCH_3, R_2/R_3 = {=}O, R_4 = R_5 = CH_3$
145 $R_1 = OCOCH_3, R_2 = H, R_3 = OH, R_4 = R_5 = CH_3$
146 $R_1 = Br, R_2/R_3 = {=}O, R_4 = R_5 = CH_3$
147 $R_1 = Br, R_2 = H, R_3 = OH, R_4 = R_5 = CH_3$
148 $R_1 = H, R_2/R_3 = {=}O, R_4 = R_5 = CH_3$, 7,8-dihydro
149 $R_1 = R_2 = R_5 = H, R_3 = OH, R_4 = CH_3$
150 $R_1 = R_2 = H, R_3 = OH, R_4 = R_5 = CH_3$
151 $R_1 = R_5 = H, R_2/R_3 = {=}O, R_4 = CH_3$
152 $R_1 = OH, R_2/R_3 = {=}O, R_4 = CH_3, R_5 = H$
153 $R_1 = H, R_2/R_3 = {=}O, R_4 = R_5 = CH_3$
154 $R_1 = OH, R_2/R_3 = {=}O, R_4 = R_5 = CH_3$
155 $R_1 = R_3 = OH, R_2 = R_5 = H, R_4 = CH_3$
156 $R_1 = R_2 = R_4 = R_5 = H, R_3 = CH_3$

ble effort has been directed towards the synthesis of analgesic derivatives from coexisting alkaloids, while metabolic studies have focused principally on mechanisms of bioinactivation.

Numerous microbial studies have been conducted on (−)-thebaine (**141**) beginning with the reports of IIZUKA et al. that extensive metabolism occurs with *Trametes sanguinea* (*Polystictus sanguineus*) and several other wood-rotting Basidomycetes (IIZUKA et al., 1960; 1962; ASAI et al., 1964). *T.sanguinea* was reported to convert thebaine (**141**) to 14-β-hydroxycodeinone (**142**) in good yield (40%), along with the further reduction product, 14β-hydroxycodeine (**143**, 8%). The former reaction presumably occurs via *O*-dealkylation, followed by allylic hydroxylation of the resultant α,β-unsaturated ketone in an analogous manner to that proposed for certain steroid hydroxylations (HOLLAND and AURET, 1975). In subsequent studies using cell suspension cultures, these authors clearly demonstrated that the source of the 14β-oxygen is molecular oxygen, not water (AIDA et al., 1966). The possibility that the 6-keto group contains the incorporated oxygen was excluded by exchange studies indicating that any incorporated oxygen at this position would not have survived incubation conditions (i. e., exchanges). In addition, cell-suspensions allowed a quantitative conversion to **142**, with no apparent formation of the reduction product **143**. Earlier studies indicated that pH controls the degree of reduction (IIZUKA et al., 1962). This reaction is also catalyzed chemically by hydrogen peroxide in acetic acid (FELDMAN, 1975).

Additional studies focused on the effect of 14β-substitution on thebaine metabolism by this organism (YAMADA et al., 1962; 1963; AIDA et al., 1966). For example, the 14β-acetoxy-derivative of codeinone (**144**) or codeine (**145**) yielded 14β-hydroxycodeine (**143**). In the case of 14β-bromocodeinone (**146**), **143** was again isolated, but found to result from solvolysis of the true enzymatic product, 14β-bromocodeine (**147**) (ABE et al., 1970).

Studies with the related organism *Trametes cinnabarina* using thebaine as the substrate yielded similar results in the formation of **142**, but also exhibited further oxidation of this metabolite to form 14β-hydroxycodeinone-*N*-oxide. This divergent pathway for **142** is presumably responsible for limiting the amount of **143** observed (GROGER and SCHUMAUDER, 1969).

Several investigations on these and related compounds have focused on the stereochemical aspects of substrates and products. Microbial reduction of the 6-keto group in 7,8-dihydrocodeinone (**148**) and related thebaine derivatives (dihydrothebainones) yielded mixtures of the 6α- and 6β-alcohols, indicating a low level of product stereospecificity using *Trametes sanguinea* (YAMADA, 1963; YAMADA et al., 1963). In addition, the unnatural (+)-thebaine and other unnatural stereoisomers yielded the same products as the natural (−)-substrates, indicating little substrate enantiospecificity (TSUDA, 1964).

LIRAS and UMBREIT examined an *Arthrobacter* species isolated by soil enrichment, which contained a constitutive, soluble hydroxylase active in hydroxylating morphine (**149**) and codeine (**150**) in the 14-position. Activity was exhibited in growing cells (LIRAS and UMBREIT, 1975a), in cell-suspension, and in cell-free systems (LIRAS and UMBREIT, 1975b). In addition, an unidentified "Substance C" produced from further metabolism of 14β-hydroxymorphine was reported. In cell-suspensions, the hydroxylation was enhanced by increasing oxygenation over a 40-fold range. In cell-free systems, the reaction was enhanced by the presence of Fe^{2+} and NADH. Later studies indicated that this organism also produces morphine-*N*-oxide and the dimerization product, pseudomorphine (ATHERHOLT, 1977). This author then explored the activity of a purified β-hydroxysteroid dehydrogenase (the α-enzyme is inactive) from *Pseudomonas testosteroni* (LIRAS et al., 1975), which metabolized morphine (**149**) to morphinone (**151**, tentative) and 14β-hydroxymorphinone (**152**), and metabolized codeine (**150**) to codeinone (**153**) and 14β-hydroxycodeinone (**154**). It was suggested that the enzyme catalyzed C-6 dehydrogenation to yield morphinone or codeinone, followed by spontaneous or enzymatic oxygenation (contaminant hydroxylase?) at C-14. This sequence is supported by the fact that 14β-hydroxymorphine (**155**) does not serve as a substrate for dehydrogenation. The further suggestion of the utility of using well characterized enzyme systems in the systematic search for metabolism in an alkaloid series was also made.

N-Dealkylation of several morphine alkaloids has also been reported using microbial systems. Morphine itself is *N*-demethylated to normorphine (**156**) by *Cladosporium cladosporioides,* with dimerization of morphine to pseudomorphine as a side-reaction (ISAKA, 1969). The highly active (analgesic) "endo-thebaines" (e. g., **157**), formed by Diels-Alder cyclization of vinylmethyl ketone with thebaine, are also attacked. For example, **157** is dealkylated to the corresponding secondary amine **158** in low yield using *Cunninghamella echinulata* NRRL A11 498 or *Xylaria* sp. (MITSCHER et al., 1968). In addition, the side-chain ketone is stereospecifically reduced to the corresponding *S*-alcohol (**159**). The cyclopropylmethyl-analog **160** was also dealkylated

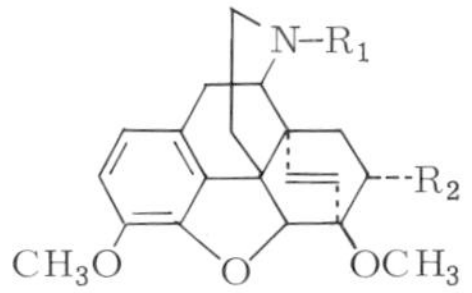

N–R_1 R_2 CH_3O O OCH_3

157 $R_1 = CH_3$, $R_2 = COCH_3$
158 $R_1 = H$, $R_2 = COCH_3$
159 $R_1 = CH_3$, $R_2 = CH(OH)CH_3$
160 $R_1 = CH_2$-cyclopropyl, $R_2 = CH(OH)CH_3$

by the *Xylaria* sp. Recent chemical methods have been developed involving the use of chloroformate reagents for the facile *N*-dealkylation of morphine derivatives (RICE and MAY, 1977; ABDEL-MONEM and PORTOGHESE, 1972). A possible advantage of microbial systems might be realized if any substrate stereoselectivity exists for this reaction, allowing for the simultaneous modification and resolution of stereoisomeric mixtures resulting from synthetic production of these alkaloids.

E. Colchicine and Related Alkaloids

The alkaloid colchicine (**161**) derived from *Colchicum autumnale,* has been utilized for centuries in the treatment of acute attacks of gouty arthritis (WOODBERRY, 1970). It is also a versatile biochemical tool in that it inhibits mitosis by binding tubulin, and thus poisons assembly of the mitotic spindle apparatus, resulting in cell division arrest (CREASEY, 1975; 1977). More recently, colchicine and its analogs have enjoyed renewed interest because of their potent antineoplastic activity (SARTORELLI and CREASEY, 1969) but which is severely limited by toxicity (LEITHER et al., 1952a; STOLINSKY et al., 1976). Studies concerning microbial bioconversions of colchicine and related compounds are of interest based on observations that certain colchicine derivatives (also potential metabolites) exhibit less toxicity than the parent alkaloid (LEITER et al., 1952a, b; HUNTER and KLASSEN, 1975). Further, an understanding of pathways leading to detoxification or the production of toxic metabolites may lead to the development of more efficaceous antitumor agents (ROSAZZA, 1978).

The microbial metabolism of colchicine was first explored by VELLUZ and BELLET (1959), who described the production of an (unspecified) *O*-dealkylated product using *Streptomyces griseus,* which was reconverted to the substrate with diazomethane. Thiocolchicine (**162**) was also dealkylated by this organism. ZEITLER and NIEMER (1969) described the ability of *Arthrobacter colchovorum,* isolated from soil near *Colchicum autumnale* by culture enrichment, to catalyze limited modification of the colchicine structure. In a medium containing a nitrogen source, but no carbon source, *N*-desacetylcolchicine (**163**) was isolated. Thus, amide cleavage to yield one equivalent of acetate is apparently sufficient to support growth. Conversely, a medium containing glucose but lacking a nitrogen source resulted in the production of 7-desacetamido-7-oxocolchicine (**164**), presumably via ace-

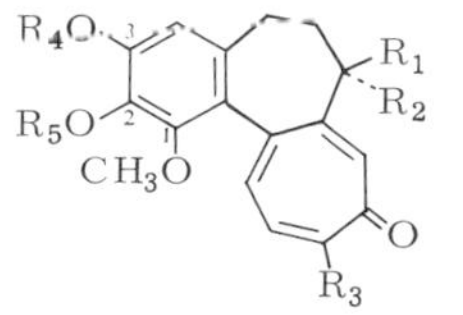

161 $R_1 = H, R_2 = NHCOCH_3, R_3 = OCH_3, R_4 = R_5 = CH_3$
162 $R_1 = H, R_2 = NHCOCH_3, R_3 = SCH_3, R_4 = R_5 = CH_3$
163 $R_1 = H, R_2 = NH_2, R_3 = OCH_3, R_4 = R_5 = CH_3$
164 $R_1/R_2 = {=}O, R_3 = OCH_3, R_4 = R_5 = CH_3$
165 $R_1 = H, R_2 = NHCOCH_3, R_3 = OCH_3, R_4 = CH_3, R_5 = H$
166 $R_1 = H, R_2 = NHCOCH_3, R_3 = OCH_3, R_4 = H, R_5 = CH_3$
167 $R_1 = H, R_2 = NHCOCH_3, R_3 = NHCH_3, R_4 = R_5 = CH_3$
168 $R_1 = H, R_2 = NHCOCH_3, R_3 = NH_2, R_4 = R_5 = CH_3$
169 $R_1 = H, R_2 = NHCOCH_3, R_3 = NHCH_3, R_4 = CH_3, R_5 = H$
170 $R_1 = H, R_2 = NHCOCH_3, R_3 = NHCH_3, R_4 = H, R_5 = CH_3$

tate cleavage and oxidative deamination to yield a source of nitrogen.

More recently, HUFFORD et al. (1979) described the production of *O*-dealkylated (phenolic) metabolites of colchicine using *Streptomyces spectabilis* ATCC 27 465 and *S.griseus* ATCC 13 968. CMR studies were diagnostic in determining the positions of dealkylation, allowing for the determination that 2-*O*-desmethylcolchicine (**165**) and 3-*O*-desmethylcolchicine (**166**) were the sole metabolites. Metabolite **165** predominates in *S.griseus,* while **166** predominates with *S.spectabilis*. It is of considerable interest that these phenolic derivatives are also produced in the modified Udenfriend redox system used as a chemical model for oxidative mammalian liver metabolism (SCHONHARTING et al., 1973), and have been observed as mammalian metabolites (SCHONHARTING et al., 1974). Further, phenolic analogs of colchicine are of interest therapeutically since they may exhibit decreased toxicity in comparison to the parent alkaloid (HUNTER and KLASSEN, 1975). For preparative scale production, however, it appears that newer chemical methods of *O*-dealkylation of colchicine would be more satisfactory (BLADE-FONT, 1979).

Colchicine chemically represents a vinylogous ester in the tropolone ring, and exhibits the appropriate reactivity for such a functional group. The vinylogous amide analog, *N*-methylcolchiceinamide (**167**) is also of interest, since such colchiceinamide analogs exhibit decreased toxicity and/or increased potency in comparison to colchicine (LEITER et al., 1952b; SCHINDLER, 1965). Initial studies on the microbial metabolism of this compound indicated that extensive metabolism occurred with *S.griseus* NRRL B-599 (DAVIS, 1981) in contrast to the low conversions reported with colchicine (HUFFORD, 1979). Preparative scale incubations yielded the *N*-demethylated product, colchiceinamide (**168**, 65%) and a minor amount of the corresponding 2- and 3-*O*-desmethyl-analogs **169** (8%) and **170** (2.5%), respectively. This major route of metabolism (*N*-dealkylation) is of therapeutic interest because, if it occurs in mammals, it may represent a form of detoxification while retaining biological activity, since **168** is less toxic, but equipotent to **167** and colchicine itself (LEITER, 1952b) (however, *vide infra*).

These studies with colchiceinamides have been extended by TUBIO and DAVIS (1983) to include mono- and dialkyl homologs. As illustrated in Fig. 5, *N,N*-dimethylcolchiceinamide (**171**) undergoes nearly complete bioconversion with *Streptomyces griseus* representing sequential and divergent pathways. Quantitatively, *N*-dealkylation predominates, yielding firstly *N*-methylcolchiceinamide (**167**) and secondly, the product of further dealkylation, colchiceinamide (**168**). *O*-Dealkylation competes to a small extent at all stages to give the corresponding 2- and 3-phenols of **171**, **167**, and **168**. Diethylcolchiceinamide (**172**) is also sequentially *N*-dealkylated to give ethylcolchiceinamide (**173**) and then colchiceinamide (**168**). *O*-Dealkylation again competes at the stage of **173**, and **168**, but not with the primary substrate **172**, which is solely *N*-dealkylated. The indicated pathways

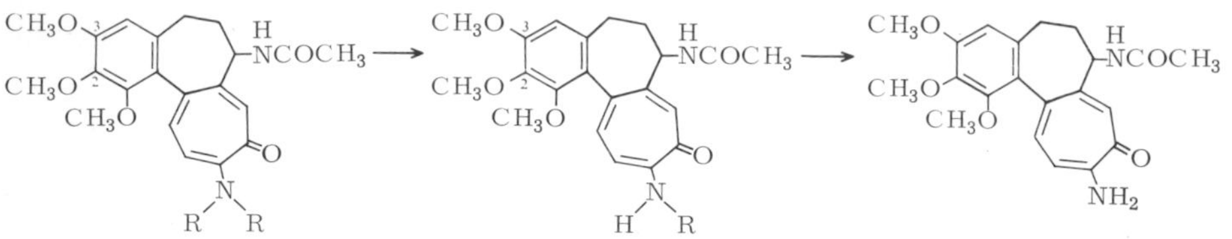

171 R = CH_3
172 R = CH_2CH_3
↓
2- and 3-Phenols
(R = CH_3 only)

167 R = CH_3
173 R = CH_2CH_3
↓
2- and 3-Phenols
(R = CH_3 and CH_2CH_3)

168
↓
2- and 3-Phenols

Figure 5. Metabolism of *N*-alkyl and *N,N*-dialkyl homologs of colchiceinamide by *Streptomyces griseus.*

were determined by chemical interconversions, isolation and structure elucidation of metabolites, and the feeding of intermediates (e. g., **173**). It would be interesting to determine if these routes of metabolism for colchiceinamides also occur in mammals. An interesting parallel can be proposed between the reported mammalian toxicity of these compounds (equated to the presence of *N*-alkyl substituents) and the metabolism by *S.griseus*. For example, colchiceinamide, which lacks any alkyl substituents cannot undergo *N*-dealkylation, is the least toxic of the series (LEITER et al., 1952b). It is conceivable that the toxicity relates to metabolic "bioactivation" such as an intermediate formed during *N*-dealkylation. An understanding of the metabolic basis for such parallels may prove useful in the design of newer therapeutic agents with decreased toxicity.

F. Steroidal Alkaloids

Interest in steroidal alkaloids is based on their potent toxicity in animals, the sensitivity or resistance of microbial parasites on plants containing these toxins, and their potential use as a source of raw materials in steroid and insecticide production (BELIC and KOMEL, 1980; PINDER, 1978; CAMBIE, 1981). For the latter application, extensive modification including side-chain and C-3 glycosidic cleavage as well as A-ring functionalization is required, and has been the major thrust of microbial transformations in this class of alkaloids. While it could be argued that these substances might be more appropriately excluded in a discussion of "alkaloids" in favor of "steroids", they are clearly alkaloidal from a chemical and toxicological standpoint. In addition, while certain organisms exhibit transformations indicative of the steroid nucleus, recent studies indicate that the presence of a nitrogen in the side-chain controls the type-reactions observed (*vide infra*).

Transformations of steroidal alkaloids can be conveniently divided into three categories, based on the position of attack, as follows: first, glycosidic cleavage at the 3-position; second, transformations of the tetracyclic steroidal nucleus (rings A-D) via specific type-reactions characteristic of usual steroidal substrates; third, transformations of the heterocyclic E- and F-rings or their seco (ring-opened) analogs.

Glycosidic cleavage of steroidal alkaloids has been investigated principally as an explanation for the detoxification (hence, resistance) of certain microbial pathogens on plants containing these substances. For example, VERHOEFF and LIEM (1975) suggested that the tomato pathogen *Botrytis cinerea* "detoxifies" tomatine (**174**) by glycoside hydrolysis to yield the aglycone, tomatidine (**175**), and demonstrated the reaction

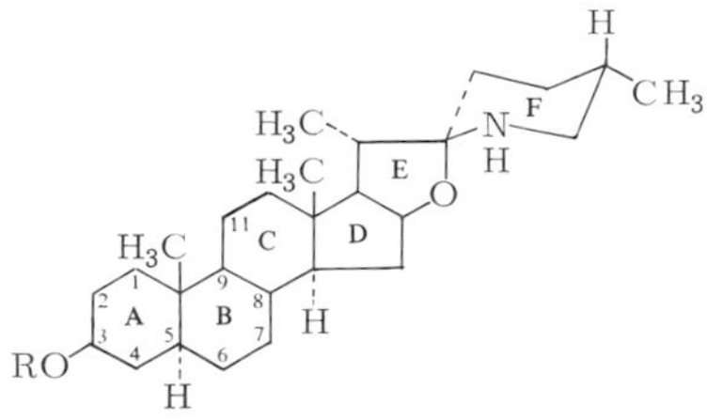

174 R = Lycotetrose
175 R = H
178 R = Triose, $\Delta^{5,6}$
179 R = H, $\Delta^{5,6}$

with enzymes from this organism. Similarly, FORD et al. (1977) determined that *Fusarium oxysporum,* the causative agent in tomato wilt, cleaves tomatine **174** to yield **175** and the sugar portion of the glycoside, lycotetrose (a tetrose composed of xylose, galactose, and two moles of glucose). ARNESON and DURBIN (1967) described partial sugar hydrolysis of α-tomatine (**174**) by *Septoria lycopersici* to yield the glycoside β-tomatine plus glucose. In analogy to studies with tomatine, recent investigations indicate that similar reactions occur with the

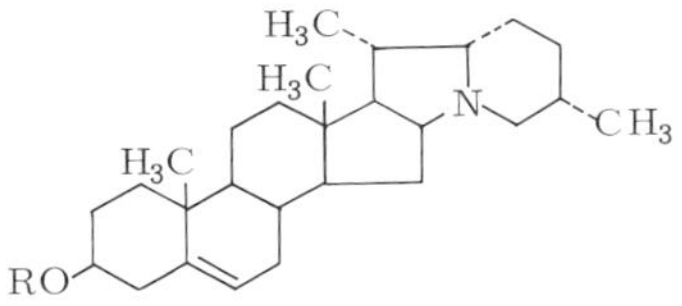

176 R = Triose
177 R = H

potato alkaloid, solanine (**176**). The potato blight fungus *Phytophthora infestans* cleaves this glycoside to solanidine (**177**) (HOLLAND and TAYLOR, 1979). RODRIGUEZ et al. (1979) expanded on the practical application of a similar reaction by exploring the utility of *Aspergillus niger* NRRL 3 in cleaving a mixture of glycoalkaloids (mostly solasonine, **178**) to yield the aglycone solasodine (**179**) in good yield. A dose-dependent inhibition of growth (and hydrolysis) was observed, and a quantitative conversion was observed by controlling substrate concentration. An advantage over chemical hydrolysis of the glycoalkaloids cited by the authors was the absence of side-products.

Specific type-reactions of the tetracyclic steroidal skeleton parallel those seen with simple steroids. SATO and HAYAKAWA (1961; 1963) observed that *Helicostylum piriforme* converts solasodine (**179**) to the 9α-hydroxy (35%), 11α-hydroxy (1%), 7β-hydroxy (1%), and 7ξ,11α-dihydroxy (tentative, 0.5%) derivatives. Likewise, this organism hydroxylates tomatidine (**175**) to the 7α,11α-dihydroxy (30%), 7α-hydroxy (5%), and 9α-hydroxy (0.5%) derivatives (SATO and HAYAKAWA, 1961; 1964). The alkaloid conessine (**180**) is similarly hydroxylated by *Aspergillus ochraceus* to the epimeric 7α- and 7β-hydroxy-analogs (11% and 4%, respectively) (KUPCHAN et al., 1963), while

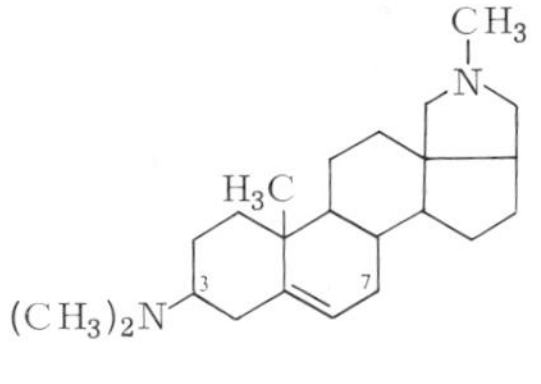

180

Gloeosporium (*Cryptoclines*) *cyclaminis* and *Hypomyces haematococcus* oxidatively deaminated this substrate to yield the 3-keto derivative in 32–64% yield (DEFLINES et al., 1962).

The most extensive studies on the transformations of steroidal alkaloids are those of BELIC and co-workers. In their studies on the metabolism of tomatidine (**175**) by *Nocardia restrictus* (and, to a lesser extent, *Mycobacterium phlei*), the alkaloid was sequentially epimerized from the 3β-ol to the 3α-ol (BELIC et al., 1977a), dehydrogenated to the 3-keto derivative, followed by dehydrogenation at the 1,2- and 4,5-positions to yield the corresponding tomata-1,4-diene-3-one, without side-chain cleavage or skeletal degradation (BELIC and SOCIC, 1971; 1972). As indicated by the authors, the latter cleavage reactions are usually observed with steroidal substrates using these organisms. The reasons for this lack of cleavage were further explored by an examination of tertiary amine analogs (*N*-methyl, **181**, and *N*-acetyl, **182**) (BELIC et al., 1977b), as well as the F-ring epimer, 5α-solasodanol (**183**)

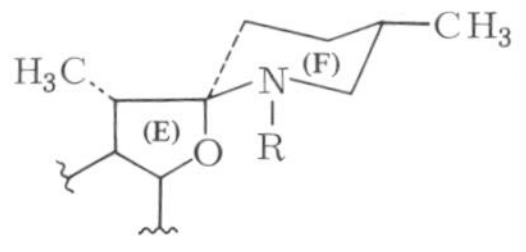

174 R = H
181 R = CH_3
182 R = $COCH_3$

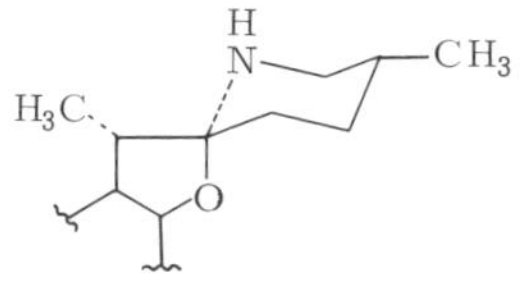

183

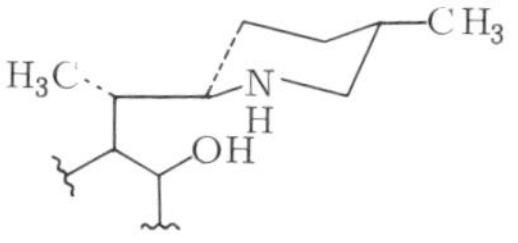

184

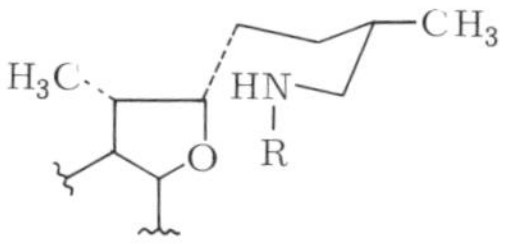

185 R = $COCH_3$
186 R = H

(BELIC et al., 1975a), all of which were similarly metabolized, i. e., by A-ring transformations only. E-Ring seco-steroids such as **184** were similarly transformed (BELIC et al., 1973), whereas the F-ring seco-analog **185** was metabolized to a very slight extent. The secondary amine **186** was only *N*-acetylated with *Nocardia restrictus*. These results indicated that transformations of steroidal alkaloids by these organisms parallel those of steroids (A-ring transformations) only if the nitrogen is heterocyclic, otherwise, they are poor substrates. In any case, they are not subject to side-chain cleavage (BELIC et al., 1975b; BELIC and SOCIC, 1975). The latter fact is now explainable based on the observation that the nitrogen in the side-chain prevents induction of the side-chain splitting enzymes (BELIC and KOMEL, 1980). If induced by cholesterol and using a chelating agent (e. g., α,α-dipyridyl) to prevent steroidal skeletal cleavage, tomatidine is converted to 1,4-androstadiene-3,17-dione in low yield. Enhancement of this process might render steroidal alkaloids as a commercially exploitable source of raw materials for steroid syntheses.

G. Miscellaneous Alkaloids

1. Nicotine and related pyridine alkaloids

Nicotine (**187**), a potent cholinergic agent, pharmacological tool, insecticide, and recreational drug (ACETO and MARTIN, 1982), as well as related pyridine alkaloids (e. g., anabasine, myosmine) have been the subject of numerous microbial biodegradation studies. Results of these investigations have indicated initial oxidative attack on the pyridine or pyrrolidine rings,

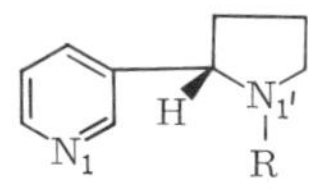

187 R = CH_3
188 R = H

ring cleavage, pigment formation, and/or total assimilation. This area has recently been reviewed by a number of authors (KIESLICH, 1976; VINING, 1969; 1980; HOLLAND and TAYLOR, 1979; IIZUKA and NAITO, 1967; DAVIS, 1982) and will not be reviewed here. Studies involving reactions of synthetic potential are more limited. However, a recent study by SINDELAR et al. (1979) describes the use of *Microsporum gypseum* ATCC 11 395 to *N*-demethylate nicotine (**187**) to nornicotine (**188**) in moderate yield. In addition, other (tentative) metabolites were suggested, including the *N*(1′)-oxide and the bis-*N*(1),*N*(1′)-oxide. Cell-free systems from this organism were capable of reducing the *N*(1′)-oxide back to the parent compound in excellent yield (80%), a reaction analogous to that observed in mammalian systems and the intestinal flora of humans. A heterocyclic (nonaromatic) amine oxide reductase was recently purified from *Streptomyces lincolnensis* ATCC 25 466 which affects the reduction of monocrotaline-*N*-oxide (THEDE et al., 1982).

2. Pyrrolizidine alkaloids

Chief interest in pyrrolizidines stems from their potent hepatotoxicity in animals grazing on *Heliotropium* species of plants. Detoxification of representative alkaloids such as heliotrine (**189**) have been shown to occur by coccal isolates of sheep rumen, involving reductive cleavage to the 1-methylene-derivative **190** (RUSSELL and SMITH, 1968; LANIGAN, 1976).

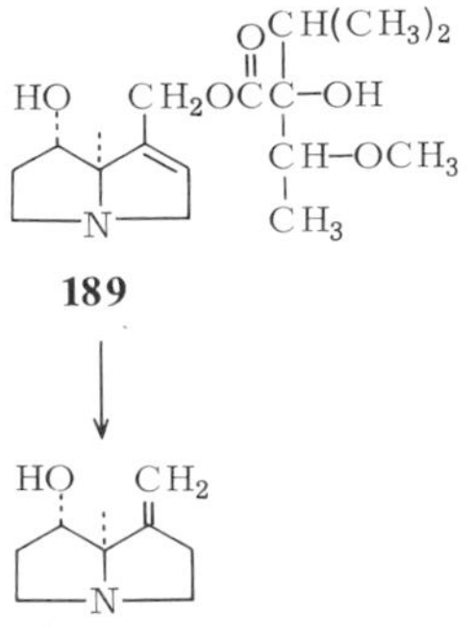

189

190

3. Quinolizidine alkaloids

Studies on the microbial metabolism of quinolizidines have focused on cultures isolated by enrichment techniques which are capable of utilizing these alkaloids as sole sources of carbon and nitrogen (MOZEJKO-TOCZKO, 1960; BRZESKI and TOCZKO, 1961; TOCZKO et al., 1961; KAKOLEWSKA-BANIUK et al., 1962). The intermediate 17-hydroxylupanine (**192**) accumulated during the course of degradation of lupanine (**191**) by *Pseudomonas lupani* (TOCZKO, 1966; TOCZKO et al., 1963). Corynebacteria are also noted for degrading related quinolizidines such as sparteine and lupinine (RYBICKA, 1964).

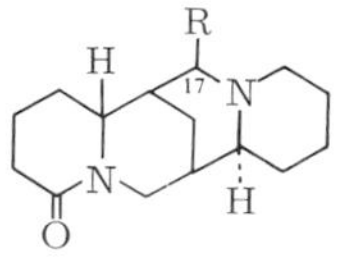

191 R = H
192 R = OH

4. Tropane alkaloids

The classic anticholinergic alkaloid atropine (**193**) and related tropane alkaloids are biodegraded by numerous cultures (KEDZIA, 1961; KACZKOWSKI, 1959; NIEMER et al., 1960; NIEMER and BUCHERER, 1961; BUCHERER, 1960; BRANTNER and HUFLER, 1970). Of potential synthetic interest is the reduction of tropinone (**194**) to the *endo*- and *exo*-alcohols tropine (**195**) and pseudotropine (**196**), respectively, by *Fusarium lini* (TAMM, 1969). Tropine is converted to pseudotropine by dehydration/hydration using a mixed culture of *Bacillus alvei* and *Diplococcus* sp. (SEILER and WERNER 1964; 1965). In addition, hydrolysis and *N*-dealkylation of atropine using *Cunninghamella echinulata* yields nortropine (**197**).

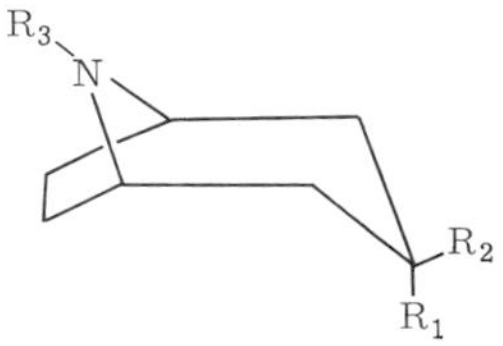

193 R_1 = O–C(=O)–CH(CH_2OH)–Ph
194 R_1/R_2 = =O, R_3 = CH_3
195 R_1 = OH, R_2 = H, R_3 = CH_3
196 R_1 = H, R_2 = OH, R_3 = CH_3
197 R_1 = OH, R_2 = H, R_3 = H

5. Ephedrine

(–)-Ephedrine (**198**) and other stereoisomers of this compound are degraded by *Arthrobacter globiformis*, an organism isolated by enrichment culture. Degradation follows a pathway involving oxidative deamination and ring cleavage (Klamann, 1976; Klamann et al., 1976). An NAD-dependent deaminating oxido-reductase was isolated from this organism which catalyzes the first degradative step (oxidative deamination) (Klamann and Lingens, 1980). *Pseudomonas putida* B-1 also degrades ephedrine by a similar, but not identical pathway.

198

III. References

Abbot, B. J. (1979). In "Developments in Industrial Microbiology", Vol. 20 (L. A. Underkofler, ed.), Chap. 32, Impressions Ltd., Garthersburg, Maryland.

Abdel-Monem, M. M., and Portoghese, P. S. (1972). J. Med. Chem. *15*, 208.

Abe, K., Onda, M., Isaka, H., and Okuda, S. (1970). Chem. Pharm. Bull. (Tokyo) *18*, 2070.

Aceto, M. D., and Martin, B. R. (1982). Med. Res. Rev. *2*, 43.

Adam, J. M., Fonzes, L., and Winternitz, I. (1973). Ann. Chim. (Paris) *8*, 71.

Ahmad, R., Saa, J. M., and Cava, M. P. (1977). J. Org. Chem. *42*, 1228.

Aida, K., Uchida, K., Iizuka, K., Okuda, S., Tsuda, K., and Uemura, T. (1966). Biochem. Biophys. Res. Commun. *22*, 13.

Aleshinskaya, E. E. (1976). Chim. Farm. Zh. *10*, 144.

Arneson, P. A., and Durbin, R. D. (1967). Phytopathol. *57*, 1358.

Asai, T., Soda, H., Tsuda, K., Okuda, S., and Tsuda, K. (1964). Japan. Patent 4 633 (64); Chem. Abstr. *61*, 6355a.

Atherholt, T. B. (1977). "Characterization of the Morphine Transformation System of *Arthrobacter* sp. 86". Thesis, Rutgers University, New Brunswick, New Jersey.

Auret, B. J., Boyd, D. R., Breen, F., and Greene, R. M. E. (1981). J. Chem. Soc. Perkin *1*, 930.

Bellet, P., and Gerard, D. (1962). Ann. Pharm. Fr. *20*, 928.

Bellet, P., and Gerard, D. (1965). Fr. Patent 1 365 278; Chem. Abstr. *62*, 1052.

Bellet, P., and VanThuong, T. (1969). Fr. Patent 1 519 524; Chem. Abstr. *71*, 70 786a.

Bellet, P., and VanThuong, T. (1970a). Ann. Pharm. Fr. *28*, 119.

Bellet, P., and VanThuong, T. (1970b). Ann. Pharm. Fr. *28*, 245.

Bellet, P., and VanThuong, T. (1972). Fr. Patent 2 068 400; Chem. Abstr. *77*, 32 721s.

Belic, I., and Komel, R. (1980). Period. Biol. *82*, 463.

Belic, I., and Socic, H. (1971). Experientia *27*, 626.

Belic, I., and Socic, H. (1972) J. Steroid Biochem. *3*, 843.

Belic, I., and Socic, H. (1975) Acta Microbiol. Acad. Sci. Hung. *22*, 389.

Belic, I., Kramer, V., and Socic, H. (1973). J. Steroid Biochem. *4*, 363.

Belic, I., Gaberc-Porrkar, V., and Socic, H. (1975a). Vestn. Slov. Kem. Drus. *22*, 49.

Belic, I., Hisl-Pintaric, V., Socic, H., and Vranjek, B. (1975b). J. Steroid Biochem. *6*, 1211.

Belic, I., Komel, R., and Socic, H. (1977a). Steroids, *29*, 271.

Belic, I., Mervic, M., Kastelic-Suhadolc, T., and Kramer, V. (1977b). J. Steroid Biochem. *8*, 311.

Berde, B., and Schild, H. O. (eds.) (1978). "Ergot Alkaloids and Related Compounds", Handbuch der Experimentellen Pharmakologie, Vol. 49. Springer-Verlag, Berlin.

Betts, R. E., Walters, D. E. and Rosazza, J. P. (1974). J. Med. Chem. *17*, 599.

Blade Font, A. (1979). Afinidad *36*, 329; Chem. Abstr. *92*, 212491k.

Borgman, R. J., Smith, R. V., and Keiser, J. E. (1975). Synthesis, 249.

Braestrup, C., Nielsen, M., and Olson, C. E. (1980). Proc. Nat. Acad. Sci. USA *77*, 2288.

Brannon, D. R., Horton, D. R., and Svoboda, G. H. (1974). J. Med. Chem. *17*, 653.

Brantner, H., and Hufler, E. (1970). Arch. Hyg. Bacteriol. *154*, 412.

BROSSI, A. (1979). Pure Appl. Chem. *51*, 681.
BROSSI, A. (1980). Trends Pharmacol. Sci. *1*, 9.
BROSSI, A. (1982). Trends Pharmacol. Sci. *3*, 239.
BRZESKI, W., and TOCZKO, M. (1961). Bull. Acad. Pol. Sci. Ser. Sci. Biol. *9*, 161.
BUCHERER, H. (1960). Zentrbl. Bakteriol. Parasitkd. Infektionskr. 2. Abtlg. *113*, 360.
BUCHERER, H. (1965). Zentralbl. Bakteriol. Parasitkd. Infektionskr. 2. Abtlg. *119*, 232.
BUCHERER, H., and NIEMER, H. (1962). Hoppe-Seyler's Z. Physiol. Chem. *328*, 108.
CAMBIE, R. C., POTTER, G. J., READ, R. W., RUTLEDGE, P. S., and WOODGATE, P. D. (1981). Aust. J. Chem. *34*, 599.
CANNON, J. G., SMITH, R. V., MODIRI, A., SOOD, S. P., BORGMAN, R. J., ALEEM, M. A., and LONG, J. P. (1972). J. Med. Chem. *15*, 273.
CAVA, M. P., VENKATESWARIA, A., SRINIVASAN, M., and EDIE, D. L. (1972). Tetrahedron *28*, 4299.
CAVA, M. P., EDIE, D. L., and SAA, J. J. (1975). J. Org. Chem. *40*, 301.
CHACHAJ, W., MALOLEPSZY, J., JANKOWSKA, R., and DRAUS-FILARSKA, M. (1972). Pol. Tyg. Lek. *27*, 2071.
CHIEN, M. M., and ROSAZZA, J. P. (1979a). Drug Metab. Dispos. *7*, 211.
CHIEN, M. M., and ROSAZZA, J. P. (1979b). J. Nat. Prod. *42*, 643.
CHIEN, M. M., and ROSAZZA, J. P. (1980). Appl. Environ. Microbiol. *40*, 741.
CHU, K.-P., LAN, L. T., PAN, W. C., CHOU, T. C., HUANG, Y. C., HSIEH, R. Y., and LIANG, S. F. (1978). K'o Hsueh T'ung Pao *23*, 761; Chem. Abstr. *90*, 119733g.
COUTTS, I. G. C., HAMBLIN, M. R., TINLEY, E. J., and BOBBITT, J. M. (1979a). J. Chem. Soc. Perkin *1*, 2744.
COUTTS, R. T., FOSTER, B. C., GRAHM, R. J., and MYERS, G. E. (1979b). Appl. Environ. Microbiol. *37*, 429.
CREASEY, W. A. (1975). "Vinca alkaloids and colchicine." In "Antineoplastic and Immunosuppressive Agents", Pt. II (A. C. SARTORELLI and D. G. JOHNS, eds.), "Handbuch der Experimentellen Pharmakologie", Vol. *38*, pp. 670–94. Springer-Verlag, Berlin.
CREASEY, W. A. (1977). "Plant alkaloids". In "Cancer 5: A Comprehensive Treatise" (F. F. BECKER, ed.), pp. 379–425. Plenum Press, New York.
DAVIS, P. J. (1980). J. Chromatogr. *193*, 170.
DAVIS, P. J. (1981). Antimicrob. Agents Chemother. *19*, 465.
DAVIS, P. J. (1982). "Microbial transformations of alkaloids." In "Microbial Transformations of Bioactive Compounds", Vol. 2 (J. P. ROSAZZA, ed.), *2*, pp. 67–90. CRC Press, Boca Raton, Florida.
DAVIS, P. J., and RIZZO, J. D. (1982). Appl. Environ. Microbiol. *43*, 884.
DAVIS, P. J., and ROSAZZA, J. P. (1976). J. Org. Chem., *41*, 2543.
DAVIS, P. J., and ROSAZZA, J. P. (1981). Bioorg. Chem. *10*, 97.
DAVIS, P. J., and TALAAT, R. E. (1981). Appl. Environ. Microbiol. *41*, 1243.
DAVIS, P. J., WIESE, D. R., and ROSAZZA, J. P. (1977a). J. Nat. Prod. *40*, 240.
DAVIS, P. J., WIESE, D. R., and ROSAZZA, J. P. (1977b). J. Chem. Soc. Perkin *1*, 1.
DAVIS, P. J., GLADE, J. C., CLARK, A. M., and SMITH, R. V. (1979). Appl. Environ. Microbiol. *38*, 891.
DEFLINES, J., MARX, A. F., VAN DER WAARD, W. F., and VAN DER SIJDE, D. (1962). Tetrahedron Lett., 1257.
DIERCKX, P., LEBLANC, G., DECOSTER, A., and CRISCUOLO, D. (1981). Int. J. Clin. Pharmacol. Ther. Toxicol. *19*, 396.
ECKENRODE, F., PECZYNSKA-CZOCH, W., and ROSAZZA, J. P. (1979). J. Nat. Prod. *42*, 690.
ECKENRODE, F., and ROSAZZA, J. P. (1982). J. Nat. Prod. *45*, 226.
FELDMAN, Kh., and LYUTENBERG, C. (1945). J. Appl. Chem. USSR *18*, 715; Chem. Abstr. *40*, 6489.
FLOSS, H. G. (1970). Naturwissenschaften *57*, 435.
FORD, J. E., MCCANCE, D. J., and DRYSDALE, R. B. (1977). Phytochem. *16*, 545.
GARNIER, J. (1975). Ann. Pharm. Fr. *33*, 183.
GODTFREDSEN, W. W. (1960). Brit. Patent 824 496; Chem. Abstr. *54*, 7768a.
GODTFREDSEN, W. W., KORSBY, G., LORCK, H., and VANGEDAL, S. (1958). Experientia *14*, 88.
GOODHUE, C. T. (1982). "The methodology of microbial transformations of organic compounds." In "Microbial Transformation of Bioreactive Compounds", Vol. 1 (J. P. ROSAZZA, ed.), p. 9. CRC Press, Boca Raton, Florida.
GOODMAN, G. G., GOODMAN, L. S., and GILMAN, A. (eds.) (1980). "The Pharmacological Basis of Therapeutics." Sixth Ed., pp. 564–67. Macmillan, New York.
GROGER, D., and SCHUMAUDER, H. P. (1969). Experientia *25*, 95.
GUSTAFSON, M. E., and ROSAZZA, J. P. (1979). J. Chem. Res., Synop. B, 166.
GUSTAFSON, M. E., ECKENRODE, F., and ROSAZZA, J. P. (1980), Planta medica *39*, 200.

HAASE-ASCHOFF, K., and LINDGENS, F. (1979). Hoppe-Seyler's Z. Physiol. Chem. *360*, 621.

HANNART, J. A. (1971). Ger. Patent 2 727 718; Chem. Abstr. *88*, 121 507w.

HANSON, K. R. (1966). J. Am. Chem. Soc. *88*, 2731.

HARTMAN, R. E., KRAUSE, E. F., ANDREWS, W. E., and PATTERSON, E. L. (1964). Appl. Microbiol. *12*, 138.

HAUER, B., HAASE-ASCHOFF, K., and LINGENS, F. (1982). Hoppe-Seyler's Z. Physiol. Chem. *363*, 499.

HOLLAND, A. H. (1981). "Enzymatic transformations of alkaloids." In "The Alkaloids; Chemistry and Physiology", Vol. *18* (R. H. F. MANSKE, founding ed., R. G. A. RODRIGO, ed., pp. 323–399. Academic Press, New York.

HOLLAND, H. L., and AURET, B. J. (1975). Can. J. Chem. *53*, 2041.

HOLLAND, H. L., and TAYLOR, G. J. (1979). Phytochem. *18*, 437.

HOWELL, C. R., STIPANOVIC, R. D., and BELL, A. A. (1972). Prestic. Biochem. Physiol. *2*, 369.

HUFFORD, C. D., COLLINS, C. C., and CLARK, A. M. (1979). J. Pharm. Sci. *68*, 1239.

HUNTER, A. L., and KLASSEN, C. D. (1975). J. Pharmacol. Exp. Ther. *192*, 605.

IIZUKA, H., and NAITO, A. (1967). "Microbial Transformations of Steroids and Alkaloids." University Park Press, State College, Pennsylvania.

IIZUKA, K., OKUDA, S., AIDA, K., ASAI, T., TSUDA, K., YAMADO, M., and SEKI, I. (1960). Chem. Pharm. Bull. (Tokyo) *8*, 1056.

IIZUKA, K., YAMADA, M., SUZUKI, J., SEKI, I., AIDA, K., OKUDA, S., ASAI, T., and TSUDA, K. (1962). Chem. Pharm. Bull. (Tokyo) *10*, 67.

ISAKA, H. (1969). Yakugaku Zasshi *89*, 1732.

ISHII, H., HAYASHI, M., NIWAGUCHI, T., and NAKAHARA, Y. (1979a). Chem. Pharm. Bull. (Tokyo) *27*, 1570.

ISHII, H., HAYASHI, M., NIWAGUCHI, T., and NAKAHARA, Y. (1979b). Chem. Pharm. Bull. (Tokyo) *27*, 3029.

ISHII, H., NIWAGUCHI, T., NAKAHARA, Y., and HAYASHI, M. (1980). J. Chem. Soc. Perkin *1*, 902.

KACZKOWSKI, J. (1959). Acta Soc. Botan. Pol. *28*, 677; Chem. Abstr. *54*, 15 515e.

KAKOLEWSKA-BANIUK, A., TOXZKO, M., and BRZESKI, W. (1962). Bull. Acad. Pol. Sci. Ser. Sci. Biol. *10*, 167.

KAMETANI, T., OHTA, Y., TAKEMURA, M., IHARA, M., and FUKUMATO, K. (1977). Biorg. Chem. *6*, 249.

KEDZIA, W. (1961). J. Pharm. Pharmacol. *13*, 614.

KERR, K. M., and DAVIS, P. J. (1983). J. Org. Chem. *48*, 928.

KIESLICH, K. (1976). "Microbial Transformations of Non-steroid Cyclic Compounds." Georg Thieme Verlag, Stuttgart.

KLAMANN, E. (1976). Zentralbl. Bakteriol. Parasitkd. Infektionskr. Hyg. Abt. 1: Orig., Reihe B. *162*, 184; Chem. Abstr. *85*, 119 400a.

KLAMANN, E., SCHROPPEL, E., BLECHER, R., and LINGENS, F. (1976). Eur. J. Appl. Microbiol. *2*, 257.

KLAMANN, E., and LINGENS, F. (1980). Z. Naturforsch. Ser. C *35*, 80.

KUPCHAN, S. M., SIH, C. J., KUBOTA, S., and RAHIM, A. M. (1963). Tetrahedron Lett., 1767.

LANIGAN, G. W. (1976). J. Gen. Microbiol. *94*, 1.

LEITER, J., DOWNING, V., HARTWELL, J. L., and SHEAR, M. S. (1952a). J. Natl. Cancer Inst. *13*, 379.

LEITER, J., HARTWELL, J. L., KLINE, I., NADKARNI, M. V., and Shear, M. J. (1952b). J. Natl. Cancer Inst. *13*, 731.

LESCA, P., RAFIDINARIVO, E., LECOINTE, P., and MANSUY, D. (1979). Chem. Biol. Interact. *24*, 189.

LIN, S. K., TIN-WA, M., and TAYLOR, E. H. (1975). J. Pharm. Sci. *64*, 2021.

LIRAS, P., and UMBREIT, W. W. (1975a). Dev. Ind. Microbiol. *16*, 401.

LIRAS, P., and UMBREIT, W. W. (1975b). Appl. Microbiol. *30*, 262.

LIRAS, P., KASPARIAN, S. S., and UMBREIT, W. W. (1975). Appl. Microbiol. *30*, 650.

LOO, Y. H., and REIDENBERG, M. (1959). Arch. Biochem. Biophys. *79*, 257.

MALLETT, G. E., FUKUDA, D. S., and GORMAN, M. (1964). Lloydia *27*, 334.

MEYERS, E., and PAN, S. C. (1961). J. Bacteriol. *81*, 504.

MITSCHER, L. A., ANDERS, W. W., MORTON, G. O., and PATTERSON, E. L. (1968). Experientia *24*, 133.

MOZEJKO-TOCZKO, M. (1960). Acta Microbiol. Pol. *9*, 157.

NABIH, T., DAVIS, P. J., CAPUTO, J. F., and ROSAZZA, J. P. (1977). J. Med. Chem. *20*, 914.

NABIH, T., YOUEL, L., and ROSAZZA, J. P. (1978). J. Chem. Soc. Perkin *1*, 757.

NEEF, G., EDER, U., PETZOLDT, K., SEEGER, A., and WIEGLEPP. (1982). J. Chem. Soc. Chem. Commun., 366.

NEUSS, N., FUKUDA, D. S., MALLETT, G. E., BRANNON, D. R., and HUCKSTEP, L. L. (1973a). Helv. Chim. Acta *56*, 248.

NEUSS, N., BRANNON, D. R., MALLETT, G. E.,

HUCKSTEP, L. L., and FUKUDA, D. S. (1973b). Lloydia *36*, 433.

NEUSS, N., FUKUDA, D. S., BRANNON, D. R., and HUCKSTEP, L. L. (1974a). Helv. Chim. Acta *57*, 1891.

NEUSS, N., MALLETT, G. E., BRANNON, D. R., MABE, J. A., HORTON, H. R., and HUCKSTEP, L. L. (1974b). Helv. Chim. Acta *57*, 1886.

NGUYEN-DANG, T., and DISSET, N. G. (1968). C. R. Acad. Sci. Ser. C *266*, 2362.

NIEMER, H., BUCHERER, H., and KOHLER, A. (1960). Hoppe-Seyler's Z. Physiol. Chem. *317*, 238.

NIEMER, H., and BUCHERER, H. (1961). Hoppe-Seyler's Z. Physiol. Chem. *319*, 161.

NIEMER, H., and BUCHERER, H. (1965). Hoppe-Seyler's Z. Physiol. Chem. *328*, 108.

NIWAGUCHI, T., NAKAHARA, Y., INOVE, T., HAYASHI, M., and ISHIJ, H. (1975). 19th Symp. Chem. Nat. Prod., Hiroshima, Japan, p. 235.

ORITANI, T., and YAMASHITA, K. (1980). Agric. Biol. Chem. *44*, 2807.

PAN, S. C., and WEISENBORN, F. L. (1958). J. Am. Chem. Soc. C, 4749.

PATTERSON, E. L., ANDRES, W. W., KRAUSE, E. F., HARTMAN, R. E., and MITSCHER, L. A. (1963). Arch. Biochem. Biophys. *103*, 117.

PETERSON, D. H., and MURRAY, H. C. (1952). J. Am. Chem. Soc. *74*, 1871.

PHILIPSON, J. D. (1971). Xenobiotica *1*, 419.

PINDER, A. R. (1978). "Steroidal alkaloids." In "Rodds's Chemistry of Carbon Compounds" (S. COFFEY and M. F. ANSELL, eds.), 2nd Ed., Vol. 4G, pp. 381–465. Elsevier, Amsterdam, Neth.

REHACEK, Z. (1980). "Biosynthesis of ergot alkaloids." In "Advances in Biochemical Engineering", Vol. *14* (A. FIECHTER, ed.), pp. 33–60. Springer-Verlag, New York.

REIGHARD, J. B., SCHIFF, P. L., SLATKIN, D. J., and KNAPP, J. E. (1981). J. Nat. Prod. *44*, 466.

RICE, K. C., and MAY, E. L. (1977). J. Heterocycl. Chem. *14*, 665.

RICE, K. C., and BROSSI, A. (1980). J. Org. Chem. *45*, 592.

RICE, K. C., RIPKA, W. C., REDEN, J., and BROSSI, A. (1980). J. Org. Chem. *45*, 601.

RODRIQUEZ, J., SEQOVIA, R., GUERREIRA, E., FERRETTI, F., ZAMARBIDE, G., and ERTOLA, R. (1979). J. Chem. Techn. Biotechnol. *29*, 525.

ROSAZZA, J. P. (1978). "Antitumor antibiotic bioactivation, biotransformation, and derivatization by microbial systems." Recent Results Cancer Res. *63*, 58.

ROSAZZA, J. P., and SMITH, R. V. (1974). Arch. Biochem. Biophys. *161*, 551.

ROSAZZA, J. P., and SMITH, R. V. (1979). Adv. Appl. Microbiol. *25*, 169.

ROSAZZA, J. P., and SMITH, R. V. (1982). "Microbial transformations as a means of preparing mammalian drug metabolites." In "Microbial Transformations of Bioactive Compounds", Vol. *2* (J. P. ROSAZZA, ed.), p. 14. CRC Press, Boca Raton, Florida.

ROSAZZA, J. P., and YANG, L. M. (1982). Abstr. of the 23rd Annual Meeting of the American Society of Pharmacognosy, Pittsburgh, PN, August 1, Abstr. 31.

ROSAZZA, J. P., STOCKLINSKI, A. W., GUSTAFSON, M. A., ADRIAN, J., and SMITH, R. V. (1975). J. Med. Chem. *18*, 791.

ROSAZZA, J. P., KAMMER, M., YOUEL, L., SMITH, R. V., ERHARDT, P. W., TRUONG, D. H., and LESLIE, S. W. (1977). Xenobiotica *7*, 133.

RUSSELL, G. R., and SMITH, R. M. (1968). Aust. J. Biol. Sci. *21*, 1277.

RYBICKA, H. (1964). Acta Agrobot. (Warsaw) *16*, 23.

SANNERSTEDT, R., and CONWAY, J. (1970). Am. Heart J. *79*, 122.

SARTORELLI, A. C., and CREASEY, W. A. (1969). Annu. Rev. Pharmacol. *9*, 51.

SATO, Y., and HAYAKAWA, S. (1961). J. Org. Chem. *26*, 4181.

SATO, Y., and HAYAKAWA, S. (1963). J. Org. Chem. *28*, 2739.

SATO, Y., and HAYAKAWA, S. (1964). J. Org. Chem. *29*, 198.

SCHAUMBERG, J. P., and ROSAZZA, J. P. (1981). J. Nat. Prod. *44*, 478.

SCHINDLER, R. (1965). J. Pharmacol. Exp. Ther. *149*, 409.

SCHOENEWALDT, E. F., and IHNEN, E. D. (1974). US Patent 3 785 927; Chem. Abstr. *81*, 11 850a.

SCHONHARTING, M., PFAENDER, P., RIEKER, A., and SIEBERT, G. (1973). Hoppe-Seyler's Z. Physiol. Chem. *354*, 421.

SCHONHARTING, M., MENDE, G., and SIEBERT, G. (1974). Hoppe-Seyler's Z. Physiol. Chem. *355*, 1391.

SEILER, N., and WERNER, G. (1964). Z. Naturforsch. A *19b*, 572.

SEILER, N., and WERNER, G. (1965) Z. Naturforsch. A *20b*, 451.

SHAMMA, M. (1972). "The Isoquinoline Alkaloids", Chap. 9. Academic Press, New York.

SIH, C. J., and ROSAZZA, J. P. (1976). "Microbial transformations in organic synthesis." In "Applications of Biochemical Systems in Organic

Chemistry" (J. B. JONES, C. J., SIH, and D. PERLMAN, eds.), Chap. 3. John Wiley & Sons, New York.

SIH, C. J., HEATHER, J. B., SOOD, R., PRICE, P., PERUZZOTTI, G., LEE, H. F. H., and LEE, S. J. (1975). J. Am. Chem. Soc. *97*, 865.

SINDELAR, R. D., ROSAZZA, J. P., and BARFKNECHT, C. F. (1979). Appl. Environ. Microbiol. *38*, 836.

SKRYABIN, G. K., GOLOVELA, L. A., TERENTJEVA, I. V., LAZURJEVSKY, G. V., NEFEDOVA, M. Y., ADANIN, V. M., SHMOTINA, G. E., and SADIRINA, G. A. (1974). Izv. Akad. Nauk SSSR, Ser. Biol., 682; Miccrobiol. Abstr. *10*, A2231.

SMISSMAN, E. S., MAKRIYANNIS, A. C., and WATASZEK, E. J. (1970). J. Med. Chem. *13*, 640.

SMITH, R. V., and DAVIS, P. J. (1978). Appl. Environ. Microbiol. *35*, 738.

SMITH, R. V., and DAVIS, P. J. (1980). "Induction of xenobiotic monooxygenases." In "Advances in Biochemical Engineering", Vol. 14 (A. FIECHTER, ed.), pp. 61–100. Springer-Verlag, New York.

SMITH, R. V., and ROSAZZA, J. P. (1975). J. Pharm. Sci. *64*, 1737.

SMITH, R. V., DAVIS, P. J., and KERR, K. M. (1983). J. Pharm. Sci., in press.

STOLINSKY, P. C., JACOBS, E. M., IRWIN, L. E., PAJAK, TH. F., and BATEMAN, J. R. (1976). Oncology *3*, 151.

TAMM, C. (1969). Planta Med. *8*, 331.

TENEN, S. S., and HIRSCH, J. D. (1980). Nature (London) *288*, 60.

TESTA, B. (1979). "Principles of Organic Stereochemistry", p. 151. Marcel Dekker, New York.

THEDE, B. M., DUFFEL, M. W., and ROSAZZA, J. P. (1982). Abstr. of the 23rd Annual Meeting of the American Society of Pharmacognosy, Pittsburgh, Pennsylvania, August 1, Abstr. 45.

TOCZKO, M. (1966). Biochim. Biophys. Acta *128*, 570.

TOCZKO, M., BRZESKI, W., and DROESE, X. (1961). Bull. Acad. Pol. Sci. Ser. Sci. Biol. *9*, 447.

TOCZKO, M., BRZESKI, W., and KAKOLEWSKA-BANIUK, A. (1963). Bull. Acad. Pol. Sci. Ser. Sci. Biol. *11*, 161.

TSUDA, K. (1964). "Chemistry of Microbiology." 6th Symposium Inst. Appl. Microbiol., University of Tokyo, Tokyo, Japan, p. 167.

TUBIO, J. T.-H., and DAVIS, P. J. (1983). Antimicrob. Agents Chemother., submitted.

VELLUZ, L., and BELLETT, P. (1959). C. R. Acad. Sci. *248*, 3453.

VERHOEFF, K., and LIEM, J. I. (1975). Phytopathol. Z. *82*, 333; Chem. Abstr. *83*, 190 398v.

VINING, L. C. (1969). "Microbial transformations of alkaloids." In "Fermentation Advances" (D. PERLMAN, ed.), p. 715. Academic Press, New York.

VINING, L. C. (1980). "Conversions of alkaloids and nitrogenous xenobiotics." In "Economic Microbiology: Microbial Enzymes and Bioconversions", Vol. 5 (A. H. ROSE, ed.), pp. 523–573. Academic Press, New York.

WANII, M. C., and WALL, M. E. (1967). J. Org. Chem. *34*, 1364.

WOODBERRY, D. (1970). In "The Pharmacological Basis of Therapeutics", 4th Ed. (L. S. GOODMAN and A. GILMAN, eds.), p. 339. Macmillan, New York.

WU, G. S., NABIH, T., YOUEL, L., PECZYNSKA-CZOCH, W., and ROSAZZA, J. P. (1978). Antimicrob. Agents Chemother. *14*, 601.

YAMADA, M. (1963). Chem. Pharm. Bull. (Tokyo) *11*, 356.

YAMADA, M., IIZUKA, K., OKUDA, S., ASAI, T., and TSUDA, K. (1962). Chem. Pharm. Bull. (Tokyo) *10*, 481.

YAMADA, M., IIZUKA, K., AKUDA, S., ASAI, T., and TSUDA, K. (1963). Chem. Pharm. Bull. (Tokyo) *11*, 206.

YAMANO, T., KISHINO, K., YAMATODANI, S., and ABE, M. (1962). Takeda Kenkyusho Nempo *21*, 83; Chem. Abstr. *59*, 3099c.

ZEITLER, H.-J., and NIEMER, H. (1969). Hoppe-Seyler's Z. Physiol. Chem. *350*, 366.

Chapter 7

Antibiotics

Oldrich K. Sebek

The Upjohn Company
Infectious Diseases Research
Kalamazoo, Michigan 49001, USA

I. General
II. Directed Biosynthesis
A. Precursor Feeding
B. Metabolic Inhibition
C. Hybrid Biosynthesis
III. Mutasynthesis
IV. Genetic Manipulation
V. Biotransformations
A. β-Lactams
B. Aminoglycosides
C. Rifamycins
VI. Concluding Remarks
VII. References

I. General

Antibiotics are among the most remarkable chemicals produced by microorganisms. Ever since the rediscovery of penicillin by CHAIN and coworkers, more than four decades ago, they have been of enormous value in human therapy, animal health care and agriculture. Their manufacture has been increasing steadily and in 1982 amounted to 14.7 million kg in the U.S. alone. Of this volume, penicillins commanded the largest share (4.3 million kg including 907000 kg of semi-synthetic penicillins) followed by 3.2 million kg of tetracyclines, 500000 kg cephalosporins, and 6.7 million kg of others (BRIGGS, 1983).

The biotechnological production of various classes of antibiotics is the subject of several chapters in Vol. 4 of this series. These chapters stress the secondary metabolism of microorganism as the source of antibiotics. The present chapter is a brief survey of some technically applied biotransformation reactions of antibiotics with the aim of producing new substances of antibiotic activity.

In view of a continuous need for new antibiotics, extensive programs have been directed to discover new ones with improved therapeutic properties, effective against resistant infections and neoplastic diseases, and also of value in other expanded areas such as low-molecular enzyme inhibitors, immunomodifiers, and others. To date, these efforts resulted in the discovery of more than 6000 antibiotic entities of many new structures. Extensive studies of those of practical interest were also carried out to develop new strains and processes in an effort to increase the economy of their production, to elucidate their biosynthesis and modes of action and to chemically synthesize new analogs with improved therapeutic properties (broader antibiotic spectra, lower toxicity, effectiveness against the continually developing antibiotic-resistant populations). The last approach was eminently successful in the chemical synthesis of clinically valuable analogs of β-lactams, aminocyclitol glycosides, rifamycins, and others. Some could also be modified by different microorganisms and inactivated by pathogenic bacteria. Such findings were not totally unexpected since various microorganisms have been described to transform compounds of many different chemical structures (CHARNEY and HERZOG, 1967; KIESLICH, 1976, IIZUKA and NAITO, 1981) and since some were also found to carry out an extensive degradation of several antibiotics: chloramphenicol (SMITH and WORRELL, 1949, 1950, 1953; LINGENS et al., 1966), streptomycin (PRAMER and STARKEY, 1951; FENTON et al., 1973), actinomycin (KATZ and PIENTA, 1957), tetracyclines (PRAMER, 1958), novobiocin (SEBEK and HOEKSEMA, 1972), penicillin (KAMEDA et al., 1961; JOHNSEN, 1977, 1981; BECKMAN and LESSIE, 1979) or streptothricin (SEBEK, unpublished).

To date, antibiotics of all major classes were found to be modified by a variety of microorganisms (by controlled fermentation, addition of specific precursors or metabolic inhibitors, use of selected strains or mutants, and more recently by application of molecular genetics). Some were antibiotic producers themselves while others were not related to the antibiotic under investigation but were selected on the basis of their ability to carry out the desired reactions of other organic compounds. In general, the changes observed have been brought about by the following manipulations which also form the structure of this chapter:

1. Directed biosynthesis
   a) Precursor feeding
   b) Metabolic inhibition
   c) Hybrid biosynthesis
2. Mutasynthesis
3. Genetic manipulation
4. Biotransformation.

Although most of the findings have been only of theoretical interest, a few processes became of considerable practical importance such as the hydrolysis of penicillins to 6-aminopenicillanic acid and of mannosidostreptomycin to streptomycin; the use of blocked mutants for the preparation of 6-demethylchlortetracycline and of rifamycin SV for the chemical synthesis of rifampicin; ribostamycin production by neomycin- and butirosin-blocked mutants; or adriamycin production by daunomycin-revertants.

II. Directed Biosynthesis

A. Precursor Feeding

Enzymes of secondary metabolism have in general broader substrate specificities and are under less stringent controls than those of the primary metabolism. Hence a fairly common characteristic of antibiotic-producing microorganisms is their ability to synthesize groups of structurally related antibiotics rather than single entities. This occurrence of product multiplicity is due to the broad specificities of the antibiotic-synthesizing systems and has been demonstrated in many cases. Thus when a precursor of an antibiotic is added to the organism which normally produces a mixture of antibiotics, such an exogenously supplied compound competes with the endogenously synthesized antibiotic precursor and may be preferentially incorporated into the corresponding antibiotic. This then results in an increased synthesis of the latter; and the formation of other antibiotics, which are produced in unsupplemented media, is partially suppressed or completely eliminated. When in turn an analog of the precursor is added, it may compete with the natural precursors and result in the formation of a new antibiotic analog.

The first observations of this kind were made during the early work on penicillin production when it was noted that in media of different composition mixtures of several natural penicillins were formed. They were found to differ from each other in the acyl side chain which is attached through an amide bond to a unit later identified as 6-aminopenicillanic acid (Fig. 1). They included penicillins G (benzylpenicillin), X (4-hydroxybenzylpenicillin), F (2-pentenylpenicillin), and K (heptylpenicillin).

When, however, the fermentation medium was supplemented with cornsteep liquor, the antibiotic titers increased considerably (MOYER and COGHILL, 1946). This increase is due to phenylacetic acid in cornsteep liquor where it was produced by spontaneous fermentation from phenylalanine (via 2-phenylethylamine) during storage. The acid in turn provides the side chain for the benzyl portion of the penicillin molecule. It directs the metabolic flow toward the synthesis of benzylpenicillin whereby the formation of the other penicillins is suppressed. With this discovery, the addition of phenylacetic acid to the fermentation media became a standard practice in the manufacture of benzylpenicillin. Since the penicillin-producing *Penicillium chrysogenum* incorporates other acids in the same way, it was then possible to prepare more than 100 new biosynthetic penicillins when various non-polar aliphatic and aryl-aliphatic carboxylic acids had been supplied to the fermentation as side chain precursors. One of them, namely penicillin V (phenoxymethylpenicillin, see Fig. 1), is formed upon the addition of phenoxyacetic acid to the fermentation.

Phenoxymethylpenicillin is of therapeutic value since it is acid-stable and hence can be administered orally but, like benzylpenicillin, is effective essentially only against Gram positive bacteria and inactivated by penicillinase-producing organisms. Beside benzylpenicillin, it is the only penicillin made by directed fermentation. Both are manufactured by more than 20 companies worldwide for direct medicinal formulation and for hydrolysis to 6-aminopenicillanic acid, a crucial intermediate for the synthesis of new generations of penicillins and cephalosporins (PERLMAN, 1977).

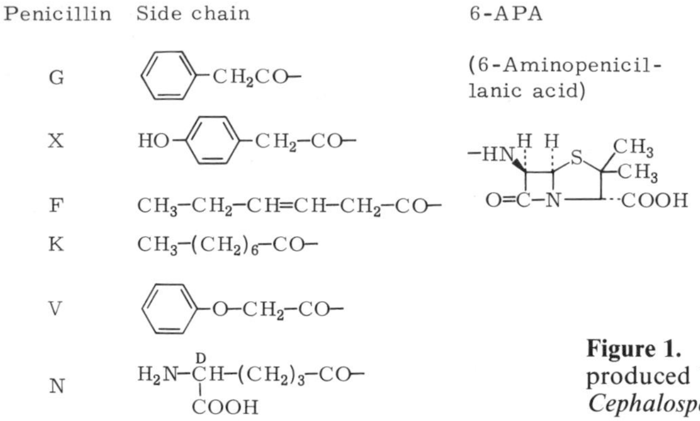

Penicillin	Side chain
G	C_6H_5–CH_2CO–
X	HO–C_6H_4–CH_2–CO–
F	CH_3–CH_2–CH=CH–CH_2–CO–
K	CH_3–$(CH_2)_6$–CO–
V	C_6H_5–O–CH_2–CO–
N	H_2N–CH(D)–$(CH_2)_3$–CO– (with COOH on the CH)

6-APA

(6-Aminopenicillanic acid)

–HN–(H H)–S–C(CH_3)(CH_3)
O=C–N–··COOH

Figure 1. Natural and biosynthetic penicillins produced by *Penicillium* sp. (G, X, F, K, V) and *Cephalosporium* sp. (N).

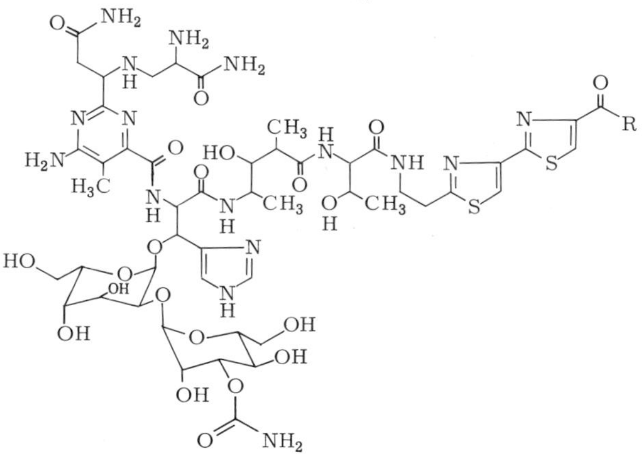

R (terminal amine)	Compound
$-\underset{H}{N}(CH_2)_3S^+(CH_3)_2X^-$	Bleomycin A_2
$-\underset{H}{N}(CH_2)_4NH\underset{NH}{\overset{\|\|}{C}}NH_2$ (Agmatine)	Bleomycin B_2
–OH	Bleomycinic acid

Figure 2. Selected bleomycins and bleomycinic acid.

These findings were also important for synthetic organic chemists. As E. CHAIN noted in an autobiographic review on penicillin shortly before his death in 1979, the idea of synthesizing new penicillin analogs by chemical means arose from the successful modification of the antibiotic by microbiological manipulation (CHAIN, 1980). As a result, thousands of novel semi-synthetic penicillins were prepared chemically and a number of them with broadened antimicrobial spectra became clinically valuable (methicillin, oxacillin and cloxacillin, nafcillin, ampicillin, carbenicillin). A great deal of work has been carried out on this aspect of penicillin research and technology and discussed in much detail (BEHRENS, 1949; COLE, 1966; DEMAIN, 1966; QUEENER and

SWARTZ, 1979; QUEENER and NEUSS, 1982).

A number of linear as well as cyclic peptide antibiotics were modified in a similar manner. Thus under normal fermentation conditions *Streptomyces verticillus* produces mixtures of 12 bleomycins. These linear glycopeptides are effective in the treatment of squamous cell carcinoma and malignant lymphoma but have a serious side effect – pulmonary fibrosis. All of them contain bleomycinic acid which consists of two tripeptides and a disaccharide (see Fig. 2) to which different C-terminal amines are attached. When the medium is supplemented with 3-aminopropyl-dimethylsulfonium chloride, an amine characteristic of bleomycin A_2, the formation of this bleomycin congener was increased from about 55% to more than 80% of the total antibiotic yield. Similar results were also obtained when natural and synthetic amines were used (agmatine, putrescine, spermidine, spermine, ethylamine). New bleomycins containing diaminoethane, 1,3-diaminopropane and their *N*-alkyl- or *N*-aminoalkyl derivatives were produced exclusively when the corresponding amines were present in the fermentation media (UMEZAWA, 1971, 1977; TAKITA and MAEDA, 1980).

Modifications were also obtained by directed biosynthesis when certain amino acid components in several cyclic homomeric peptide antibiotics were replaced by structurally similar compounds or their analogs. These modifications are possible because the biosynthesis of such peptides proceeds by a mechanism different from protein biosynthesis. It is not mediated by the ribosomal but by the thiotemplate systems, and the synthetases involved have broad substrate specificities.

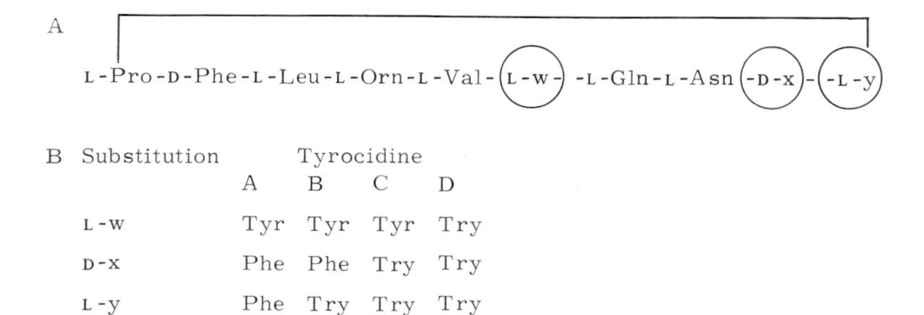

B

Substitution	Tyrocidine A	Tyrocidine B	Tyrocidine C	Tyrocidine D
L-w	Tyr	Tyr	Tyr	Try
D-x	Phe	Phe	Try	Try
L-y	Phe	Try	Try	Try

Figure 3. General structure (A) and amino acid substitutions (B) of tyrocidines A, B, C, and D.

Thus by the addition of either L- or D-tryptophan, *Bacillus brevis* which normally produces a mixture of three tyrocidines (A, B and C), was directed to an almost exclusive formation of tryptophan-containing tyrocidines C and D (a new compound, Fig. 3) while the synthesis of tyrocidines A and B was suppressed. Similar shifts in the tyrocidine ratios were also made by supplying the organism with other antibiotic precursors (namely D- and L-phenylalanine, L-isoleucine, L-alloisoleucine, and also D,L-thienylalanine, and L-pipecolic acid (MACH and TATUM, 1964; OKUDA et al., 1963; RUTTENBERG and MACH, 1966).

Actinomycins, another group of cyclic homomeric peptides, were also modified by directed biosynthesis. Their structures consist of a phenoxazinone chromophore to which two pentapeptide chains are attached and differ from each other at the amino acid sites in the chains (Fig. 4). They are produced by several streptomycetes and one *Micromonospora* sp., and are active against Gram positive bacteria but their high toxicity limits their use to certain cancer therapies. At the sites indicated in Fig. 4, the ratios of naturally produced actinomycins have been altered by the addition of certain precursors (L-valine, L-alloisoleucine, isomers of isoleucine and threonine). New analogs were formed similarly upon supplementation of the culture broths with others (such as sarcosine, 4-hydroxy-L-prol-

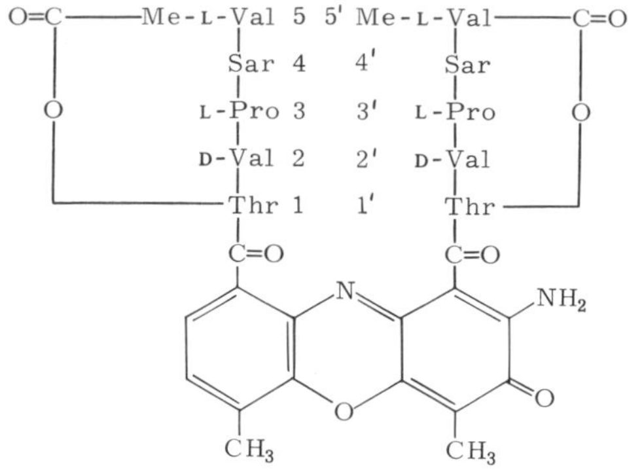

Figure 4. Structure of actinomycin D. Sar sarcosine; Me-L-Val *N*-methyl-L-valine. Amino acid substitutions at 2, 3, 2′, 3′, and 5′ yield correspondingly different actinomycins.

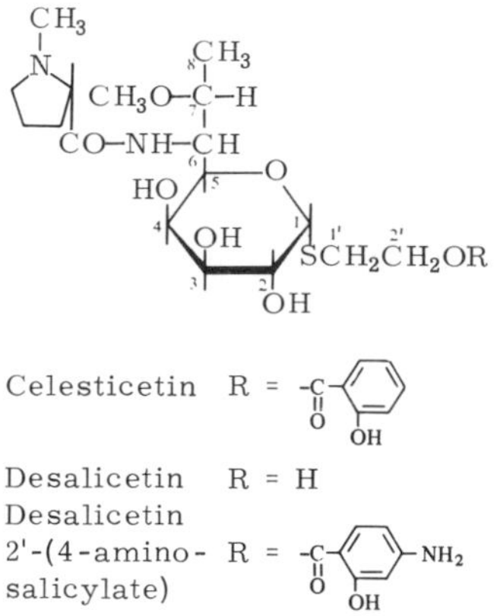

Figure 5. Celesticetin and related compounds.

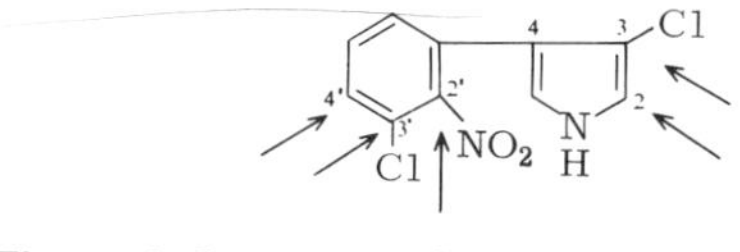

Figure 6. Structure of pyrrolnitrin. – The arrows indicate the sites of substitution.

ine, *cis*- and *trans*-4-methylproline, pipecolic, azetidine-2-carboxylic, and L-thiazolidine-4-carboxylic acids; KATZ, 1974; KATZ et al., 1977). Other antibiotics which are synthesized by non-ribosomal mechanisms, have been affected in a similar way. Thus incubation of L-α-aminobutyric acid, L-proline, and *cis*-4-methyl-D-proline with the viridogrisein-producing *Streptomyces griseoviridus* resulted in the preferential formation of new antibiotic analogs (neoviridogriseins; OKUMURA et al., 1982).

Quinomycins, cyclic homomeric peptides containing two residues of quinoxaline-2-carboxylic acid, were similarly modified. In the synthetic media and by means of cell-free extracts quinaldinic and quinazolone-3-acetic acids replaced one or both carboxylic residues and yielded new bioactive quinomycin analogs (YOSHIDA et al., 1968; KHAN et al., 1969; DHAR et al., 1971).

The addition of ethionine to the producers of certain methyl-containing antibiotics resulted in the formation of ethyl-substituted antibiotic homologs. Thus the oxytetracycline-producing *Streptomyces rimosus* yielded a mixture of the parent antibiotic and its *N*-ethyl homolog (DULANEY et al., 1962). Similarly both *S*- and *N*-methyl groups in the lincomycin molecule were replaced by the corresponding ethyl groups by means of lincomycin-producing *Streptomyces lincolnensis* (ARGOUDELIS et al., 1970).

The formation of celestosaminide antibiotics, which are structurally related to lincomycin, can be controlled by the same manipulation. Exogenous [7-^{14}C]-salicylic acid was incorporated into the celesticetin molecule without randomization to form the ester at C-2′ of desalicetin, the core moiety of celestosaminides. When in turn 4-aminosalicylic acid was added to the celesticetin-producing *Streptomyces celestis,* the corresponding desalicetin 2′-(4-aminosalicylate) was formed (see Fig. 5). New bioactive desalicetin (celesticetin) derivatives were also formed from a few related acids: anthranilic, *N*-methylanthranilic, 3- and 4-aminobenzoic, 4-methyl- and 4-dimethylaminobenzoic, and 4-acetamidobenzoic acid (ARGOUDELIS and COATS, 1974; ARGOUDELIS et al., 1974).

By the same method, some 19 different benzoic acid derivatives were similarly incorporated into new novobiocins by *Streptomyces spheroides* (WALTON et al., 1962).

Pyrrolnitrin (Fig. 6), an antifungal agent, is one of several phenylpyrrole antibiotics produced by *Pseudomonas* sp. Bromo analogs were also obtained when the chemically defined growth medium was supplied with Br-ions (AJISAKA et al., 1969; VAN PÉE et al., 1983). The 5-, 6- and 7-fluoro- as well as the 5- and 7-methyltryptophans gave rise to new corresponding bioactive aminopyrrolnitrin derivatives as did the substitution by bulkier (chloro-, bromo-, trifluoromethyl- and methoxy-) groups. Since the antibiotic is derived directly from D-tryptophan, the analog technique for derepressing the enzymes involved in the biosynthesis of amino acids (and vitamins) was applied to its production. Indeed, analog-resistant mutants were isolated and yielded increased amounts of pyrrolnitrin (HAMILL et al., 1970; ELANDER et al., 1971).

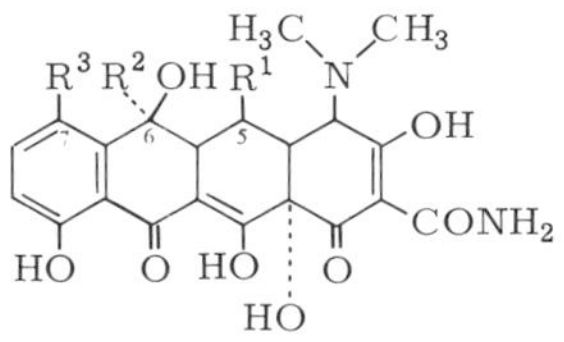

Antibiotic	Functionality R^1	R^2	R^3
Tetracycline	H	CH_3	H
Oxytetracycline	OH	CH_3	H
Chlortetracycline	H	CH_3	Cl
6-Demethylchlortetracycline	H	H	Cl

Figure 7. Structures of tetracyclines.

New derivatives of polyoxins, the pyridine nucleoside antifungal antibiotics, were similarly synthesized by *Streptomyces cacaoi* when the growth medium was supplemented with 5-fluoro-, 5-bromo- and 6-azauracil and also with 5-fluorouracil (ISONO and SUHADOLNIK, 1976).

B. Metabolic Inhibition

Modified antibiotics were also obtained when metabolic inhibitors were added to the fermentation media. Thus a number of selected compounds inhibited the chlorination step in the synthesis of chlortetracycline (Fig. 7) by *Streptomyces aureofaciens* whereby tetracycline was the predominant product formed (GOODMAN et al., 1959). The analogs of L-methionine, namely D-methionine, ethionine, and also sulfonamides, aminopterin, homocysteine derivatives, and methoximine inhibited methylation by the same chlortetracycline producer and resulted instead in the formation of 6-demethylchlortetracycline (GOODMAN and MATRISHIN, 1961; PERLMAN et al., 1961; GOODMAN and MILLER, 1962; NEIDELMAN et al., 1963 a, b).

D, L-Ethionine caused a decrease of streptomycin (Fig. 8) formation by *Strepto-*

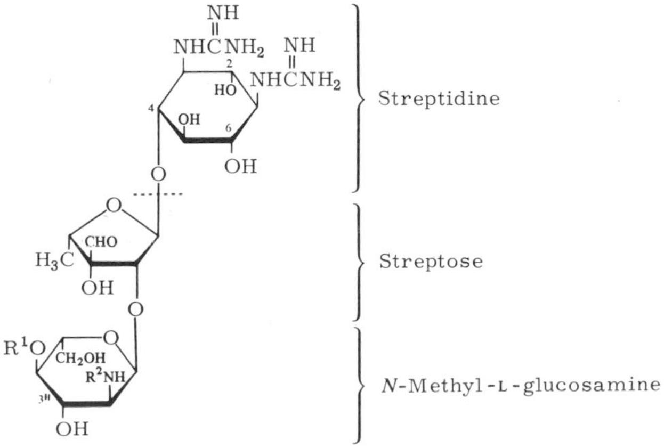

Streptidine

Streptose

N-Methyl-L-glucosamine

Figure 8. Streptomycin and related compounds. – Streptomycin: $R^1 = H$, $R^2 = CH_3$; *N*-demethylstreptomycin: $R^1 = R^2 = H$; mannosidostreptomycin: R^1 = mannose, $R^2 = CH_3$. The streptidine moiety is replaced by 2-deoxystreptidine in streptomutin A. C-3″ OH is inactivated by phosphorylation or adenylylation in streptomycin-resistant bacteria.

myces griseus with the concomitant accumulation of *N*-demethylstreptomycin (HEDING, 1968). Lincomycin biosynthesis was affected in the same way. The addition of sulfanilamide to *S. lincolnensis* fermentations resulted in the primary formation of *N*-demethyllincomycin (ARGOUDELIS et al., 1973a).

C. Hybrid Biosynthesis

This method has been used as an alternative to mutasynthesis (see below) to generate new derivatives of macrolide antibiotics such as picromycin, tylosin, platenomycin, spiramycin, or leucomycin. These antibiotics consist in general of a macrocyclic lactone ring (aglycone) linked glycosidically to amino- and/or deoxysugars. Since the aglycone moiety is synthesized from lipid precursors (acetyl-, propionyl-, butyryl-CoA) via polyketide intermediates, their formation is blocked by cerulenin, an inhibitor of fatty acid biosynthesis, while the formation of the sugar portion remains unaffected. The latter then combines with the aglycone of another macrolide antibiotic added exogenously which results in the formation of a new modified antibiotic.

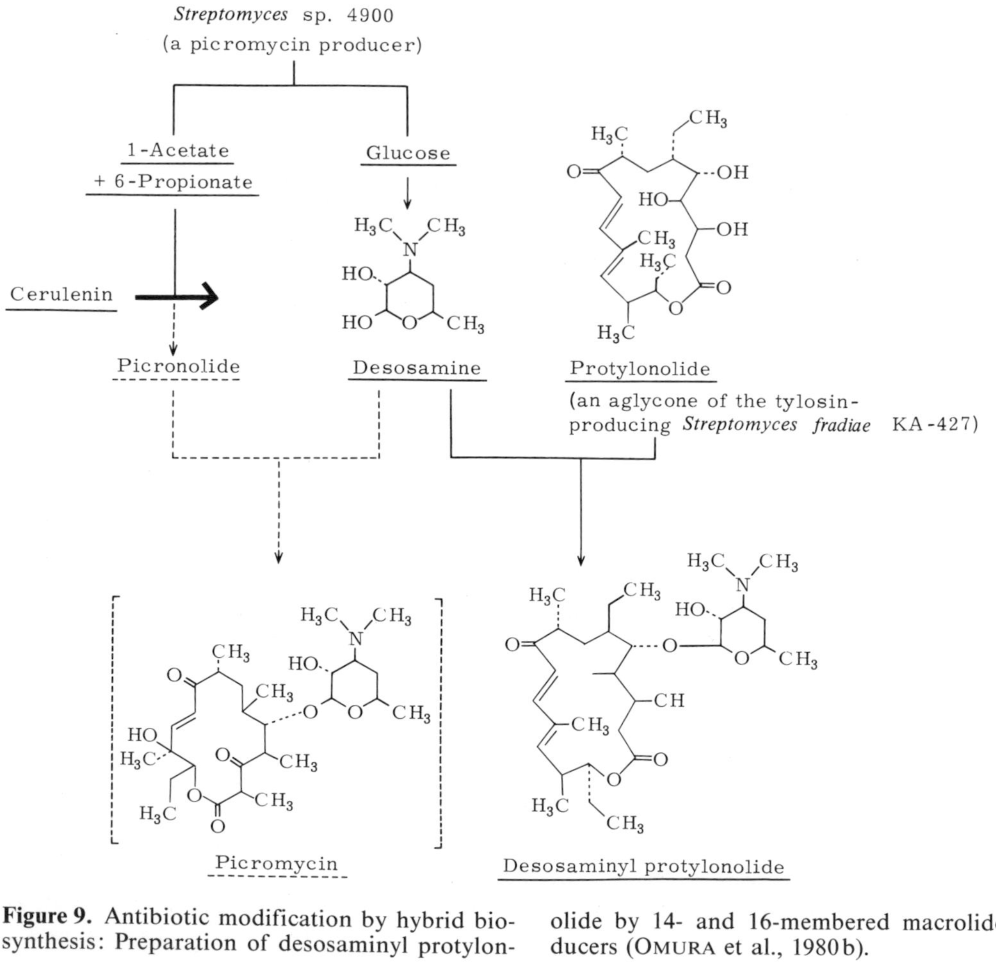

Streptomyces sp. 4900
(a picromycin producer)

1-Acetate
+ 6-Propionate

Glucose

Cerulenin

Picronolide

Desosamine

Protylonolide
(an aglycone of the tylosin-producing *Streptomyces fradiae* KA-427)

Picromycin

Desosaminyl protylonolide

Figure 9. Antibiotic modification by hybrid biosynthesis: Preparation of desosaminyl protylonolide by 14- and 16-membered macrolide producers (OMURA et al., 1980b).

Thus when *Streptomyces* sp. 4900, a producer of picromycin (a 14-membered macrolide) was grown in the presence of cerulenin, the synthesis of picronolide, the aglycone portion of the molecule, was inhibited but that of desosamine (the corresponding sugar component) was not. When in turn protylonolide, an aglycone portion of tylosin (a 16-membered macrolide) was added, it was linked to desosamine and yielded desosaminyl protylonolide, a new "hybrid" antibiotic (Fig. 9, OMURA et al., 1980b).

III. Mutasynthesis

Mutants derived from antibiotic producers have been found valuable in the preparation of new antibiotic derivatives. Thus some of those blocked in antibiotic synthesis (idiotrophs) accumulate precursors which themselves are antibiotically active as illustrated by examples listed in Table 1.

Table 1. Examples of Antibiotics Produced by Antibiotic-blocked Mutants

Antibiotic Produced	Microorganism	Reference
Adriamycin, 4-*O*-methyl-13-dihydro-daunomycin	*Streptomyces peucetius* var. *caesius*	ARCAMONE et al., 1969
Deacetoxycephalosporin	*Acremonium chrysogenum*	YOSHIDA et al., 1978; SAWADA et al., 1979; FELIX et al., 1981
9-Dehydromycarosyl-platenomycin Demycarosylplatenomycin	*Streptomyces platensis*	FURUMAI and SUZUKI, 1975
7-*O*-Demethylcelesticetin, *N*-demethylcelesticetin, *N*-demethyl-7-*O*-demethyl-celesticetin	*Streptomyces caelestis*	ARGOUDELIS et al., 1972b, 1973b
6-Demethylchlortetracycline, 6-demethyltetracycline, tetracycline	*Streptomyces aureofaciens*	MCCORMICK et al., 1951; HENDLIN et al., 1962
1-Deoxychloramphenicol	*Streptomyces venezuelae*	AKAGAWA et al., 1979
Holomycin	*Streptomyces clavuligerus*	KENIG and READING, 1979
Ribostamycin	*Bacillus circulans*	FUJIWARA et al., 1978
Rifamycin SV	*Nocardia mediterranei*	LANCINI and HENGELLER, 1969

In this way, new products were also obtained by means of the leucomycin-, spiramycin-, and carbomycin-producing strains (OMURA et al., 1980a, 1983; SADAKANE et al., 1982, 1983).

Two of them are of considerable practical importance: 1. Daunomycin 14-hydroxylase is not operative in the parent strain of *Streptomyces peucetius* which thus synthesizes daunomycin as the main product.

Table 2. Incorporation of Precursor Analogs into Mutasynthetic Antibiotics

Microorganism and Natural Antibiotic	Mutant with Blocked Synthesis of	Mutasynthetic Precursor Analog	Reference
Streptomyces fradiae -neomycin	2-Deoxystreptamine	Streptamine	Shier et al., 1969, 1973
		2-*epi*-Streptamine	Cleophax et al., 1976
		2,5-Dideoxystreptamine	Rinehart, 1977
		2,6-Dideoxystreptamine	
		2-Bromo-2-deoxystreptamine	
		6-Bromo-6-deoxystreptamine	
		3-*N*-Methyldeoxystreptamine	
		6-*O*-Methyldeoxystreptamine	
		Streptidine	
Streptomyces rimosus forma *paromomycinus* -paromomycin		Streptamine	Shier et al., 1974
		2,6-Dideoxystreptamine	
Streptomyces ribosidificus -ribostamycin		Streptamine	Kojima and Satoh, 1973
		2-*epi*-Streptamine	Oka et al., 1981
		1-*N*-Methyl-2-deoxystreptamine	
		3′,4′-Dideoxyneamine	
Streptomyces kanamyceticus -kanamycin		1-*N*-Methylstreptamine	
		2-*epi*-Streptamine	
Micromonospora purpurea -gentamicin		Streptamine	Daum et al., 1977
		2,5-Dideoxystreptamine	Rosi et al., 1977
			Testa and Tilley, 1976
Micromonospora sagamiensis -sagamicin		Streptamine	Kitamura et al., 1982
Micromonospora inyoensis -sisomicin		Streptamine	Testa and Tilley, 1975
		2-*epi*-Streptamine	Waitz et al., 1978
		5-*epi*-2-Deoxystreptamine	Testa and Tilley, 1975
		2,5-Dideoxystreptamine	
		N-Methyl-2,5-dideoxystreptamine	
Bacillus circulans -butirosin	2-Deoxystreptamine	Streptamine	Claridge et al., 1974
		Streptidine	DeFuria and Claridge, 1976
		2,5-Dideoxystreptamine	Takeda et al., 1978 a, b, c, d
		6′-*N*-Methylneamine	
		3′-4′-Dideoxyneamine	
		3′,4′-Dideoxy-6′-*N*-methylneamine	
		3′,4′-Dideoxy-6′-*C*-methylneamine	
		6′-*N*-Methylgentamine C_{1a}	
		Gentamine C_2	
	Butirosin	Gentamine C_{1a}	Cappelletti and Spagnoli, 1983
		L(−)-α-Hydroxy-γ-butyric acid	

Streptomyces griseus -streptomycin	Streptidine	Deoxystreptidine	LEMKE and DEMAIN, 1976
Streptomyces erythraeus -erythromycin	Erythronolide B	Erythronolide A, (8S)-8-Fluoroerythronolides A and B, and 3-*O*-Mycarosyl-(8S)-8-fluoroerythronolide B	SPAGNOLI and TOSCANO, 1983 TOSCANO et al., 1983
Streptomyces antibioticus -oleandomycin	Oleandomycin	Erythronolide B	SPAGNOLI et al., 1983
Micromonospora polytrota -mycinamicins	Mycinamycins	10-Deoxo-20-dihydro-12,13-de-epoxy-12,13-dehydrorosaramicin	LEE et al., 1983
Streptomyces niveus -novobiocin	Novenamine	Chlornovenamine	LEMAUX and SEBEK, 1973
Streptomyces peucetius subsp. *caesius* -adriamycin	Aklavinone	Daunomycin, 13-Dihydro-daunomycin, Carminomycin, Feudomycin A	OKI et al., 1981

When, however, the synthesis of this enzyme is restored by mutational reactivation of the "silent" gene, the main product of the resulting revertant (*S. peucetius* var. *caesius*) is adriamycin (14-hydroxydaunomycin; ARCAMONE et al., 1969). 2. A blocked mutant of the rifamycin B producing *Nocardia mediterranei* accumulates rifamycin SV, an antibiotic with improved therapeutic efficacy and low toxicity (LANCINI and HENGELLER, 1969).

Antibiotic-blocked mutants have also been useful in the preparation of new antibiotic derivatives by an approach originally proposed by BIRCH in 1963; the antibiotic-blocked mutants cannot complete the synthesis of the parent antibiotic unless the missing precursor has been supplied to them which they can no longer synthesize but which they can metabolize. When in turn a suitable analog of such a precursor is added, it may be incorporated into a new antibiotic analog (BIRCH, 1963). This concept became known as mutational biosynthesis (NAGAOKA and DEMAIN, 1975) or mutasynthesis (RINEHART, 1977). Its feasibility was first demonstrated by SHIER et al. (1969).

From the population of neomycin-producing *Streptomyces fradiae,* they isolated a mutant blocked in the synthesis of 2-deoxystreptamine (2-DOS in Fig. 10), one of the moieties of the neomycin molecule. Because of this block, the mutant synthesized neomycin only when the missing deoxystreptamine moiety had been added to the fermentation. When in turn 2-deoxystreptamine was replaced by related structures such as streptamine (see Fig. 10 and Table 2) several new neomycin analogs were formed (see SHIER et al., 1973; RINEHART, 1977).

Because of its effectiveness, this methodology was applied also to other aminoglycosides whereby corresponding analogs of the more prominent 2-DOS-containing antibiotics have been produced: namely those of paromomycin ribostamycin, kanamycin, gentamicin, sisomicin, and butirosin. Streptamine was converted to 2-hydroxysagamicin by 2-DOS-blocked mutants of a *Micromonospora sagamiensis* and similar mutants

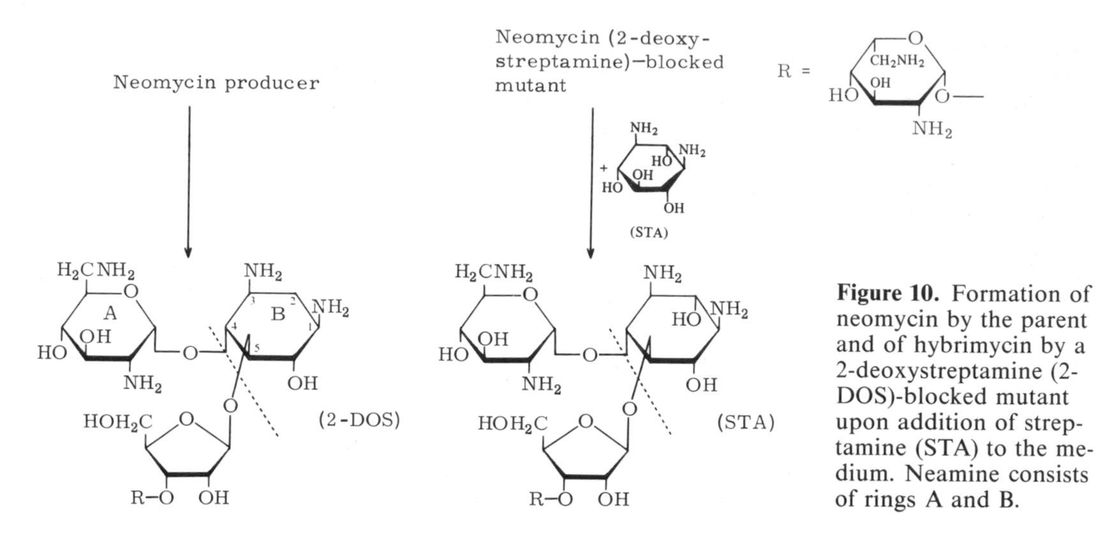

Figure 10. Formation of neomycin by the parent and of hybrimycin by a 2-deoxystreptamine (2-DOS)-blocked mutant upon addition of streptamine (STA) to the medium. Neamine consists of rings A and B.

of *Bacillus circulans* converted several neamine and paromamine analogs to the corresponding butirosin analogs. In a similar way, blocked mutants of other antibiotic classes yielded new analogs of the respective antibiotics (see Table 2).

Amikacin, one of the most effective semisynthetic aminoglycoside antibiotics, was also formed by mutasynthesis. A butirosin-blocked mutant of *Bacillus circulans* NRRL B-3313 was isolated which produced L-(−)-4-amino-2-hydroxybutyric acid (L-AHBA), a butirosin side chain. When supplied with preformed kanamycin A, the mutant synthesized amikacin (Fig. 11). Although the yields were low (3%), attributed in part to the high sensitivity of the bacterium to amikacin, this approach was viewed as a potential alternative to the expensive chemical synthesis of the antibiotic (CAPPELLETTI and SPAGNOLI, 1983).

The streptidine moiety of streptomycin (see Fig. 8) was modified in a similar way. When a streptidine idiotroph of streptomycin-producing *Streptomyces griseus* was incubated with 2-deoxystreptamine, a new streptomycin analog, streptomutin A, was formed (NAGAOKA and DEMAIN, 1975; LEMKE and DEMAIN, 1975).

Mutasynthesis of other antibiotic classes has also been investigated. Thus, for example, a new novobiocin was synthesized upon the addition of a clorinated coumarin derivative to a novobiocin-blocked *Streptomyces niveus* (BIRCH, 1972; LEMAUX and SEBEK, 1973). When erythronolide B, a 14-

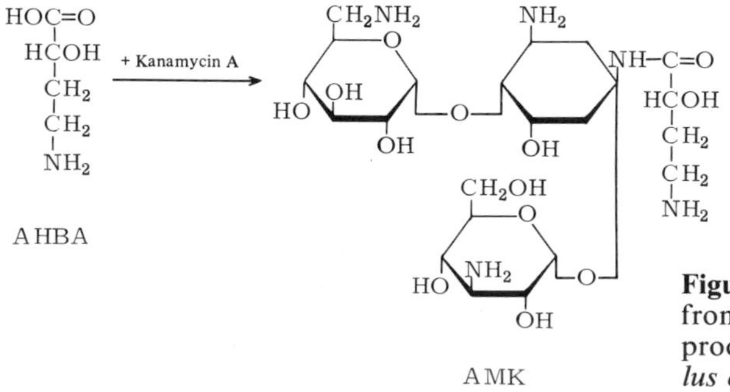

AHBA + Kanamycin A → AMK

Figure 11. Mutasynthesis of amikacin (AMK) from 4-amino-2-hydroxybutyric acid (AHBA) produced by a butirosin-blocked mutant of *Bacillus circulans*, and from added kanamycin A.

membered macrolide aglycone of erythromycins, was incubated with an oleandomycin-blocked mutant of *Streptomyces antibioticus,* four new antibiotics were formed. All of them contained the 3-*O*-oleandrosyl-5-*O*-desosaminyl portion of oleandomycin attached to erythronolide B which was modified by hydroxylation at epoxidation at C-8 and C-15. They were less active but more acid-stable than erythromycin A (Spagnoli et al., 1983). By the same technique, (8*S*)-8-fluoroerythromycins were prepared which were not only acid-stable but also showed the same antibacterial activities and spectra as the natural erythromycin. They were generated by feeding the corresponding (8*S*)-8-fluoroerythronolides to a mutant of *Streptomyces erythreus* able to convert erythronolide B to erythromycins A and B (Spagnoli and Toscano, 1983; Toscano et al., 1983). In a similar way, the aglycone of rosaramicin, a 16-membered macrolide, was incubated with a mutant of *Micromonospora polytrota* blocked in the synthesis of micamicins. The mixture yielded 6 new compounds which displayed about 50% activity of the parent antibiotics against Gram positive bacteria (Lee et al., 1983).

Adriamycin (doxorubicin) one of the most valuable antibiotics in cancer therapy, is produced by a mutant of *Streptomyces peucetius* var. *caesius.* A different mutant was also isolated which is blocked in the total synthesis of the antibiotic but retained the 13-oxidizing and 14-hydroxylating properties of the parent. Thus it was able to produce adriamycin not only from the preformed daunomycin but also from 13-dihydrodaunomycin, feudomycin A, and carminomycin I (Oki et al., 1981).

Preparation of a new penicillin by mutasynthesis is also of interest. L-α-Aminoadipic acid which is an intermediate in the L-lysine biosynthesis, is also a side-chain precursor of penicillin N (and cephalosporin C). A lysine auxotroph of *Acremonium chrysogenum* was isolated which grows and produces penicillin N only when supplied with this acid. Of several related compounds tested, it incorporated L-*S*-carboxymethylcysteine into a new semi-synthetic penicillin (RIT 2214) with the bioactivity similar to that of ampicillin. It was identified as 6-D-{[(2-amino-2-carboxy)ethylthio]acetamido}penicillanic acid (Troonen et al., 1976).

IV. Genetic Manipulation

This methodology includes the classical mutation of antibiotic producers and both intra- and interspecific recombinations. The rapidly advancing techniques of protoplast fusion (and recombinant DNA manipulation) have been already used (rifamycins, tylosin, spiramycin, cirramycin) whereby antibiotic productivity of the regenerated progeny was drastically changed (Ikeda et al., 1983). These techniques will be increasingly important in the transfer of the desired genes between divergent or wild-type microorganisms and in the construction of new recombinant strains. Mutation and intraspecific recombination have been used successfully to modify known antibiotics as illustrated by the preparation of new ansamycins by *Nocardia mediterranei.* Thus an intraspecific recombinant from a cross between two mutants blocked at different steps in the rifamycin biosynthesis, yielded seven novel ansamycins (Schupp et al., 1981; Traxler et al., 1981). Recombination between two other mutants was similarly successful (Traxler et al., 1982) and an assessment of the capability of the available technology to increase the production

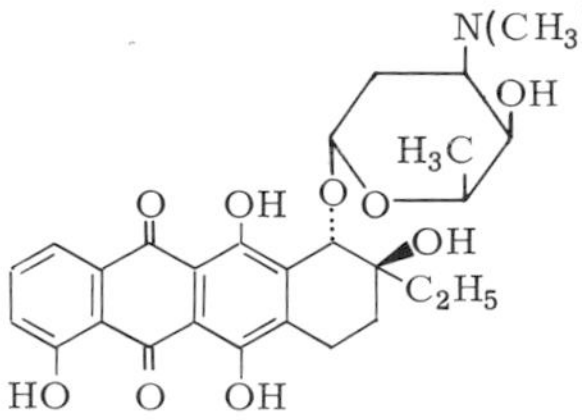

Figure 12. Structure of iremycin, an interspecific recombinant.

of aminoglycoside antibiotics was made (THOMPSON and DAVIES, 1984). An interspecific recombinant phenotype was also obtained from a cross between antibiotic-blocked mutants derived from a macrolide (turimycin)-producing *Streptomyces hygroscopicus*, and an anthracycline (violamycin)-producing *S. violaceus*. The structure of the resulting antibiotic (iremycin) is shown in Fig. 12 (SCHLEGEL and FLECK, 1980; IHN et al., 1980).

V. Biotransformations

A large number of antibiotics have been modified by various microorganisms and many of the changes found to be simple and chemically well defined reactions. Although most of these reactions have been only of academic interest, a few have become of considerable practical significance in clinical use (development of resistance to β-lactam and aminoglycoside antibiotics) and in the preparation of new semi-synthetic drugs superior to their parent compounds (semi-synthetic penicillins, cephalosporins, aminoglycosides, rifamycins, and others). The transformations have been carried out in batch cultures and more recently also by immobilized cells and enzyme technology (cf. DUNNILL, 1980).

They were divided into 11 type reactions.

- Hydrolysis and deacylation
- Acylation
- Phosphorylation
- Nucleotidylation (adenylylation)
- Oxidation
- Reduction
- Amination and deamination
- Glycosidation
- Methylation and demethylation
- Isomerization
- Hydration.

Only representative examples of these reactions are given in Tables 3–13 since a comprehensive review of this subject was published recently by MARSHALL and WILEY, 1982; see also SEBEK, 1974, 1975; SHIBATA and UYEDA, 1978). Furthermore, in view of a large amount of data that exist on the biotransformation of the therapeutically important antibiotics, only three examples dealing with the biotransformation of β-lactams, aminoglycosides, and rifamycins are given here in more details for illustration.

A. β-Lactams

As shown in Table 3, β-lactams, aminoglycosides, macrolides, and others have been hydrolyzed by different microorganisms. From the practical point of view, however, the hydrolysis of β-lactams by two types of enzymes, namely by β-lactamases and penicillin acylases, is of particular im-

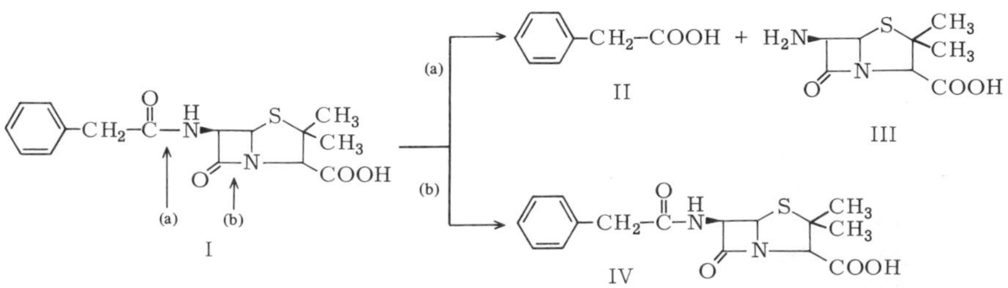

I: C_6H_5–CH_2–C(=O)–NH–[penam: S, CH_3, CH_3, COOH] —(a)→ II: C_6H_5–CH_2–COOH + III: H_2N–[penam: S, CH_3, CH_3, COOH]; —(b)→ IV: C_6H_5–CH_2–C(=O)–NH–[opened β-lactam: S, CH_3, CH_3, COOH]

Figure 13. Hydrolysis of benzylpenicillin (I) by penicillin acylases (a) to phenylacetic acid (II) and 6-aminopenicillanic acid (III), and by β-lactamases (b) to penicilloic acid (IV).

portance (Fig. 13). The former are of clinical importance as they are responsible for penicillin resistance, and the latter are used in the manufacture of 6-aminopenicillanic acid (6-APA).

The first microbial transformation of an antibiotic was that of penicillin by bacterial penicillinase (ABRAHAM and CHAIN, 1940) to antibiotically inactive penicilloic acid. This property renders penicillinase-producing bacteria resistant to the penicillin action. Since the inactivation is due to the hydrolysis of the cyclic amide bond in the β-lactam ring (see Fig. 13), the enzymes involved were named β-lactamases.

Because of their clinical importance, β-lactamases were investigated in many details. They are elaborated not only by Gram positive and Gram negative bacteria, but also by eukaryotic cells (yeasts and blue-green algae). Some are produced endocellularly and others are excreted into the medium. They are both inducible and constitutive, chromosomally controlled and also R-plasmid mediated. They differ in their substrate profiles, molecular weights, isoelectric points, immunological reactions, and in their responses to enzyme inhibitors (RICHMOND and SYKES, 1973; SYKES and MATTHEW, 1976; SLOCOMBE, 1980; LIVERMORE, 1982).

When penicillin was introduced in hospital use in the early 1940s, it proved to be of outstanding therapeutic value as most of the clinical isolates of *Staphylococcus aureus* were sensitive to it. However, the antibiotic exerted also an undesirable selective pressure on such infections. Since the growth of the penicillinase producers among staphylococcal populations were not affected by the antibiotic, the incidence of penicillin-resistant populations continued to rise and eventually reached 80% of all staphylococcal isolates by the mid 1950s (RIDLEY et al., 1970). In response to this alarming situation, thousands of new penicillin analogs were synthesized and some of them were found highly effective against such penicillin-resistant populations (cloxacillin, methicillin, nafcillin). Their administration practically eliminated the severe problem of staphylococcal resistance. In face of a similar development of penicillin resistance among other Gram positive and also Gram negative disease-causing bacteria, massive chemical and biological programs yielded new generations of β-lactams and compounds with novel β-lactam ring systems (cf. BROWN, 1982). They have low affinity for β-lactamases and broader antimicrobial spectra. Some have become of considerable therapeutical value by virtue of the antimicrobial spectra (penicillins, cephalosporins, monobactams), β-lactamase inhibitors (clavulanic acid, carbapenems; COLE, 1981) and other therapeutic advantages such as the route of administration and acid stability.

Amide hydrolysis

A wide range of microorganisms has been found to hydrolyze the amide bond of penicillins into 6-aminopenicillanic acid (6-APA) and the corresponding side chains. 6-APA has in turn served as a unique substrate for the preparation of novel semi-synthetic penicillins with improved therapeutic properties: methicillin, nafcillin, cloxacillin, ampicillin, carbenicillin, sulbenicillin, amoxicillin (cf. KATO et al., 1980), and others. They possess low affinity for β-lactamases and are therefore effective in inhibiting penicillin-resistant staphylococci, Gram negative bacteria (Enterobacteriaceae, *Pseudomonas, Haemophilus influenzae*).

6-APA was first reported in 1950 to be formed from benzylpenicillin by *Penicillium chrysogenum* Q 176 and named penicin (SAKAGUCHI and MURAO, 1950; MURAO, 1955). The compound was also formed when this penicillin producer was grown in the absence of phenylacetic acid and suggested to be the penicillin nucleus (KATO, 1953). It was identified later as 6-APA (BATCHELOR et al., 1959) and prepared microbiologically from natural penicillins (ROLINSON et al., 1960; CLARIDGE et al., 1960; HUANG et al., 1960; KAUFMANN and BAUER, 1960).

The enzyme involved was originally called "penicillin-amidase" by SAKAGUCHI

Table 3. Hydrolysis, *N*- and *O*-Deacylations

Substrate	Product	Microorganism	Reference
β-Lactams			
Hydrolysis			
Benzyl- and	Corresponding β-lactam ring cleavage products	*Escherichia coli*	Abraham and Chain, 1940
Phenoxymethylpenicillins,		*Streptomyces albus*	Johnson et al., 1973
Ampicillin, Carbenicillin		*Pseudomonas aeruginosa*	Sykes and Richmond, 1970
Cephalosporin C		*Enterobacter cloacae*	Hennessey and Richmond, 1968
Cephaloglycine, Cephaloridine, Cephalothin		*Streptomyces* sp.	Johnson et al., 1973
N-Deacylation			
Benzyl-, phenoxymethyl- and other penicillins	6-APA and the corresponding acyl side chains	*Escherichia coli*	Rolinson et al., 1960
		Fusarium semitectum	Brandl, 1965
		Penicillium chrysogenum	Sakaguchi and Murao, 1950
		Aspergillus oryzae	
		Nine different bacteria	Savidge and Cole, 1975
N-Acetyldehydroxy-thienamycin (PS-5)	Deacetylated PS-5 (NS-5)	*Streptomyces olivaceus*	Fukagawa et al., 1980
O-Deacylation			
Cephalosporin C	3-Deacetylcephalosporin	Various bacteria and actinomycetes	Demain et al., 1963
Cephalothin	3-Deacetylcephalothin	*Escherichia coli*	Nishiura et al., 1978
Nocardicin C	3-Aminonocardicinic and α-aminoadipic acids	*Pseudomonas schuylkilliensis*	Komori et al., 1978
Aminoglycosides			
Hydrolysis			
Mannosidostreptomycin	Streptomycin	*Streptomyces griseus*	Perlman and Langlykke, 1948
Validamycins	Validamycin A and Validoxylamine	Various bacteria and yeasts	Kameda et al., 1975
Validamycins A and D	Validoxylamine A	*Pseudomonas denitrificans* and other microorganisms	Kameda and Horii, 1972
Validamycin B	Hydroxyvalidamine	*Pseudomonas denitrificans*	
Validamycins C, E and F	Validamycin A	*Endomycopsis* spp. and *Candida intermedia*	

Macrolides			
Deacylation			
14-*O*-Acetyl-8-*O*-acyl-lankacidin C	Corresponding 8-*O*-Acyllankacidins	*Streptomyces rochei*	HARADA et al., 1973
Lankacidin A Lankacidinol A	Lankacidin C	*Streptomyces rochei*, *Aspergillus niger* *Aspergillus sojae* *Trametes sanguinea*	FUGONO et al., 1970
Various 8-*O*-14-*O*-di-acylated Lankacidin C derivatives	14-*O*-Acylated lankacidin C	*Streptomyces rochei*	
Leucomycins A_1 and A_3 Magnamycins A and B Middamycin	Corresponding 4″-*O*-deisovaleryl derivatives	*Cunninghamella elegans* and other fungi, *Streptomyces* sp.	THERIAULT, 1974
Leucomycin A_5	Leucomycin V (4″-debutyrylleucomycin)	*Actinoplanes missouriensis*	SINGH and RAKHIT, 1979
Maridomycin III	4″-*O*-Depropionylmari-domycin III 18-Dihydromaridomycin III, 4″-Depropionylmaridomycin III and 18-Dihydro-4″-depropionyl-maridomycin III	*Bacillus megaterium* *Streptomyces pristinaespiralis* *Streptomyces* sp.	NAKAHAMA et al., 1974a, c UYEDA et al., 1980
Others			
Chloramphenicol	1-(*p*-Nitrophenyl)-2-acetamidopropane-1,3-diol 1-(*p*-Nitrophenyl)-2-aminopropane-1,3-diol	*Streptomyces* sp. *Bacillus subtilis* *Bacillus mycoides* *Proteus vulgaris* *Escherichia coli*	MALIK and VINING, 1970 SMITH and WORRELL, 1949, 1950, 1953
Olivomycin A Chromomycins A_2 and A_3	Corresponding deacylated analogs	*Whetzelinia sclerotiorum*	SCHMITZ and CLARIDGE, 1977

Table 4. Acylation

Substrate	Product	Microorganism	Reference
***β*-Lactams**			
6-APA + phenylacetic acid	Benzylpenicillin	*Escherichia coli*	ROLINSON et al., 1960
		Alcaligenes faecalis	CLARIDGE et al., 1960
+ phenoxyacetic acid	Phenoxymethyl-penicillin	*Alcaligenes faecalis*	CLARIDGE et al., 1960
+ carboxylic acids and esters	Penicillins with the corresponding acyl side chains	*Kluyvera citrophila*	NARA et al., 1971
		Pseudomonas melanogenum	OKACHI and NARA, 1973
+ phenylglycine esters	Ampicillin	*Kluyvera citrophila*	NARA et al., 1971
7-ACA, 7-ADCA, and their organic acid esters	Corresponding cephalosporins (cephalexin, cephaloglycine)	*Kluyvera citrophila*	SHIMIZU et al., 1975
		Xanthomonas citri and other pseudomonads	TAKAHASHI et al., 1972
Aminoglycosides			
Gentamicins C, C_1 and C_{1a}	3-*N*-Acetyl derivatives of the respective substrates	*Escherichia coli*	LEGOFFIC et al., 1974
		Klebsiella pneumoniae	
		Pseudomonas aeruginosa	BIDDLECOME et al., 1976
Kanamycin A		*Escherichia coli*	LEGOFFIC et al., 1976
Kanamycin B		*Klebsiella pneumoniae*	
		Pseudomonas aeruginosa	BIDDLECOME et al., 1976
Tobramycin		*Escherichia coli*	LEGOFFIC et al., 1974
		Klebsiella pneumoniae	
Gentamicin C_1 and C_{1a}	2′-*N*-Acetyl derivatives of the respective substrates	*Providencia* spp.	CHEVEREAU et al., 1974
Sisomicin		*Providencia* spp.	
Tobramycin			
Lividomycins A and B		*Providencia* spp.	YAMAGUCHI et al., 1974
Paromomycin			
Ribostamycin			
Amikacin	6′-*N*-Acetyl derivatives of the respective substrates	*Pseudomonas aeruginosa*	HAAS et al., 1976, YAGISAWA et al., 1975
Butirosins		*Pseudomonas aeruginosa*	HAAS et al., 1976, YAGISAWA et al., 1975

Gentamicin C_{1a}		*Escherichia coli*	Benveniste and Davies, 1971
		Pseudomonas aeruginosa	Haas et al., 1976, Yagisawa et al., 1975
Gentamicin C_2		*Pseudomonas aeruginosa*	Yagisawa et al., 1975
Kanamycins A, B and C		*Moraxella* spp.	LeGoffic and Martel, 1974
		Pseudomonas aeruginosa	Haas et al., 1976, Kawabe et al., 1975, Yagisawa et al., 1975
Sisomicin		*Moraxella* spp.	LeGoffic and Martel, 1974
		Providencia spp.	
		Pseudomonas aeruginosa	Chevereau et al., 1974
Tobramycin		*Moraxella* spp.	LeGoffic and Martel, 1974
		Pseudomonas aeruginosa	Haas et al., 1976 Yagisawa et al., 1975
Neamine		*Moraxella* spp.	LeGoffic and Martel, 1974
Neomycin B		*Escherichia coli*	Benveniste and Davies, 1971
		Pseudomonas aeruginosa	Haas et al., 1976
Ribostamycin		*Pseudomonas aeruginosa*	Yagisawa et al., 1975
Macrolides			
4″-Depropionyl maridomycin III	Maridomycin V (K)	*Streptomyces* sp.	Uyeda et al., 1977
Lankacidin C	Lankacidin C butyrate, isobutyrate, valerate, and isovalerate	*Bacillus megaterium*	Nakahama et al., 1975
Tylosin	Eight derivatives acylated at C3- and C 4″	*Streptomyces thermotolerans*	Okamoto et al., 1980
Anthracyclines			
Daunomycin Adriamycin	*N*-Acetyl derivatives of the respective substrates	*Bacillus cereus* var. *mycoides*	Hamilton et al., 1977
Others			
Cycloheximide	Cycloheximide acetate	*Cunninghamella blakesleeana*	Howe and Moore, 1968
Virginiamycin (M factor)	*O*-Acetylvirginiamycin (M factor)	*Staphylococcus aureus*	deMeester and Rondelet, 1976

Table 5. Phosphorylation

Substrate	Product	Microorganism	Reference
Aminoglycosides			
Amikacin	3′-*O*-Phosphorylated derivatives of the respective substrates	*Staphylococcus aureus*	COURVALIN and DAVIES, 1977
Butirosin		*Escherichia coli* *Pseudomonas aeruginosa*	BRZEZINSKA and DAVIES, 1973
Butirosin A		*Escherichia coli* *Pseudomonas aeruginosa*	MATSUHASHI et al., 1975
Kanamycin		*Nocardia asteroides* *Escherichia coli*	SHIRAFUJII et al., 1980 YAGISAWA et al., 1972a
Kanamycin A		*Pseudomonas aeruginosa*	MATSUHASHI et al., 1975
Neamine		*Escherichia coli* *Pseudomonas aeruginosa*	BRZEZINSKA and DAVIES, 1973
Neamine Neomycin Ribostamycin Gentamicins A, B, C_{1a}, C_2 Kanamycin B Sisomicin Tobramycin		*Escherichia coli* *Pseudomonas aeruginosa* *Escherichia coli* *Staphylococcus aureus*	YAGISAWA et al., 1972a DOI et al., 1968 YAGISAWA et al., 1972a LEGOFFIC et al., 1977
Dihydrostreptomycin	3″-*O*-Phosphorylated derivatives of the respective substrates	*Pseudomonas aeruginosa*	KAWABE et al., 1971
Streptomycin		*Escherichia coli* *Pseudomonas aeruginosa*	OZANNE et al., 1969
Lincosaminides			
Celesticetin Clindamycin Lincomycin	3-*O*-Phosphorylated derivatives of the respective substrates	*Streptomyces coelicolor* *Streptomyces rochei*	COATS and ARGOUDELIS, 1971 ARGOUDELIS and COATS, 1969
Others			
Tubercidin	Tubercidin 5′-monophosphate	*Serratia marcescens*	SHIRATO et al., 1968

Table 6. Nucleotidylation (adenylylation)

Substrate	Product	Microorganism	Reference
Amikacin	4′-*O*-Adenylylated derivatives of the respective substrates	*Staphylococcus aureus*	LEGOFFIC et al., 1976
Butirosin			
Gentamicin		*Pseudomonas aeruginosa*	KABINS et al., 1974
Kanamycins A and B			
Lividomycin			
Neomycins B and C			
Paromomycin			
Ribostamycin			
Sisomicin			
Tobramycin			
3′,4′-Dideoxykanamycin B	2″-*O*-Adenylyl-, 2″-*O*-guanosylyl-, and 2″-*O*-inosylyl-3′,4′-dideoxykanamycin B	*Escherichia coli*	YAGISAWA et al., 1972b
Clindamycin	Clindamycin 3-ribonucleotides	*Streptomyces coelicolor*	ARGOUDELIS et al., 1977
Spectinomycin	9-*O*-Adenylylspectinomycin	*Acinetobacter calcoaceticus*	SHIMIZU et al., 1981

Table 7. Oxidation

Substrate	Product	Microorganism	Reference
Macrolides			
Hydroxylation			
A23187-Methyl ester and polyvinylpyrrolidone	16-Hydroxylated (and *N*-demethylated) products	*Streptomyces chartreusis*	ABBOTT et al., 1979
Josamycin	3‴-OH-Josamycin	*Streptomyces olivaceus*	NAKAHAMA et al., 1974b
Maridomycin I	3‴-OH-Maridomycin I		
Narbomycin	Picromycin	*Streptomyces narbonensis*	MAEZAWA et al., 1973
Narbonolide	Picronolide	*Streptomyces zaomyceticus*	
Ketone formation			
Lankacidinol	Lankacidin A	*Streptomyces rochei* var. *volubilis*	HARADA et al., 1973
Midecamycin A_1	Midecamycin A_3	*Streptomyes mycarofaciens*	MATSUHASHI et al., 1979
Epoxidation			
Carbomycin B	Carbomycin A	*Streptomyces hygroscopicus*	SUZUKI et al., 1977
Leucomycin A3	Maridomycin II		
cis-Propenylphosphonic acid	Fosfomycin	*Penicillium spinulosum*	WHITE et al., 1971
Anthracyclines			
Hydroxylation			
Auramycinone	11-OH-Auramycinone and 9-methyl-10-OH-daunomycin	*Streptomyces coeruleorubidus*	HOSHINO and FUJIWARA, 1983
Daunomycin	14-OH-Daunomycin (adriamycin)	*Streptomyces peucetius* var. *caesius*	ARCAMONE et al., 1969 OKI et al., 1981
ε-Isorhodomycinone	1-OH-13-Dihydrodaunomycin (and its *N*-formyl derivative)	*Streptomyces coeruleorubidus*	YOSHIMOTO et al., 1980
ε-Pyrromycinone			

Cephalosporins and lincosaminides			
Sulfoxidation			
Cephalosporins	Corresponding cephalosporin (*R*)- and (*S*)-sulfoxides	*Coriolus hirsutus* and seven other fungi (Aphyllophorales)	Torii et al., 1980
Clindamycin	Clindamycin sulfoxide	*Streptomyces armentosus*	Argoudelis et al., 1969
Lincomycin	Lincomycin sulfoxide	*Streptomyces lincolnensis*	Argoudelis and Mason, 1969
Other antibiotics			
Hydroxylation			
Ansamitocin	15-OH- and 15-*epi*-OH-Ansamitocin P-3	*Streptomyces sclerotialus*	Izawa et al., 1981a
12a-Deoxytetracycline	Tetracycline	*Curvularia lunata*	Holmlund et al., 1959
Fusidic acid	6-Keto-7α-OH-fusidic acid	*Acrocylindrium oryzae*	von Daehne et al., 1968
Griseofulvin	5′-OH- and 6′-OH-Methylgriseofulvin	*Cephalosporium curtipes*	Bod et al., 1973
Novobiocin	11-OH-Novobiocin	*Sebekia benihana*	Sebek and Dolak, 1984
Oxidation of others			
Aclacinomycin A	Aclacinomycin Y	*Streptomyces galilaeus*	Yoshimoto et al., 1979
Dihydroabikoviromycin	Abikoviromycin	*Streptomyces olivaceus*	Tsuruoka et al., 1973
Formycin B	Oxoformycin B	*Pseudomonas fluorescens* *Xanthomonas oryzae* *Nocardia interforma*	Sawa et al., 1968
Mycophenolic acid	Various oxygenated products	Different organisms	Jones et al., 1970
Rifamycin B	Rifamycin O	*Monocillium* sp.	Seong et al., 1983
Toyocamycin	Sangivamycin	*Streptomyces rimosus*	Uematsu and Suhadolnik, 1974

Table 8. Reduction

Substrate	Product	Microorganism	Reference
Anthracyclines			
Adriamycin	7-Deoxyadriamycinone and 7-Deoxyadriamycinol aglycone	*Streptomyces steffisburgensis*	MARSHALL et al., 1978
Daunomycin	Daunomycinol	*Corynebacterium equi*	ASZALOS et al., 1977
		Mucor spinosus	MARSHALL et al., 1978
	7-Deoxydaunomycinone	*Streptomyces steffisburgensis*	MARSHALL et al., 1976
Daunomycinol and its aglycone	7-Deoxydaunomycinol aglycone	*Streptomyces steffisburgensis*	MARSHALL et al., 1978
		Streptomyces coeruleorubidus	BLUMAUEROVÁ et al., 1979
		Streptomyces galilaeus	
Daunomycinone	13-Dihydrodaunomycinone	*Streptomyces aureofaciens*	KARNETOVÁ et al., 1976
Nogalamycin	7-Deoxynogalarol	*Streptomyces nogalater*	RUECKERT et al., 1979
Rubeomycin A	Rubeomycin B	*Rhodotorula glutinis*	OGAWA et al., 1983
Macrolides			
Albocycline	2,3-Dihydroalbocycline	*Streptomyces venezuelae*	SLECHTA et al., 1978
Carbomycin A	Carbomycin A P1	*Streptomyces halstedii*	FUKAGAWA et al., 1984
		Streptomyces lutea and others	
Carbomycin A	Maridomycin II	*Streptomyces hygroscopicus*	SUZUKI et al., 1977
Carbomycin B	Leucomycin A_3		
Maridomycin III	18-Dihydromaridomycin III	*Streptomyces* sp.	UYEDA et al., 1980
Midecamycin A_3	Midecamycin A_1	*Streptomyces mycarofaciens*	MATSUHASHI et al., 1979
Tylosin	Relomycin	*Nocardia corallina*	FELDMAN et al., 1963
Others			
Ascochitine	Dihydroascochitine	*Fusarium lycopersici*	OKU and NAKANISHI, 1964
Dehydrogriseofulvin	(+)-Griseofulvin and (+)-5′-Hydroxygriseofulvin	*Streptomyces cinereocrocatus*	ANDRES et al., 1969
Rifamycin S	Rifamycin SV	*Nocardia mediterranei*	GHISALBA et al., 1982
Saframycin	25-Dihydro-(and 21-decyano-25-dihydro-)saframycin	*Rhodococcus amidophilus*	TAKAHASHI et al., 1982
			YAZAWA et al., 1982

Table 9. Amination and Deamination

Substrate	Product	Microorganism	Reference
Amination			
Formycin B	Formycin	*Streptomyces* sp. *Nocardia interforma* *Xanthomonas oryzae*	OCHI et al., 1975 SAWA et al., 1968
Deamination			
Blasticidin	Deaminohydroxyblasticidin S	*Aspergillus terreus*	YAMAGUCHI et al., 1975

Table 10. Glycosidation

Substrate	Product	Microorganism	Reference
Aclacinomycin B	Aclacinomycin A	*Streptomyces galilaeus*	HOSHINO et al., 1983
Daunomycinone, Daunomycinol aglycone ε-Rhodomycinone, ε-Isorhodomycinone	Glycosides of the respective substrates	*Streptomyces coeruleorubidus*	BLUMAUEROVÁ et al., 1979
ε-Pyrromycinone		*Streptomyces galilaeus*	VANĚK et al., 1973
2-Hydroxyaclavinone	2-Hydroxyaclacinomycin A	*Streptomyces galilaeus*	MATSUZAWA et al., 1981
Erythronolide A oxime	3-*O*-Oleandrosyl-5-desosaminylerythronolide A oxime	*Streptomyces antibioticus*	LEMAHIEU et al., 1976
Erythronolide A	Erythromycin A	*Streptomyces erythreus*	SPAGNOLI and TOSCANO, 1983
Narbonolide	5-*O*-Mycaminosylnarbonolide	*Streptomyces platensis*	MAEZAWA et al., 1976
Validoxylamine A	Validamycin A	*Rhodotorula glutinis* and *Rhodotorula lactosa*	KAMEDA et al., 1975
	Validamycins A and D	*Rhodotorula marina* *Rhodotorula lactosa*	KAMEDA et al., 1980

Table 11. Methylation, Demethylation

Substrate	Product	Microorganism	Reference
Methylation			
Gentamicins C_{1a} and C_2	Gentamicins C_{2b} and C_1	*Micromonospora purpurea*	TESTA and TILLEY, 1976
N-Demethylclindamycin	*N*-Demethyl-*N*-hydroxymethyl clindamycin	*Streptomyces lincolnensis*	ARGOUDELIS et al., 1972a
Demethylation			
Celesticetin	*N*-Demethylcelesticetin, 7-*O*-demethylcelesticetin and *N*-demethyl-7-*O*-demethyl-celesticetin	*Streptomyces caelestis*	ARGOUDELIS et al., 1973b
Clindamycin	*N*′-Demethylclindamycin	*Streptomyces punipalus*	ARGOUDELIS et al., 1969
Lincomycin	*N*′-Demethylthio-1-hydroxy-lincomycin	*Streptomyces lincolnensis*	ARGOUDELIS and MASON, 1969
Ansamitocins	20-*O*-Demethyl ansamitocins	*Bacillus megaterium*	IZAWA et al., 1981b
	N-Demethylansamitocins	*Streptomyces minutiscleroticus*	
Daunomycinone derivatives	Corresponding demethylated Carminomycinone derivatives	*Sepedonium chrysospermum* *Beauveria sulfurescens*	WU et al., 1980

Table 12. Isomerization

Substrate	Product	Microorganism	Reference
Novobiocin	Isonovobiocin	*Aspergillus sphaeroides*	BATYROVA and EGOROV, 1974
Showdomycin	Isoshowdomycin	*Streptomyces* sp.	OZAKI et al., 1972

Table 13. Hydration

Substrate	Product	Microorganism	Reference
4-*epi*-Cetocycline	2-Decarboxamido-2-acetyl-4-dedimethyl-9-methyl-tetracycline	*Streptomyces rimosus*	THERIAULT et al., 1982
Anhydrotetracycline	Dehydrotetracycline	*Streptomyces aureofaciens*	BĚHAL et al., 1979

and MURAO and has since been classified as penicillin acylase. It exists in the form of different isoenzymes which display a considerable degree of substrate specificity (PLASKIE et al., 1978). Those of bacterial origin have in general higher affinity for benzylpenicillin and hence hydrolyze it more readily than phenoxymethylpenicillin while the reverse is true for acylases produced by actinomycetes and fungi (VANDAMME and VOETS, 1974; QUEENER and SWARTZ, 1979; LOWE et al., 1981; MARSHALL and WILEY, 1982). A few of them *(Escherichia coli; Kluyvera citrophila, Pseudomonas melanogenum, Bacillus megaterium)* were adapted for large-scale production of 6-APA. In such operations, suspended and immobilized cells as well as crude enzyme preparations and purified enzymes are employed (CARRINGTON, 1971; MOSS, 1977; ROLINSON, 1979; DUNNILL, 1980; VANDAMME, 1983). The efficiency of the process has been increased by strain improvement through conventional mutation, selection, and enzyme induction. More recently, such improvements were also made in *B.megaterium* through the application of recombinant DNA technology (MCCULOUGH, 1983).

Patents and publications indicate that both enzymatic and chemical hydrolysis is used for large-scale production of 6-APA (WEISSENBURGER and VAN DER HOEVEN, 1970; HUBER et al., 1972; SELLSTEDT, 1975; QUEENER and SWARTZ, 1979). Clearly, the selection of the particular process is dictated by the expertise and facilities of the individual manufacturer. Enzymatic manipulation is pH dependent. The hydrolysis occurs at an alkaline reaction (pH 7.5–8.5) while the acylation of 6-APA takes place at acidic pHs (MCDOUGALL et al., 1982).

The technical procedures mainly starting from penicillin G are nowadays carried out with immobilized whole cells of *Escherichia coli* in polyacrylamide gels (SATO et al., 1976), with immobilized penicillin acylase on Sephadex (EKSTRÖM et al., 1974; LAGERLÖF et al., 1976), or entrapped enzyme in cellulose triacetate fibers (MARCONI et al., 1973; GIACOBBE, 1977). Immobilized enzyme preparations could be used with careful operations for more than one hundred batches. Losses of enzyme activity could be avoided by a recirculation process having a higher production capacity and at lower costs, e.g., yields of 90% 6-APA at a purity of 98%. The acyl moiety, phenylacetic acid, is additionally recovered and could be reused as a precursor for penicillin G production in an integrated process.

Further processes are described with penicillin acylase from *Escherichia coli* covalently bound to polymethacrylate resin with glutaraldehyde (SAVIDGE et al., 1974), coupled to a polymer of sucrose and epichlorohydrin (CAWTHORNE, 1974), to a copolymer of acrylamide, *N,N′*-methylene-bis-acrylamide and maleic acid anhydride (HUEPER 1973a,b), or entrapped in polyacrylamide gel (MANDEL et al., 1975).

Penicillin acylase is also obtained from *Bacillus megaterium* as an extracellular enzyme by adsorption onto bentonite; the same enzyme from *Bovista plumbea* is technically used for the hydrolysis of penicillin V with a yield of 91.5% of crystallized 6-APA (BRANDL and KNAUSEDER, 1975).

Penicillin acylase from *Escherichia coli* can also perform the opposite reaction of coupling the 6-APA with an acyl moiety with its directional equilibrium controlled by pH. Because of low yields and an efficient chemical acylation, this enzymatic reaction is used only rarely. An exception is the preparation of ampicillin from 6-APA and D-phenylglycine methyl ester, and of amoxycillin from D-*p*-hydroxyphenylglycine ethyl ester at yields of less than 50% by the use of entrapped enzyme in triacetate fibers (MARCONI et al., 1973). Yields of more than 50% ampicillin could be reached by means of the enzyme obtained from *Bacillus megaterium* or *Achromobacter* sp. adsorbed to DEAE-cellulose (FUJII et al., 1973) or with succinoylation of the enzyme before adsorption (KAMOGASHIRA et al., 1972). An excellent review of all these methods is given by BRODELIUS (1978).

A two-step enzymatic process was also described for the preparation of ampicillin (D-(−)-α-aminobenzylpenicillin). Benzylpenicillin is first hydrolyzed to 6-APA with acylase of *Kluyvera citrophila* at pH 7.5. 6-APA in turn is acylated with DL-phenylglycine methyl ester at pH 5.5 by means of an acylase of *Pseudomonas melanogenum* and yields ampicillin. The feasibility of this operation is due to the high substrate specificity of the *Pseudomonas* acylase. The enzyme does not react with benzylpenicillin or phenylacetic acid which remain unaffected during the acylation step (OKACHI and NARA, 1973).

7-Aminocephalosporanic acid (7-ACA) is of similar importance in the preparation of semi-synthetic cephalosporins. The acid is obtained chemically with high efficiency using PCl_5, nitrosyl chloride, etc. to remove the D-α-aminoadipic acid side chain of cephalosporin C (see Fig. 14). Patent literature indicates that the same has been accomplished by means of *Pseudomonas putida* (NIWA et al., 1977; GOI et al., 1978). In addition, methods were described for the preparation of 7-ACA which consist of two enzymatic steps. The amino group of the aminoadipic acid side chain of cephalosporin C is first removed by D-amino acid oxidase of *Trigonopsis variabilis, Gliocladium deliquescens* or of pig kidney, to glutaryl 7-ACA (GILBERT et al., 1972; MAZZEO and ROMEO, 1972; FUJII et al., 1979). The latter is in turn deacylated by selected pseudomonads, *Comamonas* and other bacteria to 7-ACA (ICHIKAWA et al., 1981; SHIBUYIA et al.; cf. VANDAMME, 1983).

By the addition of the corresponding side chain precursors to 7-ACA, it was thus possible to prepare cephalothin, cephaloglycine, and cefamandole by means of immobilized cells and enzyme from *Bacillus megaterium* (TAKAHASHI et al., 1972; TOYO BREWING Co., 1978).

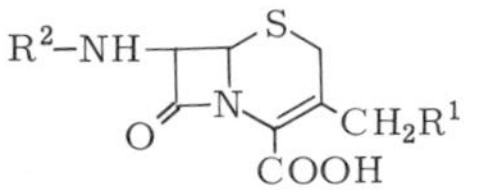

R²–NH– [cephem ring: S, N, O=, CH₂R¹, COOH]

	R^1	R^2
Cephalosporin C	$OCOCH_3$	$CO-(CH_2)_3-\overset{D}{C}H(NH_2)-COOH$
7-Aminocephalosporanic acid (7-ACA)	$OCOCH_3$	H
Desacetoxycephalosporin	H	$CO-(CH_2)_3-\overset{D}{C}H(NH_2)-COOH$
7-Aminodesacetoxy-cephalosporanic acid (7-ADCA)	H	H

Figure 14. Cephalosporin C and related compounds.

Another key intermediate for the preparation of semi-synthetic cephalosporins is 7-aminodeacetoxycephalosporanic acid (7-ADCA). This compound is obtained by chemical expansion of the five-membered thiazolidine ring of the substrate benzyl- or phenoxymethylpenicillin to the six-membered dihydrothiazine ring of cephalosporins*⁾. The subsequent deacylation of the respective penicillin side chains to 7-ADCA can be carried out in 85% yields chemically and also enzymatically with immobilized penicillin acylase or with immobilized cells of *Bacillus megaterium*. This and a few

* The same ring expansion of penicillin N to deacetoxycephalosporin C has been shown to be carried out by *Cephalosporium acremonium* (KOHSAKA and DEMAIN, 1976; YOSHIDA et al., 1978; FELIX et al., 1981).

other selected bacteria (such as *Achromobacter* sp., *Alcaligenes faecalis, Xanthomonas citri* or *Acetobacter turbidans*) in turn produce cephalexin, an orally active semi-synthetic cephalosporin, by acylation of 7-ADCA with D-α-phenylglycine methyl ester (FUJII et al., 1976; TAKAHASHI et al., 1977).

In order to synthesize cephalosporin derivatives at C-3, the original acetate group at that position must first be removed. It is therefore of practical importance that this hydrolysis is carried out specifically and more efficiently by esterases of various microorganisms, plant and mammalian tissues than by chemical methodology. Thus purified cephalosporin acetyl esterase of *Bacillus subtilis* was immobilized and re-used 20 times to deacetylate cephalosporins to deacetylcephalosporins (and acetate) in high yields (see Fig. 14; ABBOTT and FUKUDA, 1975; KONECNY and SIEBER, 1980).

Hydrolysis of other β-lactams

Of the new natural β-lactams (cephamycins, thienamycins, olivanic acids, carpetimycins, nocardicins and others) some have also been hydrolyzed by various microorganisms. Thus compound PS-5, an *N*-acetylated dehydroxythienamycin derivative and a potent β-lactamase inhibitor, was hydrolyzed by D-amino acid acylase of *Streptomyces olivaceus* and also by most of the methanol-assimilating pseudomonads (FUKAGAWA et al., 1980). Nocardicin C, a monocyclic β-lactam and a minor product in nocardicin fermentations, was found to be deacylated by *Pseudomonas schuylkilliensis* and others to 3-aminonocardicinic acid which thus may provide a substrate for new semi-synthetic nocardicins (KOMORI et al., 1978). Although nocardicin A which is the major product of nocardicin fermentation, is not subject to such hydrolysis, it can be readily converted to nocardicin C by chemical hydrogenation.

B. Aminoglycosides

Some of the natural aminoglycoside antibiotics (such as streptomycin and neomycin) have only limited application as antibacterial agents. Others, however, (such as those of the kanamycin and gentamicin families), have been of considerable therapeutic value since they are parenterally active, bactericidal and exhibit broad antibacterial spectra. Organisms resistant to aminoglycosides started to appear upon their introduction into the clinic. One kind of such resistance is due to the *N*-acetylation, *O*-adenylylation, or *O*-phosphorylation of the respective antibiotics by aminoglycoside transferases produced by the resistant strains*). They inactivate the antibiotics either partially (by *N*-acetylation) or completely (by *O*-adenylylation or *O*-phosphorylation). Once the nature and sites of these inactivations have been elucidated, a more rational design of analogs insensitive to such inactivations became possible (cf. UMEZAWA et al., 1972).

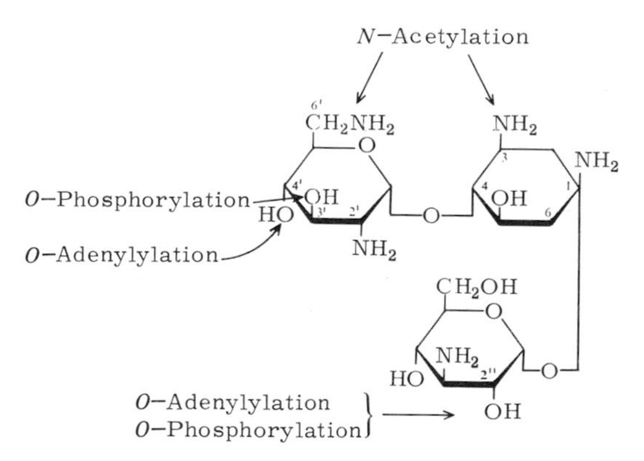

Figure 15. Structure of kanamycin B showing the sites susceptible to enzymatic modification. Kanamycin A has —OH at C′-2 instead of —NH_2.

* Two other kinds of resistance are due to (1) the mutation of the organisms to strains with altered ribosomes which no longer bind the antibiotic, and (2) the reduction or elimination of the antibiotic transport into the bacterial cell.

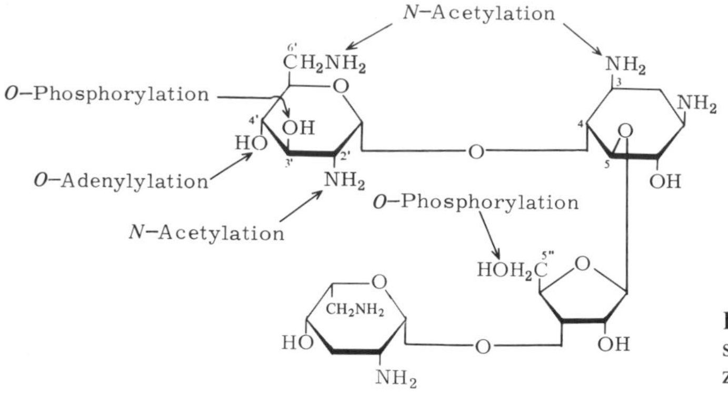

Figure 16. Structure of neomycin B showing the sites susceptible to enzymatic modification.

For purposes of illustration, Fig. 15 shows the structure of the natural aminoglycoside kanamycin B and the sites and kinds of its inactivation. Like gentamicins and sisomicins, kanamycin B consists of a 2-deoxystreptamine moiety to which two aminohexoses are attached at two non-adjacent hydroxyls (4,6-disubstituted 2-deoxystreptamine). Although its activation is mediated by all three mechanisms mentioned (*N*-acetylation, *O*-phosphorylation, and *O*-adenylylation), kanamycin B continues to be a clinically valuable antibiotic. This is due primarily to the fact that the inactivating enzymes are not uniformly distributed in all of the resistant strains and that some have been detected only in a few instances. Thus the formation of the 6′-*N*-acetylkanamycin by acetyltransferases in hospital isolates is uncommon and their levels in some cases are not clinically significant. Similarly, 2′-*N*-acetylation has been detected thus far only in *Providencia* and *Proteus*. 4′-*O*-Adenylylation and 2″-*O*-phosphorylation have been shown only in strains of *Staphylococcus aureus* which are readily controlled by penicillin therapy. Since on the other hand 3′-*O*-phosphorylation is common in many clinical bacteria including *Pseudomonas aeruginosa*, modifications at that position yielded semi-synthetic kanamycins which are highly resistant to such inactivation.

Related aminoglycosides (neomycins, Fig. 16, paromomycins, lividomycins, ribostamycin, butirosins) in which the 2-deoxystreptamine moiety is substituted in the vicinal (i.e. 4, 5) positions, are inactivated by the same three mechanisms as illustrated in Fig. 16. Streptomycin which contains streptidine instead of 2-deoxystreptamine (see Fig. 8) is inactivated by *O*-phosphorylation and *O*-adenylylation but not by *N*-acetylation. The same streptomycin 3″-adenyltransferase [AAD(3″)(9)] also inactivates spectinomycin, an aminocyclitol antibiotic. The enzymes involved are produced both by Gram positive and Gram negative bacteria. They are encoded on plasmids and located in the periplasmic space, i.e., between the inner and outer cell membrane.

An understanding of these resistance mechanisms has prompted massive programs to discover new superior semi-synthetic derivatives which would not be subject to such inactivations. Indeed some were found in which the functional groups susceptible to the inactivation were removed or sterically hindered and were found clinically valuable such as 1-*N*-ethylsisomicin (netilmicin), 3′-deoxykanamycin B (tobramycin), and 3′,4′-dideoxykanamycin B (dibekacin). An important lead for antibiotic improvement by chemical modification came also from the observation that 4-amino-2-hydroxybutyric acid (4-AHBA) in butirosin B is essential for the antipseudomonadal activity. When by analogy amikacin, the 1-*N*-AHBA derivative of kanamycin A, was prepared (Kawaguchi et al., 1972), it was shown to be clinically one of the most effective aminoglycoside antibiotics found. Amikacin was also reported to be synthesized by the mutasynthetic approach as described above and Fig. 11; see also Thompson and Davies (1984).

Finally the production of streptomycin (Fig. 8) was accompanied in earlier methods with the formation of mannosidostreptomycins in amounts up to 40% of the mixture (HOCKENHULL, 1960). Mannosidostreptomycin is 75–80% less active than streptomycin, which requires a conversion of the by-product to the main product. The streptomycin-forming actinomycete often secretes high concentrations of α-D-mannosidase, which hydrolyzes the mannosidostreptomycin (DEMAIN and INAMINE, 1970). More recent procedures of streptomycin fermentation use selected strains, which form streptomycin exclusively.

Apart from the mutasynthesis of new aminocyclitol no practical examples of biotransformation in this class are known. On the other hand, the investigation of the inactivation pathways of these antibiotics leads to the enzymatic reactions: *N*- and *O*-acylation, *O*-phosphorylation, and *O*-nucleotidylation (UMEZAWA, 1974). Although these biotransformations have no preparative value, their results suggested ways for hindering the enzymatic reaction and inactivation by corresponding chemical modification of the structures.

C. Rifamycins

Under normal fermentation conditions *Nocardia mediterranei* forms 20 different rifamycins. The addition of barbital leads to the formation of rifamycin B, with only small amounts of rifamycin Y (LANCINI and PARENTI, 1978) (Fig. 17).

If the carboxymethylation of the 4-hydroxy group is blocked by mutation, rifamycin SV is produced exclusively. This compound presents an important starting material for semi-synthetic rifamycins, of which rifampicin is an antituberculosis drug with high activity (LESTER, 1972).

Another mutant is blocked in the methylation of the hydroxy group in 27-position

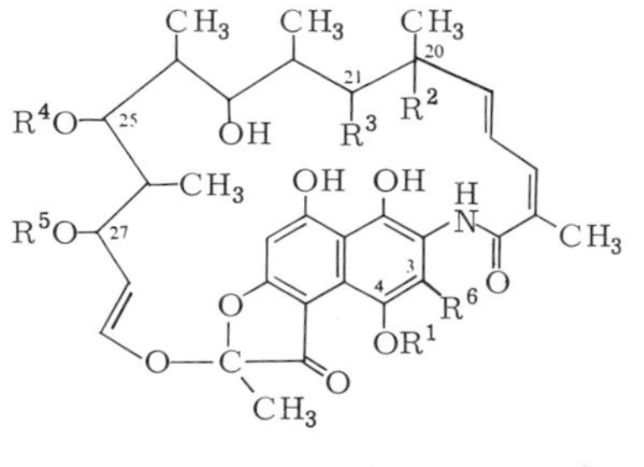

Rifamycin	R^1	R^2	R^3	R^4	R^5	R^6
B	CH_2COOH	H	OH	CH_3CO	CH_3	H
Y	CH_2COOH	OH	O=	CH_3CO	CH_3	H
SV	H	H	OH	CH_3CO	CH_3	H
Rifampicin	H	H	OH	CH_3CO	CH_3	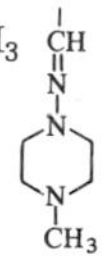 $CH{=}N{-}N\bigcirc N{-}CH_3$
27-Demethyl SV	H	H	OH	CH_3CO	H	H
27-Demethyl B	CH_2COOH	H	OH	CH_3CO	H	H
25-Deacetyl-27-Demethyl B	CH_2COOH	H	OH	H	H	H

Figure 17. Structures of rifamycin and related compounds.

and accumulates 27-demethylrifamycin SV together with the by-products 27-demethylrifamycin B and 25-deacetyl-27-demethylrifamycin B, which are further useful structures for derivatization with increased activity (LANCINI and WHITE, 1973).

Some other mutants are known to form different rifamycin intermediates, which offer further ways to new rifamycin derivatives (GHISALBA et al., 1978; MARTINELLI et al., 1978).

Although these examples do not present real biotransformations, they should be pointed out as practically used procedures within this category of compounds.

VI. Concluding Remarks

As discussed in this chapter, different microbiological methods were employed to carry out transformation of a large number of antibiotics. Microorganisms which were used for this purpose included the antibiotic producers themselves and their mutants. Others were also employed which were not related to antibiotics of interest but were selected on the basis of their known property to carry out the desired reaction of diverse structures. Some of the products were essential for the elucidation of the modes of antibiotic resistance or identification of individual steps in antibiotic biosynthesis. Others were antibiotically active or in turn served as substrates for the chemical preparation of new antibiotic derivatives. With the proper and imaginative application of the established and emerging techniques, microorganisms will continue to be of crucial importance in generating new products for human therapy and agriculture.

VII. References

ABBOTT, B. J., and FUKUDA, D. S. (1975). Appl. Microbiol. *30,* 413–419.

ABBOTT, B. J, FUKUDA, D. S., DORMAN, D. E., OCCOLOWITZ, M. D., and FAHRNER, L. (1979). Antimicrob. Ag. Chemother. *16,* 808–812.

ABRAHAM, E. P., and CHAIN, E. (1940). Nature *146,* 837.

AJISAKA, M., KARIYONE, K., JOMON, K., YAZAWA, H., and ARIMA, K. (1969). Agric. Biol. Chem. *33,* 294–295.

AKAGAWA, OKANISHI, M., and UMEZAWA, H. (1979), J. Antibiot. *32,* 610.

ANDRES, W. W., MCGAHREN, W. J., and KUNSTMANN, M. P. (1969). Tetrahedron Lett., 3777.

ARCAMONE, F., CASSINELLI, G., FANTINI, G., GRAIN, A., OREZZI, P., POL, C., and SPALA, C. (1969). Biotechnol. Bioeng. *11,* 1101–1110.

ARGOUDELIS, A. D., and COATS, J. H. (1969). J. Antibiot. *22,* 341.

ARGOUDELIS, A. D., and COATS, J. H. (1974). J. Antibiot. *27,* 674–676.

ARGOUDELIS, A. D., and MASON, D. J. (1969). J. Antibiot. *22,* 289.

ARGOUDELIS, A. D., COATS, J. H., MASON, D. J., and SEBEK, O. K. (1969). J. Antibiot. *22,* 309.

ARGOUDELIS, A. D., EBLE, T. E., and MASON, D. J. (1970). J. Antibiot. *23,* 1–8.

ARGOUDELIS, A. D., COATS, J. H., and MAGERLEIN, B. J. (1972a). J. Antibiot. *25,* 191.

ARGOUDELIS, A. D., COATS, J. H., LEMAUX, P. G., and SEBEK, O. K. (1972b). J. Antibiot. *25,* 445–455.

ARGOUDELIS, A. D., JOHNSON, L. E., and PYKE, T. R. (1973a). J. Antibiot. *26,* 429–436.

ARGOUDELIS, A. D., COATS, J. H., LEMAUX, P. G., and SEBEK, O. K. (1973b). J. Antibiot. *26,* 7–14.

ARGOUDELIS, A. D., COATS, J. H., and JOHNSON, L. E. (1974). J. Antibiot. *27,* 738–743.

ARGOUDELIS, A. D., COATS, J. H., and MIZSAK, S. A. (1977). J. Antibiot. *30,* 474.

ASZALOS, A. A. BACHUR, N. R., HAMILTON, B. K., LANGLYKKE, A. F., ROLLER, P. P., SHEIKH, M., SUTPHIN, M. S., THOMAS, M. C., WAREHEIM, D. A., and WRIGHT, L. H., (1977). J. Antibiot. *30,* 50.

BATCHELOR, F. R., DOYLE, F. P., NAYLER, J. H. C., and ROLINSON, G. N. (1959). Nature *183,* 257.

BATYROVA, A. S., and EGOROV, N. S. (1974). Antibiotiki *19,* 1089–1092.

BECKMAN, W., and LESSIE, T. G. (1979). J. Bacteriol. *140,* 1126.

BĚHAL, V., HOŠTÁLEK, Z., and VANĚK, Z. (1979). Biotechnol. Lett. *1,* 177–182.

BEHRENS, O. K. (1949). In "The Chemistry of Penicillin". (H. T. CLARKE, J. R. JOHNSON, and R. ROBINSON, eds.), pp. 657–679. Princeton University Press.

BENVENISTE, R., and DAVIES, J. (1971). Biochemistry *10,* 1787.

BIDDLECOME, S., HAAS, M., DAVIES, J., MILLER, G. H., RANE, D. F., and DANIELS, P. J. L. (1976). Antimicrob. Ag. Chemother *9,* 951.

BIRCH, A. J. (1963). Pure Appl. Chem. *7,* 527–537.

BIRCH, A. J. (1972). In "Advances in Antimicrobiol Antineoplastic Chemotherapy" (M. SEMONSKÝ and S. MASÁK, eds.), 1/2, pp. 1023–1024. Urban and Schwarzenberg, Munich.

BLUMAUEROVÁ, M., KRÁLOVCOVÁ, E., MATĚJŮ, J., and VANĚK, Z. (1979). Folia Microbiol. *24,* 117.

BOD, P., SZARKA, E., GYIMESI, J., HORVATH, G., VAJNA-MEHESFALVI, Z., and HORVATH, L. (1973). J. Antibiot. *26,* 101.

BRANDL, E. (1965). Z. Physiol. Chem. *342,* 86.

BRANDL, E., and KNAUSEDER, F. (1975). Ger. Offen. 2 503 584.

BRIGGS, T. C. (1983). In "Synthetic Organic Chemicals, U.S. Production and Sales, 1982", p. 103. U.S. Internatl. Trade Comm., Publ. No. 1422.

BRODELIUS, P. (1978). Adv. Biochem. Eng. *10,* 75–129.

BROWN, A. G. (1982). J. Antimicrob. Chemother. *10,* 365–372.

BRZEZINSKA, M., and DAVIES, J. (1973). Antimicrob. Ag. Chemother. *3,* 266.

CAPPELLETTI, L. M., and SPAGNOLI, R. (1983). J. Antibiot. *36,* 328–330.

CARRINGTON, T. R. (1971). Proc. R. Soc. London, Ser. B. *179,* 321–334.

CAWTHORNE, M. A. (1974). Ger. Offen. 2 356 630.

CHAIN, E. (1980). CHEMTECH *10,* 474–481.

CHARNEY, W., and HERZOG, H. L. (1967). "Microbial Transformations of Steroids". Academic Press, New York–London.

CHEVEREAU, M., DANIELS, P. J. L., DAVIES, J., and LEGOFFIC, F. (1974). Biochemistry *13,* 598.

CLARIDGE, C. A., GOUREVITCH, A., and LEIN, J. (1960). Nature *187,* 237.

CLARIDGE, C. A., BUSH, J. A., DEFURIA, M. D., and PRICE, K. E. (1974). Dev. Ind. Microbiol. *15,* 101–113.

CLEOPHAX, J., GERO, S. D., LEBOUL, J., AKHTAR, M., BARNETT, J. E. G., and PEARCE, C. J. (1976). J. Am. Chem. Soc. *98,* 7110–7112.

COATS, J. H., and ARGOUDELIS, A. D. (1971). J. Bacteriol. *108,* 459.

COLE, M. (1966). Process Biochem. *1,* 334–338.

COLE, M. (1981). Drugs of the Future *6,* 697–727.

COURVALIN, P., and DAVIES, J. (1977). Antimicrob. Ag. Chemother. *11,* 619.

DAUM, S. J., ROSI, D., and GOSS, W. A. (1977). J. Antibiot. *30,* 98–105.

DEMEESTER, C. and RONDELET, J. (1976), J. Antibiot. *29,* 1297.

DEFURIA, M. D., and CLARIDGE, C. A. (1976). In "Microbiology-1976", pp. 427–436 (D. SCHLESSINGER, ed.) American Society of Microbiology, 1976.

DEMAIN, A. L. (1966). In "Biosynthesis of Antibiotics", J. F. SNELL, ed. pp. 29–94. Academic Press, New York–London.

DEMAIN, A. L., and INAMINE, E. (1970). Bacteriol. Rev. *34,* 1.

DEMAIN, A. L., WALTON, R. B., NEWKIRK, J. F., and MILLER, I. M. (1963). Nature *199,* 909–910.

DHAR, M. M., SINGH, C., KHAN, A. W., ARIFF, A. J., GUPTA, C. M., and BHADURI, A. P. (1971). Pure Appl. Chem. *28,* 469–473.

DOI, O., OGURA, M., TANAKA, N., and UMEZAWA, H. (1968). Appl. Microbiol. *16,* 1276.

DULANEY, E. L., PUTTER, I., DRESCHER, D., CHAIET, L., MILLER, W. J., WOLF, F. J., and HENDLIN, D. (1962). Biochim. Biophys. Acta *60,* 447–449.

DUNNILL, P. (1980). Philos. Trans. R. Soc. London B. *290,* 409–420.

EKSTRÖM, B., LAGERLÖF, E., NATHORST-WESTFELT, L., and SJÖBERG, B. (1974). Sven. Farm. Tidskr. *78,* 531.

ELANDER, R. P., MABE, J. A., HAMILL, R. L., and GORMAN, M. (1971). Folia Microbiol. *16,* 156–165.

FELDMAN, L. I., DILL, I. K., HOLMLUND, C. E., WHALEY, H. A., PATTERSON, E. L., and BOHONOS, N. (1963). Antimicrob. Ag. Chemother., 54–57.

FELIX, H. R., PETER, H. H., and TREICHLER, H. J. (1981). J. Antibiot. *34,* 567.

FENTON, J. J., HARSCH, H. H., and KLEIN, D. (1973). J. Bacteriol. *116,* 1267–1272.

FUGONO, T., HIGASHIDE, E., SUZUKI, T., YAMAMOTO, H., HARADA, S., and KISHI, T. (1970). Experientia *26,* 26.

FUJII, T., HANAMITSU, K., IZUMI, R., YAMAGU-

CHI, T., and WATANABE, T. (1973). Japan. Kokai 73: 99393.

FUJII, T., MATSUMOTO, K., and WATANABE, T. (1976). Process Biochem. *11* (10), 21-24.

FUJII, T., SHIBUYA, T., and MATSUMOTO, K. (1979). Proc. Annual Meetg. Agric. Chem. Soc. Japan, April 1-4.

FUJIWARA, T., TANIMOTO, T., MATSUMOTO, K., and KONDO, E. (1978). J. Antibiot. *31,* 966.

FUKAGAWA, Y., KUBO, K., ISHIKURA, T., and KOUNO, K. (1980). J. Antibiot. *33,* 543-549.

FUKAGAWA, Y., MUTOH, Y., ISHIKURA, T., and LEIN, J. (1984). J. Antibiot. *37,* 118-126.

FURUMAI, T., and SUZUKI, M. (1975). J. Antibiot. *28,* 770, 775, and 783.

GHISALBA, O., TRAXLER, P., and NÜESCH, J. (1978). J. Antibiot. *32,* 1124.

GHISALBA, O., ROOS, R., SCHUPP, T., and NÜESCH, J. (1982). J. Antibiot. *35,* 74.

GIACOBBE, F. (1977). 4th Enzyme Eng. Conf., Bad Neuenahr, W.-Germany.

GILBERT, D. A., ARNOLD, B. H., and FIELDES, R. A. (1972). Brit. Patent 1272769.

GOI, H., NIWA, T., NOJIRI, C., MIYADO, S., SEKI, M., and YAMADA, Y. (1978). Japan. Patent 53-94093.

GOODMAN, J. J., and MATRISHIN, M. (1961). J. Bacteriol. *82,* 615.

GOODMAN, J. J., and MILLER, P. A. (1962). Biotechnol. Bioeng. *4,* 391-402.

GOODMAN, J. J., MATRISHIN, M., YOUNG, R. W., and MCCORMICK, J. R. D. (1959). J. Bacteriol. *78,* 492.

HAAS, M., BIDDLECOME, S., DAVIES, J., LUCE, C. E., and DANIELS, P. J. L. (1976). Antimicrob. Ag. Chemother. *9,* 945.

HAMILL, R. L., ELANDER, R. P., MABE, J. A., and GORMAN, M. (1970). Appl. Microbiol. *19,* 721-725.

HAMILTON, B. K., SUTPHIN, M. S., THOMAS, M. C., WAREHEIM, D. A., and ASZALOS, A. A. (1977). J. Antibiot. *30,* 425.

HARADA, S., YAMAZAKI, T., HATANO, K., TSUCHIYA, K., and KISHI, T. (1973). J. Antibiot. *26,* 647.

HEDING, H. (1968). Acta Chem. Scand. *22,* 1649-1954.

HENDLIN, D., DULANEY, E. L., DRESCHER, D., COOK, T., and CHAIET, L. (1962). Biochim. Biophys. Acta *58,* 635-636.

HENNESSEY, T. D., and RICHMOND, M. H. (1968). Biochem. J. *109,* 469.

HOCKENHULL, D. J. D. (1960). In "Progress in Industrial Microbiology" (HOCKENHULL, ed.), Vol. 2, p. 133. Haywood & Co. Ltd., London.

HOLMLUND, C. E., ANDRES, W. W., and SHAY, A. J. (1959). J. Am. Chem. Soc. *81,* 4750.

HOSHINO, T., and FUJIWARA, A. (1983). J. Antibiot. *36,* 1463-1467.

HOSHINO, T., SEKINE, Y., and FUJIWARA, A. (1983). J. Antibiot. *36,* 1458.

HOWE, R., and MOORE, R. H. (1968). Experientia *24,* 904.

HUANG, H. T., ENGLISH, A. R., SETO, T. A., SHULL, G. M., and SOBIN, B. A. (1960). J. Am. Chem. Soc. *82,* 3790.

HUBER, F. M., CHAUVETTE, R. R., and JACKSON, B. G. (1972). In "Cephalosporins and Penicillins" (E. H. FLYNN, ed.), p. 27. Academic Press, New York.

HUEPER, F. (1973a). Ger. Offen. 2157972.

HUEPER, F. (1973b). Ger. Offen. 2157970.

ICHIKAWA, S., SHIBUYA, Y., MATSUMOTO, K. FUJII, T., and KOMATSU, K. (1981). Agric. Biol. Chem. *45,* 2231.

IHN, W., SCHLEGEL, B., FLECK, W. F., and SEDMERA, P. (1980). J. Antibiot. *33,* 1457-1461.

IIZUKA, H., and NAITO, A. (1981). "Microbial Conversion of Steroids and Alkaloids". University Tokyo Press and Springer-Verlag, Berlin, Heidelberg, New York.

IKEDA, H., INOUE, M., and OMURA, S. (1983). J. Antibiot. *36,* 283-288.

ISONO, K., and SUHADOLNIK, R. J. (1976). Arch. Biochem. Biophys. *173,* 141-153.

IZAWA, M., NAKAHAMA, K., KASAHARA, F., ASAI, M., and KISHI, T. (1981a). J. Antibiot. *34,* 1587.

IZAWA, M., WADA, Y., KASAHARA, F., ASAI, M., and KISHI, T. (1981b). J. Antibiot. *34,* 1591.

JOHNSEN, J. (1977). Microbiol. *115,* 271.

JOHNSEN, J. (1981). J. Gen. Appl. Microbiol. *27,* 499-503.

JOHNSON, K. DUSART, J., CAMPBELL, J. N., and GHUYSEN, J. M. (1973). Antimicrob. Ag. Chemother. *3,* 289.

JONES, D. F., MOORE, R. H., and CRAWLEY, G. C. (1970). J. Chem. Soc. Sect. C., 1725.

KABINS, S., NATHAN, C., and COHEN, S. (1974). Antimicrob. Ag. Chemother. *5,* 565.

KAMEDA, Y., KIMURA, Y., TOYOURA, E., and OMORI, T. (1961). Nature *191,* 1122.

KAMEDA, Y., and HORII, S. (1972). J. Chem. Soc., Chem. Commun., 746.

KAMEDA, Y., HORII, S., and YAMANO, T. (1975). J. Antibiot. *28,* 298.

KAMEDA, Y., ASANO, N., WAKAE, O., and IWASA, T. (1980). J. Antibiot. *33,* 764.

KAMOGASHIRA, T., KAWAGUCHI, T., MIYAZAKI, W., and DOI, T. (1972). Japan. Kokai 72:28190.

KARNETOVÁ, J., MATĚJŮ, J., SEDMERA, P., VOKOUN, J., and VANĚK, Z. (1976). J. Antibiot. *29,*1199.

KATO, K. (1953). J. Antibiot. *6*, 130, 184.

KATO, K., KAWAHARA, K., TAKAHASHI, T., and IGARASI, S. (1980). Agric. Biol. Chem. *44*, 821.

KATZ, E. (1974). Cancer Chemother. Rep. Part 1, *58*, 83–91.

KATZ, E., and PIENTA, P. (1957). Science *126*, 402–403.

KATZ, E., WILLIAMS, W. K., MASON, K. T., and MAUGER, A. B. (1977). Antimicrob. Ag. Chemother. *11*, 1056–1063.

KAUFMANN, W., and BAUER, K. (1960). Naturwissenschaften *47*, 474.

KAWABE, H., KOBAYASHI, F., YAMAGUCHI, M., UTAHARA, R., and MITSUBASHI, S. (1971). J. Antibiot. *24*, 651.

KAWABE, H., KONDO, S., UMEZAWA, H., and MITSUBASHI, S. (1975). Antimicrob. Ag. Chemother. *7*, 494.

KAWAGUCHI, T., NAITO, S., NAKAGAWA, S., and FUJISAWA, K. (1972). J. Antibiot. *25*, 695.

KENIG, M., and READING, C. (1979). J. Antibiot. *32*, 549.

KHAN, A. W., BHADURI, A. P., GUPTA, C. M., and DHAR, M. M. (1969). Indian J. Biochem. *6*, 220–221.

KIESLICH, K. (1976). "Microbial Transformations of Non-steroid Compounds". G. Thieme Publ., Stuttgart.

KITAMURA, S., KASE, H., ODAKURA, Y., IIDA, T., SHIRABARA, K., and NAKAYAMA, K. (1982). J. Antibiot. *35*, 94.

KOHSAKA, M., and DEMAIN, A. L. (1976). Biochem. Biophys. Res. Commun. *70*, 465.

KOJIMA, M., and SATOH, A. (1973). J. Antibiot. *26*, 784.

KOMORI, T., KUNUGITA, K., NAKAHARA, K., AOKI, H., and IMANAKA, H. (1978). Agric. Biol. Chem. *42*, 1439.

KONECNY, J., and SIEBER, M. (1980). Biotechnol. Bioeng. *22*, 2013.

LAGERLÖF, E., NATHORST-WELTFELT, L., EKSTRÖM, B., and SJÖBERG, B. (1976). Methods Enzymol. *44*, 759–768.

LANCINI, G. (1983). In "Biochemistry and Genetic Regulation of Commercially Important Antibiotics" (L. C. VINING, ed.), Addison-Wesley Publ. Co., London, pp. 231–254.

LANCINI, G. C., and HENGELLER, C. (1969). J. Antibiot. *22*, 637.

LANCINI, G. C., and PARENTI, F. (1978). In "Antibiotics and Other Secondary Metabolites" (R. HÜTTER, T. LEISINGER, J. NÜESCH, and W. WEHRLI, eds.), p. 129. Academic Press, New York.

LANCINI, G. C., and WHITE, R. J. (1973). Process Biochem. *8*, 14.

LEE, B. K., PUAR, M. S., PATEL, M., BARTNER, P., LOTVIN, J., MUNAYYER, H., and WAITZ, J. A. (1983). J. Antibiot. *36*, 742.

LEGOFFIC, F., and MARTEL, A. (1974). Biochimie *56*, 893.

LEGOFFIC, F., MARTEL, A., and WITCHITZ, J. (1974). Antimicrob. Ag. Chemother. *6*, 680.

LEGOFFIC, F., MARTEL, A., CAPMAU, M. L., BOCA, B., GOEBEL, P., CHARDEN, H., SOUSSY, C. J., DUVAL, J., and BOUANCHAUD, D. H. (1976). Antimicrob. Ag. Chemother. *10*, 258.

LEGOFFIC, F., MARTEL, A., MOREAU, N., CAPMAU, M. L., SOUSSY, C. J., and DUVAL, J. (1977). Antimicrob. Ag. Chemother. *12*, 26.

LEMAHIEU, R. A., AX, H. A., BLOUNT, J. F., CARSON, M., DESPREAUX, C. W., PRUESS, D. L., SCANNELL, J. P., WEISS, F., and KIERSTEAD, R. W. (1976). J. Antibiot. *29*, 728.

LEMAUX, P. G., and SEBEK, O. K. (1973). Abstr. 13th Intersci. Conf. Antimicrob. Agents Chemother. Abstr. No. 49.

LEMKE, J. R., and DEMAIN, A. L. (1976). Eur. Appl. Microbiol. *2*, 91–94.

LESTER, W. (1972). Annu. Rev. Microbiol. *26*, 85.

LINGENS, F., EBERHARDT, H., and OLTMANNS, O. (1966). Biochim. Biophys. Acta *130*, 345–354.

LIVERMORE, D. M. (1982). J. Antimicrob. Chemother. *10*, 168.

LOWE, D. A., ROMANCIK, G., and ELANDER, R. P. (1981). Dev. Ind. Microbiol. *22*, 163.

MACH, B., and TATUM, E. L. (1964). Proc. Natl. Acad. Sci. U.S.A. *52*, 876–884.

MAEZAWA, I., HORI, T., KINUMAKI, A., and SUZUKI, M. (1973). J. Antibiot. *26*, 771.

MAEZAWA, I., I., KINUMAKI, A., and SUZUKI, M. (1976). J. Antibiot. *29*, 1203–1208.

MALIK, V. S., and VINING, L. C. (1970). Can. J. Microbiol. *16*, 173.

MANDEL, M. O., KÖSTNER, A. J., SIMER, E. KH., KLEINER, G. J., ELIZAROVSKAJA, L. M., and SHTAMER, V. YA. (1975). Appl. Biochem. Microbiol. *11*, 197.

MARCONI, W., CECERE, F., MORISI, F., DELLA PENUA, G., and RAPPUOLI, B. (1973). J. Antibiot. *26*, 228.

MARSHALL, V. P., and WILEY, P. F. (1982). In "Microbial Transformations of Bioactive Compounds" (J. P. ROSAZZA, ed.), Vol. 1, pp. 45–80. CRC Press, Boca Raton, Florida.

MARSHALL, V. P., REISENDER, E. A., and WILEY, P. F. (1976). J. Antibiot. *29*, 966.

MARSHALL, V. P., MCGOVREN, J. P., RICHARD, F. A., RICHARD, R. E., and WILEY, P. F. (1978). J. Antibiot. *31*, 336.

MARTINELLI, E., ANTONINI, P., CRICCHIO, R.,

LANCINI, G., and WHITE, R. J. (1978). J. Antibiot. *31,* 949.
MATSUHASHI, Y., YAGISAWA, M., KONDO, S., TAKENCHI, T., and UMEZAWA, H. (1975). J. Antibiot. *28,* 442.
MATSUHASHI, Y., OGAWA, H., and NAGAOKA, K. (1979). J. Antibiot. *32,* 777.
MATSUZAWA, Y., YOSHIMOTO, H., SHIBAMOTO, N., TOBE, H., OKI, T., NAGANAWA, H., TAKEUCHI, T., and UMEZAWA, H. (1981). J. Antibiot. *34,* 959.
MAZZEO, P., and ROMEO, A. (1972). J. Chem. Soc. Perkin *1,* 2532.
MCCORMICK, J. R. D., SJOLANDER, N. O., HIRSCH, U., JENSEN, E. R., and DOERSCHUK, A. P. (1951). J. Am. Chem. Soc. *79,* 4561.
MCCULLOUGH, J. E. (1983). Bio/Technology *1,* 879.
MCDOUGALL, B., DUNNILL, P., and LILLY, M. D. (1982). Enzyme Microb. Technol., 114.
MOSS, M. O. (1977). In "Topics in Enzyme and Fermentation Biotechnology" (A. WISEMAN, ed.), p. 111. Ellis Horwood Ltd., Chichester.
MOYER, A. J., and COGHILL, R. D. (1946). J. Bacteriol. *51,* 57, 79.
MURAO, S. (1955). J. Agric. Chem. Soc. Jpn. *28,* 400, 404.
NAGAOKA, K., and DEMAIN, A. L. (1975). J. Antibiot. *28,* 627-635.
NAKAHAMA, K., ISAWA, M., MUROI, M., KISHI, T., USHIDA, M., and IGARASI, S. (1974a). J. Antibiot. *27,* 425.
NAKAHAMA, K., KISHI, T., and IGARASI, S. (1974b). J. Antibiot. *27,* 433.
NAKAHAMA, K., KISHI, T., and IGARASI, S. (1974c). J. Antibiot. *27,* 487.
NAKAHAMA, K., HARADA, S., and IGARASI, S. (1975). J. Antibiot. *28,* 390.
NARA, T., OKACHI, R., and MISAWA, M. (1971). J. Antibiot. *24,* 321.
NEIDELMAN, S. L., ALBU, F., and BIENSTOCK, E. (1963a). Biotechnol. Bioeng. *5,* 87-89.
NEIDELMAN, S. L., BIENSTOCK, E., and BENNETT, R. C. (1963b). Biochim. Biophys. Acta *71,* 199-201.
NISHIURA, T., KAWADA, Y., SHIOMI, Y., O'HARA, K., and KONO, M. (1978). Antimicrob. Ag. Chemother. *13,* 1036.
NIWA, T., NOJIRI, C., GOI, H., MIYADO, S., KAI, F., SEKI, M., YAMADA, Y., and NIIDA, T. (1977). Japan. Patent 52-143259.
OCHI, K., YOSHIMA, S., and EGUCHI, Y. (1975). J. Antibiot. *28,* 965.
OGAWA, Y., MIZUKOSHI, S., and MORI, H. (1983). J. Antibiot. *36,* 1561.
OKA, Y., ISHIDA, H., MORIOKA, M., NUMASAKI, Y., YAMAFUGI, T., and OSONO, T. (1981). J. Antibiot. *34,* 777.
OKACHI, R., and NARA, T. (1973). Agric. Biol. Chem. *37,* 2797.
OKAMOTO, R., FUKUMOTO, T., NOMURA, H., KIYOSHIMA, K., NAKAMURA, K., and TAKAMATSU, A. (1980). J. Antibiot. *33,* 1300.
OKI, T., TAKATSUKI, Y., TOBE, H., YOSHIMOTO, A., TAKEUCHI, T., and UMEZAWA, H. (1981). J. Antibiot. *34,* 1229.
OKU, H., and NAKANISHI, T. (1964). Naturwissenschaften *51,* 538.
OKUDA, K., EDWARDS, G. C., and WINNICK, T. (1963). J. Bacteriol. *85,* 329-338.
OKUMURA, Y., FUKAGAWA, Y., OKAMOTO, R., and ISHIKURA, T. (1982). Agric. Biol. Chem. *46,* 731-737.
OMURA, S., KITAO, C., and SADAKANE, N. (1980a). J. Antibiot. *33,* 911-912.
OMURA, S., IKEDA, H., MATSUBARA, H., and SADAKANE, N. (1980b). J. Antibiot. *33,* 1570.
OMURA, S., SADAKANE, N., TANAKA, Y., and MATSUBARA, H. (1983). J. Antibiot. *36,* 927-930.
OZAKI, M., KARIYA, T., KATO, H., and KIMURA, T. (1972). Agric. Biol. Chem. *36,* 451.
OZANNE, B., BENVENISTE, R., TIPPER, D., and DAVIES, J. (1969). J. Bacteriol. *100,* 1144.
PERLMAN, D. (1977). CHEMTECH *7,* 434-443.
PERLMAN, D., and LANGLYKKE, A. F. (1948). J. Am. Chem. Soc. *70,* 3968.
PERLMAN, D., HEUSER, L. J., SEMAR, J. B., FRAZIER, W. R., and BOSKA, J. A. (1961). J. Am. Chem. Soc. *83,* 4481.
PLASKIE, A., ROETS, E., and VANDERHAEGE, H. (1978). J. Antibiot. *31,* 783.
PRAMER, D. (1958). Appl. Microbiol. *6,* 221.
PRAMER, D., and STARKEY, R. L. (1951). Science *113,* 127.
QUEENER, S. W., and SWARTZ, R. (1979). In "Economic Microbiology, Vol. 3, Secondary Products of Metabolism" (A. H. ROSE, ed.), pp. 35-122. Academic Press, London.
QUEENER, S. W., and NEUSS, N. (1982). In "Chemistry and Biology of β-Lactam Antibiotics", Vol. 3 (R. B. MORIN, and M. GORMAN, eds.), pp. 1-81. Academic Press, New York.
RICHMOND, M. H., and SYKES, R. B. (1973). Adv. Microbiol. Physiol. *9,* 31.
RIDLEY, M., BARRIE, D., LYNN, R., and STEAD, K. C. (1970). Lancet *269*/1, 230.
RINEHART, K. L., Jr. (1977). Pure Appl. Chem. *49,* 1361-1384.
ROLINSON, G. N. (1979). J. Antimicrob. Chemother. *5,* 7.
ROLINSON, G. N., BATCHELOR, F. R., BUTTERWORTH, D., CAMERON-WOOD, J., COLE, M., EUSTACE, G. C., HART, M. V., RICHARDS, M., and CHAIN, E. B. (1960). Nature *187,* 236.

ROSI, D., GOSS, W. A., and DAUM, S. J. (1977). J. Antibiot. *30,* 88–97.
RUECKERT, P. W., WILEY, P. F., MCGOVREN, J. P., and MARSHALL, V. P. (1979). J. Antibiot. *32,* 141.
RUTTENBERG, M. A., and MACH, B. (1966). Biochemistry *5,* 2864–2869.
SADAKANE, N., TANAKA, Y., and OMURA, S. (1982). J. Antibiot. *35,* 680–687.
SADAKANE, N., TANAKA, Y., and OMURA, S. (1983). J. Antibiot. *36,* 921.
SAKAGUCHI, K., and MURAO, S. (1950). J. Agric. Chem. Soc. Jpn. *23,* 411.
SATO, T., TOSA, T., and CHIBATA, I. (1976). Eur. J. Appl. Microbiol. *2,* 153.
SAVIDGE, T. A., and COLE, M. (1975). Methods Enzymol *43,* 705.
SAVIDGE, T. A., POWELL, L. W., and WARREN, K. B. (1974). Ger. Offen. 2336829.
SAWA, T., FUKAGAWA, Y., HOMMA, I., WAKASHIRO, T., TAKENCHI, T., and HORI, M. (1968). J. Antibiot. *21,* 334.
SAWADA, Y., HUNT, N. A., and DEMAIN, A. L. (1979). J. Antibiot. *32,* 1303.
SCHLEGEL, B., and FLECK, W. F. (1980). Z. Allg. Mikrobiol. *20,* 527.
SCHMITZ, H., and CLARIDGE, C. A. (1977). J. Antibiot. *30,* 635.
SCHUPP, T., TRAXLER, P., and AUDEN, J. A. L. (1981). J. Antibiot. *34,* 965.
SEBEK, O. K. (1974). Lloydia *37,* 115–133.
SEBEK, O. K. (1975). Acta Microbiol. Acad. Sci. Hung. *22,* 381–388.
SEBEK, O. K., and DOLAK, L. A. (1984). J. Antibiot. *37,* 136.
SEBEK, O. K., and HOEKSEMA, H. (1972). J. Antibiot. *25,* 434–436.
SELLSTEDT, J. H. (1975). U.S. Patent 3896110.
SEONG, B. L., SON, H. J., MHEEN, T.-I., and HAN, M. H. (1983). J. Antibiot. *36,* 1402.
SHIBATA, M., and UYEDA, M. (1978). Annu. Rep. Ferment. Proc. *2,* 267–303.
SHIBUYA, Y., MATSUMOTO, K., and FUJII, T. (1981). Agric. Biol. Chem. *45,* 1561.
SHIER, W. T., RINEHART, K. L., Jr., and GOTTLIEB, D. (1969). Proc. Natl. Acad. Sci. U.S.A. *63,* 198–204.
SHIER, W. T., OGAWA, S., HITCHENS, M., and RINEHART, K. L., Jr. (1973). J. Antibiot. *26,* 551.
SHIER, W. T., SHAEFER, P. C., GOTTLIEB, D., and RINEHART, K. L., Jr., (1974). Biochemistry *13,* 5073.
SHIMIZU, M., MASUIKE, T., FUJITA, H., KIMURA, K., OKACHI, R., and NARA, T. (1975). Agric. Biol. Chem. *39,* 1225.
SHIMIZU, S., INOUE, M., and MITSUHASHI, S. (1981). J. Antibiot. *34,* 869.
SHIRAFUJI, H., KIDA, M., NOGAMI, I., and YONEDA, M. (1980). Agric. Biol. Chem. *44,* 1561.
SHIRATO, S., YOSHIDA, K., and MIYAZAKI, Y. (1968). J. Ferment. Technol. *46,* 233.
SINGH, K., and RAKHIT, S. (1979). J. Antibiot. *32,* 78.
SLECHTA, L., CIALDELLA, J., and HOEKSEMA, H. (1978). J. Antibiot. *31,* 319.
SLOCOMBE, B. (1980). In "Augmentin, Proc. 1st Symp." (G. N. ROLINSON and A. WATSON, eds.), pp. 8–18. Excerpta Medica, Amsterdam.
SMITH, G. N., and WORREL, C. S. (1949). Arch. Biochem. *24,* 216.
SMITH, G. N., and WORREL, C. S. (1950). Arch. Biochem. *28,* 232.
SMITH, G. N., and WORREL, C. S. (1953). J. Bacteriol. *65,* 313.
SPAGNOLI, R., and TOSCANO, L. (1983). J. Antibiot. *36,* 435.
SPAGNOLI, R., CAPPELLETTI, L., and TOSCANO, L. (1983). J. Antibiot. *36,* 365–375.
SUZUKI, M., TAKAMAKI, T., MIYAGAWA, K., ONO, H., HIGASHIDE, E., and UCHIDA, M. (1977). Agric. Biol. Chem. *41,* 419.
SYKES, R. B., and MATTHEW, M. (1976). J. Antimicrob. Chemother. *2,* 115.
SYKES, R. B., and RICHMOND, M. H. (1970). Nature *226,* 952.
TAKAHASHI, T., YAMAZAKI, Y., KATO, K., and ISONO, M. (1972). J. Am. Chem. Soc. *94,* 4035.
TAKAHASHI, T., KATO, K. YAMASAKI, Y., and ISONO, M. (1977). Japan. J. Antibiot. Suppl. *30,* S-230.
TAKAHASHI, K., YAZAWA, K., KISHI, K., MIKAMI, Y., ARAI, T., and KUBO, A. (1982). J. Antibiot. *35,* 196.
TAKEDA, K., KINUMAKI, A., OKUNO, S., MATSUSHITA, T., and ITO, Y. (1978a). J. Antibiot. *31,* 1039.
TAKEDA, K., KINUMAKI, A., FURUMAI, T., YAMAGUCHI, T., OSHIMA, S., and ITO, Y. (1978b). J. Antibiot. *31,* 247.
TAKEDA, K., KINUMAKI, A., HAYASAKA, H., YAMAGUCHI, T., and ITO, Y. (1978c). J. Antibiot. *31,* 1031.
TAKEDA, K., OKUNO, S., OHASHI, Y., and FURUMAI, T. (1978d). J. Antibiot. *31,* 1023.
TAKITA, T., and MAEDA, K. (1980). J. Heterocycl. Chem. *17,* 1799–1802.
TESTA, R. T., and TILLEY, B. C. (1975). J. Antibiot. *28,* 573.
TESTA, R. L., and TILLEY, B. C. (1976). J. Antibiot. *29,* 140.

THERIAULT, R. J. (1974). U.S. Patent 3784447.
THERIAULT, R. J., GUAGLIARDI, L., HUDSON, P. B., MITSCHER, L. A., PARK, Y.-H., and SWAYZE, J. K. (1982). J. Antibiot. *35,* 364–366.
THOMPSON, C. J., and DAVIES, J. A. (1984). Trends Biotechnol. *2,* 43.
TORII, H., ASANO, T., MATSUMOTO, N., KATO, K., TSUCHIMA, S., and KAKINUMA, A. (1980). Agric. Biol. Chem. *44,* 1431.
TOSCANO, L., FIORIELLO, G., SPAGNOLI, R., CAPPELLETTI, L., and ZANUSO, G. (1983). J. Antibiot. *36,* 1439.
TOYO BREWING Co. (1978). Japan. Patent J5-3118-591.
TRAXLER, P., SCHUPP, T., FUHRER, H., and RICHTER, W. J. (1981). J. Antibiot. *34,* 971.
TRAXLER, P., SCHUPP, T., and WEHRLI, W. (1982). J. Antibiot. *35,* 594.
TROONEN, H., ROELANTS, R., and BOON, B. (1976). J. Antibiot. *29,* 1258–1267.
TSURUOKA, T., SHOMURA, T., OGAWA, Y., EZOKI, N., WATANABE, H., AMANO, S., INOUYE, S., and NIIDA, T. (1973). J. Antibiot. *26,* 168.
UEMATSU, T., and SUHADOLNIK, R. J. (1974). Arch. Biochem. Biophys. *162,* 614.
UMEZAWA, H. (1971). Pure Appl. Chem. *28,* 665–680.
UMEZAWA, H. (1974). Adv. Carbohydr. Chem. Biochem. *30,* 183.
UMEZAWA, H. (1977). Lloydia, J. Nature Prod. *40,* 67.
UMEZAWA, H., NISHMURA, Y., TSUCHIYA, T., and UMEZAWA, S. (1972). J. Antibiot. *25,* 743.
UYEDA, M., MORI, S., MORITA, M., OGATA, T., MORI, M., and SHIBATA, M. (1977). J. Antibiot. *30,* 1130.
UYEDA, M., NAKAMICHI, K., SHIGEMI, K., and SHIBATA, M. (1980). Agric. Biol. Chem. *44,* 1399.
VAN PÉE, K.-H., SALCHER, O., FISCHER, P., BOKEL, M., and LINGENS, F. (1983). J. Antibiot. *36,* 1735.
VANDAMME, E. J. (1983). Enzyme Microbiol. Technol. *5,* 403.
VANDAMME, E. J., and VOETS, J. P. (1974). Adv. Appl. Microbiol. *17,* 311.
VANĚK, Z., TAX, J., KOMERSOVÁ, I., and ECKARDT, K. (1973). Folia Microbiol. *18,* 524.
VON DAEHNE, W., LORCH, H., and GODTFREDSEN, W. O. (1968). Tetrahedron Lett. *47,* 4843.
WAITZ, J. A., MILLER, G. H., MOSS, JR. E., and CHIN, P. J. S. (1978). Antimicrob. Ag. Chemother. *13,* 41.
WALTON, R. B., MCDANIEL, L. E., and WOODRUFF, H. B. (1962). Dev. Ind. Microbiol. *3,* 370–375.
WEISSENBURGER, H. W. O., and VAN DER HOEVEN (1970). Reç. Trav. Chim. Pays-Bas *89,* 1081.
WHITE, R. F., BIRNBAUM, J., MEYER, R. T., TEN BROCKE, J., CHEMERDA, J. M., and DEMAIN, A. L. (1971). Appl. Microbiol. *22,* 55.
WU, C S., GARD, A., and ROSAZZA, J. P. (1980). J. Antibiot. *33,* 705.
YAGISAWA, M., NAGANAWA, H., KONDO, S., TAKEUCHI, T., and UMEZAWA, H. (1972a). J. Antibiot. *25,* 492.
YAGISAWA, M., YAMAMOTO, H., NAGANAWA, H., KONDO, S., TAKEUCHI, T., and UMEZAWA, H. (1972b). J. Antibiot. *25,* 748.
YAGISAWA, M., KONDO, S., TAKEUCHI, T., and UMEZAWA, H. (1975). J. Antibiot. *28,* 486.
YAMAGUCHI, M., MITSUHASHI, S., KOBAYASHI, F., and ZENDA, H. (1974). J. Antibiot. *27,* 507.
YAMAGUCHI, I., SHIBATA, H., SETO, H., and MISATO, T. (1975). J. Antibiot. *28,* 7.
YAZAWA, K., ASAOKA, T., TAKAHASHI, K., MIKAMI, Y., and ARAI, T. (1982). J. Antibiot. *35,* 915.
YOSHIDA, T., KIMURA, Y., and KATAGIRI, K. (1968). J. Antibiot. *21,* 465–467.
YOSHIDA, M., KONOMI, T., KOHSAKA, M., BALDWIN, J. E., HERCHEN, S., SINGH, P., HUNT, N. A., and DEMAIN, A. L. (1978). Proc. Natl. Acad. Sci. U.S.A. *75,* 6253.
YOSHIMOTO, A., OGASAWARA, T., KITAMURA, I., OKI, T., INUI, T., TAKEUCHI, T., and UMEZAWA, H. (1979). J. Antibiot. *32,* 472.
YOSHIMOTO, A., MATSUZAWA, Y., OKI, T., NAGANAWA, H., TAKEUCHI, T., and UMEZAWA, H. (1980). J. Antibiot. *33,* 1150.

Chapter 8

Aromatic and Heterocyclic Structures

Peter R. Wallnöfer
Gabriele Engelhardt

Bayerische Landesanstalt für Bodenkultur und Pflanzenbau, Abt. Pflanzenschutz
München 19, Federal Republic of Germany

I. General and Historical
II. Important Microorganisms
  A. Use of Microorganisms in Mixed Cultures
  B. Manipulation of Degradative Genes of Microorganisms
III. Biochemistry
  A. Common Pathways of Aromatic Metabolism
    1. Aerobic attack on the aromatic ring
      a) Entry into the cell
      b) Transformation of side chains and modification of substituents
      c) Reactions converting aromatic compounds into ring fission substrates
      d) Pathways of aromatic ring cleavage and subsequent reactions
    2. Anaerobic degradation of aromatic compounds
      a) Anaerobic photometabolism
      b) Metabolism of aromatic compounds through anaerobic nitrate respiration
      c) Anaerobic degradation of aromatic compounds during methane fermentation
      d) The fate of benzene ring substituents under anaerobic conditions
  B. Common Reactions in the Metabolism of Heterocyclic Compounds
    1. Five membered ring systems
    2. Six membered heterocyclic compounds
IV. Aromatic and Heterocyclic Compounds with Economical and Ecotoxicological Significance
  A. Aromatic Compounds
    1. Phenolic pesticides and terminal aromatic metabolites of pesticides
    2. Industrial pollutants
      a) Phthalic acid esters
      b) Lignosulfonates

c) Surfactants
d) Dyes
e) Aromatics released during combustion
B. Heterocyclic Compounds
V. Application and Economical Importance
A. Use of Microorganisms and Microbial Enzymes for the Degradation of Xenobiotics
B. Use of Microorganisms for Bioconversion of Wastes
C. Use of Microorganisms for the Preparation of Important Intermediates in the Degradation of Aromatic and Heterocyclic Structures
1. Intermediates of industrial or pharmaceutical relevance
a) Microbial hydroxylation of naphthyridines
b) Microbial preparation of 4-hydroxymethyl-3-chloroanilines
c) Microbial oxidation of polynuclear aromatic hydrocarbons
2. Intermediates important for studies of selected biochemical reactions
a) (+)*cis*-1,2-Dihydroxy-1,2-dihydronaphthalenecarboxylic acid
b) Substituted 5-oxo-furan-2-acetic acid compounds
VI. References

I. General and Historical

Increasing numbers and amounts of aromatic and heterocyclic compounds are being produced industrially. Furthermore, benzenoid residues such as those found in lignin are among the most common monomeric units found in nature. It may sometimes be difficult to determine to what extent compounds are derived from natural sources such as vegetation, forest fires, and volcanoes and to what extent they are produced by manufacturing industries, agriculture, and other human activities. Aromatic and heterocyclic compounds may have diverse effects on the environment. Serious pollutants are those compounds which are present in high enough quantities to cause health hazards. Examples of such aromatic compounds are benzene, toluene, diisocyanate, and various pesticides. Most dangerous of all are those compounds, like the polycyclic aromatic hydrocarbons, which may become metabolized into highly mutagenic, carcinogenic, and teratogenic derivatives.

Pollution by aromatic and heterocyclic chemicals has been controlled to a considerable extent in recent years, often by using the metabolic capabilities of microorganisms. The complications arising from political, legal, and economic factors are well documented in the controversies surrounding the use of benzene and 2,4,5-T-(2,4,5-trichlorophenoxyacetate), a herbicide which formerly was often contaminated by the much more teratogenic 2,3,7,8-tetrachlorodibenzo-*p*-dioxin. A further problem is that abolition of one source of pollution can lead to the creation of another: for instance, liquid effluents which can cause water pollution are often derived from scrubbers installed to abate air pollution.

Despite these complications, there is no doubt that there is an enormous increase of man-made aromatic compounds and that many industrial processes produce large amounts of synthetic aromatics either as the main products or as wastes. Phthalic acid esters are particularly good examples, being frequently used as plasticizers but having a variety of other applications ranging from antifoaming agents in the paper industry to perfume vehicles in cosmetic production; after serving their purpose they have to be disposed of. The formation of aromatic compounds as waste or by-products is exemplified by the production of coke-oven liquors which sometimes may contain up to 2 g of different phenols per liter.

In addition to other industrial advances that can be expected in the next few decades, a shortage of oil in many parts of the world will almost certainly lead to increased exploitation of resources such as coal, oil shales, and lignin. This will mean a challenge for biochemists and microbiologists, many of whom are concerned with the manipulation and the metabolism of aromatic compounds. It may be interesting to note that the related but much older problem of gasworks' liquors helped to increase the knowledge of the biochemical mechanisms by which aromatic compounds are degraded by microorganisms. It had been known for many years that phenols were gradually destroyed when spent gasworks' liquor was mixed with sewage and then passed through filter beds. During a study of this process a Gram negative bacterium, most likely *Acinetobacter calcoaceticus*, was isolated which could grow on phenol. This organism can degrade many aromatic compounds and was used in some of the earliest studies on the elucidation of the metabolic pathway by which dihydric phenols are subjected to intra-diol ring-cleavage leading to 3-oxoadipic acid (3-oxohexanedioic acid).

II. Important Microorganisms

The present state of knowledge of the microbiological breakdown of aromatic and heterocyclic compounds has already been summarized in a number of detailed reviews (KIESLICH, 1976; DAGLEY, 1978a, b, c; EVANS, 1977; CALLELY, 1978; FEWSON, 1981; HILL, 1978). Because the breakdown of aromatic and heterocyclic compounds is a vital biochemical step in the natural carbon cycle, many microorganisms are capable of cleaving the aromatic ring. Eubacteria, yeasts, higher fungi, and one or two photosynthetic bacteria which perform transformations under anaerobic conditions, are capable of breaking down aromatic substrates. Up to now, the following species have been reported to carry out these cleavages (KIESLICH, 1976):

Prokaryota		
Rhodospirillaceae	*Rhodospirillum*	*capsulata*
	Rhodopseudomonas	*palustris*
		gelatinosa
Pseudomonadaceae	*Pseudomonas*	*acidovorans*
		aeruginosa
		arvilla
		boreopolis
		candatus
		convexa
		cruciviae
		dacunhae
		desmolyticum
		fluorescens
		multivorans
		oleovorans
		ovalis
		phendis
		pictorum
		putida
		rhatonis
		salopium
		stutzeri
		subcreta
		sp.
		testosteroni
genera of uncertain affiliation	*Alcaligenes*	*faecalis*
		sp.
	Achromobacter	*jophagum*
		cycloclastes
		sp.
Enterobacteriaceae	*Escherichia*	*coli*
	Klebsiella	sp.
	Enterobacter	*aerogenes*
Vibrionaceae	*Vibrio*	*cuneata*
		cyclosites
		neocistes
		tyrosinatica
		sp.
genus of uncertain affiliation	*Flavobacterium*	*aquatile*
		pelegrinum
		helvolum
		resinovorum
		sp.
Neisseriaceae	*Moraxella*	*calcoacetica*
		lwoffii

Micrococcaceae	*Micrococcus*	*chinicus*
		pittonensis
		pyogenes var *sphaericus*
		sphaeroides
		ureae
		sp.
Bacillaceae	*Bacillus*	*albolactis*
		alcalescens
		benzoicus
		cereus
		closteroides
		hexacarbovorum
		natans
		phenanthrenicus bakiensis
		phenanthrenicus guricus
		platychoma
		subtilis
		thermophenolicus
		spp.
Coryneform group of bacteria	*Clostridium*	sp.
	Corynebacterium	*simplex*
		equi
		renale
		sp.
	Arthrobacter	*globiformis*
		simplex
		sp.
Mycobacteriaceae	*Mycobacterium*	*actinomorphum*
		agreste
		brevicale
		butyricum
		coeliacum
		convolutus
		crystallophagum
		erythropolis
		flavum
		hyalinum
		lacticola
		phlei
		rhodochrous
		smegmatis
		tuberculosis
Nocardiaceae		*vadosum*
	Nocardia	*equi*
		coeliaca
		corallina
		convoluta
		erythropolis
		opaca
		restricta
		rubra
		sp.
Streptomycetaceae	*Rhodococcus*	sp.
Eukaryota	*Streptomyces*	sp.
Ascomycetes		
Endomycetaceae	*Debaryomyces*	sp.
	Pichia	sp.

Saccharomycetaceae	*Saccharomyces*	*cerevisiae*
		sp.
Fimetariaceae	*Neurospora*	*crassa*
		sp.
Basidiomycetes		
Agaricaceae	*Russula*	*nigricans*
Polyporaceae	*Lenzites*	*trabea*
	Polyporus	*versicolor*
Fungi imperfecti		
Cryptococcaceae	*Candida*	*brumphi*
		pulcherima
		sp.
Dermatiaceae	*Hormodendrum*	sp.
Moniliaceae	*Aspergillus*	*niger*
		sp.
	Botrytis	sp.
	Oospora	sp.
	Penicillium	*chrysogenum*
		notatum
		sp.
Rhodotorulaceae	*Rhodotorula*	*glutinis*
Sporobolomycetaceae	*Sporobolomyces*	sp.
Torulopsidaceae	*Torulopsis*	*dattila*
		incospicua
		utilis
		sp.

A. Use of Microorganisms in Mixed Cultures

The majority of laboratory-based studies of biodegradation have focused on the metabolism of single compounds by pure cultures. There are, however, various drawbacks to the pure-culture approach which may lead to an incorrect assessment of the potential of microorganisms in nature to degrade aromatic and heterocyclic compounds.

In recent years it became evident that biodegradation of such compounds often requires the presence of a microbial community consisting of several different kinds of microorganisms. Application of continuous-flow enrichment methods has led to the isolation of various types of such communities and has enabled the study of the interactions between the different organisms. Continuous cultures also allow an analysis of mixed culture/mixed substrate systems in respect to biodegradations. Furthermore, such systems have allowed the study of adaptations of the organisms to novel substrates under conditions of continuous and strong selective pressure. Thus, studies using the continuous culture as one of the possible environmentally relevant laboratory model systems have led to a more complete understanding of the biodegradation of aromatic and heterocyclic compounds in nature.

Microbial interactions

Microbial interactions play an important role in the biodegradation of aromatic and heterocyclic compounds in nature. Among the interactions that have been encountered are various forms of mutualism (all members of the mixture benefit from each other's presence) and of commensalism (one member of a community benefits from the presence of a second population which itself neither stimulates nor inhibits the activity of the first organism).

Recent studies on the utilization of mixed substrates by a variety of different types of microbes have shown that under growth limitation by multiple sources of carbon and/or energy these compounds are simultaneously used rather than in a diauxic pattern as frequently observed in batch culture (HARDER and DIJKHUIZEN, 1976). Mixed-substrate utilization by pure and mixed cultures is of considerable importance in the biodegradation of aromatic and heterocyclic compounds in nature.

B. Manipulation of Degradative Genes of Microorganisms

The biological degradation of aromatics and heterocyclics is a result of catabolism by individual microbial members of bacterial communities or the concerted catabolism by consortia of different microbes. The net result can be the total conversion of a compound to CO_2 and biomass at one extreme or its partial metabolism to less toxic materials at the other. However, there are many compounds which are too recalcitrant to be attacked by natural microorganisms and it is difficult, if not impossible, to enrich cultures which have any effect upon them. The theoretical possibility exists to degrade these compounds by constructing strains through the introduction of a number of enzyme activities from different bacteria into a certain chosen bacterium which under natural conditions might have little or no chance to exchange genetic information. Such hybrid metabolic pathways could be constructed in a number of ways, but the two most promising methods are: (a) the cloning of catabolic enzymes onto plasmid or bacteriophage vectors by the techniques of recombinant DNA technology, and (b) the use of naturally cloned catabolic DNA located on degradative plasmids (FRANKLIN et al., 1981).

The use of degradative plasmids will probably prove most effective, at least in the near future. This statement is based upon the requirement for a fairly complex sequence of reactions for the complete conversion of the kind of molecules which cause environmental problems, many of which are heterocyclic aromatic compounds with a number of constituent groups. Recombinant DNA technology is still somewhat limited by the size of the DNA fragments which can be inserted into a vector, and this obviously limits the number of genes which can be interdependently cloned.

Some plasmids have been found to contain genes for integrated and regulated degradative pathways in a form which can be transferred or mobilized between bacterial hosts. As microbiologists are becoming aware of the existence of such extrachromosomal catabolic genes, more are being discovered. Most degradative plasmid genes found in saprophytic cultures are of esoteric biochemical nature and are likely to be used to degrade the more recalcitrant chemicals.

In order to optimize the construction of strains by these methods a complete understanding of the biochemistry of these pathways, of the regulation and chemical nature of the inducers, the number of operons, and the location of all the relevant genes on the vector genome, is necessary (KNACKMUSS, 1981).

III. Biochemistry

A. Common Pathways of Aromatic Metabolism

The microbial degradation of aromatic compounds by ring fission is predominantly an aerobic process. Some aromatic compounds, however, can be mineralized in different bacteria also in the absence of molecular oxygen by reaction sequences in which the benzene nucleus is first reduced and then cleaved by hydrolysis.

The chemicals may serve as a carbon, nitrogen, or energy source for the microorganisms or may only be a substrate for cometabolism. There is now considerable evidence that the microbial metabolism of some aromatic xenobiotics involves co-oxidative stages. However, there is little direct proof of co-oxidation under field conditions.

1. Aerobic attack on the aromatic ring

Aromatic compounds can be either totally or partly degraded by microorganisms depending on the number of rings and especially on the type of substituents. The basic reaction sequences involved have been clearly described by DAGLEY (1978a, b, c). They include the following reaction steps: entry into the cell (a), modification of side chains (b), formation of ring-fission substrates (c), ring-cleavage, and transformation of the products of ring-fission into common compounds of intermediary metabolism (d).

a) Entry into the cell

Entry of aromatic compounds into the cell has often been assumed to occur by diffusion, but there is evidence that specific transport mechanisms exist as was shown for benzoic acid and mandelic acid (HIGGINS and MANDELSTAM, 1972; COOK and FEWSON, 1972; THAYER and WHEELIS, 1976).

b) Transformation of side chains and modification of substituents before ring cleavage

The principal reactions involved in the transformation of benzene ring substituents include β-oxidation, oxidative dealkylation, thioether oxidation, phosphorothionate oxidation, epoxidation of carbon-carbon double bonds, hydroxylation, hydrolysis, dehalogenation, and nitro-reduction. The number of reactions is few compared to the number of compounds metabolized, reflecting the relatively small number of types of chemicals. Various correlations between the chemical structures of different compounds and their routes of metabolism have been noted.

On the other hand, it strongly depends on the bacterial species whether a substituent remains intact or is transformed or eliminated before ring cleavage.

Alkyl, alkoxy, and carboxyl substituents

Methyl groups may either be oxidized or remain intact during hydroxylation of the aromatic ring. Thus, *m*- and *p*-cresol or *p*-cymene can be transformed to hydroxymethyl derivatives by an oxygen-dependent hydroxylation and by successive dehydrogenations forming the carboxylic acid via the aldehyde. However, *o*-, *m*-, and *p*-cresol can also be directly hydroxylated to give 3-methyl- and 4-methylcatechol intermediates, respectively (CHAPMAN, 1972; DEFRANK and RIBBONS, 1977). Similar pathways are described for the degradation of toluene. Some *Pseudomonas* and *Achromobacter* sp. oxidize toluene to 3-methylcatechol, whereas a strain of *Pseudomonas aeruginosa* oxidizes the methyl group of toluene to benzylalcohol and further to benzaldehyde and benzoic acid (GIBSON et al., 1972).

Larger alkyl side chains may also remain intact during the formation of the corresponding catechol or undergo oxidation. Carboxylic acids are then formed by oxidation of the terminal methyl group. A number of alkylbenzenes or alkylbenzene sulfonates are oxidized in this manner. Provided that extensive branching does not exist, the longer carboxyalkyl substituents can undergo β-oxidation forming either phenylacetic or benzoic acids. Phenoxyalkanoic acids can also undergo β-oxidation, releasing either phenoxyacetic or phenoxypropionic acids. The carboxymethyl substituent remains intact before ring cleavage occurs as shown by the conversion of phenylacetic acid to homogentisic acid (2,5-dihydroxyphenylacetic acid) and of 4-hydroxyphenylacetic acid to homoprotocatechuic acid (3,4-dihydroxyphenylacetic acid).

The carboxyethyl substituent is not always degraded by β-oxidation but may occasionally remain intact in the substrate for ring cleavage. In this manner a strain of *Achromobacter* uses β-phenylpropionic acid for growth by forming 2,3-dihydroxyphenylpropionic acid, which is also formed from *trans*-cinnamic acid by *Pseudomonas* sp. Another three-carbon side chain – the carboxyethylene substituent of *p*-hydroxy-*trans*-cinnamic acid – apparently is preserved when *Pseudomonas fluorescens* cleaves the intermediate caffeic acid. The same group in *trans*-ferulic acid is degraded by *Pseudomonas acidovorans* under formation of vanillaldehyde and acetic acid (CHAPMAN, 1972).

Although generally remaining unaltered, carboxyl groups can also be eliminated. This may either occur via oxidation as in the conversion of anthranilic acid, salicylic acid, and benzoic acid to catechol, or nonoxidatively as in the conversion of 4,5-dihydroxyphthalic acid to protocatechuic acid in *Pseudomonas* and *Nocardia* sp., and of protocatechuic acid to catechol in *Klebsiella aerogenes* (CHAPMAN, 1972; ENGELHARDT et al., 1976).

Alkoxy substituents generally are dealkylated to give the parent phenol, with the concomitant liberation of the alkyl moiety as an aldehyde. Vanillic acid (3-methoxy-4-hydroxybenzoic acid), for example, is converted to protocatechuic acid by loss of its methoxy group in the form of formaldehyde. As in the hydroxylation of methyl groups, these reactions are catalyzed by monooxygenases which introduce hydroxyl groups forming semi-acetals. Hydrolysis of the latter (probably a rapid non-enzymic process) yields the corresponding phenol plus an aldehyde. Higher ethers such as *p*-ethoxybenzoate and *p*-*n*-propoxybenzoate are similarly metabolized. Degradation of the herbicides 4-chloro-2-methylphenoxyacetic acid and 2,4-dichlorophenoxyacetic acid by bacteria is also initiated by such an attack to give substituted phenols and glyoxylic acid.

Dealkylation of methoxy groups, however, does not always occur before ring cleavage: *Pseudomonas putida* may degrade syringic acid (3,5-dimethoxy-4-hydroxybenzoic acid) via 3-*o*-methyl gallic acid, which then undergoes ring cleavage (CHAPMAN, 1972).

Thioalkyl substituents

Methylthio- and methylsulfinyl groups are not removed or dealkylated before ring cleavage to form the respective muconic semialdehyde derivatives as shown for the metabolism of 4-(methylthio)- and 4-(methylsulfinyl)-substituted phenols by a *Norcardia* species (see Table 2) (ENGELHARDT et al., 1977; RAST et al., 1979). The limited number of observations of *S*-dealkylations may be due to the greater tendency for oxidation of the sulfur to the sulfone and sulfoxide.

Alkamino groups

The microbially mediated removal of an alkyl group from a nitrogen atom has been recorded for numerous pesticides and apparently proceeds via unstable hydroxy-methyl intermediates (HILL, 1978).

Fatty acid side chains

Many aromatic pesticides and some industrial pollutants have fatty acid side chains, or substituents which readily are converted into the latter. These units are microbially metabolized by β-oxidation (HILL, 1978).

Esters, amides, nitriles

Numerous pesticides and some industrial pollutants as, e. g. phthalic acid esters contain ester amide or nitrile moieties which may undergo hydrolysis to the corresponding acid and an alcohol or amine.

The incorporation of a molecule of water into a substrate (hydrolysis) is catalyzed by a large group of commonly occurring enzymes, including esterases, amidases, nitrilases, phosphatases, and chitinases.

The ability of microorganisms to enzymatically hydrolyze pesticides has been proven by the isolation and partial or complete purification of active enzyme systems (HILL, 1978).

Of the numerous pesticides possessing potential sites for hydrolytic attack, the carbamate and organophosphorous compounds have probably received the greatest attention.

Nitro, amino, sulfonic acid, and halogen substituents

Although the presence of nitro, amino, halogen, and sulfonic acid substituents is known to render benzenoid compounds more resistant towards microbial degradation, a sufficient number of examples can be cited to illustrate the metabolic fate of such groups.

Although nitro substituents are frequently reduced to amines, probably via nitroso- and hydroxylamino intermediates, these are occasionally found as by-products rather than as obligatory intermediates. Thus, aminobenzoic acids are present in cultures of bacteria growing at the expense of different nitrobenzoic acids. The latter acids, however, are not degraded by reduction but by replacement of the nitro substituents by hydroxyl groups with the concomitant formation of nitrite. Similar pathways operate for the degradation of 3,5-dinitro-*o*-cresol by *Arthrobacter simplex,* of 2- and 4-nitrophenol by *Pseudomonas, Moraxella,* and *Bacillus* sp. (CHAPMAN, 1972; SPAIN et al., 1979; SUDHAKAR et al., 1976), of 3-nitrophenol by the same *Bacillus* sp., and of 2,4-dinitrophenol by a *Pseudomonas* sp. (SUDHAKAR et al., 1976).

The nitro group can also remain unchanged, however, as in the formation of 4-nitrocatechol from 4-nitrophenol by *Pseudomonas* sp. ATCC 29 358 (SUDHAKAR et al., 1978).

Arylamines are intermediates in the microbial degradation of phenylamide type pesticides (WALLNÖFER et al., 1983) as well as in the catabolism of 3,5-dinitro-*o*-cresol by *Pseudomonas* species. In these cases, it is the amino substituent that is replaced by hydroxyl. Therefore, *p*-aminobenzoic acid is converted to *p*-hydroxybenzoic acid and 5-hydroxyanthranilic acid to gentisic acid (CHAPMAN, 1972). In the microbial metabolism of aniline by *Pseudomonas* and *Nocardia* species, of 2-, 3-, and 4-chloroaniline by *Alcaligenes faecalis* and a *Pseudomonas multivorans* strain, and of 3,4-dichloroaniline by *A.faecalis* and *P.putida,* however, the deamination step is shown to occur by dioxygenation forming catechol (WALKER and HARRIS, 1969; BACHOFER et al., 1975), 4-chlorocatechol (SUROVTSEVA et al., 1980; REBER et al., 1979; LATORRE, 1982), or 3,4-dichlorocatechol (SUROVTSEVA et al., 1981; YOU and BARTHA, 1982), respectively. Aniline, but not the halogen anilines, is used as a carbon source for growth.

Contrarily, the amino group may also remain unchanged during ring cleavage. Thus, *o*-aminophenol is cleaved by the intradiol dioxygenase, pyrocatechase of a *Pseudomonas arvilla* strain, in an extradiol manner to give picolinic acid as the major product (QUE, 1978).

There appear to be at least two different ways in which sulfonic acid substituents are metabolized by bacteria. Benzenesulfonate is oxidized by a *Bacillus* species that is able to degrade detergents of the alkylbenzene-sulfonate type. It appears that a mixed-function oxygenase forms catechol and simultaneously releases the sulfonic acid substituent as sulfite. However, results with another *Pseudomonas* sp. strongly suggested that the sulfonic acid group of *p*-toluenesulfonate is not removed as a sulfite but as a sulfate (CHAPMAN, 1972).

Elimination of halogen from aromatic compounds is of special interest because of its relevance to the widely used chloroaromatic compounds which are known to decompose very slowly. Direct dechlorination of aromatic compounds is very rarely described in the literature. The mechanism of such dechlorinations from the aromatic ring by microbial monooxygenases is suggested to follow the so-called "NIH-shift" in which the *p*-substituent migrates to the *m*-position or is eliminated. Dechlorination can also occur by direct hydroxyl replacement as shown in the transformation of 4-chlorobenzoic acid to 4-hydroxybenzoic acid by an *Arthrobacter* species (JANKE and FRITSCHE, 1978). Additionally halogen substituents can be eliminated when aromaticity is lost during dioxygenation (KNACKMUSS, 1981).

Much more common among microorganisms than direct elimination of chlorine from the aromatic ring, however, is the dechlorination of aliphatic intermediates formed after ring cleavage. Ring cleavage of halocatechols is considered to be the crucial step for the total degradation of these compounds because of steric hindrance and inductive and mesomeric effects of the substituents. Dehalogenation of halocatechols always occurs after ring cleavage during one of the isomerization steps between muconic acid and the tricarboxylic acid cycle intermediates. Thus, 2,4-dichlorophenoxy acetic acid (2,4-D) is degraded by an *Arthrobacter* sp. via 2,4-dichlorophenol to 3,5-dichlorocatechol which undergoes intradiol ring cleavage to give the dichlorosubstituted muconic acid. Elimination of the *p*-chlorosubstituent occurs during lactonization of this muconic acid, whereas it is suggested that the *o*-chlorosubstituent is eliminated from the chlorosuccinic acid formed after cleavage of the halogenated 3-oxoadipic acid (JANKE and FRITSCHE, 1978; KNACKMUSS, 1981).

Fluorine can also be eliminated from the aromatic ring during initial dioxygenation (ENGESSER et al., 1980; KNACKMUSS, 1981). Catechol is formed if the 2-position is involved in the formation of the intermediate *cis*-diol; fluorine is, however, retained in the product 3-fluorocatechol, if the 6-position is occupied. Ring cleavage of 3-fluorocatechol is recognized as a critical step in 2- and 3-fluorobenzoic acid degradation forming 2-fluoro-*cis,cis*-muconic acid as the final metabolite (SCHREIBER et al., 1980). The failure of many fluorosubstituted aromatic compounds to support growth has frequently been attributed to their breakdown to fluoroacetic acid with subsequent poisoning of the tricarboxylic acid cycle.

c) Reactions converting aromatic compounds into ring fission substrates

All aromatic compounds must be transformed into *ortho*- or *para*-dihydroxybenzenes before ring cleavage can occur. Converging metabolic pathways are used to transform the substrates into relatively few of those di- or trihydric phenolic key intermediates of which the most important are catechol, protocatechuic acid, homogentisic acid, homoprotocatechuic acid, gentisic acid, and gallic acid (see DAGLEY, 1978a, b, c).

Catechol, for example, is formed from a variety of mono- and 1,2-disubstituted aromatic compounds such as mandelic acid, benzoic acid, anthracene, phenanthrene, naphthalene, anthranilic acid, salicylic acid, phenol, and benzene (Fig. 1).

Protocatechuic acid is formed by different reactions from 1,2-, 1,3-, and 1,4-substituted aromatic compounds as, e. g., 3- and 4-hydroxybenzoic acid, phthalic acid, 3- and 4-nitrobenzoic acid, and *m*- and *p*-cresol (Fig. 2).

Among the dihydric phenols which are substrates for ring cleavage gentisic acid is the most prominent one. It is formed from anthranilic acid, salicylic acid, *m*-cresol, and β-naphthol (Fig. 3).

Trihydric phenols may result from the bacterial oxidation of *meta*-dihydric phenols such as orcinol, resorcinol, or thymol (CHAPMAN, 1972). Since much has been published in this field the reader may refer to reviews containing more detailed information on the various reactions leading to substrates for ring fission enzymes by CHAPMAN (1972), DAGLEY (1978b), GIBSON (1971), KIESLICH (1976), and to all the original work.

Microbial hydroxylation of the benzene nucleus involves the incorporation of molecular oxygen into the substrate and is accomplished by two different mechanisms. In the first, the introduction of a single hydroxyl group is catalyzed by a monoxygenase (hydroxylase) according to the following reaction type in Eq. (1):

$$R{-}H + NADH + H^+ + O_2 \rightarrow R{-}OH + NAD^+ + H_2O \quad (1)$$

Many of the enzymes are flavoproteins, although in a few cases the first hydroxylation involves a prosthetic group other than flavin. For instance, a cytochrome P-450 is used by a species of *Pseudomonas* for hydroxylation of D-camphor and both mandelate 4-hydroxylase from *Pseudomonas convexa,* and benzoate 4-hydroxylase from *Aspergillus niger* have an absolute requirement

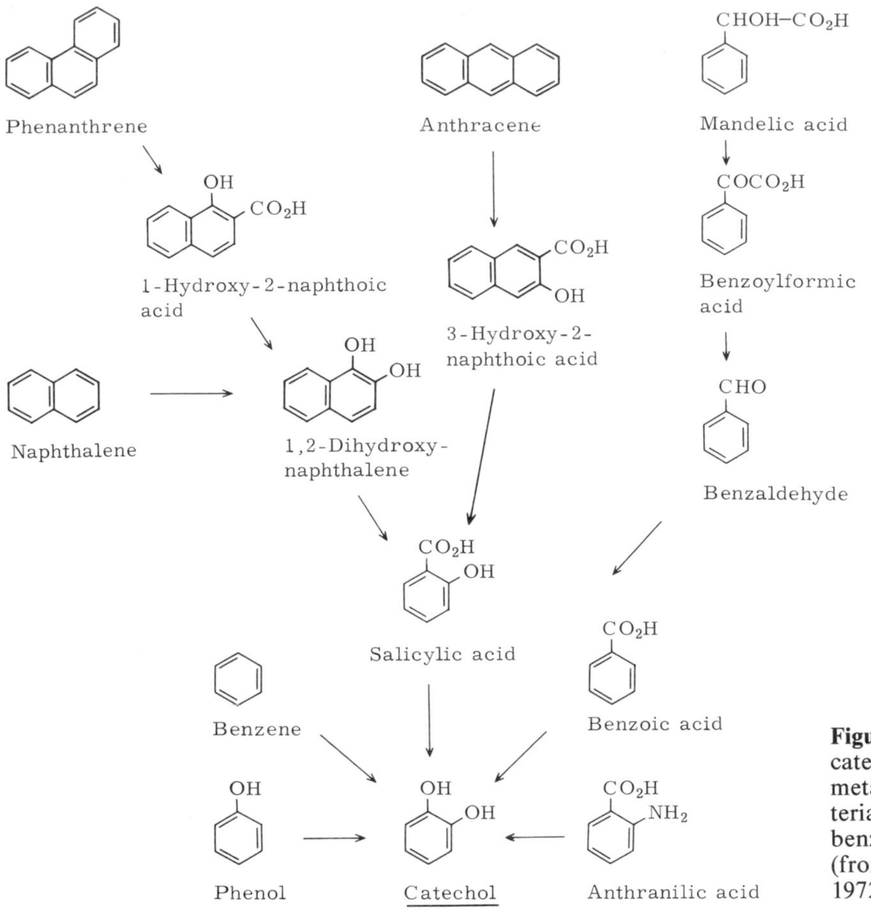

Figure 1. The role of catechol as a central metabolite in the bacterial degradation of benzenoid compounds (from CHAPMAN, 1972).

for tetrahydropteridine (DAGLEY, 1978b). Characteristic of the microbial monooxygenases is their stringent substrate specificity as compared to the broad substrate specificity of, e. g., liver cytochrome P-450 hydroxylase. Thus, bacterial hydroxylases direct the catabolism of different aromatic substrates to the specific degradative sequences. Reactions of the monooxygenase type are exemplified by the bacterial oxidation of a number of phenols and phenolic acids such as phenol, the cresols, *m*- and *p*-hydroxybenzoic acid, or salicylic acid.

The second mechanism of microbial hydroxylation forms *ortho*-dihydric phenols. In bacteria it is accomplished by the simultaneous incorporation of both atoms of molecular oxygen with the intermediate formation of a dihydrodiol which is invariably in the *cis*-configuration and requires the participation of both a dioxygenase and a dehydrogenase. These enzymes which are suggested to exist as a complex (ROGERS and GIBSON, 1977) are involved in the bacterial oxidation of benzoic acid, anthranilic acid, phthalic acid, and some mononuclear and polynuclear hydrocarbons. *Trans*-dihydrodiols of aromatic compounds can also be formed in bacteria, but probably only as anabolic intermediates.

Another mechanism of double hydroxylation of aromatic compounds is carried out by fungi when grown in the presence of a second carbon source. The reaction leads

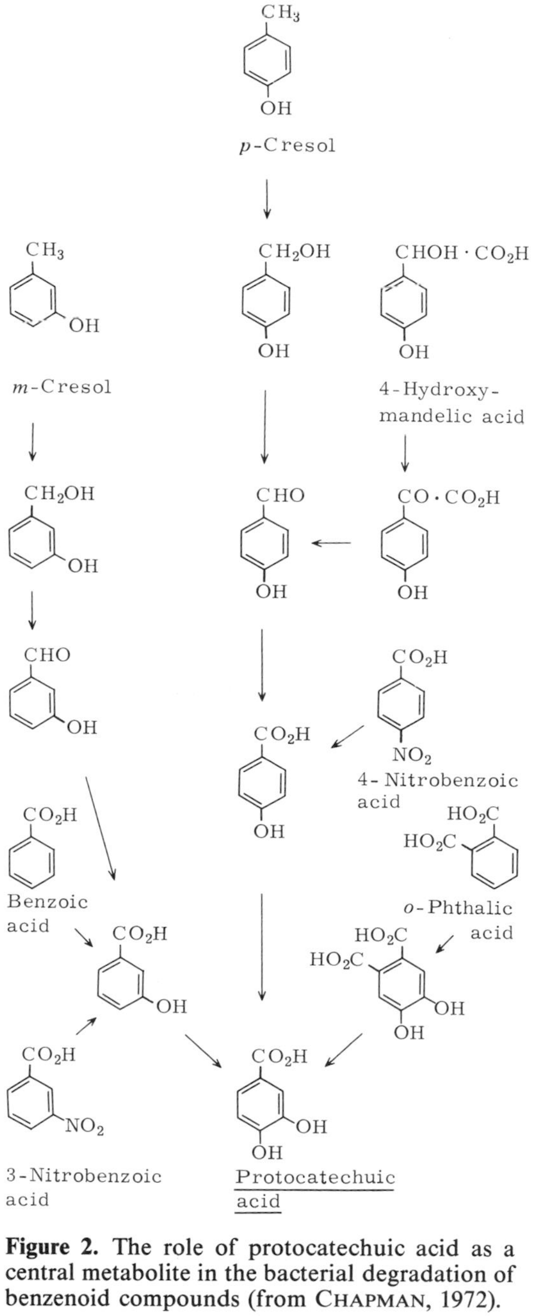

Figure 2. The role of protocatechuic acid as a central metabolite in the bacterial degradation of benzenoid compounds (from CHAPMAN, 1972).

via the formation of arene oxide (epoxide) intermediates which can be hydrated to a *trans*-dihydrodiol as in other eukaryotes (GIBSON, 1971; GIBSON et al., 1975).

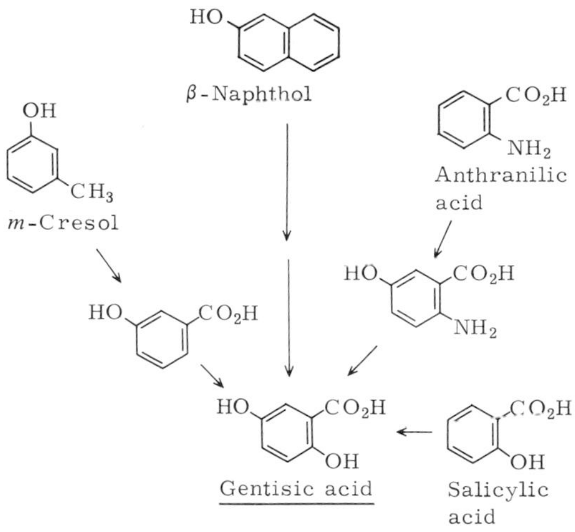

Figure 3. The role of gentisic acid as a central metabolite in the degradation of benzenoid compounds (from CHAPMAN, 1972).

d) Pathways of aromatic ring cleavage and subsequent reactions

Ortho-fission pathways of catechol and protocatechuic acid

Ring cleavage is achieved by reactions in which both atoms of molecular oxygen are introduced into the substrate molecule by dioxygenases. In the case of the *ortho*-dihydric phenols catechol, protocatechuic acid, and homoprotocatechuic acid, intradiol (*"ortho"*) cleavage (reactions a and c in Fig. 4) or extradiol (*"meta"*) cleavage (reactions b, d, e, f in Fig. 4) may occur. Extradiol fission may be distal (reaction d in Fig. 4) or proximal (reaction e in Fig. 4) to the primary substituent. Cleavage of catechol by catechol 1,2-dioxygenase and of protocatechuic acid by protocatechuate 3,4-dioxygenase leads to *cis,cis*-muconate (*cis,cis*-2,4-hexadienedioic acid) or its 3-carboxy derivative. Alternatively, the products of the extradiol mode of ring fission by catechol 2,3-dioxygenase or protocatechuate 4,5-dioxygenase are 2-hydroxymuconic semialdehyde (2-hydroxy-6-oxo-2,4-hexadienoic acid) and 2-hydroxy-4-carboxymuconic semialdehyde (4-carboxy-2-hydroxy-6-oxo-hexadienoic acid). The *para*-dihydric phenols gentisic and homogentisic acid

cis,cis-Muconic acid ← (a, O_2) Catechol → (b, O_2) 2-Hydroxymuconic semialdehyde

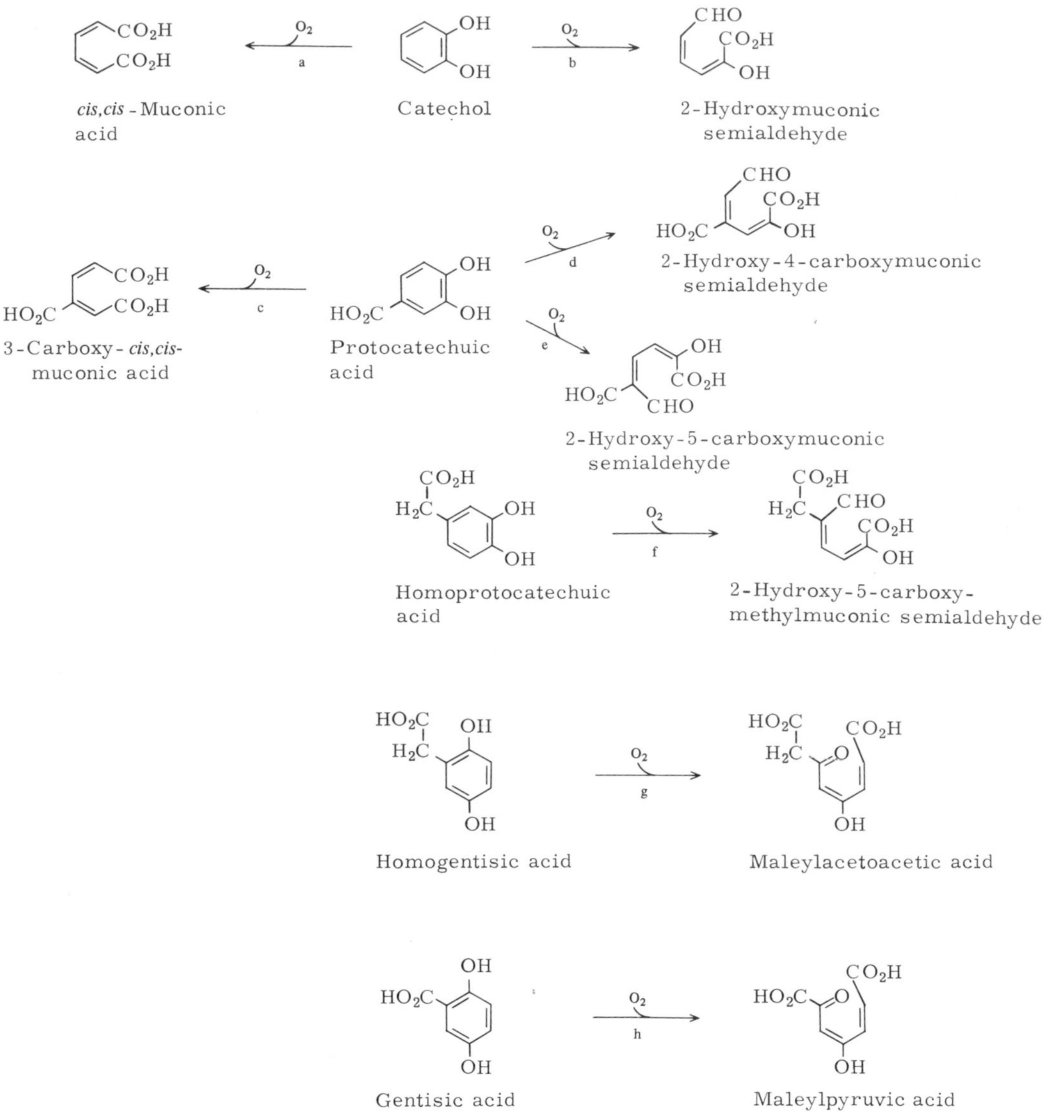

3-Carboxy-*cis,cis*-muconic acid ← (c, O_2) Protocatechuic acid → (d, O_2) 2-Hydroxy-4-carboxymuconic semialdehyde

Protocatechuic acid → (e, O_2) 2-Hydroxy-5-carboxymuconic semialdehyde

Homoprotocatechuic acid → (f, O_2) 2-Hydroxy-5-carboxymethylmuconic semialdehyde

Homogentisic acid → (g, O_2) Maleylacetoacetic acid

Gentisic acid → (h, O_2) Maleylpyruvic acid

Figure 4. Main enzymatic reactions involved in cleavage of dihydroxy benzenes (from FEWSON, 1981).

(reactions h and g in Fig. 4) are split by gentisate 1,2-dioxygenase and homogentisate 1,2-dioxygenase forming maleylpyruvic (3-hydroxy-1-oxo-2,4-pentadienedioic acid) and maleylacetoacetic acid (4-hydroxy-6-oxo-2,4-octadienedioic acid), respectively, as ring fission products (Fig. 4). Many of the dioxygenases have been crystallized and extensively characterized (reviews in HAYAISHI, 1974; BOYER, 1975). They have a fairly strict but not an absolute specificity for one substrate: thus, catechol 1,2-dioxygenases from various organisms will tolerate the presence of methyl and ether alkyl substituents at C-3 or C-4, the rate of attack depending upon the source of the enzyme. This is particularly important in the metabolism of xenobiotics. In addition, catechol

1,2-dioxygenase can catalyze a limited degree of extradiol cleavage (HOU et al., 1977).

The mode of ring-fission strongly depends on the structure of the 1,2-dihydric phenol, on the bacterial species, and on the substrate used for growth. Hence, a catechol 1,2-dioxygenase is induced in *Moraxella lwoffii* growing with benzoic acid, whereas a catechol 2,3-oxygenase is induced when the same organism grows with naphthalene (CHAPMAN, 1972). Similarly, the kind of substituent on the benzene nucleus influences the method of ring fission used. In *Nocardia* sp. DSM 43 251, for example, phenols substituted with electron-donating substituents as dimethyl, methoxy, and methylthio groups elicit the induction of the enzymes of the 2,3-cleavage pathway of catechol. When grown with unsubstituted phenols or phenols substituted by electron-withdrawing groups, however, the enzymes of the 1,2-cleavage pathway of catechol are induced (ENGELHARDT et al., 1979).

After ring fission the resulting open chemical structures enter the main channels of metabolism, such as the tricarboxylic acid cycle. The reactions involved include hydrations, hydrolyses, and aldolase fissions. Thus, the products of intradiol cleavage of catechol and protocatechuic acid are converted to succinic acid and acetyl-CoA by the 3-oxoadipate pathway. The pathways of both compounds converge at 5-carboxymethyl-2-oxo-2,3-dihydrofuran (formerly 3-oxoadipate enol lactone) beyond which the reaction steps are common to both pathways (Fig. 5). The enzymes involved are highly specific. Their properties, regulation, and possible evolutionary relationships are reviewed in a number of papers (STANIER and ORNSTON, 1973; ORNSTON and PARKE, 1977).

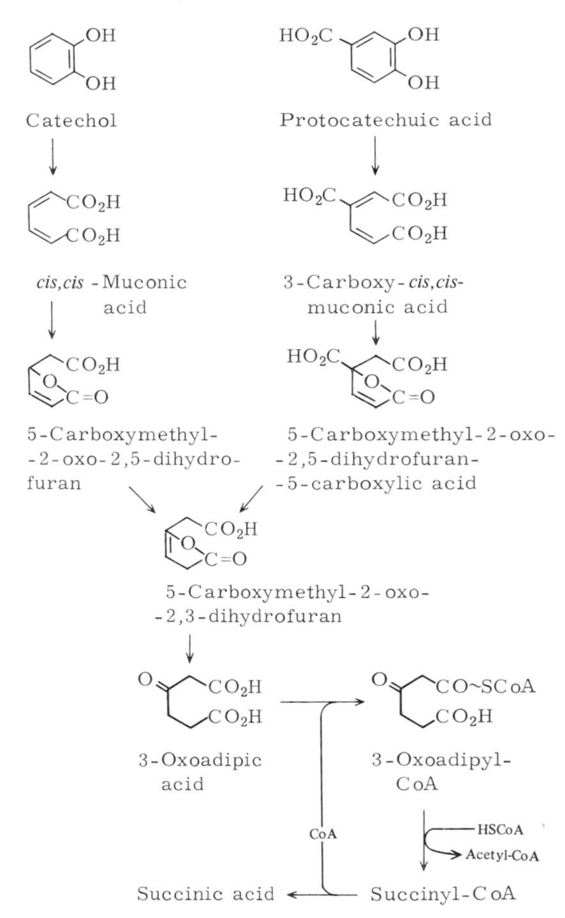

Figure 5. The *ortho*-fission pathways of catechol and protocatechuic acid in bacteria.

Meta-fission pathways of catechol and protocatechuic acid

By extradiol fission of 1,2-dihydric phenols an entirely different sequence of reactions is initiated (reviews by DAGLEY, 1978a, b, c). The final products are amphibolic intermediates such as pyruvic acid and fumaric acid, or closely related compounds such as acetaldehyde, succinic semialdehyde, propionaldehyde, or acetoacetic acid. Therefore, catechol and its analogs are cleaved by a 2,3-dioxygenase to 2-hydroxymuconic semialdehyde or its derivatives which may then either undergo hydrolysis or NAD^+-dependent dehydrogenation. The reactions from 2-keto-pent-4-enoic acid are common to both pathways (CHAPMAN, 1972) (Fig. 6). In 3-methylcatechol only the hydrolytic route exists, since after ring cleavage a keto acid and not a semialdehyde is formed (Fig. 6).

The bacterial metabolism of protocatechuic acid via *meta*-fission is a distinctive property of the non-fluorescent species of

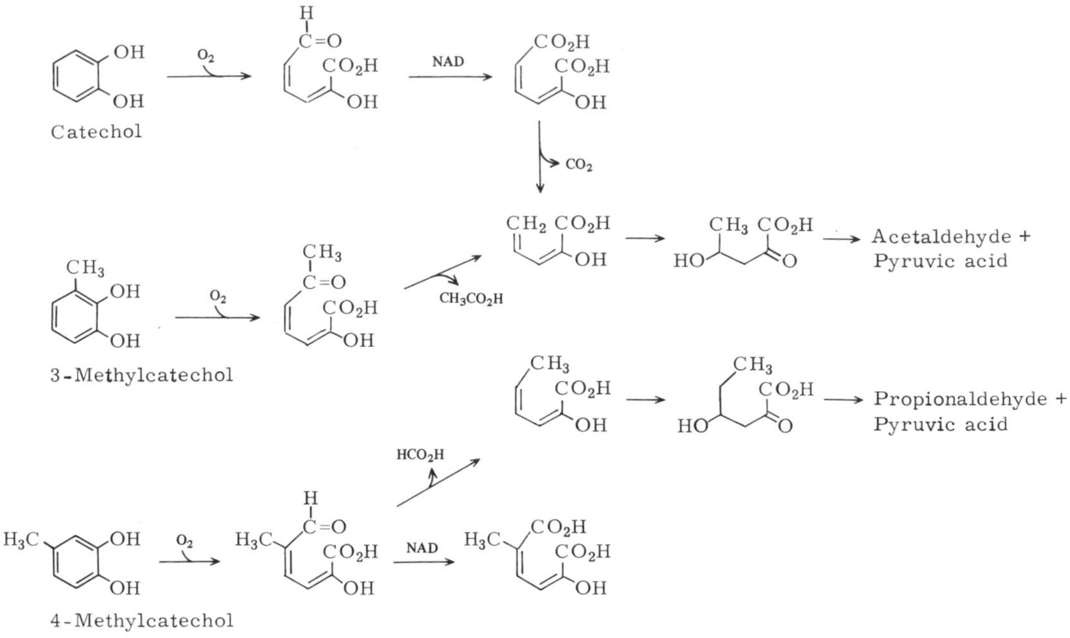

Figure 6. The *meta*-fission pathways of catechol and its homologs in bacteria (from CHAPMAN, 1972).

Pseudomonas, namely *P.acidovorans* and *P.testosteroni* (STANIER et al., 1966) and of the micrococci (KEYSER et al., 1976). Analogous to the hydrolytic cleavage of catechol, after ring fission at the 4,5-position both hydrolytic elimination of formic acid and NAD^+-dependent dehydrogenation occur, followed by hydration to the 2-oxo-4-hydroxy-4-methylglutaric acid and by aldolase cleavage to pyruvate. The alternative pathway via dehydrogenation, however, is different from the catechol *meta*-cleavage pathway since instead of the decarboxylation step the tricarboxylic acid is reduced to the hydroxy keto acid followed by aldolase cleavage to pyruvic acid and oxaloacetic acid (Fig. 7) (CHAPMAN, 1972).

The ring of homoprotocatechuic acid is opened at the 2,3-position by *meta*-cleavage (CHAPMAN, 1972). The intermediates formed, however, have not yet been identified. 2,3-Dihydroxy-*p*-cumic acid (4-isopropyl-2,3-dihydroxybenzoic acid) is an intermediate in the catabolism of *p*-cymene in *P.putida* PL-W. It is also cleaved by a 2,3-dioxygenase forming isopropylpyruvic acid and oxaloacetaldehyde. Metabolism of the latter via a series of degradation steps leads to acetaldehyde (DEFRANK and RIBBONS, 1977).

Degradation of gentisic and homogentisic acid

Gentisic acid is cleaved by a dioxygenase at the carbon-carbon bond between the carboxyl group and the neighboring hydroxyl group to give maleylpyruvic acid followed by a glutathione-dependent isomerization to fumarylpyruvic acid which is then hydrolyzed to fumaric acid and pyruvic acid (Fig. 8). In some organisms, however, hydrolysis occurs without isomerization, as e. g., in the degradation of xylenols by *Pseudomonas* sp. The pathway of homogentisic acid metabolism is analogous (Fig. 8).

Quinol is degraded by dioxygenative ring cleavage at the 1,2-position to give 4-hydroxymuconic semialdehyde, further metabolism of which leads to the formation of 2-oxoadipic acid (CHAPMAN, 1972).

Degradation of trihydric phenols

Metabolism of 1,2,4-trihydroxybenzene involves *ortho*-fission yielding maleylacetic acid.

The two trihydric phenols 2,3,5-trihydroxytoluene and 3-hydroxythymoquinol can be considered as substituted catechols which undergo *meta*-cleavage forming 2,4,6-triketo acids. Similarly, alkylcatechols are substrates for *meta*-fission while alkyl-substituted quinols are not.

Gallic acid is cleaved at the 3,4-bond forming a tricarboxylic acid which is metabolized further in analogy to one of the *meta*-fission pathways of protocatechuic acid (CHAPMAN, 1972) (Fig. 7).

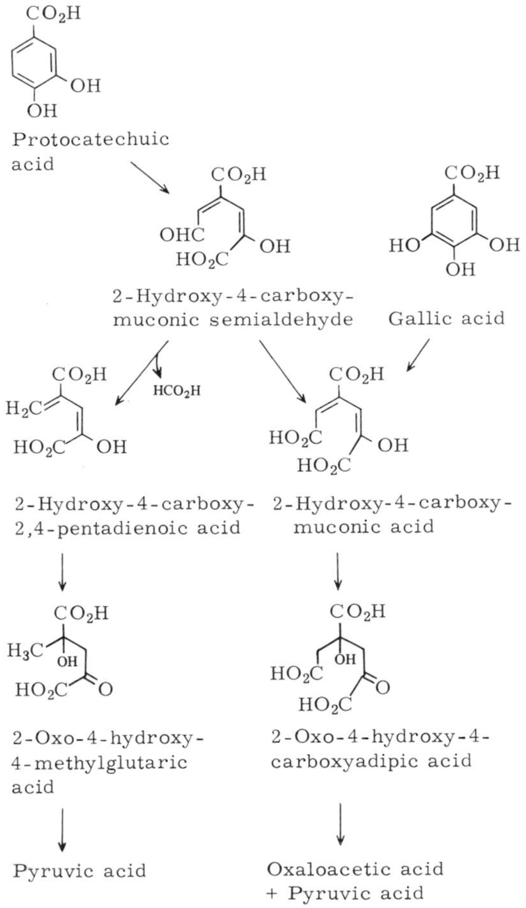

Protocatechuic acid

2-Hydroxy-4-carboxy-muconic semialdehyde

Gallic acid

HCO_2H

2-Hydroxy-4-carboxy-2,4-pentadienoic acid

2-Hydroxy-4-carboxy-muconic acid

2-Oxo-4-hydroxy-4-methylglutaric acid

2-Oxo-4-hydroxy-4-carboxyadipic acid

Pyruvic acid

Oxaloacetic acid + Pyruvic acid

Figure 7. The *meta*-fission pathways of procatechuic acid metabolism and catabolism of gallic acid (from CHAPMAN, 1972).

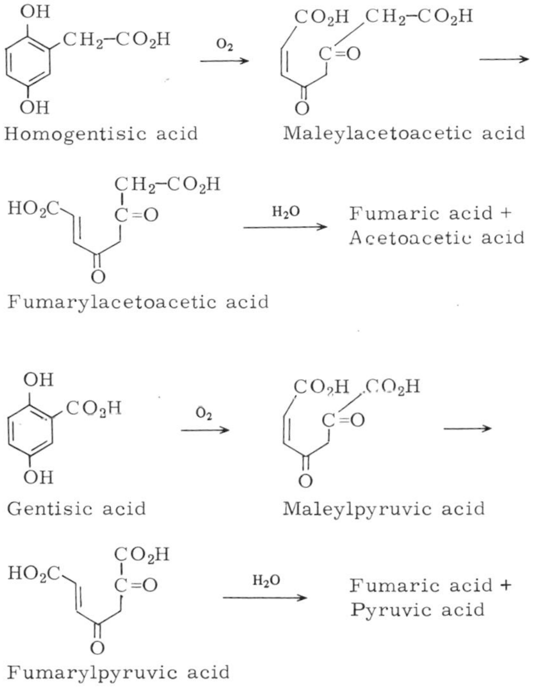

Homogentisic acid $\xrightarrow{O_2}$ Maleylacetoacetic acid $\longrightarrow$

Fumarylacetoacetic acid $\xrightarrow{H_2O}$ Fumaric acid + Acetoacetic acid

Gentisic acid $\xrightarrow{O_2}$ Maleylpyruvic acid $\longrightarrow$

Fumarylpyruvic acid $\xrightarrow{H_2O}$ Fumaric acid + Pyruvic acid

Figure 8. The pathways of gentisic acid and homogentisic acid metabolism in bacteria (from CHAPMAN, 1972).

2. Anaerobic degradation of aromatic compounds

Degradation of aromatic compounds during methane fermentation was reported as early as 1934 by TARVIN and BUSWELL but it has been shown only fairly recently how the benzene nucleus is opened and degraded by bacteria in the absence of molecular oxygen; this may be accomplished by photometabolism, nitrate respiration, and methanogenic fermentation (review by EVANS, 1977). In these pathways, which may also be used for the catabolism of xenobiotics, the benzene ring is first reduced and then cleaved by hydrolysis to give aliphatic acids for cell growth.

a) Anaerobic photometabolism

Different species of the purple non-sulfur bacteria, the Rhodospirillaceae, can use simple aromatic carboxylic acids such as benzoic acid and the hydroxybenzoic acids as the sole source of carbon both aerobically by respiration as well as anaerobically in the presence of light (EVANS, 1977). Anaerobically grown cells were devoid of the enzymes of the aerobic pathways and produced none of the known intermediates of aerobic aromatic metabolism. In cell suspensions of *Rhodopseudomonas palustris* photometabolizing (U-^{14}C) benzoic acid, cyclohexanecarboxylic acid, cyclohex-1-enecarboxylic acid, 2-hydroxycyclohexanecarboxylic acid, 2-oxocylohexanecarboxylic acid, and pimelic acid became labelled. Therefore, DUTTON and EVANS (1970) proposed a new method of aromatic ring metabolism which includes successive reduction steps of benzoic acid (or a derivative) to give cyclohexanecarboxylic acid followed by a coenzyme-A mediated β-oxidation phase (pathway A in Fig. 9). The same pathway for the photometabolism of benzoic acid has been proposed independently by GUYER and HEGEMAN (1969) from a totally different experimental approach. Using a washed chromatophore suspension of *Rhodopseudomonas palustris,* EVANS (1977) obtained strong evidence for the presence of a light-dependent membrane-bound proton-translocating redox system. The low potential reductant was suggested to be a ferredoxin. Furthermore, an appropriate β-oxidation suite of enzymes exists in these cells which is responsible for the sequence of β-oxidations resulting in ring cleavage.

Besides the simple aromatic acids which are metabolized photosynthetically by all species of the Athiorhodaceae, some can also degrade phloroglucinol by photometabolism. Thus, cultures of *Rhodopseudomonas gelatinosa* grown on phloroglucinol in the presence of light formed dihydrophloroglucinol and 2-oxo-4-hydroxyadipic acid (EVANS, 1977).

b) Metabolism of aromatic compounds through anaerobic nitrate respiration

Anaerobic metabolism of benzoic and hydroxybenzoic acids in the presence of nitrate is reported for different soil microorganisms (OSHIMA, 1965; TAYLOR et al., 1970). A reductive pathway is shown to be operating in the decomposition of benzoic acid to adipic acid by a *Moraxella* sp. as well as in the anaerobic decomposition of

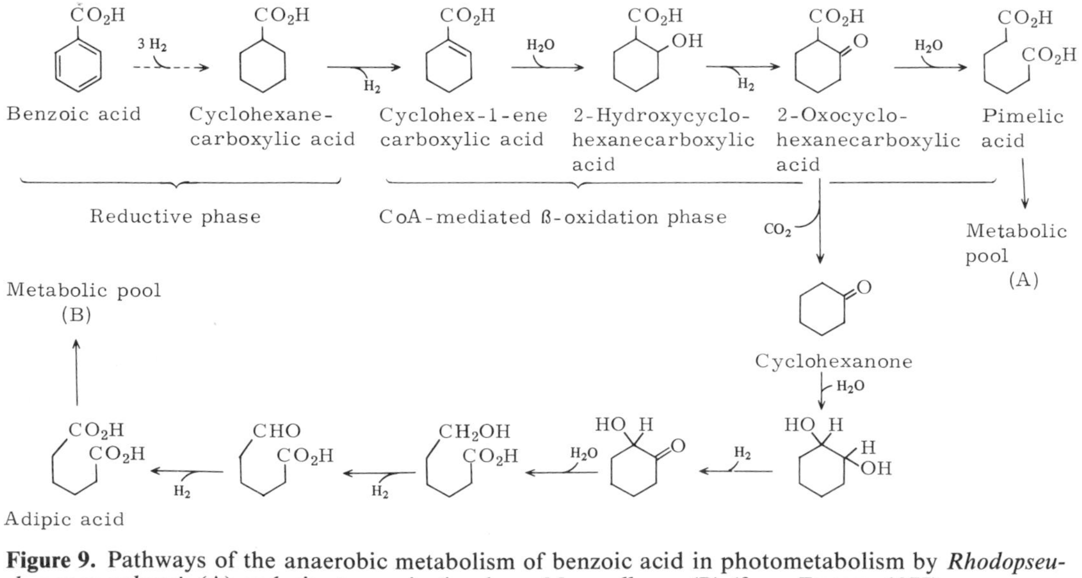

Figure 9. Pathways of the anaerobic metabolism of benzoic acid in photometabolism by *Rhodopseudomonas palustris* (A) and nitrate respiration by a *Moraxella* sp. (B) (from EVANS, 1977).

phenol and other aromatic compounds by a mixed bacterial population in the presence of nitrate (BAKKER, 1977; WILLIAMS and EVANS, 1975). Concomitant with the disappearance of the aromatic substrate, nitrate is reduced mainly to nitrogen gas. Cyclohexanecarboxylic acid, cyclohex-1-enecarboxylic acid, 2-hydroxycyclohexanecarboxylic acid, and adipic acid are identified in the nitrate-dependent metabolism of benzoic acid by the *Moraxella* sp. as well as by a *Pseudomonas* sp. strain PN1. The production of adipic instead of pimelic acid is explained by a divergence at the 2-oxocyclohexanecarboxylic acid stage, with its decarboxylation to cyclohexanone followed by alicyclic ring cleavage by an unknown mechanism (pathway B in Fig. 9). In the mixed culture growing on (ring-U-^{14}C) phenol $^{14}CO_2$-labelled *n*-caproic acid and acetic acid are detected.

Recently, evidence was obtained that phthalic acids are also biodegradable by denitrifying, mixed cultures of bacteria (AFTRING et al., 1981).

A probable interpretation of the events in the anaerobic nitrate-dependent catabolism of aromatic compounds is given by EVANS (1977): the reductive phase is accomplished by a ferredoxin-type reductant followed by the β-oxidation sequence and ring cleavage to aliphatic acids. Since these have to serve as the carbon and energy source, a part must be oxidized and the resulting reduced coenzymes re-oxidized via a membrane bound proton-translocating redox system which is coupled by the electron transport chain to nitrate through nitrate reductase.

c) Anaerobic degradation of aromatic compounds during methane fermentation

The formation of methane and CO_2 from simple aromatic compounds occurs in the absence of nitrate, sulfate, and light by a consortium of microorganisms. The composition of the fermentation gases is shown to be in good accordance with the following equation (NOTTINGHAM and HUNGATE, 1969):

$$4\ {}^{14}C_6H_5COOH + 18\ H_2O \rightarrow 15\ {}^{14}CH_4 + 9\ {}^{14}CO_2 + 4\ CO_2 \quad (2)$$

Using isotopic trapping experiments with benzoate-methanogenic cultures cyclohexanecarboxylic acid, cyclohex-1-enecarboxylic acid, heptanoic acid, valeric acid, butyric acid, propionic acid, and acetic acid were identified as intermediates (KEITH, 1972). However, only cyclohexanecarboxylic acid, cyclohex-1-enecarboxylic acid, propionic acid, and acetic acid are detected during methanogenic fermentation of benzoic acid by both rumen and sewage sludge cultures, from which the latter also contain adipic acid (BALBA and EVANS, 1977). From quantitative studies on the rate of formation and degradation of intermediates and products the following steps [Eqs. (3)–(7)] in the overall conversion of benzoic acid to methane are proposed (FERRY and WOLFE, 1976):

$$4\ C_6H_5COOH + 24\ H_2O \rightarrow 12\ CH_3COOH + 4\ HCOOH + 8\ H_2 \quad (3)$$
$$12\ CH_3COOH \rightarrow 12\ CH_4 + 12\ CO_2 \quad (4)$$
$$4\ HCOOH \rightarrow 4\ CO_2 + 4\ H_2 \quad (5)$$
$$3\ CO_2 + 12\ H_2 \rightarrow 3\ CH_4 + 6\ H_2O \quad (6)$$
$$\text{Net: } 4\ C_6H_5COOH + 18\ H_2O \rightarrow 15\ CH_4 + 13\ CO_2 \quad (7)$$

It is unlikely that the microbial communities studied by different workers are identical in the composition of their bacterial species, although they produce the same intermediates during fermentation of benzoic acid. In all these cultures the aromatic substrate is first reduced with the intermediate formation of cyclohexanecarboxylic acid and cyclohex-1-enecarboxylic acid and then cleaved to form volatile aliphatic acids (heptanoic, valeric, butyric, propanoic, and acetic acid) which are degraded further to substrates for methane bacteria. The electrons generated are assumed to be excreted as H_2 which is necessary for the reduction of CO_2 to CH_4. During the anaerobic degradation of benzoic acid C-1 is converted to methane and C-4 and C-7 primarily to CO_2. The carboxyl carbon of propanoic acid, a possible intermediate, is derived from C-4 of benzoic acid (FINA et al., 1978). The formation of heptanoic acid necessitates a break between the annular C-1 and C-6, and the formation of acetic acid from C-1 and C-7 necessitates a break between the annular C-1 and C-2 (KEITH et al., 1978). Previously breaks have been indicated between C-1 and C-6 and between C-3 and C-4 (FINA et al., 1978). Including these breaks and the identified intermediates a possible pathway is proposed by KEITH et al. (1978) for the methanogenic fermentation of benzoic acid by a microbial consortium (Fig. 10). The acetic acid is suggested to arise by β-oxidation of heptanoic and valeric acid. The reactions leading to butyric acid, however, could not yet be determined.

Besides benzoic acid different ligno-aromatic compounds are biodegradable to methane, as recently shown for eleven simple aromatic lignin derivatives (HEALY and YOUNG, 1979). Methanogenic enrichment cultures can be obtained which grow anaerobically on vanillin, vanillic acid, ferulic acid, cinnamic acid, benzoic acid, catechol, protocatechuic acid, phenol, *p*-hydroxybenzoic acid, syringic acid, and syringaldehyde. Microbial communities acclimated to a particular aromatic substrate are simultaneously adapted to other selected aromatic substrates. From carbon balance measurements with vanillic and ferulic acid evidence is obtained that the aromatic ring is cleaved and that the amounts of methane produced closely agree with the calculated values. Thus, nearly stoichiometric amounts of the substrate are converted to CO_2 and CH_4 which indicates that half or more of the organic carbon in aromatic ring derivatives can be potentially converted to methane gas. These results also suggest that aromatic carbon derivatives can be mineralized to CO_2 and CH_4 in highly anaerobic environments.

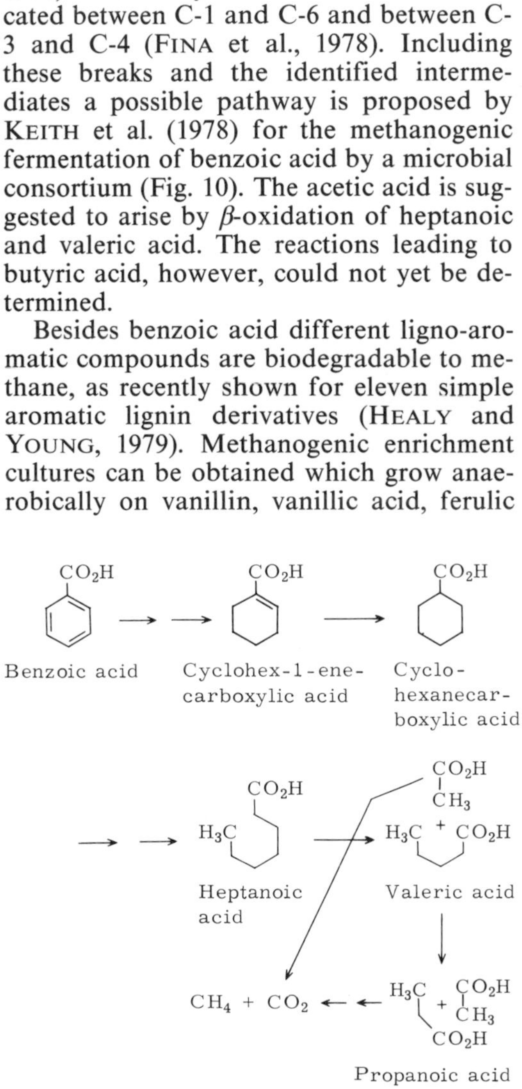

d) The fate of benzene ring substituents under anaerobic conditions

Halogen substituents are removed in anaerobic environments from organic substrates by two different reaction types: (1) Dehydrohalogenation has been observed with both aromatic and aliphatic halides (although relatively infrequently for the latter). It involves the removal of adjacent halogen and hydrogen atoms and the formation of a C,C double bond. (2) Reductive dehalogenation of aliphatic halides is a major degradation route, favored by anaerobic conditions and like dehydrohalogenation, has been reported for many of the chlorinated pesticides. It probably involves carriers such as reduced cytochrome oxidase and FAD (HILL, 1978).

Those persistent environmental contaminants as DDT, the isomers of hexachlorocyclohexane, and cyclodiene insecticides are more rapidly transformed under anaerobic than under aerobic conditions. Strict anaerobes such as different species of *Clostridium* as well as other facultative anaerobes are very active in the reductive dechlorination of chlorinated hydrocarbons (JAGNOW et al., 1977; OHISA, 1980).

Recent studies on the metabolism of monofluoro- and monochlorobenzoic acids by the denitrifying bacterium *Pseudomonas*

Figure 10. Suggested pathway from benzoic acid to methane and carbon dioxide by a microbial consortium.

PN-1 (TAYLOR et al., 1979) showed that these compounds do not support growth of the bacterium, neither aerobically nor anaerobically (nitrate respiration). Only *o*- and *p*-fluorobenzoic acid, of the monohalogenated benzoic acids tested, are definitely degraded by *Pseudomonas* PN-1. Anaerobic growth rates on non-halogenated substrates are increased by *p*-fluorobenzoic acid under metabolization of the compound and release of F^-. Cells grown anaerobically on *p*-hydroxybenzoic acid catabolize *o*- and *p*-fluorobenzoic acid with a release of F^-. Degradation of *p*-fluorobenzoic acid, but not of *o*-fluorobenzoic acid, occurs only after a lag-phase and is inhibited by chloramphenicol thereby indicating the need for additional enzyme(s) to attack the *para*-isomer.

The inability of *Pseudomonas* PN-1 to grow at the expense of *o*-fluorobenzoate is suggested to be due to the production of a toxic fluoro compound during its degradation analogous to the formation of fluorocitrate in the aerobic metabolism of fluorinated aromatic compounds.

Nitro substituents of aromatic compounds have been shown to be reduced to amino groups in anaerobic environments as in the presence of oxygen. Thus, a strain of *Pseudomonas aeruginosa* reduces 2,4,6-trinitrophenol (picric acid) to the mutagen 2-amino-4,6-dinitrophenol under anaerobic conditions (WYMAN et al., 1979).

B. Common Reactions in the Metabolism of Heterocyclic Compounds

More than one third of all known organic compounds have heterocyclic structures. They include many biologically important compounds as metabolites, coenzymes, or components of macromolecules and other natural products as well as substances with pharmaceutical or pesticidal properties. On the other hand, many heterocyclic compounds are present in industrial effluents and before discharge have to be removed or degraded to harmless products to avoid pollution.

The hetero atom(s) can be the nutritional source for biosynthetic purposes. This is especially the case with many heterocyclic compounds containing nitrogen or sulfur. Microorganisms can use these compounds as sole sources of carbon and energy as well as sole nitrogen or sulfur sources (COOK and HÜTTER, 1981). Though occasionally excreted as methylamine, nitrogen is nearly always eliminated as ammonia. Sulfur finally yields sulfate or hydrogen sulfide or mercaptans of low molecular weight. Oxygen is finally incorporated into hydroxyl groups, frequently the —OH in water, or in carbon dioxide.

Heterocyclic compounds vary considerably in stability. Some can be easily cleaved through hydrolysis. Thus, 5-valerolactone and 6-capronolactone are hydrolyzed rapidly to 5-hydroxyvaleric acid and 6-hydroxycaproic acid, respectively (CALLELY, 1978). Similar to the benzene derivatives other heterocyclic compounds must first be transformed to hydroxyl derivatives by introduction of one or more hydroxyl groups before ring cleavage can occur.

These quite stable ring systems, however, are hydroxylated without the participation of oxygen and without the expenditure of reducing power. The sequences shown in Fig. 11 demonstrate that the oxygen atom of water is incorporated into the pyridine rings of nicotinic acid and picolinic acid. Similarly, a species of *Pseudomonas* uses water to hydroxylate the furan ring of furan-2-carboxylic acid, converting the substrate first into its coenzyme A ester. This initial activation involves an investment of energy, but the actual hydroxylation reaction does not. In each of these systems, in contrast to hydroxylation in the benzene ring, electrons are not supplied to the enzymes but are removed through the cytochrome chain.

The conversion of nicotinic acid to 6-hydroxynicotinic acid has also been demonstrated for strictly anaerobic systems. Thus a FAD-containing (and non-heme iron) enzyme from a *Clostridium* species inserts ox-

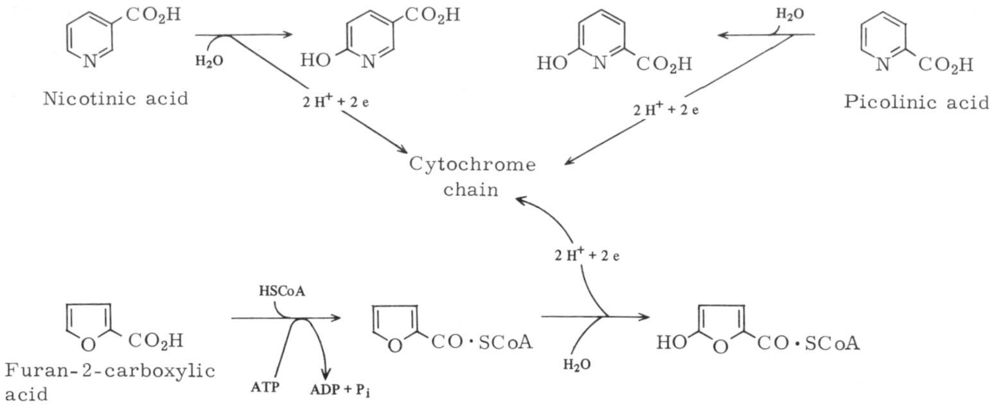

Nicotinic acid
Picolinic acid
Cytochrome chain
Furan-2-carboxylic acid

Figure 11. Reaction sequences in the hydroxylation of nicotinic acid, picolinic acid, and furan-2-carboxylic acid (from DAGLEY, 1972).

ygen derived from water into nicotinic acid, and, in this case, NADP serves as the necessary acceptor of electrons. 6-Hydroxynicotinic acid is therefore the initial product formed from nicotinic acid by both aerobic and anaerobic bacteria.

The ability to use the oxygen of water for hydroxylation may be due to the fact that electrons are more readily localized in the pyridine than the benzene ring (see DAGLEY, 1972).

In the following the degradation pathways of some structurally simpler heterocyclic compounds of both biological and industrial importance are presented.

1. Five membered ring systems

Many biologically and ecotoxicologically important compounds possess one or more five membered heterocyclic ring systems or one such ring condensed to a benzene ring or another heterocyclic one.

In the heterocyclic rings of pyrrol, furan, and thiophene the heteroatoms can each contribute a lone pair of electrons into the unsaturated ring resulting in an aromatic character of these substances.

Furan-2-carboxylic acid (formerly called furoic or pyromuconic acid) is metabolized to 2-oxoglutaric acid which enters the tricarboxylic acid cycle (CALLELY, 1978). The first reaction involved was shown to be the "activation" of furan-2-carboxylic acid with coenzyme A and ATP giving the coenzyme A derivative which is then hydroxylated as described above. Keto-enol tautomerism of the 5-hydroxy derivative gives an unsaturated lactone which is hydrolyzed to the enol form of 2-oxoglutaryl-CoA, from which 2-oxoglutaric acid is finally derived (Fig. 12). Since the introduction of the hydroxyl group results at least in the formation of a lactone these compounds are easily cleaved by hydrolysis as demonstrated

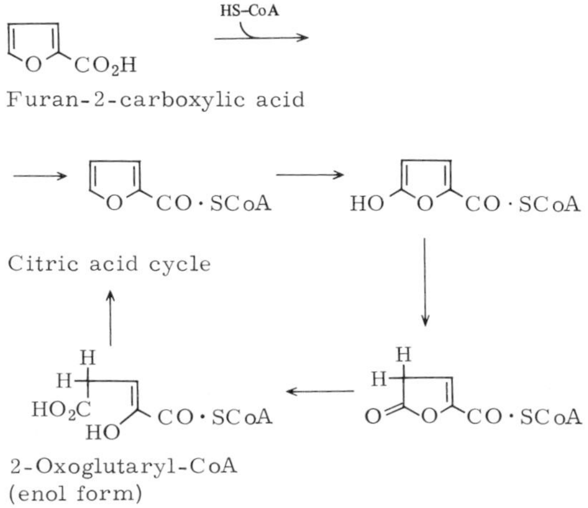

Furan-2-carboxylic acid
Citric acid cycle
2-Oxoglutaryl-CoA (enol form)

Figure 12. Pathway for the degradation of furan-2-carboxylic acid (also called 2-furoic acid) by bacteria (from CALLELY, 1978).

earlier for 5-valerolactone and 6-caprolactone.

Thiophene-2-carboxylic acid seems to be degraded by an analogous reaction sequence as was shown for furan-2-carboxylic acid (CALLELY, 1978).

The amino acid tryptophan which possesses a pyrrol ring condensed to a benzene ring is metabolized by a variety of pseudomonads as the sole source of carbon, energy, and nitrogen under aerobic conditions. Tryptophan is degraded via two alternative pathways, the so-called "aromatic" pathway and the "quinoline" pathway (KIESLICH, 1976; CALLELY, 1978). The latter is employed by pseudomonads of the acidovorans group. In both pathways the first reaction is the fission of the pyrrol ring by tryptophan oxygenase forming *N*-formyl kynurenine. The formyl group is then cleaved from the molecule by formyl kynurenine formamidase yielding formate and kynurenine. In the aromatic pathway kynurenine is converted to catechol via anthranilic acid, whereas in the quinoline pathway kynurenic acid is formed by condensation to give a new heterocyclic ring.

In addition to the microbial degradation of tryptophan by ring cleavage this compound is transformed by different bacteria and fungi to a variety of products via deamination, decarboxylation, acetylation, or hydroxylation (KIESLICH, 1976).

Thiophene and benzothiophene which is a contaminant of crude oils and coal-tar distillates together with other condensed thiophenes are degraded by an anaerobic bacterium under formation of hydrogen sulfide (CALLELY, 1978).

The imidazole ring which contains two nitrogen atoms seems to be easily opened by bacteria. Hence, L-histidine is degraded by a reaction sequence very similar to that employed in the degradation of furan- and thiophene-2-carboxylic acid. The imidazole ring is transformed to give an imidazolone which is then cleaved hydrolytically (CALLELY, 1978).

The microbial degradation of nitrogeneous compounds containing a five membered saturated heterocyclic ring system has been investigated by several authors. 2-Pyrrolidone is hydrolyzed by *Pseudomonas aeruginosa* into γ-aminobutyric acid, and pyrrolidine is shown to be degraded by *P. fluorescens* and an *Arthrobacter* sp. to succinic acid via γ-aminobutyric acid (GUPTA et al., 1975).

2. Six membered heterocyclic compounds

Among the heterocyclic compounds the pyridine ring is considered to be the most important from the view of environmental and industrial chemicals. Pyridine and its alkyl derivatives are widely used as industrial solvents. They have to be removed from industrial effluents before such effluents can be safely discharged. In addition, pyridine compounds find their way into the environment, e. g., by application of the pyridine herbicides diquat and parquat.

Pyridine is degraded by comparatively few bacteria, actinomycetes being predominating, when the compound is supplied as the sole source of carbon, nitrogen, and energy. In sewage, pyridine is rapidly removed by activated sludge. Degradation of pyridine by a *Bacillus* and a *Nocardia* sp. has been shown to occur via two distinct metabolic routes (WATSON and CAIN, 1975). Both organisms grow rapidly on this compound as the sole C, N, and energy source. The monohydroxypyridines, tetrahydropyridine, piperidine, and some other analogs are not utilized for growth by these bacteria. In both organisms an initial reduction of pyridine produces 1,4-dihydropyridine which, in the case of *Nocardia* Z1 (route A, Fig. 13) undergoes a hydrolytic N-C-2 ring cleavage and subsequent deamination to glutaraldehyde followed by successive oxidations to glutaric semialdehyde, glutaric acid, and glutaryl-CoA. In the *Bacillus* sp. (route B), on the other hand, hydrolysis or a dioxygenase attack at the C-2, C-3 double bond forms 4-(formylamino)-but-3-en-1-al which undergoes immediate oxidation to the corresponding semialdehyde, 4-(formylamino)-but-3-enoic acid. Hydrolysis of the C-6-N bond liberates formamide and

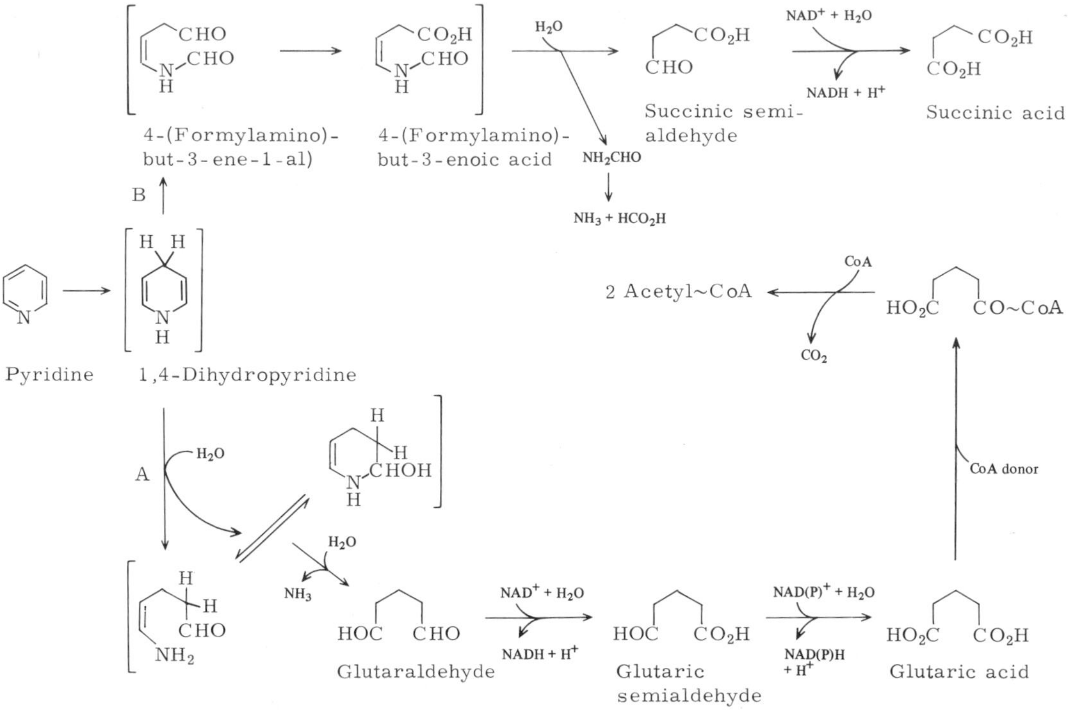

Figure 13. Metabolic routes for the biodegradation of pyridine by a *Nocardia* (A) and a *Bacillus* sp. (B). Compounds in brackets are hypothetical (from WATSON and CHAIN, 1975).

succinic semialdehyde, the former of which is attacked by a specific formamide amidohydrolase to produce formate and NH_3, whereas succinic semialdehyde is further oxidized to succinic acid by an equally specific dehydrogenase (Fig. 13) (WATSON and CAIN, 1975).

Nicotinic acid is degraded by pseudomonads via 6-hydroxynicotinic acid via two successive hydroxylations and decarboxylation to give the paradiol 2,5-dihydroxypyridine. This is then cleaved by an oxygenase to formate and maleamic acid from which maleic and finally fumaric acid is formed. Alternatively, a *Bacillus* species produces 2,6-dihydroxynicotinic acid from the 6-hydroxy compound. It is suggested to be decomposed via pyridine 2,3,6-triol yielding maleamic acid and formic acid.

In these two pathways and those describing the breakdown of other pyridine-ring compounds (also route B of pyridine degradation) the overall strategy is broadly the same, i. e., the pyridine ring is normally cleaved between C-2 and C-3 to give a C—N—C—C—C—C open-chain compound. The isolated carbon is disposed of (usually as formic acid) and then the amino group is lost, leaving a four carbon compound which is converted to a tricarboxylic acid cycle intermediate.

As shown with monohydroxy pyridines the pyridine paradiol is not always the substrate for ring cleavage. Thus, the 2- and 3-hydroxy derivatives are both converted to the paradiol 2,5-dihydroxy pyridine, whereas the orthodiol 3,4-dihydroxypyridine is formed from the 4-hydroxy isomer. This pyridine orthodiol is analogous to the ortho-dihydroxy compounds commonly encountered as ring-cleavage substrates in the degradative pathways of benzene-ring compounds. This orthodiol pathway, however, is suggested to be not as common as the paradiol route for pyridine degradation (CALLELY, 1978).

The fermentation of nicotinic acid by a *Clostridium* has been reported, the end products being propionic acid, acetic acid, carbon dioxide, and ammonia (CALLELY, 1978).

1-Methyl-4-carboxypyridinium ion, the photolytic decomposition product of the herbicide paraquat, can also be degraded by soil bacteria via two different degradation pathways. In the first which is performed by an *Achromobacter* species *N*-methylisonicotinic acid is hydroxylated on C-2, followed by demethylation of the ring nitrogen atom. The compound is then hydroxylated further to the 2,6-dihydroxy derivative before ring cleavage. In the alternative pathway in an *Arthrobacter* strain the *N*-methyl group is released as methylamine very late in the metabolic sequence after ring fission has occurred (see Tab. 7). In the latter pathway no hydroxylated heterocyclic compounds are involved. In contrast, the pyridine ring is first reduced followed by a direct oxygenative cleavage of the link between carbon atoms C-2 and C-3 analogous to route B of pyridine degradation performed by a *Nocardia* sp. (Fig. 13). This direct oxygenase cleavage of a double bond without previous hydroxylation represents a new mode of ring fission performed by microorganisms.

The degradation of the pyridine alkaloid nicotine which contains a heteroparaffinic pyrrolidine ring substituted with a *N*-methyl group linked by a carbon-carbon bond to pyridine by *Pseudomonas* and *Arthrobacter* species is well documented (KIESLICH, 1976, CALLELY, 1978). It is presented in Chapter 6 of this volume.

Six membered heterocyclic rings containing two nitrogen atoms are very common in nature. Four types of reactions on the pyrimidine molecule are known: hydroxylation, ribosidation, ring cleavage, and hydrogenation (KIESLICH, 1976). Ring cleavage of these compounds occurs by similar mechanisms as described for pyridine. Thus, uracil is first hydroxylated to give barbituric acid which is then cleaved hydrolytically yielding urea and malonic acid.

Utilization of pyrimidines as a nitrogen source for growth is a common property of the Rhodospirillaceae. Thus, *R.capsulata* R10 utilizes cytidine, uracil, and thymine, evidently incorporating the total pyrimidine nitrogen into biomass, whereas cytosine does not support growth. Degradation of the pyrimidines apparently follows the reductive pattern with release of CO_2. This is accompanied by synthesis of the key enzymes of pyrimidine catabolism, dihydrouracil dehydrogenase and dihydropyrimidine amidohydrolase which splits dihydrouracil yielding *N*-carbamyl-β-alanine (KASPARI, 1979).

IV. Aromatic and Heterocyclic Compounds with Economical and Ecotoxicological Significance

Microorganisms must be considered in any discussion of the behavior of pesticides and industrial pollutants in natural environments, both because of what they are able and unable to do. Microorganisms are frequently the major and sometimes the only means by which these chemicals are eliminated from a variety of ecosystems, and they are therefore important in governing persistence.

Microorganisms differ widely in their ability to metabolize "foreign" chemicals, the so-called xenobiotics. While a large number of reactions of pesticides have been observed in the presence of microorganisms, many workers have failed to differentiate enzymatic processes from those resulting from the extra- or intracellular conditions, of redox potential and pH, or from the presence of heavy metals, etc.

Although details of many enzyme reactions with pesticides are still lacking the volume of knowledge is steadily increasing. Studies of the oxygenases, for example, are well advanced and many of these enzymes have been isolated, purified, and prepared in crystalline form. Investigations of the requirements for enzyme activity have mostly progressed simultaneously. Most of the data result from laboratory studies with microbes grown under controlled culture conditions and from studies of the isolated and purified enzyme systems from these organisms. Information on microbially mediated reactions of pesticides and industrial pollutants under natural environmental conditions has, as of necessity, been mostly extracted from literature on soil studies (HILL, 1978).

A. Aromatic Compounds

1. Phenolic pesticides and terminal aromatic metabolites of pesticides

Phenolic pesticides include fungicides as well as herbicides. Of these the nitrophenols are of considerable interest with respect to their application as herbicides.

Degradation of many pesticides in the environment, e. g., in soils, is often incomplete leaving behind metabolites such as the aromatic moieties of the starting compounds which are more polar and recalcitrant to microbial attack (WALLNÖFER et al., 1984). Such polar metabolites are mainly anilines and halogenoanilines, released from phenylamide-type herbicides and fungicides, as well as substituted phenols derived from aroxyalkanoic acid herbicides, and organophosphorus and methylcarbamate insecticides. Halogenated phenols, like 2,4-dichlorophenol, are the predominating metabolites with respect to the economic importance of the starting pesticides. In Tables 1, 2 and 3 the results obtained on the microbial degradation of phenolic pesticides and aromatic metabolites of pesticides by isolated microorganisms are presented.

In soils the bulk of these residues become immobilized by chemical reaction to humus compounds. Thus, pesticide-derived phenols are able to form polymeric materials by reactions with monomeric lignin units catalyzed by microbial oxidases and peroxidases.

For studying the biodegradation of such humus-bound metabolites some veratrylglycerol-β-arylethers have been synthesized as model compounds, since β-arylether structures represent the most abundant bond type in the peroxidase-catalyzed polymerization of coniferyl alcohol (ENGELHARDT et al., 1981a). A model compound containing a 2,4-dichlorophenylether moiety, is degraded by different bacteria of the genera *Corynebacterium* and *Rhodococcus*. Degradation obviously occurs after oxidation to 2-aroxy-3-hydroxy-3-(3,4-dimethoxyphenyl)-propionic acid via cleavage of the aliphatic chain between C-2 and C-3. The corresponding aroxyacetic acid, 3,4-dimethoxybenzaldehyde, and 3,4-dimethoxybenzoic acid could be identified as degradation products (Fig. 14). This result at least indicates that humus-bound phenolic pesticide metabolites may be released from the humic core by biochemical reactions different from those responsible for their incorporation, e. g., the laccase reaction. The potential liberation of aroxyacetic acids from soil-bound substituted phenols might be of considerable ecotoxicological significance since herbicidal activities of such compounds cannot be excluded.

2. Industrial pollutants

From the industrial aromatic pollutants which cause problems of disposal or increase environmental pollution the phthalic acid esters, lignosulfonates, surfactants, dyes, and different chlorinated hydrocarbons play a major role.

a) Phthalic acid esters

Phthalic acid esters have been synthesized on a massive scale for the last two to

Table 1. Degradation of Phenolic Pesticides (WALLNÖFER et al., 1984)

Compound	Transformation Product	Reaction Type	Microorganism
(OH; R_1, R_2, R_3, R_4, R_5) $R_1=R_3=NO_2$; $R_5=CH_3$ Dinitro-*o*-cresol (DNOC, H[a)])	2,3,5-Trihydroxytoluene	n. d.[b)]	*Pseudomonas* *Arthrobacter* *Azotobacter* sp.
	2-Hydroxy-3-methyl-5-nitroacetanilide	Reduction, acetylation	
$R_1=R_3=NO_2$; $R_5=C_2H_5CH(CH_3)—$ 6-(2-Butyl)-2,4-dinitrophenol (Dinoseb, H)	2-Hydroxy-3-(2-butyl)-5-nitroacetanilide	Reduction, acetylation	*Azotobacter* sp.
$R_1=R_5=I$; $R_3=CN$ 3,5-Diiodo-4-hydroxybenzonitrile (Ioxynil, H)	3,5-Diiodo-4-hydroxybenzamide (A)	Addition of water	*Fusarium solani*
			Flexibacterium BR 4
	3,5-Diiodo-4-hydroxybenzoic acid	Hydrolysis of A	*Fusarium solani*
$R_1=R_5=Br$; $R_3=CN$ 3,5-Dibromo-4-hydroxybenzonitrile (Bromoxynil, H)	3,5-Dibromo-4-hydroxybenzoic acid	Hydrolysis	*Fusarium solani*
$R_1=R_2=R_3=R_4=R_5=Cl$ (Pentachlorophenol, H)	Pentachloroanisole	Methylation	*Trichoderma virgatum*

a H Herbicide
b n. d. not determined

three decades, mainly for the formation of plastics. Considerable attention has been directed to the possible toxicity of these compounds. In recent years, several reports on the biodegradation of phthalic acid esters have been published. These reports have established that phthalic acid esters are easily biodegradable in the laboratory with pure cultures of bacteria (KEYSER et al., 1976), as well as in activated-sludge digestion and hydrosoils (Tab. 4).

b) Lignosulfonates

The two most common ways of manufacturing paper are the "kraft" or sulfate pulping and the acidic sulfite cooking process.

Chemical pulping is performed in order to remove hemicelluloses and the encrusting lignin from cellulose. The effluents resulting from the acidic sulfite process are called "sulfite spent liquors" which, beside lignosulfonates, contain considerable amounts of hexoses, pentoses, and organic

Table 2. Degradation of Phenols Derived from Pesticides (Starting Compound)

Compound	Transformation Product	Reaction Type	Microorganism	Reference
OH; R_1, R_5, R_2, R_4, R_3 $R_3 = NO_2$ 4-Nitrophenol (Parathion, I[a])	4-Nitrocatechol	Monooxygenation	*Flavobacterium* sp.	WALLNÖFER et al., 1984
$R_3 = SOCH_3$ 4-(Methylsulfinyl)-phenol (Fensulfothion, I)	4-(Methylsulfinyl)-catechol (A)	Monooxygenation	*Nocardia* sp. DSM 43 252	ENGELHARDT et al., 1977
	6-Oxo-2-hydroxy-5-(methylsulfinyl)-*cis,cis*-2,4-hexadienoic acid	*m*-Cleavage (2,3) of A	*Nocardia* sp. DSM 43 252	
$R_2 = CH_3$, $R_3 = SCH_3$ 3-Methyl-4-(methylthio)-phenol (Fenthion, I)	3-Methyl-4-(methylthio)-catechol (A)	Monooxygenation	*Nocardia* sp. DSM 43 252	RAST et al., 1979
	4-Methyl-5-(methylthio)-catechol (B)	Monooxygenation		
	2,5-Dihydro-4-methyl-3-(methylthio)-5-oxo-2-furanyl acetic acid	*o*-Cleavage and lactonization of A	*Nocardia* sp. DSM 43 252	
	2,5-Dihydro-2-methyl-3-(methylthio)-5-oxo-2-furanyl acetic acid	*o*-Cleavage and lactonization of B		

$R_1 = R_3 = Cl$ 2,4-Dichlorophenol (2,4-D, H[b])	3,5-Dichlorocatechol (A)	Monooxygenation	*Nocardia* sp. DSM 43 252	ENGELHARDT et al., 1979
	1,3-Dichloro-*cis,cis*-1,3-butadiene-1,4-dicarboxylic acid (B)	*o*-Cleavage of A	*Arthrobacter* sp.	SHARPEE et al., 1973
	5-Carboxymethylene-2-oxo-3-chloro-2,5-dihydrofuran	Lactonization of B with HCl elimination	*Arthrobacter* sp.	
$R_1 = CH_3$; $R_3 = Cl$ 2-Methyl-4-chlorophenol (MCPA, H)	3-Methyl-5-chlorocatechol (A)	Monooxygenation	*Nocardia* sp. DSM 43 252	ENGELHARDT et al., 1979
	5-Carboxymethylene-2-oxo-3-methyl-2,5-dihydrofuran	*o*-Cleavage and lactonization with HCl elimination of A		
$R_1 = R_3 = R_4 = Cl$ 2,4,5-Trichlorophenol (2,4,5-T, H)	3,4,6-Trichlorocatechol	Monooxygenation	*Pseudomonas* sp.	WALLNÖFER et al., 1984
			Achromobacter	
			Nocardia restricta DSM 43 199	ENGELHARDT et al., 1979

[a] I Insecticide
[b] H Herbicide

Table 3. Degradation of Anilines Derived from Pesticides (Starting Compounds)

Compound	Transformation Product	Reaction Type	Microorganism	Reference
R_1, NH_2, R_2	Catechol (A)	Dioxygenative deamination	*Nocardia* sp. AM 44	BACHOFER et al., 1975
			Pseudomonas sp.	WALKER and HARRIS, 1969
$R_1=R_2=H$ Aniline (Carboxin, F[a]) Propachlor, H[b])	Catechol (A)		*Pseudomonas* sp. S 9	LATORRE, 1982
	2-Hydroxy-6-oxo-*cis,cis*-2,4-hexadienoic acid	*m*-Cleavage of A	*Pseudomonas* sp. S 9	LATORRE, 1982
$R_1=Cl$ 4-Chloroaniline (Monuron, Monolinuron, H)	4-Chlorocatechol (A)	Dioxygenative deamination	*Pseudomonas multivorans* An 1	REBER et al., 1979
			Pseudomonas sp. S 9	LATORRE, 1982
	4-Chlorocatechol (A)		*Pseudomonas* OCA-1	
	2-Chloro-*cis,cis*-1,3-butadiene-1,4-carboxylic acid (B)	*o*-Cleavage of A	*Pseudomonas* OCA-1	
	5-Carboxymethylene-2-oxo-2,5-dihydrofuran	Lactonization of B with HCl elimination	*Pseudomonas* OCA-1	
	4-Chloroacetanilide	Acetylation	*Bacillus firmus*	WALLNÖFER et al., 1984
	4-Chloropropionanilide	Propionylation	*Bacillus firmus*	
	7-Chloro-2-amino-3H phenoxazine-3-one	n. d.[c]	*Bacillus firmus*	
	4,4′-Dichloroazobenzene	Oxidation	*Fusarium oxysporum* Schlecht	WALLNÖFER et al., 1984
	4,4′-Dichloroazoxybenzene	Oxidation	*Fusarium oxysporum* Schlecht	
	4,4′-Dichlorodiazoaminobenzene	n. d.	*Paracoccus* sp.	WALLNÖFER et al., 1984

$R_1=Br$ 4-Bromoaniline (Metobromuron, H)	4-Bromoacetanilide	Acetylation	*Talaromyces wortmannii*	WALLNÖFER et al., 1984
			Fusarium oxysporum Schlecht	
$R_1=R_2=Cl$ 3,4-Dichloroaniline (Diuron, Linuron, Propanil, H)	2,3-Dichloro-*cis,cis*-butadiene-1,4-dicarboxylic acid (A)	Dioxygenative deamination and *o*-cleavage	*Pseudomonas putida*	YOU and BARTHA, 1982
	5-Carboxymethylene-2-oxo-4-chloro-2,5-dihydrofuran (B)	Lactonization of A with HCl elimination	*Pseudomonas putida*	
	4-Oxo-3-chloropentanoic acid	Further degradation of B	*Pseudomonas putida*	
	3,3′,4,4′-Tetrachloroazobenzene	Oxidation	*Escherichia coli*	CORKE et al., 1979
	[3,4-Dichlorobenzenediazonium ion] (A)	Reaction with nitrite	*Escherichia coli*	
	1,3-Bis(3,4-dichlorophenyl)-triazene	Reaction of A with 3,4-dichloroaniline		
	3,3′,4,4′-Tetrachlorobiphenyl	From A		

[a] F Fungicide
[b] H Herbicide
[c] n.d. not determined

Table 4. Degradation of Phthalates

Compound	Transformation Product	Reaction Type	Microorganism	Reference
(benzene ring with CO_2R_1 and CO_2R_2) $R_1 = R_2 = n$-octyl; 2-ethylhexyl; *n*-butyl; methyl Dialkylphthalates	Monoalkylphthalate(s)	Hydrolysis	*Corynebacterium petrophilum* ATCC 19080 *Arthrobacter hydrocarboglutamicus* ATCC 15583 *Mycobacterium phlei* *Penicillium lilacinum*	ENGELHARDT et al., 1975
$R_1 = R_2 =$ 2-ethylhexyl; *n*-butyl; methyl	Phthalic acid (A)	Hydrolysis	*Nocardia erythropolis* S 1	KURANE et al., 1980
	3,4-Dihydroxybenzoic acid (B)	Dioxygenation and decarboxylation of A		
	3-Oxoadipic acid	*o*-Cleavage of B		
$R_1 = R_2 =$ methyl	3,4-Dihydroxybenzoic acid (A)	Hydrolysis, dioxygenation and decarboxylation	*Micrococcus* 12 B	KEYSER et al., 1976
	4-Oxo-2-carboxy-1-butene-1,4-dicarboxylic acid	*m*-Cleavage (4,5-) of A and oxidation		
$R_1 =$ H; $R_2 =$ butyl	Phthalic acid (A)	Hydrolysis	*Arthrobacter* sp. DSM 20389	ENGELHARDT and WALLNÖFER, 1978
	3,4-Dihydroxybenzoic acid	Dioxygenation and decarboxylation of A		
$R_1 = R_2 =$ H Phthalic acid	4,5-Dihydroxyphthalic acid (A)	Dioxygenation	*Pseudomonas testosteroni* NH 1000	NAKAZAWA and HAYASHI, 1978
	3,4-Dihydroxybenzoic acid (B)	Decarboxylation of A		
	6-Oxo-2-hydroxy-4-carboxy-2,4-hexadienoic acid	*m*-Cleavage (4,5-) of B		
	3-Hydroxyphthalic acid	Monooxygenation	*Corynebacterium* I P 4	HARADA and KOIWA, 1977
	3,4-Dihydroxybenzoic acid	Dioxygenation and decarboxylation		
	3,4-Dihydroxybenzoic acid (A)	Dioxygenation and decarboxylation	*Pseudomonas* sp.	RIBBONS and EVANS, 1960

	2-Carboxy-*cis,cis*-1,3-butadiene-1,4-dicarboxylic acid (B)	*o*-Cleavage of A		
	3-Oxoadipic acid	From B by *o*-cleavage pathway		
	3,4-Dihydroxybenzoic acid (A)	Dioxygenation and decarboxylation	*Nocardia* sp. DSM 43250	ENGELHARDT et al., 1976
	3-Oxoadipic acid	From A via *o*-cleavage pathway	*Nocardia* sp. DSM 43251 *Nocardia* sp. DSM 43252 *Arthrobacter* sp. DSM 20389 *Arthrobacter* sp. DSM 20390 *Alcaligenes* sp. DSM 30128	
HOOC COOH 1,3-Benzenedicarboxylic acid	3,4-Dihydroxybenzoic acid	Oxygenation, decarboxylation	*Pseudomonas testosteroni*	KEYSER et al., 1976
COOH HOOC 1,4-Benzenedicarboxylic acid	3,4-Dihydroxybenzoic acid	Oxygenation, decarboxylation	*Pseudomonas testosteroni*	KEYSER et al., 1976
			Nocardia sp. DSM 43250 *Nocardia* sp. DSM 43251 *Nocardia* sp. DSM 43252 *Arthrobacter* sp. DSM 20389 *Arthrobacter* sp. DSM 20390 *Alcaligenes* sp. DSM 30128	ENGELHARDT et al., 1976

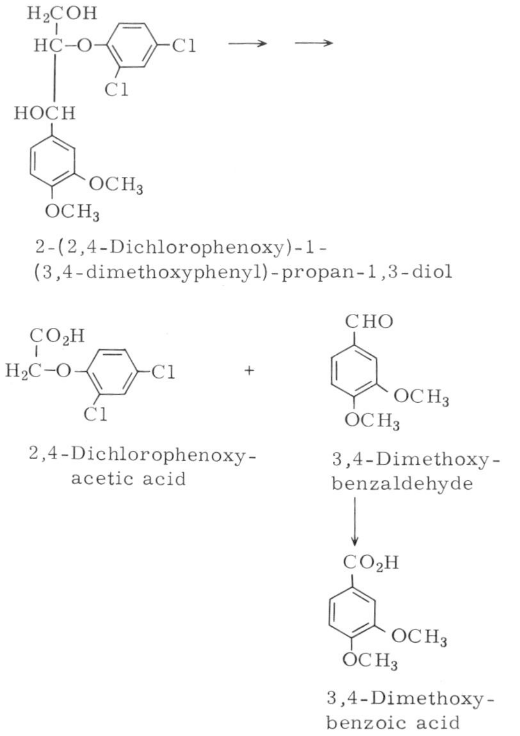

2-(2,4-Dichlorophenoxy)-1-(3,4-dimethoxyphenyl)-propan-1,3-diol

2,4-Dichlorophenoxy-acetic acid + 3,4-Dimethoxy-benzaldehyde

3,4-Dimethoxy-benzoic acid

Figure 14. Degradation of 2-(2,4-dichlorophenoxy)-1-(3,4-dimethoxyphenyl)-propan-1,3-diol by *Corynebacterium equi* ATCC 6939.

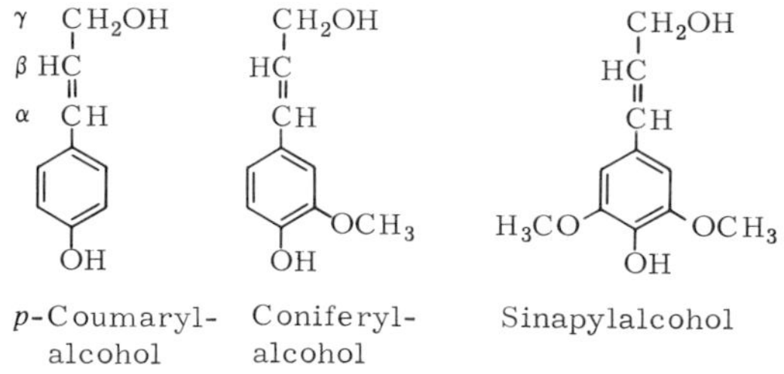

p-Coumaryl-alcohol Coniferyl-alcohol Sinapylalcohol

Figure 15. Lignin precursors.

acids. In principle, the sulfite cooking consists of subjecting wood to the action of sulfurous acid at elevated temperatures for a sufficiently long time.

Because of the very complex structure of lignin, the lack of regularity in structural units, and reactions and interactions between different components of the wood during cooking, a detailed knowledge of the structure of lignosulfonates (LS) is far from complete. In spite of this situation, at least the main routes of chemical alterations of lignin are known. These have been elucidated by use of adequate model substances representing characteristic bonding types of lignin. Today it is generally accepted that lignin results from enzymatic dehydrogenation of a mixture of three *p*-hydroxycinnamyl alcohols (Fig. 15). Therefore, phenylpropane units are considered as the basic elements in lignin structure.

All investigations on the microbial degradation of lignosulfonates clearly show that the acidic sulfite process converts lignin into derivatives which show increased resistance against microbial attack compared to lignin itself. Obviously there must be a correlation between biological availability and structural changes in the lignin. Because the introduction of the sulfonic group into the polymeric lignin is the main chemical reaction, it is reasonable to consider sulfonation as an important hindrance to biological degradation.

Single strains of microorganisms or mixed cultures have been isolated by enrichment techniques which very often need long periods to adapt to LS. According to PANDILA (1973) microorganisms are able to remove about 15–20% of LS through assimilation or merely by adsorption on the surface of the organism within a reasonable time. It also becomes clear that low molecular weight LS are more susceptible to microbial attack. Soil fungi and bacteria appear to be more effective in degrading LS than white rot fungi. Because in some of the investigations LS have been used as the sole carbon source, many fungi seemed to be unable to degrade them. Until now it is generally believed that most of the fungi need an additional carbon source which enables them to degrade lignin via co-oxidation (HÜTTERMANN et al., 1977).

It may be assumed that most of the microorganisms known to degrade lignin might also be able to degrade LS to some extent. As has been elucidated by CRAWFORD and CRAWFORD (1980) there are differences in lignin-degrading abilities observed for different organisms which could reflect differences in the completeness of their lignolytic enzyme system and could also indicate differences in the mode of the attack on lignin. It is reasonable to expect mixed cultures or natural microfloras to bring about most effective degradation of LS. This will be profitable for practical applications.

c) Surfactants

In spite of large production and extensive use by industrialized societies modern formulations of surfactants rarely resist biodegradation (CAIN, 1981). This phenomenon invites explanations of how the competent microflora of soils and natural waters have acquired the ability to break down these xenobiotic materials.

In the case of the alkyl sulfates and linear alkylbenzene sulfonates the available evidence would suggest that the organisms use their enzymes for lipid catabolism to affect ω- and β-oxidation (and in some rarer cases α-oxidation) of the alkyl chains. Naturally occurring sulfate esters have been shown to act as inducers of the sulfatases involved in catabolism of detergent alkyl sulfates. There is also recent evidence to support the view that desulfonation of the aromatic ring in aryl sulfonates may be effected by modified aryl dioxygenases which normally attack an unsubstituted aromatic nucleus; this probably occurs with alkylbenzene sulfonates, as well. The enzymes of desulfonation and ring fission are sometimes plasmid-encoded.

d) Dyes

In all areas the work on biodegradation of dyes has concentrated on azo dyes. With a few exceptions these chemicals do not appear to constitute an environmental hazard. Biodegradation is considered from various points of view: metabolism in mammals in the context of food and drug colorants, the fate of color released into nature, and removal of dyes from colored industrial effluents in water-purification plants. Mammalian metabolism of the dyes was found to take place in the liver as well as in microorganisms of the gastro-intestinal tract. Experiments with sewage sludge and other mixed cultures revealed that a reduction of the azo structures to the corresponding amines is relatively easily achieved under anaerobic conditions. The same is true for a number of pure cultures isolated from sludge and other sources. With oxygen present, degradation does not spontaneously occur. Long adaptation periods are necessary and enzymes which develop under these conditions are very specific.

Biodegradation of azo compounds can occur in an aerobic and in an anaerobic system (Fig. 16). Anaerobic decoloration is relatively easy to achieve and is accomplished by a number of microorganisms with rather non-specific enzymes. The first step of degradation in both systems is a reductive fission of the azo group.

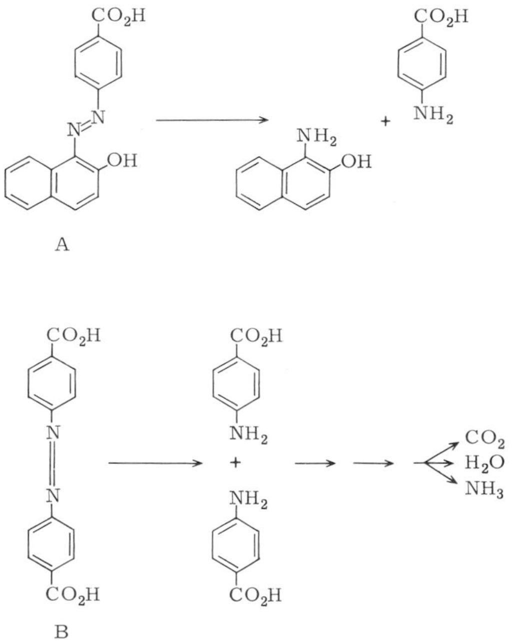

Figure 16. Microbial degradation af azo dyes. Anaerobic degradation of orange II by *Bacillus cereus* (A); aerobic degradation of 4,4′-dicarboxyazobenzene by *Flavobacterium* sp. (B).

Under anaerobic conditions no further decomposition was observed, whereas in the presence of oxygen the aromatic metabolites formed are metabolized (KULLA, 1981).

Table 5. Degradation of Biphenyl, Polychlorinated Biphenyls, and Polycyclic Hydrocarbons

Compound	Transformation Product	Reaction Type	Microorganism	Reference
Biphenyl	2,3-Dihydroxybiphenyl (A)	Dioxygenation	*Arthrobacter simplex*	TITTMANN and LINGENS, 1980
	2-Hydroxy-6-oxo-6-phenyl-2,4-hexadienoic acid (B)	*m*-Cleavage (2,3) of A	*Arthrobacter simplex*	
	Benzoic acid	Further oxidation of B	*Arthrobacter simplex*	
	4-Hydroxybiphenyl	Monooxygenation	*Oszillatoria* sp. strain JCM	CERNIGLIA et al., 1980
$R_3 = Cl$ 4-Chlorobiphenyl	4-Chloro-4′-hydroxybiphenyl	Monoxygenation	*Rhizopus japonicus* ATCC 24794	WALLNÖFER et al., 1973
	4-Chlorobenzoic acid	n. d.[a)]	*Achromobacter* sp.	AHMED and FOCHT, 1973
$R_3 = R_3' = Cl$ 4,4′-Dichlorobiphenyl	3-Chloro-2-hydroxy-6-oxo-6-(4-chlorophenyl)-2,4-hexadienoic acid	Dioxygenation and *m*-cleavage (2,3)	*Alcaligenes* sp.	FURUKAWA and MATSUMURA, 1976
$R_1 = R_3 = R_3' = Cl$ 2,4,4′-Trichlorobiphenyl	3-Chloro-2-hydroxy-6-oxo-6-(2,4-dichlorophenyl)-2,4-hexadienoic acid	Dioxygenation and *m*-cleavage (2,3)	*Alcaligenes* sp.	FURUKAWA and MATSUMURA, 1976

$R_1 = R_4 = R'_3 = Cl$ 2,5,4'-Trichlorobiphenyl	3-Chloro-2-hydroxy-6-oxo-6-(2,5-dichlorophenyl)-2,4-hexadienoic acid	Dioxygenation and *m*-cleavage (2,3)	*Alcaligenes* sp.	FURUKAWA and MATSUMURA, 1976
$R_1 = R_2 = R_3 = R_4 = R'_2 = Cl$ 2,3,3',4,5-Pentachlorobiphenyl	2,3,4,5-Tetrachlorobenzoic acid	n.d.	gram⁻ Bacteria gram⁺ Cocci from soil	BALLSCHMITTER et al., 1977
$R_1 = R_2 = R_3 = R_4 = R'_3 = Cl$ 2,3,4,4',5-Pentachlorobiphenyl	2,3,4,5-Tetrachlorobenzoic acid	n.d.	gram⁻ Bacteria gram⁺ Cocci from soil	BALLSCHMITTER et al., 1977
Benzo[*a*]anthracene	*cis*-1,2-Dihydroxy-1,2-dihydrobenzo[*a*]anthracene	Dioxygenation and hydrogenation	*Beijerinckia* sp.	GIBSON et al., 1975
Benzo[*a*]pyrene	*cis*-9,10-Dihydroxy-9,10-dihydrobenzo[*a*]pyrene	Dioxygenation and hydrogenation	*Beijerinckia* sp.	GIBSON et al., 1975

[a] n.d. not determined

e) Aromatics released during combustion

Waste products from chemical manufacture or technological processes involving combustion (burning of coal, gasoline, or diesel engines) usually contain large numbers of different organic compounds, most of which have not yet been identified. But even "defined" industrial products are often complex mixtures. For example, polychlorinated biphenyls (PCBs) consist of at least a hundred chlorobiphenyls (of the 209 possible products). In addition, a number of impurities have been found. Furthermore, conversion products in the environment (photoproducts, metabolites) from the PCB components raise the total number of known products associated with PCB in the environment to almost 200.

For PCBs the biodegradation studies produced relatively clear-cut results. The rate of biodegradation decreases with increasing chlorine content of the preparation; it is very low for most tetrachlorobiphenyls; pentachlorobiphenyls and more chlorinated biphenyls are practically nonbiodegradable.

Table 5 presents the different reactions of microorganisms on aromatic industrial pollutants.

B. Heterocyclic Compounds

Since an excellent review on the microbial transformation of heterocyclic compounds, particularly of natural origin, is presented by KIESLICH (1976), the Tables 6, 7, and 8 show only some examples for the degradation of heterocyclic compounds of both industrial and ecotoxicological importance.

Among the five membered heterocyclic substances especially furan-, benzimidazole-, benzthiazole, and triazole derivatives are of interest, since these chemicals are used as pesticides in agricultural environments (Table 6).

Six membered heterocyclic compounds with special ecotoxicological relevance are the chlorinated dibenzo-*p*-dioxins which may be released accidently during the production of chlorophenols (Table 7).

Another class of chemicals showing six membered heterocyclic ring systems are the symmetrical and asymmetrical triazines which constitute an important group of pesticides, widely applied as soil herbicides. Ring cleavage of these compounds which are commonly very resistant to microbial attack has recently been described (Table 8). The heterocyclic ring of the 1,2,4-triazinone herbicide metamitron is thought to be opened hydrolytically between the carbonyl group and the substituted amino group by a reaction similar to the hydrolytic cleavage of uracil.

V. Application and Economical Importance

A. Use of Microorganisms and Microbial Enzymes for the Degradation of Xenobiotics

It has long been recognized that with an increasing production of chemicals there is also the problem of environmental pollution. Unwanted discharge of xenobiotics containing aromatic or heteroaromatic structures into the environment can occur from numerous sources. Primarily, xenobiotic compounds applied, e. g., to agricultural areas which through drift, water runoff, or soil erosion do not remain at the target site undesirably enter the environment and cause pollution. The amount of xenobiotics thus annually entering the environment is difficult to quantify and nothing can be done microbiologically to reduce this adverse input.

Another major source of undesired chemical discharge occurs at the pesticide production and formulation plants as well as in any other plant producing commercial chemicals or drugs.

With regard to controlling discharges from production plants, chemical procedures such as solvent extraction or distillation coupled to conventional biological treatment processes are generally not able to meet zero-discharge requirements. To meet this level or low-ppb discharge new technologies are required. Biological systems involving cell-free enzymatic degradation of undesirable chemicals may represent a promising new technology for disposal of discharges.

The ability of particular bacteria and fungi to change the parent molecule into less complex metabolites which may then be further metabolized by the same organism or by a wider range of secondary microorganisms existing in the soil or water environment, plays an important role in the degradation of xenobiotic chemicals. Generally, once the initial enzymatic degradative step has occurred, the substance loses its extreme toxicity and the metabolites formed can be degraded by a wider range of microorganisms.

In research involving development of cell-free enzyme systems for the detoxification of xenobiotics, the initial enzymatic reaction with the parent molecule is of great importance. If enzymes can be found which can greatly decrease the toxicity of the starting compound by a simple hydrolysis or other enzymatic reaction, then these enzymes could possibly be used as potent tools for the detoxification of xenobiotics.

The possibility of obtaining cell-free enzymes capable of detoxifying various classes of aromatic and heterocyclic compounds has been described (MUNNECKE, 1981). Many enzymes responsible for aromatic metabolism are encoded on plasmids as mentioned above. Research can thus make use of genetic engineering techniques: Multiple gene copies, multiple expression, and movement of enzymes to a more suitable industrial strain may lead to a potential application of aromatic and heteroaromatic degrading enzymes, e. g., in soil-spill cleanup and waste water treatment.

B. Use of Microorganisms for Bioconversion of Wastes

In view of the energy consumption in industry and the expected aggravation of the energy situation, the great potential of dissolved wood substance and solid lignocellulosic wastes in the pulp industry may be used for the generation of energy.

Industry expects progress in microbiology to contribute to waste-water cleanup. On a global scale, this is a matter of some 4 million tons of organic degradation products of lignin that cannot be eliminated by degradation or adsorption, even if the latest technological achievements, including biological waste-water treatment, were introduced all over the world.

Very positive activities have been taken up in the past decade on the potential applications of ligninolytic systems for treatment of wastes and effluents from the pulp industry.

In the past years, intensive work has been conducted on ligninolytic systems. However, surveys by KIRK et al. (1980), and CRAWFORD and CRAWFORD (1980) indicate that the biochemistry of lignin degradation still remains largely unexplained. These surveys show that biodegradation of lignin is probably an aerobic process. Furthermore, it seems that a group of soil bacteria, including actinomycetes, will degrade lignin to a limited extent. Thus, it is possible that the pulp industry will soon have a new group of microorganisms at its disposal for special purposes.

Table 6. Degradation of 5-Membered Heterocyclic Compounds with one or two Heteroatoms

Compound	Transformation Product	Reaction Type	Microorganism	Reference
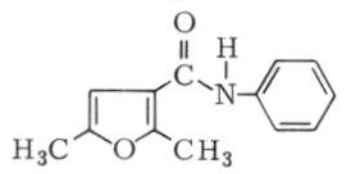 2,5-Dimethyl-3-furancarboxanilide, F[a]	Aniline (A)	Hydrolysis	*Bacillus sphaericus* ATCC 12123	WALLNÖFER et al., 1984
	2,5-Dimethyl-3-furancarboxylic acid	Hydrolysis		
	Acetanilide	Acetylation of A	*Aspergillus niger* ATCC 36782	WALLNÖFER et al., 1984
	2-Hydroxymethyl-5-methyl-3-furancarboxanilide	Monooxygenation	*Rhizopus japonicus* ATCC 24794; *Rhizopus peka* ATCC 24796; *Rhizopus nigricans* ATCC 24795	WALLNÖFER et al., 1984
	5-Hydroxymethyl-2-methyl-3-furancarboxanilide	Monooxygenation		
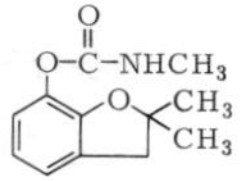 (2,3-Dihydro-2,2-dimethyl-7-benzofuranyl)-methylcarbamate (Carbofuran, I[b])	(3-Hydroxy-2,3-dihydro-2,2-dimethyl-7-benzofuranyl)-methylcarbamate	Monooxygenation	*Penicillium* sp.	WALLNÖFER et al., 1984
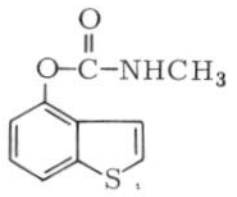 (Benzo[*b*]thiophen-2-yl)-methylcarbamate (Mobam, I)	4-Hydroxy-benzo[*b*]thiophene	Hydrolysis	Rumen bacteria	WALLNÖFER et al., 1984

Dibenzofuran	2-Hydroxydibenzofuran	Monooxygenation	*Cunnighamella elegans*	CERNIGLIA et al., 1979
	3-Hydroxydibenzofuran	Monooxygenation	*Cunnighamella elegans*	
	trans-2,3-Dihydroxy-2,3-dihydro-dibenzofuran (A_1)	Monooxygenation to form arene oxide and subsequent addition of water	*Cunnighamella elegans*	
	2,3-Dihydroxydibenzofuran	Dehydrogenation of A_1	*Cunnighamella elegans*	
	cis-1,2-Dihydroxy-1,2-dihydrodibenzofuran (A_2)	Dioxygenation and hydrogenation	*Beijerinckia* sp.	CERNIGLIA et al., 1979
	cis-2,3-Dihydroxy-2,3-dihydrodibenzofuran (A_3)	Dioxygenation and hydrogenation	*Beijerinckia* sp.	
	1,2-Dihydroxydibenzofuran	Dehydrogenation of A_2	*Beijerinckia* sp.	
	2,3-Dihydroxydibenzofuran	Dehydrogenation of A_3	*Beijerinckia* sp.	
Methyl-(benzimidazole-2-yl)-carbamate (Carbendazim, F)	(5-Hydroxybenzimidazole-2-yl)-methylcarbamate	Monooxygenation	*Aspergillus nidulans*	WALLNÖFER et al., 1984
3-(Benzo-1,3-thiazole-2-yl)-1,1-dimethylurea (Metabenzthiazuron, H[c])	3-(Benzo-1,3-thiazole-2-yl)-1-methylurea	Demethylation	*Cunninghamella echinulata* Thaxter ATCC 38447; *Hypocrea Cf. pilulifera*	WALLNÖFER et al., 1984
	3-(6-Hydroxybenzo-1,3-thiazole-2-yl)-1,3-dimethylurea	Monooxygenation	*Cunnighamella echinulata* Thaxter ATCC 38447	WALLNÖFER et al., 1984
	3-(Benzo-1,3-thiazole-2-yl)-3-(hydroxymethyl)-1-methylurea	Monooxygenation	*Hypocrea Cf.pilulifera*	WALLNÖFER et al., 1984
	1-(Benzo-1,3-thiazole-2-yl)-1-methylurea	Demethylation	*Hypocrea Cf.pilulifera*	WALLNÖFER et al., 1984
3-Amino-1,2,4-triazole (Amitrol, H)	3-(3-Amino-1,2,4-triazole-1-yl)-alanine	n. d.[d]	*Escherichia coli*	WALLNÖFER et al., 1984

[a] F Fungicide [b] I Insecticide [c] H Herbicide [d] n. d. not determined

Table 7. Degradation of 6-Membered Heterocyclic Compounds with one or two Heteroatoms

Compound	Transformation Product	Reaction Type	Microorganism	Reference
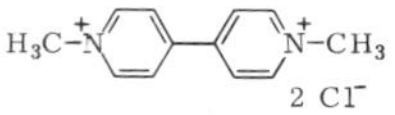 1,1′-Dimethyl-4,4′-bipyridylium dichloride (Paraquat, H[a])	1-Methyl-4-carboxypyridinium ion	n. d.[b]	*Pseudomonas fluorescens* *Aerobacter aerogenes* *Streptomyces* *Nocardia*	WALLNÖFER et al., 1984
	1-Methyl-4,4′-bipyridylium ion	n. d.[b]	Soil bacteria	WALLNÖFER et al., 1984
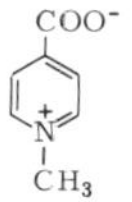 1-Methyl-4-carboxypyridinium ion	2-Hydroxy-1-methyl-4-carboxy-pyridinium ion (A)	Monoxygenation	*Achromobacter*	CALLELY, 1978
	2-Hydroxy-4-pyridinecarboxylic acid (B)	Demethylation of A		
	2,6-Dihydroxy-4-pyridinecarboxylic acid (C)	Monooxygenation of B		
	Malic acid monoamide	Ring cleavage of C		
	1-Methyl-1,4-dihydro-4-pyridinecarboxylic acid (A)	Reduction	*Arthrobacter*	CALLELY, 1978
	1-(Formyl-methylamino)-4-oxo-1-butene (B)	Ring cleavage of B		
	1-Methylamino-4-oxo-1-butene	Further degradation of B		
	4-Methylamino-3-butenoic acid			
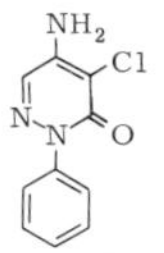 1-Phenyl-4-amino-5-chloropyridazone-(6) (Chloridazon, H)	1-(2,3-Dihydroxy-2,3-dihydrophenyl)-4-amino-5-chloropyridazone-(6) (A)	Dioxygenation and hydrogenation	"Chloridazon degrading bacteria"	EBERSPÄCHER and LINGENS, 1981
	1-(2,3-Dihydroxyphenyl)-4-amino-5-chloro-pyridazone-(6) (B)	Dehydrogenation of A	"Chloridazon degrading bacteria"	
	1-[4-Amino-5-chloropyridazone-(6)-1-yl]-*cis,cis*-1,3-butadiene-1,4-dicarboxylic acid (C)	*o*-Cleavage of B	"Chloridazon degrading bacteria"	

Compound	Metabolite	Reaction	Organism	Reference
	4-Amino-5-chloro-1H-pyridazone-(6)	Further degradation of C	"Chloridazon degrading bacteria"	
$R_n = H$ Dibenzo-*p*-dioxin	*cis*-1,2-Dihydroxy-1,2-dihydrobenzo-*p*-dioxin (A)	Dioxygenation	*Pseudomonas* sp. NCIB 9816 *Beijerinckia* B 8/36	KLEČKA and GIBSON, 1980
	1,2-Dihydroxydibenzo-*p*-dioxin	Dehydrogenation of A	*Beijerinckia* B 8/36	KLEČKA and GIBSON, 1980
$R_1 = Cl$ 1-Chlorodibenzo-*p*-dioxin	*cis*-1,2-Dihydroxy-1,2-dihydro-1-chlorodibenzo-*p*-dioxin	Dioxygenation and hydrogenation	*Beijerinckia* B 8/36	KLEČKA and GIBSON, 1980
$R_2 = Cl$ 2-Chlorodibenzo-*p*-dioxin	*cis*-1,2-Dihydroxy-1,2-dihydro-2-chlorodibenzo-*p*-dioxin	Dioxygenation and hydrogenation	*Beijerinckia* B 8/36	KLEČKA and GIBSON, 1980
$R_2 = R_3 = R_7 = R_8 = Cl$ 2,3,7,8-Tetrachlorodibenzo-*p*-dioxin (TCDD)	"Phenolic" metabolite	n. d.	Soil bacteria	PHILIPP et al., 1981
2,3-Dihydro-5-carboxanilido-6-methyl-1,4-oxathiine (Carboxin, F[c])	4-Oxo-2,3-dihydro-5-carboxanilido-6-methyl-1,4-oxathiine (A)	Monooxygenation	*Rhizopus japonicus* ATCC 24794	WALLNÖFER et al., 1984
	4,4-Dioxo-2,3-dihydro-5-carboxanilido-6-methyl-1,4-oxathiine	Monooxygenation of A		
	Aniline	Hydrolysis	*Bacillus sphaericus* ATCC 12123	WALLNÖFER et al., 1984

[a] H Herbicide
[b] n. d. not determined
[c] F Fungicide

Table 8. Degradation of 6-Membered Heterocyclic Compounds with three Heteroatoms

Compound	Transformation Product	Reaction Type	Microorganism	Reference
[triazine ring structure with R1, R2, R3] $R_1=Cl$; $R_2=NHC_2H_5$; $R_3=NHCH(CH_3)_2$ 2-Chloro-4-ethylamino-6-isopropylamino-1,3,5-triazine (Atrazine, H[a])	2-Hydroxy-4-ethylamino-6-isopropylamino-1,3,5-triazine	Hydrolysis	*Fusarium roseum*	WALLNÖFER et al., 1984
	2-Chloro-4-amino-6-ethylamino-1,3,5-triazine	Dealkylation	*Aspergillus fumigatus*	
$R_1=OH$; $R_2=NH_2$; $R_3=NHCH(CH_3)_2$ 2-Hydroxy-4-amino-6-isopropylamino-1,3,5-triazine	2,4-Dihydroxy-6-isopropylamino-1,3,5-triazine	n. d.[b]	*Pseudomonas* sp. strain A	COOK and HÜTTER, 1981
$R_1=OH$; $R_2=NH_2$; $R_3=NHC_2H_5$ 2-Hydroxy-4-amino-6-ethylamino-1,3,5-triazine	2,4-Dihydroxy-6-(ethylamino)-1,3,5-triazine	n. d.	*Pseudomonas* sp. strain A	COOK and HÜTTER, 1981
$R_1=R_2=OH$; $R_3=NHCH(CH_3)_2$ 2,4-Dihydroxy-6-isopropylamino-1,3,5-triazine	CO_2; NH_3	n. d.	*Pseudomonas* sp. strain D	COOK and HÜTTER, 1981
$R_1=OH$; $R_2=R_3=NH_2$ 2-Hydroxy-4,6-diamino-1,3,5-triazine	CO_2; NH_3	n. d.	*Pseudomonas* sp. strain D, A	COOK and HÜTTER, 1981
$R_1=R_2=OH$; $R_3=NH_2$ 2,4-Dihydroxy-6-amino-1,3,5-triazine	CO_2; NH_3	n. d.	*Klebsiella pneumoniae* strain 90	COOK and HÜTTER, 1981
$R_1=R_2=R_3=OH$ 2,4,6-Trihydroxy-1,3,5-triazine	CO_2; NH_3	n. d.	*Pseudomonas* sp. strain A *Klebsiella pneunomiae* strain 90	COOK and HÜTTER, 1981

Compound	Metabolite	Reaction	Organism	Reference
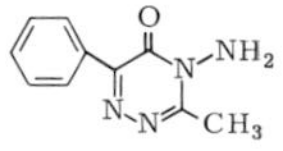 3-Methyl-4-amino-6-phenyl-1,2,4-triazine-5(4H)-one (Metamitron, H)	3-Methyl-4-amino-6-(2,3-dihydro-2,3-dihydroxyphenyl)-1,2,4-triazine-5(4H)-one (A)	Dioxygenation and hydrogenation	"Chloridazon degrading bacteria"	BLECHER et al., 1979
	3-Methyl-4-amino-6-(2,3-dihydroxyphenyl)-1,2,4-triazine-5(4H)-one (B)	Dehydrogenation of A	"Chloridazon degrading bacteria"	
	4-Amino-3-methyl-1,2,4-triazine-5(4H)-one	Further degration of B	"Chloridazon degrading bacteria"	
	3-Methyl-6-phenyl-1,2,4-triazine-5(4H)-one	Deamination	*Pseudomonas* sp.	WALLNÖFER et al., 1984
			Rhizopus japonicus ATCC 24794	
			Cunninghamella echinulata Th.	
	3-Methyl-6-phenyl-2,3-dihydro-1,2,4,5-tetrazine-2-carboxylic acid	n. d.	*Arthrobacter* sp. DSM 20389	ENGELHARDT et al., 1982
	Benzoylformic acid acetylhydrazone	Ring cleavage by hydrolysis	*Arthrobacter* sp. DSM 20389	
	Benzoylformic acid	Further degradation	*Arthrobacter* sp. DSM 20389	
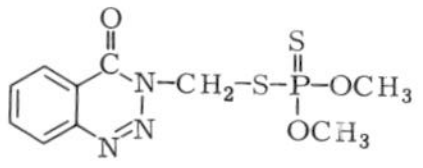 *O,O*-Dimethyl-*S*-(4-oxo-1,2,3-benzotriazine-3[4H]-yl-methyl)-phosphoro-dithioate (Azinphos Methyl, I[c])	Bis-(benzazimidyl)-methyl-disulfide	Hydrolysis and oxidative dimerization	*Pseudomonas fluorescens* DSM 1976	ENGELHARDT et al., 1981b
	Benzazimide	n. d.	*Pseudomonas fluorescens* DSM 1976	ENGELHARDT et al., 1981b
	Anthranilic acid	n. d.	*Pseudomonas fluorescens* DSM 1976	ENGELHARDT et al., 1981b
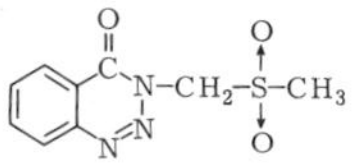 Methyl-(4-oxo-1,2,3-benzotriazine-3[4H]-yl-methyl)-sulfone	Anthranilic acid	n. d.	*Pseudomonas fluorescens* DSM 1976	ENGELHARDT et al., 1984
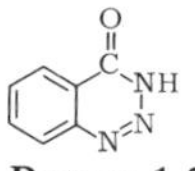 Benzo-1,2,3-triazine-4[3H]-one (Benzazimide)	5-Hydroxy-benzo-1,2,3-triazine-4[3H]-one	Monooxygenation	*Pseudomonas* sp. DSM 5030	ENGELHARDT et al., 1984

[a] Herbicide [b] n. d. not determined [c] I Insecticide

C. Use of Microorganisms for the Preparation of Important Intermediates in the Degradation of Aromatic and Heterocyclic Structures

1. Intermediates of industrial or pharmaceutical relevance

The catabolism of various aromatic and heterocyclic compounds by microorganisms often leads to "dead end" metabolites which accumulate in the culture medium. Such experiments can provide suitable tools for large scale preparation of industrially or pharmaceutically interesting compounds as for instance hydroxylated compounds, aromatic acids, etc. (Fig. 17).

a) Microbial hydroxylation of naphthyridines

Penicillium adametzi and seven other species convert nalidixic acid (1,4-dihydro-1-ethyl-7-methyl-4-oxo-1,8-naphthyridine-3-carboxylic acid) to 1,4-dihydro-1-ethyl-7-hydroxymethyl-4-oxo-1,8-naphthyridine-3-carboxylic acid (Fig. 17) (HAMILTON et al., 1969). This conversion by fungi is a finding with considerable potential importance, since the hydroxylated product has been reported to be of the same order of activity as the parent drug against Meningococci resistant to sulfonamide. Undoubtedly, some of the chemotherapeutic activity of nalidixic acid is attributed to its conversion to this active metabolite. Although this important metabolite can be synthesized chemically, the process is difficult and the yield is low. Therefore, a microbial transformation can be of considerable utility.

Large quantities of culture for the isolation of the 7-hydroxy metabolite are obtained by inoculating 1.0 liter of 48 h cultures in soy-dextrose medium into a 14-liter stirred jar fermenter. The fermenter contains 10.0 L of soy-dextrose medium autoclaved for 45 min at 121 °C. The inoculated medium is incubated at 30 °C with a filtered airflow of 2 L per min and with agitation of 450 r.p.m. with a double impellor. After 48 h 1.0 g of nalidixic acid in 50 mL of dimethylformamide is added, and the fermentation is continued for 24 h. The course of the fermentation is followed by chromatography of samples (HAMILTON et al., 1969). The fermentation is stopped by adding 200 mL of 10 N HCl.

The acidified fermentation beer is extracted three times with 4 L of dichloromethane. The extracts are combined and concentrated to about 100 mL. The crystalline product that separates is collected by filtration. This product is recrystallized three times from ethyl acetate and once from acetone to yield 0.6 g of chromatographically pure material melting at 256–257 °C.

b) Microbial preparation of 4-hydroxymethyl-3-chloroanilines

The aniline derivatives *N*-(2-diethylaminoethyl)-3-chloro-4-methyl-aniline and 1-(3-chloro-4-methylphenyl)-piperazine show a schistosomicidal effect in mice but apparently are quite inactive in monkeys and man. Their 4-hydroxymethyl derivatives, however, seem to be the therapeutically active forms as demonstrated by ROSI et al. (1967).

The special transformation reaction can be carried out by microorganisms as follows:

Microbial oxidation of 1-(3-chloro-4-methylphenyl)-piperazine hydrochloride (I)

Ten liters of a soy-dextrose medium of the following composition is used for fermentation: 1 kg of cerelose, 150 g of soybean meal, 50 g of yeast, 50 g of salt, 2.5 g of $MgSO_4 \cdot 7\,H_2O$, 13.8 g of $NaH_2PO_4 \cdot H_2O$, 301 g of $Na_2HPO_4 \cdot 12\,H_2O$, and tap water to make the final volume 10 L. The solution at pH 7.3 is sterilized. One seed flask containing *Aspergillus sclerotiorum* is added to the above solution which is stirred in a fermenter kept in a water bath at 28 °C. After 24 h the addition of I starts. About 10 g/day is added in two portions over a period of 5 days. The total amount of substrate consumed is 53 g. To the fermentation vessel 130 mL of 10 N NaOH is ad-

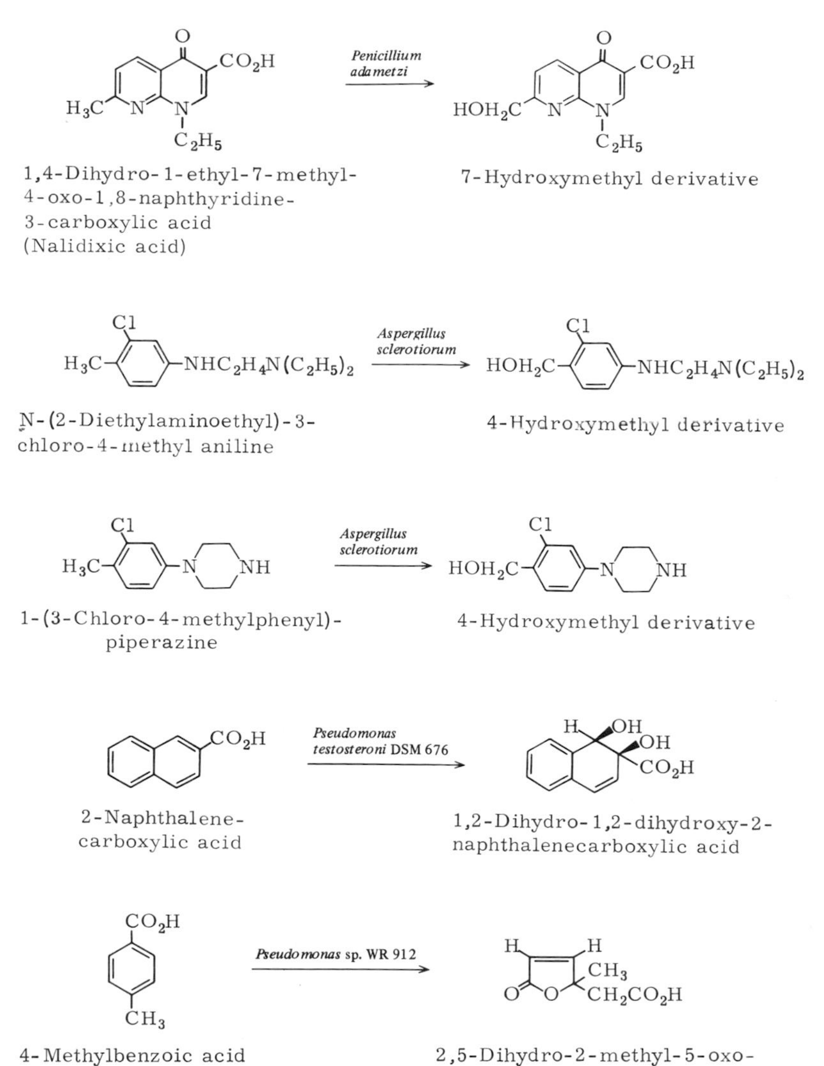

Figure 17. Important microbial intermediates in the degradation pathways of aromatic and heterocyclic structures.

ded to stop further growth of the organism. The vessel is extracted with two 20 L portions of CH_2Cl_2, and the extracts are partially concentrated *in vacuo* and evaporated to furnish a crystalline residue. This is recrystallized from 200 mL of ethyl acetate after the insoluble fraction is filtered off. The total yield of the transformation product 1-(3-chloro-4-hydroxymethylphenyl)-piperazine (m. p. 122 °C) is 73% of the theoretical estimate.

Microbial oxidation of N-(2-diethylaminoethyl)-3-chloro-4-methyl-aniline hydrochloride (II)

Essentially the same procedure is used here. A total of 161 g (0.58 mol) of substrate is converted over an 8-day period. The substance is extracted and concentrated as in the above procedure. About 93 g of *N*-(2-diethylaminoethyl)-3-chloro-4-hydroxymethyl-aniline (m. p. 66–67.5 °C) are obtained as white crystals.

c) Microbial oxidation of polynuclear aromatic hydrocarbons

The fermentation of polynuclear aromatic hydrocarbon substrates can be achieved with suitable oxidative ring splitting microorganisms in the presence of a carbohydrate co-substrate which is capable of supporting growth: This method results in the yield of a greater amount of oxidation products, e. g., yields up to 5 g per liter of broth. Compounds containing at least two condensed rings, including the halo and nitro derivatives thereof, are oxidized. Thus, for example, with glucose as co-substrate *Corynebacterium novum* ATCC 15 570 converts naphthalene into salicylic acid, phenanthrene into 1-hydroxy-2-naphthalenecarboxylic acid, and anthracene into 2-hydroxy-3-naphthalenecarboxylic acid (SUN OIL, 1967).

Since salicylic acid may be toxic to the microbial cells themselves, the production and accumulation of this product is limited by its own inhibitory effect on cell growth. A simple method is to remove the acid during fermentation by adsorption to an ion exchange resin added to the solution without giving any negative effect on the fermentation itself, a method which is possibly applicable to other similar fermentations.

Thus, cells of *Pseudomonas aeruginosa* B-344 are incubated in a medium containing NH_4Cl, 0.25%; K_2HPO_4, 2.0%; $MgSO_4 \cdot 7\,H_2O$, 0.05%; $MgCl_2 \cdot 5\,H_2O$, 0.02%; $FeSO_4 \cdot 7\,H_2O$, 0.005%; $CaCl_2 \cdot 2\,H_2O$, 0.05%; and a trace of KJ and $CuSO_4 \cdot 5\,H_2O$. Powdered naphthalene is sterilized by UV radiation and added at the time of inoculation in 2% concentration. All fermentations are conducted in a 500 mL flask containing 25–40 mL of the medium which is inoculated with 1 mL of a 24-h seed culture of *Pseudomonas aeruginosa* B-344. An aerobic condition is obtained by shaking reciprocally at a rate of 150 strokes/min with a 7 cm stroke. About 10 mL anion exchange resin (Amberlite IRA-400, 20–50 mesh), directly or packed in a cellophane tube (Visking. Co 20/32) in the shape of a sausage is autoclaved for 3 min at 105 °C and thrown into the culture 20 h after inoculation. At the same time 1.0 g of naphthalene is added. Culture shaking is continued for 48 h. Elution of salicylic acid, after washing the resin with distilled water, is carried out with a mixture of equal volumes of 2 N HCl and ethanol. Recovery of the acid by this elution should be more than 95%. From the directly applied resin 558.0 mg of salicylic acid is obtained which is 3.75 times the control levels (148.6 mg). When the resin is wrapped with a cellophane membrane, 817.2 mg of salicylic acid is recovered which is 5.47 times the control and indicates that direct contact of the resins with growing cells should be avoided (TONE et al., 1968).

2. Intermediates important for studies of selected biochemical reactions

a) (+)*cis*-1,2-Dihydroxy-1,2-dihydronaphthalenecarboxylic acid

During degradation of naphthalene by *Pseudomonas testosteroni* DSM 676 the intermediate *cis,cis*-dihydroxy-1,2-dihydronaphthalene is dehydrated enzymatically to

1,2-naphthalenediol which is further metabolized. The analogous oxidation of 2-naphthol and naphthalene-2-sulfonic acid leads to unstable intermediates from which 1,2-naphthalenediol is formed without dehydration by spontaneous hydroxy and hydrogen sulfite-elimination, respectively. 2-Naphthalenecarboxylic acid is also easily hydroxylated to the corresponding (+)*cis*-1,2-dihydroxy-1,2-dihydro-2-naphthalenecarboxylic acid. Since from this compound no spontaneous cleavage of formiate can take place, it accumulates in the culture medium (Fig. 17). For large scale preparation of this intermediate, salicylic acid is used as carbon source which does not inhibit the hydroxylation and serves as an inductor of all enzymes of the naphthalene degradation pathway as, e. g., the naphthalene dioxygenase.

Salicylic acid degrading cultures of *Pseudomonas testosteroni* DSM 676 co-oxidize 2 g of 2-naphthalenecarboxylic acid almost quantitatively to the corresponding (+)-*cis,*-1,2-dihydroxy-1,2-dihydronaphthalenecarboxylic acid. The acidic extraction of the culture medium with ethyl acetate, followed by crystallization out of ethylmethylketone results in colorless crystals with a yield of 74% and a melting point of 147 °C (KNACKMUSS et al., 1976).

b) Substituted 5-oxo-furan-2-acetic acid compounds

3,5-Dichlorobenzoate grown cells of *Pseudomonas* sp. WR 912 readily co-metabolize 3-methyl-, 4-methyl-, and 3,5-dimethylbenzoate. During incubation, the "dead-end" metabolites 2,5-dihydro-4-methyl-, 2,5-dihydro-2-methyl- (Fig. 17), and 2,5-dihydro-2,4-dimethyl-5-oxo-furan-2-acetic acid accumulate quantitatively in the culture fluid (HARTMANN et al., 1979).

VI. References

AFTRING, R. P., CHALKER, B. E., TAYLOR, B. F. (1981). Appl. Environ. Microbiol. *41*, 1177.

AHMED, M., and FOCHT, D. D. (1973). Can. J. Microbiol. *19*, 47.

BACHOFER, R., LINGENS, F., and SCHÄFER, W. (1975). FEBS Lett. *50*, 288.

BAKKER, G. (1977). FEMS Microbiol. Lett. *1*, 103.

BALBA, M. T., and EVANS, W. C. (1977). Biochem. Soc. Trans. *5*, 302.

BALLSCHMITTER, K., UNGLERT, CH., and NEU, H. J. (1977). Chemosphere *6*, 51.

BLECHER, R., KOCH, U., BALLHAUSE, B., and LINGENS, F. (1979). Z. Pflanzenkr. Pflanzenschutz *86*, 93.

BOYER, P. D. (Ed.) (1975). "The Enzymes", Vol. XII, 3rd Ed. Academic Press, New York–London.

CAIN, R. B. (1981). In "Microbial Degradation of Xenobiotics and Recalcitrant Compounds" (T. LEISINGER, A. M. COOK, R. HÜTTER, and J. NÜESCH, eds.), p. 325. Academic Press, London.

CALLELY, A. G. (1978). In "Progress of Industrial Microbiology" *14* (M. J. BULL, ed.), p. 205. Elsevier, Amsterdam–Oxford–New York.

CERNIGLIA, C. E., VANBAALEN, CH., and GIBSON, D. T. (1980). Arch. Microbiol. *125*, 203.

CERNIGLIA, C. E., MORGAN, J. C., and GIBSON, D. T. (1979). Biochem. J. *180*, 175.

CHAPMAN, P. J. (1972). In "Degradation of Synthetic Organic Molecules in the Biosphere". Proc. Conference, Natl. Acad. Sci., Washington, D. C., p. 17.

COOK, A. M., and FEWSON, C. A. (1972). Biochem. Biophys. Acta *290*, 384.

COOK, A. M., and HÜTTER, R. (1981). J. Agric. Food Chem. *29*, 1135.

CORKE, C. T., BUNCE, N. J., BEAUMONT, A.-L., and MERRICK, R. L. (1979). J. Agric. Food Chem. *27*, 644.

CRAWFORD, D. L., and CRAWFORD, R. L. (1980). Enzyme Microbiol. Technol. *2*, 11.

DAGLEY, S. (1972). In "Degradation of Synthetic Organic Molecules in the Biosphere". Proc. Conference, Natl. Acad. Sci., Washington, D. C., p. 1.

DAGLEY, S. (1978a). In "The Bacteria" (L. N. ORNSTON, J. R. SOKATCH, eds.), Vol. VI, p. 305. Academic Press, London–New York.

DAGLEY, S. (1978b). Naturwissenschaften *65*, 85.

DAGLEY, S. (1978c). Q. Rev. Biophys. *11*, 577.

DEFRANK, J. J., and RIBBONS, D. W. (1977). J. Bacteriol. *129*, 1365.

DUTTON, P. L., and EVANS, W. C. (1970). Arch. Biochem. Biophys. *136*, 288.

EBERSPÄCHER, J., and LINGENS, F. (1981). In "Microbial Degradation of Xenobiotics and

Recalcitrant Compounds" (T. LEISINGER, A. M. COOK, R. HÜTTER, and J. NÜESCH, eds.), p. 271. Academic Press, London.

ENGELHARDT, G., WALLNÖFER, P. R., and HUTZINGER, O. (1975). Bull. Environ. Cont. *3*, 342.

ENGELHARDT, G., WALLNÖFER, P. R., RAST, H. G., and FIEDLER, F. (1976). Arch. Microbiol. *109*, 109.

ENGELHARDT, G., RAST, H. G., and WALLNÖFER, P. R. (1977). Arch. Microbiol. *114*, 25.

ENGELHARDT, G., and WALLNÖFER, P. R. (1978). Appl. Environ. Microbiol. *35*, 243.

ENGELHARDT, G., RAST, H. G., and WALLNÖFER, P. R. (1979). FEMS Microbiol. Lett. *5*, 377.

ENGELHARDT, G., WALLNÖFER, P. R., and RAST, H. (1981a). In: "Microbial Degradation of Xenobiotics and Recalcitrant Compounds" (T. LEISINGER, A. M. COOK, R. HÜTTER, and J. NÜESCH, eds.), p. 293. Academic Press, London.

ENGELHARDT, G., ZIEGLER, W., and WALLNÖFER, P. R. (1981b). FEMS Microbiol. Lett. *11*, 165.

ENGELHARDT, G., ZIEGLER, W., WALLNÖFER, P. R., JARCZYK, H. J., and OEHLMANN, L. (1982). J. Agric. Food Chem. *30*, 278.

ENGELHARDT, G., OEHLMANN, L., WAGNER, K., WIEDEMANN, M., and WALLNÖFER, P. R. (1984). J. Agric. Food. Chem., in press.

ENGESSER, K.-H., SCHMIDT, E., and KNACKMUSS, H.-J. (1980). Appl. Environ. Microbiol. *39*, 68.

EVANS, W. C. (1977). Nature *270*, 17.

FERRY, J. G., and WOLFE, R. S. (1976). Arch. Microbiol. *107*, 33.

FEWSON, C. A. (1981). In "Microbial Degradation of Xenobiotics and Recalcitrant Compounds" (T. LEISINGER, A. M. COOK, R. HÜTTER, and J. NÜESCH, eds.), p. 141. Academic Press, London.

FINA, L. R., BRIDGES, R. L., COBLENTZ, T. H., and ROBERTS, F. F. (1978). Arch. Microbiol. *118*, 169.

FRANKLIN, F. G. H. (1981). In "Microbial Degradation of Xenobiotics and Recalcitrant Compounds" (T. LEISINGER, A. M. COOK, R. HÜTTER, and J. NÜESCH, eds.), p. 109. Academic Press, London.

FURUKAWA, K., and MATSUMURA, F. (1976). J. Agric. Food Chem. *24*, 251.

GIBSON, D. T. (1971). CRC Crit. Rev. Microbiol. *1*, 199.

GIBSON, D. T. (1972). In "Degradation of Synthetic Organic Molecules in the Biosphere". Proc. Conference, Natl. Acad. Sci., Washington, D. C., p. 116.

GIBSON, D. T., MAHADEVAN, V., JERINA, D. M., YAGI, H., and YEH, H. J. C. (1975). Science *189*, 295.

GUPTA, R. C., KAUL, S. M., and SHUKLA, O. P. (1975). Ind. J. Biochem. Biophys. *12*, 263.

GUYER, M., and HEGEMAN, G. (1969). J. Bacteriol. *99*, 906.

HAMILTON, P. B., ROSI, D., PERUZZOTTI, G. P., and NIELSON, E. D. (1969). Appl. Microbiol. *17*, 237.

HARADA, T., and KOIWA, S. (1977). J. Ferment. Technol. *55*, 97.

HARDER, W., and DIJKHUIZEN, L. (1976). In "Continuous Culture 6: Applications and New Fields" (A. C. R. DEAN, D. C. ELLWOOD, C. G. T. EVANS, and J. MELLING, eds.), p. 297. Ellis Harwood, Chichester–Oxford.

HARTMANN, J., REINEKE, W., and KNACKMUSS, H.-J. (1979). Appl. Environ. Microbiol. *37*, 421.

HAYAISHI, O. (ed.) (1974). "Molecular Mechanisms of Oxygen Activation". Academic Press, New York–London.

HEALY, J. B., JR., and YOUNG, L. Y. (1979). Appl. Environ. Microbiol. *38*, 84.

HIGGINS, S. J., and MANDELSTAM, J. (1972). Biochem. J. *126*, 917.

HILL, I. R. (1978). In "Pesticide Microbiology" (I. R. HILL and S. J. L. WRIGHT, eds.), p. 137. Academic Press, London.

HOU, C. T., PATEL, R., and LILLARD, M. O. (1977). Appl. Environ. Microbiol. *33*, 725.

HÜTTERMANN, A., GEBAUER, M., VOLGER, CH., and RÖSGER, CH. (1977). Holzforschung *31*, 83.

JAGNOW, G., HAIDER, K., and ELLWARDT, P.-CHR. (1977). Arch. Mikrobiol. *115*, 285.

JANKE, D., and FRITSCHE, W. (1978). Z. Allg. Mikrobiol. *18*, 365.

KASPARI, H. (1979). Microbiologica (Bologna) *2*, 231.

KEITH, C. L. (1972). Diss. Abstr. Intern. B *33*, 3214.

KEITH, C. L., BRIDGES, R. L., FINA, L. R., IVERSON, K. L., and CLORAN, J. A. (1978). Arch. Microbiol. *118*, 173.

KEYSER, P., PUJAR, B. G., EATON, R. W., and RIBBONS, D. W. (1976). Environ. Health Perspectives *18*, 159.

KIESLICH, K. (1976). "Microbial Transformations of Non-Steroid Cyclic Compounds". Georg Thieme Verlag, Stuttgart.

KIRK, T. K., HIGUCHI, T., and CHANG, H. M. (eds.) (1980). In "Lignin Degradation". Vol. *2*, p. 235. CRC-Press, Boca Raton, Florida.

KLEČKA, G. M., and GIBSON, D. T. (1980). Appl. Environ. Microbiol. *39*, 288.

KNACKMUSS, H.-J., BECKMANN, W., and OTTING, W. (1976). Angew. Chem. *88*, 581.

KNACKMUSS, H.-J. (1981). In "Microbial Degradation of Xenobiotics and Recalcitrant Compounds" (T. LEISINGER, A. M. COOK, R. HÜTTER, and J. NÜESCH, eds.), p. 189. Academic Press, London.

KULLA, H. G. (1981). In "Microbial Degradation of Xenobiotics and Recalcitrant Compounds" (T. LEISINGER, A. M. COOK, R. HÜTTER, and J. NÜESCH, eds.), p. 387. Academic Press, London.

KURANE, R., SUZUKI, T., and TAKAHARA, Y. (1980). Agric. Biol. Chem. *44*, 523.

LATORRE, I. (1982). Ph. Doctoral Thesis, Kiel, W. Germany.

MUNNECKE, D. M. (1981). In "Microbial Degradation of Xenobiotics and Recalcitrant Compounds" (T. LEISINGER, A. M. COOK, R. HÜTTER, and J. NÜESCH, eds.), p. 251. Academic Press, London.

NAKAZAWA, T., and HAYASHI, E. (1978). Appl. Environ. Microbiol. *36*, 264.

NOTTINGHAM, P. M., and HUNGATE, R. E. (1969). J. Bacteriol. *98*, 1170.

OHISA, N., YAMAGUCHI, M., and KURIHARA, N. (1980). Arch. Microbiol. *125*, 221.

ORNSTON, L. N., and PARKE, D. (1977). Curr. Top. Cell. Regul. *12*, 210.

OSHIMA, T. (1965). Z. Allg. Mikrobiol. *5*, 386.

PANDILA, M. M. (1973). Paper Magazine Can. *74*, 80.

PHILIPP, M., KRASNOBAJEW, Y., ZEYER, J., and HÜTTER, R. (1981). In "Microbial Degradation of Xenobiotics and Recalcitrant Compounds" (T. LEISINGER, A. M. COOK, R. HÜTTER, and J. NÜESCH, eds.), p. 221. Academic Press, London.

QUE, L., JR. (1978). Biochem. Biophys. Res. Commun. *84*, 123.

RAST, H. G., ENGELHARDT, G., WALLNÖFER, P. R., OEHLMANN, L., and WAGNER, K. (1979). J. Agric. Food Chem. *27*, 699.

REBER, H., HELM, V. KARANTH, N. G. K. (1979). Eur. J. Appl. Microbiol. Biotechnol. *7*, 181.

RIBBONS, D. W., and EVANS, W. C. (1960). Biochem. J. *76*, 310.

ROGERS, J. E., and GIBSON D. T. (1977). J. Bacteriol. *130*, 1117.

ROSI, D., PERUZZOTTI, G., DENNIS, E. W., BERBERIAN, D. A., FREELE, H., TULLAR, B. F., and ARCHER, S. (1967). J. Med. Chem. *10*, 867.

SCHREIBER, A., HELLWIG, M., DORN, E., REINECKE, W., and KNACKMUSS, H.-J. (1980). Appl. Environ. Microbiol. *39*, 58.

SHARPEE, K. W., DUXBURY, I. M., and ALEXANDER, M. (1973). Appl. Microbiol. *26*, 445.

SPAIN, J., WYSS, O., and GIBSON, D. T. (1979). Biochem. Biophys. Res. Commun. *88*, 634.

STANIER, R. Y., PALLERONI, N. J., and DOUDOROFF, M. (1966). J. Gen. Microbiol. *43*, 159.

STANIER, R. Y., and ORNSTON, L. N. (1973). Adv. Microbiol. Physiol. *9*, 89.

SUDHAKAR, B., SIDDARAMAPPA, R., and SETHUNATHAN, N. (1976). Antonie van Leeuwenhoek J. Microbiol. Serol. *42*, 461.

SUDHAKAR, B., SIDDARAMAPPA, R., WAHID, P. A., and SETHUNATHAN, N. (1978). Antonie van Leeuwenhoek J. Microbiol. Serol. *44*, 171.

SUN OIL (1967). Brit. Patent 1 056 729.

SUROVTSEVA, E. G., VASIL'EVA, G. K., VOL'NOVA, A. I., and BASKUNOV, B. P. (1980). Dokl. Akad. Nauk SSSR *254*, 226.

SUROVTSEVA, E. G., VASIL'EVA, G. K., BASKUNOV, B. P., and VOL'NOVA, A. I. (1981). Mikrobiologiya *50*, 740.

TARVIN, D., and BUSWELL, A. M. (1934). J. Am. Chem. Soc. *56*, 1751.

TAYLOR, B. F., CAMPBELL, W. L., and CHINOY, I. (1970). J. Bacteriol. *102*, 430.

TAYLOR, B. F., HEARN, W. L., and PINKUS, S. (1979). Arch. Microbiol. *122*, 301.

THAYER, J. R., and WHEELIS, M. L. (1976). Arch. Microbiol. *110*, 37.

TITTMANN, U., and LINGENS, F. (1980). FEMS Microbiol. Lett. *8*, 255.

TONE, H., KITAI, A., and OZAKI, A. (1968). Biotechnol. Bioeng. *10*, 689.

WALLNÖFER, P. R., ENGELHARDT, G., SAFE, S., and HUTZINGER, O. (1973). Chemosphere *2*, 69.

WALLNÖFER, P. R., HUTZINGER, O., ENGELHARDT, G., and ZIEGLER, W. (1984). In "Handbook of Microbiology" (A. J., LASKIN, and H. A. LECHAVELIER, eds.) Vol. VI b, CRC-Press, Cleveland, in press.

WALKER, N., and HARRIS, D. (1969). J. Appl. Bacteriol. *32*, 457.

WATSON, G. K., and CAIN, R. B. (1975). Biochem. J. *146*, 157.

WILLIAMS, R. J., and EVANS, W. C. (1975). Biochem. J. *148*, 1.

WYMAN, J. F., GUARD, H. E., WON, W. D., QUAY, J. H. (1979). Appl. Environ. Microbiol. *37*, 222.

YOU, J.-S., and BARTHA, R. (1982). J. Agric. Food Chem. *30*, 274.

Chapter 9

Aliphatic Hydrocarbons

Matthias Bühler
Joachim Schindler

Abteilung Biotechnologie
Henkel KGaA, Düsseldorf
Federal Republic of Germany

I. Introduction
II. Hydrocarbon-degrading Microorganisms
III. Uptake of *n*-Alkanes
IV. Transformation of Short-chain Alkanes
  A. Introduction
  B. Metabolic Pathways
  C. Methane Monooxygenase
  D. Methanol Dehydrogenase
  E. Formaldehyde Oxidation
  F. Formate Oxidation
  G. Some Technical Aspects
V. Transformation of Higher Aliphatic Hydrocarbons
  A. Introduction
    1. Survey
    2. Criteria for metabolic degradation pathways in microorganisms
  B. Initial Steps of Alkane Oxidation
    1. Hydroxylation by mixed-function monooxygenase systems
      a) Cytochrome P-450-dependent systems
      b) Monooxygenase systems without cytochrome P-450
    2. Dehydrogenation mechanism
    3. Hydroperoxide mechanism
  C. Terminal Oxidation Pathways to Fatty Acids
    1. Monoterminal oxidation
    2. Diterminal oxidation
  D. Subterminal Oxidation Pathways
  E. Oxidation of Alkenes

F. Branched-chain Alkanes and Alkenes
G. Co-oxidation
VI. Further Metabolism of Fatty Acids
A. Survey
B. β-Oxidation
C. α-Oxidation
D. Lipids
VII. Genetics and Strain Improvement
A. Methylotrophic Organisms
B. Microorganisms Degrading Higher Alkanes
VIII. Industrial Application: A Survey of Patents
IX. References

I. Introduction

Since the oil crisis in 1973 it has increasingly become obvious that petroleum - besides providing energy - is a highly important and non-renewable feedstock for chemistry. This way of thinking has also strongly penetrated the field of hydrocarbon biotransformation - as reflected by the turbulence encountered in single cell protein production from alkanes (EINSELE, 1983).

This review deals with products obtained by microbial or enzymatic transformation of aliphatic hydrocarbons limited to branched- and straight-chain alkanes as well as alkenes with methane and related short-chain paraffins.

The origins of this field date back to MIYOSHI (1895) and SÖHNGEN (1906), who first reported on hydrocarbon-utilizing microorganisms. It was only during the early 1960s, however, that hydrocarbon biochemistry entered industrial research, when mainly petroleum companies became interested in topics such as bacterial oil prospecting, refinery waste disposal, or deparaffinization of kerosine and gas oil. Since then, the use of inexpensive petroleum fractions as a carbon source in fermentations substituting for carbohydrates has been widely investigated for cell mass production as well as for biochemical synthesis of amino acids, fatty acids, steroids, vitamins, and other substances of commercial interest (ABBOTT and GLEDHILL, 1971; BIRD and MOULTON, 1972; FUKUI and TANAKA, 1981). In spite of the critical economics of hydrocarbons, the potential of their microbial transformations is reflected by the increasing number of publications and patents on promising products such as dicarboxylic acids, ω-hydroxy fatty acids, or bioemulsifiers.

It should, however, be recognized that hydrocarbon substrates deserve academic interest in their own right. Compared with conventional carbohydrate substrates, the water insoluble hydrocarbons as well as other lipophilic compounds reveal new challenges concerning the transformation processes in an aqueous environment (Table 1).

The mechanisms of alkane emulsification, uptake, and primary oxidation as well as their further breakdown and the fundamental genetics involved in these steps will be discussed as far as these have been elucidated; a graphical outline of the contents is given in Fig. 1.

Several reviews covering parts of these topics have recently been published (FUKUI and TANAKA, 1981; HOU, 1982; MIALL,

Table 1. Transformations of Hydrocarbons: The Problems and their Metabolic Solutions by n-Alkane-degrading Microorganisms

Property of Hydrocarbon	Resulting Problem	Solution by Alkane-degrading Microorganisms	Possible Application
Insolubility in water	How to dissolve/ emulsify the substrate	Several uptake mechanisms	Bioemulsifiers for various applications
Chemically unreactive (kinetically inert)	How to activate the substrate	Several oxidation mechanisms catalyzed by specialized enzymes	Transformation and degradation of hydrocarbons and crude oil (oil spills)
Lipophilic	Deprivation of hydrophilic metabolites	Reversal of metabolism: glycolytic/lipogenic to lipolytic/glycogenic or to lipogenic/glycogenic (for short-chain alkanes)	Synthesis of lipophilic substances

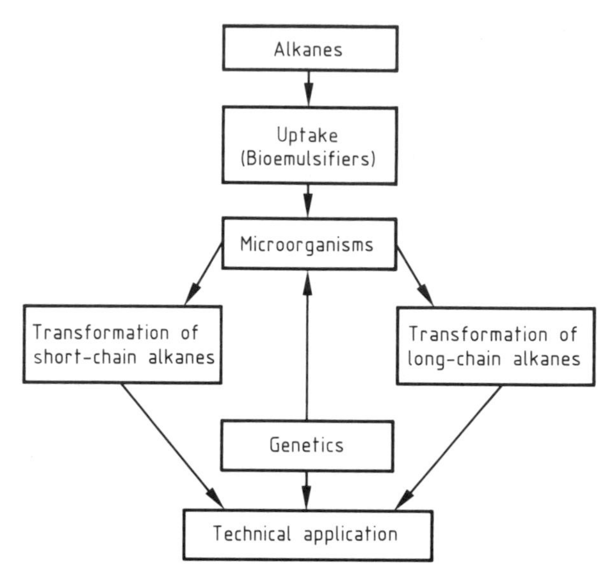

Figure 1. Graphical outline of the topics described in this chapter.

1980; RATLEDGE, 1978, 1980; REHM and REIFF, 1981, 1982; SEBEK and KIESLICH, 1977). While we have tried to help the interested reader to proceed beyond this handbook article by indicating pertinent references and patents, the citation of original literature is certainly not complete; we apologize to those authors whose valuable contributions we may have inadvertently omitted.

II. Hydrocarbon-degrading Microorganisms

Hydrocarbons are not only fossil relicts withdrawn from the biosphere while stored as oil deposits inside the earth, but also metabolites permanently resynthesized in nature by numerous plants and microorganisms (ALBRO, 1976; BACHOFEN, 1982; TORNABENE, 1976, 1978, 1981, 1982; WAYMAN and WHITELEY, 1979; WEETE, 1976). The latter ones serve as growth substrates for other microorganisms in the course of recycling processes in nature and explain why microbial strains capable of degrading hydrocarbons are ubiquitous.

The occurrence of hydrocarbon oxidizing microorganisms depends on the environmental conditions. The population of those specialists increases with the availability of the respective substrates as the enumeration of petroleum-degrading microorganisms in various biotopes has shown. Oil spills caused by accidents at sea or on land demonstrated impressively the hydrocarbon-degrading potential of microorganisms (BARTHA and ATLAS, 1977; BERDICHEVSKAYA, 1982; COLWELL and WALKER, 1977; GRIFFITHS et al., 1982; GUTNICK and ROSENBERG, 1977; MCKENZIE and HUGHES, 1976; REISFELD et al., 1972; WOLFE, 1977).

Numerous species belonging to bacteria, yeasts, and fungi have been described as utilizers of paraffins. Representatives have been found in various taxonomic groups so that only the major important genera may be mentioned here. (ATLAS, 1981; AUSTIN et al., 1977; BEMMANN and TRÖGER, 1975; BOS and DEBRUYN, 1973; DAVIES and WESTLAKE, 1979; GRANGE, 1974; KIYOHARA et al., 1982; KLUG and MARKOVETZ, 1971; LEVI et al., 1979; RATLEDGE, 1978; SHENNAN and LEVI, 1974; VUILLEMIN et al., 1981; WALKER and COLWELL, 1976a, b).

Bacteria

Achromobacter, Acinetobacter, Actinomyces, Alcaligenes, Arthrobacter, Aeromonas, Bacillus, Brevibacterium, Corynebacterium, Flavobacterium, Micrococcus, Mycobacterium, Nocardia, Pseudomonas, Streptomyces, and *Vibrio.*

Yeasts

Candida, Debaryomyces, Hansenula, Pichia, Rhodosporidium, Rhodotorula, Saccharomyces, Sporobolomyces, Torulopsis, Trichosporium.

Fungi

Aspergillus, Aureobasidium, Cladosporium, Cunninghamella, Fusarium, Mortierella, Mucor, Paecilomyces, Penicillium, Sporotrichum, and *Verticillium.*

Crude oils differ widely in their composition and one single species is not able to degrade the entire spectrum of constituents. The *n*-alkanes or paraffins - one group of these components - are saturated, straight-chain hydrocarbons and usually more readily attacked by microorganisms than the other constituents. The diverse components are degraded differently. This was formulated already by ZOBELL (1946) and confirmed or modified by other authors (EINSELE and FIECHTER, 1971; SHENNAN and LEVI, 1974; LEVI et al., 1979; SCHAEFFER et al., 1979; WALKER and COLWELL, 1976a):

1. Aliphatic paraffins are more readily degraded than aromatic hydrocarbons.
2. Long-chain paraffins are degraded preferentially to short-chain paraffins.
3. Saturated compounds are degraded more readily than unsaturated compounds.
4. Branched chains are decomposed less readily than straight-chain compounds.

This review is focused on the microbial degradation of aliphatic hydrocarbons. Most research has been done with *n*-alkanes with a chain length of 9 to 18 carbon atoms because of their better degradability compared to lower *n*-alkanes. This behavior has been ascribed to the higher toxicity of the short-chain *n*-alkanes due to their better solubility in aqueous solutions (KLUG and MARKOVETZ, 1971; GILL and RATLEDGE, 1973).

In nature biodegradation of hydrocarbons predominantly occurs by mixed populations and in presence of other organic compounds. In laboratories, however, degradation is examined mainly in pure cultures and in defined nutrition media. Strains initially known to be unable to utilize *n*-alkanes do oxidize them, when cultivated in mixed cultures or in the presence of other carbon sources. These culture conditions are known as co-metabolism or co-oxidation and may preferentially lead to the accumulation of intermediate metabolites (ALEXANDER, 1981; HARRISON, 1978; KACHHOLZ and REHM, 1977, 1978; KIM and REHM, 1982; PERRY and GIBBSON, 1977; REHM and REIFF, 1982; REISFELD et al., 1972).

III. Uptake of n-Alkanes

n-Alkanes with chain lengths up to C_{44} in the gaseous, liquid or solid state are utilized by microorganisms. The catabolism of *n*-alkanes occurs intracellularly and the degree of their metabolization depends on the accessibility of these substances to the microbial cells.

The mechanism of alkane uptake has been reviewed by several authors (KNAPP and HOWELL, 1980; MIURA, 1978; REDDY et al., 1982; ROY et al., 1979; VERKOOYEN and RIETEMA, 1980a, b; VERKOOYEN et al., 1980; WATKINSON, 1980). While gaseous alkanes are soluble to some extent in aqueous solution, the interactions of the almost water insoluble liquid or solid hydrocarbons with microbial cells have been the subject of various theories. The microbial species and the chemical nature of the substrate play an important role in deciding which way of interaction is used. Mainly three different modes of uptake have been discussed:

1. Direct contact through attachment of the cells to large oil drops.
2. Direct contact through accommodation of submicron oil droplets to the cell surface.
3. Uptake of hydrocarbons dissolved in the aqueous phase.

The solubility of *n*-alkanes in the aqueous phase decreases as the molecular weight increases. In media containing mixtures of paraffins the low molecular weight substrates were first consumed.

The degradation of hydrocarbons by microorganisms has often been observed to be accompanied by the excretion of emulsifying agents and the microbial formation of hydrocarbon dispersions. In general, formation of these compounds is induced when the organisms grow on alkanes. However, constitutive formation of biosurfactants has also been described. Thus, *Corynebacterium lepus* grown on glucose as its only carbon source synthesized and stored glycolipids intracellularly; if alkenes were added, the glycolipid was excreted into the fermentation medium (DUVNJAK and KOSARIC, 1981). Externally added surfactants, fatty acids, or lipids stimulate the assimilation of alkanes by *Acinetobacter* and *Pseudomonas* (BREUIL and KUSHNER, 1980).

A hydrocarbon-solubilizing factor from *Pseudomonas aeruginosa* was identified as a rhamnolipid (HISATSUKA et al., 1971). The same authors isolated a "protein-like-activator" which was formed in the presence of *n*-alkanes but not of glucose. This component was released from the cells after treatment with ethylene diamine tetraacetate, had emulsifying abilities and stimulated the oxidation of alkanes (HISATSUKA et al., 1972, 1975, 1977). In the culture medium of a hydrocarbon-utilizing yeast *Candida petrophilum* an emulsifying factor has been detected, too (IGUCHI et al., 1969).

Candida tropicalis was shown to build up microemulsions of *n*-hexadecane by synthesis of surface active substances (EINSELE et al., 1975). The yeast *Torulopsis bombicola* produced extracellular sophorolipids (Fig. 2). These emulsifiers stimulate the growth on alkanes and turned out to be specific only for this *Torulopsis* species (ITO and INOUE, 1982). A mutant capable of excreting a surface active metabolite when growing on *n*-alkanes was isolated from a yeast strain which originally lacked this ability (NAKAHARA et al., 1981). Polymeric extracellular emulsifying agents called "Emulsans" have been isolated from *Acinetobacter* and subsequently characterized. Oil-in-water emulsions were stabilized by even small quantities of these biosurfactants (BAYER et al., 1981; ROSENBERG et al., 1979; ZOSIM et al., 1982).

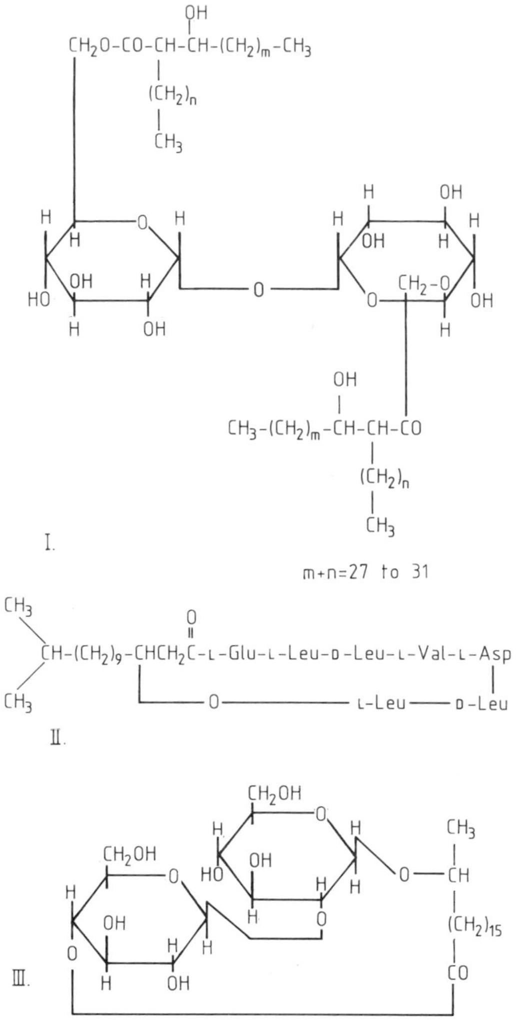

Figure 2. Some structures of bioemulsifiers. – I. Trehalose dimycolate from *Rhodococcus erythropolis* (RAPP et al., 1979). II. Surfactin or Subtilysin, a lipopeptide isolated from *Bacillus subtilis* (KAKINUMA, 1969). III. Deacetylated sophoroselipid from *Torulopsis* sp. (TULLOCH et al., 1967).

The number of microorganisms which have been found to synthesize biosurfactants have increased especially during re-

Table 2. Biosurfactants – Some Chemical Structures and Their Microbial Origin

Classification	Organism	Reference
Trehaloselipid (e. g., "cord factor")	*Arthrobacter paraffineus*	SUZUKI et al. (1969, 1974)
	Brevibacterium vitarumen	LANEELLE and ASSELINEAU (1977)
	Corynebacterium ovis	IONEDA and SILVA (1979)
	Corynebacterium diphteriae	THOMAS et al. (1979)
	Mycobacterium spp.	BATRAKOV et al. (1981)
	Micromonospora sp.	TABAUD et al. (1971)
	Nocardia asteroides	IONEDA et al. (1970)
	Propionibacterium shermanii	PROTTEY and BALLOU (1968)
	Rhodococcus erythropolis	RAPP et al. (1979)
with substituted sugar moiety	*Arthrobacter* sp.	ITOH and SUZUKI (1974)
	Corynebacterium sp.	SUZUKI ct al. (1974)
	Nocardia sp.	
Rhamnolipid	*Pseudomonas aeruginosa*	EDWARDS and HAYASHI (1965)
		ITOH et al. (1971)
		HISATSUKA et al. (1971)
	Pseudomonas sp.	YAMAGUCHI et al. (1976)
Sophoroselipid	*Torulopsis magnoliae*	GORIN et al. (1961)
	Torulopsis gropengiesseri	TULLOCH et al. (1967)
		JONES and HOWE (1968a)
	Torulopsis bombicola	ITO and INOUE (1982)
		ITO et al. (1980)
Diglycosyl diglyceride	*Lactobacillus fermenti*	WICKEN and KNOX (1970)
		BRUNDISH et al. (1967)
Polysaccharide-lipid (-protein) complex	*Arthrobacter (Acinetobacter) calcoaceticus*	ZUCKERBERG et al. (1979)
		KAPLAN and ROSENBERG (1982)
	Pseudomonas sp. PG-1	BRADE and GALANOS (1982)
	Candida lipolytica	REDDY et al. (1983)
	Candida tropicalis	PAREILLEUX (1979)
		KÄPPELI and FIECHTER (1976, 1977)
Lipo-peptide (e. g., "Subtilysin" or "Surfactin")	*Bacillus subtilis*	ARIMA et al. (1968)
		BERNHEIMER and AVIGAD (1970)
		TAKAHARA et al. (1976)
		Besson et al. (1977)
	Corynebacterium lepus	COOPER et al. (1979)
	Candida petrophilum	IGUCHI et al. (1969)
	Streptomyces canus	HEINEMANN et al. (1953)
Ornithine/lysine lipid (e. g., "Cerelipin")	*Agrobacterium tumefaciens*	TAHARA et al. (1976a, b)
	Gluconobacter cerinus	KAWANAMI (1971)
	Thiobacillus thiooxidans	KNOCHE and SHIVELEY (1972)
	Pseudomonas rubescens	WILKINSON (1972)
Protein ("protein-like activator")	*Pseudomonas aeruginosa*	HISATSUKA et al. (1977)
Phospholipid	*Thiobacillus thiooxidans*	BEEBE and UMBREIT (1971)
	Corynebacterium lepus	COOPER et al. (1979)
	Corynebacterium alkanolyticum	KIKUCHI et al. (1973)
Fatty acids	*Micrococcus cerificans*	MAKULA and FINNERTY (1972)
Corynomycolic acid	*Corynebacterium lepus*	COOPER et al. (1981)

Table 2. Continued

Classification	Organism	Reference
Fatty alcohols	*Arthrobacter paraffineus*	SUZUKI and OGAWA (1972)
	Mycobacterium lacticolum	MILKO et al. (1976)
Glycerides and esters	*Acinetobacter* sp.	MAKULA et al. (1975)
	Mycobacterium rhodochrous	HOLDORN and TURNER (1969)
	Corynebacterium fascicans	COOPER et al. (1982)
	Clostridium pasteurianum	COOPER et al. (1980)

cent years (Table 2) and also interesting industrial applications have been developed. The chemical structures of some biosurfactants are depicted in Fig. 2.

Surfactants decrease the interfacial tension and increase the interfacial area. Improved hydrocarbon transport and good growth were correlated in most cases with conditions which provided greater interfacial area (ERICKSON and NAKAHARA, 1975).

Maximum specific growth rates were only obtained with submicron droplets of hydrocarbons in the culture solution (YOSHIDA et al., 1973; YOSHIDA and YAMAME, 1974). The phenomenon of "pseudosolubilization" – the accommodation ability of oil drops – has been described by GOMA et al. (1973).

The uptake of liquid hydrocarbons by microbial cells occurs predominantly from the solubilized or accommodated form (REDDY et al., 1982). Microorganisms with a high affinity for hydrocarbons utilize them as oil drops or in the form of submicron accommodation. Microorganisms with low affinity degrade the accommodation-form more effectively than the drop-form. Specific properties of the organism and the kind of hydrocarbon to be metabolized determines which mechanism of uptake a hydrocarbon-degrading microorganism uses (MIURA, 1978; MIURA et al., 1977a, b).

Most of the *Candida* cells grown on hexadecane were adsorbed to the drops of the hydrocarbon (GUTIERREZ and ERICKSON, 1977). Agglomerations between microorganisms and oil droplets have been observed, as well as the formation of microdroplets around yeast cells (BLANCH and EINSELE, 1973; EINSELE et al., 1975; VERKOOYEN and RIETEMA, 1980a, b; VERKOOYEN et al., 1980).

In addition to extracellular biosurfactants, appropriate structures on the cell surface may participate in stabilizing alkane-water emulsions. Strong evidence for such a mechanism comes from immunological studies on *Acinetobacter* RAG-1 where a cell-associated form of biosurfactant was found to accumulate on the surface of the cell before its release to the medium (GOLDMANN et al., 1982). Furthermore, it was shown for *Corynebacterium fascians* CF 15 that lowering of surface and interfacial tension was mainly due to a mixture of extracellular neutral lipids, whereas emulsion stabilization (and the de-emulsification properties) were primarily associated with the bacterial cells. This organism produced much larger amounts of extracellular neutral lipids when the substrate was an alkane, but the ability to emulsify alkane-water mixtures was similar for cells grown on carbohydrates or alkanes, respectively (COOPER et al., 1982).

Cells of *Candida tropicalis* grown on glucose exhibited a 25% lower adsorption capacity than those grown on *n*-alkane. The induced binding affinity was brought about by a polysaccharide-fatty acid complex at the cell surface (KÄPPELI and FIECHTER, 1976, 1977). These cells contained twice as many lipids than those grown on glucose. The lipids may provide a more hydrophobic region through which the lipophilic substrates may permeate more easily (HUG et al., 1974).

The transport of hydrocarbons into the cells of *Acinetobacter* species was mediated by the accumulation of extracellular vesicular particles rich in phospholipids and lipopolysaccharides (KÄPPELI and FIECHTER, 1980; KÄPPELI et al., 1978, 1981; KÄPPELI and FINNERTY, 1979, 1980).

Cells of *Acinetobacter calcoaceticus* which adhered strongly to hydrocarbons possessed numerous thin fimbriae on the cell surface (ROSENBERG et al., 1982). A rearrangement of the cell wall structures, such as the formation of canals, pores, protrusions, membrane vesicles, and intricate membrane complexes have been observed in various microorganisms grown on hydrocarbons (DMITRIEV et al., 1980; FISCHER et al., 1982; GULEVSKAYA and SHISHKANOVA, 1982; KOZLOVA et al., 1973; MEIZEL et al., 1976; OSUMI et al., 1975).

After penetrating the cell wall the hydrocarbons accumulate at the interfaces between hydrophobic and hydrophilic regions. The ultrastructure of hydrocarbon-grown cells of *Acinetobacter* showed intracytoplasmic, membrane-bound hexadecane inclusions. Physiological relationships exist between those membranes and the intracellular hydrocarbon inclusions (AURICH, 1979; ERICKSON and NAKAHARA, 1975; KENNEDY and FINNERTY, 1975; PETRIKEVICH and DOVGUN, 1980; SCOTT and FINNERTY, 1976; SCOTT et al., 1976).

IV. Transformation of Short-chain Alkanes

A. Introduction

Microorganisms oxidizing short-chain *n*-alkanes (C_2–C_8) have been studied to a minor extent so far (CERNIGLIA et al., 1976; MCLEE et al., 1972; PERRY, 1980). However, methane-utilizing organisms receive increasing attention due to their important ecological functions and their metabolic features. Enormous quantities of organic material are anaerobically degraded and converted to methane. The fact that only minor amounts are released into the atmosphere is ascribed mainly to the activities of aerobic methanotrophic bacteria. The total heterotrophic population in sediments of certain biotopes has been found to consist of up to 8% methanotrophs (HANSON, 1980; HIGGINS et al., 1980a, 1981a; QUAYLE, 1972; SCHLEGEL et al., 1976; WHITTENBURY et al., 1970a, b).

The literature on methylotrophic microorganisms has grown exponentially in recent years comprising monographs and review articles (ANTHONY, 1980, 1982; BEST and HIGGINS, 1983; COLBY et al., 1979; DALTON, 1980a, b; HAMER, 1979; HANSON, 1980; HARDER and VAN DIJKEN, 1976; HIGGINS et al., 1981a; HOU et al., 1980a; QUAYLE, 1972, 1976; SAHM, 1977; TANI et al., 1978; TANI and YAMADA, 1980).

Most representatives of this interesting group of microorganisms are obligate methane or methanol utilizers, growing only on C_1-compounds such as methane, methanol, or methylamine. Also various facultative methylotrophs have been described which grow not only on C_1-compounds but also on conventional carbon sources like glucose or other carbohydrates. Although methylotrophs are able to grow only on C_1-compounds, some enzymes of the methane oxidation pathway in these organisms transform or oxidize hydrocarbons of a chain length up to C_{16}.

Besides these metabolic properties the characteristic DNA base composition, the lipid content, and the special intracellular membranes (BRYOM, 1981) also justify a consideration of this group separately from microorganisms utilizing higher *n*-alkanes (Table 3).

Table 3. Properties Characterizing the Main Groups of Methanotrophs (ANTHONY, 1982)

	Morphology	Resting Stage[b)]	Rosette Formation	DNA Base Ratio (G+C)	Major Carbon Pathway	Motility[c)]
Type I[a)] (obligate)						
Methylococcus	Coccus	*Azotobacter*-type cyst (immature)	—	62–64%	RuMP[d)]	—
Methylomonas	Rod	*Azotobacter*-type cyst (immature)	—	50–54%	RuMP[d)]	+P
Methylobacter	Rod	*Azotobacter*-type cyst	—	50–54%	RuMP[d)]	±P
Type II[a)] (obligate)						
Methylosinus	Rod or pearshaped	Exospore	+	62.5%	Serine	+PT
Methylocystis	Rod or vibroid	Lipid cyst	+	62.5%	Serine	—
Type II[a)] (facultative)						
Methylobacterium	Rod	Exospore	+	58–62%	Serine	+P

[a] Type I bacteria have internal membranes arranged as bundles of vesicular discs; Type II bacteria have membranes arranged in pairs around the cell periphery.
[b] Not all organisms form an identifiable resting stage.
[c] P single polar flagellum; PT polar tufts of flagella; ± indicates that some strains are motile; all *Methylomonas* strains were motile except *M.streptobacterium*.
[d] RuMP ribulose monophosphate.

B. Metabolic Pathways

The oxidation route of methane in methylotrophic bacteria is initiated by a mixed function monooxygenase (Fig. 3):

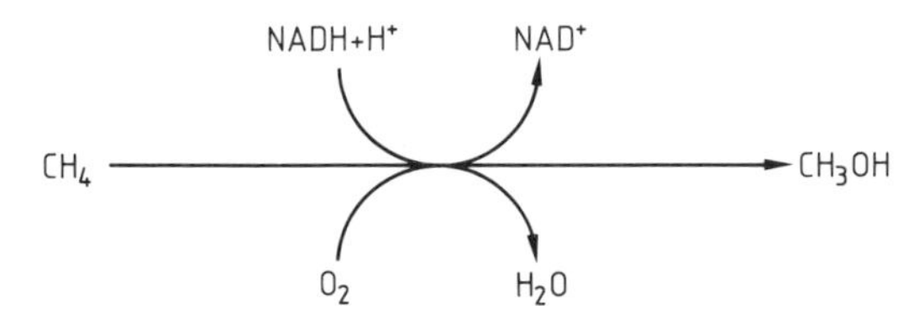

Figure 3. Methane monooxygenase reaction.

The methanol thus produced is further oxidized via formaldehyde and formate to carbon dioxide (Fig. 4).

The oxidation of methanol is catalyzed by a methanol dehydrogenase possessing a novel prosthetic group as only recently detected. Formaldehyde may be oxidized by a special NAD^+-formaldehyde dehydrogenase, by the unspecific methanol dehydrogenase, or by other dye-linked dehydrogenases. This pathway ends with the oxidation of formate by a NAD^+-linked formate dehydrogenase regenerating NADH.

The cellular biomass is synthesized by the assimilation of formaldehyde via the ribulose monophosphate pathway (Quayle cycle, type I group) or via the serine pathway (type II group). Additionally, the autotrophic fixation of CO_2 into phosphorylated sugar has been observed (Calvin cycle). A further possibility of formaldehyde fixation exists in the dihydroxyacetone pathway which has been found only in me-

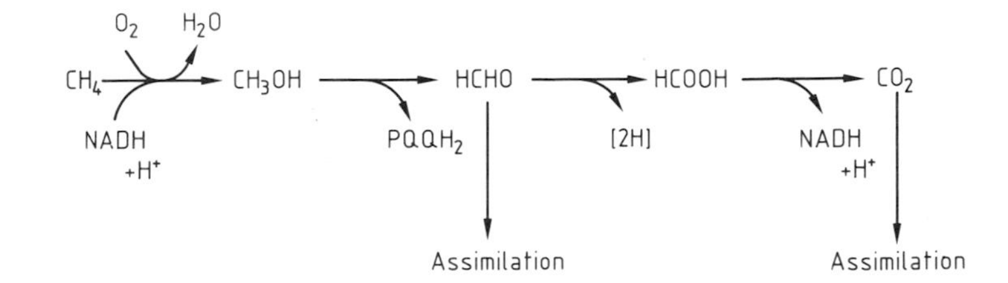

Figure 4. Oxidation pathway of methane to carbon dioxide (ANTHONY, 1982). - NADH Nicotinamide adenine dinucleotide, reduced; $PQQH_2$ Pyrroloquinoline-quinone, reduced.

thylotrophic yeasts (ANTHONY, 1980, 1982; BABEL and HOFMANN, 1982; KATO et al., 1977; WAITES and QUAYLE, 1980).

C. Methane Monooxygenase

In most bacterial species so far examined the oxidative enzymes exhibit a surprisingly broad substrate specificity. Whole cells of *Methylosinus trichosporium* or other methylotrophic bacteria oxidize *n*-alkanes, alkenes, aromatic, alicyclic, and terpenoid hydrocarbons, alcohols, phenol, pyridine, and ammonia (HIGGINS et al., 1979; HOU et al., 1980b; PATEL et al., 1978, 1980a, b).

The methane oxidation catalyzed by cell-free systems has been studied rather extensively in the three strains *Methylococcus capsulatus* (Bath), *Methylosinus trichosporium,* and *Methylomonas methanica* (DALTON, 1981). The soluble methane monooxygenase (MMO) of *Methylococcus capsulatus* consists of three protein components, it requires NADH and exhibits almost the same broad substrate specificity as the enzyme of

Table 4. Oxidation of Alkanes by Extracts of *Methylococcus capsulatus, Methylosinus trichosporium,* and *Methylomonas methanica* (DALTON, 1980b)

Substrate	Products	*M.capsulatus* Amount (mol)	*M.capsulatus* Specific Activity	*M.trichosporium* Amount (mol)	*M.trichosporium* Specific Activity	*M.methanica* Amount (mol)	*M.methanica* Specific Activity
Methane	Methanol	2.02	84	1.88	51	2.79	73
Ethane	Ethanol	1.64	68	0.16		1.68	87
	Ethanal			0.25	33	1.65	
	Acetate			0.83			
Propane	1-Propanol	0.65	69	0.04			
	2-Propanol	1.00		0.66	29	1.27	44
	Propanal			0.42		0.41	
Butane	1-Butanol	1.10	77				
	2-Butanol	0.92		0.60	33	0.54	17
	n-Butanal			0.42		0.13	
Pentane	1-Pentanol	0.49	73				
	2-Pentanol	1.26		1.12	35	0.38	17
	n-Pentanal			0.3		0.28	
Hexane	1-Hexanol	0.60	40	0.17			
	2-Hexanol	0.36		0.28	25		
	n-Hexanal			0.48			
Heptane	1-Heptanol	0.14	27				
	2-Heptanol	0.51		0.68	19		
	n-Heptanal			0.05			
Octane	1-Octanol	0.04	9				
	2-Octanol	0.39					

Methylosinus trichosporium. The MMO of *Methylomonas methanica* showed a somewhat higher specificity (Table 4).

The following transformations are catalyzed (Table 5): the hydroxylation of primary and secondary alkyl C—H bonds (up to hexadecane), the formation of epoxides from internal and terminal alkenes, the hydroxylation of aromatic compounds, the *N*-oxidation of pyridine, the oxidation of CO to CO_2, and the oxidation of methanol to formaldehyde (COLBY and DALTON, 1978; COLBY et al., 1977; STIRLING and DALTON, 1979, 1980; STIRLING et al., 1979; TONGE et al., 1977).

The MMO of *Methylosinus trichosporium* has been described to be particulate. The formation of soluble or particulate MMO seems to depend on the organization of the intracytoplasmic membranes (BEST and HIGGINS, 1983).

Epoxides are of high industrial importance because of their ability to polymerize, thus forming epoxy homopolymers and copolymers. They may also be formed from olefins by the MMO of methanotrophs. HOU et al. (1979a) examined the capability of *M.trichosporium, Methylococcus capsulatus,* and *Methylobacterium organophilum* to oxidize C_2-C_4 *n*-alkenes to their corresponding 1,2-epoxides using intact cells. The metabolites are not further oxidized and accumulate extracellularly. Only methane-grown cells show hydroxylation or epoxidation activity. The facultative methanotrophic *Methylobacterium* sp. CRL-26 ox-

Table 5. Compounds Oxidized by Crude Extracts of Both *Methylosinus trichosporium* and *Methylomonas methanica* (STIRLING et al., 1979)

Substrate	Products	*Methylosinus trichosporium* Amount	*Methylosinus trichosporium* Specific Activity	*Methylomonas methanica* Amount	*Methylomonas methanica* Specific Activity
Methane	Methanol	1.88	50.8	2.79	73.3
	Ethanol	0.16		1.68	
Ethane	Ethanal	0.25	32.6	1.65	87.5
	Acetate	0.83		0	
	1-Propanol	0.04		0	
Propane	2-Propanol	0.66	29.4	1.27	44.2
	Propanal	0.42		0.41	
n-Butane	2-Butanol	0.60	33.5	0.54	17.5
	n-Butanal	0.42		0.13	
n-Pentane	2-Pentanol	1.12	34.7	0.38	17.5
	n-Pentanal	0.30		0.28	
Ethene	Epoxyethane	2.20	59.0	6.02	158.3
Propene	1,2-Epoxypropane	2.03	53.5	3.10	81.4
1-Butene	1,2-Epoxybutane	1.69	44.4	3.83	100.4
	trans-2,3-Epoxybutane	0.24		0.82	
trans-2-Butene	*trans*-2-Buten-1-ol	0	40.5	0.38	68.3
	trans-2-Buten-1-al	1.30		1.39	
	cis-2,3-Epoxybutane	0.52		2.34	
cis-2-Butene	*cis*-2-Buten-1-al	0.63	37.4	0.28	70.8
	Butanone	0.27		0.06	
Dimethyl ether	n. d.[a]	n. d.[a]	25.8	n. d.[a]	65.0
Diethyl ether	Ethanol	0.74	144.4	0	27.2
	Ethanal	4.60		1.04	
Chloromethane	n. d.[a]	n. d.[a]	29.0	n. d.[a]	120.9
Dichloromethane	n. d.[a]	n. d.[a]	26.3	n. d.[a]	43.2

[a] n. d. not determined

idizes *n*-alkenes to primary and secondary alcohols and to the corresponding epoxides. Substituted alkenes are also oxidized (Table 6). Hydroxylation and epoxidation reactions are catalyzed in the presence of oxygen and $NADH_2$, with the cofactor regenerated by the NAD^+-linked secondary alcohol dehydrogenase or formate dehydrogenase (Table 7) (PATEL et al., 1982a, b).

The cofactor reduction for the NADH-dependent, soluble MMO of *Methylosinus* sp. CRL 31 can also be achieved by means of the alcohol dehydrogenase, formate dehydrogenase, diol dehydrogenase, or secondary alcohol dehydrogenase with ethanol, formate, 1,2-propane diol, or 2-propanol as co-substrate (HOU et al., 1982).

D. Methanol Dehydrogenase

Methanol is oxidized to formaldehyde by a $NAD(P)^+$-independent dehydrogenase exhibiting a broad substrate specificity. This enzyme was finally identified as an unusual quinoprotein dehydrogenase. Elec-

Table 6. Epoxidation of Alkenes by Soluble Methane Monooxygenase from *Methylobacterium* sp. CRL-26 (PATEL et al., 1982b)

Substrate	Product	Specific Activity (nmol/min/mg Protein)
Ethylene	Ethylene oxide	55
Propylene	Propylene oxide	100
1-Butene	1,2-Epoxybutane	87
Butadiene	1,2-Epoxybutene	75
Isobutylene	1,2-Epoxyisobutene	95
cis-2-Butene	*cis*-2,3-Epoxybutane	22
	cis-2-Buten-1-ol	15
trans-2-Butene	*trans*-2,3-Epoxybutane	25
	trans-2-Buten-1-ol	18
2-Methyl-1-butene	n. d.[a]	42
2-Methyl-2-butene	n. d.[a]	16
1-Bromo-1-butene	n. d.[a]	83
2-Bromo-2-butene	n. d.[a]	30
Isoprene	1,2-Epoxyisoprene	38

[a] n. d. Product not determined

Table 7. Epoxidation of Propylene by Soluble Methane Monooxygenase from *Methylosinus* sp. CRL 31 Using Cofactor Regeneration by Various Dehydrogenases and Their Substrates (HOU et al., 1982)

Cofactor Generation System	Rate of Epoxidation of Propylene (nmol/min/mg Protein)
Formate dehydrogenase, formate, and NAD^+	70
Diol dehydrogenase, 1,2-propanediol, and NAD^+	60
Secondary alcohol dehydrogenase, 2-propanol, and NAD^+	60
Primary alcohol dehydrogenase, ethanol, and NAD^+	70
Control 1 NADH	75
Control 2 NAD^+	0

trons seem to be transferred to the electron transport chain at the level of cytochrome c. The dehydrogenase and cytochrome c of the obligate methylotroph *Methylophilus methylotrophus* interact in a similar way to that of the facultative methylotroph *Pseudomonas* AM 1 (ANTHONY, 1982; BEARDMORE-GRAY et al., 1983; DE BEER et al., 1983; DUINE and FRANK, 1981; OHTA and TOBARI, 1981). The primary alcohol dehydrogenases in *Hyphomicrobium* WC, *Pseudomonas* tp-1, and *Pseudomonas* W1 are active against a series of aliphatic alcohols, formaldehyde, and to a limited extent against acetaldehyde; higher aldehydes are not oxidized (SPERL et al., 1974).

A similarly wide substrate specificity could be demonstrated for the primary alcohol dehydrogenase of three methanol-utilizing *Pseudomonas* strains capable of oxidizing methanol, ethanol, 1-propanol, 1-butanol, 1-pentanol, methoxyethanol, and formaldehyde (YAMANAKA, 1981; YAMANAKA and MATSUMOTO, 1977) and for the *Methylosinus sporium* enzyme (PATEL and FELIX, 1978).

The MMO may produce primary and secondary alcohols from *n*-alkanes. A series of newly isolated obligate and facultative methylotrophic bacteria produce methylketones from secondary alcohols by a NAD^+-linked secondary alcohol dehydrogenase (HOU et al., 1979b, 1980a; PATEL et al., 1981).

E. Formaldehyde Oxidation

Formaldehyde is further oxidized to formate. Various types of dehydrogenases may be involved in this dehydrogenation. Of the NAD^+-linked dehydrogenases only a few have been fully characterized. Formaldehyde may also be oxidized by the methanol dehydrogenase as mentioned above. Various strains with high levels of methanol dehydrogenase activity possess only low or no activity of NAD^+-dependent formaldehyde dehydrogenase. Most dehydrogenases show high substrate specificity. Only the dye-linked formaldehyde dehydrogenase of *Methylosinus trichosporium* and *Methylomonas methylovora* oxidizes a wide range of aliphatic and aromatic aldehydes, methylglyoxal, glyoxal, glyoxalate, glyceraldehyde, glycolaldehyde, and glutaraldehyde (Table 8) (BEST and HIGGINS, 1983; MARISON and ATTWOOD, 1980; PATEL et al., 1980a).

Table 8. Substrate Specificity of Aldehyde Dehydrogenase of *Methylosinus trichosporium* (PATEL et al., 1980a)

Substrate	Oxidation Rate (%)
Formaldehyde	100
Acetaldehyde	80
Propionaldehyde	100
Butyraldehyde	85
Pentaldehyde	105
Hexaldehyde	95
Heptaldehyde	50
Octaldehyde	43
Nonaldehyde	35
Decaldehyde	15
Benzaldehyde	20
Salicylaldehyde	5
Glyoxylate	25
Glyceraldehyde	20

F. Formate Oxidation

In most strains the oxidation of formate to carbon dioxide is mediated by a NAD^+-formate dehydrogenase. These enzymes were often found to be relatively unstable. Beside the soluble NAD^+-linked enzyme a membrane-bound dehydrogenase also exists which transfers electrons at the level of cytochrome b. This final enzyme of the oxidation route plays an important role in providing reducing power for the MMO (ANTHONY, 1982; ZATMAN, 1981).

The broad substrate specificity of the MMO allows a more distinctive explanation of the co-oxidation phenomenon mainly detected in obligate methylotrophs.

Organisms, while growing on utilizable substrates like methane, can concomitantly oxidize other compounds like ethane, propane, butane, or others present in the medium, even though those substances cannot serve as growth substrates (LEADBETTER and FOSTER, 1960; WHITTENBURY et al., 1970a; PERRY, 1979).

G. Some Technical Aspects

These various transformation possibilities qualify the oxidative enzymes of methylotrophs in a special way for applications in industrial processes. These enzymes may offer new alternatives for the production of interesting chemicals which is impressively demonstrated by the increasing number of patents in this field (Section VIII). Advantages and disadvantages of biocatalytic systems applied to *in vitro* processes have extensively been discussed (ANTHONY, 1982; BEST and HIGGINS, 1983).

To overcome the problems of multiple products formation isolated enzymes should be used. But the MMO can only catalyze oxidation reactions at the expense of reduced pyridine nucleotides. These coenzymes are expensive and unstable. Therefore, such processes are only economical if the reducing power is continuously regenerated. This may be effected by the use of whole cells (DROZD, 1980; IZUMI et al., 1983; SOMERVILLE and MASON, 1979) or by the input of electrical energy (HIGGINS et al., 1980b; HILL and HIGGINS, 1980; SIMON et al., 1981).

The catalytic stability of enzymes and cofactors used in bioreactors may be enhanced by immobilization, and reduced cofactors can be supplied by recycling systems. Product yields may be increased by high concentrations of biomass or catalysts and by avoiding product inhibition.

The MMO of *Methylobacterium, Methylococcus,* or *Methylosinus,* which catalyzes hydroxylations and epoxidations, can be regenerated by the dehydrogenases included in the reaction route, as exemplified in Fig. 5 (HIGGINS et al., 1980b; HOU et al., 1982; PATEL et al., 1982a, b; WISEMAN and KING, 1982). A continuous high productivity process was studied in which ethane was co-oxidized to extracellular acetate during growth on methane (DROZD, 1980). Cells of *Methylomonas* sp. catalyzing the oxidation of methane to methanol by a MMO were immobilized on acetate cellulose filters and were stable for at least twenty days. This system provided the basis for a methane sensor containing immobilized cells and an oxygen electrode (OKADA et al., 1981). The propylene-utilizing *Mycobacterium* PY1 oxidizes gaseous alkenes including ethylene to the corresponding epoxides by a NAD(P)H-dependent MMO. The prolonged formation of ethylene oxide was achieved by using immobilized cells in a gas-solid reactor in the presence of the co-substrate propionaldehyde for the regeneration of NAD(P)H (DE BONT and VAN GINKEL, 1983).

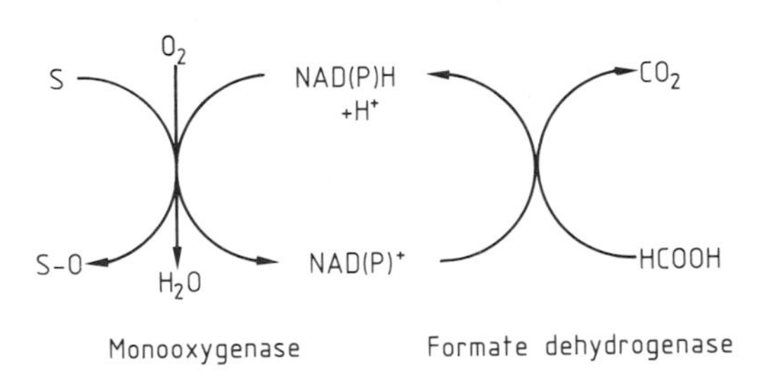

Figure 5. NAD(P)H_2-regeneration in a two-enzyme system (HIGGINS et al., 1980b). S Substrate; S-O Substrate oxidized.

The *in vitro* transfer of electrons to proteins can be mediated in several ways: firstly, by using electron carriers of low molecular weight like ascorbic acid, methylene blue, or viologen dyes, or secondly, electrochemically by using modified electrodes (HIGGINS et al., 1980b; KELLY and KIRWAN, 1977). The electrochemical reduction of the cofactor NAD^+ has been shown for the cytochrome P-450-dependent monooxygenase of rabbit liver (MOHR et al., 1982; SCHELLER et al., 1977). The prosthetic groups of redox proteins like monooxygenases are directly reducible by means of electrochemical methods (Cass et al., 1980; Higgins et al., 1980b; Hill and Higgins, 1980).

V. Transformation of Higher Aliphatic Hydrocarbons

A. Introduction

1. Survey

A wide range of microorganisms possesses the enzymatic capacity to assimilate alkanes with a chain length greater than C_9. On the other hand the metabolism of short-chain paraffins is restricted to a lesser number of species obviously due to the toxicity of these substances. In the case of co-oxidation particularly short-chain hydrocarbons are oxidized but not assimilated. Generally, unsaturated or branched aliphatic hydrocarbons are degraded more slowly or to an incomplete extent.

From the isolated or detected products of alkane oxidation several possible degradation pathways have been derived (Fig. 6), some of which have not yet been fully proved.

After uptake of the hydrocarbon substrate by microbial cells, the primary steps in alkane oxidation form the corresponding alcohols as the first well established intermediates. They are subsequently oxidized to the corresponding fatty acids. Such oxidations can occur at one or at both terminals, or at internal positions of the carbon chain leading to mono-, di-, or subterminal reaction sequences. The fatty acids thus formed undergo various metabolic alterations such as

- incorporation into cellular lipids
- elongation or shortening (β-oxidation) of the chain
- formation of derivatives (α- or ω-oxidation).

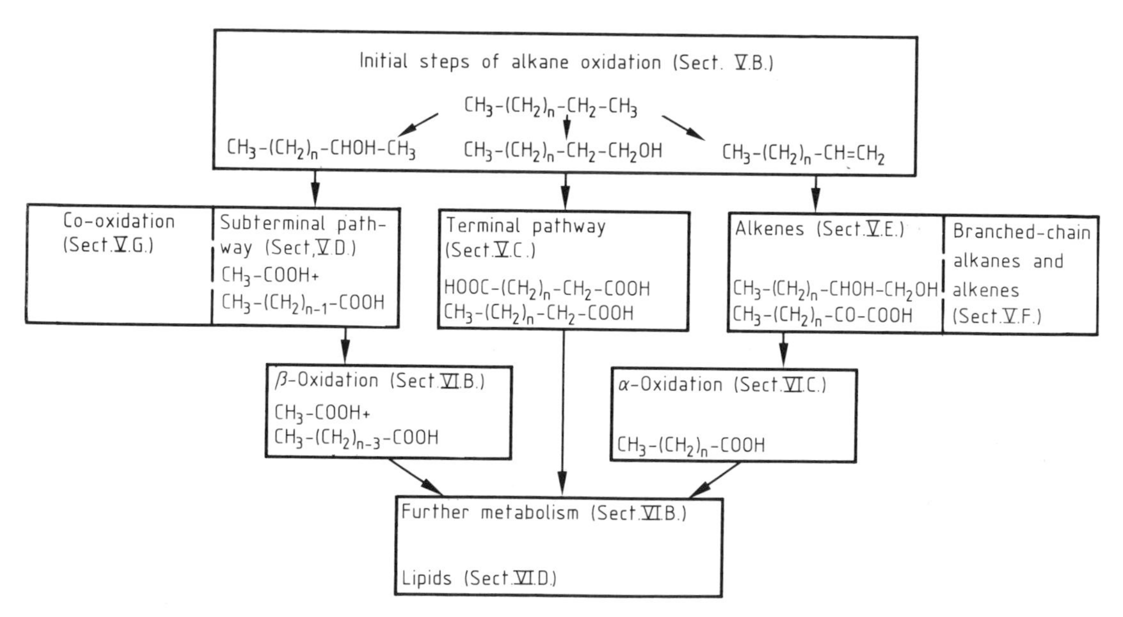

Figure 6. Survey of the metabolism of higher aliphatic hydrocarbons.

Besides their role in providing energy via β-oxidation, the fatty acids can deliver acetyl-CoA for the anabolic pathways and induce special metabolic pathways such as the glyoxylate cycle or the methylcitric acid cycle.

2. Criteria for metabolic degradation pathways in microorganisms

An important criterion in the evaluation of a degradation pathway is the isolation of intermediates and products. However, the amounts of a detected substance can only permit qualitative conclusions.

Another criterion is the determination and isolation of enzymes involved in a certain pathway, although unambiguous results on enzymes specific for a certain pathway are not yet available. On the other hand, interpretation of oxidation studies with whole cells, or even with crude extracts, is always open to criticism as alternative explanations can often be found to account for the experimental results.

A third criterion is the detection of fatty acids in intra- and extracellular lipids of alkane-grown microorganisms. Clear evidence for a correlation between the *n*-alkane chain length and the fatty acid pattern in lipids has only been found for the monoterminal oxidation pathway.

These criteria are also due to co-oxidative degradation. In this case one should be aware that the products are formed specifically from *n*-alkanes by alkane oxidation and not also from other carbon sources such as glucose.

B. Initial Steps of Alkane Oxidation

Three different mechanisms have been reported for the initial attack on the *n*-alkane molecule:

- Hydroxylation
  by a monooxygenase system (mixed-function oxidase)
  with and without involvement of cytochrome P-450
- Dehydrogenation
  to the corresponding alkene followed by a hydration
- Hydroperoxidation
  via a free-radical mechanism and subsequent reduction to alcohols.

1. Hydroxylation by mixed-function monooxygenase systems

Molecular oxygen is incorporated into *n*-alkanes to give the corresponding primary alcohol and the monocarboxylic acid of the same chain length as the substrate (STEWART et al., 1959; IMADA, 1967). This has been demonstrated for bacteria (e.g., NIEDER and SHAPIRO, 1975), for yeasts (e. g., SOUW et al., 1977), and for molds (e. g., HOFFMANN and REHM, 1976a, b).

The formation of the primary alcohol is catalyzed by a complex monooxygenase (hydroxylase), which may be linked to one of several possible electron-carrier systems. The two systems which are known involve either cytochrome P-450 (Table 9) or rubredoxin (Table 10), but in neither case has the reaction mechanism been worked out unambiguously.

a) Cytochrome P-450-dependent systems

Hydroxylase systems of the cytochrome P-450 coupled monooxygenase type are broadly distributed in both eukaryotic and prokaryotic cells (ULLRICH, 1979). The involvement of cytochrome P-450 is easily detected by the characteristic spectra of its oxidized and reduced state, its sensitivity to carbon monoxide, and its insensitivity to cyanide (e. g., FERRIS et al., 1976). The function of P-450 is specifically inhibited by piperonyl butoxide (JAFFE et al., 1969).

In *Candida tropicalis* the P-450-linked hydroxylase system is located within the mi-

Table 9. Cytochrome-linked Monooxygenase Systems in Hydrocarbon-utilizing Microorganisms

Microorganisms	Remarks[a)]	References
Bacteria:		
Bacillus megaterium sp. ATCC 14581	P-450 (detec.), inhib. (CO), (ω-2)-fatty acid hydroxylase	MATSON and FULCO (1981)
Corynebacterium sp. 7E1C	P-450 (detec.), C_8, NADH-dependent	CARDINI and JURTSCHUK (1970)
Pseudomonas aeruginosa sp. S7B1	cyt c, NHI, flavoprotein (pur.), reconst. system	MATSUYAMA et al. (1981)
Pseudomonas oleovorans	cyt o induc. (C_8), ω-hydroxylase system	PETERSON (1970)
Pseudomonas putida sp. PpG 786	P-450 (pur.), camphor	YU and GUNSALUS (1974)
Pseudomonas putida sp. PpG 6	hydroxylase (isol.)	FISH et al. (1983)
Yeasts:		
Candida guilliermondii	inhib. (CO)	TITTELBACH et al. (1976)
sp. H 17	P-450 induc. (C_{16}),	MAUERSBERGER et al. (1980)
	P-450 (pur.), reductase (fract.)	MÜLLER et al. (1979)
Candida lipolytica	P-450 induc. (C_{16}),	JL'CHENKO et al. (1980)
	cyt o (pur.)	BARONCELLI et al. (1979)
Candida maltosa	P-450 induc. (C_{14}),	TAKAGI et al. (1979)
Candida tropicalis	P-450 (pur.), reconst. system,	BERTRAND et al. (1979)
sp. CBS 6947	reductase (pur.)	BERTRAND et al. (1980)
sp. LM7	P-450 (fract.), reductase (fract.), lipid (fract.)	DUPPEL et al. (1973)
sp. IFO 0589	P-450 induc. (C_{14})	TAKAGI et al. (1979)
sp. 101	P-450 and reductase induc. (C_{16})	GILEWICZ et al. (1979)
ATCC 32113	P-450 induc. (C_{16}), under O_2-limitation enhanced	GMÜNDER et al. (1981)
Lodderomyces elongisporus	P-450 induc. (C_{14})	MAUERSBERGER et al. (1980)
Saccharomyces lipolytica	P-450 (pur.), anaerob	YOSHIDA et al. (1977)
Saccharomycopsis lipolytica	P-450, reductase (fract.),	DELAISSE et al. (1981)
	system induc. (C_{16}), P-450 (detec.)	MARCHAL et al. (1982)
Torulopsis candida	P-450 induc. (C_{16})	IL'CHENKO et al. (1980)
Fungi:		
Cunninghamella bainieri	inhib. (CO not CN), P-450 (detec.)	FERRIS et al. (1976)

[a] Abbreviations used: cyt o cytochrome o; detec. detected; fract. fractionated into subcellular fractions; induc. (C_{14}) induced by tetradecane; inhib. (CO not CN) inhibited by carbon monooxide and not by cyanide; pur. (partly) purified; NHI non-heme-iron (protein); reconst. reconstituted

crosomes and consists of cytochrome P-450, $NADPH_2$-cytochrome c (cytochrome P-450) reductase, and a heat-stable phospholipid (GALLO et al., 1971; DUPPEL et al., 1973). Only long-chain alkanes and their metabolic derivatives with more than 13 carbon atoms induce the synthesis of the P-450-dependent hydroxylation system or at least increase the concentrations of its components considerably compared to cells grown on

Table 10. Monooxygenase Systems Independent of Cytochromes in Hydrocarbon-utilizing Microorganisms

Microorganisms	Remarks[a)]	References
Acinetobacter calcoaceticus	rubredoxin (pur.), reductase (pur.)	AURICH et al. (1976) CLAUS et al. (1978)
Pseudomonas aeruginosa	3 comp. (fract.), C_7, no P-450 (detec.)	VAN RAVENSWAY CLAASEN and VAN DER LINDEN (1971)
Pseudomonas desmolytica	3 comp. (pur.), rubredoxin (detec.), C_{14}	KUSUNOSE et al. (1967)
Pseudomonas (marine sp.)	3 comp. (fract.), C_{10}, no P-450 (detec.), inhib. (CN)	HAMMER and LIEMANN (1976)
Pseudomonas putida var. *oleovorans*	3 comp. (fract.), C_8, rubredoxin (pur.), ω-hydroxylase (pur.)	PETERSON and COON (1968) LODE and COON (1971) RUETTINGER et al. (1977)
Cladosporium resinae	no P-450 (detec.), no inhib. (PB[b)])	WALKER and COONEY (1973)

[a] Abbreviations used: cyt o cytochrome o; detec. detected; fract. fractionated into subcellular fractions; induc. (C_{14}) induced by tetradecane; inhib. (CN) inhibited by cyanide; pur. (partly) purified

[b] PB piperonyl butoxide

non-hydrocarbon substrates (GALLO et al., 1973, 1976).

There is no evidence that the microbodies formed in alkane-grown yeasts are associated with the described system (KAWAMOTO et al., 1977). The catalytic mechanism of the hydroxylating system in yeast (Fig. 7) has not been fully proved yet, but it seems to be similar to the microsomal hydroxylase system from liver (STROBEL and COON, 1971). Another three-component enzyme system that catalyzes the $NADPH_2$-dependent oxidation of *n*-hexadecane to 1-hexadecanol has been purified and reconstituted from *Pseudomonas aeruginosa*. It consists of a non-heme iron protein, a heme protein with the spectral characteristics of cytochrome c, and a flavoprotein of low molecular weight (MATSUYAMA et al., 1981).

Though the P-450-linked monooxygenase system is specifically $NADPH_2$-dependent, some exceptions have been observed (Table 9). $NADH_2$ is found to be the cofactor of the alkane monooxygenase from a *Corynebacterium* sp. (CARDINI and JURTSHUK, 1970). This might be due to transhydrogen-

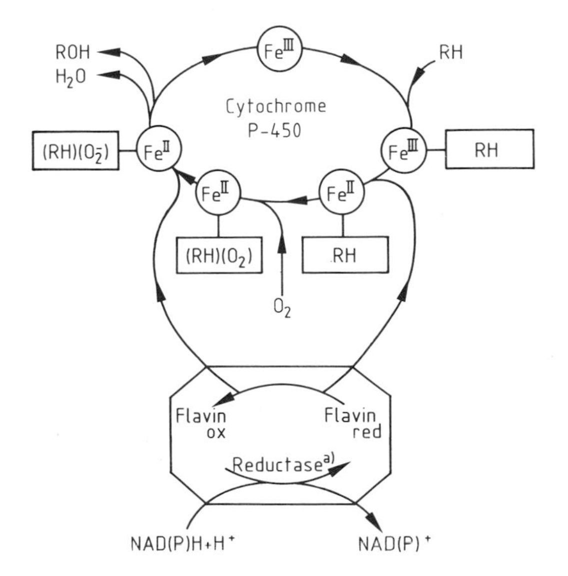

Figure 7. P-450-dependent hydroxylation system. – Proposed mechanism according to that in liver microsomes.

[a] NAD(P)H: cytochrome P-450 (cytochrome c) oxidoreductase (lipid associated)

ase activity in the crude enzyme system or due to impurities in the substrate. Additionally, it may be concluded that there is a number of interlinked electron transport chains involved with varying sensitivities to activators and inhibitors. Thus, a critical evaluation of the details will be extremely difficult especially for the industrially important yeasts, because of the hindrance by their cell wall. However, the ability to prepare protoplasts is yielding new perspectives for the future (PEPERDY, 1980; YAMAMURA et al., 1975).

b) Monooxygenase systems without cytochrome P-450

There is considerably less information on hemoprotein-independent hydroxylating systems for alkanes than on monooxygenases containing cytochrome P-450.

The system of *Pseudomonas oleovorans* var. *putida* involves three proteins (Fig. 8): a rubredoxin, a non-heme iron protein, a NADH-rubredoxin reductase, and an alkane hydroxylase (MCKENNA and COON, 1970).

The rubredoxin can serve as an electron carrier in the *in vitro* reduction of alkyl hydroperoxides (BOYER et al., 1971) to the corresponding alcohols by the reductase, but no intermediate hydroperoxide has yet been isolated.

The system catalyzes the ω-oxidation of long-chain fatty acids with more than twelve C-atoms as well as the epoxidation of 1-alkenes (MAY and ABBOTT, 1973).

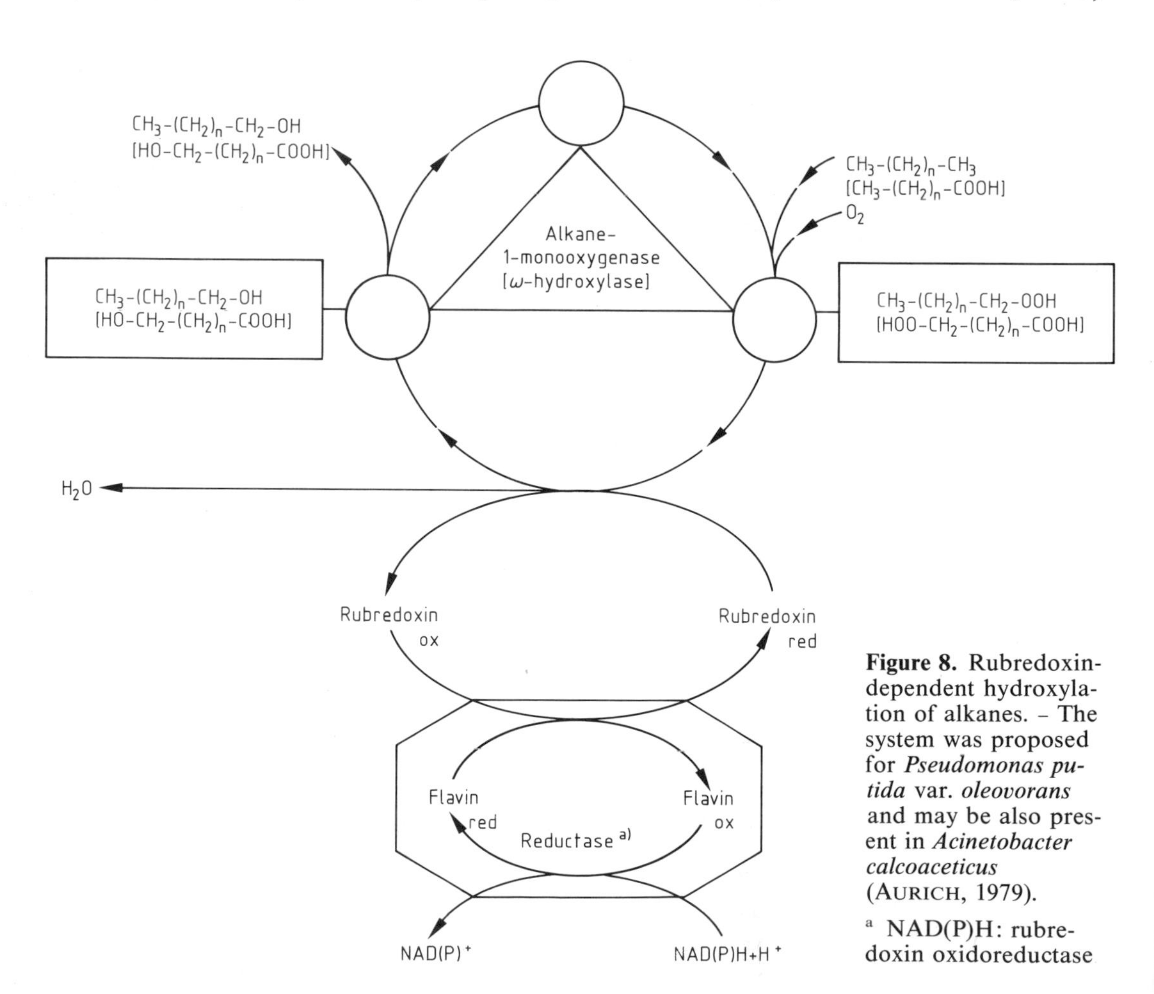

Figure 8. Rubredoxin-dependent hydroxylation of alkanes. – The system was proposed for *Pseudomonas putida* var. *oleovorans* and may be also present in *Acinetobacter calcoaceticus* (AURICH, 1979).

[a] NAD(P)H: rubredoxin oxidoreductase

These properties may be due to the broad substrate specificity of the monooxygenase system. Similar hydroxylation systems, which contain no heme iron and which are sensitive to cyanide but not to carbon monoxide, have been reported for several other *Pseudomonas* spp., *Acinetobacter calcoaceticus,* and *Cladosporium resinae* (Table 10).

2. Dehydrogenation mechanism

Following this reaction pathway alkanes are reduced by an NAD^+-dependent dehydrogenase. The 1-alkenes thus formed can be hydrated to alcohols via an epoxide or by the addition of water.

There have been several reports on the detection and isolation of 1-alkenes after microbial growth on *n*-alkanes (Table 11). Additionally, a NAD^+-linked alkane hydroxylase and a $NADPH_2$-linked alkene hydroxylase have been isolated from a Pseudomonas species grown anaerobically on hexadecane with nitrate as terminal electron acceptor. Thereby, the hydroxylase converts 1-decene to 1-decanol, and the dehydrogenase produces 1-decene from *n*-decane (PAREKH et al., 1977).

On the other hand, there is considerable evidence against 1-alkene being an intermediate. Firstly, the dehydrogenation of *n*-alkane to 1-alkene by NAD^+ is thermodynamically unfavorable (MCKENNA and KALLIO, 1965). Furthermore, different patterns of oxidation products were found with *n*-alkanes or the corresponding 1-alkenes as nutrients (IIZUKA et al., 1969). Therefore, impurities in the substrate or traces of oxygen in the reaction mixture have been made responsible for the oxidation via alkenes (KLUG and MARKOVETZ, 1971; GALLO et al., 1973).

Table 11. Alkenes as Possible Intermediates in Microorganisms Degrading Long-chain Alkanes (modified from REHM and REIFF, 1981). – The significance of the pathway in the organism is arbitrarily symbolized by +, + +, or + + +.

Microorganisms		Remarks[a)]	References
Bacteria:			
Pseudomonas aeruginosa	+ + +	anaerob; C_7; cfe	SENEZ and AZOULAY (1961)
—	+	anaerob; C_7; rc	CHOUTEAU et al. (1962)
Pseudomonas sp.	+ + +	anaerob; C_8; C_{10}; C_{16} purified enzyme	PAREKH et al. (1977)
Achromobacter sp.	+ + +	C_{10}–C_{14}	KESTER and FOSTER (1963)
Micrococcus cerificans	+	anaerob; C_{16}; gc	WAGNER et al. (1967)
Mycobacterium phlei	+	anaerob; C_{16}; gc	WAGNER et al. (1967)
Nocardia sp.	+	anaerob; C_{16}; gc	WAGNER et al. (1967)
Nocardia salmonicolor	+	anaerob; C_{16}; rc (glucose)	ABBOTT and CASIDA (1968)
Yeasts:			
Candida rugosa	+ + +	aerob; C_{10}; rc	IIZUKA et al. (1968)
	+ + +	anaerob; C_{10}; rc	IIDA and IIZUKA (1970)
Candida tropicalis	+	anaerob; C_{10}; cfe	LEBEAULT and AZOULAY (1971)
Candida parapsilosis	+	C_{14}; gc, cfe	SOUW et al. (1978)
Rhodotorula sp.	+	anaerob; C_{16}; gc	WAGNER et al. (1967)

[a] Abbreviations used: cfe cell-free extract, rc resting cells, gc growing cells

Finally, most microbial systems assimilating hydrocarbons appear obligately dependent on molecular oxygen which is not required for this mechanism. Thus, there seems to be only minor evidence which would favor a general oxidation pathway of *n*-alkanes proceeding to the 1-alcohol via the 1-alkene as a free intermediate.

Furthermore, however, there is only little support in literature for the formation of intermediary hydroperoxides in *n*-alkane oxidation.

3. Hydroperoxide mechanism

A free-radical mechanism in the primary oxidation of *n*-alkanes (Fig. 9) has been suggested by LEADBETTER and FOSTER (1960) and further evaluated by FINNERTY (1977). This step might be classified as a dioxygenase reaction because of the incorporation of molecular dioxygen into the alkane molecule as was shown by STEWART et al. (1959).

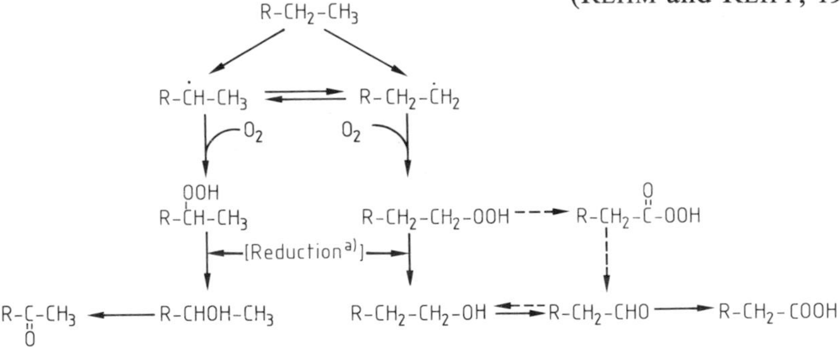

Figure 9. Formation of hydroperoxides via free-radical mechanism and their further metabolism. – Suggestions by LEADBETTER and FOSTER (1960), full line, and FINNERTY (1977), dotted line.
[a] by bacteria and fungi (UPDEGRAFF and BOVEY, 1958)

Thereby, the formation of primary and secondary alkyl hydroperoxides is followed by the reduction to the corresponding alcohols and the further oxidation to aldehydes or ketones. For *Acinetobacter* sp. peroxy acids have been proposed as an alternative intermediate (FINNERTY, 1977).

The oxidation of alkyl-1-hydroperoxides to fatty acid esters could be detected in *Micrococcus cerificans* (FINNERTY et al., 1962). The involvement of free radicals in the oxidation of methane by *Methylococcus capsulatus* was also proposed (HUTCHINSON et al., 1976).

C. Terminal Oxidation Pathways to Fatty Acids

By the terminal oxidation pathways one or both of the terminal methyl groups are oxidized to the corresponding fatty acids, thus classified as the mono- and diterminal pathway, respectively. In Table 12 the occurrence of the terminal pathways among microorganisms has been summarized (REHM and REIFF, 1981).

1. Monoterminal oxidation

In contrast to the exact mechanism for the oxidation of an alkane to the corresponding alcohol, the oxidative pathway from the primary alcohol via the aldehyde to the 1-monocarboxylic acid is unambiguous (Fig. 10) and has been reviewed by KLUG and MARKOVETZ (1971), EINSELE and FIECHTER (1971), as well as by REHM and REIFF (1981).

Sufficient literature has been provided on the isolation of the corresponding 1-alkanols and fatty acids as a proof for this alkane oxidation pathway. The isolation of an intermediary aldehyde was first achieved by accumulating 1-alkanals as semicarbazones in the reaction mixture (BÜNING-PFAUE and REHM, 1972).

In the two oxidative steps leading from alcohol via aldehyde to fatty acid an alcohol dehydrogenase and an aldehyde dehydrogenase are involved, several examples of which have been characterized and isolated (Tables 13 and 14).

There are two types of alcohol dehydrogenases with respect to alkane degradation (Table 13). The pyridine nucleotide-dependent enzymes appear to be cytoplasmatic and resemble in some respect the alcohol dehydrogenase from baker's yeast (AURICH, 1979). The second type is generally a membrane-bound flavoenzyme coupled to an electron transport system with a cytochrome as the possible primary acceptor *in vivo*. As an exception, such an enzyme, but without a detectable flavin, has been found in *Pseudomonas aeruginosa* (TASSIN et al., 1973).

In *Acinetobacter* as well as in *Pseudomonas putida* or *Pseudomonas aeruginosa* both types of alcohol dehydrogenase are present. But only the particular enzyme is directly connected to the assimilation of alkanes (TAUCHERT et al., 1978a, b). At least in bacteria only the $NAD(P)^+$-dependent alcohol dehydrogenase is constitutive, whereas the acceptor-dependent enzyme as well as the aldehyde dehydrogenase and the alkane monooxygenase are inducible (AURICH, 1979).

There are examples of microbial long-chain alcohol dehydrogenases with rather broad substrate specificity concerning the chain length of the 1-alkanols, such as in *Candida tropicalis* (LEBEAULT and AZOULAY, 1971) and in *Cladosporium resinae* (WALKER and COONEY, 1973). The capability of the alcohol dehydrogenase to oxidize hydroxy compounds other than 1-alkanols, e. g., 2-alkanols, 1,2-alkanediols, 1,ω-alkanediols, and ω-hydroxyalkanoic acids into the corresponding carbonyl compounds is essential for the diterminal and the subterminal pathways as described later.

The thermodynamic equilibrium is unfavorable for the transformation of alcohol to aldehyde. The conditions for product for-

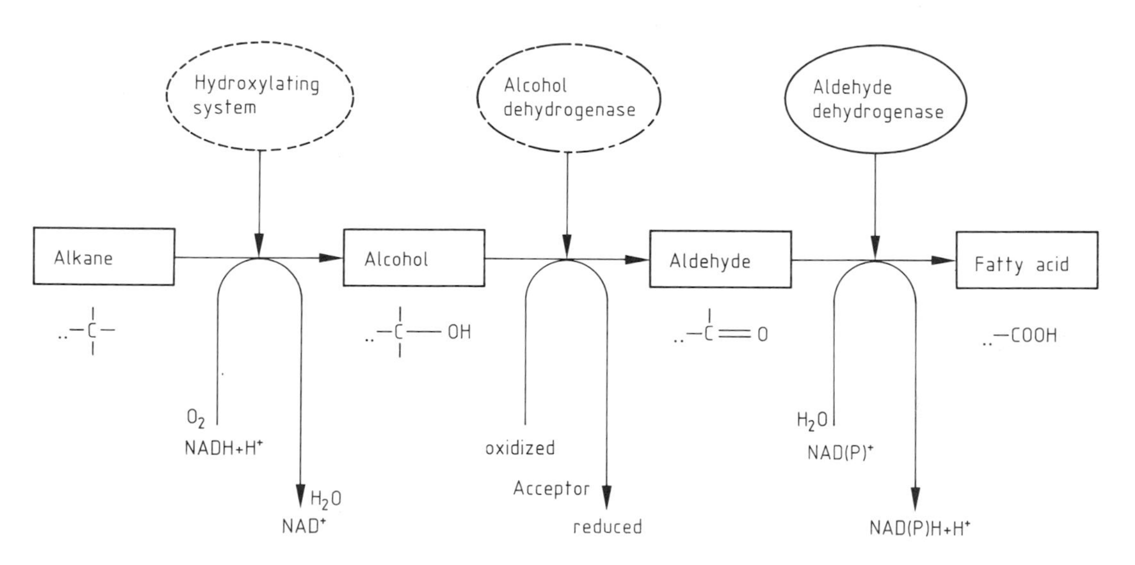

Figure 10. Monoterminal oxidation pathway.

Table 12. Occurrence of Terminal Pathways among Microorganisms Degrading Long-chain *n*-Alkanes (modified from REHM and REIFF, 1981). – The significance of the pathway in the organism is arbitrarily symbolized by +, + +, or + + +.

Microorganisms	mono-terminal	di-terminal	References
Pseudomonas aeruginosa	+		FREDERICKS (1967)
Pseudomonas aeruginosa	+ + +		VAN RAVENSWAY CLAASEN and VAN DER LINDEN (1971)
Pseudomonas aeruginosa	+ +		SCHNABEL and REHM (1971)
Pseudomonas fluorescens	+ +		SCHNABEL and REHM (1971)
Pseudomonas oleovorans	+ +		PETERSON (1970)
Pseudomonas putida	+ + +		NIEDER and SHAPIRO (1975)
Pseudomonas sp.	+ +	+	KILLINGER (1970)
Acetobacter peroxydans	+ +		SCHNABEL and REHM (1971)
Acinetobacter calcoaceticus	+ +		SORGER and AURICH (1978a)
Acinetobacter sp.	+ + +		MAKULA et al. (1975)
Micrococcus cerificans	+		WAGNER et al. (1967)
Micrococcus cerificans	+ +		FINNERTY et al. (1962)
Bacillus coagulans	+		KACHHOLZ and REHM (1977)
Bacillus lentus	+		KACHHOLZ and REHM (1977)
Bacillus macerans	+		KACHHOLZ and REHM (1977)
Bacillus stearothermophilus	+		KACHHOLZ and REHM (1978)
Corynebacterium simplex	+ + +	+ +	KESTER and FOSTER (1960)
Corynebacterium simplex	+ +	+ +	BACCHIN et al. (1974)
Arthrobacter paraffineus	+ + +		SUZUKI and OGAWA (1972)
Brevibacterium erythrogenes	+ + +		PIRNIK et al. (1974)
Mycobacterium phlei	+	+	WAGNER et al. (1967)
Mycobacterium rhodochrous	+		FREDERICKS (1967)
Mycobacterium smegmatis	+	+ +	LUKINS and FOSTER (1963)
Nocardia hydrocarbonoxydans	+ +		NOLOF and HIRSCH (1962)
Nocardia petroleophila	+ +		SEELER (1962)
Nocardia sp.	+		WAGNER et al. (1967)
Streptomyces eurythermus	+		GROSSEBÜTER et al. (1979)
Streptomyces griseus	+		GROSSEBÜTER et al. (1979)
Pichia sp.	+ +	+ +	OGINO et al. (1965)
Pichia haplophila		+ +	SHIIO and UCHIO (1971)
Candida albicans		+ +	SHIIO and UCHIO (1971)
Candida cloacae		+ +	UCHIO and SHIIO (1971)
Candida guilliermondii		+ +	KRAUEL and WEIDE (1978)
Candida guilliermondii	+ + +	+	SCHUNCK et al. (1978)
Candida intermedia		+ + +	SHIIO and UCHIO (1971)
Candida lipolytica		+ +	SHIIO and SERIZAWA (1975)
Candida lipolytica	+ + +		TABUCHI and UCHIO (1971)
Candida maltosa		+ +	SHIIO and UCHIO (1971)
Candida parapsilosis		+ + +	SHIIO and UCHIO (1971)
Candida parapsilosis	+ + +		SOUW et al. (1977)
Candida rugosa		+ +	IIZUKA et al. (1966)
Candida tropicalis	+ + +	+	LEBEAULT et al. (1970a)
Candida tropicalis		+	OKUHARA et al. (1971)
Candida tropicalis	+ + +		DUPPEL et al. (1973)
Candida zeylanoides		+	SHIIO and UCHIO (1971)
Rhodutorula sp.	+		WAGNER et al. (1967)
Torulopsis gropengiesseri	+ + +		JONES and HOWE (1968a, c)
Absidia spinosa	+ +		HOFFMANN and REHM (1976a, b)
Cunninghamella bainieri	+ + +		FERRIS et al. (1976)

Table 12. Continued

Microorganisms	mono-terminal	di-terminal	References
Cunninghamella blakesleeana	+++		ALLEN and MARKOVETZ (1970)
Cunninghamella echinulata	+		HOFFMANN and REHM (1976a, b)
Mortierella isabellina	+		HOFFMANN and REHM (1976a, b)
Rhizopus nigricans	+		HOFFMANN and REHM (1976a, b)
Aspergillus versicolor	++	++	LIN et al. (1971)
Penicillium lilacinum	++	++	LIN et al. (1971)
Botrytis sp.	++	++	YAMADA and TORIGOE (1966)
Cladosporium resinae	++		WALKER and COONEY (1973)

Table 13. Some Characteristics of Long-chain Alcohol Dehydrogenases from Alkane-assimilating Microorganisms

Microorganisms	Performance	Electron Acceptor	Induced by	References
Acetobacter calcoaceticus	particulate	DCPIP[c]	C_{13}–C_{18}	TAUCHERT et al. (1975)
	soluble	$NADP^+$	const.[b]	TAUCHERT et al. (1976)
Pseudomonas aeruginosa	particulate	PMS[d]	C_{16}	TASSIN et al. (1973)
	soluble	$NADP^+$	const.[b]	TASSIN and VANDECASTEELE (1972)
Pseudomonas putida var. *oleovorans* PpG6	particulate	DCPIP[c]	C_8	TAUCHERT et al. (1978a, b)
	soluble	NAD^+	const.[b]	
Pseudomonas sp. *marine*	particulate[a]	NAD^+	C_7	HAMMER and LIEMANN (1976)
Pseudomonas sp. strain 196Aa	soluble	NAD^+/FAD	C_{16}	PAREKH et al. (1977)
Candida intermedia	particulate	NAD^+	C_{10}	LIU and JOHNSON (1971)
Candida lipolytica	particulate	NAD^+	C_{10}–C_{13}	YAMADA et al. (1980)
Candida tropicalis	particulate	NAD^+	C_{14}	GALLO et al. (1974)
sp. 101	soluble	NAD^+	C_{14}	LEBEAULT and AZOULAY (1971)
sp. ATCC 20336	particulate	NAD^+	C_{10}–C_{13}	YAMADA et al. (1980)
Saccharomyces cerevisiae	particulate	NAD^+	C_{14}	ROCHE and AZOULAY (1969)
Torulopsis candida	particulate	NAD^+	C_{16}	KRANZOVA and SAPOZHIKOVA (1979)

[a] associated with monooxygenase
[b] constitutive enzyme
[c] 2,6-dichlorophenol-indophenol
[d] phenazine-methosulfate

mation might be improved by a high affinity (low K_m-value) of the enzyme for the long-chain alcohol being present at probably very low concentrations and by a natural electron acceptor with a high redox potential. Such conditions have been found in *Pseudomonas aeruginosa* (TASSIN et al., 1973). On the other hand, the formation of carboxylic acids from the corresponding aldehydes are thermodynamically more fa-

Table 14. Some Characteristics of Long-chain Aldehyde Dehydrogenases from Alkane-assimilating Microorganisms

Microorganisms	Performance	Acceptor	Inductor	References
Acinetobacter calcoaceticus	particulate	$NADP^+$	C_{13}–C_{18}	AURICH and EITNER (1977) SORGER and AURICH (1978b)
Pseudomonas aeruginosa	soluble	$NAD(P)^+$	C_7	BERTRAND et al. (1973)
	particulate	NAD^+	C_7	
Pseudomonas sp. strain 196Aa	soluble	NAD^+	C_{16}	PAREKH et al. (1977)
Candida intermedia	particulate	NAD^+	C_{14}	LIU and JOHNSON (1971)
Candida lipolytica	particulate	NAD^+	C_{10}–C_{13}	YAMADA et al. (1980)
Candida tropicalis	particulate	NAD^+	C_{14}	LEBEAULT et al. (1970b)

vorable than the oxidation of an alcohol to an aldehyde, thus giving an additional reason for the difficulties in isolating intermediary aldehydes.

All the aldehyde dehydrogenases found so far in alkane-utilizing microorganisms are $NAD(P)^+$-dependent (Table 14). They appear either membrane-bound as in *Acinetobacter calcoaceticus* or cytoplasmatic as in *Pseudomonas putida* (SORGER and AURICH, 1978a, b). Both locations have been evaluated for the aldehyde dehydrogenases of *Pseudomonas aeruginosa* (GUERRILLOT and VANDECASTEELE, 1977). The membrane-bound enzyme could be induced only when the organism was grown on *n*-paraffin.

The aldehyde dehydrogenases also reveal a relatively broad substrate specificity (LEBEAULT et al., 1970a, b) and some of them show a pronounced substrate inhibition (SORGER and AURICH, 1978b).

The activities of long-chain alcohol dehydrogenase and aldehyde dehydrogenase have been detected in microsomes, mitochondria, and peroxisomes of alkane-grown *Candida tropicalis* cells. The possible functions of these yeast organelles with respect to *n*-alkane metabolism are summarized in Fig. 11 (FUKUI and TANAKA, 1981).

2. Diterminal oxidation

The diterminal oxidation of *n*-alkanes to dioic acids (DCs) are found mainly in yeasts and to a minor extent in bacteria. The formation of α,ω-DCs occurs via the monoacids, as recently confirmed by YI and REHM (1982a). They demonstrated that a mutant of *Candida tropicalis* is capable of converting *n*-dodecane, *n*-dodecanol, *n*-dodecanoic acid, and 12-hydroxydodecanoic acid to α,ω-DC-12. The mutant produces DC-12, DC-10, DC-8, and DC-6 from dodecane. Resting cells of the same strain also excrete α,ω-dodecanediol from dodecane or dodecanol. The existence of this intermediate together with the fact that ω-hydroxydodecanoic acid and DC-12 are synthesized from α,ω-dodecanediol gives evidence for an alternative pathway leading to DCs via α,ω-diols (Fig. 12; YI and REHM, 1982b, c).

The formation of DCs from *n*-alkanes may result from different types of oxidation processes. One of them leads to products without a shortening of the carbon chain and is considered to be a co-oxidation reaction. The other type produces DCs shorter than the original alkanes due to degradation by β-oxidation. The DCs thereby arising are of odd chain length if the substrate consists of a chain with odd carbon numbers and are even-chained if the *n*-alkane is even-numbered. The transformation of *n*-decane to decanoic acid, 10-hydroxydecanoic acid, and DC-10 by *Rhodococcus rhodochrous* (formerly *Corynebacterium* 7E1C) has been already described by KESTER and FOSTER (1963). Decanoic acid does not serve as a growth substrate. The same organism accumulates consistent amounts of

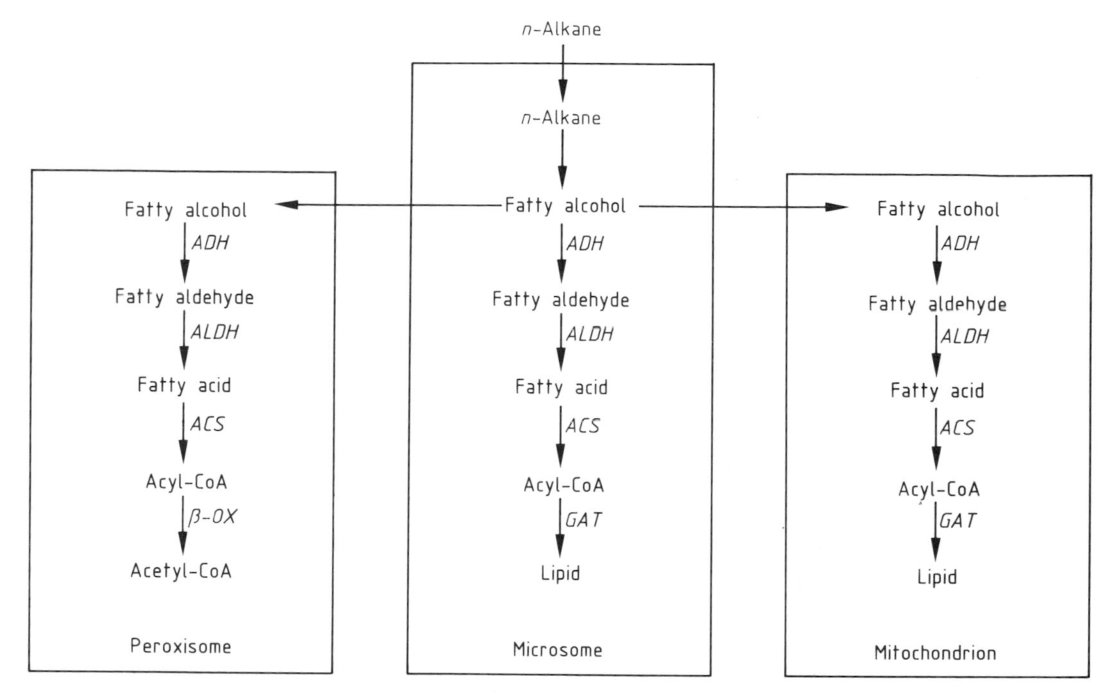

Figure 11. Possible functions of microsomes, mitochondria, and peroxisomes in alkane-utilizing yeasts (YAMADA et al., 1980). – Abbreviations: ACS acyl-CoA synthetase; ADH alcohol dehydrogenase; ALDH aldehyde dehydrogenase; GAT glycerophosphate acetyltransferase; β-OX fatty acid β-oxidation system.

esters of long-chain acids with long-chain alcohols such as decyldecanoate (BACCHIN et al., 1974).

Mutant strains were isolated from the alkane utilizing yeast *Candida cloacae* which have a defect within their degradation system (SHIIO and UCHIO, 1971; UCHIO and SHIIO, 1972a, b). These mutants are unable to assimilate the DCs to the original extent but transform *n*-alkanes to the homologous DCs in the presence of acetate as a co-substrate. The active mutant of the first generation produces up to 29.3 g/L of DC-16 from *n*-hexadecane but still produces small amounts of shorter DCs. A further mutant MR-12 of the second generation shows only 10% of the degradation capacity of the original strain and has already produced 42.7

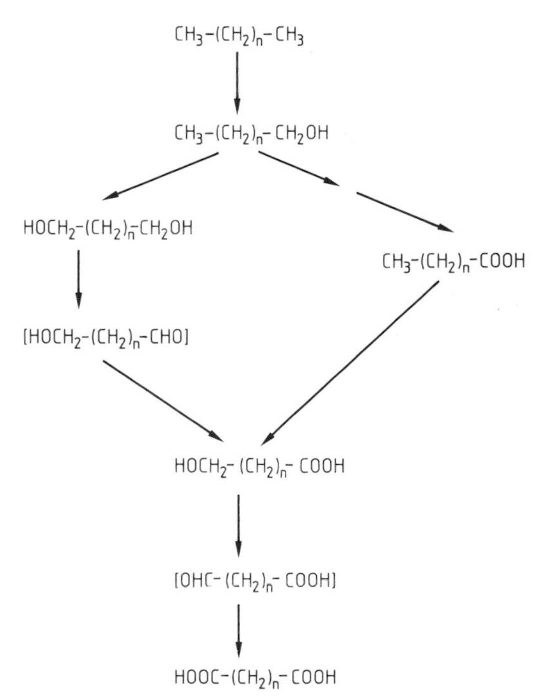

Figure 12. Metabolic pathways of α,ω-dodecane dioic acid formation from *n*-dodecane (YI and REHM, 1982c). – [] products not isolated.

Table 15. DC Production from *n*-Alkanes and their Oxidized Derivatives by Resting Cells of Strain M-1 and MR-12 (UCHIO and SHIIO, 1972b)

Substrate Added		Strains M-1		MR-12
		DC Produced Mainly (g/L)	Other DC Produced (g/L)	DC Produced (g/L)
n-C_{12}	10% (v/v)	DC-12, 8.20	DC-10, 0.01	DC-12, 9.50
Laurylalcohol	10% (v/v)	DC-12, 5.20	DC-10, 0.02	DC-12, 9.20
Laurylaldehyde	10% (v/v)	DC-12, 4.40	DC-10, 0.01	DC-12, 9.20
1,12-Dodecanediol	3.3% (w/v)	DC-12, 8.20	0	DC-12, 10.10
MC-12	3.3% (w/v)	DC-12, 0	0	DC-12, 4.20
n-C_{14}	10% (v/v)	DC-14, 12.4	DC-12, 0.01	DC-14, 24.00
1-Tetradecene	10% (v/v)	DC-12, 2.65	DC-14, 1.11	—
Myristylalcohol	1% (w/v)	DC-14, 9.80	DC-12, 0.91	—
Myristylaldehyde	1% (w/v)	DC-14, 6.70	DC-12, 0.04	—
MC-14	3.3% (w/v)	DC-14, 1.90	DC-12, 0.02	DC-14, 11.20,42
n-C_{16}	10% (v/v)	DC-16, 18.46	DC-14, 0.01	DC-16, 42.70
1-Hexadecene	10% (v/v)	DC-14, 4.60	DC-16, 1.29	—
Palmitylalcohol	1% (w/v)	DC-16, 0.30	DC-8, 0.08	—
MC-16	3.3% (w/v)	DC-16, 0.08	DC-14, 0.01	DC-16, 1.20
Methyl palmitate	10% (v/v)	DC-16, 1.20	0	DC-16, 7.20
Ethyl palmitate	10% (v/v)	DC-16, 0.08	0	DC-16, 6.00

Abbreviations: *n*-C *n*-Alkane, DC Dicarboxylic acid, MC Monocarboxylic acid

g/L of DC-16. This yield could finally be increased to 61.5 g/L by optimization of the culture conditions (UCHIO and SHIIO, 1972c). The strain oxidizes various *n*-alkane derivatives such as alcohols, diols, aldehydes, fatty acids, and methyl- or ethylesters of fatty acids showing the highest affinity for *n*-alkanes (Table 15).

A mutant strain of *Torulopsis candida* 99 produces 33.1 g/L of DC-10 from *n*-decane under optimal conditions after 102 hours of fermentation while the parent strain yields only 5.82 g/L. The mutant can still assimilate DC-10 to some extent (KANEYUKI et al., 1980; OGATA et al., 1973).

A Chinese group obtained 86.6 g/L of DC-15 in batch fermentation with resting cells of a *Candida* mutant, which equals a transformation rate of 67.7% (SHEN et al., 1977).

Microorganisms carrying out the second type of reaction synthesize mixtures of DCs in far lower yields after incubation with *n*-alkanes. In this case, the arising DCs are mainly of heterogenous chain length. A yeast strain Y-3 capable of growing on hydrocarbons as sole carbon source produces a series of dioic acids from *n*-undecane: DC-11, DC-9, DC-7, and DC-5 (OGINO et al., 1965).

Resting cells of *Candida rugosa* oxidize *n*-decane to DC-10, DC-8, DC-6, and DC-4 (IIZUKA et al., 1966). When *Torulopsis candida* 99 was cultured in a medium containing *n*-decane, DC-10, DC-8, DC-6, and DC-4 could be detected. About 5 g/L of DC-10 were extracellularly accumulated. The strain can utilize *n*-alkanes with a chain length between C_9 and C_{18}. DCs with the same number of carbons are synthesized from C_9- and C_{10}-alkanes, whereas carbon chains longer than C_{11} result mainly in DC-7 or DC-6 metabolites (OGATA et al., 1973). The transformation of tridecane by *Candida guilliermondii* yielded about 60 mg/L of DC-5 and 75 mg/L of DC-7 beside several other DCs of odd chain length. The strain also oxidizes monocarboxylic acids at the ω-position (KRAUEL et al., 1973; KRAUEL and WEIDE, 1978).

YI and REHM (1982d) succeeded in producing DCs with κ-carrageenan entrapped cells of *Candida tropicalis*. The synthesis of DCs, monocarboxylic acids, ω-hydroxyacids, and α,ω-diols by the immobilized mutant turned out to be a function of the incubation time and happened to operate better with alkanols and alkanoic acids than with *n*-alkanes.

Since the chemical synthesis of long-chain DCs is more difficult to perform, the microbial transformation of *n*-alkanes may be of special industrial interest. The co-oxidation process may be of particular value since high product yields might be reached with relatively cheap co-substrates such as methane, methanol, acetate, or others. DCs are basic substrates for the synthesis of perfumes, plastisizers, lubricants, polyurethanes, and polyamides. Therefore, fermentation processes for the microbial production of DCs have been the subject of a series of patent applications, as can be seen in Table 25.

D. Subterminal Oxidation Pathways

By the subterminal oxidation pathways one or more methylene groups of the *n*-alkane are oxidized to secondary alcohols and the corresponding ketones, which are further metabolized to fatty acids (Fig. 13).

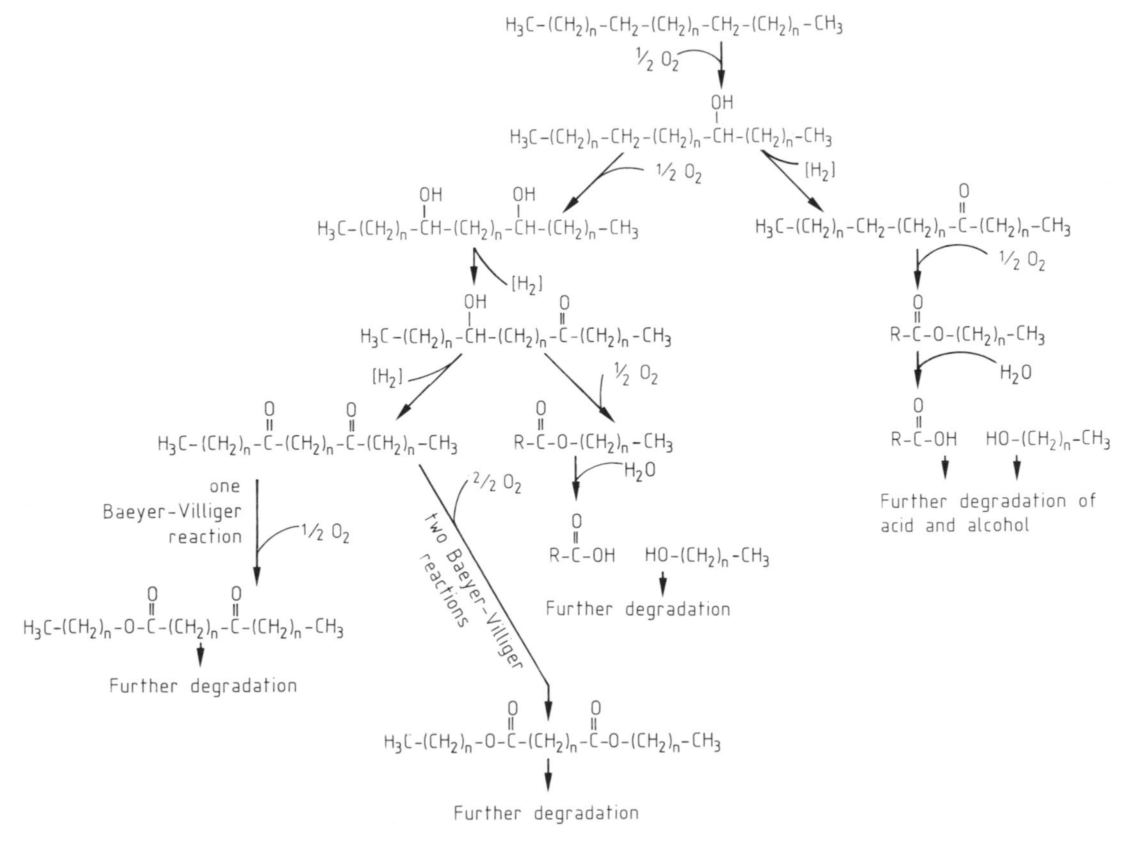

Figure 13. Subterminal degradation pathways of long-chain alkanes (REHM and REIFF, 1981).

The initial attack on the *n*-alkane molecule leading to secondary alcohols occurs by similar mechanisms as already described (see Sect. V.B.). The subsequent step yielding the corresponding ketones may be catalyzed by the same alcohol dehydrogenase as described earlier for the terminal pathways (AURICH, 1979). However, no clear evidence has yet been reported.

The following transformation of the ketones is classified to be a Baeyer-Villiger type reaction. Thus, e. g., 2-tridecanone in *Pseudomonas aeruginosa* is oxidized and rearranged into undecyl acetate, which is split into 1-undecanol and acetate (FORNEY and MARKOVETZ, 1970). The ester formation requires the incorporation of molecular oxygen (BRITTON et al., 1974); this is catalyzed by a $NADPH_2$-dependent monooxygenase, as shown for *Pseudomonas* (MARKOVETZ, 1978).

The occurrence of undecyl acetate as an intermediate in this pathway has been confirmed by a detailed study of the undecyl acetate esterase from *Pseudomonas cepacia (multivorans)* (SHUM and MARKOVETZ, 1974a, b) and by the characterization of other alkyl acetate esterases from *Nocardia* sp. (EUBANKS et al., 1974) as well as from *Cladosporium cladosporioides* (YAMAKAWA et al., 1978).

The oxidation of *n*-alkanes at C-3, C-4, C-5, and higher positions has been found in various molds and bacteria (Table 16). It could be demonstrated for *Bacillus* spp. (KACHHOLZ and REHM, 1978), *Fusarium lini* (THIELE and REHM, 1979), and *Mortierella isabellina* (HOFFMANN and REHM, 1978) that a cleavage of the long-chain alkane molecule may occur by a Baeyer-Villiger type rearrangement even if the position of the oxidative attack is different from C-2.

A unique oxidation pathway of secondary alcohols has been proposed for a *Pseudomonas* sp. The oxidation via ketone, hy-

Table 16. Occurrence of Subterminal Pathways among Microorganisms Degrading Long-chain *n*-Alkanes (modified from REHM and REIFF, 1981). – The significance of the pathway in the organism is arbitrarily symbolized by +, ++, or +++.

Microorganisms	2-Position	2-, 3-, 4-Position	4-, 5-, 6-Position	References
Pseudomonas aeruginosa		+		SCHNABEL and REHM (1971)
Pseudomonas cepaciacens	++			BRITTON et al. (1974)
Pseudomonas fluorescens	+			SCHNABEL and REHM (1971)
Pseudomonas sp.		+	+	KILLINGER (1970)
Acetobacter peroxydans	+			SCHNABEL and REHM (1971)
Micrococcus smegmatis	+			LUKINS and FOSTER (1963)
Nocardia sp.	++			BRITTON et al. (1974)
Candida intermedia	+			KLUG and MARKOVETZ (1967)
Candida parapsilosis	++			SOUW et al. (1977)
Torulopsis gropengiesseri	+			JONES and HOWE (1968a, c)
Absidia spinosa		++	+++	HOFFMANN and REHM (1976a, b)
Cunninghamella echinulata		++	+	HOFFMANN and REHM (1976a, b)
Mortierella isabellina		++	+++	HOFFMANN and REHM (1976a, b)
Aspergillus flavus		+	+++	PELZ and REHM (1973)
Aspergillus ochraceus		+++	++	PELZ and REHM (1973)
Aspergillus niger		++	++	PELZ and REHM (1973)
Penicillium javanicum		++	+++	PELZ and REHM (1973)
Penicillium sp.		+		ALLEN and MARKOVETZ (1970)
Penicillium sp.		++	+++	PELZ and REHM (1973)
Verticillum sp.		+	+++	PELZ and REHM (1973)
Fusarium lini		++	+++	THIELE and REHM (1976)

droxykctone, and diketone is followed by a cleavage into aldehyde and fatty acid according to the findings with 4-decanol (LIJMBACH and BRINKHUIS, 1973). However, further data supporting this pathway in other microorganisms are not yet available (KACHHOLZ and REHM, 1978).

A variety of compounds resulting from subterminal pathways are found in co-oxidation cultures especially of *Bacillus* spp. and *Streptomyces* spp., but not in cultures with long-chain hydrocarbons as the sole carbon source (see Sect. V.G.).

E. Oxidation of Alkenes

As already quoted, there is some evidence against 1-alkenes as intermediates in the oxidation of *n*-alkanes to alcohols, because different patterns of oxidation products arise from alkanes or alkenes. Otherwise, alkenes are closely related to alkanes with respect to their chemical constitution so that they act as metabolic analogs and can be readily oxidized by hydrocarbon-utilizing cells.

The initial attack may occur either at the terminal methyl group or at the double bond; a variety of corresponding products has been isolated (JONES and HOWE, 1968a). The oxidation of 1-hexadecene by *Candida lipolytica* leads to the 1,2-diol, the ω-unsaturated primary and secondary alcohols, the 1,2-epoxide, and the 2-hydroxy carboxylic acid, each having the same carbon skeleton as the alkenc substrate (KLUG and MARKOVETZ, 1968). The alkane hydroxylating system of *Pseudomonas oleovorans* also catalyzes the epoxidation of 1-alkenes as well as the hydroxylation of the terminal methyl group, thus accounting for most of the compounds depicted in Fig. 14 (MAY and ABBOT, 1972, 1973).

The metabolic relevance of the epoxide has not yet been clarified. It might be an in-

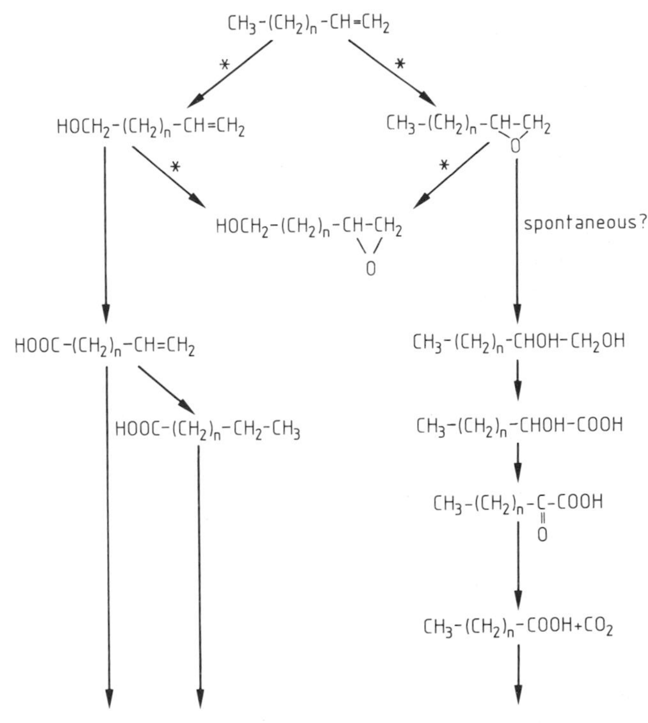

Figure 14. Oxidation of 1-alkenes. – Reactions indicated by an asterisk are catalyzed by the same hydroxylating system which attacks the terminal methyl group of *n*-alkanes (modified from MAY and ABBOTT, 1972, 1973).

termediate in the formation of the 1,2-diol by *Candida lipolytica* (KLUG and MARKOVETZ, 1968). In *Pseudomonas oleovorans*, however, there is apparently no enzymatic transformation at the epoxide function, but further oxidation occurs on the methyl group at the opposite end of the molecule (ABBOTT and HOU, 1973). Slow and perhaps spontaneous hydrolysis of the epoxide can lead to the corresponding diol which in turn is oxidized to the 2-hydroxy acid as was shown for *Pseudomonas aeruginosa* (VAN DER LINDEN and THIJSSE, 1965; HUYBREGTSE and VAN DER LINDEN, 1964).

The decarboxylation product of the 2-hydroxy acid has been recovered for example after growth of *Torulopsis gropengiesseri* (JONES and HOWE, 1968a) and *Mycobacterium vaccae* (KING and PERRY, 1975) on 1-alkenes with 14 to 18 carbon atoms. In the latter organism, ω-unsaturated fatty acids of the same and of shorter chain length as the alkene substrate also occur, indicating a simultaneous attack at the methyl group including β-oxidation. This latter pathway seems to be favored in *Cunninghamella elegans* (CERNIGLIA and PERRY, 1974).

Saturated fatty acids with the same carbon chain as the alkene substrate have been recovered in several organisms. This indicates either the presence of *n*-alkane in the substrate or an enzyme system for the reduction of the double bond probably at the level of ω-unsaturated fatty acids (KING and PERRY, 1975; YANAGAWA et al., 1972).

Subterminal and diterminal oxidations also occur during the growth of microorganisms on 1-alkenes. This has been derived from the recovered secondary alcohols (STEWART et al., 1959) or ω- and $(\omega-1)$-hydroxy fatty acids esterified as glycolipids in *Torulopsis gropengiesseri* (JONES and HOWE, 1968b).

Only little is known about the oxidation of internal alkenes although these substances are produced from *n*-alkanes by *Nocardia salmonicolor* (ABBOTT and CASIDA, 1968). For *Mycobacterium vaccae* some shifting of the internal double bond has been observed. These primary products have been methylated leading to the corresponding branched-chain saturated fatty acids (KING and PERRY, 1975). However, such reactions may be peculiar to genera related to *Mycobacterium* spp. (RATLEDGE, 1976).

F. Branched-chain Alkanes and Alkenes

There are only a few microorganisms which are able to utilize branched-chain hydrocarbons. The reason for this phenomenon may be that either the alkyl branches hinder the uptake of the hydrocarbon into the cell or the branched-chain hydrocarbons are less susceptible to the degradative enzymes of the β-oxidation pathway (SCHAEFFER et al., 1979).

Single-branched alkanes are preferentially oxidized at the terminus most distant from the branching point. The resulting fatty acid is incorporated into the cell lipids (JONES and HOWE, 1968b; KING and PERRY, 1975) or further metabolized. Subsequent degradation via β-oxidation would finally lead to either α-methyl butyric acid and/or β-methyl butyric acid, the latter of which is not oxidizable to the keto acid which would be necessary for cleavage into the acids. Such difficulties for further metabolism may be overcome by ω-oxidation to finally produce β-methylsuccinic acid as postulated for *Brevibacterium erythrogenes* (PIRNIK et al., 1974).

For a multiple-branched alkane such as pristane (2,3,10,14-tetramethylpentadecane) the mechanism of oxidation has been elucidated using *Corynebacterium* sp. (MCKENNA and KALLIO, 1971) and *Brevibacterium erythrogenes* (PIRNIK et al., 1974) as shown in Fig. 15.

The oxidative degradation of squalene, as a methyl-branched, unsaturated hydrocarbon (C_{30}-triterpene), by an *Arthrobacter* sp. yields geranyl acetone (not further metabolized) and a series of degradation products,

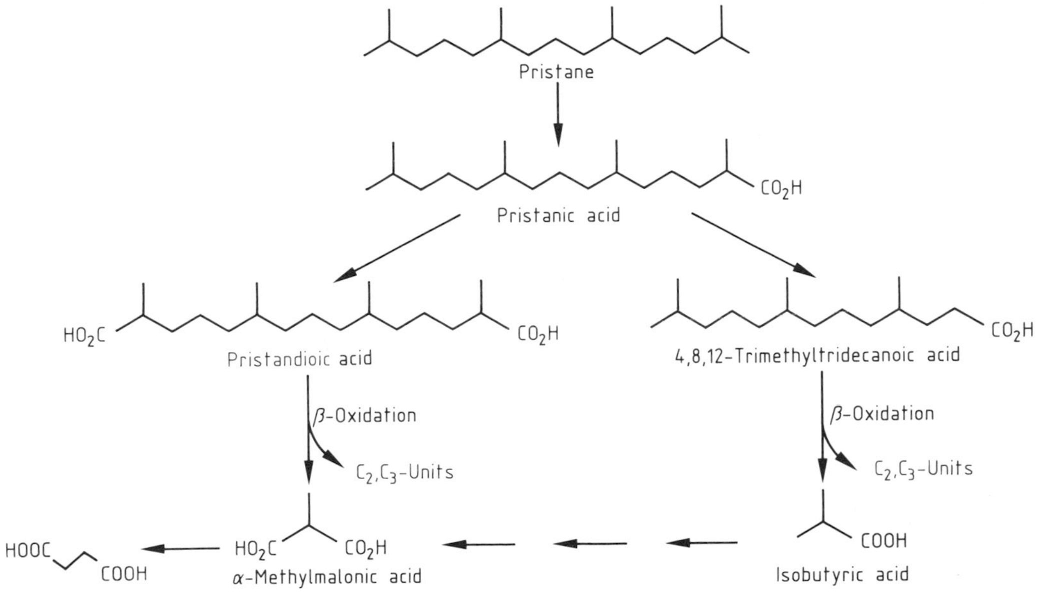

Pristane

Pristanic acid

HO_2C Pristandioic acid CO_2H

4,8,12-Trimethyltridecanoic acid CO_2H

β-Oxidation

C_2,C_3-Units

HOOC COOH

HO_2C CO_2H α-Methylmalonic acid

COOH Isobutyric acid

Figure 15. Pathway of oxidation of pristane (PIRNIK et al., 1974; MCKENNA and KALLIO, 1971).

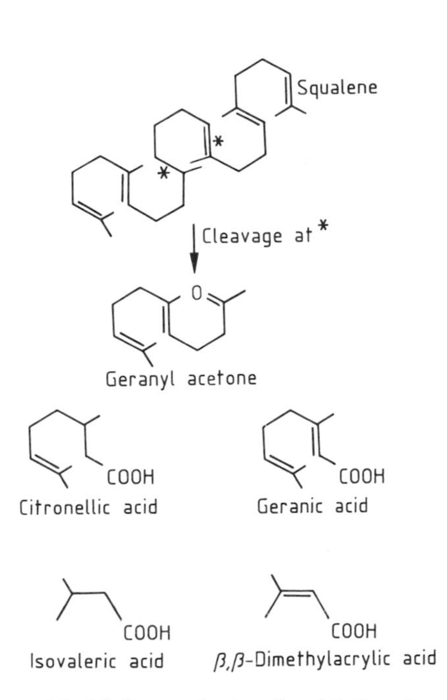

Figure 16. Main products of oxidative degradation of squalene by *Arthrobacter* sp. (YAMADA et al., 1975).

indicating the connection to intermediates of the tricarboxylic acid cycle (YAMADA et al., 1975, 1977). As demonstrated in Fig. 16 the initial oxidations take place at the double bonds.

G. Co-oxidation

Some microorganisms are able to transform long-chain alkanes only in the presence of another carbon source such as glucose or even another hydrocarbon (see Sect. IV.G.). Such a coupled reaction is classified as co-oxidation. This phenomenon should be clearly distinguished from bioconversion, i. e., transformation by non-proliferating (resting) cells and from co-metabolism which is appropriately applied to the conversion of pesticides (HORVATH, 1972; RAYMOND et al., 1971; LEADBETTER and FOSTER, 1960).

The failure of microorganisms to grow on a particular hydrocarbon is not only because they are not capable of attacking the

Table 17. Co-oxidation of Long-chain *n*-Alkanes (modified from REHM and REIFF, 1981). – The significance of the pathway in the organism is arbitrarily symbolized by +, ++, or +++.

Microorganisms	2-, 3-, 4-Position	4-, 5-, 6-Position	References
Bacillus coagulans	++	+	KACHHOLZ and REHM (1977)
Bacillus lentus	+	+	KACHHOLZ and REHM (1977)
Bacillus macerans	+	+++	KACHHOLZ and REHM (1977)
Bacillus stearothermophilius	+	+++	KACHHOLZ and REHM (1978)
Arthrobacter paraffineus	+		KLEIN and HENNING (1969)
Streptomyces eurythermus	+	++	GROSSEBÜTER et al. (1979)
Streptomyces griseus	+	++	GROSSEBÜTER et al. (1979)
Streptomyces violaceoruber	+	+	GROSSEBÜTER et al. (1979)
Chlorella vulgaris	+	+	SCHRÖDER and REHM (1981)

substrate but also due to the inability to use the oxidation products as growth substrates.

So far co-oxidation of long-chain alkanes has been formed predominantly for the genera of *Bacillus, Streptomyces,* and *Chlorella* (Table 17). In all examples the patterns of degradation products are consistent with subterminal pathways.

In addition, disubterminal intermediates such as diols, ketols, and diketones have been found in co-oxidation cultures of *Bacillus* spp. Here the two functional groups are located at the primary and a secondary C-atom on opposite ends of the molecule (see Fig. 13). Such substances have hitherto been unknown for microorganisms which grow on long-chain alkanes as the sole carbon source.

VI. Further Metabolism of Fatty Acids

A. Survey

Fatty acids produced by microbial alkane oxidation can be further metabolized via different pathways such as

1. β-oxidation yielding acetyl-CoA and carbon chains shortened by C-2-units (Sect. VI.B.),
2. α-oxidation leading to fatty acid carbon chains diminished by a C-1-unit (Sect. VI.C.),
3. ω-oxidation producing intact or degraded carbon chains, both terminals being oxidized (Sect. V.C.2.),
4. chain elongation and *de novo* synthesis by acetyl units (Sect. VI.B.),
5. incorporation of intact or altered fatty acid carbon skeletons into lipids (Sect. VI.D.).

B. β-Oxidation

The degradation of fatty acids in microorganisms follows the β-oxidation pathway as elucidated in animal tissues. This aspect of microbial lipid metabolism has been reviewed by FINNERTY and MAKULA (1975); a discussion on the energetics and localization of this pathway is given by RATLEDGE (1978).

The key step for β-oxidation and other metabolic transformations of fatty acids is the synthesis of the acyl-CoA thioester at the expense of adenosine triphosphate catalyzed by acyl-CoA synthetase. This enzyme has been studied in several bacteria (CALMES and DEAL, 1973; TRUST and MILLIS, 1971) and yeasts (DUVNJAK et al.,

1970; TRUST and MILLIS, 1970) grown on *n*-alkanes. Recently, in oleate-grown *Candida lipolytica* the existence of two acyl-CoA synthetases have been shown (Table 18) differing in subcellular localization, control mechanism, and metabolic functions (MISHINA et al., 1978a, b).

of acetyl-CoA by condensation with oxalacetate and glyoxylate. Results predominantly achieved with yeasts suggest that in alkane-utilizing cells the glyoxylate cycle shows a higher activity than the TCA cycle (HILDEBRAND and WEIDE, 1974; NABESHIMA et al., 1977; TANAKA et al., 1977).

Table 18. Comparison of the Properties of Acyl-CoA Synthetase from *Candida lipolytica* (MISHINA et al., 1978a, b)

Properties	Synthetase I	Synthetase II
Induction by fatty acid	No	Yes
Phosphatidylcholine dependency	No	Yes
Stability	High	Low
Substrate specificity	Narrow	Wide
Solubilization by Titron X-100	Easy	Difficult
Subcellular localization	Microsomes, Mitochondria, etc.	Peroxisomes
Function	Lipid synthesis	Fatty acid degradation

The β-oxidation system of yeasts is exclusively localized in peroxisomes, the development of which is induced by alkanes and fatty acids (TERANISHI et al., 1974; OSUMI et al., 1975). The mechanism has been found to be similar to that in castor beans (COOPER and BEEVERS, 1969) and rat liver (LAZAROW and DE DUVE, 1976). This means that acyl-CoA is oxidized by acyl-CoA oxidase (FAD-containing) yielding enoyl-CoA with concomitant consumption of molecular oxygen and formation of hydrogen peroxide. Catalase induced by alkanes participates in the degradation of the hydrogen peroxide thus formed. In the presence of NAD^+ and coenzyme A enoyl-CoA is further metabolized finally yielding acetyl-CoA (from even- and odd-chain alkanes) and propionyl-CoA (from odd-chain alkanes).

Acetyl-CoA, thus formed, can be further processed by the tricarboxylic acid cycle (TCA) which produces carbon dioxide and the reducing power linked to a respiratory system to yield energy. In addition, the glyoxylate cycle plays an important role in the biosynthesis of cellular components by the production of one C_4-compound, such as malate or succinate, from two molecules

C. α-Oxidation

The key reaction of this pathway is the transformation of the fatty acid to the corresponding α-hydroxy derivative which will then undergo oxidative decarboxylation:

$$R{-}CH_2{-}COOH \longrightarrow RCHOH{-}COOH \longrightarrow CO_2 + RCHO \longrightarrow RCOOH$$

The detection of 2-alkanones in an alkane-grown microorganism is no conclusive evidence for the occurrence of the α-oxidation pathway. These ketones are probably split into acetic acid and a primary alcohol for further oxidation.

However, firm evidence for α-oxidation has been provided for *Arthrobacter (Corynebacterium) simplex*. Palmitic acid is oxidized to α-hydroxypalmitic acid by a washed cell suspension and further to pentadecanoic acid when cells are used to grow on pentadecane (YANO et al., 1969, 1971a, b). In *Candida lipolytica* and probably some other yeasts propionyl-CoA derived from odd-chain alkanes is metabolized via unique ancillary C_7 tricarboxylic acids in

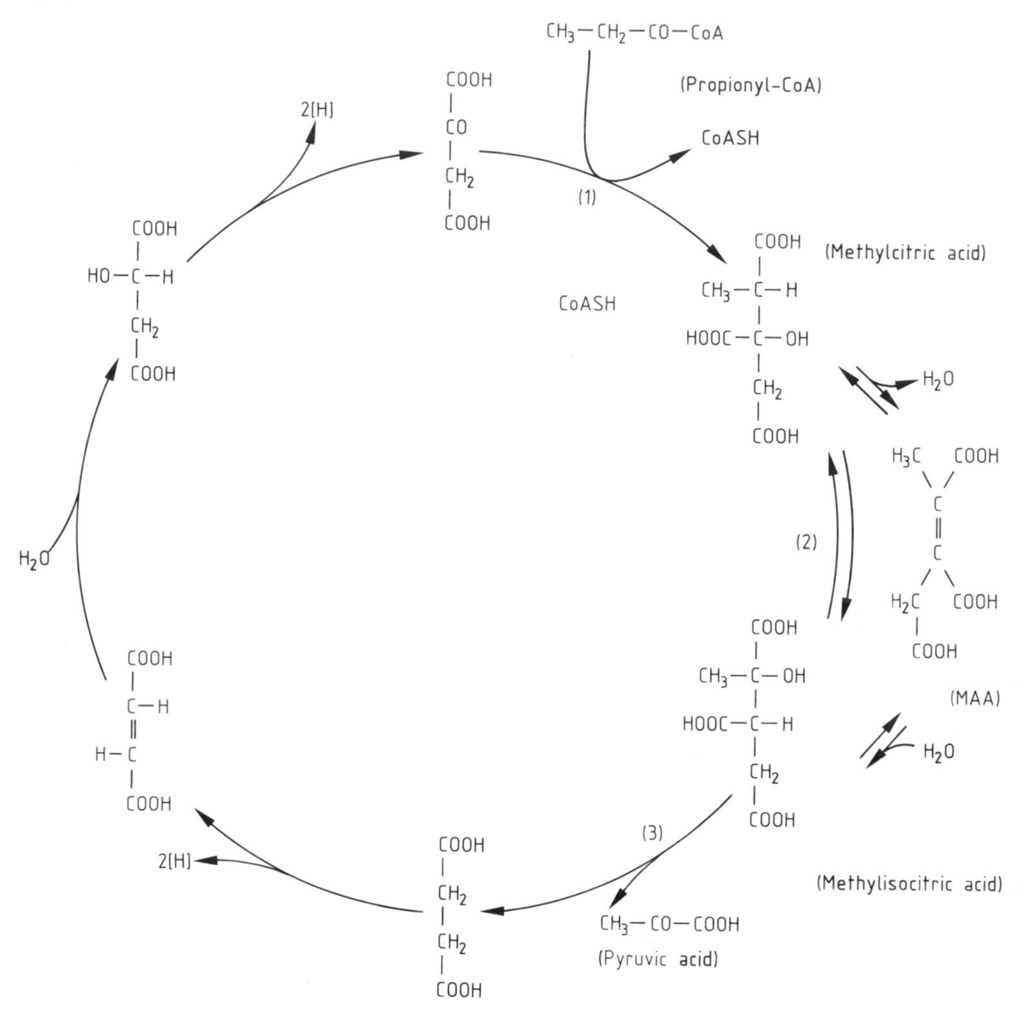

Figure 17. Methylcitric acid cycle in *Candida lipolytica* (TABUCHI and UCHIYAMA, 1975). – The intermediates are shown as their free acids. – MAA methylaconitic acid. Enzymes: (1) methylcitrate synthase; (2) aconitase; (3) methylisocitrate lyase.

the culture broth (TABUCHI and SERIZAWA, 1975). The key enzymes of the methylcitric acid cycle, the methylcitrate synthase and methylisocitrate lyase, seem to be constitutive as the enzymes of the TCA cycle, while the key enzymes of the glyoxylate cycle are inducible (TABUCHI and IGOSHI, 1978) (Fig. 17).

The existence of α-oxidation also in other organisms could be the reason for the high levels of even-chain fatty acids in the lipid of organisms grown on odd-chain alkanes (GILL and RATLEDGE, 1973). To account for these findings, either an unusual elongation system operates using, e. g., propionate, or α-oxidation occurs to be followed by conventional chain elongation by C_2-units. Another possible pathway may be the *de novo* synthesis yielding even-numbered cellular fatty acid even from odd-chain alkanes, as was shown for yeasts (TANAKA et al., 1978).

D. Lipids

It is well established that the hydrocarbon used as substrate determines the fatty acid composition within the lipid fraction of the organism; this has been extensively reviewed (BIRD and MOULTON, 1972; RATTREY et al., 1975; RATLEDGE, 1980; REHM and REIFF, 1981). The fatty acid composition may also depend on the growth phase during batch cultivation (JWANNY, 1975; MISHINA et al., 1973), though there are also examples of only minor variations (HUG and FIECHTER, 1973).

Alkanes and 1-alkenes of a chain length from C_{14} to C_{18} are usually accommodated with the fewest changes in the chain length no matter whether there is an even or an odd number of carbon atoms (REHM and REIFF, 1981).

Fatty acids corresponding to hydrocarbons with chain lengths below C_{14} obviously cannot be used to produce cytoplasmic membranes or storage lipids with satisfactory physical properties. Chain elongation of such acids frequently occurs as has been demonstrated for *Pseudomonas* sp. (BIRD and YEONG, 1974) and *Candida lipolytica* (TANAKA et al., 1976). At least in the latter organism the elongation mechanism is different from the *de novo* synthesis system. This has been concluded from inhibition studies with cerulenin, an anti-lipogenic antibiotic, which inhibits *de novo* synthesis of fatty acids but not the intact incorporation into lipids nor the elongation system (Fig. 18).

Microorganisms grown on odd-numbered alkanes of shorter chain length (C_{11} and C_{13}) also contain substantial proportions of even-chain fatty acids with 16 and 18 carbon atoms. Exemplarily this has been shown for *Candida* spp. (MISHINA et al., 1973; SKIPTON et al., 1974), for *Cunninghamella elegans,* and *Penicillium zonatum* (CERNIGLIA and PERRY, 1974). It has not been clarified yet whether these findings are due to the *de novo* synthesis of fatty acids from acetyl-CoA or to another mechanism (see Sect. VI.C.), for there are antithetical results on the alkane repression of acetyl-CoA carboxylase, the key enzyme of the *de novo* synthesis in yeast (GILL and RATLEDGE, 1973; MISHINA et al., 1976; FINNERTY, 1977; FUKUI and TANAKA, 1981; RATLEDGE, 1978).

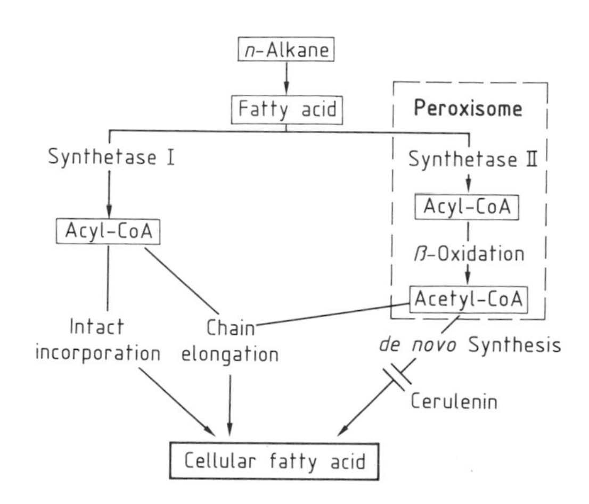

Figure 18. Proposed scheme of fatty acid metabolism in alkane-utilizing yeasts (FUKUI and TANAKA, 1981).

Alkanes longer than C_{18} are oxidized to fatty acids by the same mechanism like the alkanes with shorter carbon chains. However, microorganisms in general cannot grow on these acids which, therefore, remain mainly unmodified in large amounts, e. g., in the bound lipid fraction. Thus, the cells of *Mycobacterium convolutum* grown on *n*-alkanes with chains up to C_{28} contain at most only 12% of nonadecanoic acid as the longest acid (HALLAS and VESTAL, 1978). The presence of high proportions of pentadecanoic acid up to 45% indicates that a mid-chain cleavage possibly occurs, related to the various subterminal oxidation pathways (Sects. V.D. and V.E.).

Fatty acids are highly toxic to the cell although their toxicity may be reduced by partition of the acids into the *n*-alkane used as substrate (HUNKOVA and FENCL, 1977). Therefore, the vast majority of fatty acids are esterified to glycerol, to phosphoglyceric acid derivatives, to carbohydrates, or to fatty alcohols. The resulting triacylglycerols, phospholipids, glycolipids, or waxes may serve as structural components like phospholipids for the membranes or as storage products like the triacylglycerols within lipid vacuols.

The lipid content of microorganisms is increased when grown on alkanes instead of non-hydrocarbons. This might be due to a slow growth rate which can also enforce the lipid accumulation, because the rate of lipid synthesis is not controlled by the growth of the organism (GILL et al., 1977; HALL and RATLEDGE, 1977). It has been demonstrated for a continuous culture of *Candida tropicalis,* maintaining the same growth rate on hexadecane as on glucose, that the extractable lipid content doubles to 12% of the biomass with the hydrocarbon as substrate (HUG et al., 1974).

Table 19. Lipid Contents of Some Microorganisms Grown on *n*-Alkanes (RATLEDGE, 1980)

Organisms	Lipid Content of Biomass[a)]	References
Nocardia sp.	56% (crude lipid)	RAYMOND and DAVIES (1960)
Candida 107	37% (crude), 26% (corr.)[b)]	THORPE and RATLEDGE (1972)
Candida lipolytica	27% (fatty acids)	PELECHOVA et al. (1971)
Candida lipolytica	17%	NYNS et al. (1968)
Candida lipolytica	47% (corr.)[b)]	JWANNY(1975)
Candida lipolytica	4% (corr.)[b)]	MISHINA et al. (1977)
Candida tropicalis	17% (purified)	MISHINA et al. (1977)
Candida tropicalis	18% (crude), 6% (corr.)[b)]	THORPE and RATLEDGE (1972)
Candida tropicalis	17% (crude), 10% (corr.)[b)]	HUG and FIECHTER (1973)
Candida tropicalis	12% (crude?)	HUG et al. (1974)
Mycotorula japonica	15%	YAMAGUCHI and KURASAVA (1976)
Pichia vanriji	20% (fatty acids)	PELECHOVA et al. (1971)
Rhodotorula glutinis	35% (fatty acids)	PELECHOVA et al. (1971)
Rhodotorula gracilis	32% (fatty acids)	PELECHOVA et al. (1971)

[a] Values given for extracted lipid which if unfractionated may contain residual hydrocarbon substrate
[b] After subtraction of residual intracellular hydrocarbon content

Table 20. Cellular and Extracellular Lipids of *Acinetobacter* sp. HO1-N (FINNERTY, 1977)

Lipid Class	Cellular Lipid (μmol/g dry cell wt)		Extracellular Lipid (μmol/L culture medium)	
	NBYE[a)]	C_{16}[b)]	NBYE[a)]	C_{16}[b)]
Phospholipids	46.0	129	0	0
Mono- and diacylglycerol	0.38	6.8	0	410
Triacylglycerol	1.78	2.5	2.4	25.6
Fatty acid	7.5	8.2	4.0	60.0
Fatty alcohol	trace	2.6	0	0.5
Wax ester	11.5	18.0	0	280
Hexadecane	0	360	0	n. d.[c)]

[a] Cells grown on nutrient broth/yeast extract
[b] Cells grown on hexadecane as sole carbon source
[c] not determined

Generally, yeasts and molds exhibit the highest lipid contents among microorganisms, whereas only a few bacteria have a natural tendency to accumulate lipids (Table 19). The quantity of lipids within an organism can change during growth and may decline from its maximum value during late stages of growth and in the stationary phase (JWANNY, 1975).

Besides the total amount, the nature of the lipid components is also modified by growth on hydrocarbons as documented for an *Acinetobacter* sp. in Table 20 (FINNERTY, 1977). The increase in total phospholipid is related to the induction of extensive cytoplasmic membrane systems in response to growth on alkanes (KENNEDY and FINNERTY, 1975). The intracellular alkane inclusions amount to about 8% of the dry weight of hexadecane-grown cells (MAKULA et al., 1975). The accumulation of extracellular lipids in hydrocarbon-grown cultures emphasizes the physiological and biochemical differences induced by hydrocarbons, especially in terms of surfactant characteristics (Section III.).

Condensation of a long-chain alcohol and a long-chain fatty acid produces a wax. Microbial production of such substances has been known since cetyl palmitate was isolated from *Acinetobacter* sp. HO1-N *(Micrococcus cerificans)* growing on hexadecane (STEWART et al., 1959). Several other saturated wax esters are produced by *Acinetobacter* sp. (MAKULA et al., 1975) and *Nocardia* spp. (RAYMOND and JAMISON, 1971). As a less conventional type the unusual diester didecyldecane-1,10-dioate is formed by *Corynebacterium* 7E1C from *n*-decane (BACCHIN et al., 1974). Recently wax esters containing a large percentage of mono- and di-unsaturated components have been described from *Acinetobacter* sp. HO1-N grown on *n*-alkanes (C_{16}–C_{20}). These wax esters show a close chemical similarity with those found in sperm whale and jojoba oils (DEWITT et al., 1982). Only little is known of the enzymatic mechanism of wax formation, though the condensing activity from *Mycobacterium cereformans* has been sedimented (KRASSILNIKOV et al., 1973).

VII. Genetics and Strain Improvement

Microorganisms with improved properties may be gained by the application of genetic engineering techniques such as the development of recombinant DNA systems. These improvements may include the utilization of a broader substrate spectrum by one species, the synthesis of new products, the accumulation of special intermediate metabolites due to blocked pathways, or the increase of productivity by the enhanced synthesis of special enzymes (SHAPIRO et al., 1979, 1980, 1981; WILLIAMS and FRANKLIN, 1980).

A. Methylotrophic Microorganisms

Methylotrophic bacteria exhibit special potentials in synthesizing valuable chemicals while growing on cheap substrates such as methane or methanol. Methylotrophs already show broad substrate specificities but as soon as gene transfer systems are developed other metabolites may also be producible provided the suitable genes become available. The methane monooxygenase of the obligate methane utilizer *Methylococcus* NCJB 11083 has been shown to be constitutive. In the facultative methane utilizer *Methylobacterium organophilum* XX and *M.ethanolicum* the ability to oxidize methane is inducible (Table 21).

The enzymes necessary for methanol oxidation and assimilation in facultative methylotrophic bacteria are in most cases inducible by C_1-compounds. In *M.organophilum* XX and *Hyphomicrobium* X all genes coding for C_1-enzymes are proposed to be grouped in one operon with the exception of the gene for formate dehydrogenase. This enzyme is not coordinatedly controlled

Table 21. The Regulation of Methane Oxidation in Obligate and Facultative Methane Utilizers (O'CONNOR, 1981a)

Organisms	Growth Substrate	Presence of Methane Oxidation	Presence of Internal Membrane
Methylococcus NCIB 11083	Methane	+	+
	Methanol	+	+
Methylobacterium organophilum XX	Methane[a)]	+	+
	Methanol[a)]	–	–
	Glucose[a)]	–	–
	Methane + glucose[a)]	–	–
Methylobacterium ethanolicum H4.14	Methane[a)]	+	+
	Methanol[a)]	– or low	–
	Ethanol[a)]	– or low	–
	Succinate[a)]	– or low	–
	Methane + ethanol[a)]	+	+
	Methane + succinate[a)]	+	+
Methylobacterium organophilum R6	Methane[a)]	+	+
	Methanol	–	n. t.
	Succinate	–	n. t.

[a] Oxygen in the atmosphere was maintained at 0.10 atm or less
n. t. not tested

with the other C_1-enzymes (O'CONNOR, 1981a, b; O'CONNOR and HANSON, 1978).

Many facultative methylotrophs are unstable concerning the ability to oxidize methane. This behavior gave rise to the suggestion that methane utilization might be plasmid-encoded. Strains of *M.organophilum* contain a single large plasmid. Spontaneous mutants which had lost their ability to grow on methane no longer contained this plasmid. In obligate methylotrophs such plasmids could not be detected (HANSON, 1980; O'CONNOR, 1981b). The construction of plasmid-chromosome hybrids for genetic studies has been successful in several species and may be useful in developing strains with new desired qualities. Plasmids have been detected in different strains of *Methylomonas clara*. They are supposed not to be involved in methanol metabolism, but would be useful as vectors for gene cloning (STAHL and ESSER, 1982). The P1 group plasmids carrying R-factors have a wide host range. Thus, they could also be transferred to the facultative methylotroph *Pseudomonas extorquens* and to the obligate methylotroph *Methylosinus trichosporium*. They may be used for mapping of genetic markers including genes for methylotrophy or other desirable properties (WARNER et al., 1980). Enhanced chromosome mobilizing (ECM) plasmids such as R68.45 have been successfully transferred to a variety of facultative methylotrophs and can also promote chromosome transfer to *Methylophilus methylotrophus* (HOLLOWAY, 1981). A hybrid plasmid consisting of R1162-DNA and chromosomal *Pseudomonas* AM1-DNA, which contained the genetic information for methanol dehydrogenase, was transferred to a methanol dehydrogenase negative strain which regained the capability to grow on methanol (GAUTIER and BONEWALD, 1980).

A preliminary map of the *M.methylotrophus* AS1 genome has been obtained by the isolation of prime plasmids which can complement mutant functions when transferred to appropriate strains of *Pseudomonas aeruginosa* PAO (MOORE et al., 1983).

The transfer of the glutamate dehydrogenase of *Escherichia coli* by the plasmid vector pTB70 to *M.methylotrophus* increased the efficiency of the conversion rate from methanol to biomass (SCP) and has been the item of a patent application (WINDASS et al., 1980; ICI, 1980).

The improvement of carbon conversion in the SCP production process has also been achieved by modification of the genes specifying the oxidative transformation of methanol in *M.methylotrophus.* Increased energetic efficiency in the utilization of methanol could be accomplished by the replacement of genes coding for methanol oxidase and methanol dehydrogenase by an adh gene being capable of expressing an alcohol dehydrogenase enzyme. By means of the plasmid pTB107 as vector, the adh gene has been transferred from *Bacillus stearothermophilus* to the methylotrophic organism. The new strain as well as the methods leading to it have been claimed in a recent patent (ICI, 1981).

B. Microorganisms Degrading Higher Alkanes

Man-modified life is patentable as ruled for the first time by the US Supreme Court in favor of the patent application of CHAKRABARTY (1981). The techniques for developing multi-plasmid strains of *Pseudomonas* have been claimed therein. These strains combine the capabilities in metabolizing camphor, octane, salicylate, and naphthalene.

The conversion of higher alkanes is not only determined chromosomally but also by plasmids. Especially strains of *Pseudomonas* have been shown to contain plasmids for the degradation of aliphatic or aromatic hydrocarbons. The octane- and camphor-degradative plasmids OCT and CAM, originally isolated from *Pseudomonas putida,* belong to the IncP-2 group. The oxidation of longer alkanes seemed to be determined chromosomally in most strains of *Pseudomonas. P.aeruginosa* and *P.putida* are not able to grow on *n*-alkanes shorter than undecane unless they carry an IncP-2 plasmid which contains the alk genes. Strains of both species were able to oxidize chain lengths of 6 to 10 carbons as soon as they contained alk genes located on OCT, CAM-OCT, or other IncP-2 plasmids (FENNEWALD et al., 1979; FENNEWALD and SHAPIRO, 1977, 1979; SHAPIRO et al., 1980).

The OCT plasmid has been transferred from *Pseudomonas oleovorans* to several other species of *Pseudomonas* (CHAKRABARTY et al., 1973, 1978). High copy-number Inc Q plasmids such as RSF1010 were introduced into *P.aeruginosa* and *P.putida* and maintained stable. This suggests the potential usefulness of these plasmids as vectors in *Pseudomonas* strains (NAGAHARI and SAKAGUCHI, 1978).

In alk strains of *P.putida* the first two enzymes of the alkane oxidation pathway (alkane hydroxylase and alcohol dehydrogenase) are inducible while the subsequent steps of the oxidative route are constitutive (GRUND et al., 1975). Regulatory and structural genes encoding the alkane oxidizing

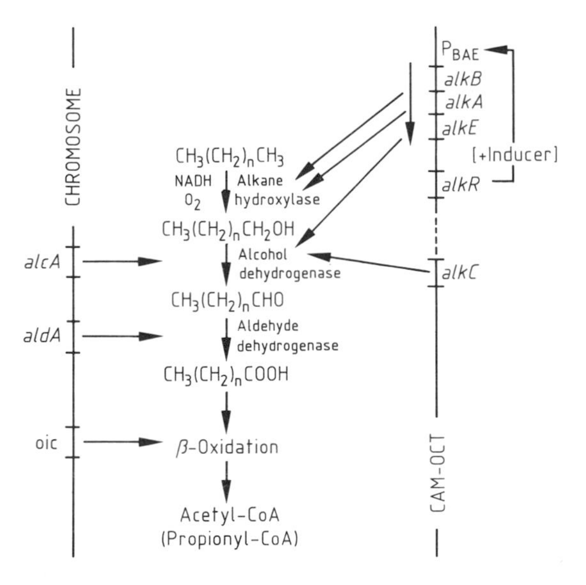

Figure 19. Genetic control of *n*-alkane oxidation by plasmid-bearing *Pseudomonas putida* (SHAPIRO et al., 1981).

enzymes have been localized on the IncP-2 plasmid of *P.putida* (Fig. 19).

The alkane hydroxylase is expressed by the alkA and alkB loci which are responsible for the soluble and membrane associated components, respectively. The unlinked alkC gene determines the inducible membrane alcohol dehydrogenase. The alkD locus appears to encode a product necessary for the synthesis of a membrane alkane hydroxylase component and of membrane alcohol dehydrogenase. The alkE gene controls the synthesis of membrane alcohol dehydrogenase, as well as the ability to grow on alcohols. The alkBAE operon is controlled by the alkR locus encoding one or more positive regulatory proteins. While the first two enzymatic steps of *n*-alkane oxidation originate from plasmid genes, most of the genes for further oxidation are located on chromosomes. The dehydrogenation step from alcohol to aldehyde is accomplished by an overlap between plasmid and chromosomal genes (BENSON, 1979; BENSON et al., 1977; BENSON and SHAPIRO, 1975, 1976; FENNEWALD et al., 1979; NIEDER and SHAPIRO, 1975; SHAPIRO et al., 1981).

VIII. Industrial Applications: A Survey of Patents

The patent literature of 1972 to 1982 has been reviewed for microbial processes utilizing alkanes as substrates, excluding the production of biomass. This (incomplete) summary will provide some material to outline the economical trends in fermentation technology of alkanes. However, the chances for an industrial realization of these processes are difficult to predict.

The references are listed in tables according to the products described therein: alkanols and alkane oxides (Table 22), ketones (Table 23), alkanoic acids and their derivatives (Table 24), alkane dioic acids (Table 25), and biosurfactants (Table 26).

For convenience the educts and products are characterized by chain length (e.g., C_6 for hexane), functional group (e.g., C_6-2-one), and degree of saturation (e.g., C_4H_6 for butadiene).

Table 22. Production of Alkanols and Alkyl Oxides from Hydrocarbons (Patents)

Products	Educts	Biocatalysts	Documents
C_6-C_{16}—OH	C_6-C_{16}	*Methylosinus trichosporium* *Methylococcus*	ICI, UK Pat. Appl. 2081306 (1982)
C_8—OH C_{10}—OH C_{12}—OH C_{14}—OH C_{16}—OH	C_8 C_{10} C_{12} C_{14} C_{16}	*Acinetobacter* NCIB 11613	ICI, Eur. Pat. Appl. 0062492 (1982)
Pristanol	Pristane	*Nocardia* BPM 1613	Agency of Industrial Science and Technology, Jpn. Kokai Tokkyo Koho 8199793 (1981)
C_1—OH C_2-oxide C_3-oxide C_4-oxide C_4H_6-oxide	C_1 C_3H_6 C_4H_8 C_4H_6	*Methylosinus* *Methylocystis* *Methylomonas* *Methylobacter* *Methylococcus*	Exxon, US Pat. 4269940 (1981)
C_1—OH C_3-oxide C_4-oxide C_4H_6-oxide	C_1 C_3H_6 C_4H_8 C_4H_6	*Methylosinus trichosporium*	Exxon, US Pat. 4266034 (1981)

Table 22. Continued

Products	Educts	Biocatalysts	Documents
C_6-C_{16}—OH C_3-oxide	C_6-C_{16}	*Methylosinus trichosporium*	ICI, UP Pat. Appl. 2024205 (1980)
C_{16}—OH	C_{16}	*Candida guilliermondii*	Akademie der Wissenschaften der DDR Ger. Democrat. Republ. Pat. 134639 (1979)
C_1—OH C_2—OH C_2-oxide C_3-oxide	C_2H_4-C_4H_8	*Methylosinus trichosporium*, *Methylocystis pouvus*, *Methylomonas* sp.	Exxon, UK Pat. Appl. GB 2019390 A (1979)
C_4-oxide C_4H_6-oxide	C_4H_6	*Methylococcus* sp.	
C_3-C_6—OH	C_3-C_6	*Methylosinus* sp. *Methylococcus* sp. *Methylobacter* sp. *Methylobacterium* sp. *Methylomonas* sp.	Exxon UK Pat. Appl. 2019772 (1979)
C_1—OH C_2-C_4-oxides	C_2H_4-C_4H_8 C_4H_8	*Methylosinus* sp. *Methylocystis* sp. *Methylomonas* sp. *Methylococcus* sp. *Methylobacterium* sp.	Exxon, DE 2915108 (1979)
C_2-C_4-oxides C_4H_6-oxide	C_2H_4-C_4H_8 C_4H_6	*Methylosinus* sp. *Methylocystis* sp. *Methylomonas* sp.	Exxon, UK Pat. Appl. 2018822 A (1979)
C_6—OH C_{16}—OH C_8—OH	C_6 C_{16} C_8	*Methylosinus trichosporium*	ICI, UK Pat. Appl. 2024205 (1979)
C_x-oxides	C_xH_{2x}		Bio Research Center Comp., Ltd., US Pat. 4106986 (1978)
C_8H_{16}-oxide C_8-dioxide	C_8H_{14}	*Pseudomonas oleovorans* ATCC 29347	Exxon, DE 2756287 (1976)
C_{14}—OH	C_{14}	*Arthrobacter* *Brevibacterium* *Micrococcus* *Corynebacterium* *Nocardia* *Pseudomonas*	Kyowa Hakko Kogyo Co., Ltd., Jap. Pat. 7928708 (1976)
C_{11}-C_{20}—OH	C_{11}-C_{20}	*Arthrobacter* sp. *Brevibacterium* sp. *Corynebacterium* sp. *Nocardia* sp.	Kyowa KK, Jap. Pat. 76028708 (1976)
C_{11}-C_{20}—OH	C_{11}-C_{20}	*Arthrobacter* sp. *Brevibacterium* sp. *Micrococcus* sp. *Corynebacterium* sp. *Nocardia* sp. *Pseudomonas* sp.	Kyowa Hakko Kogyo Co. Ltd., Jap. Pat. 7084057 (1976)
C_{14}—OH	C_{14}	*Pichia* yeast	Mobil Oil Corp., US Pat. 3880739 (1975)
C_4—OH	C_{12}-C_{18}	*Candida lipolytica* ATCC 20341	Pfizer, US Pat. 3756917 (1973)
C_6-C_{20}—OH	C_6-C_{20}	*Nocardia*	Phillips Petroleum Company, US Pat. 3510401 (1970)

Table 23. Production of Ketones from Hydrocarbons (Patents)

Products	Educts	Biocatalysts	Documents
C_3-C_6-ones C_3-C_6—OH C_3-C_6-oxides	C_1-C_6	*Methylosinus sporium* *Methylocystis parvus* *Methylomonas* sp. *Methylobacter* sp. *Methylococcus* sp.	Exxon, US Pat. 4269940 (1981)
C_3-C_6-ones	C_3-C_6	*Methylosinus* sp. *Methylomonas* sp. *Methylobacter* sp. *Methylococcus* sp.	Exxon, US Pat. 4286630 (1981)
C_3-C_6-ones	C_3-C_6—OH	*Methylosinus* sp. *Methylomonas* sp. *Methylobacter* sp. *Methylococcus* sp.	Exxon, US Pat. 4241184 (1980)
C_3-C_6-ones	C_3-C_6	*Methylosinus* sp. *Methylobacter* sp. *Methylococcus* sp. *Methylobacterium* sp. *Methylomonas* sp. *Methylocystis* sp.	Exxon, UK Pat. Appl. 2018822 (1979)
C_3-C_6-ones	C_3-C_6	*Methylosinus* sp. *Methylomonas* sp. *Methylobacter* sp. *Methylococcus* sp. *Methylobacterium* sp.	Exxon, UK Pat. Appl. 2018772A (1979)
C_3-C_6-ones	C_3-C_6	*Methylotrophic microorganisms*	Exxon, DE 2915106 (1979)
Cycl. ketones	C_6-C_{22}	*Torulopsis bombicola*	Phillips Petroleum Co., US Pat. 3963571 (1976)
C_{13}—CHO	C_{14}	*Pichia* sp.	Mobil Oil Corp., US Pat. 3880739 (1975)
C_6-2-one C_6-3-one C_6-4-one	C_{16} C_6H_{14}	*Arthrobacter* ATCC 21237	Texaco Inc., US Pat. 3625824 (1971)
C_6-C_{20}-ones	C_6-C_{20} (Iso-)	*Nocardia*	Phillips Petroleum Company, US Pat. 3510401 (1970)

Table 24. Production of Alkanoic Acids and Their Derivatives from Hydrocarbons (Patents)

Products	Educts	Biocatalysts	Documents
Chloropalmitate	C_{12}-C_{18} Alkylchlorides	*Arthrobacter* *Nocardia* *Corynebacterium*	Kao Soap Co. Ltd., Jpn. Kokai Tokkyo Koho 8250893 (1982)
C_{13}-oic acid	C_{13}	not specified	Nippon Mining Co. Ltd., Jpn. Kokai Tokkyo Koho 82105193 (1982)
C_{12}-oic acid C_{12}-dioic acid	C_{11}-C_{18}	*Candida tropicalis* 1098 (FERM-P3291)	Bio Research Center Co. Ltd., Jpn. Kokai Tokkyo Koho 81154993 (1981)
C_x-oic acids	C_x		Bio Research Center Co. Ltd., Jap. Pat. 56082095 (1981)
ω, (ω-1)-2-OH C_{12}-C_{18}-oic acid	$C_{12}H_{24}$ $C_{18}H_{36}$	*Candida tropicalis* BR-254	Bio Research Center Co. Ltd., DE 2937292 (1980)
C_{10}-oic acid C_{11}-oic acid C_{12}-oic acid C_{13}-oic acid C_{14}-oic acid C_{15}-oic acid C_{16}-oic acid	C_{10} C_{11} C_{12} C_{13} C_{14} C_{15} C_{16}	*Debaryomyces phaffii* ATCC 20499	Bio Research Center Co. Ltd., DE 2853847 (1979)
1-OH—C_{10}-oic acid	$C_{10}H_{20}$	*Candida lipolytica*	Bio Research Center Co. Ltd., Jpn. Kokai Tokyo Koho 78130489 (1978)
1-OH—C_{10}-C_{18}-oic acids	$C_{10}H_{20}$-$C_{18}H_{36}$	*Candida lipolytica* ATCC 20496	Bio Research Center Co. Ltd., US Pat. 4059488 (1977)
1-OH—C_6-C_{22}-oic acids	C_6-C_{22}	*Torulopsis bombicola*	Phillips Petroleum Co., US Pat. 3963571 (1976)
C_{14}-C_{20}-oic	C_{14}-C_{20}	Immobilized *Saccharomycopsis lipolytica* in gels	Brit. Petroleum Co. Ltd., Belg. Pat. 841057 (1976)
C_{18}-oic-Me C_{16}-oic-Me		*Torulopsis bombicola* PRL 319-67	Phillips Petroleum Co., US Pat. 3796630 (1974)
13-OH—C_{13}-oic acid 12-OH—C_{13}-oic acid	C_{13}	*Corynebacterium dioxydans* MC-1-1 (FERM-P No 690 ATCC 21766)	Hasegawa Co., Ltd., DE 214626 (1972)
C_{15}-C_{20}-oic acids	C_{15}-C_{20}	*Arthrobacter paraffineus* ATCC 15591	Kyowa Hakko Kogyo Co. Ltd., DE 1903335 (1970)

Table 25. Production of Alkane Dioic Acids from Alkanes (Patents)

Products	Educts	Biocatalysts	Documents
C_{14}-dioic acid C_{12}-dioic acid C_8-dioic acid C_6-dioic acid	n-C_x	*Pichia carboniferus*	Dainippon Ink Chemicals, Inc., Jpn. Kokai Tokkyo Koho 82 129 694 (1982)
C_{13}-dioic acid	n-C_{13}	*Brettanomyces petrophilum* ATCC 20 224	Mitsubishi Petrochemical Co., Ltd., Jpn. Kokai Tokkyo Koho 8 279 889 (1982)
C_{13}-dioic acid	n-C_{13}	*Torulopsis candida*	Mitsubishi Petrochemical Co. Ltd., Jpn. Kokai Tokkyo Koho 82 102 191 (1982)
C_{13}-dioic acid	n-C_{13}	*Brettanomyces petrophilum* ATCC 20 224	Mitsui Petrochemical Ind. Ltd., Japan Pat. 7 079 889 (1982)
C_x-dioic acid	n-C_x	*Torulopsis candida*	Mitsui Petrochemical Ind. Ltd., Japan Pat. 5 726 394 (1982)
C_{13}-dioic acid	n-C_{13}	—	Nippon Mining Co. Ltd., Jpn. Kokai Tokkyo Koho 82 105 193 (1982)
C_{13}-dioic acid	n-C_{13}	*Candida tropicalis*	Nippon Mining Co. Ltd., US Pat. 4 339 536 (1982)
C_{12}-dioic acid C_{12}-dioic acid	n-C_{11}-n-C_{18}	*Candida tropicalis* 1098 (FERM-P 3291)	Bio-Research Center Co. Ltd., Jpn. Kokai Tokkyo Koho 81 154 993 (1981)
C_{11}-C_{18}-dioic acids	n-C_{11}-C_{18}	*Candida* sp.	Bio Research Center Co. Ltd., Japan Pat. 56 154 993 (1981)
C_{10}-dioic acid	n-C_{10}	*Candida* Thus, *Candida tropicalis* B-70-3	Ashai Denka Kogyo K.K., J 7 903 950 (1979)
C_4-dioic acid	C_{16}	*Brettanomyces petrophilum* (ATCC 20 224)	Ashai Chemical Ind. Co. Ltd., Japan Pat. 7 749 066 (1977)
C_{10}-dioic acid	n-C_{10}	*Torulopsis* TIT-99	Mitsui Petrochemical Ind. Ltd., Jpn. Kokai Tokkyo Koho 7 425 186 (1974)
C_{18}-dioic-di-Me C_{16}-dioic-Me C_{18}-oic-Me	n-C_{18}	*Torulopsis bombicola* PRL 319-67	Phillips Petroleum Co., US Pat. 3 796 630 (1974)
C_{10}-C_{16}-dioic acids	C_{10}-C_{16}	*Corynebacterium* 7EIC (ATCC 19 067)	DuPont, US Pat. 3 784 445 (1974)
C_{16}-dioic acid	n-C_{16}	*Candida* sp.	Ajinomoto Co. Inc., DE 2 140 133 (1973)
C_{12}-C_{22}-dioic acids	n-C_{12}-C_{22}	*Corynebacterium* sp. ATCC 21 745	DuPont, US Pat. 3 773 621 (1973)
C_{10}-dioic acid C_8-dioic acid	n-C_{10}	*Candida lopsta* MF1 (FERM-P1040)	Mitsui Shipbuilding and Engineering Co. Ltd., Jpn. Kokai Tokkyo Koho 7 339 690 (1973)
C_{13}-dioic acid	n-C_{13}	*Corynebacterium dioxydans* MC-1-1 (FERM-P No690 ATCC 21 766)	Hasegawa Co., DE 216 426 (1972)

Table 26. Production of Biosurfactants from Hydrocarbons (Patents)

Products	Educts	Microorganisms	Documents
Biosurfactant	Hexadecane 3%	*Corynebacterium salvinicum* SFC	J. E. ZAJIC and R. K. GERSON, US Pat. 4355109 (1982)
Demulsifier	Hexadecane 4%	*Nocardia amarea* *Rhodococcus aurantiacus* *Rhodococcus rubropertinctus*	J. E. ZAJIC and D. G. COOPER, Can. Pat. 1133840 (1982)
Biosurfactant	Hexadecane	*Corynebacterium salvinicum* SFC	J. E. ZAJIC et al., Can. Pat. 1125688 (1982)
Biosurfactant	Hexadecane	*Corynebacterium salvinicum* SFC	J. E. ZAJIC et al., SFC Can. Pat. 1125683 (1982)
Emulsan	Hexadecane 0.2%	*Acinetobacter* ATCC 31012	Petroleum Fermentation N.V., US Pat. 4311830 (1982)
Biosurfactant	Paraffins (C_6-C_8)	*Corynebacterium lepus* OSGB-1 (UWO-OSGB1) *Arthrobacter terregens* 5A (UWO-5A)	Canadian Patents and Development Ltd., Can. Pat. 1114759 (1981)
2,6,10-Trimethyl-1-pentadecanol	Norpristane	*Nocardia* BPM 1613 (FERM-P1609)	Agency of Industrial Sciences and Technology, Jap. Pat. 57150391 (1981)
2,6,10-Trimethyl-1-pentadecanol 2,6,10,14-Tetramethyl-1-pentadecanol (Pristanol) 2,6,10,14-Tetramethyl-14-pentadecene-1-ol 2,6,10,14-Tetramethyl-1-hexadecanol	Isoprenoid hydrocarbons	*Nocardia* BPM 1613 (FERM-1609)	Agency of Industrial Sciences and Technology, Jap. Pat. 57150392 (1980)
Trehalose lipid	Hydrocarbons (Oil spill)	Yeast sp.	F. WAGNER et al. DE 2843685 (1980)
Glycolipids	*n*-Alkanes C_8–C_{24} (2%) Crude oil (9%)	*Nocardia rhodochrous*	F. WAGNER et al. DE 2805823 (1979)
Biosurfactant	*n*-Alkanes	*Candida* sp.	VEB Petrol-chemisches Kombinat Schwedt, Ger. Democr. Republ. Pat. 139069 (1979)
Biosurfactant	Paraffins 1.5%	*Corynebacterium hydrocarboclastus*	Canadian Patents and Development Ltd., Can. Pat. 990668 (1976)

Acknowledgements
The authors appreciate stimulating discussions with Dr. A. Einsele, Prof. Dr. H.-J. Rehm, and Dr. R. Schmid. The skillful help of Mrs. U. Griesbach and Mr. K. Siekmann in editing the manuscript is gratefully acknowledged.

IX. References

ABBOTT, B. J., and CASIDA, L. E. (1968). J. Bacteriol. *96,* 925.

ABBOTT, J. B., and GLEDHILL, W. E. (1971). Adv. Appl. Microbiol. *14,* 249.

ABBOTT, B. J., and HOU, C. T. (1973). Appl. Microbiol. *26,* 86.

ALBRO, P. W. (1976). In "Chemistry and Biochemistry of Natural Walls" (P. E. KOLATTUKUDY, ed.), pp. 419–443. Elsevier, Amsterdam.

ALEXANDER, M. (1981). Science *211,* 132.

ALLEN, J. E., and MARKOVETZ, A. J. (1970). J. Bacteriol. *103,* 426.

ANTHONY, C. (1980). In "Hydrocarbons in Biotechnology" (D. E. F. HARRISON, I. J. HIGGINS, and R. WATKINSON, eds.), pp. 35–57. The Institute of Petroleum, London.

ANTHONY, C. (1982). "The Biochemistry of Methylotrophs". Academic Press, London.

ARIMA, K., KAKINUMA, A., and TAMURAA, G. (1968). Biochem. Biophys. Res. Commun. *31,* 488.

ATLAS, R. M. (1981). Microbiol. Rev. *45,* 180.

AURICH, H. (1979). Sitzungsber. Akad. d. Wiss. DDR, Math., Naturwiss., Tchn. (16N), 3.

AURICH, H., and EITNER, G. (1977). Z. Allg. Mikrobiol. *17,* 203.

AURICH, H., SORGER, D., and ASPERGER, O. (1976). Acta Biol. Med. Germ. *35,* 443.

AURICH, H., BRÜCKNER, A., ASPERGER, O., BEHRENDS, B., and FUTTIG, A. (1977). Z. Allg. Mikrobiol. *17,* 249.

AUSTIN, B., CALOMIRIS, J. J., WALKER, J. D., and COLWELL, R. R. (1977). Appl. Environ. Microbiol. *34,* 60.

BABEL, W., and HOFMANN, K. H. (1982). Arch. Microbiol. *132,* 179.

BACCHIN, P., ROBERTIELLO, A., and VIGLIA, A. (1974). Appl. Microbiol. *28,* 737.

BACHOFEN, R. (1982). Experientia *38,* 47.

BARONCELLI, V., BOCCALON, C., GIANNINI, J., and RENZI, P. (1979). Mol. Cell Biochem. *28,* 3.

BARTHA, R., and ATLAS, R. M. (1977). Adv. Appl. Microbiol. *22,* 225.

BASSEL, J., and ORGYDZIAK, D. M. (1979). In "Genetics of Industrial Microorganisms" (O. K. SEBEK and A. I. LASKIN, eds.), pp. 160–165. Am. Soc. Microbiology, Washington.

BATRAKOV, S. G., ROZYNOV, B. V., KORONELLI, T. V., and BERGELSON, L. D. (1981). Chem. Phys. Lipids *29,* 241.

BAYER, E. A., ROSENBERG, E., and GUTNICK, D. (1981). J. Gen. Microbiol. *127,* 295.

BEARDMORE-GRAY, M., O'KEEFFE, D. T., and ANTHONY, C. (1983). J. Gen. Microbiol. *129,* 923.

BEEBE, J. L., and UMBREIT, W. W. (1971). J. Bacteriol. *108,* 612.

BEMMANN, W., and TRÖGER, R. (1975). Zentralbl. Bakteriol. Parasitenkd. Infektionskr. Hyg. Abt. 2 *129,* 742.

BENSON, S. A. (1979). J. Bacteriol. *140,* 1123.

BENSON, S. A., and SHAPIRO, J. (1975). J. Bacteriol. *123,* 759.

BENSON, S. A., and SHAPIRO, J. (1976). J. Bacteriol. *126,* 794.

BENSON, S. A., FENNEWALD, M., SHAPIRO, J., and HUETTNER, C. (1977). J. Bacteriol. *132,* 614.

BERDICHEVSKAYA, M. V. (1982). Mikrobiologiya *51,* 146.

BERNHEIMER, A. W., and AVIGAD, L. S. (1970). J. Gen. Microbiol. *61,* 361.

BERTRAND, J. C., GALLO, M., and AZOULAY, E. (1973). Biochemie *55,* 343.

BERTRAND, J. C., GILEWICZ, M., BAZIN, H., ZACEK, M., and AZOULAY, E. (1979). FEBS Lett. *105,* 143.

BERTRAND, J. C., GILEWICZ, M., BAZIN, H., and AZOULAY, E. (1980). Biochem. Biophys. Res. Commun. *94,* 889.

BESSON, F., PEYPOUX, F., MICHEL, G., and DECAMBE, L. (1977). Eur. J. Biochem. *77,* 61.

BEST, D. J., and HIGGINS, I. J. (1983). In "Topics in Enzyme and Fermentation Biotechnology", Vol. 7 (A. WISEMAN, ed.), pp. 38–75, John Wiley & Sons, New York.

BIRD, C. W., and MOULTON, P. (1972). Top. Lipid Chem. *3,* 125.

BIRD, C. W., and YEONG, Y. C. (1974). Chem. Ind. (London), 459.

BLANCH, H. W., and EINSELE, A. (1973). Biotechnol. Bioeng. *15,* 861.

BOS, P., and DEBRUYN, J. C. (1973). Antonie van Leeuwenhoek. J. Microbiol. Serol. *39,* 99.

BOYER, R. F., LODE, E. T., and COON, M. J. (1971). Biochem. Biophys. Res. Commun. *44,* 925.

BRADE, H., and GALANOS, C. (1982). Eur. J. Biochem. *122,* 233.

BREUIL, C., and KUSHNER, D. J. (1980). Can. J. Microbiol. *26,* 223.

BRITTON, L. N., BRAND, J. M., and MARKOVETZ, A. J. (1974). Biochim. Biophys. Acta *369,* 45.

BRUNDISH D. E., SHAW, N., and BADDILEY, J. (1967). Biochem. J. *105,* 885.

BRYOM, D. (1981). In "Taxonomy of Methylo-

trophs: a Reappraisal" (H. DALTON, ed.), pp. 278–284. Heyden, London.
BÜNING-PFAUE, H., and REHM, H.-J. (1972). Arch. Microbiol. *86,* 231.
CALMES, R., and DEAL, S. J. (1973). J. Bacteriol. *114,* 249.
CARDINI, G., and JURTSHUK, P. (1970). J. Biol. Chem. *245,* 2789.
CASS, A. E. G., EDDOWES, M. J., HILL, H. A. O., and UOSAKI, K. (1980). Nature *285,* 673.
CERNIGLIA, C. E., and PERRY, J. J. (1974). J. Bacteriol. *118,* 844.
CERNIGLIA, C. E., BLEVINS, W. T., and PERRY, J. J. (1976). Appl. Environ. Microbiol. *32,* 764.
CHAKRABARTY, A. M. (1981). US Patent 4259444.
CHAKRABARTY, A. M., CHOU, G., and GUNSALUS, I. C. (1973). Proc. Natl. Acad. Sci. USA *70,* 1137.
CHAKRABARTY, A. M., FRIELLO, D. A., and BOPP, L. M. (1978). Proc. Natl. Acad. Sci. USA *75,* 3109.
CHOUTEAU, J., AZOULAY, E., and SENEZ, J. C. (1962). Nature *194,* 576.
CLAUS, R., ASPERGER, O., and KLEBER, H.-P. (1978). Wiss. Ztschr. Karl-Marx-Univ., Math.-Nat. Reihe *27,* 17.
COLBY, J., and DALTON, H. (1978). Biochem. J. *171,* 461.
COLBY, J., STIRLING, D. I., and DALTON, H. (1977). Biochem. J. *165,* 395.
COLBY, J., DALTON, H., and WITTENBURY, R. (1979). Annu. Rev. Microbiol. *33,* 481.
COLWELL, R. R., and WALKER, J. D. (1977). CRC Crit. Rev. Microbiol. *5,* 423.
COOPER, D. G., and BEEVERS, H. (1969). J. Biol. Chem. *244,* 3514.
COOPER, D. G., ZAJIC, J. E., and GERSON, D. F. (1979). Appl. Environ. Microbiol. *37,* 4.
COOPER, D. G., ZAJIC, J. E., GERSON, D. F., and MANNINEN, K. J. (1980). J. Ferment. Technol. *58,* 83.
COOPER, D. G., ZAJIC, J. E., and DENIS, C. (1981). J. Am. Oil Chem. Soc. *58,* 77.
COOPER, D. G., AKIT, J., and KOSARIC, N. (1982). J. Ferment. Technol. *60,* 19.
DALTON, H. (1980a). In "Hydrocarbons in Biotechnology" (D. E. F. HARRISON, I. J. HIGGINS, and R. WATKINSON, eds.), pp. 85–97. Heyden and Son, The Institute of Petroleum, London.
DALTON, H. (1980b). Adv. Appl. Microbiol. *26,* 71.
DALTON, H. (1981). In "Microbial Growth on C_1 Compounds" (H. DALTON, ed.), pp. 1–10. Heyden, London.
DAVIES, J. S., and WESTLAKE, D. W. S. (1979). Can. J. Microbiol. *25,* 146.
DE BEER, R., DUINE, J. A., FRANK, J., and WESTERLING, J. (1983). Eur. J. Biochem. *130,* 105.
DE BONT, J. A. M., and VAN GINKEL, D. G. (1983). Enzym. Microbiol. Technol. *5,* 55.
DELAISSE, J. M., MARTIN, P., VERHEYEN-BOUVY, M. F., and NYNS, E. J. (1981). Biochim. Biophys. Acta *676,* 77.
DEWITT, S., ERVIN, J. L., HOWES-ORCHISON, D., DALIETOS, D., NEIDLEMAN, S. L., and GEIGERT, J. (1982). J. Am. Oil Chem. Soc. *59,* 69.
DMITRIEV, V., TSIOMENKO, A., KULAEV, I., and FIKHTE, B. (1980). Eur. J. Appl. Microbiol. Biotechnol. *9,* 211.
DROZD, J. W. (1980). In "Hydrocarbons in Biotechnology" (D. E. F. HARRISON, I. J. HIGGINS, and R. J. WATKINSON, eds.), pp. 75–83. Heyden, London.
DUINE, J. A., and FRANK, J. (1981). In "Microbial Growth on C_1 Compounds" (H. DALTON, ed.), pp. 31–41. Heyden, London.
DUPPEL, W., LEBEAULT, J. M., and COON, M. J. (1973). Eur. J. Biochem. *36,* 583.
DUVNJAK, Z., and KOSARIC, N. (1981). Biotechnol. Lett. *3,* 583.
DUVNJAK, Z., LEBEAULT, J. M., ROCHE, B., and AZOULAY, E. (1970). Biochim. Biophys. Acta *202,* 447.
EDWARDS, J. R., and HAYASHI, J. A. (1965). Arch. Biochem. Biophys. *11,* 415.
EINSELE, A. (1983). In "Biotechnology" Vol. 3 (H.-J. REHM and G. REED, eds.), pp. 43–81. Verlag Chemie, Weinheim - Deerfield Beach/Florida - Basel.
EINSELE, A., and FIECHTER, A. (1971). Adv. Biochem. Eng. *1,* 169.
EINSELE, A., SCHNEIDER, H., and FIECHTER, A. (1975). J. Ferment. Technol. *53,* 241.
ERICKSON, L. E., and NAKAHARA, T. (1975). Proc. Biochem. *10* (5), 9.
EUBANKS, E. F., FORNEY, F. W., and LARSON, A. D. (1974). J. Bacteriol. *120,* 1133.
FENNEWALD, M., and SHAPIRO, J. (1977). J. Bacteriol. *132,* 622.
FENNEWALD, M. A., and SHAPIRO, J. A. (1979). J. Bacteriol. *139,* 264.
FENNEWALD, M., BENSON, S., OPPICI, M., and SHAPIRO, J. (1979). J. Bacteriol. *133,* 940.
FERRIS, J. P., MACDONALD, L. H., PARIE, M. A., and MARTIN, M. A. (1976). Arch. Biochem. Biophys. *175,* 443.
FINNERTY, W. R. (1977). TIBS II, 73.
FINNERTY, W. R., and MAKULA, R. A. (1975). CRC Crit. Rev. Microbiol. *4,* 1.

FINNERTY, W. R., HAWTREY, E., and KALLIO, R. E. (1962). Z. Allg. Mikrobiol. *2,* 169.

FISCHER, W., BRÜCKNER, B., and MEYER, H. W. (1982). Z. Allg. Mikrobiol. *22,* 227.

FISH, N. M., HARBRON, S., ALLENBY, D., and LILLY, M. D. (1983). Eur. J. Appl. Microbiol. Biotechnol. *17,* 57.

FORNEY, F. W., and MARKOVETZ, A. J. (1970). J. Bacteriol. *102,* 281.

FREDERICKS, K. (1967). Antonie van Leeuwenhoek J. Microbiol. Serol. *33,* 41.

FUKUI, S., and TANAKA, A. (1981). Adv. Biochem. Eng. *19,* 217.

GALLO, M., BERTRAND, J. C., and AZOULAY, E. (1971). FEBS Lett. *19,* 45.

GALLO, M., BERTRAND, J. C., ROCHE, B., and AZOULAY, E. (1973). Biochim. Biophys. Acta *294,* 624.

GALLO, M., ROCHE, B., and AZOULAY, E. (1976). Biochim. Biophys. Acta *419,* 425.

GAUTIER, F., and BONEWALD, R. (1980). Mol. Gen. Genet. *178,* 375.

GILEWICZ, M., ZACEK, M., BERTRAND, J.-C., and AZOULAY, E. (1979). Can. J. Microbiol. *25,* 201.

GILL, C. O., and RATLEDGE, C. (1973). J. Gen. Microbiol. *78,* 337.

GILL, C. O., HALL, M. J., and RATLEDGE, C. (1977). Appl. Environ. Microbiol. *33,* 231.

GMÜNDER, F. K., KÄPPELI, O., and FIECHTER, A. (1981). Eur. J. Appl. Microbiol. Biotechnol. *12,* 129.

GOLDMANN, S., SHABTAI, Y., RUBINOVITZ, C., ROSENBERG, E., and GUTNICK, D. L. (1982). Appl. Environ. Microbiol. *44,* 165.

GOMA, G., PAREILLEUX, A., and DURAND, G. (1973). Arch. Microbiol. *88,* 97.

GORIN, P. A. J., SPENCER, J. F. T., and TULLOCH, A. P. (1961). Can. J. Chem. *39,* 846.

GRANGE, J. M. (1974). J. Appl. Bacteriol. *37,* 465.

GRIFFITHS, R. P., CALDWELL, B. A., CLINE, J. D., BROICH, W. A., and MORITA, R. Y. (1982). Appl. Environ. Microbiol. *44,* 435.

GROSSEBÜTER, W., REIFF, I., and REHM, H.-J. (1979). Eur. J. Appl. Microbiol. Biotechnol. *8,* 139.

GRUND, A., SHAPIRO, J., FENNEWALD, M., BACHA, P., LEAHY, J., MARKBREITER, K., NIEDER, M., and TOEPFER, M. (1975). J. Bacteriol. *123,* 546.

GUERRILLOT, L., and VANDECASTEELE, J. P. (1977). Eur. J. Biochem. *81,* 185.

GULEVSKAYA, S. A., and SHISHKANOVA, N. V. (1982). Mikrobiologiya *51,* 82.

GUTIERREZ, J. R., and ERICKSON, L. E. (1977). Biotechnol. Bioeng. *19,* 1331.

GUTNICK, D. L., and ROSENBERG, E. (1977). Annu. Rev. Microbiol. *31,* 379.

HALL, M. J., and RATLEDGE, C. (1977). Appl. Environ. Microbiol. *33,* 577.

HALLAS, L. E., and VESTAL, J. B. (1978). Can. J. Microbiol. *24,* 1197.

HAMER, G. (1979). In "Economic Microbiology". Vol. 4 (A. H. ROSE, ed.), pp. 315-360. Academic Press, London.

HAMMER, K. D., and LIEMANN, F. (1976). Zentralbl. Bakteriol. Hyg., I. Abt. Orig. B *162,* 169.

HANSON, R. S. (1980). Adv. Appl. Microbiol. *26,* 3.

HARDER, W., and VAN DIJKEN, J. P. (1976). In "Microbial Production and Utilization of Gases" (G. SCHLEGEL, G. GOTTSCHALK, and N. PFENNIG, eds.), pp. 403-418. E. Goltze KG, Göttingen.

HARRISON, D. E. F. (1978). Adv. Appl. Microbiol. *24,* 129.

HEINEMANN, B., KAPLAN, M. A., MUIR, R. D., and HOOPER, I. R. (1953). Antibiot. Chemother. (Washington, D. C.) *3,* 1239.

HIGGINS, I. J., HAMMOND, R. C., SARIASLANI, F. S., BEST, D., DAVIES, M. M., TRYHORN, S. E., and TAYLOR, F. (1979). Biochem. Biophys. Res. Comm. *89,* 671.

HIGGINS, I. J., BEST, D. J., and HAMMOND, R. C. (1980a). Nature *286,* 561.

HIGGINS, I. J., HAMMOND, R. C., PLOTKIN, E., HILL, H. A. O., UOSAKI, K., EDDOWES, M. J., and CASS, A. E. G. (1980b). In "Hydrocarbons in Biotechnology" (D. E. F. HARRISON, I. J. HIGGINS, and WATKINSON, eds.), pp. 181-193. Heyden, London.

HIGGINS, I. J., BEST, D. J., HAMMOND, R. C., and SCOTT, D. (1981a). Microbiol. Rev. *45,* 556.

HIGGINS, I. J., BEST, D. J., and SCOTT, D. (1981b). In "Microbial Growth on C_1 Compounds" (H. DALTON, ed.), pp. 11-20. Heyden, London.

HILDEBRAND, W., and WEIDE, H. (1974). Z. Allg. Mikrobiol. *14,* 47.

HILL, A. O., and HIGGINS, I. J. (1980). Brit. Patent 2 033 428.

HISATSUKA, K., NAKAHARA, T., and YAMADA, K. (1972). Agric. Biol. Chem. *36,* 1361.

HISATSUKA, K., NAKAHARA, T., SANO, N., and YAMADA, K. (1971). Agric. Biol. Chem. *35,* 686.

HISATSUKA, K., NAKAHARA, T., MINODA, Y., and YAMADA, K. (1975). Agric. Biol. Chem. *39,* 999.

HISATSUKA, K., NAKAHARA, T., MINODA, Y., and YAMADA, K. (1977). Agric. Biol. Chem. *41,* 445.

HOFFMANN, B., and REHM, H.-J. (1976a). Eur. J. Appl. Microbiol. *3,* 19.
HOFFMANN, B., and REHM, H.-J. (1976b). Eur. J. Appl. Microbiol. *3,* 31.
HOFFMANN, B., and REHM, H.-J. (1978). Eur. J. Appl. Microbiol. *5,* 189.
HOLDORN, R. S., and TURNER, A. G. (1969). J. Appl. Microbiol. *34,* 448.
HOLLOWAY, B. W. (1981). In "Microbial Growth on C_1 Compounds" (H. DALTON, ed.), pp. 317–324. Heyden & Son, London.
HORVATH, R. S. (1972). Bacteriol. Rev. *36,* 146.
HOU, C. T. (1982). In "Microbial Transformations of Bioactive Compounds", Vol. I (J. P. ROSAZZA, ed.), pp. 21–107. CRC Press, Boca Raton, Florida.
HOU, C. T., PATEL, R., LASKIN, A. J., and BARNABE, N. (1979a). Appl. Environ. Microbiol. *38,* 127.
HOU, C. T., PATEL, R., LASKIN, A. J., BARNABE, N., and MARCZAK, I. (1979b). Appl. Environ. Microbiol. *38,* 135.
HOU, C. T., PATEL, R. N., and LASKIN, A. I. (1980a). Adv. Appl. Microbiol. *26,* 41.
HOU, C. T., PATEL, R. N., LASKIN, A. I., and BARNABE, N. (1980b). FEMS Microbiol. Lett. *9,* 267.
HOU, C. T., PATEL, R. N., LASKIN, A. I., and BARNABE, N. (1982). J. Appl. Biochem. *4,* 379.
HUG, H., and FIECHTER, A. (1973). Arch. Mikrobiol. *88,* 87.
HUG, H., BLANCH, H. W., and FIECHTER, A. (1974). Biotechnol. Bioeng. *16,* 965.
HUNKOVA Z., AND FENCL, Z. (1977). Biotech. Bioeng. *16,* 965.
HUTCHINSON, D. W., WHITTENBURY, R., and DALTON, H. (1976). J. Theor. Biol. *58,* 325.
HUYBREGTSE, R., and VAN DER LINDEN, A. C. (1964). Antonie van Leeuwenhoek J. Microbiol. Serol. *30,* 185.
ICI (Imperial Chemical Industries, Ltd.) (1980). Brit. Patent 8 007 849.
ICI (Imperial Chemical Industries, Ltd.) (1981). Brit. Patent 8 117 191.
IGUCHI, T., TAKEDA, I., and OHSAWA, H. (1969). Agric. Biol. Chem. *33,* 1657.
IIDA, M., and IIZUKA, H. (1970). Z. Allg. Mikrobiol. *10,* 245.
IIZUKA, H., IIDA, M., and TOYODA, S. (1966). Z. Allg. Mikrobiol. *6,* 335.
IIZUKA, H., IIDA, M., UNAMI, Y., and HOSHINO, Y. (1968). Z. Allg. Mikrobiol. *8,* 145.
IIZUKA, H., IIDA, M., and FUJITA, S. (1969). Z. Allg. Mikrobiol. *9,* 223.
ILCHENKO, A. P., MAUERSBERGER, S., MATYASHOVA, R. N., and LOSINOV, A. B. (1980). Mikrobiologiya *49,* 452.
IMADA, Y. (1967). Biotechnol. Bioeng. *9,* 45.
INOUE, S., and ITO, S. (1982). Biotechnol. Lett. *4,* 3.
IONEDA, T., and SILVA, C. L. (1979). Chem. Phys. Lipids *23,* 63.
IONEDA, T., LEDERER, E., and ROZANIS, J. (1970). Chem. Phys. Lipids *4,* 375.
ITO, S., and INOUE, S. (1982). Appl. Environ. Microbiol. *43,* 1278.
ITO, S., KINATA, M., and INOUE, S. (1980). Agric. Biol. Chem. *44,* 2221.
ITOH, S., and SUZUKI, T. (1974). Agric. Biol. Chem. *38,* 1443.
ITOH, S., HONDA, H., TOMITA, F., and SUZUKI, T. (1971). J. Antibiot. *24,* 855.
IZUMI, Y., MISHRA, S. K., GOSH, B. S., TANI, Y., and YAMADA, H. (1983). J. Ferment. Technol. *61,* 135.
JAFFE, H., FUJII, K., GUERIN, H., SANGUPIA, M., and EPSTEIN, S. S. (1969). Biochem. Pharmacol. *18,* 1045.
JONES, D. F., and HOWE, R. (1968a). J. Chem. Soc. Commun., 2801.
JONES, D. F., and HOWE, R. (1968b). J. Chem. Soc. Commun., 2809.
JWANNY, E. W. (1975). Z. Allg. Mikrobiol. *15,* 423.
KACHHOLZ, T., and REHM, H.-J. (1977). Eur. J. Appl. Microbiol. *4,* 101.
KACHHOLZ, T., and REHM, H.-J. (1978). Eur. J. Appl. Microbiol. *6,* 39.
KÄPPELI, O., and FIECHTER, A. (1976). Biotechnol. Bioeng. *18,* 967.
KÄPPELI, O., and FIECHTER, A. (1977). J. Bacteriol. *131,* 917.
KÄPPELI, O., and FIECHTER, A. (1980). Biotechnol. Bioeng. *22,* 1829.
KÄPPELI, O., and FINNERTY, W. R. (1979). J. Bacteriol. *140,* 707.
KÄPPELI, O., and FINNERTY, W. R. (1980). Biotechnol. Bioeng. *22,* 495.
KÄPPELI, O., MÜLLER, M., and FIECHTER, A. (1978). J. Bacteriol. *133,* 952.
KÄPPELI, O., FIECHTER, A., and FINNERTY, W. R. (1981). Adv. Biotechnol. (Proc. Int. Ferment. Symp.), 6th 1980, *1,* 177.
KAKINUMA, A., LUGINO, H., ISONO, M., TAMURA, G., and ARIMA, K. (1969). Agric. Biol. Chem. *33,* 973.
KANEYUKI, H., DENO, H., HIRATSUKA, J., MATSUYOSHI, T., and FURUKAWA, T. (1980). J. Ferment. Technol. *58,* 405.
KAPLAN, N., and ROSENBERG, E. (1982). Appl. Environ. Microbiol. *44,* 1335.
KATO, N., TSUJI, K., OHASHI, H., TANI, Y., and OGATA, K. (1977). Agric. Biol. Chem. *41,* 29.
KAWAMOTO, S., TANAKA, A., YAMAMURA, M., TERANISHI, Y., FUKUI, S., and OSUMI, M. (1977). Arch. Microbiol. *112,* 1.

KAWAMOTO, S., NOZAKI, C., TANAKA, A., and FUKUI, S. (1978). Eur. J. Biochem. *83,* 609.

KAWANAMI, J. (1971). Chem. Phys. Lipids *7,* 159–172.

KELLY, R. N., and KIRWAN, D. J. (1977). Biotechnol. Bioeng. *19,* 1215.

KENNEDY, R. S., and FINNERTY, W. R. (1975). Arch. Microbiol. *102,* 85.

KESTER, A. S., and FOSTER, J. W. (1960). Bacteriol. Proc., 168.

KESTER, A. S., and FOSTER, J. W. (1963). J. Bacteriol. *85,* 859.

KIKUCHI, M., KANAMARU, T., and NAKAO, Y. (1973). Agric. Biol. Chem. *37,* 2405.

KILLINGER, A. (1970). Arch. Microbiol. *73,* 153.

KIM, Y. B., and REHM, H.-J. (1982). Eur. J. Appl. Microbiol. Biotechnol. *14,* 105.

KING, D. H., and PERRY, J. J. (1975). Can. J. Microbiol. *21,* 85.

KIYOHARA, H., NAGAO, K., and YANO, K. (1982). Appl. Environ. Microbiol. *43,* 454.

KLEIN, D. A., and HENNING, F. A. (1969). Appl. Microbiol. *17,* 676.

KLUG, M. J., and MARKOVETZ, A. J. (1967). J. Bacteriol. *93,* 1847.

KLUG, M. J., and MARKOVETZ, A. J. (1968). J. Bacteriol. *96,* 1115.

KLUG, M. J., and MARKOVETZ, A. J. (1971). Adv. Microbiol. Physiol. *5,* 1.

KNAPP, J. S., and HOWELL, J. A. (1980). In "Topics in Enzyme and Fermentation Biotechnology" (A. WISEMAN, ed.), Vol. 4, pp. 85–143. John Wiley & Son, New York.

KNOCHE, H. W., and SHIVELEY, J. M. (1972). J. Biol. Chem. *247,* 170.

KOZLOVA, T. M., MEDVEDEVA, G. A., MEIZEL, M. N., RYLKIN, S. S., and SHUL'GA, A. V. (1973). Mikrobiologiya *42,* 937.

KRANZOWA, V. I., and SAPOZHNIKOVA, G. P. (1979). Mikrobiologiya *48,* 145.

KRASSILNIKOV, N. A., KORONELLI, T. V., OSTROVSKY, D. N., and BISKO, N. A. (1973). Mikrobiologiya *42,* 779.

KRAUEL, H., and WEIDE, H. (1978). Z. Allg. Mikrobiol. *18,* 47.

KRAUEL, H., KUNZE, R., and WEIDE, H. (1973). Z. Allg. Mikrobiol. *13,* 55.

KRETSCHMER, A., BOCK, H., and WAGNER, F. (1982). Appl. Environ. Microbiol. *44,* 864.

KUSUNOSE, M., MATSUMOTO, J., ICHIHARA, K., KUSUNOSE, E., and NOZAKA, J. (1967). J. Biochem. *61,* 665.

LANEELLE, M.-A., and ASSELINEAU, J. (1977). Biochim. Biophys. Acta *486,* 205.

LAZAROW, P. B., and DE DUVE, C. (1976). Proc. Natl. Acad. Sci. USA *73,* 2043.

LEADBETTER, E. F., and FOSTER, J. W. (1960). Arch. Mikrobiol. *35,* 92.

LEBEAULT, J. M., and AZOULAY, E. (1971). Lipids *6,* 444.

LEBEAULT, J. M., ROCHE, B., DUVNJAK Z., and AZOULAY, E. (1970a). Arch. Microbiol. *72,* 140.

LEBEAULT, J. M., ROCHE, B., DUVNJAK, Z., and AZOULAY, E. (1970b). Biochim. Biophys. Acta *220,* 373.

LEDERER, E. (1967). Chem. Phys. Lipids *1,* 294.

LEVI, J. D., SHENNAN, J. L., and EBBON, G. P. (1979). In "Economic Microbiology" (A. H. ROSE, ed.), Vol. 4, pp. 361–419. Academic Press, London.

LIJMBACH, G. W. M., and BRINKHUIS, E. (1973). Antonie van Leeuwenhoek J. Microbiol. Serol. *39,* 415.

LIN, H. T., IIDA, M., and IIZUKA, H. (1971). J. Ferment. Technol. *49,* 206.

LIU, C.-M., and JOHNSON, M. (1971). J. Bacteriol. *106,* 830.

LODE, E. T., and COON, M. J. (1971). J. Biol. Chem. *246,* 791.

LUKINS, H. B., and FOSTER, J. W. (1963). J. Bacteriol. *85,* 1174.

MAKULA, R. A., and FINNERTY, W. R. (1972). J. Bacteriol. *112,* 398.

MAKULA, R. A., LOOKWOOD, P. J., and FINNERTY, W. R. (1975). J. Bacteriol. *121,* 250.

MARCHAL, R., METCHE, M., and VAN DECASTEELE, J.-P. (1982). J. Gen. Microbiol. *128,* 1125.

MARISON, J. W., and ATTWOOD, M. M. (1980). J. Gen. Microbiol. *117,* 305.

MARKOVETZ, A. J. (1982). Abstr. XII Intern. Congr. Microbiology, München, 03–08 Sept., p. 18.

MATSON, R. S., and FULCO, A. J. (1981). Biochem. Biophys. Res. Commun. *103,* 531.

MATSUYANA, H., NAKAHARA, T., and MINODA, Y. (1981). Agric. Biol. Chem. *45,* 9.

MAUERSBERGER, S., MATYASHOVA, R. N., MÜLLER, H.-G., and LUSINOV, A. B. (1980). Eur. J. Appl. Microbiol. Biotechnol. *9,* 285.

MAY, S. W., and ABBOTT, B. J. (1972). Biochem. Biophys. Res. Commun. *48,* 1230.

MAY, S. W., and ABBOTT, B. J. (1973). J. Biol. Chem. *248,* 1725.

MCKENNA, E. J., and COON, M. J. (1970). J. Biol. Chem. *245,* 3882.

MCKENNA, E. J., and KALLIO, R. E. (1965). Annu. Rev. Microbiol. *19,* 183.

MCKENNA, E. J., and KALLIO, R. E. (1971). Proc. Natl. Acad. Sci. USA *68,* 1552.

MCKENZIE, P., and HUGHES, D. E. (1976). In "Microbiology in Agriculture, Fisheries and

Food". Soc. Appl. Bacteriol. (Symp. Ser. No. 4), p. 91.

McLee, A. G., Kormendy, A. C., and Wayman, M. (1972). Can. J. Microbiol. *18,* 1191.

Meizel, M. N., Medvedeva, G. A., and Kozlova, T. M. (1976). Mikrobiologiya *45,* 844.

Miall, L. M. (1980). In "Hydrocarbons in Biotechnology" (D. E. F. Harrison, I. J. Higgins, and R. Watkinson, eds.), pp. 25–34. The Institute of Petroleum, London.

Milko, E. S., Egorov, N. N., and Revenko, A. A. (1976). Mikrobiologiya *45,* 808.

Mishina, M., Yanagawa, S., Tanaka, A., and Fukui, S. (1973). Agric. Biol. Chem. *37,* 863.

Mishina, M., Kamiryo, T., Tanaka, A., Fukui, S., and Numa, S. (1976). Eur. J. Biochem. *71,* 301.

Mishina, M., Isurgi, M., Tanaka, A., and Fukui, S. (1977). Agric. Biol. Chem. *41,* 517.

Mishina, M., Kamiryo, T., Tashiro, S., Hagihara, T., Tanaka, A., Fukui, S., Osumi, M., and Numa, S. (1978a). Eur. J. Biochem. *89,* 321.

Mishina, M., Kamiryo, T., Tashiro, S., and Numa, S. (1978b). Eur. J. Biochem. *82,* 347.

Miura, Y. (1978). Adv. Biochem. Eng. *9,* 31.

Miura, Y., Okazaki, M., Hamada, S.-J., Murakawa, S.-J., and Yugen, R. (1977a). Biotechnol. Bioeng. *19,* 701.

Miura, Y., Okazaki, M., Murakawa, S., Hamada, S.-J., and Ohno, K. (1977b). Biotechnol. Bioeng. *19,* 715.

Miyoshi, M. (1895). Jahrb. Wiss. Bot. *28,* 269.

Mohr, P., Scheller, F., and Kühn, M. (1982). Prikl. Biokhim. Mikrobiol. *18,* 481.

Moore, A. T., Nayudu, M., and Holloway, B. W. (1983). J. Gen. Microbiol. *129,* 785.

Müller, H. G., Schunck, W.-H., Riege, P., and Honeck, H. (1979). Acta Biol. Med. Ger. *38,* 345.

Nabeshina, W., Tanaka, A., and Fukui, S. (1977). Agric. Biol. Chem. *41,* 275.

Nagahari, K., and Sakaguchi, K. (1978). J. Bacteriol. *133,* 1527.

Nakahara, T., Kawashima, H., and Tabuchi, T. (1981). Adv. Biotechnol. (Proc. Int. Ferment. Symp.), 6th 1980 (Pub. 1981) *1,* 149.

Nieder, M., and Shapiro, J. (1975). J. Bacteriol. *122,* 93.

O'Connor, M. L. (1981a). Appl. Environ. Microbiol. *41,* 437.

O'Connor, M. L. (1981b). In "Microbial Growth on C_1 Compounds" (H. Dalton, ed.), pp. 294–300. Heyden, London.

O'Connor, M. L., and Hanson, R. (1977). J. Gen. Microbiol. *101,* 327.

O'Connor, M. L., and Hanson, R. (1978). J. Gen. Microbiol. *104,* 105.

Ogata, K., Kaneyuki, H., Kato, N., Tani, Y., and Yamada, H. (1973). J. Ferment. Technol. *51,* 227.

Ogino, S., Yano, K., Tamura, G., and Arima, K. (1965). Agric. Biol. Chem. *29,* 1009.

Ohta, S., and Tobari, J. (1981). J. Biochem. *90,* 215.

Okada, T., Karube, I., and Suzuki, S. (1981). Eur. J. Appl. Microbiol. Biotechnol. *12,* 102.

Okuhara, M., Kubochi, Y., and Harada, T. (1971). Agric. Biol. Chem. *35,* 1367.

Osumi, M., Fukuzumi, F., Teranishi, Y., Tanaka, A., and Fukui, S. (1975). Arch. Microbiol. *103,* 1.

Pareilleux, A. (1979). Eur. J. Appl. Microbiol. Biotechnol. *8,* 91.

Parekh, V. R., Traxler, R. W., and Sobek, J. M. (1977). Appl. Environ. Microbiol. *33,* 881.

Patel, R. N., and Felix, A. (1978). J. Bacteriol. *128,* 413.

Patel, R. N., Hou, C. T., and Felix, A. (1978). J. Bacteriol. *136,* 352.

Patel, R. N., Hou, C. T., Derelanko, P., and Felix, A. (1980a). Arch. Biochem. Biophys. *203,* 654.

Patel, R. N., Hou, C. T., Laskin, A. J., Felix, A., and Derelanko, P. (1980b). Appl. Environ. Microbiol. *39,* 720.

Patel, R. N., Hou, C. T., Laskin, A. J., Felix, A., and Derelanko, P. (1980c). Appl. Environ. Microbiol. *39,* 727.

Patel, R. N., Hou, C. T., Laskin, A. J., and Derelanko, P. (1981). J. Appl. Biochem. *3,* 218.

Patel, R. N., Hou, C. T., Laskin, A. J., and Felix, A. (1982a). J. Appl. Biochem. *4,* 175.

Patel, R. N., Hou, C. T., Laskin, A. J., and Felix, A. (1982b). Appl. Environ. Microbiol. *44,* 1130.

Peberdy, J. F. (1980). Enzyme Microbiol. Technol. *2,* 23.

Pelechova, J., Krumphanzl, V., Uher, J., and Dyr, J. (1971). Folia Mikrobiol. *16,* 103.

Pelz, B. F., and Rehm, H.-J. (1973). Arch. Mikrobiol. *92,* 153.

Perry, J. J. (1979). Microbiol. Rev. *43,* 59.

Perry, J. J. (1980). Adv. Appl. Microbiol. *26,* 89.

Perry, J. J., and Gibbson, D. T. (1977). CRC Crit. Rev. Microbiol. *5,* 387.

Peterson, J. A. (1970). J. Bacteriol. *103,* 714.

Peterson, J. A., and Coon, M. J. (1968). J. Biol. Chem. *243,* 329.

PETRIKEVICH, S. B., and DOVGUN, L. J. (1980). Mikrobiologiya *49,* 78.

PIRNIK, M. P., and MCKENNA, E. J. (1977). CRC Crit. Rev. Microbiol. *5,* 413.

PIRNIK, M. P., ATLAS, R. M., and BARTHA, R. (1974). J. Bacteriol. *119,* 868.

PROTTEY, C., and BALLOU, C. E. (1968). J. Biol. Chem. *243,* 6196.

QUAYLE, J. R. (1972). Adv. Microbiol. Physiol. *7,* 119.

QUAYLE, J. R. (1976). In "Microbial Production and Utilization of Gases" (G. SCHLEGEL, G. GOTTSCHALK, and N. PFENNIG, eds.), pp. 353–357. E. E. Goltze KG, Göttingen.

RAPP, P., BOCK, H., WRAY, V., and WAGNER, F. (1979). J. Gen. Microbiol. *115,* 491.

RATLEDGE, C. (1976). Adv. Microbiol. Physiol. *13,* 115.

RATLEDGE, C. (1978). In "Developments in Biodegradation of Hydrocarbons" (R. J. WATKINSON, ed.), pp. 1–46. Applied Science Publishers, London.

RATLEDGE, C. (1980). In "Hydrocarbons in Biotechnology". (D. E. F. HARRISON, I. J. HIGGINS, and R. WATKINSON, eds.), pp. 133–153. The Institute of Petroleum, London.

RATTREY, J. B. M., SCHIBECI, A., and KIDBY, D. K. (1975). Bacteriol. Rev. *39,* 191.

RAYMOND, R. L., and DAVIES, J. B. (1960). Appl. Microbiol. *8,* 329.

RAYMOND, R. L., and JAMISON, V. W. (1971). Adv. Appl. Microbiol. *14,* 93.

RAYMOND, R. L., JAMISON, V. W., and HUDSON, J. O. (1971). Lipids *6,* 453.

REDDY, P. J., SINGH, H. D., ROY, P. K., and BARUAH, J. N. (1982). Biotechnol. Bioeng. *24,* 1241.

REDDY, P. G., SINGH, H. D., PATHAK, M. G., BHAGAT, S. D., and BARUAH, J. N. (1983). Biotechnol. Bioeng. *25,* 387.

REHM, H.-J., and REIFF, I. (1981). Adv. Biochem. Eng. *19,* 175.

REHM, H.-J., and REIFF, I. (1982). Acta Biotechnol. *2,* 127.

REISFELD, A., ROSENBERG, E., and GUTNICK, D. (1972). Appl. Microbiol. *24,* 363.

ROCHE, B., and AZOULAY, E. (1969). Eur. J. Biochem. *8,* 426.

ROHDE, H. G., SCHRÖDER, S., SCHIRPKE, B., and WEIDE, H. (1975). Z. Allg. Mikrobiol. *15,* 195.

ROSENBERG, E., PERRY, A., GIBSON, D. T., and GUTNICK, D. L. (1979). Appl. Environ. Microbiol. *37,* 409.

ROSENBERG, M., BAYER, E. A., DELAREA, J., and ROSENBERG, E. (1982). Appl. Environ. Microbiol. *44,* 929.

ROY, P. K., SINGH, H. D., BHAGAT, S. D., and BARUAH, J. N. (1979). Biotechnol. Bioeng. *21,* 955.

RUETTINGER, R. T., GRIFFITH, G. R., and COON, M. J. (1977). Arch. Biochem. Biophys. *183,* 528.

SAHM, H. (1977). Adv. Biochem. Eng. *6,* 77.

SCHAEFFER, T. L., CANTWELL, S. G., BROWN, J. L., WATT, D. S., and FALL, R. R. (1979). Appl. Environ. Microbiol. *38,* 742.

SCHELLER, F., RENNEBERG, R., STRUAD, P., POMMERENNING, K., and MOHR, P. (1977). Bioelectrochem. Bioeng. *4,* 500.

SCHLEGEL, H. G., GOTTSCHALK, G., and PFENNIG, N. (1976). In "Microbial Production and Utilization of Gases". E. Goltze KG, Göttingen.

SCHNABEL, H., and REHM, H.-J. (1971). Naturwissenschaften *58,* 55.

SCHRÖDER, E., and REHM, H.-J. (1981). Eur. J. Appl. Microbiol. Biotechnol. *12,* 36.

SCHUNCK, W. H., RIEGE, P., BLASIG, R., HONECK, H., and MUELLER, H. G. (1978). Acta Biol. Med. Ger. *37,* K3.

SCOTT, C. C. L., and FINNERTY, W. R. (1976). J. Bacteriol. *127,* 481.

SCOTT, C. C. L., MAKULA, R. A., and FINNERTY, W. R. (1976). J. Bacteriol. *127,* 469.

SEBEK, O. K., and KIESLICH, K. (1977). In "Annu. Rep. Ferment. Proc.", Vol. 1 (D. PERLMAN and G. T. TSAO, eds.), pp. 267–297. Academic Press, London.

SEELER, G. (1962). Arch. Mikrobiol. *43,* 213.

SENEZ, J. C., and AZOULAY, E. (1961). Biochim. Biophys. Acta *47,* 306.

SHAPIRO, J. A., FENNEWALD, M., and BENSON, S. (1979). In "Genetics of Industrial Microorganisms" (O. K. SEBEK and A. I. LASKIN, eds.), pp. 147–153. Am. Soc. Microbiol., Washington.

SHAPIRO, J. A., BENSON, S., and FENNEWALD, M. (1980). Plasmids Transposons, Proc. Annu. Symp. Sci. Basis Med. 4th, p. 1.

SHAPIRO, J. A., CHARBIT, A., BENSON, S., CARUSO, M., LAUX, R., MEYER, R., and BANUETT, F. (1981). In "Basic Life Science" 18 (Trends in the Biology of Fermentations for Fuels and Chemicals). (A. HOLLAENDER, ed.), pp. 243–269. Plenum Press, New York.

SHEN, Y.-CH., LON, S. C., XU, K. R., XIA, G. X., and CHIAO, J.-S. (1977). Chem. Abstr. *91,* 3902k.

SHENNAN, J. L., and LEVI, J. D. (1974). In "Progress in Industrial Microbiology" (D. J. D. HOCKENHULL, ed.), Vol. 13, pp. 1–57. Churchill, Livingstone.

SHIIO, J., and UCHIO, R. (1971). Agric. Biol. Chem. *35,* 2033.
SHUM, A. C., and MARKOVETZ, A. J. (1974a). J. Bacteriol. *118,* 880.
SHUM, A. C., and MARKOVETZ, A. J. (1974b). J. Bacteriol. *118,* 890.
SIMON, H., GÜNTER, H., BADER, J., and TISCHER, W. (1981). Angew. Chem. Int. Ed. Engl. *20,* 861.
SKIPTON, M. D., WATSON, K., HOUGHTON, R. L., and GRIFFITHS, D. E. (1974). J. Gen. Microbiol. *84,* 94.
SÖHNGEN, M. L. (1906). Zentralbl. Bakteriol. Parasitenkd. Infektionskr. Hyg. Abt. 2 *15,* 513.
SOMERVILLE, H. J., and MASON, J. R. (1979). Biochem. Soc. Trans. *7,* 85.
SORGER, H., and AURICH, H. (1978a). Wiss. Z. Univers. Leipzig, Math.-Naturwiss. Reihe *27,* 35.
SORGER, H., and AURICH, H. (1978b). Z. Allg. Microbiol. *18,* 587.
SOUW, P., LUFTMANN, H., and REHM, H.-J. (1977). Eur. J. Appl. Microbiol. *3,* 289.
SOUW, P., SCHULTE, E., and REHM, H.-J. (1978). XII. Intern. Congr. Microbiology, München, 0.3–08. Sept. 1978, Abstr. No. 8, p. 145.
SPERL, G. T., FORREST, H. S., and GIBBSON, D. T. (1974). J. Bacteriol. *118,* 541.
STAHL, U., and ESSER, K. (1982). Eur. J. Appl. Microbiol. Technol. *15,* 223.
STEWART, J. E., KALLIO, R. E., STEVENSON, D. P., JONES, A. C., and SCHISSLER, D. O. (1959). J. Bacteriol. *78,* 441.
STIRLING, D. J., and DALTON, H. (1979). Eur. J. Biochem. *96,* 205.
STIRLING, D. J., and DALTON, H. (1980). J. Gen. Microbiol. *116,* 277.
STIRLING, D. J., COLBY, J., and DALTON, H. (1979). Biochem. J. *177,* 361.
STROBEL, H. W., and COON, M. J. (1971). J. Biol. Chem. *246,* 7826.
SUZUKI, T., and OGAWA, K. (1972). Agric. Biol. Chem. *36,* 457.
SUZUKI, T., TANAKA, K., MATSURBA, I., and KINOSHITA, S. (1969). Agric. Biol. Chem. *33,* 1619.
SUZUKI, T., TANAKA, H., and ITOH, S. (1974). Agric. Biol. Chem. *38,* 557.
TABAUD, M., TISNOVSKA, H., and VILKAS, E. (1971). Biochemistry *53,* 55.
TABUCHI, T., and IGOSHI, K. (1978). Agric. Biol. Chem. *42,* 2381.
TABUCHI, T., and SERIZAWA, N. (1975). Agric. Biol. Chem. *39,* 1055.
TABUCHI, T., and UCHIYAMA, H. (1975). Agric. Biol. Chem. *39,* 2035.
TAHARA, Y., KAMEDA, M., YAMADA, Y., and KONDO, K. (1976a). Agric. Biol. Chem. *40,* 243.
TAHARA, Y., YAMADA, Y., and KONDO, K. (1976b). Agric. Biol. Chem. *40,* 1449.
TAKAGI, M., MORIYA, K., and YANO, K. (1979). Cell Mol. Biol. *25,* 371.
TAKAHARA, Y., HIROSE, Y., YASUDA, N., MITSUGI, K., and MURAO, S. (1976). Agric. Biol. Chem. *40,* 1901.
TANAKA, A., HAGIHARA, T., NISHIKAWA, Y., MISHINA, M., and FUKUI, S. (1976). Eur. J. Appl. Microbiol. *3,* 115.
TANAKA, A., NABESHINA, S., TOKUDA, M., and FUKUI, S. (1977). Agric. Biol. Chem. *41,* 795.
TANAKA, A., HAGIHARA, T., KAMIRYO, T., MISHINA, M., TASHIRO, S.-I., NUMA, S., and FUKUI, S. (1978). Eur. J. Appl. Microbiol. Biotechnol. *5,* 79.
TANI, Y., and YAMADA, H. (1980). Biotechnol. Bioeng. *22,* Suppl. 1, 163.
TANI, Y., KATO, N., and YAMADA, H. (1978). Adv. Appl. Microbiol. *24,* 165.
TASSIN, J.-P., and VANDECASTELE, J.-P. (1972). Biochim. Biophys. Acta *276,* 31.
TASSIN, J. P., CELIER, C., and VANDECASTEELE, J. P. (1973). Biochim. Biophys. Acta *315,* 220.
TAUCHERT, H., ROY, M., SCHÖPP, W., and AURICH, H. (1975). Z. Allg. Mikrobiol. *15,* 457.
TAUCHERT, H., GRUNOW, M., HARNISCH, H., and AURICH, H. (1976). Acta Biol. Med. Ger. *35,* 1267.
TAUCHERT, H., GRUNOW, M. and AURICH, H. (1978a). Z. Allg. Mikrobiol. *18,* 675.
TAUCHERT, H., SCHÖPP, W., and AURICH, H. (1978b). Wiss. Z. Univers. Leipzig, Math.-Naturwiss. Reihe *27,* 25.
TERANISHI, Y., KAWAMOTO, S., TANAKA, A., USUMI, M., and FUKUI, S. (1974). Agric. Biol. Chem. *38,* 1221.
THIELE, H., and REHM, H.-J. (1976). Eur. J. Appl. Microbiol. Biotechnol. *6,* 361.
THOMAS, D. W., MATIDA, A. K., SILVA, C. L., and JONEDA, T. (1979). Chem. Phys. Lipids *23,* 267.
THORPE, R. F., and RATLEDGE, C. (1972). J. Gen. Microbiol. *72,* 151.
TITTELBACH, M., RHODE, H. G., and WEIDE, H. (1976). Z. Allg. Mikrobiol. *16,* 155.
TONGE, G. M., HARRISON, D. E. F., and HIGGINS, I. J. (1977). Biochem. J. *161,* 333.
TORNABENE, T. G. (1976). In "Microbial Energy Conversion" (H. G. SCHLEGEL and J. BARNEA, eds.), pp. 281–299. E. Goltze KG, Göttingen.
TORNABENE, T. G. (1978). Natl. Meet. Am. Chem. Soc. Div. Environ. Chem. *18,* 99.

TORNABENE, T. G. (1981). In "Trends in the Biology of Fermentations for Fuels and Chemicals." (HOLLAENDER et al., eds.), pp. 421–438. Plenum Press, New York.

TORNABENE, T. G. (1982). Experientia *38,* 43.

TRUST, T. J., and MILLIS, N. F. (1970). J. Bacteriol. *104,* 1397.

TRUST, T. J., and MILLIS, N. F. (1971). J. Bacteriol. *105,* 1216.

TULLOCH, A. P., HILL, A., and SPENCER, J. F. T. (1967). Chem. Commun., 584.

UCHIO, R., and SHIIO, I. (1972a). Agric. Biol. Chem. *36,* 426.

UCHIO, R., and SHIIO, I. (1972b). Agric. Biol. Chem. *36,* 1169.

UCHIO, R., and SHIIO, I. (1972c). Agric. Biol. Chem. *36,* 1389.

UCHIO, R., and SHIIO, I. (1972d). Agric. Biol. Chem. *36,* 1395.

ULLRICH, V. (1979). Top. Curr. Chem. *83,* 68.

UPDEGRAFF, D. M., and BOVEY, F. A. (1958). Nature *181,* 890.

VAN DER LINDEN, A. C., and THIJSSE, G. J. E. (1965). Adv. Enzymol. *27,* 469.

VAN RAVENSWAY CLAASEN, J. C., and VAN DER LINDEN, A. C. (1971). Antonie van Leeuwenhoek J. Microbiol. Serol. *37,* 339.

VERKOOYEN, A. H. M., and RIETEMA, K. (1980a). Biotechnol. Bioeng. *22,* 571.

VERKOOYEN, A. H. M., and RIETEMA, K. (1980b). Biotechnol. Bioeng. *22,* 615.

VERKOOYEN, A. H. M., VAN DEN OEVER, A. M. C., and RIETEMA, K. (1980). Biotechnol. Bioeng. *22,* 597.

VUILLEMIN, N., DUPEYRON, C., LELUAN, G., and BORY, J. (1981). Ann. Pharm. Fr. *39,* 155.

WAGNER, F., ZAHN, W., and BÜHRING, U. (1967). Angew. Chem. *79,* 314.

WAITES, M. J., and QUAYLE, J. R. (1980). J. Gen. Microbiol. *118,* 321.

WALKER, J. D., and COLWELL, R. R. (1976a). Appl. Environ. Microbiol. *31,* 189.

WALKER, J. D., and COLWELL, R. R. (1976b). Appl. Environ. Microbiol. *31,* 198.

WALKER, J. D., and COONEY, J. J. (1973). J. Bacteriol. *115,* 635.

WARNER, P. J., HIGGINS, I. J., and DROZD, J. W. (1980). FEMS Microbiol. Lett. *7,* 181.

WATKINSON, R. J. (1980). In "Hydrocarbons in Biotechnology" (D. E. F. HARRISON, I. J. HIGGINS, and J. R. WATKINSON, eds.), pp. 11–24. The Institute of Petroleum, London.

WAYMAN, M., and WHITELEY, M. (1979). In "Liquid Fuels from Carbonates by a Microbial System" (H. TOMLENSON, ed.), pp. 120–132. American Chemical Society, Washington.

WEETE, J. D. (1976). In "Chemistry and Biochemistry of Natural Waxes". (P. E. KOLATTUKUDY, ed.), pp. 350–418. Elsevier, Amsterdam.

WHITTENBURY, R., DAVIES, S. L., and DAVEY, J. F. (1970a). J. Gen. Microbiol. *61,* 219.

WHITTENBURY, R., PHILLIPS, K. C., and WILKINSON, J. F. (1970b). J. Gen. Microbiol. *61,* 205.

WICKEN, A. J., and KNOX, K. W. (1970). J. Gen. Microbiol. *60,* 293.

WILKINSON, S. G. (1972). Biochim. Biophys. Acta *270,* 1.

WILLIAMS, P. A. (1978). In "Developments in Biodegradation of Hydrocarbons - 1". (J. R. WATKINSON, ed.), pp. 135–164. Appl. Science Publishers, London.

WILLIAMS, P. A. (1979). In "Genetics of Industrial Microorganisms" (O. K. SEBEK, and A. J. LASKIN, eds.), pp. 154–159. Am. Soc. Microbiology, Washington.

WILLIAMS, P. A., and FRANKLIN, F. C. H. (1980). In "Hydrocarbons in Biotechnology" (D. C. H. HARRISON, eds.), pp. 155–180. Heyden & Son, The Institute of Petroleum, London.

WINDASS, J. D., WORSEY, M. J., PIOLI, E. M., PIOLI, D., BARTH, P. T., ATHERTON, K. T., DART, E. C., BYROM, D., POWELL, K., and SENIOR, P. J. (1980). Nature *287,* 396.

WISEMAN, A., and KING, D. J. (1982). Top. Enzyme Ferment. Biotechnol. (A. WISEMAN, ed.) *6,* 151.

WOLFE, D. A. (1977). In "Fate and Effects of Petroleum Hydrocarbons in Marine Ecosystems and Organisms." Pergamon Press, Oxford.

YAMADA, Y., and TORIGOE, Y. (1966). J. Agric. Chem. Soc. *40,* 364.

YAMADA, Y., MOTOI, H., KINOSHITA, S., TANAKA, N., and OKADA, H. (1975). Appl. Microbiol. *29,* 400.

YAMADA, Y., KUSUHARA, N., and OKADA, H. (1977). Appl. Environ. Microbiol. *33,* 771.

YAMADA, T., NAWA, H., KAWAMOTO, S., TANAKA, A., and FUKUI, S. (1980). Arch. Microbiol. *128,* 145.

YAMAGUCHI, K., and KURASAWA, M. (1976). Agric. Biol. Chem. *40,* 719.

YAMAGUCHI, M., SATO, A., and YUKUYAMA, A. (1976). Chem. Ind. (London), 741.

YAMAMURA, M., TERANISHI, Y., TANAKA, A., and FUKUI, S. (1975). Agric. Biol. Chem. *39,* 13.

YAMANAKA, K. (1981). In "Microbial Growth on C_1 Compounds" (H. DALTON, ed.), pp. 21–30. Heyden, London.

YAMANAKA, K., and MATSUMOTO, K. (1977). Agric. Biol. Chem. *41,* 467.

YAMAKAWA, Y., GOTO, S., and YOKOTSUKA, J. (1978). Agric. Biol. Chem. *42,* 269.

YANAGAWA, S., TANAKA, A., and FUKUI, S. (1972). Agric. Biol. Chem. *36,* 2129.

YANO, I., FURUKAWA, Y., and KUSUNOSE, M. (1969). Biochim. Biophys. Acta *187,* 166.

YANO, I., FURUKAWA, Y., and KUSUNOSE, M. (1971a). J. Gen. Appl. Microbiol. *17,* 429.

YANO, J., FURUKAWA, Y., and KUSUNOSE, M. (1971b). Eur. J. Biochem. *23,* 220.

YI, Z.-H., and REHM, H.-J. (1982a). Eur. J. Appl. Microbiol. Biotechnol. *14,* 254.

YI, Z.-H., and REHM, H.-J. (1982b). Eur. J. Appl. Microbiol. Biotechnol. *15,* 144.

YI, Z.-H., and REHM, H.-J. (1982c). Eur. J. Appl. Microbiol. Biotechnol. *15,* 175.

YI, Z.-H., and REHM, H.-J. (1982d). Eur. J. Appl. Microbiol. Biotechnol. *16,* 1.

YOSHIDA, F., and YAMAME, T. (1974). Biotechnol. Bioeng. *16,* 635.

YOSHIDA, F., YAMANE, T., and NAKAMOTO, K. (1973). Biotechnol. Bioeng. *15,* 257.

YOSHIDA, Y., AOYAMA, Y., KUMAOKA, H., and KUBOTA, S. (1977). Biochem. Biophys. Res. Commun. *78,* 1005.

YU, C.-A., and GUNSALUS, I. C. (1974). J. Biol. Chem. *249,* 107.

ZATMAN, L. J. (1981). In "Microbial Growth on C_1 Compounds" (H. DALTON, ed.), pp. 42–54. Heyden, London.

ZOBEL, C. E. (1946). Bacteriol. Rev. *10,* 1.

ZOSIM, Z., GUTNICK, D., and ROSENBERG, E. (1982). Biotechnol. Bioeng. *24,* 281.

ZUCKERBERG, A., DIVER, A., PEERI, Z., GUTNICK, D. L., and ROSENBERG, E. (1979). Appl. Environ. Microbiol. *37,* 414.

Chapter 10

Amino Acids and Peptides

Günter Schmidt-Kastner
Peter Egerer
BAYER AG
Verfahrensentwicklung Biochemie
Wuppertal 1, Federal Republic of Germany

I. Biotransformation – an Introduction
II. Biotransformation of Amino Acids
  A. Aliphatic Amino Acids
  B. Aromatic Amino Acids
  C. Basic Amino Acids
  D. Acidic Amino Acids
  E. Hydroxyl- and Sulfur-containing Amino Acids
  F. Side Chain Amino Acids of Semi-synthetic β-Lactams
  G. Miscellaneous
III. Biotransformation of Peptides
  A. Low Molecular Weight Peptides
  B. Insulin
IV. Biotransformation – Economic Aspects
V. References

I. Biotransformation – an Introduction

Biotransformation is the transformation of a substance A (substrate) to a substance B (product) using a biocatalyst. There are no principal differences in the outcome of a chemical reaction whether it was carried out with an inorganic, an organic, or a biocatalyst. However, one advantage of the biocatalyst is its high stereospecificity and regiospecificity. Biocatalysts are enzymes produced by and isolated from microorganisms, plant cells, and animal cells. Biocatalysts for biotransformations can either be used (see Chapter 1) as:

- suspended growing/resting cells,
- immobilized cells,
- enzymes in solution, or
- immobilized enzymes

Immobilization is the fixation of biocatalysts to high molecular weight carriers by covalent bonding, by adsorption, or by physical containment, e.g., as discussed at the Henniker Conference on Enzyme Engineering in 1971 (WINGARD, 1972).

The purpose of immobilization is to increase the stability of the biocatalyst, to optimize the economy of a process by the repeated use of the biocatalyst, and to simplify product isolation. In multistep biotransformations a substance A will be transformed in several steps to a product by two or more biocatalysts (A→B→C→...). Two biocatalysts can be coimmobilized on one carrier or can be immobilized on separate carriers. Some biotransformations are based on the use of the bare enzyme, others additionally require coenzymes (see Fig. 4). Enzyme and coenzyme can be coimmobilized or can be combined with chemical processes to form a multistep chemoenzymatic process. This review will comprise the biotransformation of amino acids and peptides, but will not cover fermentation, biosynthesis, or precursor-directed biosynthesis.

II. Biotransformation of Amino Acids

For the preparation of enantiomerically pure L-amino acids by biotransformation there are several, generally applicable methods, such as:

- racemate resolution of D,L-amino acid derivatives by hydrolytic enzymes (see Figs. 2, 6, 7, 9, 10, 11, 12)
- stereospecific condensation of achiral compounds (see Figs. 3, 8)
- reductive amination of α-keto acids (see Fig. 4)
- modification of chiral precursors, e.g., by hydroxylation (see Fig. 5).

As depicted in Fig. 1, racemate resolution can be subclassified into

- esterase-catalyzed resolution of D,L-amino acid esters
- esterase-catalyzed resolution of *N*-acyl-D,L-amino acid esters
- acylase-catalyzed resolution of *N*-acyl-D,L-amino acids
- amidase- or nitrilase-catalyzed resolution of D,L-amino amides or of D,L-amino nitriles, and
- resolution of 5-substituted D,L-hydantoins.

The enzymes are strictly L-specific with the exception of the D-specific hydantoinases.

About 80 years ago, WARBURG (1906) prepared L-leucine in about 83% yield from D,L-leucine propylester by biotransformation using a pancreas extract (Fig. 2). With the increasing demand for highly pure L-amino acids as food and feed additives as well as for L- and D-amino acids as bulk chemicals, used for instance for the synthesis of pharmaceuticals (KLEEMANN, 1982), economically feasible synthetic routes had to be investigated. Racemate resolution through formation of diastereomeric salts, e.g., with camphorsulfonic acid, and subsequent fractionation by crystallization is ex-

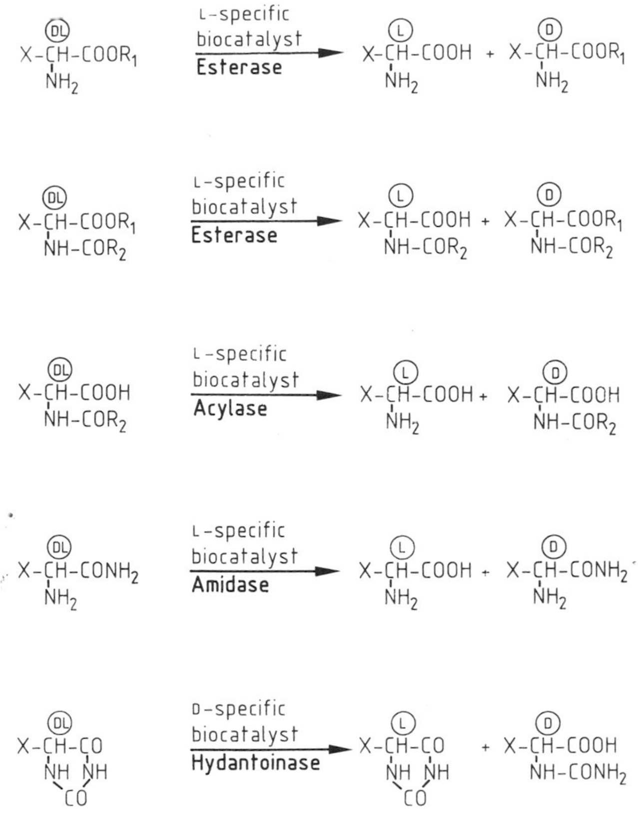

(DL) X-CH(NH$_2$)-COOR$_1$ —L-specific biocatalyst, Esterase→ (L) X-CH(NH$_2$)-COOH + (D) X-CH(NH$_2$)-COOR$_1$

(DL) X-CH(NH-COR$_2$)-COOR$_1$ —L-specific biocatalyst, Esterase→ (L) X-CH(NH-COR$_2$)-COOH + (D) X-CH(NH-COR$_2$)-COOR$_1$

(DL) X-CH(NH-COR$_2$)-COOH —L-specific biocatalyst, Acylase→ (L) X-CH(NH$_2$)-COOH + (D) X-CH(NH-COR$_2$)-COOH

(DL) X-CH(NH$_2$)-CONH$_2$ —L-specific biocatalyst, Amidase→ (L) X-CH(NH$_2$)-COOH + (D) X-CH(NH$_2$)-CONH$_2$

(DL) X-CH-CO-NH-CO-NH (hydantoin) —D-specific biocatalyst, Hydantoinase→ (L) X-CH-CO-NH-CO-NH (hydantoin) + (D) X-CH(NH-CONH$_2$)-COOH

Figure 1. Subclassification of racemate resolution.

pensive in most cases. Racemate resolution by biotransformation using immobilized biocatalysts is an economic alternative for the production of amino acids (CHIBATA, 1979). After having studied a variety of methods suitable for immobilization, ionic adsorption of aminoacylase to DEAE-Sephadex has been chosen for continuous industrial production of several L-amino acids from chemically synthesized *N*-acyl-D,L-amino acids in a column reactor (TOSA et al., 1967). It is necessary for economic reasons to recycle the D- or the L-isomer by racemization through heating in the presence of strong acid or base, or under mild conditions in acetic acid in the presence of catalytic amounts of aldehydes (TANABE SEIYAKU, 1982a; YAMADA et al., 1983, and literature cited therein ...).

Racemization can be greatly enhanced by enzyme catalysis, e.g., by the various amino

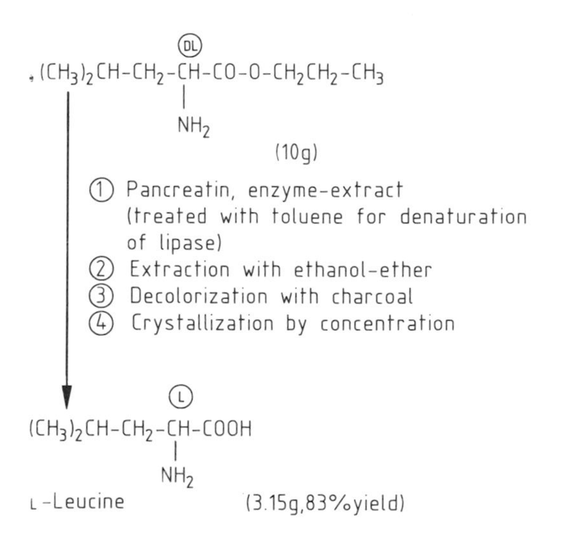

Figure 2. Racemate resolution of D,L-leucine propylester by pancreatin (O. WARBURG, 1906). – First published paper on biotransformation.

acid racemases. A special case of L- or D-amino acid synthesis by racemate resolution is the route via chemically synthesized 5-substituted D,L-hydantoins (see Fig. 10). At higher pH, the 5-phenyl-substituted hydantoin derivatives racemize spontaneously and rapidly, and the desired D-amino acid can be prepared quantitatively (OLIVIERI et al., 1981). A further improvement is the resolution of 5-substituted D,L-hydantoin derivatives applying thermophilic microorganisms (BASF AG, 1980). At elevated temperatures up to 90 °C the solubility of the hydantoins, the spontaneous re-racemization of the undesired L-isomer, and the enzymatic cleavage to the D-carbamoyl amino acids is enhanced to yields of 85–95%.

An enzyme membrane reactor has been developed for D,L-amino acid racemate resolution allowing a high enzyme density and thus a high conversion ratio at high flow rates (REHOVOT RESEARCH PRODUCTS, 1978).

For example, bacterial amino acid ester hydrolase, subtilisin, and chymotrypsin have been coupled to the polyvinylchloride-silica membrane surface and employed for the resolution of *N*-acetyl-D,L-methionine methylester, D,L-phenylglycine methylester, and D,L-tryptophan methylester.

In a typical experiment, the space-time yield of *N*-acetyl-L-methionine is about 27 kg $L^{-1}d^{-1}$ at a 76% conversion efficiency (REHOVOT RESEARCH PRODUCTS, 1978).

Chymotrypsin which has been immobilized in a liquid membrane of kerosene or cyclohexane can be used for resolution of D,L-amino acid esters in an emulsion-type reactor (SCHEPER et al., 1982; 1983). D- and L-Amino acid esters, especially the hydrophobic D,L-phenylalanine esters, are readily soluble in the liquid membrane. The formed L-amino acids are insoluble in the liquid membrane and are transported through the membrane by a hydrophobic quarternary ammonium salt. This emulsion-type enzyme membrane reactor can be principally used for continuous racemate resolution (SCHÜGERL et al., 1983).

A very recent route to L-amino acids of high optical purity is the reductive amination of a series of aliphatic α-keto acids. The enzymes catalyzing this reaction generally require cofactors, e.g., NAD(P). High costs of NAD(P), however, prevent an equimolar use, and therefore inexpensive methods for continuous cofactor regeneration are necessary. By means of an ultrafiltration reactor unit, a continuous process has been developed by WANDREY et al. (1981) based on L-leucine dehydrogenase, formate dehydrogenase, and NAD with an enlarged molecular weight. Formic acid is an inexpensive electron source. CO_2 is the only coproduct, and separation of product and coproduct is unnecessary. As has been demonstrated by SIMON et al. (1981), NAD(P) can also be regenerated by electroenzymatic or electromicrobial processes using methylviologen as an artificial electron mediator. With whole cells of *Clostridium sporogenes* it has been possible to transform α-keto acids to L-amino acids in ammonium buffer solutions (SIMON et al., 1981). Future industrial application will depend on an ecconomical way to regenerate the necessary cofactor.

For a comparison of various processes for the production of L- or D-amino acids by biotransformation, fermentation, or by precursor-directed fermentation, the reviews by ABBOTT (1976), ENEI et al. (1982), LEUENBERGER and KIESLICH (1982), and by SODA et al. (1983) are recommended. FUKUI and TANAKA (1980) have reviewed the production of amino acids from alkane media in Japan.

A. Aliphatic Amino Acids

Alanine. Two alternate basic methods have been developed for the preparation of L-alanine by biotransformation:

- racemate resolution of chemically synthesized alanine derivatives by hydrolytic enzymes, or
- synthesis from fumarate via aspartate through the action of specific lyases.

The enantioselective hydrolysis of D,L-5-methyl-hydantoin by dihydropyrimidinase produces exclusively *N*-carbamoyl-L-alanine which can be easily cleaved to L-alanine. D-5-Methylhydantoin is recycled by chemical racemization. Dihydropyrimidinase can be isolated from calf liver and entrapped in nylon fibers, or can be extracted from microorganisms (SNAM PROGETTI, 1973; KANEGAFUCHI CHEM. KK, 1976b). Hydrolysis of *N*-acetyl-D,L-amino acids by aminoacylase from hog kidney has also been applied to prepare L-alanine. The acylase is adsorbed to DEAE-Sephadex and operated continuously in a column reactor (CHIBATA et al., 1976).

0.04 L/h. The system is operated at a 70–80% conversion rate during the first stage and between 40% and 75% during the second stage. In order to maintain a constant conversion rate under continuous operation conditions, addition of fresh enzyme is necessary to compensate for inactivated enzymes. Formation of L-malic acid can be completely avoided (JANDEL et al., 1982). Acid treatment of the microorganisms to destroy alanine racemase activity improves the yield of L-alanine (TANABE SEIYAKU, 1981b).

Reductive amination of pyruvate using L-alanine dehydrogenase and formate dehydrogenase was carried out in an ultrafiltra-

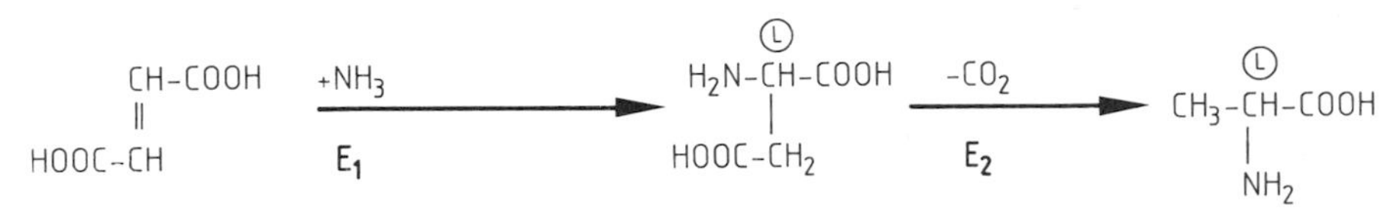

Figure 3. Multistep-biotransformation of fumarate: stereospecific amination to L-aspartic acid using aspartase (E_1) and decarboxylation of L-aspartic acid to L-alanine using L-aspartate-β-decarboxylase (E_2).

The enzymatic synthesis of L-alanine has been extensively investigated and developed by Tanabe Seiyaku. About two decades ago, the β-decarboxylation of L-aspartate to L-alanine with *Pseudomonas dacunhae* has been published (CHIBATA et al., 1965). A selected *E.coli* strain with a high aspartase activity (see Sect. II.D.) can be used for L-alanine production in a two-step biotransformation using two immobilized microorganisms (TAKAMATSU et al., 1982). For the same route, an alternate process in a two-stage enzyme membrane reactor has been designed employing both soluble aspartase from *Escherichia coli* ATCC 11303 and L-aspartate-β-decarboxylase from *Pseudomonas dacunhae* (JANDEL et al., 1982) (Fig. 3). It is possible to perform the conversion of fumarate to L-aspartate at pH 8.5 and 20 °C and the decarboxylation of L-aspartic acid to L-alanine at pH 7.0 and 30 °C starting from 0.4 mol/L fumarate and 0.44 mol/L ammonium ions at a feed rate of

tion reactor unit with polyethyleneglycol-coupled NAD^+ (WANDREY et al., 1978).

Formate dehydrogenase from *Candida boidinii* is used as the NADH-regenerating enzyme producing easily removable CO_2 and protons.

Similarly, L-alanine has been synthesized from pyruvate by using L-alanine dehydrogenase and hydrogenase (DANIELSSON et al., 1982). The NAD^+-dependent hydrogenase from *Alcaligenes eutrophus* (SCHNEIDER and SCHLEGEL, 1976) is used for NADH-regeneration.

Leucine, isoleucine, valine. The first biotransformation of an amino acid reported in the literature was the preparation of L-leucine (83% yield) from racemic leucine propylester by means of a pancreas extract (WARBURG, 1906) (see Fig. 2).

L-Leucine can be prepared from the racemic methylester by pronase B from *Streptomyces griseus* (YAMSKOV et al., 1981). For the preparation of L-valine, however, pronase B is not very well suited since L-valine methylester hydrolysis proceeds about 10^2 to 10^3 times slower. Ca^{2+} increases the enantioselectivity of pronase B by a factor of 2 (YAMSKOV et al., 1981).

A convenient route to the aliphatic hydrophobic L-amino acids leucine, isoleucine, and valine is based on the L-leucine dehydrogenase from *Bacillus sphaericus* (Fig. 4) (DEGUSSA, 1979; WICHMANN et al., 1981). In a small ultrafiltration reactor, L-leucine dehydrogenase, formate dehydrogenase, and PEG-10000-NAD^+ give a yield of 42.5 g $L^{-1}d^{-1}$ of L-leucine. The applicability of the process seems to be limited by the coenzyme stability: the coenzyme inactivation rate was calculated to be 16.8% per day compared to 0.83% for formate dehydrogenase and 1.35% for L-leucine dehydrogenase (WICHMANN et al., 1981). L-Leucine dehydrogenase has been purified, characterized, and found to be active not only with 2-oxo-4-methylpentanoic acid but also with 2-oxo-3-methylpentanoic acid and with 2-oxo-3-methylbutyric acid (OHSHIMA et al., 1978) thus being useful for the production of L-isoleucine and L-valine by the same process.

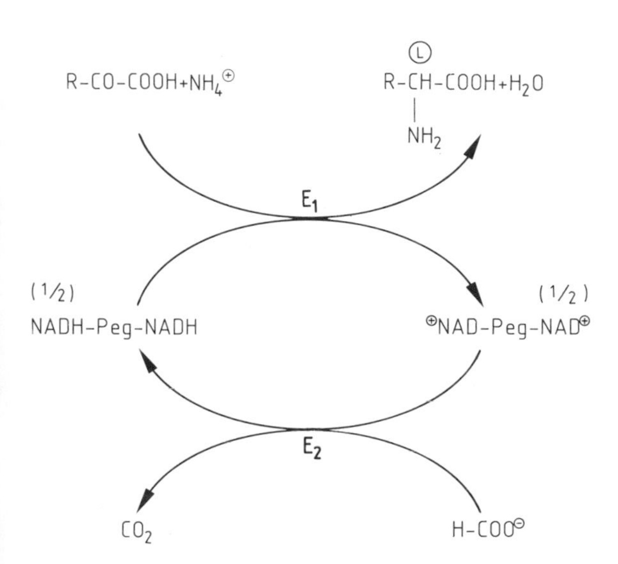

Figure 4. Coenzyme-dependent enzymatic synthesis of L-amino acids from α-keto acids with ammonia by amino acid dehydrogenases (E_1) and coenzyme regeneration by formate dehydrogenase (E_2) with formate as cosubstrate.

B. Aromatic Amino Acids

Phenylalanine. A variety of different approaches have been reported for the synthesis of L-phenylalanine by biotransformation. L- and D-Phenylalanine were prepared starting from the racemic *N*-acetyl-D,L-phenylalanine mixture via the *p*-toluidyl-L-amide through the enantioselective action of papain in 73% and 77–84 % overall yields, respectively (HUANG and NIEMANN, 1951). Racemic *N*-benzyloxycarbonyl-D,L-phenylalanine methylester was resolved by chymotrypsin, and 72% of L-phenylalanine and 90% of D-phenylalanine methylester were recovered by crystallization and extraction after subsequent hydrogenation of the *N*-protected L-acid and the *N*-protected D-ester (BERGER et al., 1973).

L-Phenylalanine has been resolved also on an industrial scale from *N*-acetyl-D,L-phenylalanine by acylase. The enzyme, however, was discarded after every batch. After adsorption of acylase to DEAE-Sephadex the enzyme could be employed in a continuous process (TOSA et al., 1967). Problems may arise from the desorption of acylase by the ionic strength of the substrate solution. An alternate process has been developed with free acylase in a semicontinuously operated ultrafiltration reactor (KITAHARA and ASAI, 1983). This process starts from a saturated solution of a racemic mixture of the ammonium salts of *N*-acetyl-D- and L-phenylalanine and is based on the separation of L-phenylalanine and *N*-Ac-D-PheNH_4 by crystallization. Saturation conditions are re-established by addition of the racemic salt mixture. Thus, after the sixth run, 79% of the totally liberated L-phenylalanine and 99% *N*-Ac-D-PheNH_4 were recovered by the alternate process, the purity being 98.2% and 93.9%, respectively. By the simultaneous process, the separated crystal fraction contained 33.6% L-phenylalanine, 57.2% N-Ac-D-PheNH_4, and 9.2% N-Ac-D,L-PheNH_4 after the third run. This technique has turned out to be suitable also for the production of L-tryptophan, L-valine, and L-phenylglycine and their D-enantiomers (KITAHARA and ASAI, 1983).

Incubation of 1–3% solutions of 2-acetamidocinnamate with resting cells of *Achromobacter liquidum, Flavobacter flavescens, Arthrobacter paraffineus, Corynebacterium hydrocarboclastus,* or *Micrococcus denitrificans* gives a yield of 0.3–1.5 g $L^{-1}d^{-1}$ of L-phenylalanine (TANABE SEIYAKU, 1980).

The same research group investigated the possibility of L-phenylalanine synthesis from *trans*-cinnamic acid catalyzed by L-phenylalanine ammonia lyase from *Rhodotorula glutinis* (YAMADA et al., 1981). 70% of the *trans*-cinnamic acid had been converted to L-phenylalanine after 24 h at pH 10. With *Rhodotorula* cells entrapped in κ-carrageenan, the enzymatic activity dropped to about 30% of the original value after the first use. The commercialization of the process is limited by the low content of L-phenylalanine ammonia lyase in the immobilized cells and by the low stability of the enzyme.

Tyrosine and derivatives. Enzymatic synthesis of L-tyrosine can be achieved by condensation starting from phenol, pyruvate, and ammonia, or from phenol and D,L-serine. L-Tyrosine formation is catalyzed by β-tyrosinase from several microorganisms, for example, *Escherichia* or *Erwinia herbicola* (KUMAGAI et al., 1970a,b; KUMAGAI et al., 1972). By the same method a series of L-tyrosine derivatives, e.g., L-DOPA, and methyl- or Cl-substituted L-tyrosine are obtained (YAMADA and KUMAGAI, 1978). L-Tyrosine or L-DOPA can be obtained also by a β-replacement from D,L-serine and phenol or pyrocatechol using *Erwinia herbicola* ATCC 21434 as the source of β-tyrosinase. Within 48 h, 60 g L^{-1} of L-tyrosine and 55 g L^{-1} of L-DOPA were prepared (ENEI et al., 1973a,b). β-Tyrosinase was covalently immobilized through pyridoxal derivatives on sepharose. The immobilized enzyme lost about 30% of its activity during 7 days of operation (FUKUI et al., 1975b). In addition, L-tyrosine can be formed from glycine, formaldehyde, and phenol by a partially purified tyrosine phenol lyase (β-tyrosinase) isolated from *Escherichia* A-21 (FALEEV et al., 1980). Thereby L-tyrosine was formed in a two-step enzymatic process catalyzed by a serine transhydroxymethylase and by the tyrosine phenol lyase.

L-Tyrosine can be produced from phenylalanine by aromatic hydroxylation (FRIEDRICH and SCHLEGEL, 1972; HANEDA et al., 1973).

Several methods for preparing L-DOPA by racemate resolution have been patented: e.g., the treatment of D,L-N-benzoyl-3-(4-hydroxy-3-methoxyphenyl)alanine and aniline with papain yielding 87.4% L-DOPA after enantioselective hydrolysis of the L-anilide (BOEHRINGER C. H., SOHN, 1971), or the resolution of 3,4-dihydroxy-D,L-phenylalanine esters by α-chymotrypsin to yield about 70% L-DOPA having an optical purity exceeding 99.8% (CANADIAN PATENTS AND DEVELOPMENT Ltd., 1971).

SIH et al. have demonstrated that a variety of fungi, especially *Stemphylium consortiale,* can efficiently hydroxylate *N*-acetyl and *N*-formyl derivatives of L-tyrosine to the corresponding L-DOPA derivatives. The amino group was derivatized to prevent microbial deamination and further degradation of L-tyrosine (SIH, 1977).

In a different way L-tyrosine was shown to be hydroxylated regiospecifically to L-DOPA by horseradish peroxidase using dihydroxyfumaric acid as an electron source (Fig. 5) (KLIBANOV et al., 1981).

Tryptophan and derivatives. As in the case of L-tyrosine enzymatic preparation of L-tryptophan is mainly focused on the condensation method with indole and D,L-serine or with indole, pyruvate, and ammonia as substrates. SNAM Progetti developed both a batchwise and a continuously fed recycle reactor with partially purified *E.coli* tryptophanase entrapped in cellulose triacetate fibers (SNAM PROGETTI, 1971; DINELLI, 1972; ZAFFARONI et al., 1974). The applicability was limited because of the strong adsorption of indole on the fiber matrix.

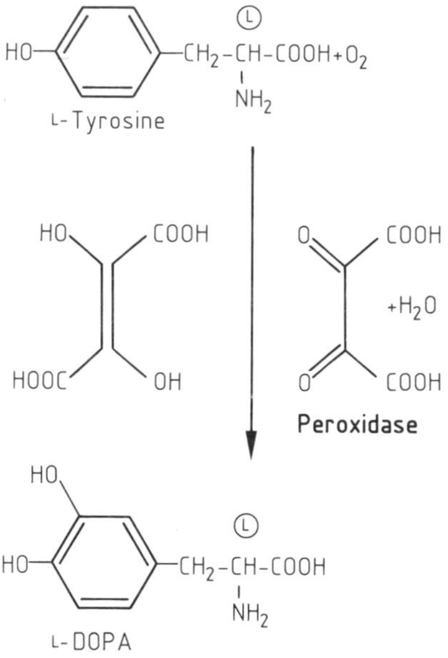

Figure 5. Regiospecific hydroxylation of L-tyrosine with O_2 to L-DOPA by horseradish peroxidase and dihydroxyfumaric acid as the electron source.

Resting cells of *E.coli* entrapped in polyacrylamide gel were used for L-tryptophan and 5-hydroxy-L-tryptophan production (TANABE SEIYAKU, 1972a,b). Yields were 1.5 g L-tryptophan and 1.1 g 5-hydroxy-L-tryptophan both from 2.0 g D,L-serine and a serine/indole (5-hydroxyindole) ratio of 2:1. Similarly to β-tyrosinase, tryptophanase was covalently bound to pyridoxal-modified sepharose by several methods (IKEDA and FUKUI, 1973; IKEDA et al., 1974; FUKUI et al., 1975a,b). During five days of batchwise operation L-tryptophan was synthesized with no significant loss of activity by an immobilized preparation of apo-tryptophanase bound to pyridoxal-phosphate sepharose (FUKUI et al., 1975b). 5-Hydroxy-L-tryptophan can be also prepared in a modified version using tryptophanase and an unspecific amino acid racemase from *Pseudomonas putida* (MITSUI TOATSU, 1980). In this process D-serine is continuously racemized. The production of L-tryptophan from indole and L- or D,L-serine by *E.coli* B 10 with high tryptophanase activity can be optimized up to yields of 14.2 g/100 mL (BANG et al., 1983a). The activity yield after immobilization of the *Escherichia coli* cells in polyacrylamide was 56%. In a batch reactor the remaining activity after 30 runs was 76–79%, in a continuous, stirred tank reactor after 50 days about 80%. The maximum yield of the continuously operated system was 0.12 g L^{-1} h^{-1} of L-tryptophan (BANG et al., 1983b).

5-Hydroxy-L-tryptophan, a precursor of the neurotransmitter serotonin (5-hydroxytryptamine), can be synthesized through microbial or enzymatic regioselective hydroxylation of L-tryptophan (SCHERING, 1971; DAUM and KIESLICH, 1974). Preferably, conversion yields up to 100% can be achieved incubating 10 h cultures of *Bacillus subtilis* ATCC 21733 with 0.2 g/L L-tryptophan for a period of about 60 h. After isolation of 5-hydroxytryptophan through adsorption to cation-exchange resin and through chromatography on Sephadex G 15, the overall yield, however, drops to about 35%. Similarly, *N*-acyl-L-tryptophans have turned out to be suitable substrates (SCHERING, 1971; DAUM and KIESLICH, 1974).

By the combined catalytic action of acylase from *Aspergillus* (10 units) and resting cells of *Proteus vulgaris, E.coli, Aerobacter aerogenes, Klebsiella pneumoniae,* or of *Bacillus alvei,* 0.9–2.7 g L^{-1} of L-tryptophan were obtained from 0.3% indole and 0.4% α-acetamidoacrylate (MITSUI TOATSU, 1978). In addition to the generally applicable racemate resolution with acylase (see: TOSA et al., 1967; KITAHARA and ASAI, 1983) L-tryptophan derivatives were claimed to be prepared from D,L-tryptophanamide through enantioselective hydrolysis of the L-amide by amidase-containing microorganisms (UBE INDUSTRIES KK, 1980).

C. Basic Amino Acids

Lysine. A combination of chemical and microbiological methods led to the preparation of L-lysine from a chemically synthesized mixture of *meso*-diaminopimelic acid and the D,L-isomers (GORTON et al., 1963) (Fig. 6). 2-Ethoxy-3,4-dihydropyran was converted in a four-step chemical synthesis to the diaminopimelic acid mixture in 80–90% overall yield (E. I. DU PONT DE NEMOURS, 1961a). Subsequent enzymatic decarboxylation of the *meso*-isomer by resting cells or cell extracts of *Bacillus sphaericus* 7054 gave 52% L-lysine calculated on the total amount of diaminopimelic acid (E. I. DU PONT DE NEMOURS, 1961b). Recovered diaminopimelic acid (41%) was epimerized by treatment with Dowex-50 ion exchanger at 180 °C.

A milestone in the industrial production of L-lysine was the process developed by Toray Industries (TORAY IND., 1970a, b), a combination of chemical and enzymatic synthesis. In a three-step chemical process, D,L-α-amino-ε-caprolactam is obtained from cyclohexene. Enantioselective hydrolysis of the L-isomer by cells of *Candida humicola* containing L-specific α-amino-ε-caprolactamase produces L-lysine at 98–100% optical purity. The D-α-amino-ε-caprolactam is re-racemized by *Alcaligenes faecalis* cells containing an α-amino-ε-caprolactam racemase (Fig. 7). For the same purpose a combination of *Achromobacter obae,* which produces high cellular levels of α-amino-ε-caprolactam racemase, and of *Cryptococcus laurentii,* an L-α-amino-ε-caprolactamase producing yeast, is utilized in industry to manufacture L-lysine with high yields (FU-

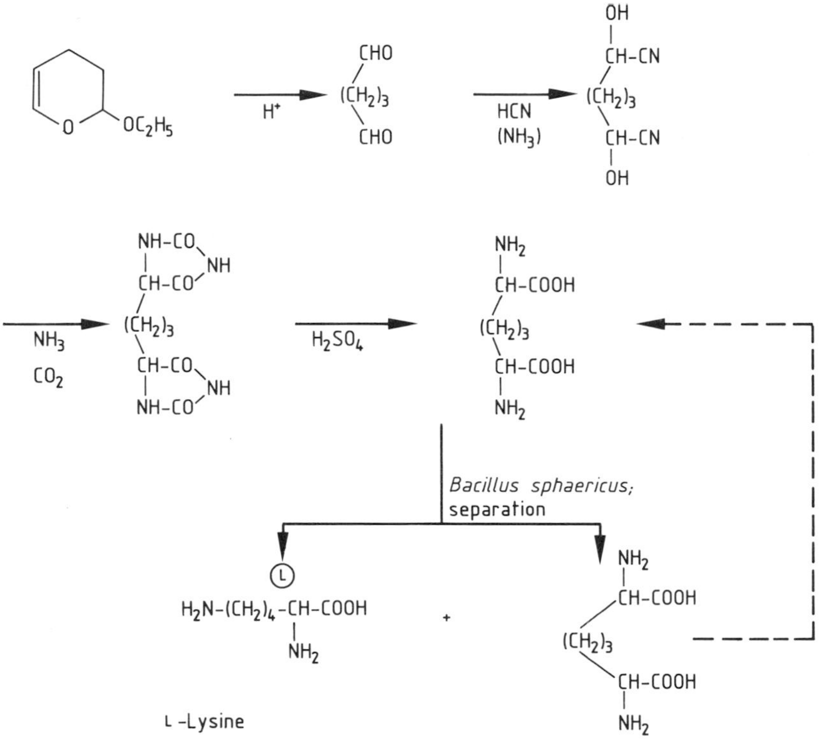

Figure 6. Chemoenzymatic synthesis of L-lysine.

KUMURA, 1977a, b). The respective enzymes have been purified and characterized (FUKUMURA et al., 1978; AHMED et al., 1982; 1983).

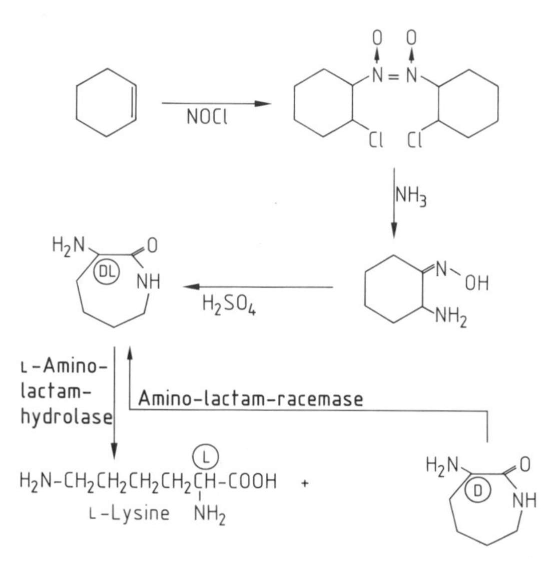

Figure 7. Chemoenzymatic synthesis of L-lysine (Toray-process). Racemate resolution of D,L-α-aminocaprolactam by L-α-amino-ε-caprolactamase and racemization of the D-enantiomer by α-amino-ε-caprolactam racemase in a multistep-biotransformation.

D. Acidic Amino Acids

Aspartic Acid. Racemate resolution of diethyl-*N*-acetyl-D,L-aspartate to the β-ethyl-α-*N*-acetyl-L-aspartate on a laboratory scale was shown to be effected by α-chymotrypsin catalysis in 74% yield (COHEN et al., 1963).

At present, only the biotransformation of fumarate to L-aspartate by *Escherichia coli* aspartase is of industrial importance. The applicability of this enzymatic reaction for production of L-aspartic acid has been investigated by KITAHARA et al. (1959). Aspartase preparations can be contaminated with fumarase leading to malate production in an unwanted side reaction. Heat treatment selectively inactivates fumarase activity (TOSA et al., 1973). By entrapment in polyacrylamide aspartase converts up to 95 % of the fumarate in a continuously operating column reactor. The production rate was about 18 g L^{-1} h^{-1} and the half-life of the enzyme about 20 days (TOSA et al., 1973). Aspartase could be further stabilized by entrapping whole cells of *E.coli* (TANABE SEIYAKU, 1974; CHIBATA et al., 1974). *E.coli* cells entrapped in polyacrylamide gel continuously transformed 1 M ammonium fumarate to L-aspartate at a yield of 95% with no loss of activity during 40 days (TOSA et al., 1974).

Modified versions were developed, as for instance: the continuous production of L-aspartic acid with duolite-ADS-aspartase at a conversion rate of higher than 99% during 3 months on a production scale (YOKOTE et al., 1978), and entrapment of *E.coli* cells in κ-carrageenan and subsequent hardening of the gel by crosslinking with glutaraldehyde and hexamethylenediamine. The half-life of the biocatalyst was increased up to more than 2 years under continuous turnover (SATO et al., 1979). Together with L-alanine (see Sect. II.A.), D-aspartic acid can be prepared by racemate resolution of D,L-aspartic acid by L-aspartic acid β-decarboxylase according to a method described by TANABE SEIYAKU (1981b).

Glutamic acid. L-Glutamic acid is mainly produced by fermentation. About 120000 tons of Na-glutamate have been produced solely in Japan in 1979 (JAPAN. CHEM. WEEK, 1983). An example of a biotransformation is the synthesis of L-glutamate from α-ketoglutarate in 96% yield (DI COSIMO et al., 1981). Glutamate dehydrogenase requires NAD(P)H as a cofactor, which can be continuously regenerated by electrochemical and enzymatic methods using methylviologen as a mediator together with ferredoxin reductase. After 7 days the residual activities of the polyacrylamide entrapped enzymes were 92% for glutamate dehydrogenase and 80% for ferredoxin reductase. The ferrodoxin reductase had a half-life of about 16 days (DI COSIMO et al., 1981).

E. Hydroxyl- and Sulfur-containing Amino Acids

Serine and threonine. L-Serine is produced upon addition of methanol and glycine at the end of the exponential growth phase of *Pseudomonas* species with a maximum yield of 4.7 g/L (WAGNER et al., 1975; KEUNE et al., 1976). The enzyme responsible for L-serine formation is serine hydroxymethyltransferase (see also Sect. II.B. for tryptophan).

Similarly, L-serine can be synthesized from glycine by resting cells of *Sarcina albida* (EMA et al., 1979; OMORI et al., 1983). With about 100 mg/mL glycine, 22 mg/mL of L-serine accumulated within 4 days of incubation of the cells. It is pointed out by the authors that strains with low serine-degrading activity should be used, and that high amounts of residual glycine might hamper L-serine recovery. L-Serine isolation can be improved by employing L-serine dehydratase and a special crystallization method using *m*-xylene-4-sulfonic acid (OMORI et al., 1983).

L-Threonine can be produced by enantioselective hydrolysis of 2-oxazoline derivatives using a variety of microorganisms (DENKI KAGAKU KOGYO CO., 1981). With cells of *Kurthia zopfi* 100 mg/mL D,L-5-methyl-2-oxazoline-4-carboxylic acid can be converted to 6 mg/mL of L-threonine (DENKI KAGAKU KOGYO CO., 1981).

Cysteine and methionine. A stereospecific condensation of allylthiol, pyruvate, and ammonia catalyzed by *Aerobacter aerogenes* cells has been patented for the formation of L-cysteine vie *S*-allyl-L-cysteine (MITSUBISHI KASEI, 1974) (Fig. 8). This condensation reaction is catalyzed by the enzyme cysteine desulfhydrase (KUMAGAI et al., 1975). Similarly, *Enterobacter cloacae* with high cysteine desulfhydrase activity was used to convert 3-chloro-L-alanine and sodium sulfide to L-cysteine with over 80% yield (YAMADA and KUMAGAI, 1978).

After screening about 500 stock cultures and about 400 soil samples, *Pseudomonas* sp. AJ 3854 was isolated as the most active bacterium forming L-cysteine and L-cystine from the chemically synthesized precursor D,L-2-amino-Δ^2-thiazoline-4-carboxylic acid. From 10 mg/mL of the D,L-racemate 6.1 mg/mL L-cysteine and/or L-cystine were produced corresponding to a 100% conversion. Thus, both hydrolysis and racemization are catalyzed by one strain (SANO et al., 1977).

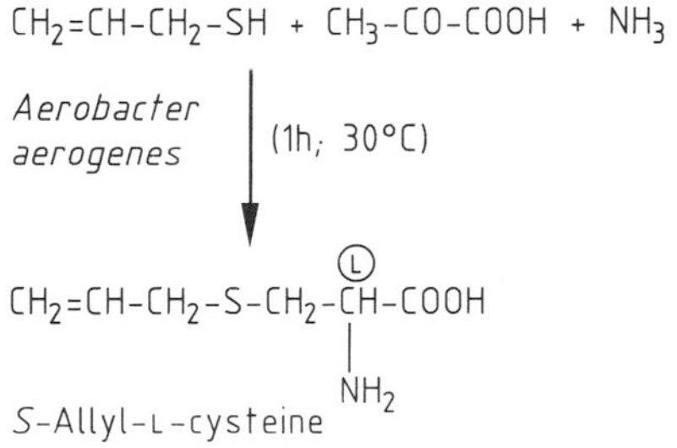

$CH_2{=}CH{-}CH_2{-}SH + CH_3{-}CO{-}COOH + NH_3$

Aerobacter aerogenes ↓ (1h; 30°C)

$CH_2{=}CH{-}CH_2{-}S{-}CH_2{-}\overset{(L)}{C}H(NH_2){-}COOH$

S-Allyl-L-cysteine

Figure 8. Stereospecific condensation of allylthiol, pyruvate, and ammonia to *S*-allyl-L-cysteine.

Several methods for L-methionine preparation have been accomplished through racemate resolution. *N*-Acetyl-D,L-methionine was resolved to L-methionine and *N*-acetyl-D-methionine using a kidney extract containing acylase (FODOR et al., 1949). Similarly, the D,L-isopropylester led to L-methionine and D-methionine isopropylester catalyzed by a pancreas extract (BRENNER and KOCHER, 1949).

Biotransformation of nitriles was shown to be another possibility to manufacture L- and D-amino acids (JALLAGEAS et al., 1980). Racemic α-amino-γ-methylthiobutyronitrile hydrochloride in a 6% solution was hydrolyzed by 20–40 g/L (dry weight) of a *Brevibacterium* mutant to 50% D-methionine amide and 50% L-methionine. After separation of the L-methionine, the D-amide is hydrolyzed to D-methionine by the wildtype *Brevibacterium* R 312 (CNRS-ANVAR, 1979; JALLAGEAS et al., 1980; 1981).

F. Side Chain Amino Acids of Semi-synthetic β-Lactams

D-Phenylglycine and derivatives. D-Phenylglycine and D-*p*-hydroxyphenylglycine are valuable synthons for the preparation of semisynthetic penicillins and cephalosporins. On an industrial scale, D,L-camphorsulfonic acid is used for racemate resolution of D,L-phenylglycine and derivatives (GREENSTEIN and WINITZ, 1961). D,L-Camphorsulfonic acid, however, is relatively expensive. An early approach by an enzymatic method was the hydrolysis of *N*-chloroacetyl-D,L-phenylglycine yielding 91% L-phenylglycine and 88% *N*-chloroacetyl-D-phenylglycine through incubation with *Azotobacter chroococcum* for 70 h (BANYU PHARMACEUTICALS, 1966). The corresponding resolution of *N*-chloroacetyl-D,L-4-methoxyphenylglycine with an aminoacylase from *Alternaria kikuchiana* was described by NIPPON SODA KK (1976). In combination with the Strecker amino acid synthesis, amidation and subsequent racemate resolution of the D,L-phenylglycine amide derivatives by aminopeptidase from *Pseudomonas putida* led to an efficient production of L- and D-phenylglycines by the DSM Geelen-NOVO process (STAMICARBON B.V., 1974; 1976).

To overcome the low solubility of *N*-acetyl-D,L-phenylglycine esters in aqueous systems, immobilized subtilisin or other proteases are used for racemate resolution in dioxane giving D-phenylglycine derivatives in high yield (BAYER AG, 1978).

Additionally, enzymatic racemate resolution of *N*-acetyl-D,L-phenylglycine methylester and its derivatives has been demonstrated in biphasic solvents, e.g., water and methylisobutyl ketone, by means of immobilized subtilisin yielding 90–99% *N*-acetyl-D-phenylglycine of high optical purity (BAYER AG, 1979; SCHUTT et al., 1984, in press) (Fig. 9).

An aminoacylase from *Micrococcus agilis* immobilized on DEAE-cellulose by adsorption was applied for stereospecific hydrolysis of *N*-acetyl-enantiomers of phenylglycine. The half-life of the biocatalyst was about 30 days (SZWAJCER et al., 1981).

However, the economically most promising way to obtain the D-phenylglycine derivatives is via the hydantoin intermediate; several processes were developed in Italy and Japan (SNAM PROGETTI, 1975; KANEGAFUCHI, 1976a, b; 1979b). D,L-5-Phenyl-

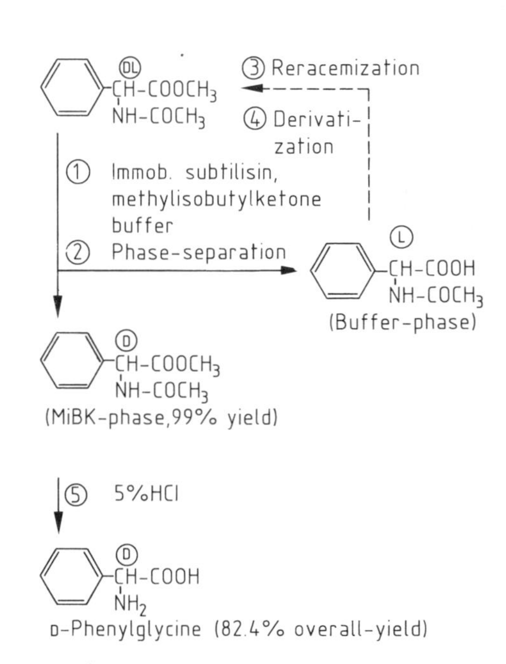

Figure 9. Racemate resolution of *N*-acetyl-D,L-phenylglycine methylester by immobilized subtilisin in a two-phase system.

hydantoin and the phenyl substituted derivatives can be synthesized by the Bucherer method from the corresponding benzaldehydes. By animal dihydropyrimidinase or by microbial hydantoinase the D,L-hydantoins were enantioselectively hydrolyzed to the *N*-carbamoyl-D-amino acid (YAMADA et al., 1980, and literature cited therein; OLIVIERI et al., 1979; 1981).

The *N*-carbamoyl-D-amino acids are converted to free D-amino acids either by chemical treatment with HNO_2 or HCl (YAMADA et al., 1980) or by a *N*-carbamoyl-D-amino acid amidohydrolase from *Agrobac-*

terium radiobacter (OLIVIERI et al., 1979; 1981). *A.radiobacter* offers the advantage to produce both *N*-carbamoyl-D-amino acid amidohydrolase and hydantoinase with activities of 11.5 and 17.1 U/g of dried cells (Fig. 10). This two-step enzymatic biotransformation which employs only one cell species in immobilized form seems to be most promising for industrial application (OLIVIERI et al., 1981).

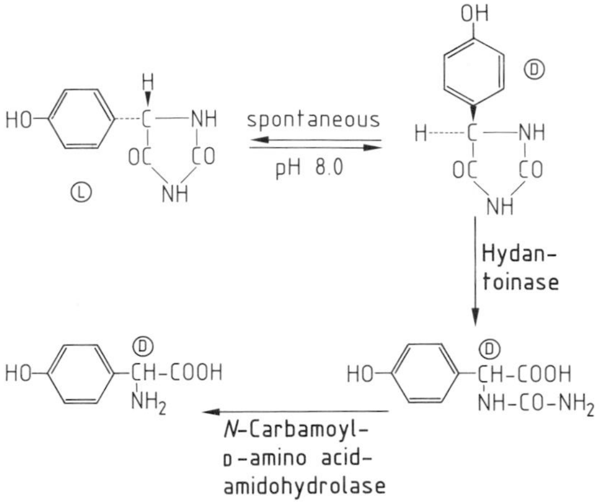

Figure 10. Racemate resolution of *p*-hydroxy-D,L-5-phenylhydantoin by hydantoinase, spontaneous reracemization of the L-isomer, and hydrolysis of the D-carbamoyl derivative to *p*-hydroxy-D-phenylglycine by *N*-carbamoyl-D-amino acid amidohydrolase in a chemoenzymatic multistep-biotransformation.

Kanegafuchi has developed a multistep-chemoenzymatic process for the racemate resolution of hydantoins (Fig. 11). The process is operated on an industrial scale with an output of 200 tons/year at their Singapore subsidiary. The process consists of a Mannich-type condensation of phenol, glyoxylic acid, and urea to the D,L-5-*p*-hydroxyphenylhydantoins followed by racemate resolution using microbial hydantoinase to *N*-carbamoyl-D-phenylglycine. The hydantoinase is D-specific, thus D-hydantoin is hydrolyzed. The L-hydantoin is spontaneously re-racemized to the D,L-hydantoins at pH 8.0. In the last step the *N*-carbamoyl-D-phenylglycine is hydrolyzed by HNO_2 to D-*p*-hydroxy-phenylglycine (KANEGAFUCHI, 1977; 1979a; YAMADA et al., 1980).

3-(R)- and 3-(S)-Aminoglutaric acid. 3-Aminoglutaric acid is an important intermediate for the preparation of a variety of physiologically active substances, such as a series of carbapenem β-lactam type antibiotics like thienamycin. Hydrolysis of diethyl-3-acetamidoglutarate by α-chymotrypsin proceeds slowly but with high optical purity to the ethyl-3-(*R*)-acetamido-glutarate, the prochiral β-carbon becoming chiral (COHEN and KHEDOURI, 1961a, b).

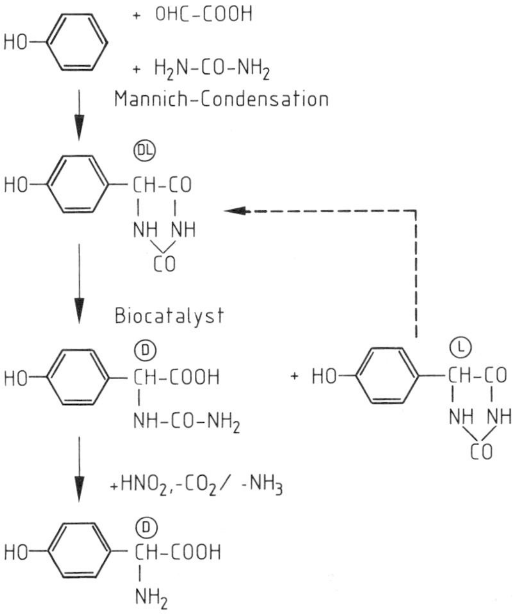

Figure 11. Production of *p*-hydroxy-D-phenylglycine by racemate resolution of D,L-*p*-hydroxyphenylhydantoin using D-specific hydantoinase as biocatalyst and spontaneous re-racemization of the L-isomer in a multistep-chemoenzymatic process (Kanegafuchi-process).

Large-scale production of 3-(*S*)-protected amino-glutarate monoalkylesters is possible by stereoselective hydrolysis of the achiral dialkylester using *Pichia farinosa* IFO 0534, especially if the amino-protective group is a bulky substituent (Kanegafuchi, 1980). The protective group can be removed by catalytic hydrogenation or by mild acid hydrolysis. The stereoselective hydrolysis of the 3-aminoglutaric acid dialkylester by esterase can be decisively influenced by the *N*-protective group. Aryloxy- and alkyloxycarbonyl groups with 2–7 carbon atoms yield 3-(*S*)-protected aminoglutaric acid monoester, whereas the unprotected diester yields 3-(*R*)-aminoglutaric acid monoester. The 3-(*R*)-glutaric acid was obtained with low optical purity due to nonenzymatic hydrolysis, but the 3-(*S*)-derivative is formed in high optical yield (Zaidan Hojin Biseibutsu Kagaku, 1980; 1981a, b; Ohno et al., 1981). The overall chemicoenzymatic synthesis starts from citric acid which is converted to 3-ketoglutarate dialkylester and further to 3-amino glutaric acid dialkylester by reductive amination.

G. Miscellaneous

L-2-Amino-4-methylphosphinobutyric acid. D,L-2-amino-4-methylphosphinobutyric acid, also called phosphotricin, is a bactericide and herbicide, however, only the L-enantiomer is active. The racemic mixture can be easily synthesized by chemical methods. The racemate resolution of the D,L-mixture was attempted through the *N*-acyl-derivatives using acylases from various species of the genera *Pseudomonas, Streptomyces,* and *Aspergillus*. The *N*-phenylacetyl derivatives can be enantioselectively hydrolized using immobilized penicillin G acylase from *E.coli* (Meiji Seika Kaisha, 1979; Hoechst AG, 1980) (Fig. 12).

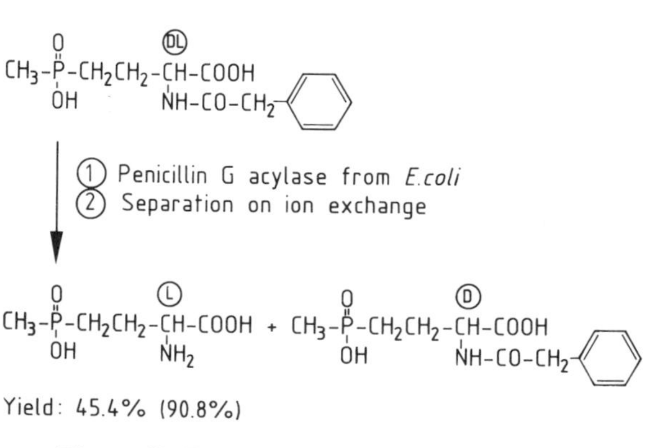

(DL) $CH_3-P(=O)(OH)-CH_2CH_2-CH(NH-CO-CH_2-C_6H_5)-COOH$

① Penicillin G acylase from *E.coli*
② Separation on ion exchange

(L) $CH_3-P(=O)(OH)-CH_2CH_2-CH(NH_2)-COOH$ + (D) $CH_3-P(=O)(OH)-CH_2CH_2-CH(NH-CO-CH_2-C_6H_5)-COOH$

Yield: 45.4% (90.8%)

Figure 12. Racemate resolution of phosphotricin by penicillin G acylase.

III. Biotransformation of Peptides

Based on earlier observations that proteases catalyze peptide bond formation, Bergmann and Fraenkel-Conrat (1937) investigated the specificity of papain, bromelin, and cathepsin for peptide synthesis. For example carbobenzoxy glycine anilide, carbobenzoxy glycine phenylhydrazide, and benzoyl-L-leucine anilide were prepared in 61.5%, 84%, and 64% yield, respectively, by this method.

In order to reverse proteolysis, thermodynamics and kinetics have to be taken into account. The equilibrium constant can be influenced favoring peptide-bond formation by using mixtures of water and water-miscible solvents like dimethylsulfoxide, dimethylformamide, or triethylene glycol. As demonstrated by Homandberg et al. (1978) the equilibrium constant of the condensation between carbobenzoxy tryptophan and glycine in 60% aqueous triethylene glycol increases by a factor of almost 20 towards peptide synthesis compared to pure aqueous solution. Shifting the pH-value may also drastically influence the equilibrium constant. Thus, either proteolysis or peptide bond formation can be effected by trypsin just by lowering the pH from 9.5 to 6.5. Peptide bond formation is enhanced by a high concentration of the nucleophilic free amino acid. Thus, the cleavage of the acylenzyme intermediate by water will be suppressed and the intermediate will preferentially react with the nucleophilic amino group to form a peptide bond. The transpeptidation by carboxypeptidase Y is assumed to proceed via an acylenzyme intermediate. This protease has been successfully employed in insulin semisynthesis and in enkephalin preparation (WIDMER et al., 1981; BREDDAM et al., 1981).

The removal of the ethyl ester protecting group has been demonstrated by treatment with immobilized carboxypeptidase Y at pH 8.5 (ROYER and ANANTHARMAIAH, 1979). The selective cleavage of the ester bond is possible because carboxypeptidase Y possesses quite different pH-optima for esterase and peptidase activity.

A detailed review on the enzymatic manipulation of protecting groups is given by GLASS (1981). For another review on the use of enzymes in peptide synthesis the reader is referred to CHAIKEN et al. (1982).

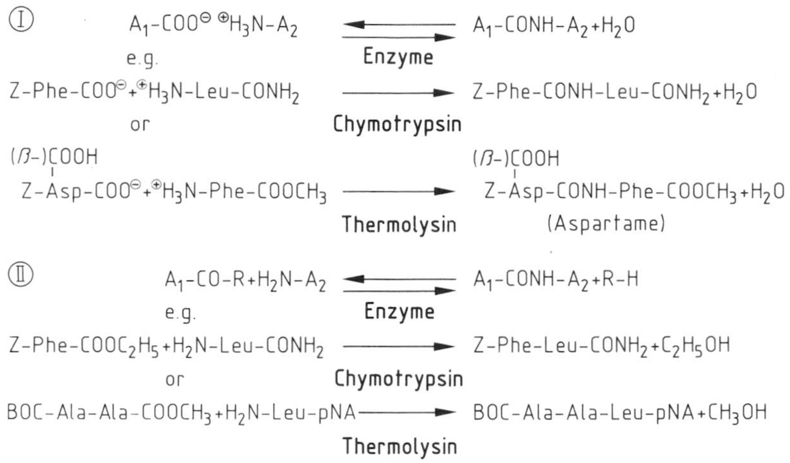

In the following, enzymatic synthesis of biologically active peptides including insulin is reviewed with special emphasis on possible industrial application.

(I) $A_1\text{-COO}^{\ominus}\ {}^{\oplus}H_3N\text{-}A_2 \rightleftharpoons A_1\text{-CONH-}A_2 + H_2O$ (Enzyme)

e.g.

$\text{Z-Phe-COO}^{\ominus} + {}^{\oplus}H_3N\text{-Leu-CONH}_2 \longrightarrow \text{Z-Phe-CONH-Leu-CONH}_2 + H_2O$ (Chymotrypsin)

or

$(\beta\text{-})\text{COOH}$ | $\text{Z-Asp-COO}^{\ominus} + {}^{\oplus}H_3N\text{-Phe-COOCH}_3 \longrightarrow$ $(\beta\text{-})\text{COOH}$ | $\text{Z-Asp-CONH-Phe-COOCH}_3 + H_2O$ (Aspartame) (Thermolysin)

(II) $A_1\text{-CO-R} + H_2N\text{-}A_2 \rightleftharpoons A_1\text{-CONH-}A_2 + \text{R-H}$ (Enzyme)

e.g.

$\text{Z-Phe-COOC}_2H_5 + H_2N\text{-Leu-CONH}_2 \longrightarrow \text{Z-Phe-Leu-CONH}_2 + C_2H_5OH$ (Chymotrypsin)

or

$\text{BOC-Ala-Ala-COOCH}_3 + H_2N\text{-Leu-pNA} \longrightarrow \text{BOC-Ala-Ala-Leu-pNA} + CH_3OH$ (Thermolysin)

Figure 13. The two major routes of peptide synthesis.

The fundamental types of peptide bond formation are shown in Fig. 13.

The carboxyl group can be used for synthesis in a non-activated or activated form (e.g., ester or amide). The amino group is protected. On the other hand, the amino group can be used for enzymatic synthesis. In this case the carboxyl group has to be protected. The carboxyl group can be protected, e.g., enzymatically with phenylhydrazin using papain (refer to KULLMANN, 1982) (see Fig. 15). A selective removal of protecting groups can be achieved using penicillin acylase, which, e.g., cleaves the phenylacetyl protecting group from the ε-amino group of lysine in the synthesis of vasopressin derivatives (BRTNIK et al., 1981).

A. Low Molecular Weight Peptides

Aspartame. α-L-Aspartyl-L-phenylalanine methyl ester was only accidentally discovered. It turned out to be about 100–200 times sweeter than sucrose. Structural requirements for the sweet taste were found to be rather rigid necessitating an unchanged α-L-aspartic acid residue but allowing modification of the L-phenylalanine moiety of the molecule as long as the ester

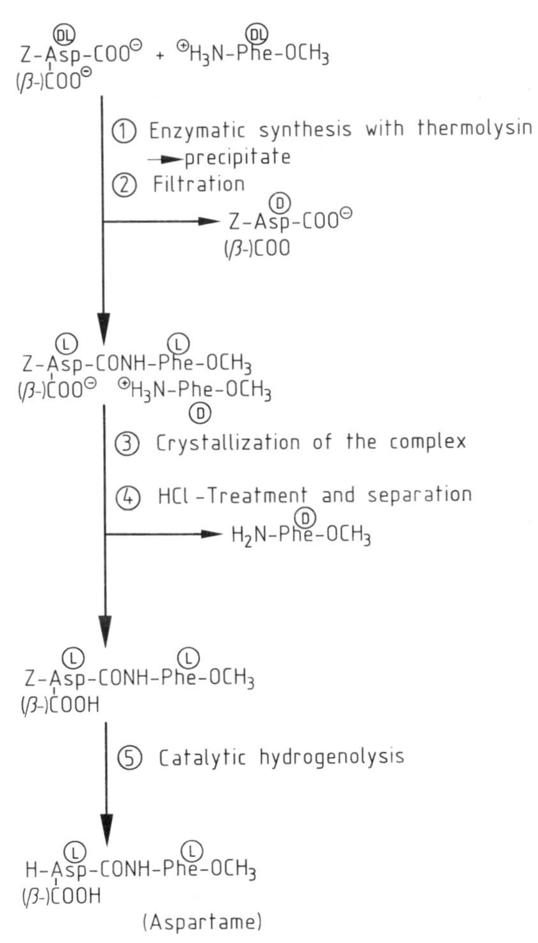

Figure 14. Stereospecific enzymatic synthesis of L-α-aspartame from benzyloxycarbonyl-D,L-aspartic acid and D,L-phenylalanine methylester by thermolysin in a chemoenzymatic process.

group and the L-configuration is retained (MAZUR et al., 1969).

Aspartame can be synthesized by chemical methods, e. g., by condensing α-amino- and β-carboxyl-protected L-aspartic acid or unprotected L-aspartic acid anhydride and L-phenylalanine methyl ester (see MAZUR et al., 1969; ARIYOSHI et al., 1973). However, all of these methods either require two protective groups or a separation of the α- and β-isomers. By enzymatic synthesis with whole cells or cell extracts, the condensation gives exclusively the α-isomer and the difficult task to protect the β-carboxylate can be avoided (SEARLE, 1981). By the chemicoenzymatic process the aminoprotected L-aspartic acid is condensed with L-phenylalanine alkyl ester hydrochloride to the *N*-protected dipeptide ester using a protease; after separation of the salt complex the protecting group can be removed by catalytic hydrogenation (SEARLE, 1981).

To avoid the production of expensive pure L-amino acid derivatives, it was attempted to start from racemic raw materials, and crude thermolysin was used as biocatalyst (ISOWA et al., 1979). As can be seen in Fig. 14, the salt of amino-protected α-L-aspartame and D-phenylalanine methylester which is easily separable into its components by HCl treatment is obtained in 95% yield by precipitation whereas the unreacted Z-D-aspartic acid can be recovered from the aqueous solution. This is an interesting example of an enantioselective and a regioselective biocatalyst that is of industrial importance. In order to reuse the thermolysin, the enzyme can be immobilized either by adsorption or by covalent bonding (OYAMA et al., 1981). It was discussed, that the application of water-saturated ethylacetate can keep the enzyme adsorbed in the inner sphere of the porous carrier to avoid its leakage. Compared to the results in aqueous media, however, the reaction rates are rather slow (OYAMA et al., 1981).

A further biotransformational approach to the synthesis of α-L-aspartame is the enzymatic esterification of α-L-aspartyl-L-phenylalanine or of the *N*-protected derivatives thereof in aqueous alcohol by proteases (SEARLE, 1980). Again, this biotransformation has the advantage of being highly selective for the α-carboxy group and does not esterify the β-carboxylate of the aspartyl part of the dipeptide. Aspartame is expected on the market as a sweetener.

Glutathione. Glutathione is a physiologically active tripeptide (γ-glutamylcysteinylglycine), which serves in animal cells as an activator of certain enzymes, as an antioxidant, and as a component of the γ-glutamyl cycle responsible for the transport of amino acids (FLOHE et al., 1974). The biosynthesis in animal cells proceeds from 5-oxoproline via L-glutamate and via two peptide condensation reactions to glutathione. All three enzymatic steps require ATP. Glutathione can be produced from animal cells by extraction or by chemical synthesis. The biotransformational procedure is based on γ-glutamylcysteine synthetase and glutathione synthetase from *Escherichia coli* or *Saccharomyces cerevisiae*, and additionally requires an intact ATP-regenerating system. In a series of experiments MURATA et al. (1979; 1980) applied immobilized *E.coli* cells containing acetate kinase as well as the glutathione-synthesizing enzymes together with co-immobilized dextran-bound ATP or, alternately, the glycolytic pathway of *S.cerevisiae* cells to produce glutathione. Using the glycolytic pathway with glucose for ATP-regeneration, the polyacrylamide-entrapped cells formed glutathione at a rate of 0.65 µmol h^{-1} mL^{-1} gel in a batch experiment (MURATA et al., 1980). The activity of the immobilized biocatalyst significantly decreased during continuous operation in a column reactor: the half-life was calculated to be 32 days (MURATA et al., 1980). Co-immobilization of a mixture of two of the cell species increased the effi-

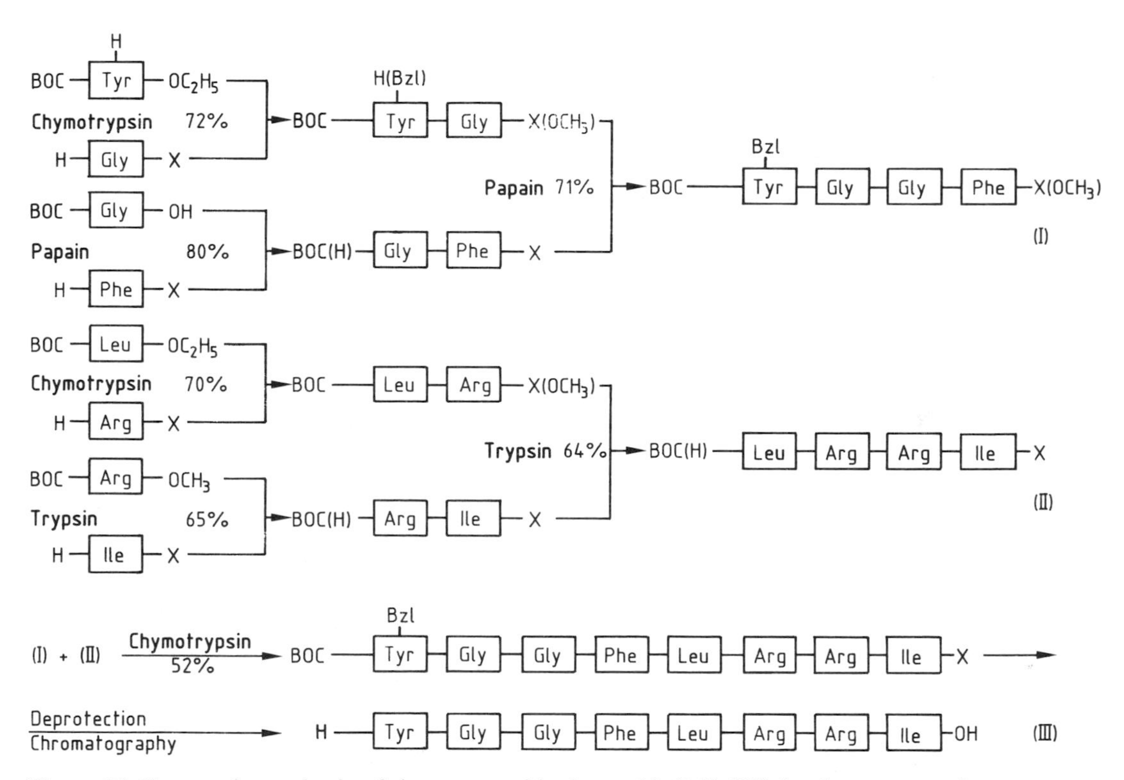

Figure 15. Enzymatic synthesis of the octapeptide dynorphin(1-8) (III) by the enzymes chymotrypsin, papain, and trypsin.

ciency of the ATP utilization through concomitant production of NADP (MURATA et al., 1981). Maximum rates of glutathione production are about 0.8 μmol h^{-1} mL^{-1}, but there is no significant improvement compared to the use of bare *S.cerevisiae* cells (0.6 μmol h^{-1} mL^{-1}). Feedback inhibition of γ-glutamylcysteine synthetase by glutathione and by ATP, glutathione degradation as well as transport limitation across the cell membrane are discussed to be the main obstacles that have to be overcome in order to attain economical production rates of glutathione (MURATA et al., 1981).

Dynorphin(1-8). Dynorphin(1-8), a biologically active opioid octapeptide, has been synthesized forming all the peptide bonds by subsequent catalysis using three different proteases (KULLMANN, 1982). As depicted in Fig. 15, α-chymotrypsin catalyzes the formation of the two dipeptides Tyr-Gly (72% yield) and Leu-Arg (70% yield). Papain catalyzes the formation of Gly-Phe (80% yield) and trypsin the enzymatic synthesis of Arg-Ile (65%). Condensation of Tyr-Gly and Gly-Phe is achieved with papain (71%), that of Leu-Arg and Arg-Ile with trypsin (64%). From the two tetrapeptides, the octapeptide Tyr-Gly-Gly-Phe-Leu-Arg-Arg-Ile [dynorphin(1-8)] is obtained with α-chymotrypsin (52%). Appropriate protecting groups are used throughout the entire synthesis. Papain is employed to catalyze the protection of α-carboxyl residues with phenylhydrazine. The advantage of this totally enzymatic procedure is its regio- and stereospecificity. Especially racemization can be avoided, and the guanidino group of arginine does not have to be protected as in chemical synthesis (KULLMANN, 1982).

Leu- and met-enkephalin. Two separate procedures have been described to prepare Leu- and Met-enkephalin in high yield. Starting from Arg, the hexapeptide Arg-Tyr-Gly-Gly-Phe-Met can be enzymatically synthesized step by step using carboxypeptidase Y (WIDMER et al., 1981). Therefrom, Met-enkephalin is easily obtained by trypsin-catalyzed fission of the N-terminal Arg-Tyr bond. As is illustrated in Fig. 16, all the single enzymatic steps give good yields (60–95%). Compared to other carboxypeptidases, the carboxypeptidase Y does not contain essential metals but has both an active-site serine and an active-site histidine (HAYASHI et al., 1976).

Carboxypeptidase Y has been also applied successfully for insulin synthesis (see Sect. III.B.).

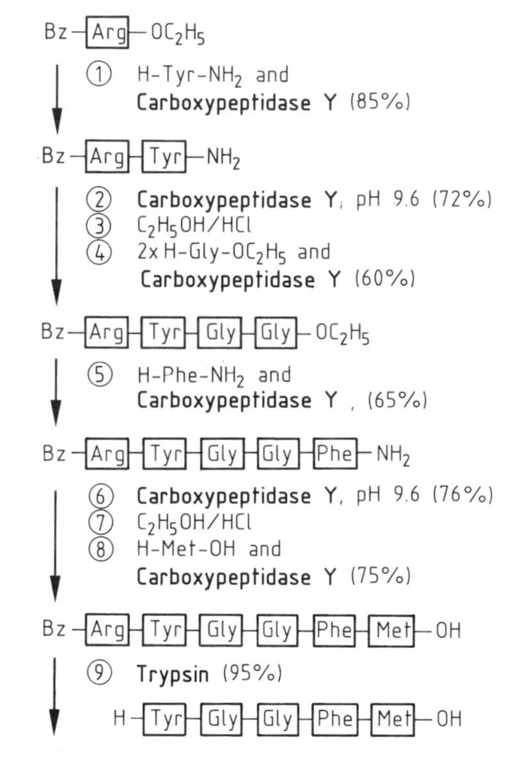

Figure 16. Stepwise enzymatic synthesis of the pentapeptide Met-enkephalin by carboxypeptidase Y using benzyl-arginine as *N*-terminal-protecting group and deprotection of the hexapeptide intermediate by trypsin (yields of single steps in parentheses).

Corresponding to dynorphin(1-8) which contains the Leu-enkephalin sequence, Met- and Leu-enkephalin are also accessible either through papain or α-chymotrypsin catalysis (KULLMANN, 1980; 1982). The amino- and α-carboxyl-protected pentapeptides can be obtained from the protected tetrapeptides and with protected Leu or Met using papain or α-chymotrypsin. It should be noted that the incubation time influences the coupling yields. Optimal incubation times vary considerably between 10 min and 40 h depending on the enzyme and on the substrates (KULLMANN, 1980). Leu- and Met-enkephalin are neurotransmitters.

Angiotensin. In comparison to the enzymatic synthesis of dynorphin(1-8) and of Met- and Leu-enkephalin, the preparation of the peptide hormone valine-5-angiotensin II amide (Asn-Arg-Val-Tyr-Val-His-Pro-Phe-NH_2) has been accomplished by enzymatic condensation of the chemically synthesized peptide fragments (ISOWA et al., 1977). The enzymatic condensation of chemically pre-synthesized Val-Tyr and Val-His-Pro-Phe by papain, nagarse, or by microbial metallo-enzymes proceeds with 57%, 74%, or 41% yield. Accordingly, the condensation of the tetrapeptides Asn-Arg-Val-Tyr and Val-His-Pro-Phe to the octapeptide Val-5-angiotensin II amide is effected by papain in 78% yield (ISOWA et al., 1977). Val-5-angiotensin II amide is a vasopressor which is related to the angiotensins, a class of peptide hormones derived from plasma angiotensinogen.

B. Insulin

Insulin is the most important peptide hormone essential for diabetics. Insulins from animals and humans differ in their amino acid composition. Porcine insulin and human insulin differ only in the C-terminal amino acid of the B-chain (Fig. 17).

Insulin is extracted from animal pancreas and purified by crystallization. Crystallized insulin can be further purified by chromatographic methods to eliminate immunological side effects (monocomponent insulin). Alternately, human insulin has to be used. Human insulin can be produced by recombinant DNA-technology (Humulin, Eli Lilly). Human insulin can be also produced by biotransformation of porcine insulin (BROMER et al., 1967; RUTTENBERG, 1972; SHANGHAI INSULIN RESEARCH GROUP, 1973; HOECHST AG, 1974).

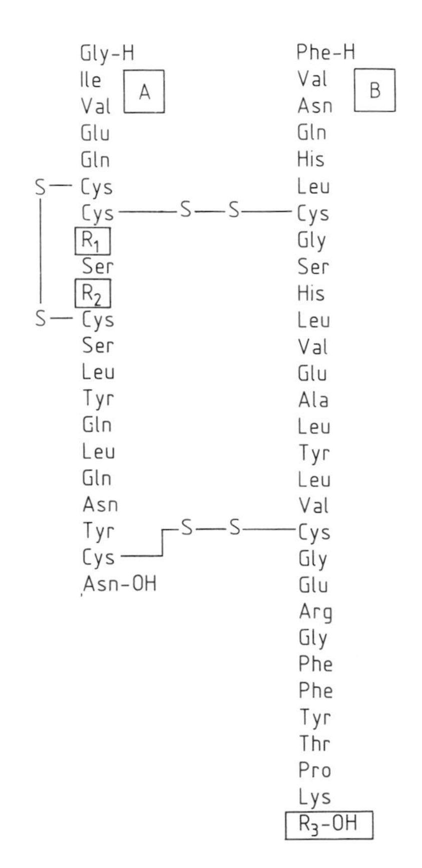

	Insulin from Bovine	Porcine	Human
R_1	Ala	Thr	Thr
R_2	Val	Ileu	Ileu
R_3	Ala	Ala	Thr

Figure 17. Amino acid composition of bovine, porcine, and human insulin.

Two different routes have been designed. According to the specificity of trypsin (Arg), porcine insulin can be hydrolyzed to des-(B23-30)octapeptide insulin (see BROMER et al., 1967). The desoctapeptide can be subsequently coupled with the human (B23-30)octapeptide by chemical methods (Fig. 18) or by biotransformation using enzymes (RUTTENBERG, 1972; INOUYE et al., 1979).

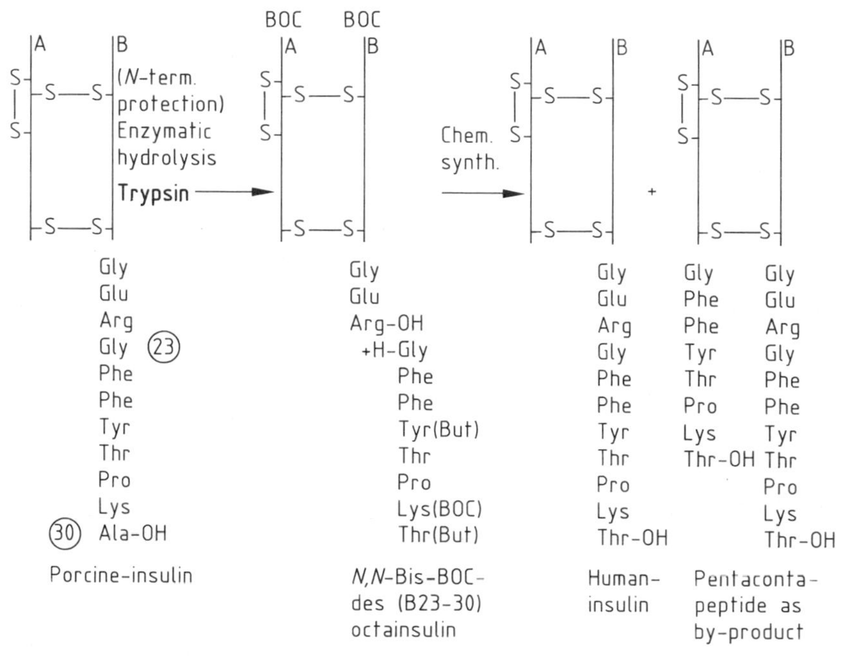

Figure 18. Enzymatic hydrolysis of porcine-insulin by trypsin to *N,N*-bis-BOC-des(B23-30)octainsulin and resynthesis using a protected octapeptide to human-insulin by a chemical process with a pentacontapeptide as by-product.

The second route is the exchange of only the C-terminal amino acid of the B-chain (alanine by threonine) in two steps by the action of one or two different enzymes (MORIHARA et al., 1979; 1980; SHIONOGI KK, 1980a) or the direct exchange by transpeptidation using carboxypeptidase Y (BREDDAM et al., 1981; DE FORENEDE BRYGGERIER A/S, 1980; JOHANSON, 1983) or using trypsin (HOECHST AG, 1981; NORDISK INSULIN LAB., 1981).

RUTTENBERG (1972) developed an easy and inexpensive route from porcine to human insulin. After reversible blocking of the α- and γ-carboxyl groups, porcine insulin is digested with trypsin to yield des(B23-30)octapeptide insulin. After reversible blocking of the free amino groups with BOC-azide, the human desoctapeptide is coupled to des(B23-30)octapeptide insulin by dicyclohexylcarbodiimide. Removal of all the protecting groups and purification on carboxymethyl cellulose leads to human insulin in an overall yield of about 70% – e.g., 2.7 g from 4.0 g porcine insulin.

By a similar approach, the SHANGHAI INSULIN RESEARCH GROUP (1973) also prepared human insulin from porcine insulin and synthetic human-type octapeptide via desoctapeptide insulin. Desoctapeptide insulin is obtained in about 45% yield from porcine insulin by trypsin catalysis. *N*-protected desoctapeptide insulin pentamethylester can be isolated in about 23% overall yield in a three-step chemical synthesis. Final condensation between the protected desoctapeptide insulin and high molar excess of the protected octapeptide in the presence of carbodiimide yields crude human insulin. After deprotection and purification on Sephadex G-50 and/or extraction with aqueous acidic 2-butanol, 4.5 mg of pure human insulin were obtained from 30 mg crude insulin.

In a modified version, des(B23-30)octapeptide insulin having free carboxyl groups is condensed with the human (B23-30)octapeptide at Arg B22 by cyclohexylcarbodiimide. However, the yields are low and purification steps are necessary to eliminate by-products (OBERMEIER and GEIGER, 1976).

Compared to the aforementioned methods, an important improvement is achieved using trypsin not only for the hydrolysis of the porcine insulin but also for the peptide bond formation (INOUYE et al., 1979). In order to favor peptide bond formation, the coupling reaction is performed at pH 6.5 with a large molar excess of human-type (B23-30)octapeptide. Hydrolysis with trypsin is optimal at pH 9–9.5. Based on porcine insulin the overall yields were 56%. For a more economical way most of the surplus of octapeptide is recovered and reused (INOUYE et al., 1979). Following the procedure of INOUYE et al. (1979) two mutant forms of human insulin, (Leu B24)insulin and (Leu B25)insulin in which B24 or B25 phenylalanine are replaced by leucine, can be prepared by trypsin-assisted synthesis from the desoctapeptide and from the (Leu B24) and (Leu B25) human-type octapeptides (TAGER et al., 1980; GATTNER et al., 1980). (Leu B24) and (Leu B25)insulin are obtained in 25% and 10% yield, respectively. The coupling yields depend strongly on the amino acid composition and sequence of the octapeptide (GATTNER et al., 1980).

Further insulin analogs have been prepared. Des(B24-30)heptapeptide insulin was prepared from des(B26-30)pentapeptide insulin by limited hydrolysis with carboxypeptidase A (LU and YU, 1980) and B23-D-Ala-des(B25-30)hexapeptide via des-(B22-30)nonapeptide insulin by trypsin and carboxypeptidase B (ZHU et al., 1981). Crystalline des(B25-B30)hexapeptide insulin was prepared by enzymatic condensation of the des(B23-30)octapeptide insulin with glycylphenylalanine in 60-fold molar excess using trypsin in 74% yield. The pH-optimum for the trypsin-catalyzed condensation is at pH 8.0 (CAO et al., 1981) compared to INOUYE et al. (1979) who have found pH 6.5 to be optimal.

A more economical route from porcine to human insulin is possible by selectively cleaving the B-chain C-terminal alanine residue of porcine insulin and subsequently condensing a threonine ester derivative and the des(Ala B30)insulin (SHIONOGI, 1980a). The selective removal of the C-terminal alanine by carboxypeptidase A has been successfully demonstrated by SCHMITT and GATTNER (1978) in an ammonium bicarbonate buffer solution which is necessary to hinder the simultaneous release of C-terminal asparagine from the A-chain. Additionally, an improved procedure has been used with a serine protease from *Achromobacter lyticus* which is specific for the carboxyl side of lysine (MORIHARA et al., 1980). However, the protease has to be employed twice at pH 8.3 and pH 6.5 with intermediate isolation and lyophilization of the des(Ala B30)insulin. Overall yields of 52% for porcine insulin and of 58% for bovine insulin have been obtained. Unreacted des(Ala B30)insulin (about 25%) can be reused (MORIHARA et al., 1980).

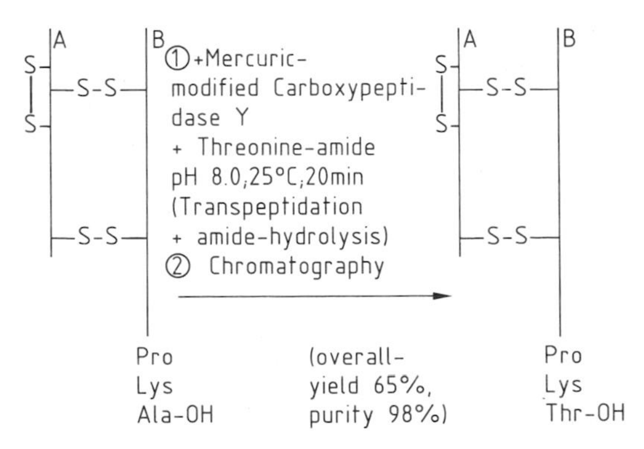

Figure 19. Enzymatic synthesis of human insulin from porcine insulin by carboxypeptidase Y modified with mercurials: transpeptidation and subsequent deamidation.

Carboxypeptidase Y catalyzes the removal of B-chain C-terminal alanine and the condensation of the des(Ala B30)insulin with threonine ester or threonine amide in one step at the same pH 7.5 (BREDDAM et al., 1981; DE FORENEDE BRYGGERIER A/S, 1980). The authors stated that the concomitant enzymatic deamidation of the insulin amide (Thr B30) by carboxypeptidase Y at pH 9.5 gives better yields of pure human insulin than acid-catalyzed deesterification (DE FORENEDE BRYGGERIER A/S, 1980) (Fig. 19). Thus, at pH 9.5, not only transpeptidation of the C-terminal alanine and threonine amide but also the removal of the amido group form C-terminal threonine amide is effected by carboxypeptidase Y in one step. Overall yield of pure human insulin (peak II) after DEAE-chromatography is about 20–30% (BREDDAM et al., 1981).

Recently, it has been demonstrated that selective modification of carboxypeptidase Y with mercuric ions increases the specificity of the transpeptidase activity towards substrates with a penultimate lysyl residue (BREDDAM, 1983). The overall yield of homogeneous human insulin can be increased to 50–65% applying mercury halide derivatives of carboxypeptidase Y (X-Hg-CPD-Y) and carboxypeptidase Y modified with methylmercuric iodide (Me-Hg-CPD-Y) for the transpeptidation and for the deamidation reaction, respectively (BREDDAM and JOHANSEN, 1983).

This increase is mainly due to a change in the product spectrum of the transpeptidase reaction, e.g., peak I comprises 52% of the total area of peaks I, II, III after ion exchange chromatography compared to only 21% with unmodified carboxypeptidase Y. Additionally, the composition of peak I which consists of human insulin amide and of the condensation products INS-Pro-Lys-Thr-Thr-NH_2 and INS-Pro-Lys-Ala-Thr-NH_2 can be markedly influenced by the pH of the reaction solution, by the nature of the halide anion of Hg-CPD-Y, and by the reaction time. In the final deamidation step, human insulin of highest purity (98.2%) is obtained from a peak I preparation containing 54% INS-Pro-Lys-Thr-Thr-NH_2 and 46% human insulin amide, at 90–95% deamidation (BREDDAM and JOHANSEN, 1983).

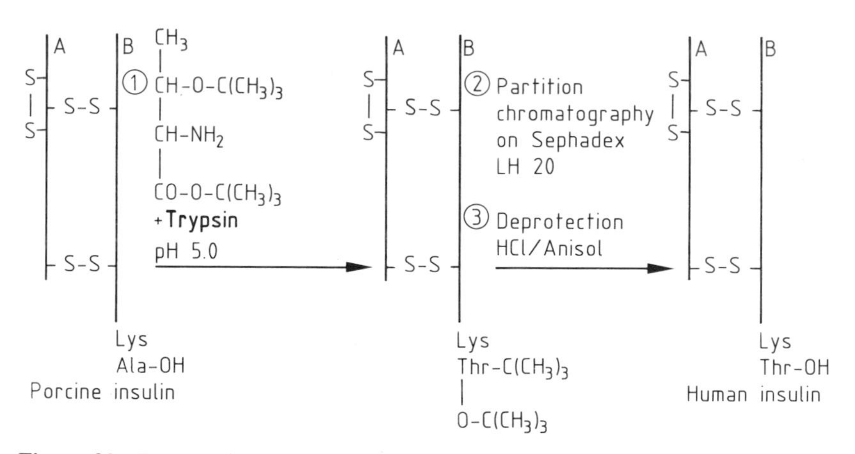

Figure 20. Enzymatic synthesis of human insulin from porcine insulin and an excess of *O-tert*-butyl-L-threonine-*tert*-butylester catalyzed by trypsin.

A new semi-synthesis of human insulin by trypsin-catalyzed transpeptidation of porcine insulin in the presence of about 30–130 fold molar excess of L-threonine tert-

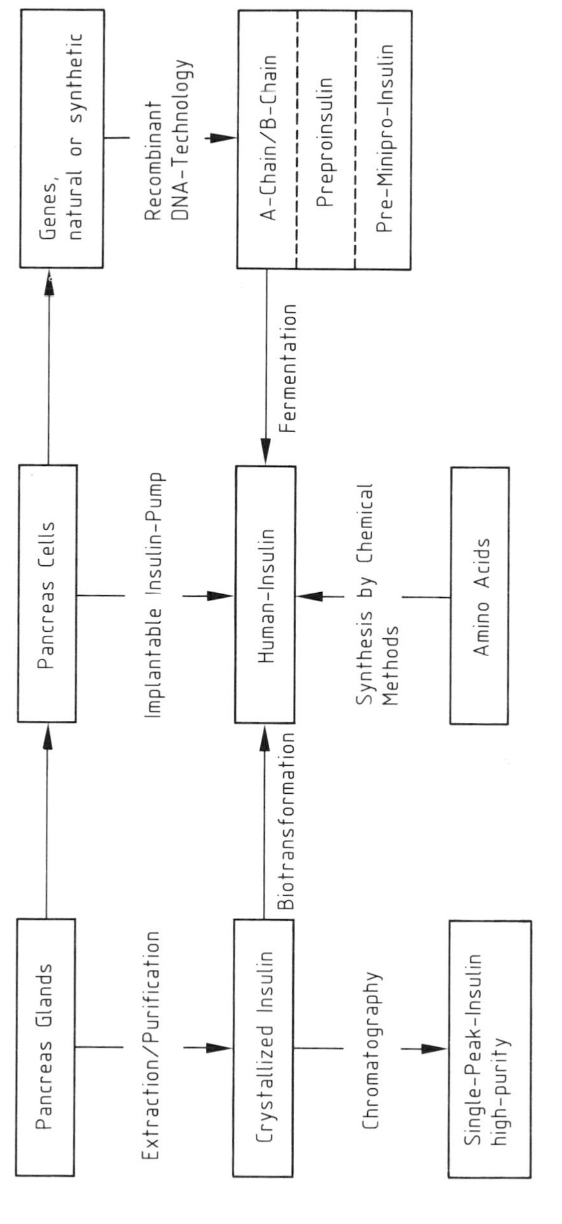

Figure 21. Alternative processes for the production of human insulin.

butylester has been described by JONCZYK and GATTNER (1981). The formation of side products, especially of des(B23-30)insulin, can be decisively reduced in 70–80% dimethylformamide, below pH 6.5, and in phosphate buffer of high ionic strength. Yields of human insulin ester are 50–70%, however, purification by HPLC or by ion exchange chromatography and hydrolysis of the human insulin ester to the human insulin further reduce the overall yield to less than 30% (JONCZYK and GATTNER, 1981).

According to the Hoechst procedure it is possible to transform porcine insulin to human insulin in the presence of a high molar excess of substituted threonine ester by trypsin. During this one-step transpeptidation (Fig. 20) the pH of the reaction solution must be kept under the isoelectric point of trypsin (pH 5.4). Yields vary between 40% and 90% depending on pH, type of threonine ester, and purity of the final product (HOECHST AG, 1981). Additionally, application has been filed for a patent using trypsin or an Achromobacter protease to prepare insulin derivatives in one step in the presence of 30–70% organic solvent (NORDISK INSULIN LABORATORIUM, 1981).

Several porcine insulin analogs like (B30-Phe) and (B30-Leu)insulin can be prepared in a two-step reaction using carboxypeptidase A, trypsin, or other proteases and may be useful for treating insulin-resistant diabetes mellitus (SHIONOGI KK, 1980b,c).

For all the enzymatic methods cited relatively high amounts of proteolytic enzymes are needed. The used enzyme has to be of sufficient purity to avoid proteolytic side reactions.

Human insulin can be economically produced either from pancreas insulin through biotransformation or by recombinant DNA-technology – see Fig. 21. Only the product with the highest purity will have the best chances to be the market leader in the future.

L-Glutamate	340 000 t
DL-Methionine	120 000 t
L-Lysine	34 000 t
	494 000 t
Glycine	6 000 t
DL-Alanine	2 000 t
L-Cysteine/L-Cystine	700 t
L-Arginine	500 t
L-Aspartic acid	450 t
L-DOPA	200 t
L-Tryptophan	200 t
L-Histidine	200 t
L-Threonine	160 t
L-Phenylalanine	150 t
L-Leucine	150 t
L-Isoleucine	150 t
L-Valine	150 t
L-Methionine	150 t
L-Alanine	130 t
L-Proline	100 t
	11 390 t
L-Serine	50 t
L-Ornithine	50 t
L-Citrulline	50 t
L-Aspargine	50 t
L-Tyrosine	50 t
L-Hydroxyproline	50 t
	300 t
in total	505 690 t

Figure 22. World supply of amino acids in 1980.

IV. Biotransformation – Economic Aspects

Amino acids are utilized by the feed and food industries and to a minor extent by the pharmaceutical industry. The world supply of amino acids in 1980 is depicted in Fig. 22 (AKASHI, MANILA-CONFERENCE, 1982). Accordingly, a total of 506000 t were produced that year, 97.7% of which consisted of the three amino acids L-glutamic acid (67.2%), D,L-methionine (23.7%), and L-lysine (6.8%). A second, smaller fraction totaling 1.7% is composed of glycine (1.2%), D,L-alanine (0.4%), and L-cysteine/cystine (0.1%). All other amino acids merely amount to 0.6% of the total production. L-Glutamic acid, D,L-methionine, and L-lysine are produced by large-scale procedures and are available at a price of ca. 5–10 US $/kg. Most other amino acids are ten times as expensive ranging from 50–100 US $/kg. Relations between prices (US $/kg) and supplies in 1980 are presented in Fig. 23. In reference to these data it can be assumed that prices of presently expensive amino acids could be lowered in case of increasing demand.

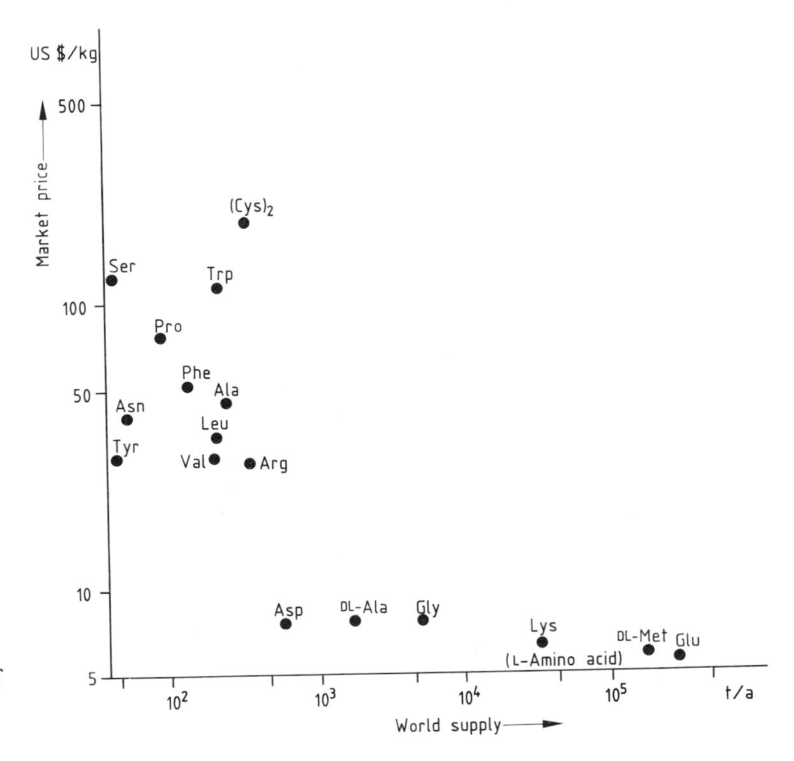

Figure 23. Market prices of amino acids relative to the world supply in 1980.

L-Glutamic acid is produced in form of its sodium salt (340 000 t in 1980), which is used to enhance the taste of foods. D,L-Methionine and L-lysine are supplements to animal feed. Main producers of D,L-methionine are the Degussa AG/Federal Republic of Germany (capacity of 75 000 t/a), Rhône-Poulenc AEC SA/France, and Nippon Soda Co. Ltd./Japan. Large plants are under construction in the US (Monsanto, capacity of ca. 40 000 t) and in Brazil (Unirhodic SA, capacity of ca. 20 000 t). Main producers of L-lysine are Eurolysin SA/France, Ajinomoto Co. Ltd./Japan, Miwan Ltd./Korea, Toray Ind. Inc./Japan, IQT SA/Spain, and Fermex SA/Mexico. A plant with a capacity of ca. 10 000 t of L-lysine is being constructed by Kyowa Hakko Koygo Co. Ltd./Japan. In comparison to the three aforementioned amino acids only small amounts of the expensive L-tryptophan are being produced (Diamalt AG/Federal Republic of Germany; Showa Denko Co. Ltd./Japan).

Assumably, utilization of amino acids by the pharmaceutical industries amounts to only 1–2% of the total world supply; however, the value of this material amounts to 18%. Mostly these amino acids are used for infusions and special diets. Thus, comparably high demands are made on quality and notably on the optical purity of the L-amino acids used for pharmaceutical purposes.

The Degussa plant in Konstanz/Federal Republic of Germany produces the optically pure L-amino acids L-methionine, L-valine, L-phenylalanine, and L-tryptophan via racemate resolution in an enzyme-membrane-reactor (Fig. 24, partial view of the production machinery at Degussa AG/Konstanz). The capacity of the plant is 5 t of L-amino acids per month. Soluble proteases accomplishing racemate resolution and separation of the products by means of membranes yield pyrogen-free L-amino acids suitable for the production of pharmaceuticals.

Further examples of pharmaceutical applications of amino acids are the use of: *N*-acetyl-L-hydroxyproline (oxacepol) as an antiarthritic, of captopril, a derivative of L-proline, in therapy of hypertension, of L-tryptophan as a tranquilizer, and of L-DOPA against the symptoms of Parkinson's disease.

The synthesis of the sweetener aspartame (L-aspartyl-L-phenylalanine-methylester; Searle Corp./USA) involves L-phenylalanine. The cosmetics industry utilizes L-cysteine, available from Diamalt AG and

Figure 24. Enzyme-membrane-reactor (see arrow) for the production of L-amino acids by racemate resolution of *N*-acyl-D,L-amino acids at the DEGUSSA plant in Konstanz, Federal Republic of Germany.

Degussa AG/Federal Republic of Germany; small amounts are also produced by Il Shin Co./Korea and Showa Denko/Japan. Glycine is used as a buffering substance in pharmaceutical and cosmetic industries.

Amino acids can be obtained by the following four known methods:

- hydrolysis/extraction of proteins
- fermentation
- biotransformation
- chemical synthesis

Largely, amino acids are produced via fermentation. Only low amounts are formed through biotransformations. However, the production of optically pure amino acids through biotransformation is constantly expanding.

It should be mentioned that certain unnatural side-chain amino acids, such as D-phenylglycine and *p*-hydroxy-D-phenylglycine also have a considerable economic importance, as they are used in semi-synthesis of penicillins and cephalosporins. Their economic value can be estimated from the 1982 turnover rate of β-lactam antibiotics which amounted to a total of 5 billion US $.

The turnover of insulin in 1983 ranged in the category of 400 million US $, the USA accounting for ca. 150 million US $ and Europe for ca. 130 million US $ (HEPNER, 1983).

V. References

In this chapter the priority date of patents is expressed in the German sequence: day, month, year.

ABBOTT, B. J. (1976). "Preparation of pharmaceutical compounds by immobilized enzymes and cells." Adv. Appl. Microbiol. *20,* 203–257.

AHMED, S. A., ESAKI, N., and SODA, K. (1982). "Purification and properties of α-amino-ε-caprolactam racemase from *Achromobacter obae.*" FEBS Lett. *150,* 370–374.

AHMED, S. A., ESAKI, N., TANAKA, H., and SODA, K. (1983). "Racemization of α-amino-δ-valerolactam catalyzed by α-amino-ε-caprolactam racemase from *Achromobacter obae.*" Agric. Biol. Chem. *47,* 1149–1150.

AKASHI, T. (1982). "World Supply of Amino Acid." Chemrawn II, Manila Conference 1982.

ARIYOSHI, Y., YAMATANI, T., UCHIYAMA, N., ADACHI, Y., and SATO, N. (1973), "The synthesis of a sweet peptide, α-aspartyl-L-phenylalanine methylester, without the use of protecting groups." Bull. Chem. Soc. Jpn. *46,* 1893–1895.

BANG, W. G., LANG, S., SAHM, H., and WAGNER, F. (1983a). "Production of L-tryptophan by *Escherichia coli* cells." Biotechnol. Bioeng. *25,* 999–1011.

BANG, W. G., BEHRENDT, U., LANG, S., and WAGNER, F. (1983b). "Continuous production of L-tryptophan from indole and L-serine by immobilized *Escherichia coli* cells." Biotechnol. Bioeng. *25,* 1013–1025.

BANYU PHARMACEUTICALS (1966). "Process for Producing Phenylglycine Having Optical Activity or Its Derivatives." Japan. Patent 45-08635 (8. 10. 1966).

BASF AG (1980). „Verfahren zur Herstellung von D-*N*-Carbamoyl-*alpha*-aminosäuren und Mikroorganismen dafür." EP 46186 (18. 08. 80).

BAYER AG (1978). "Stereoselective Resolution of Phenylglycine Derivatives and 4-Hydroxyphenylglycine Derivatives with Enzyme Resins." US Patent 4260684 (21. 02. 78).

BAYER AG (1979). „Stereoselektive Spaltung von Phenylglycinderivaten mit Enzymharzen." EP 22492 (07. 07. 79).

BERGER, A., SMOLARSKY, M., KURN, N., and BOSSHARD, H. R. (1973). "A new method for the synthesis of optically active α-amino acids and their *N(α)*-derivatives via acylamino malonates." J. Org. Chem. *38,* 457–460.

BERGMANN, M., and FRAENKEL-CONRAT, H. (1937). "The role of specificity in the enzymatic synthesis of proteins. Synthesis with intracellular enzymes." J. Biol. Chem. *119,* 707–720.

BOEHRINGER, C. H., SOHN (1971). „Verfahren zur Herstellung von L-DOPA". DOS (Ger. Offen.) 2103245 (25. 1. 1971).

BREDDAM, K., WIDMER, F., and JOHANSEN, J. T. (1981). "Carboxypeptidase Y catalyzed C-terminal modification in the B-chain of porcine insulin." Carlsberg Res. Commun. *46*, 361–372.

BREDDAM, K. (1983). "Modification of the single sulfhydryl group of carboxypeptidase Y with mercurials. Influence on enzyme specificity." Carlsberg Res. Commun. *48*, 9–19.

BREDDAM, K., and JOHANSEN, J. T. (1983). "Semisynthesis of human insulin utilizing chemically modified carboxypeptidase Y." Carlsberg Res. Commun., in press.

BRENNER, M., and KOCHER, V. (1949). „Eine einfache enzymatische Methode zur Herstellung von D- und L-Methionin." Helv. Chim. Acta *32*, 333–337.

BROMER, W. W., and CHANCE, R. E. (1967). "Preparation and characterization of desoctapeptide-insulin." Biochim. Biophys. Acta *133*, 219–223.

BRTNIK, F., BARTH, T., and JOST, K. (1981). "Use of the phenylacetyl group for protection of the lysine *N(ε)*-amino group in synthesis of peptides." Coll. Czech. Chem. Commun. *46*, 1983–1989.

CANADIAN PATENTS AND DEVELOPMENT Ltd. (1971). "Resolution of Racemates of Ring-Substituted Phenylalanines." US Patent 3813317 (02. 04. 1971).

CAO, Q. P., CUI, D. F., and ZHANG, Y. S. (1981). "Enzymatic synthesis of deshexapeptide insulin." Nature *292*, 774–775.

CHAIKEN, I. M., KOMORIYA, A., OHNO, M., and WIDMER, F. (1982). "Use of enzymes in peptide synthesis." Appl. Biochem. Biotechnol. *7*, 385–399.

CHIBATA, I., KAKIMOTO, T., and KATO, J. (1965). "Enzymatic production of L-alanine by *Pseudomonas dacunhae*." J. Appl. Microbiol. *13*, 638–645.

CHIBATA, I., TOSA, T., and SATO, T. (1974). "Immobilized aspartase-containing microbial cells: Preparation and enzymatic properties." Appl. Microbiol. *27*, 878–885.

CHIBATA, I., TOSA, T., SATO, T., and MORI, T. (1976). "Production of L-amino acids by aminoacylase adsorbed on DEAE-Sephadex." In "Methods in Enzymology", Vol. 44 (K. MOSBACH, ed.), pp. 746–759. Academic Press, New York.

CHIBATA, I. (1979) "Development of enzyme engineering – application of immobilized cell systems." Kemia-Kemi *6*, 705–714.

CNRS-ANVAR (1979). "Procédé de Préparation D'acides α-Aminés Optiquement Actifs par Hydrolyse Biologique de Nitriles α-Aminés Racémiques." Fr. Patent 2447359 (24. 01. 1979).

COHEN, S. G., and KHEDOURI, E. (1961a). "Requirements for stereospecificity in hydrolysis by α-chymotrypsin. Diethyl-β-acetamidoglutarate." J. Am. Chem. Soc. *83*, 1093–1096.

COHEN, S. G., and KHEDOURI, E. (1961b). "Requirements for stereospecificity in hydrolysis by α-chymotrypsin. IV. The hydroxyl substituent. Absolute configurations." J. Am. Chem. Soc. *83*, 4228–4233.

COHEN, S. G., CROSSLEY, J., and KHEDOURI, E. (1963). "Action of α-chymotrypsin on diethyl-*N*-acetylaspartate and on diethyl-*N*-methyl-*N*-acetylaspartate." Biochemistry *2*, 820–823.

DANIELSSON, B., WINQUIST, F., MALPOTE, J. Y., and MOSBACH, K. (1982). "Regeneration of NADH with immobilized systems of alanine dehydrogenase and hydrogen dehydrogenase." Biotechnol. Lett. *4*, 673–678.

DAUM, J., and KIESLICH, K. (1974). „Darstellung von 5-Hydroxytryptophan durch mikrobiologische Hydroxylierung von L-Tryptophan." Naturwissenschaften *61*, 167–168.

DE FORENEDE BRYGGERIER A/S (1980). "A Process for Enzymatic Replacement of the B-30 Amino Acid in Insulins." EP 45187 (24. 07. 80).

DEGUSSA AG (1979). „Verfahren zur kontinuierlichen enzymatischen Umwandlung von wasserlöslichen α-Ketocarbonsäuren in die entsprechenden Aminosäuren." DE 2930070 (25. 07. 79).

DENKI KAGAKU KOGYO (1981). "L-Threonine no Seihō." Jap. Patent 58-023790 (04. 08. 81).

DI COSIMO, R., WONG, C. H., DANIELS, L., and WHITESIDES, G. M. (1981). "Enzyme-catalyzed organic synthesis: Electrochemical regeneration of NAD(P)H from NAD(P) using methylviologen and flavoenzymes." J. Org. Chem. *46*, 4622–4623.

DINELLI, D. (1972). "Fibre-entrapped enzymes." Process Biochem. *7*, Aug., 9–12.

E. I. DU PONT DE NEMOURS (1961a). "Production of L-Lysine from Synthetic Diaminopimelic Acid." US Patent 2976218 (21. 03. 61).

E. I. DU PONT DE NEMOURS (1961b). "Diaminopimelic Acid." Brit. Patent 875353 (16. 08. 1961).

EMA, M., KAKIMOTO, T., and CHIBATA, I. (1979). "Production of L-serine by *Sarcina albida*." Appl. Environ. Microbiol. *37*, 1053–1058.

ENEI, H., MATSUI, H., NAKAZAWA, H., OKUMURA, S., and YAMADA, H. (1973a). "Synthesis of L-tyrosine or 3,4-dihydroxy-L-phenylalanine from D,L-serine and phenol or pyrocatechol." Agric. Biol. Chem. *37*, 493–499.

ENEI, H., NAKAZAWA, H., OKUMURA, S., and

YAMADA, H. (1973b). "Synthesis of L-tyrosine or 3,4-dihydroxy-L-phenylalanine from pyruvic acid, ammonia, and phenol or pyrocatechol." Agric. Biol. Chem. *37,* 725–735.

ENEI, H., SHIBAI, H., and HIROSE, Y. (1982). "Amino acids and nucleic acid-related compounds." In "Annual Reports on Fermentation Processes", Vol. 5 (G. T. TSAO, ed.) pp. 79–100. Academic Press, New York.

FALEEV, N. G., SADOVNIKOVA, M. S., MARTINKOVA, N. S., and BELIKOV, V. M. (1980). "Formation of tyrosine from glycine, formaldehyde, and phenol under the action of a partially purified preparation of tyrosine phenol lyase: the participation of a different enzyme." Enzyme Microb. Technol. *2,* 305–308.

FLOHE, L. (1974). "Glutathione." Academic Press, New York.

FODOR, P. J., PRICE, V. E., and GREENSTEIN, J. P. (1949). "Preparation of L- and D-alanine by enzymatic resolution of acetyl-D,L-alanine." J. Biol. Chem. *178,* 503–509.

FRIEDRICH, B., and SCHLEGEL, H. G. (1972). „Die Hydroxylierung von Phenylalanin durch *Hydrogenomonas eutropha* H16." Arch. Mikrobiol. *83,* 17–31.

FUKUI, S., IKEDA, S., FUJIMURA, M., YAMADA, H., and KUMAGAI, H. (1975a). "Production of L-tryptophan, L-tyrosine and their analogues by use of immobilized tryptophanase and immobilized β-tyrosinase." Eur. J. Appl. Microbiol. *1,* 25–39.

FUKUI, S., IKEDA, S., FUJIMURA, M., YAMADA, H., and KUMAGAI, H. (1975b). "Comparative studies of the properties of tryptophanase and tyrosine phenol lyase immobilized directly on sepharose or by use of sepharose-bound pyridoxal-5′-phosphate." Eur. J. Biochem. *51,* 155–164.

FUKUI, S., and TANAKA, A. (1980). "Production of useful compounds from alkane media in Japan." Adv. Biochem. Eng. *17,* 1–35.

FUKUMURA, T. (1977a). "Bacterial racemization of α-amino-ε-caprolactam." Agric. Biol. Chem. *41,* 1321–1325.

FUKUMURA, T. (1977b). "Conversion of D- and D,L-α-amino-ε-caprolactam into L-lysine using both yeast cells and bacterial cells." Agric. Biol. Chem. *41,* 1327–1330.

FUKUMURA, T., TALBOT, G., MISONO, H., TERAMURA, Y., KATO, K., and SODA, K. (1978). "Purification and properties of a novel enzyme L-α-amino-ε-caprolactamase from *Cryptococcus laurentii.*" FEBS Lett. *89,* 298–300.

GATTNER, H. G., DANHO, W., BEHN, C., and ZAHN, H. (1980). "The preparation of two mutant forms of human insulin, containing leucine in positions B24 or B25, by enzyme-assisted synthesis." Hoppe-Seyler's Z. Physiol. Chem. *361,* 1135–1138.

GLASS, J. D. (1981). "Enzymes as reagents in the synthesis of peptides." Enzyme Microbiol. Technol. *3,* 2–8.

GORTON, B. S., COKER, J. N., BROWDER, H. P., and DEFIEBRE, C. W. (1963). "A process for the production of lysine by chemical and microbiological synthesis." Ind. Eng. Chem. Prod. Res. Dev. *2,* 308–314.

GREENSTEIN, J. P., and WINITZ, M. (1961). "Chemistry of the Amino Acids." Vol. 1, p. 658. John Wiley & Sons, New York.

HANEDA, K., WATANABE, S., and TAKEDA, I. (1973). "Production of 3,4-dihydroxy-L-phenylalanine from l-tyrosine by microorganisms." J. Ferment. Technol. *51,* 398–406.

HAYASHI, R. (1976). "Carboxypeptidase Y." In "Methods in Enzymology", Vol. 45 (L. LORAND, ed.), pp. 568–587. Academic Press, New York.

HEPNER, L., and ASSOC. (1983). "Product opportunities in fermentation." Personal communication.

HOECHST AG (1974). „Verfahren zur Herstellung von Humaninsulin." DE 2460753 (21. 12. 74).

HOECHST AG (1980). „Verfahren zur enzymatischen Herstellung von L-2-Amino-4-methylphosphinobuttersäure." EP 54897 (23. 12. 80).

HOECHST AG (1981). „Verfahren zur Herstellung von Humaninsulin." DE 2460753 (21. 12. 74). Schweineinsulin oder dessen Derivaten." EP 56951 (17. 01. 81).

HOMANDBERG, G. A., MATTIS, J. A., and LASKOWSKI, M., Jr. (1978). "Synthesis of peptide bonds by proteinases. Addition of organic cosolvents shifts peptide bond equilibrium toward synthesis." Biochemistry *17,* 5220–5227.

HUANG, H. T., and NIEMANN, C. (1951). "The enzymatic resolution of D,L-phenylalanine." J. Am. Chem. Soc. *73,* 475–476.

IKEDA, S., and FUKUI, S. (1973). "Preparation of pyridoxal-5′-phosphate-bound sepharose and its use for immobilization of tryptophanase." Biochem. Biophys. Res. Commun. *52,* 482–488.

IKEDA, S., HARA, H., and FUKUI, S. (1974). "Preparation and characterization of several new immobilized derivatives of pyridoxal-5′-phosphate." Biochim. Biophys. Acta *372,* 400–406.

INOUYE, K., WATANABE, K., MORIHARA, K., TOCHINO, Y., KANAYA, T., EMURA, J., and SAKAKIBARA, S. (1979). "Enzyme assisted semisynthesis of human insulin." J. Am. Chem. Soc. *101,* 751–752.

ISOWA, Y., OHMORI, M., SATO, M., and MORI, K. (1977). "The enzymatic synthesis of protected valine-5-angiotensin II amide-1." Bull. Chem. Soc. Jpn. *50,* 2766–2772.

ISOWA, Y., OHMORI, M., ICHIKAWA, T., MORI, K., NONAKA, Y., KIHARA, K., OYAMA, K., SATOH, H., and NISHIMURA, S. (1979). "The thermolysin-catalyzed condensation reactions of *N*-substituted aspartic and glutamic acids with phenylalanine alkyl esters." Tetrahedron Lett. *28,* 2611–2612.

JALLAGEAS, J. C., ARNAUD, A., and GALZY, P. (1980). "Bioconversion of nitriles and their applications." Adv. Biochem. Eng. *14,* 1–32.

JALLAGEAS, J. C., ARNAUD, A., and GALZY, P. (1981). "Utilisation de l'activité α-aminoamidasique d'une bacterie nitrilasique pour la production d'acides α-amines stereospecifiques.". In "Advances in Biotechnology", Vol. 3 (M. MOO-YOUNG, ed.), pp. 227–233. Pergamon Press, Toronto.

JANDEL, A. S., HUSTEDT, H., and WANDREY, C. (1982). "Continuous production of L-alanine from fumarate in a two-stage membrane reactor." Eur. J. Appl. Microbiol. Biotechnol. *15,* 59–63.

JAPAN CHEMICAL WEEK (1983). "Development of large-scale outlet for amino acid urgently needed." Vol. *24,* No. 1196 (March 17th), p. 12.

JOHANSEN, J. T. (1983). Personal communication.

JONCZYK, A., and GATTNER, H. G. (1981). „Eine neue Semisynthese des Humaninsulins. Tryptisch-katalysierte Transpeptidierung von Schweineinsulin mit L-Threonin-*tert*-butylester." Hoppe-Seyler's Z. Physiol. Chem. *362,* 1591–1598.

KANEGAFUCHI CHEM. KK (1976a). "Process for Preparing D-(−)-*N*-Carbamoyl-2-(Phenyl or Substituted Phenyl)Glycines." Japan. Patent 53-69884, US Patent 4094741 (04. 02. 1976; 03. 12. 1976).

KANEGAFUCHI CHEM. KK (1976b). „Verfahren zur Herstellung von D-*N*-Carbamoyl-aminosäuren." DOS (Ger. Offen.) 2757980 (30. 12. 1976).

KANEGAFUCHI CHEM. KK (1977). "Optically Active Phenylglycines." Japan. Patent 53-103441 (21. 02. 1977).

KANEGAFUCHI CHEM. KK (1979a). "D-Amino Acids." Japan. Patent 55-104890 (6. 2. 1979).

KANEGAFUCHI CHEM. KK (1979b). "Production of *N*-Carbamoyl-D-Amino Acids." Japan. Patent 56-058493 (16. 10. 1979).

KANEGAFUCHI CHEM. KK (1980). "Optically Active β-(*S*)-Aminoglutarate Monoalkyl Esters." Fr. Patent 2497230 (30. 12. 1980).

KEUNE, H., SAHM, H., and WAGNER, F. (1976). "Production of L-serine by the methanol utilizing bacterium *Pseudomonas* 3ab." Eur. J. Appl. Microbiol. *2,* 175–184.

KITAHARA, T., FUKUI, S., and MISAWA, M. (1959). "Preparation of L-aspartic acid by bacterial aspartase." J. Gen. Appl. Microbiol. (Tokyo) *5,* 74–77.

KITAHARA, T., and ASAI, S. (1983). "Optical resolution of D,L-phenylalanine with acylase." Agric. Biol. Chem. *47,* 991–996.

KLEEMANN, A. (1982). "α-Aminosäuren als Bausteine für Heterocyclen." Chem. Ztg. *106,* 151–167.

KLIBANOV, A. M., BERMAN, Z., and ALBERTI, B. N. (1981). "Preparative hydroxylation of aromatic compounds catalyzed by peroxidase." J. Am. Chem. Soc. *103,* 6263–6264.

KULLMANN, W. (1980). "Proteases as catalysts for enzymatic synthesis of opioid peptides." J. Biol. Chem. *255,* 8234–8238.

KULLMANN, W. (1982). "Enzymatic synthesis of dynorphin(1-8)." J. Org. Chem. *47,* 5300–5303.

KUMAGAI, H., YAMADA, H., MATSUI, H., OHKISHI, H., and OGATA, K. (1970a). "Tyrosine phenol lyase. I. Purification, crystallization, and properties." J. Biol. Chem. *245,* 1767–1772.

KUMAGAI, H., YAMADA, H., MATSUI, H., OHKISHI, H., and OGATA, K. (1970b). "Tyrosine phenol lyase. II. Cofactor requirements." J. Biol. Chem. *245,* 1773–1777.

KUMAGAI, H., KASHIMA, N., YAMADA, H., ENEI, H., and OKUMURA, S. (1972). "Purification, crystallization, and properties of tyrosine phenol lyase from *Erwinia herbicola.*" Agric. Biol. Chem. *36,* 472–482.

KUMAGAI, H., SEJIMA, S., CHOI, Y. S., TANAKA, H., and YAMADA, H. (1975). "Crystallization and properties of cysteine desulfhydrase from *Aerobacter aerogenes.*" FEBS Lett. *52,* 304–307.

LEUENBERGER, H. G. W., and KIESLICH, K. (1982). „Biotransformationen." In „Handbuch der Biotechnologie". (P. PRÄVE, U. FAUST, W. SITTIG, D. A. SUKATSCH, eds.), pp. 453–482. Akademische Verlagsgesellschaft, Wiesbaden.

LU, Z., and YU, R. (1989). "Preparation and crystallization of des(B-chain C-terminal) heptapeptide insulin." Sci. Sin. *23,* 1592–1598.

MAZUR, R. H., SCHLATTER, J. M., and GOLDKAMP, A. H. (1969). "Structure-taste relationship of some dipeptides." J. Am. Chem. Soc. *91,* 2684–2691.

MEIJI SEIKA KAISHA (1979). "Process for the Optical Resolution of D,L-2-Amino-4-Methylphosphinobutyric Acid." US Patent 4226941 (27. 09. 79).

MITSUI TOATSU CHEM INC. (1978). "L-Tryptophan no Seizō Hōhō." Japan. Patent 55-88698 (26. 12. 78).

MITSUI TOATSU CHEM INC. (1980). "5-Hydroxy-Tryptophan no Seizōhō." Japan. Patent 57-083288 (07. 11. 1980).

MITSUBISHI KASEI (1974). "Enzymatic Production of *S*-Allyl-L-Cysteine." Japan. Patent 50-132178 (10. 04. 1974).

MORIHARA, K., OKA, T., and TSUZUKI, H. (1979). "Semi-synthesis of human insulin by trypsin-catalyzed replacement of Ala-B30 by Thr in porcine insulin." Nature *280,* 412–413.

MORIHARA, K., OKA, T., TSUZUKI, H., TOCHINO, Y., and KANAYA, T. (1980). "Achromobacter protease I-catalyzed conversion of porcine insulin into human insulin." Biochem. Biophys. Res. Commun. *92,* 396–402.

MURATA, K., TANI, K., KATO, J., and CHIBATA, I. (1979). "Application of immobilized ATP in the production of glutathione by a multienzyme system." J. Appl. Biochem. *1,* 283–290.

MURATA, K., TANI, K., KATO, J., and CHIBATA, I. (1980). "Continuous production of glutathione using immobilized microbial cells containing ATP-generating system." Biochimie *62,* 347–352.

MURATA, K., TANI, K., KATO, J., and CHIBATA, I. (1981). "Glycolytic pathway as an ATP generation system and its application to the production of glutathione and NADP." Enzyme Microbiol. Technol. *3,* 233–242.

NIPPON SODA KK (1976). "Optical Resolution of *N*-Acyl-D.L-Amino Acid." Japan. Patent 53-006489 (05. 07. 1976).

NORDISK INSULIN LABORATORIUM (1981). "A Process for the Preparation of Insulin Derivatives." WO 82/4069 (20. 05. 81).

OBERMEIER, R., and GEIGER, R. (1976). "A new semisynthesis of human insulin." Hoppe-Seyler's Z. Physiol. Chem. *357,* 759–767.

OHNO, M., KOBAYASHI, S., IIMORI, T., WANG, Y. F., and IZAWA, T. (1981). "Synthesis of (*S*)- and (*R*)-4-(methoxycarbonyl)methyl-2-azetidinone by chemico-enzymatic approach." J. Am. Chem. Soc. *103,* 2405–2406.

OHSHIMA, T., MISONO, H., and SODA, K. (1978). "Properties of crystalline leucine dehydrogenase from *Bacillus sphaericus*." J. Biol. Chem. *253,* 5719–5725.

OLIVIERI, R., FASCETTI, E., ANGELINI, L., and DEGEN, L. (1979). "Enzymatic conversion of *N*-carbamoyl-D-amino acids to D-amino acids." Enzyme Microb. Technol. *1,* 201–204.

OLIVIERI, R., FASCETTI, E., ANGELINI, L., and DEGEN, L. (1981). "Microbial transformation of racemic hydantoins to D-amino acids." Biotechnol. Bioeng. *23,* 2173–2183.

OMORI, K., KAKIMOTO, T., and CHIBATA, I. (1983). "L-Serine production by mutant of *Sarcina albida* defective in L-serine degradation." Appl. Environ. Microbiol. *45,* 1722–1726.

OYAMA, K., NISHIMURA, S., NONAKA, Y., KIHARA, K., and HASHIMOTO, T. (1981). "Synthesis of an aspartame precursor by immobilized thermolysin in an organic solvents." J. Org. Chem. *46,* 5242–5244.

REHOVOT RESEARCH PRODUCTS Ltd. (1978). "A Method for the Performance of Enzymatic Reactions and Reactors Therefore." Israel. Patent 54116 (23. 02. 1978).

ROYER, G. P., and ANANTHARMAIAH, G. M. (1979). "Peptide synthesis in water and the use of immobilized carboxypeptidase Y for deprotection." J. Am. Chem. Soc. *101,* 3394–3395.

RUTTENBERG, M. A. (1972). "Human insulin: Facile synthesis by modification of porcine insulin." Science *177,* 623–626.

SANO, K., YOKOZEKI, K., TAMURA, T., YASUDA, N., NODA, I., and MITSUGI, K. (1977). "Microbial conversion of D,L-2-amino-Δ^2-thiazoline-4-carboxylic acid to L-cysteine and L-cystine: Screening of microorganisms and identification of products." Appl. Environ. Microbiol. *34.* 806–810.

SATO, T., NISHIDA, Y., TOSA, T., and CHIBATA, I. (1979). "Immobilization of *Escherichia coli* cells containing aspartase activity with κ-carrageenan. Enzymic properties and application for L-aspartic acid production." Biochim. Biophys. Acta *570,* 179–186.

SCHEPER, T., HALWACHS, W., and SCHÜGERL, K. (1982). „Darstellung von L-Aminosäuren durch kontinuierliche, enzymkatalysierte Aminosäureester-Hydrolyse mittels Flüssigmembran-Emulsionen.“ Chem. Ing. Tech. *54,* 696–697.

SCHEPER, T., HALWACHS, W., and SCHÜGERL, K. (1983). "Preparation of L-Amino Acids by means of Continuous Enzyme-catalyzed D,L-Amino Acid Ester Hydrolysis Inside Liquid Surfactant Membranes." Int. Solvent Extraction Conference ISEC, Denver, Colorado.

SCHERING AG (1971). "Verfahren zur Herstellung von 5-Hydroxy-L-tryptophan." DOS (Ger. Offen.) 2150535 (06. 10. 1971).

SCHMITT, E. W., and GATTNER, H. G. (1978). „Verbesserte Darstellung von Des-alanyl(B30)-insulin." Hoppe-Seyler's Z. Physiol. Chem. *359.* 799–802.

SCHNEIDER, K., and SCHLEGEL, H. G. (1976). "Purification and properties of soluble hydrogenase from *Alcaligenes eutrophus.*" Biochim. Biophys. Acta *452,* 66–80.

SCHÜGERL, K., HALWACHS, W., POPPE, W., MELZNER, D., MOHRMANN, A., and SCHEPER, T. (1983). "New Extraction Applications of Liquid Membranes." Summer National Meeting AIChE, Denver, Colorado.

SCHUTT, H., SCHMIDT-KASTNER, G., ARENS, A., and PREISS, M. (1984). "Preparation of optically active aryl-D-glycine side-chain synthons for semisynthetic penicillins and cephalosporins by immobilized subtilisins in two-phase systems." Biotechnol. Bioeng., in press.

SEARLE (1980). "Process for Esterification of α-L-Aspartyl-L-Phenylalanine." US Patent 4293648 (02. 09. 80).

SEARLE (1981). "Preparation of Amino Protected L-Aspartyl-L-Phenylalanine Alkyl Ester." Brit. Patent 2092161 (02. 02. 81).

SIH, C. J. (1977). "The Volwiler Lecture: Microbes in synthesis." Am. J. Pharm. Educ. *41,* 432–437.

SHANGHAI INSULIN RESEARCH GROUP (1973). "Studies on the structure-function relationships of insulin. I. The relationship of the C-terminal peptide sequence of B-chain to the activity of insulin." Sci. Sin. *16,* 61–70.

SHIONOGI (1980a). "Semi-Synthesis of Human Insulin." US Patent 4320196 (09. 04. 80).

SHIONOGI (1980b). "Insulin-Related Derivatives by Action of Protease." Japan. Patent 57-050940 (10. 09. 80).

SHIONOGI (1980c). "Preparation of Insulin Analogues by Action of Protease." Japan. Patent 57-067548 (14. 10. 80).

SIMON, H., GÜNTHER, H., BADER, J., and TISCHER, W. (1981). "Electro-enzymatic and electro-microbial stereospecific reductions." Angew. Chem. Int. Ed. Eng. *20,* 861–863.

SNAM PROGETTI (1971). "Verfahren zur enzymatischen Herstellung von L-Tryptophan." DOS (Ger. Offen.) 2219671 (23. 04. 71).

SNAM PROGETTI (1973). "Verfahren zur Herstellung von L-Carbamylaminosäuren und der entsprechenden L-Aminosäuren." DOS (Ger. Offen.) 2422737 (11. 05. 73).

SNAM PROGETTI (1975). "Verfahren zur Herstellung von D-Carbamylaminsäuren und der entsprechenden D-Aminosäuren." DOS (Ger. Offen.) 2621076 (12. 05. 75).

SODA, K., TANAKA, H., and ESAKI, N. (1983). "Amino acids." In "Biotechnology, Vol. 3: Biomass, Microorganisms for Special Applications, Microbial Products I, Energy from Renewable Ressources" (H. J. REHM and G. Reed, eds.; H. DELLWEG, Vol. ed.), pp. 479–530. Verlag Chemie, Weinheim – Deerfield Beach/Florida – Basel.

STAMICARBON B. V. (1974). "Process for the Enzymatic Resolution of D,L-Phenyl Glycine Amide into Its Optically Active Antipodes." US Patent 3971700 (14. 06. 1974).

STAMICARBON B. V. (1976). "Verfahren zur Herstellung eines Enzympräparates mit Aminopeptidase-Aktivität." Luxemburg. Patent 74142 (08. 01. 1976).

SZWAICER, E., SZEWCZUK, A., and MORDARSKI, M. (1981). "Application of immobilized aminoacylase from *Micrococcus agilis* for isolation of D-phenylglycine from its racemic mixture." Biotechnol. Bioeng. *23,* 1675–1681.

TAGER, H., THOMAS, N., ASSOIAN, R., RUBENSTEIN, A., SAEKOW, M., OLEFSKY, J., and KAISER, E. T. (1980). "Semisynthesis and biological activity of porcine (Leu B24)insulin and (Leu B25)insulin." Proc. Natl. Acad. Sci. USA *77,* 3181–3185.

TAKAMATSU, S., UMEMURA, J., YAMAMOTO, K., SATO, T., TOSA, T., and CHIBATA, I. (1982). "Production of L-alanine from ammonium fumarate using two immobilized microorganisms." Eur. J. Appl. Biotechnol. *15,* 147–152.

TANABE SEIYAKU, KK. (1972a). "5-Hydroxy-L-Tryptophan Production by an Immobilized Microbe." Japan. Patent 49-81590 (15. 12. 1972).

TANABE SEIYAKU KK. (1972b) "L-Tryptophan Production by an Immobilized Microbe." Japan. Patent 49-81591 (15. 12. 1972).

TANABE SEIYAKU KK. (1974). "Process for the Production of L-Aspartic Acid." Japan. Patent 49-25189 (06. 03. 1974).

TANABE SEIYAKU KK. (1980). "L-Phenylalanine no Seizōhō." Japan. Patent 57-099198 (05. 12. 80).

TANABE SEIYAKU KK. (1981a). "Process for Racemizing an Optically Active *Alpha*-Amino Acid or a Salt Thereof." EP 57092 (23. 01. 81).

TANABE SEIYAKU KK. (1981b). "L-Alanine oyobi D-Asparagine San no Seihō." Japan. Patent 57-132882 (13. 02. 81).

TORAY INDUSTRIES KK. (1970a). "Verfahren zur Herstellung von L-Lysin." DOS (Ger. Offen.) 2157171 (19. 11. 70).

TORAY INDUSTRIES KK (1970b). "Verfahren zur Racemisierung von α-Amino-ε-caprolactam." DOS (Ger. Offen.) 2163018 (23. 12. 70).

TOSA, T., MORI, T., FUSE, N., and CHIBATA, I. (1967). "Studies on continuous enzyme reactions. IV. Preparation of a DEAE-Sephadex-aminoacylase column and continuous optical resolution of acyl-D,L-amino acids." Biotechnol. Bioeng. *9,* 603–615.

TOSA, T., SATO, T., MORI, T., MATSUO, Y., and CHIBATA, I. (1973). "Continuous production of L-aspartic acid by immobilized aspartase." Biotechnol. Bioeng. *15,* 69–84.

TOSA, T., SATO, T., MORI, T., and CHIBATA, I. (1974). "Basic studies for continuous production of L-aspartic acid by immobilized *Escherichia coli* cells." Appl. Microbiol. *27,* 886–889.

UBE INDUSTRIES KK (1980). "Optically active L-tryptophan derivatives from D.L-tryptophan amide." EP 43211 (24. 06. 80).

WAGNER, F., SAHM, H., and KEUNE, H. (1975). "Microbial production of L-serine." DE 2554530 (04. 12. 1975).

WANDREY, C., WICHMANN, R., BÜCKMANN, A. F., and KULA, M. R. (1978). "Immobilization of biocatalysts using ultrafiltration techniques." In "First European Congress on Biotechnology, Interlaken", pp. 44–47.

WANDREY, C., WICHMANN, R., and JANDEL, A. S. (1981). In "Abstracts of the 6th Enzyme Engineering Conference, Kashikojima", p. 13.

WARBURG, O. (1906). "Spaltung des Leucinesters durch Pankreasferment." Hoppe-Seyler's Z. Physiol. Chem. *48,* 205–213.

WICHMANN, R., WANDREY, C., BÜCKMANN, A. F., and KULA, M. R. (1981). "Continuous enzymatic transformation in an enzyme membrane reactor with simultaneous NAD(H)-regeneration." Biotechnol. Bioeng. *23,* 2789–2802.

WIDMER, F., BREDDAM, K., and JOHANSEN, J. T. (1981). "Carboxypeptidase Y as a catalyst for peptide synthesis in aqueous phase with minimal protection." In "Proc. 16th European Peptide Symp. 1980, Helsinki" (K. BRUNFELDT, ed.), pp. 46–55. Scriptor, Copenhagen.

WINGARD, L. B., JR. (1972). "Enzyme engineering." In "Biotechnology and Bioengineering Symposium", No. 3. John Wiley & Sons, New York.

YAMADA, H., and KUMAGAI, H. (1978). "Microbial and enzymatic processes for amino acid production." Pure Appl. Chem. *50,* 1117–1127.

YAMADA, H., SHIMIZU, S., and YONEDA, K. (1980). "Biseibutsu no Hydantoinase o Mochiiru D-Amino San no Gosei-*p*-D-Hydroxyphenylglycine no Seisan o Chushin Toshite." Hakko To Kogyo *38,* 27–36.

YAMADA, S., NABE, K., IZUO, N., NAKAMICHI, K., and CHIBATA, I. (1981). "Production of L-phenylalanine from *trans*-cinnamic acid with *Rhodotorula glutinis* containing L-phenylalanine ammonia lyase activity." Appl. Environ. Microbiol. *42,* 773–778.

YAMADA, S., HONGO, C., YOSHIOKA, R., and CHIBATA, I. (1983). "Method for the racemization of optically active amino acids." J. Org. Chem. *48,* 843–846.

YAMSKOV, I. A., TIKHONOVA, T. V., and DAVANKOV, V. A. (1981). "Proteases as catalysts of enantioselective aminoester hydrolysis. 3. Pronase-catalyzed hydrolysis of methylesters of leucine, phenylalanine, and valine." Enzyme Microb. Technol. *3,* 141–143.

YOKOTE, Y., MAEDA, S., YABUSHITA, H., NOGUCHI, S., KIMURA, K., and SAMEJIMA, H. (1978). "Production of L-aspartic acid by *Escherichia coli* aspartase immobilized on phenol-formaldehyde resin." J. Solid Phase Biochem. *3,* 247–261.

ZAFFARONI, P., VITOBELLO, V., CECERE, F., GIACOMOZZI, E., and MORISI, F. (1974). "Synthesis of L-tryptophan from indole and D,L-serine by tryptophan synthetase entrapped in fibres. Preparation and properties of free and entrapped enzyme." Agric. Biol. Chem. *38,* 1315–1342.

ZAIDAN HOJIN BISEIBUTSU KAGAKU KEN (1980). "New Processes for the Production of Di-Alkyl Esters and Optically Active Mono-Alkyl Esters of 3-Aminoglutaric Acid." EP 50799 (21. 10. 80).

ZAIDAN HOJIN BISEIBUTSU KAGAKU KEN (1981a). "Kōgaku Kassei β-(*R* mata wa *S*)-Aminoglutaric San Mono Alkyl Ester no Seizō Hōhō." Japan. Patent 57-159495 (30. 03. 81).

ZAIDAN HOJIN BISEIBUTSU KAGAKU KEN (1981b). "Kōgaku Kassei β-(*S*)-Aminoglutaric San Mono Alkyl Ester no Seizō Hō." Japan. Patent 57-177696 (07. 04. 81).

ZHU, S., LI, T., CUI, D., CAO, Q., and ZHANG, Y. (1981b). "Kōgaku Kassei β-(*S*)-Aminogluon the biological activity of insulin." Sci. Sin. *34,* 264–271.

Chapter 11

Carbohydrates

Anneliese Crueger
Verfahrensentwicklung Biochemie

Wulf Crueger
Biotechnikum Mikrobiologie

BAYER AG
Wuppertal 1
Federal Republic of Germany

I. Introduction
II. Types of Reactions
III. Fructose Production by Glucose Isomerase
  A. Introduction
  B. Glucose Isomerase
    1. Historic perspective
    2. Enzyme sources for production
    3. Fermentation of glucose isomerase
    4. Immobilization of glucose isomerase
  C. Fructose Production Process
  D. Economics
IV. Bioconversion in L-Ascorbic Acid Production
  A. Introduction
  B. Reichstein's Synthesis
    1. Microorganisms used for the conversion of D-sorbitol to L-sorbose
    2. Process
  C. New Bioconversion Steps in L-Ascorbic Acid Synthesis
    1. Process via L-sorbosone
    2. Process via 5-keto-D-gluconate
    3. Process via 2,5-diketo-D-gluconate

V. Gluconic Acid
A. Introduction
B. Microorganisms
C. Processes
D. Further Developments in Gluconic Acid Production
E. Application
VI. Kojic Acid
VII. Dihydroxyacetone
VIII. Isomaltulose
IX. Raffinose Conversion by α-Galactoside-galactohydrolase
A. Introduction
B. α-Galactosidase Producing Microorganisms
C. Industrial Processing
1. Production of mycelial pellets
2. Hydrolysis of raffinose
D. Economics
X. Lactose Hydrolysis by β-Lactase
A. Introduction
B. Microorganisms and Their Lactases
C. Fermentation for Lactase Production
1. *Aspergillus oryzae*
2. *Escherichia coli*
3. *Bacillus stearothermophilus*
D. Immobilization of Lactase
E. Economics
XI. Sucrose Conversion by Invertase
A. Introduction
B. Invertase Producing Microorganisms
C. Inversion Processes
XII. References

I. Introduction

The metabolism of carbohydrates is an important part of the primary metabolism in microorganisms. Therefore, the attack of microorganisms on a certain carbohydrate results in a broad range of reactions. A complete utilization may be observed resulting in the production of CO_2, H_2O, and cell mass. In other cases only slight alterations of the molecular skeleton may occur. These latter reactions, such as incomplete oxidation, reduction, aldose-ketose isomerization, or glycosyl transfer may also be used for industrial sugar transformations.

This review is chiefly concerned with these reactions and will not deal with the many intermediary steps of primary metabolism. The most important sugar transformation reactions are given in Tables 1 through 4. Most of these reactions cannot compete with chemical processes because of their low yields. Only few microbiological transformation procedures, which will be described in detail in Sect. III to XI, have been optimized and are used for production on a technical scale.

Detailed reviews on carbohydrate transformation are by SPENCER and GORIN (1965), HERP et al. (1970), and KIESLICH (1976).

II. Types of Reactions

Transformation of the carbohydrate molecule may be achieved by oxidation, reduction, isomerization, esterification, and transglycosidation. Many different reactions have been described in the literature, of which the most important transformation steps are oxidation, reduction, isomerization, and glycoside hydrolysis.

Oxidation, reduction and isomerization reactions are summarized with some examples in Tables 1–3, following the classification of KIESLICH (1976).

Glycosyl transfer reactions are reviewed in detail by SPENCER and GORIN (1965); Table 4 lists several disaccharide molecules obtained via transglucosidation with fructose as the acceptor molecule. The reactions of glycoside hydrolases, together with their glycosyl transferase activities, e. g.:

α-D-glucosidases
β-D-glucosidases
α-D-galactosidases
β-D-galactosidases
α,β-D-mannosidases
β-D-fructofuranosidases

and others are described in detail by NISIZAWA and HASHIMOTO (1970).

III. Fructose Production by Glucose Isomerase

A. Introduction

Until 1976 the main sweetener used in the world was sucrose which came from sugar beets (40%) and sugar cane (60%). A total of 81 $\times$ 10^6 tons of sucrose were produced that year. In the U.S. 3.6 $\times$ 10^6 tons of sucrose are used by consumers per year, the soft drink and food industry utilizing 6.3 $\times$ 10^6 tons.

D-Glucose (dextrose) has 70–75% the sweetening ability of sucrose. However, D-fructose (levulose), the other monosaccharide moiety of sucrose, has twice the sweetening power of sucrose (BARKER, 1976). In addition, fructose plays an important role in the diet of diabetics as it is only slowly reabsorbed by the stomach and intestinal tract and does not influence the blood glucose level.

Table 1. Microbial Oxidations of Carbohydrates (SPENCER and GORIN, 1965; KIESLICH, 1976)

Substrate	Product	Microorganism
Oxidation at C-1; —CHO → —COOH		
D-Ribose	D-Ribonic acid	*Pseudomonas* spp.
D-Xylose	D-Xylonic acid	*Penicillium corylophilum*
D-Arabinose	D-Arabonic acid	*Pseudomonas* spp.
L-Arabinose	L-Arabonic acid	*Pseudomonas* spp., *Acetobacter* spp.
D-Glucose	D-Gluconic acid	*Aspergillus niger*
2-Deoxy-D-glucose	2-Deoxy-D-gluconic acid	*Pseudomonas aeruginosa*
6-Deoxy-6-fluoro-D-Glucose	6-Deoxy-6-fluoro-2-keto-D-gluconic acid	*Aerobacter aerogenes*
D-Mannose	D-Mannonic acid	*Acetobacter gluconicum*
D-Galactose	D-Galactonic acid	*Acetobacter aceti*, *Pseudomonas fluorescens*
Lactose	Lactobionic acid	*Pseudomonas calco-acetica*
Maltose	Maltobionic acid	*Pseudomonas quercito-gallica*
Oxidation at C-1; —CH_2OH → —CHO		
Fructose	Glucosone	*Gluconoacetobacter roseus*
L-Sorbosone	L-Sorbose	*Gluconobacter melanogenus*
Oxidation at C-1; —CH_2OH → COOH		
L-Sorbose	2-Keto-L-gulonic acid	*Acetobacter suboxydans*
Oxidation at C-2; Polyalcohols		
Glycerol	Dihydroxyacetone	*Acetobacter suboxydans*, *Gluconobacter* spp.
D-Mannitol	D-Fructose	*Gluconobacter suboxydans*
Volemitol (D-glycero-D-talo-heptitol)	D-Mannoheptulose	*Acetobacter suboxydans*
	D-Sedoheptulose	*Acetobacter xylinum*
Oxidation at C-2; Onic acids		
Gluconic acid	2-Ketogluconic acid	*Acetobacter dioxyacetonicum*
L-Gulonic acid	2-Keto-L-gulonic acid	*Xanthomonas translucens*, *Pseudomonas* spp.
D-Galactonic acid	2-Keto-D-galactonic acid	*Pseudomonas aeruginosa*
Oxidation at C-3		
Lactose	3-Ketolactose	*Alcaligenes* sp.
Maltose	3-Ketomaltose	*Alcaligenes faecalis*
L-Gulonic acid	3-Keto-L-gulonic acid	*Schwanniomyces occidentalis*
Oxidation at C-4		
D-Arabonic acid	4-Keto-D-arabonic acid	*Acetobacter suboxydans*
Oxidation at C-5		
D-Sorbitol	L-Sorbose	*Acetobacter xylinum* *Acetobacter suboxydans*
Oxidation at C-6		
Gluconic acid	6-Ketogluconic acid	*Acetobacter xylinum*
Oxidation at Several C		
Glucose	Kojic acid	*Aspergillus flavus*, *A.oryzae*

Table 1. (continued)

Substrate	Product	Microorganism
Glucose (oxidation at C-1, C-5)	5-Ketogluconic acid	*Acetobacter suboxydans*
D-Sorbitol (C-2, C-5)	5-Ketofructose	*Gluconobacter suboxydans*
Gluconic acid (C-2, C-5)	2,5-Diketogluconic acid	*Pseudomonas albosesamae, Gluconobacter rubiginosus*
Glucose (C-1, C-2, C-5)	2,5-Diketogluconic acid	*Acetobacter melanogenum, Pseudomonas albosesamae*
Glucose (C-1, C-6)	Saccharic acid	*Aspergillus niger*

Table 2. Microbial Reductions of Carbohydrates (SPENCER and GORIN, 1965; KIESLICH, 1976)

Substrate	Product	Microorganism
Reduction of $-CHO \longrightarrow -CH_2OH$		
D-Ribose	Ribitol	*Candida polymorpha*
L-Arabinose	L-Arabitol	*Candida polymorpha*
D-Xylose	Xylitol	*Candida polymorpha*
D-Mannose	Mannitol	*Torulopsis* spp.
Galactose	Dulcitol	Yeast strains
Reduction of $> C{=}O \longrightarrow > CHOH$		
Fructose	Mannitol	*Lactobacillus brevis*
5-Keto-D-gluconic acid	L-Gulonic acid	*Fusarium* spp.
5-Ketofructose	Fructose	*Acetobacter albidus*
Reductions of Other Aldols or Ketols		
Diacetyl	Acetoin	Yeast
Diacetyl	2,3-Butanediol	Yeast
Glycerolaldehyde	1,2-Ethanediol	Yeast

Table 3. Microbial Transformation of Carbohydrates
Aldose-Ketose Isomerizations (SPENCER and GORIN, 1965; KIESLICH, 1976)

Substrate	Product	Microorganism
D-Glucose	D-Fructose	*Bacillus coagulans* *Actinoplanes missouriensis* *Streptomyces olivaceus*
D-Xylose	D-Xylulose	*Lactobacillus xylosus*
L-Arabinose	L-Ribulose + L-Xylulose	*Lactobacillus plantarum*
D-Mannose	D-Fructose	*Xanthomonas phaseoli* *X.rubrilineans*
L-Rhamnose	L-Rhamnulose	*Escherichia coli, Lactobacillus plantarum*
D-Galacturonic acid	D-Glucuronic acid	*Serratia marcescens*

Table 4. Synthesis of different α-D-Glucopyranosidofructoses by Transglucosidation with Fructose as the Acceptor. (For detailed listing of glycosyl transfer reactions see SPENCER and GORIN, 1965)

Systematic Name	Trivial Name
α-D-Glucopyranosido-1,1-fructose	Glycosylfructose
α-D-Glucopyranosido-1,2-fructose	Sucrose
α-D-Glucopyranosido-1,3-fructose	Turanose
α-D-Glucopyranosido-1,4-fructose	Maltulose
α-D-Glucopyranosido-1,5-fructose	Leucrose
α-D-Glucopyranosido-1,6-fructose	Isomaltulose (Palatinose)

Fructose was originally produced from invert sugar solutions using the calcium fructonate method. This involves the mixing of calcium hydroxide with an invert sugar solution, treatment with CO_2, separation of the $CaCO_3$, followed by vacuum evaporation and crystallization. Since 1964 fructose has been produced on the industrial scale by using cationic resins (hydrolysis and separation, LAUER, 1980).

Glucose is produced from starch by means of acid hydrolysis or enzymatic hydrolysis. As sugar beets are available only 100 days per year, wheat, corn or mannitol are used as starting materials. Studies have been carried out on the chemical conversion of glucose to fructose using alkaline conditions and high temperatures. However, these methods turned out to be non-selective, and non-metabolizable sugars such as psicose (Fig. 1) are formed in addition to formate and various colored substances. Therefore, chemically obtained fructose syrup has not been employed commercially.

Fructose is produced biochemically from inulin, sucrose, or glucose from starch (Fig. 2). Today, however, due to inulin shortage and high prices, the inulin technique is no longer used (KIERSTAN, 1980).

Isomerization of glucose to fructose by a microbial enzyme was first reported by MARSHALL and KOOI, 1957. They discovered that in *Pseudomonas hydrophila* isomerization takes place without phosphorylation. In a cell with "normal" metabolism carbohydrate isomerization occurs following a phosphorylation step.

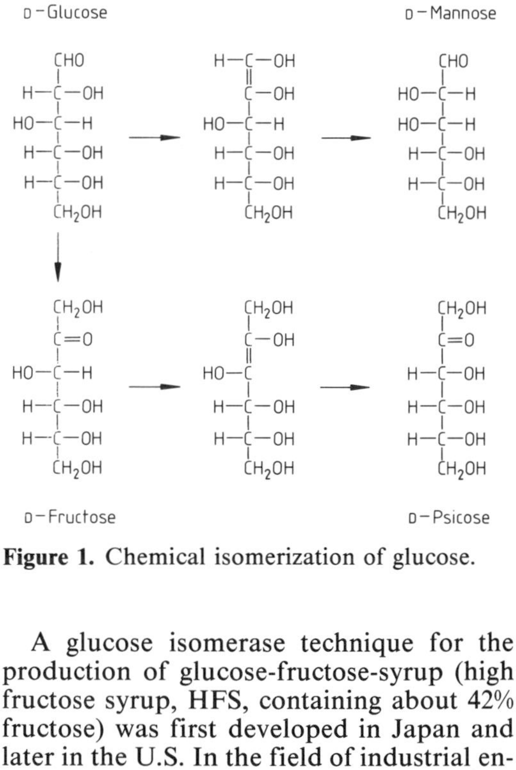

Figure 1. Chemical isomerization of glucose.

A glucose isomerase technique for the production of glucose-fructose-syrup (high fructose syrup, HFS, containing about 42% fructose) was first developed in Japan and later in the U.S. In the field of industrial enzymes the glucose isomerase process shows the largest expansion of the market today. Initially soluble enzymes were added to each batch until the process became optimized by using immobilized enzymes either in a continuous operation or reusing the immobilized material. This optimized technique is the largest commercial application of an immobilized enzyme. Latest developments have been an increase of fructose

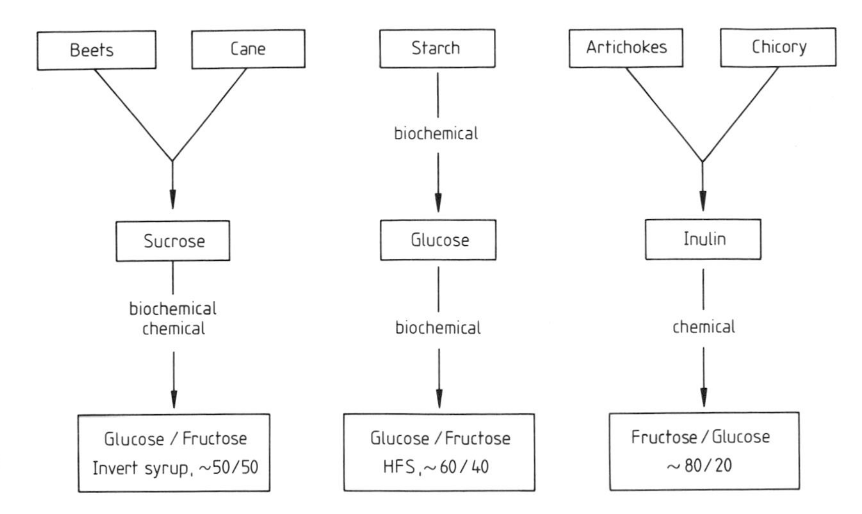

Figure 2. Production of fructose syrup.

concentration in solution and on the other hand, separation of fructose from fructose-bearing solutions. In 1978 the U.S. produced 22% of its sweeteners from corn; until 1983 the percentage increased to 33%. Reviews on this topic were written by BUCKE (1977), ANTRIM et al. (1979), BARKER and SHIRLEY (1980) and CHEN (1980a,b).

B. Glucose Isomerase

1. Historic perspective

Various types of enzymes are able to convert glucose to fructose (Fig 3). The first

CHO, H—C—OH, HO—C—H, H—C—OH, H—C—OH, CH_2OH (D-Glucose) —Glucose isomerase→ CH_2OH, C=O, HO—C—H, H—C—OH, H—C—OH, CH_2OH (D-Fructose)

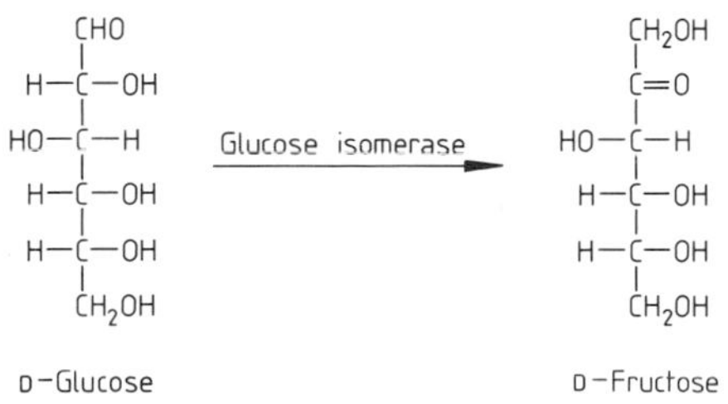

Figure 3. Reaction of glucose isomerase.

type is a glucose-phosphate isomerase (D-glucose-6-phosphate-ketol-isomerase, EC 5.3.1.9). Producers of this enzyme are *Escherichia intermedia, E.freundii, Aerobacter aerogenes,* and *A.cloacae.* The enzymes need arsenate to form a glucose-arsenate complex which is isomerized:

glucose + arsenate ↔ glucose-arsenate
glucose-arsenate + enzyme ↔ glucose-arsenate-enzyme
glucose-arsenate-enzyme ↔ fructose + arsenate + enzyme

Only for some of the enzymes is xylose needed as an inducer (NATAKE and YOSHIMURA, 1963, 1964; NATAKE, 1966, 1968). The enzymes have a pH optimum at 7.0 and a temperature optimum at 50 °C. However, due to their requirement of arsenate these enzymes are not used in production processes.

A glucose isomerase (D-glucose ketol-isomerase, EC 5.3.1.18) has been characterized which is linked to NAD^+ and produced by *Bacillus megaterium* (TAKASAKI and TANABE, 1962, 1963). Its pH optimum is at 7.8 and the temperature optimum at 35 °C.

Various heterolactic acid bacteria (*Lactobacillus brevis, L.fermenti, L.pentoaceticus, L.buchneri, L.mannitopolus, L.gayonii*) produce glucose isomerases (YAMANAKA, 1962, 1963a, 1963b, 1965, 1968; BARKER and SHIRLEY, 1980). These enzymes require

D-xylose as an inducer as well as manganese ions. In Japan these enzymes have been thoroughly studied for their utility in production. A disadvantage of these enzymes is their relative instability at higher temperatures compared to others.

2. Enzyme sources for production

The only commercially employed enzymes are D-xylose-isomerases (D-xylose ketol-isomerase, EC 5.3.1.5). Their advantages are:

- a low pH optimum (less secondary reactions)
- high specific activity
- a high temperature optimum (~ 80°C; less contaminations)
- cofactors (ATP, NAD^+) are not needed.

Many microorganisms are able to produce glucose isomerase induced by xylose activity. SUEKANTE and IIZUKA (1981) tested 603 strains of bacteria on xylose media (Table 5). The results show that isomerase activity may be found widely among the various microorganisms (Table 6). 70% of the tested actinomycetes, 13 of 16 strains of *Escherichia,* and all 14 *Aerobacter* strains showed isomerase activity. On the other hand, only 80 of 210 tested *Pseudomonas* strains were able to grow on xylose showing only minor isomerase activities. 40 yeast strains out of 56 tested ones, gave a positive result: 24 strains showed medium and 16 strains fairly weak activities.

Among the strains with xylose isomerase activity mainly the following are of commercial interest, today:

- *Bacillus coagulans* (Novo Industry A/S)
- *Arthrobacter* sp. (Reynolds Tobacco Comp., ICI Americas Inc.)
- *Actinoplanes missouriensis* (Gist-Brocades NV, Anheuser-Busch Inc.)
- *Streptomyces olivaceus* (Miles Laboratories Inc.)
- *Streptomyces olivochromogenes* (CPC International Inc.)
- *Streptomyces phaeochromogenes* (Nagase)
- *Streptomyces albus* (Miles - Kali Chemie).

3. Fermentation of glucose isomerase

In order to reach optimal yields the inducer xylose has to be added (Fig. 4). Xylose is very expensive and so it has been re-

Table 5. Composition (w/v%) of Media Used for Screening for Glucose Isomerase (SUEKANTE and IIZUKA, 1981)

Bacteria, Yeasts, and Actinomycetes		Acetic Acid Bacteria		Lactic Acid Bacteria	
D-Xylose	2	D-Xylose	1	D-Xylose	1
$MgSO_4 \cdot 7H_2O$	0.025	D-Glucose	0.3	D-Glucose	0.1
$(NH_4)_2HPO_4$	0.6	Autolyzed yeast solution	20	Yeast extract	0.4
KH_2PO_4	0.2	$MgSO_4 \cdot 7H_2O$		Nutrient broth	1
pH 6.8		$(NH_4)_2HPO_4$		CH_3COONa	1
		KH_2PO_4		$MgSO_4 \cdot 7H_2O$	0.02
		pH 6.8		NaCl	0.001
				$FeSO_4 \cdot 7H_2O$	0.001
				$MnSO_4 \cdot 4H_2O$	0.001
				pH not adjusted	

Table 6. Distribution of Glucose Isomerase Activity (SUEKANTE and IIZUKA, 1981)

Microorganism (Genera)	Number of Tested Strains	Number of Strains to Grow on Xylose	Number of Strains with Glucose Isomerase Activity	
			medium	minimal
Pseudomonas	210	80	–	80
Xanthomonas	7	2	1	1
Acetobacter	30	10	7	3
Gluconobacter	10	4	3	1
Aeromonas	3	–		
Protaminobacter	1	–		
Bacterium	5	4	1	3
Kluyvera	5	2	2	
Vibrio	2	–		
Agrobacterium	3	2	2	
Alcaligenes	4	–		
Achromobacter	14	4	2	2
Flavobacterium	14	6	4	2
Escherichia	16	13	13	
Aerobacter	14	14	13	1
Erwinia	7	6	6	
Serratia	23	3	3	
Proteus	1	–		
Micrococcus	31	10	8	2
Staphylococcus	2	–		
Sarcina	17	9	4	5
Brevibacterium	29	7	3	4
Streptococcus	10	6	5	1
Leuconostoc	7	5	4	1
Lactobacillus	21	14	12	2
Propionibacterium	2	2	2	
Corynebacterium	16	–		
Microbacterium	2	–		
Arthrobacter	2	–		
Bacillus	95	13	13	
Streptomyces	124	88	63	25

placed by xylan hydrolysate and wheat bran.

A typical medium consists of (PARK et al., 1980):

- 3.0% wheat bran acid hydrolysate
- 2.0% corn steep liquor
- 0.1% $MgSO_4 \cdot 7\ H_2O$
- 0.01% $CoCl_2 \cdot 6\ H_2O$

On the other hand, today there are constitutive mutants being used for production purposes. SHIEH (1977) described a process with *Actinoplanes missouriensis;* similarly, LONG (1978) has worked with *Arthrobacter.* Table 7 presents data from a fermentation with *Arthrobacter* nov. sp. NRRL B-3728.

The regulation of glucose isomerase production is influenced by catabolite repression. In batch and continuous fermentation glucose acts as a repressor. *Bacillus coagulans* does not produce glucose isomerase during the log-phase. When glucose is used up in the medium, growth comes to a halt. Then, following a typical diauxic pattern, catabolism of other carbohydrates starts

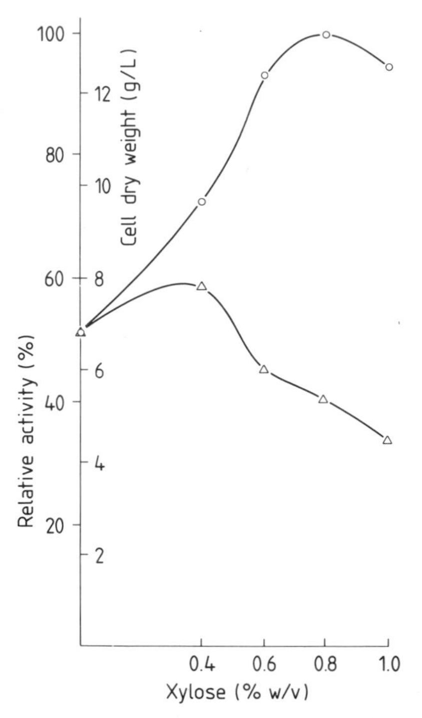

Figure 4. Induction of glucose isomerase formation by xylose (VAHERI and KAUPPINEN, 1977). – o—o cell dry weight, Δ—Δ glucose isomerase activity.

Table 7. *Arthrobacter*-Production Medium (LONG, 1978)

Components	Seed Culture	Production Culture
Dextrose	2.0 %	2.0 %[a)]
Meat protein	0.3 %	0.6 %
Yeast extract	0.15%	0.15%
$(NH_4)_2HPO_4$	0.6 %	0.6 %
KH_2PO_4	0.2 %	0.2 %
$MgSO_4 \cdot 7\,H_2O$	0.01%	0.01%[a)]

[a] Separate sterilization

and enzyme production increases. In batch fermentations with *Bacillus* strains maximum enzyme titers are obtained within 24 hours. Today the *Bacillus coagulans* enzyme may be produced in a continuous fermentation as well. In this case a specific growth rate of $\mu = 0.1 - 0.4\ h^{-1}$ is achieved. Not only glucose limitation but also oxygen limitation is optimal at the production scale.

Table 8. Fermentation Conditions for Glucose Isomerase Production (Batch Fermentation)

Strain	pH	Time (h)	Temp. (°C)	Activity (U/mL)[a)]
Bacillus coagulans	6.8	24	40	
Streptomyces olivaceus	7.0	24	30	
Streptomyces albus	7.5	40	30	33.3
Arthrobacter sp.	6.9	55	30	
Actinoplanes missouriensis	7.0	72	30	67.5

[a] U/mL amount of enzyme which produces 1 μmol/min of fructose

Table 9. Influence of the Nitrogen Source on Glucose Isomerase Production

Nitrogen Source	Activity (%)
Corn steep liquor	100
Casein hydrolysate	73
Soya peptone	66
Yeast extract	58
Malt extract	24

Microanaerobic conditions within the cells stabilize the system. *Streptomyces* strains only produce isomerase under batch conditions; continuous processes have never reached the production scale.

Table 8 shows the fermentation conditions up to a 120 m^3 scale.

Not only yeast extract and meat protein but other cheaper protein sources may replace those which were originally used in the production medium. The nitrogen source is a critical point which still has to be optimized for the different strains. Table 9 shows a comparison of the different organic nitrogen sources. For many strains corn steep liquor seems to be most efficient and therefore is most commonly used. Some bacteria are able to utilize inorganic nitrogen [NH_4Cl, $(NH_4)_2HPO_4$, or $(NH_4)_2SO_4$]; however, these are not optimally suitable on a production scale.

In most cases mineral salts such as $CoCl_2 \cdot 6\ H_2O$, $MgSO_4 \cdot 7\ H_2O$, and $MnSO_4 \cdot 4\ H_2O$ had to be added to the production culture medium. It is important to reduce the cobalt component because of its potential to cause environmental hazards. *Arthrobacter* and *Streptomyces olivaceus* (REYNOLDS, 1973) as well as some mutants of *Streptomyces olivochromogenes* (CPC INTERNATIONAL INC., 1975b) do not require cobalt for optimal production.

Very little research has been published on genetic improvement for isomerase production (Table 10). The amplification of the glucose isomerase gene in *Escherichia coli* has meant significant progress; here the activity could be increased four-fold (WOVCHA et al., 1983).

The first strain isolated for technical use was *Bacillus coagulans* (YOSHIMURA et al., 1966). The extraction and characterization of the enzyme was carried out by DANNO et al., 1967. Glucose isomerase from *Actinoplanes missouriensis* was described by GONG et al., 1980.

4. Immobilization of glucose isomerase

Enzyme immobilization can either be accomplished with the enzyme or with the glucose isomerase-bearing cell by the following procedures:

- covalent binding to an insoluble carrier
- adsorption to an insoluble carrier
- entrapping in a matrix
- immobilization within the cells

The immobilization of *Arthrobacter* on a technical scale starts with a flocculation process in order to separate the cells from the medium and to form the support matrix. By adding cationic and anionic polyelectrolytes a polysalt complex which includes bacterial cells is formed. Then, the immobilized cells are dried and homogenized (BUNGARD et al., 1979). Two immobilization techniques for *Actinoplanes missouriensis* and *Bacillus coagulans* are shown in Figs. 5 and 6. The diameter of immobilized cell particles is about 0.3–2 mm; fibrous material is on the market as well.

Table 11 shows a comparison of commercial systems based on complete cells.

C. Fructose Production Process

Glucose isomerization with immobilized cell systems is achieved in batch reactors; since 1977 it has also been carried out in continuously stirred and in feed tank reactors as well as in fixed bed continuous column operation with a height of up to 5 m. The advantages of a continuous process are: smaller reactor volumes, no necessity for the use of cobalt, better and easier process control, and lower purification costs (Table 12).

The dextrose syrup obtained by saccharification of starch has to be purified by filtration, carbon treatment and ion exchange. Since the dry substance has an influence on contact time, enzyme activity, and viscosity a 35–45% dextrose syrup concentration is used.

Table 10. Increase of Isomerase Activity by Genetic Manipulation

Strain	Mutagenic Agent	Increase of Productivity (%)	Reference
Streptomyces ATCC 21175	Ethylene imine	62	BENGTSON and LAMM (1974)
S. olivaceus NRRL 3583	UV	16	MILES LABORATORIES INC. (1974)
S. olivochromogenes ATCC 21114	UV	50	CPC INTERNATIONAL INC. (1975 a, b, c)
S. nigrificans	UV	198	DEMNEROVA et al. (1979)

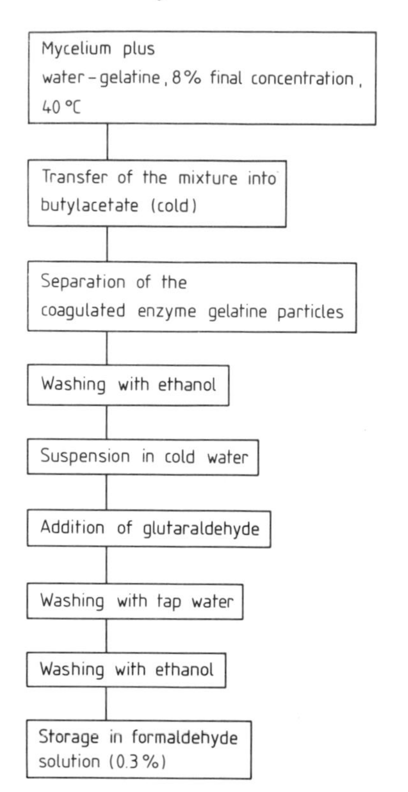

Figure 5. Immobilization steps for *Actinoplanes missouriensis* (HUPKES and VAN TILBERG, 1976).

Oxygen inactivates the enzyme and is responsible for an increased formation of secondary products. Therefore, a low oxygen tension is obtained by means of evaporation of the glucose solution. An industrial process (NOVO, 1976) follows the conditions shown in Fig. 7.

The required contact time depends upon enzyme activity, feed concentration, and the desired conversion rate. For color- and acid formation the contact time is a critical factor. A mathematical model to describe the relationship between production rate, enzyme purity, and temperature was developed by ROELS and VAN TILBERG (1979). The working temperature necessitates a compromise between high activity, high stability, and low contamination.

The pH which is adjusted to 8.5 de-

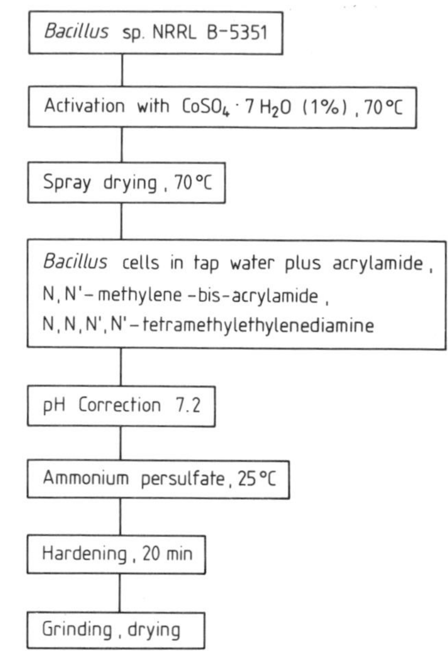

Figure 6. Production of immobilized *Bacillus coagulans* (AMOTZ and THIESEN, 1973).

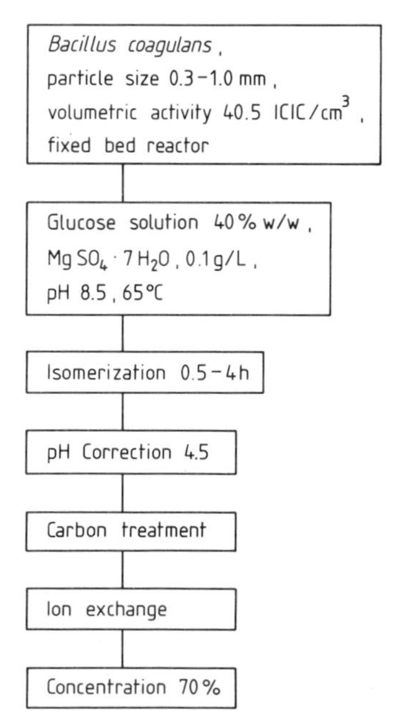

Figure 7. Fixed bed continuous isomerization process.

Table 11. Comparison of Commercial Glucose Isomerase Preparations

Microorganism	pH Optimum	Temperature Optimum (°C)	Metal Requirement		
			Cobalt (M)	Magnesium (M)	Manganese (M)
Bacillus coagulans (ANTRIM et al. 1979)	7.0	75	$1 \cdot 10^{-3}$	$1 \cdot 10^{-1}$	+
Actinoplanes missouriensis (ANHEUSER-BUSCH INC. 1975; SCALLET et al., 1974)	7.0–7.2	80–85	$3 \cdot 10^{-4}$	$3 \cdot 10^{-3}$	–
Streptomyces olivaceus (MILES LAB. INC., 1972)	8.5	60–70	+	+	
Streptomyces olivochromogenes (SUEKANTE et al., 1978)	8.0–9.0	80–100	+	+	
Arthrobacter sp. (LEE et al., 1972)	8.0	50–90	–		
Streptomyces albus (VAHERI and KAUPPINEN, 1977)	7.0	70	–	$5 \cdot 10^{-3}$	

Table 12. Processes for Batch and Continuous Production of High Fructose Syrup (ZITTAN et al., 1975)

Process Parameter	Batch Process	Continuous Process
Reactor volume (m^3)	1–50	30
Enzyme consumption (t)	17–20	10
$MgSO_4 \cdot 7\,H_2O$ (t)	4.3	2.2
$CoSO_4 \cdot 7\,H_2O$ (t)	2.2	–
Pigment synthesis (%)	0.05–0.1	0.0–0.02
Psicose production (%)	0.1	0.1
Purification	Carbon treatment, ion exchange	Carbon treatment

creases within the enzyme bed. If it drops below a critical point cobalt must be added in order to assure an acceptable stability.

Calcium has an inhibitory function. Therefore, the feed solution should be ion exchanged to obtain a calcium concentration below 1 ppm. For optimal productivity the concentration of magnesium as activator and stabilizer should be kept at 0.4 mM. A plant with a capacity of 100 t per day using two parallel lines with three reactors in series is shown in Fig. 8 (OESTERGAARD and KNUDSEN, 1976). One plant with a capacity of 120 000 t/a has been installed.

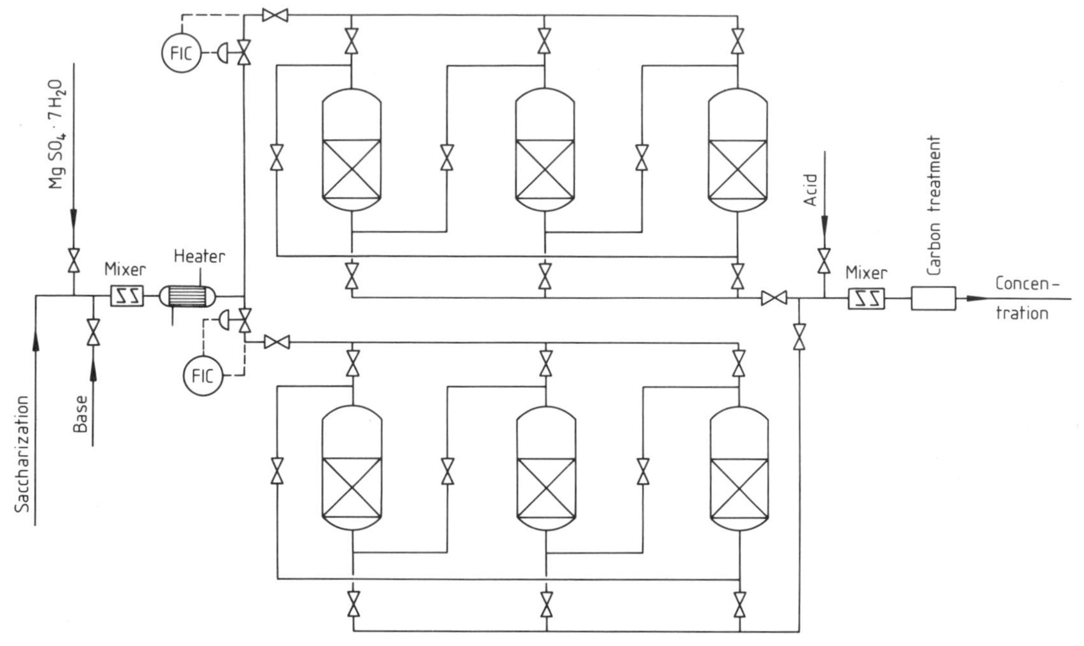

Figure 8. Reactor arrangement for a 100 t/d production (OESTERGAARD and KNUDSEN, 1976). – Enzyme bed diameter: 1.32 m, height: 1.60 m.

On an activity basis the half life for the *Actinoplanes missouriensis* process is estimated at 1 200 h (GIST BROCADES, 1979). The productivity (kg glucose in syrup converted/kg enzyme) is between 1 000–9 000.

After isomerization the high fructose syrup is treated with carbon, ion exchanged with strong acid cation-exchange resins and slightly alkaline anion-exchange resins, and then concentrated in order to reach a stable state (~ 71% dry weight). The following characteristics are typical for a high fructose syrup (HFS):

Dry substance	71%
pH	4–5
Ash	0.05–0.1%
Fructose	42% (in dry substance)
Glucose	53%
Oligosaccharide	5%
Psicose	< 0.1%

The high fructose syrup is directly sold to the food industry. In the U.S. it is less expensive than sucrose or invert syrup.

D. Economics

In 1980 the world production of HFS was about 3.7×10^6 t. Mostly it was offered for sale in the U.S. Due to free competition conditions HFS is sold 10–15% cheaper than sugar.

In the U.S. fructose corn syrup is produced by the following companies: American Maize Products Comp., Amstar Corp., Archer Daniels Midland Corn Sweeteners, Cargill Inc., Clinton Corn Processing Comp., CPC International Inc., The Hubinger Comp., and A. E. Stayley Manufacturing Co.

In Europe, especially in the EEC, the sugar industry has a strong political influence. The result is a production rate of no more than 300 000 t. There are plants in Germany, Spain, United Kingdom, Netherlands, Belgium, France, Italy, Ireland, Yugoslavia, and Hungary.

In Japan 17 plants have a capacity of 500 000 t/a.

The syrup is a direct substitute for sucrose in carbonated and still soft drinks, canned fruits, lactic acid beverages, juice, bread, ice-cream, frozen candies, and processed milk.

In addition to an increase in HFS-production developments in this field will most likely proceed along the following lines:

(1) Reduction of isomerase costs
Amplification of the glucose isomerase genes in other organisms will lead to an increase of fermentation productivity.

(2) Higher fructose levels
Higher isomerization yields may be achieved by raising the working temperature (ANTRIM et al., 1979). Temperatures above 70°C lead to an increase in fructose concentration up to 50% and more. Another possibility to increase the fructose concentration is to work with resinous molecular exclusions (KELLER et al., 1981). BARKER et al. (1983) produced a fructose syrup containing more than 90% fructose by forming fructose-oxyanion complexes with germanate.

(3) Combination of saccharification and isomerization
Several production processes exist which use starch as the initial substrate (NAARDEN INTERNATIONAL, 1976; HEBEDA and LEACH, 1975; WALON, 1977; KATWA and RAO, 1983). A combination of α-amylase, glucoamylase, and glucose isomerase may yield a 39–45% fructose syrup in a single step. Gunei Chemical Co. has installed a 70 000 t/a production plant.

(4) Large scale production of fructose
Modern chromatography techniques using ion-exchange resins to separate fructose and glucose have proved the best methods on a technical scale, to date (LAUER, 1980). In France a product with the following specifications is on the market:

Fructose syrup (levulys 70/95®)	
Dry weight (%)	70–70.5
Fructose (%)	95–95.5
Glucose (%)	4.5–5.0
$(\alpha)_D^{20}$	−85°
Calcium (mg/kg)	100
pH	4.0

This material is sold in crystalline form.

IV. Bioconversion in L-Ascorbic Acid Production

A. Introduction

L-Ascorbic acid (vitamin C) was first isolated from lemons. In 1933 the synthesis of L-ascorbic acid was published independently by REICHSTEIN's and by HAWORTH's group. Industrial production of L-ascorbic acid today still follows REICHSTEIN's procedure (REICHSTEIN and GRÜSSNER, 1934) including a highly selective biochemical dehydrogenation from D-sorbitol to L-sorbose (historical review by HUGHES, 1983).

Ascorbic acid is used in vitamin preparations or as an antioxidant in food processing; about 35 000 tons are produced annually.

Several processes with further bioconversion steps were developed (review by KULHANEK, 1970), but because of low yields, these procedures have not yet attained industrial significance.

B. Reichstein's Synthesis

D-Glucose is catalytically hydrogenated to D-sorbitol. In modern processes an enzymatically hydrolyzed starch syrup can be used as source of glucose. The next step is the regiospecific microbial dehydrogenation of D-sorbitol to L-sorbose. For protective purposes the crystalline L-sorbose is condensed with acetone to diacetonesorbose which is chemically oxidized to diacetone-2-keto-L-gulonic acid. High temperature cleavage leads to 2-keto-L-gulonic acid which further reacts by esterification, enolization, and lactonization to yield vitamin C (Fig. 9). The economic advantage of this process is demonstrated by the fact that the

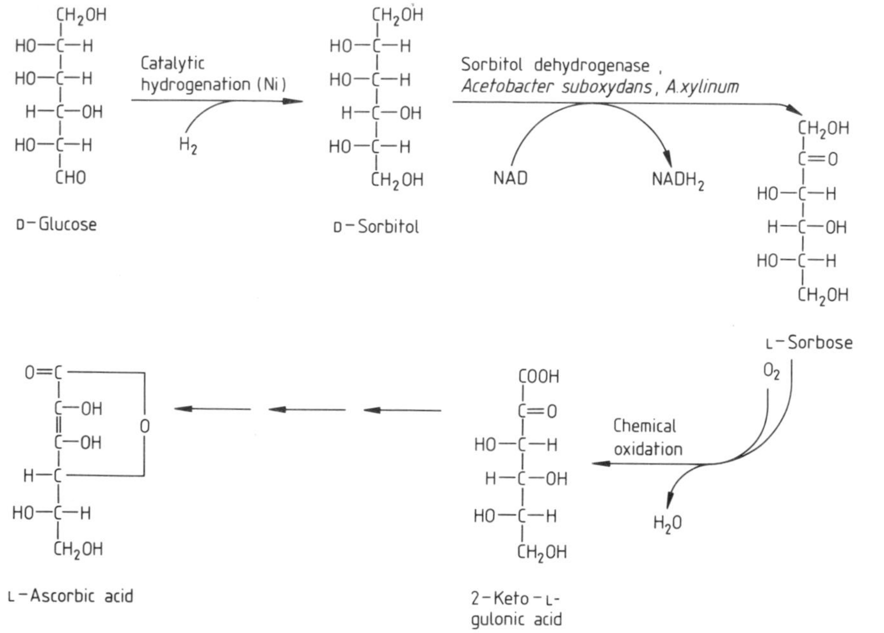

Figure 9. Microbial dehydrogenation of D-sorbitol to L-sorbose during L-ascorbic acid production.

formation of 1 kg of ascorbic acid merely requires 2 kg of glucose.

1. Microorganisms used for the conversion of D-sorbitol to L-sorbose

The most frequently used organisms for the D-sorbitol conversion to L-sorbose are *Acetobacter xylinum* and *A.suboxydans;* the latter requires pantothenic, *p*-aminobenzoic, and nicotinic acids as growth factors.

One disadvantage of the commonly used *Acetobacter* strains is their nickel sensitivity. Presently the nickel associated with sorbitol, which is produced via catalytic hydrogenation using Raney nickel catalysts, is removed by cation-exchange resins. The nickel content in medium additives, e.g., in corn steep liquor is reduced by heat precipitation. Another possibility to solve this problem would be to adapt the cultures to higher nickel concentrations. *Acetobacter melanogenum* tolerates a concentration of up to 10 mg of nickel per liter. Today industrial-scale fermentations with *A.suboxydans* run with a 20% sorbitol solution containing 24 mg/L of nickel.

2. Process

Initially sorbitol conversions were carried out with surface cultures of *Acetobacter xylinum* yielding 40–60% sorbose after 6 weeks, or with *A.suboxydans* yielding 80–90% sorbose after 7 days (FULMER et al., 1936).

The first submerged process in a rotary drum, using a medium containing 20% sorbitol and 5% yeast extract, leads to a 58% yield of sorbose after 33–45 hours. More than 80% of the resulting sorbose in the broth could be isolated (WELLS et al., 1937).

LOCKWOOD (1954) described an industrial process with *A.suboxydans* where high quality aluminum or nickel-free stainless steel bioreactors were used which were agitated by air.

Today fermentations are carried out in stirred and aerated fermenters similar to those used for antibiotics production. The most important parameters are the following:

Medium. The medium consists of sorbitol, a nitrogen source, and $CaCO_3$. At the production scale usually a 20% (w/v) solution of technical grade sorbitol is used. Higher sorbitol concentrations are inhibitory to growth. Therefore, it is advantageous to start the fermentation with 10–20% sorbitol and to add sorbitol up to the final concentration of 28–35%. Yeast extract (0.1–0.5%) may be used as a nitrogen source. Another more frequently used nitrogen source is corn steep liquor (0.1–0.3%). The pH is adjusted to 5–6.

Inoculum. If the sorbitol concentration does not exceed 20%, the inoculum is cultivated in the same medium for about 20 h. Good results are obtained if the seed is taken from a running batch. The amount of inoculum varies between 3–10%.

Aeration and agitation. Effective aeration is essential for the seed culture as well as for the sorbitol conversion step. With *Acetobacter xylinum* and *A.suboxydans* the air can be substituted by oxygen under increased pressure resulting in a decrease of fermentation time, whereas *A.melanogenum* is inhibited by an elevated oxygen content.

After about 24 h at 30–35 °C the fermentation is stopped: The content of reducing sugars reaches about 96–99%. After filtration or centrifugation and deionization of the clear liquid the solution is concentrated to induce crystallization. Data published for isolated sorbose gave yields of about 87% (LIEBSTER et al., 1956).

Towards the end of a fermentation, depending on fermentation conditions, small amounts of sorbose may be converted to D-fructose and 5-keto-D-fructose as minor components.

A continuous fermentation was developed on a technical scale. A method using polyacrylamide-immobilized cells has been described.

C. New Bioconversion Steps in L-Ascorbic Acid Synthesis

Efforts were made to develop alternative pathways to 2-keto-L-gulonic acid (KGA), but as yet none of the following techniques can compete with REICHSTEIN's procedure.

1. Process via L-sorbosone

L-Sorbose is oxidized via L-sorbosone (L-xylo hexogulose) to KGA (Fig. 10). For the first step to sorbosone resting or immobilized cells of *Gluconobacter melanogenus* IFO 3 293 are used (MARTIN and PERLMAN, 1975, 1976). The oxidation of L-sorbosone to KGA is accomplished by strains or mutants of, e.g., *Pseudomonas putida* ATCC 21 812, *Gluconobacter melanogenus* IFO 3 293, *Bacillus subtilis* NRRL 558, *Candida albicans* NRRL 477, or *Penicillium digitatum* ATCC 10 030. With cells or cell-free extracts of the preferred strain *Pseudomonas putida* ATCC 21 812 the conversion of 5 g/L gave yields of 80% KGA (procedure described by MAKOVER and PRUESS, 1975).

2. Process via 5-keto-D-gluconate

This method follows the pathway shown in Fig. 11 (for details and further references see: KULHANEK, 1970). The dehydrogenation process (A) with *Acetobacter suboxydans* from glucose to 5-keto-D-gluconate proceeds in two steps. A 10% solution of glucose in an aerated fermenter had been converted in 33 h at 25 °C when $CaCO_3$ was added; the yield of 5-keto-D-gluconate reached 90% of the expected theoretical value.

Calcium 5-keto-D-gluconate is catalytically hydrogenated (B) to a mixture of calcium D-gluconate with calcium L-gulonate in a 1:1 ratio.

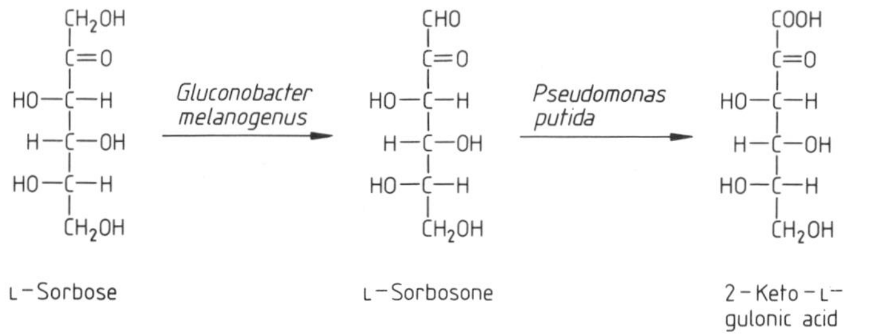

Figure 10. Conversion of L-sorbose to 2-keto-L-gulonic acid via L-sorbosone.

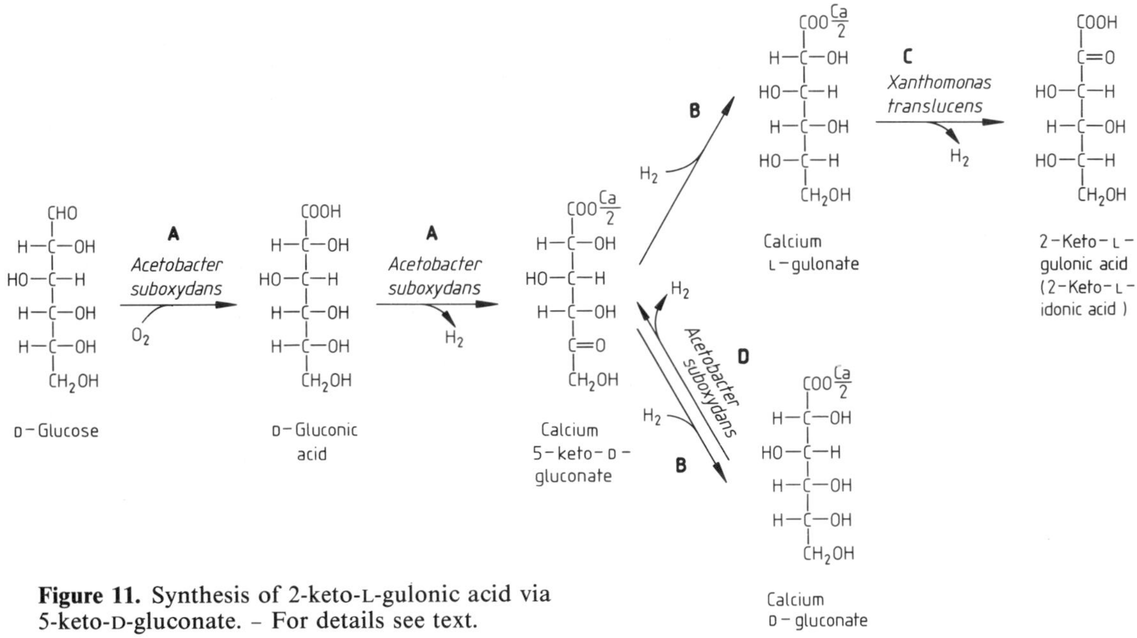

Figure 11. Synthesis of 2-keto-L-gulonic acid via 5-keto-D-gluconate. – For details see text.

Of the above mixtures only calcium L-gulonate can be converted to KGA (C). Several microorganisms are suitable for this conversion. KITA (1979) described a technique with growing cultures of *Xanthomonas translucens* ATCC 10 768 where yields of 90% or more were obtained with up to 15% L-gulonate in the form of a non-inhibitory calcium or sodium salt.

The reconversion of calcium D-gluconate to 5-keto-D-gluconate (D) by *Acetobacter suboxydans* occurs at a yield of about 70%. Further chemical or microbial methods are known.

3. Process via 2,5-diketo-D-gluconate

2,5-Diketo-D-gluconate is formed from glucose via D-gluconate and 2-keto-D-gluconate. Strains suitable for this reaction are *Pseudomonas albosesamae* ATCC 21 998, *Gluconobacter rubiginosus* IFO 3 244 (SONOYAMA et al., 1976a), or *Acetobacter melanogenus* ATCC 9 937 (STROSHANE and PERLMAN, 1977). With *Acetobacter cerinum* the 2,5-diketo acid is not further oxidized. After a fermentation period of about 40 h yields of 95% are achieved by starting with 11% glucose and adding further glucose (5.5%) after 20 h of fermentation (KITA and HALL, 1979).

2,5-Diketo-D-gluconate is converted to KGA by *Brevibacterium ketosoreductum* ATCC 21 914 and several strains of *Brevibacterium* sp. (ATCC 31 083, 31 082) or *Corynebacterium* sp. (ATCC 31 081, 31 090, 31 089, 31 088; SONOYAMA et al., 1975, 1976b). For example, *B.ketosoreductum* was cultivated in a 1.5 L fermenter containing 600 mL of nutrient broth (glycerol 0.3%, polypeptone 0.1%, yeast extract 0.1%, KH_2PO_4 0.1%, $MgSO_4 \cdot 7\ H_2O$ 0.02%; pH 6.3–7.0) to which 1.5% calcium 2,5-diketo-D-gluconate was added. Fermentation occurred at 30°C with an aeration of 1 vvm and an agitation of 300 rpm. After 72 h 1.89 g/L of KGA had been produced.

Mixed cultures with strains able to carry out both reaction steps may convert glucose directly into KGA (SONOYAMA et al., 1976a).

V. Gluconic Acid

A. Introduction

Gluconic acid and its δ-lactone are produced by dehydrogenation of D-glucose (Fig. 12). The conversion of glucose to δ-gluconolactone is catalyzed by glucose oxidase which is highly specific for β-D-glucose; the rate of oxidation of β-glucose is about 157 times faster than that of α-D-glucose. The developing H_2O_2 is cleaved by catalase to H_2O. The δ-lactone hydrolyzes spontaneously to gluconic acid.

B. Microorganisms

The dehydrogenation step can be performed by many microorganisms. Reports mostly indicate the production of gluconic acid with *Aspergillus niger, Penicillium chrysogenum, P.luteum-purpurogenum* and other Penicillia, several *Acetobacter* spp., in particular *A.suboxydans,* various strains of *Pseudomonas, Pullularia pullulans,* species of *Moraxella, Endomycopsis,* and others (MIALL, 1978). Unfavorably many bacterial strains oxidize gluconate further to ketogluconic acid.

C. Processes

Gluconic acid, δ-gluconolactone, calcium or sodium gluconate, and glucose oxidase are mainly produced through batch fermentation, with *Aspergillus niger,* in some cases also with *Acetobacter suboxydans.* A detailed description of the production procedures was made in Volume 3, Chapter 3e of this series.

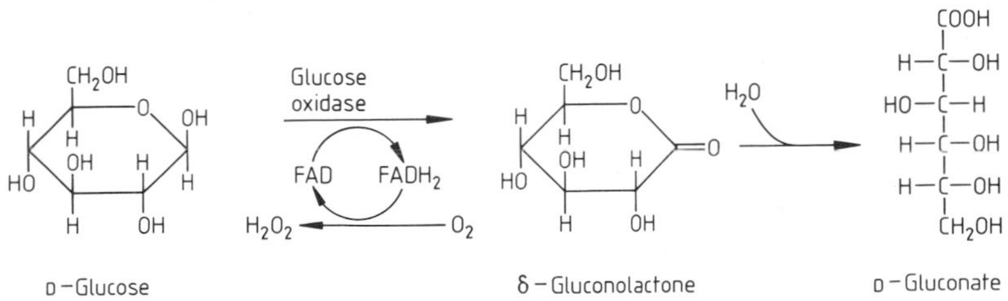

D-Glucose — Glucose oxidase (FAD ⇄ $FADH_2$; O_2 → H_2O_2) → δ-Gluconolactone — H_2O → D-Gluconate

Figure 12. Dehydrogenation of glucose to gluconic acid by glucose oxidase.

D. Further Developments in Gluconic Acid Production

Fermentation processes must compete with several high yielding chemical processes such as electrolytic oxidation in the presence of bromine or various palladium-catalyzed air oxidation processes (MIALL, 1978).

Future procedures in gluconate production may possibly be continuous methods (YAMADA, 1977) or will make use of immobilized cells or immobilized enzymes. Significant achievements have been made to develop reactor systems employing this technology. To date these procedures have not attained commercial standards.

Immobilized cells

Cells of *Aspergillus niger* NRRL 3, immobilized in calcium alginate gel, have yielded a 44% conversion of a 15% glucose solution within 24 h; using pure oxygen 93% conversions have been achieved (LINKO, 1981).

Glucose oxidase has been immobilized with glutaraldehyde and ovalbumin on the mycelium pellets of *Aspergillus* sp. Complete oxidation of glucose was attained in a batch system only under sufficient aeration (Fig. 13; KARUBE et al., 1977).

TRAMPER et al. (1983) used cells of *Gluconobacter oxydans* subsp. *suboxydans* (ATCC 621 H), which were entrapped in various alginate gels. The aim of this work was to evaluate the conditions for a continuous process of gluconic acid production.

Immobilized enzymes

Immobilized GOD (glucose oxidase) is commercially available and can be purchased, e.g., from Boehringer or Aldrich. GOD has been immobilized on various inorganic and organic support materials and on molecular sieves (for references see: PIFFERI et al., 1982). When GOD is immobilized on sepharose by photochemically (UV) initiated craft copolymerization with $FeCl_3$ as photocatalyst, the kinetic properties of the immobilized enzyme are close to those of the free enzyme ($K_m{}' = 2.0 \times 10^{-2}$ M for the immobilized enzyme; $K_m{}' = 1.93 \times 10^{-2}$ M for the free enzyme; D'ANGIURO and CREMONESI, 1982).

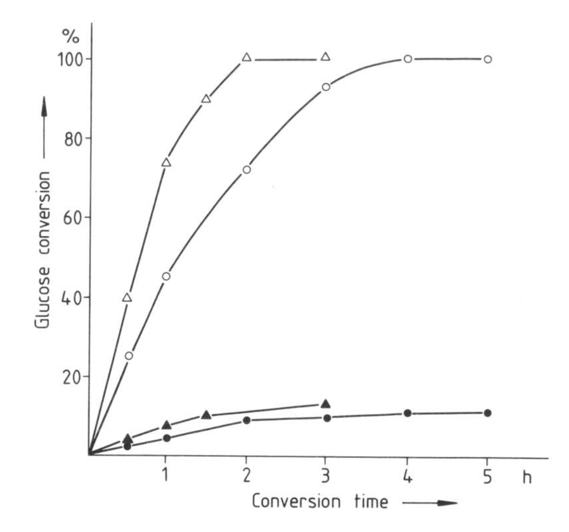

Figure 13. Oxidation of glucose by glucose oxidase pellets in a batch system. – Glucose oxidase was immobilized on the mycelium pellets of *Aspergillus* sp. No. 319 by glutaraldehyde and ovalbumin, with an activity yield of immobilization of 95% (KARUBE et al., 1977). For glucose conversion 1 g of wet glucose oxidase pellets was added to 50 mL of glucose solution at pH 5.0; aeration was achieved by bubbling air through the system,

with aeration △—△ 0.5% glucose,
○—○ 1.0% glucose,

without aeration ▲—▲ 0.5% glucose,
●—● 1.0% glucose.

Multi-enzyme systems

These systems are being developed mainly in order to use more complex systems than glucose, e.g., hydrolyzed starch. As a model, maltose was converted to gluconic acid using a glucoamylase of *Rhizopus delemar* in combination with glucose oxidase and gluconolactonase of *Aspergillus niger* (CHO and BAILEY, 1977).

A second approach is the conversion of sucrose to fructose and gluconate, both of which can easily be separated from the mixture. D'SOUZA and NADKARNI (1980) described a continuous conversion process carried out in a continuous-flow stirred tank reactor system for over 20 days. They used a multi-enzyme complex consisting of an invertase- and catalase-producing *Saccharomyces cerevisiae,* where glucose oxidase was fixed to the cell walls by concanavalin A. The cells were entrapped in polyacrylamid gel.

E. Application

Glucose oxidases (β-D-glucose: O_2 oxidoreductase, GOD; EC 1.1.3.4) from the cultures of *Penicillium notatum* and other Penicillia were first described as antibiotics (Penicillin B, Notatin) because of their inhibitory effect on bacterial growth. This however is due to the formation of H_2O_2 by GOD. In the meantime this enzyme has been applied for various practical purposes.

Technical grade preparations of GOD with varying catalase activities have been widely used in food processing to remove glucose or to prevent oxidative changes in, e.g., beer, juice, mayonnaise, or milk powder.

High amounts of purified GOD are used in medicine for quantitative determinations of glucose in body fluids.

Another analytical application of immobilized GOD is the glucose electrode. In a system developed by CLARK (1970) glucose oxidase is held on a filter trap and the hydrogen peroxide produced in this enzymatic reaction is measured with a platinum electrode. The stability of this electrode exceeded 14 months, the range of determined glucose was 2×10^{-2} to 10^{-4} M. A glucose electrode based on this reaction type is sold by Yellow Springs Instr. Co. (Yellow Springs, Ohio) and by Radiometer (Copenhagen, Denmark). Various other types of biosensors using GOD are presently being developed (GONDO et al., 1980; SUZUKI et al., 1982; ENFORS, 1982).

Free gluconic acid is mainly used in the dairy industry for cleaning purposes. Its salts are used in iron therapy as iron gluconate and iron phosphogluconate.

Calcium gluconate serves as a source of calcium in medicine and in veterinary practice. Further applications are in the tannery, metal, and dye industries.

Sodium gluconate is the main product on the market because of its metal sequestering qualities. δ-Gluconolactone is mainly used in the food industry, e. g., in baking powders, in sausage manufacturing, or in Japan as a coagulant of soybean protein.

The annual production of gluconic acid and its salts is in the range of 45 000 tons (BRONN, 1976). In Japan about 3 000 tons of δ-gluconolactone are produced for soybean curd manufacture and 8 500 tons of gluconic acid are produced for industrial purposes; producers are Kyowa Hakko and Fujisawa (YAMADA, 1977). U.S. producers are: Grain Processing, Pfizer, Stauffer, and the Premier Malt Division of Pabst Brewery. Main European manufacturers are Pfizer (Eire), Roquette Frères (France), Benckiser (Germany), and Akzo (Netherlands) (MIALL, 1978).

VI. Kojic Acid

The production of kojic acid (5-hydroxy-2-hydroxymethyl-4-pyrone) has only been of minor commercial interest; for several years it had been produced by bioconversion of glucose with *Aspergillus flavus* or *A.oryzae* (Fig. 14). Glycerol, arabinose, or ethanol may also be used as a substrate. Kojic acid was employed as a chelating

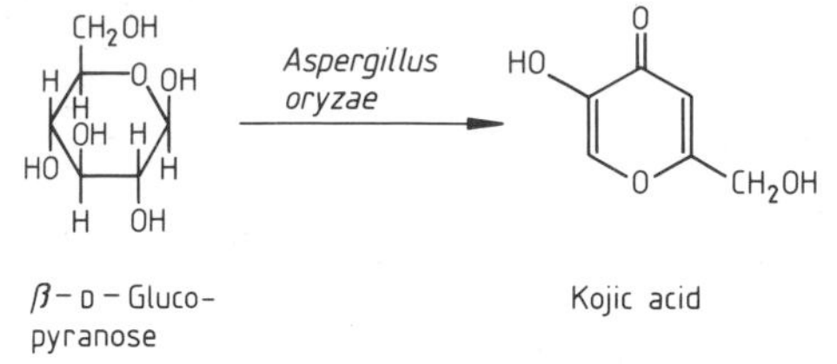

Figure 14. Kojic acid production by bioconversion of glucose.

agent for the production of insecticides, and furthermore, for analytical iron determinations as kojic acid gives a deep red color with as little as 0.1 ppm of ferric iron.

The fermentation conditions with glucose as substrate are similar to the citric acid or itaconic acid processes. By varying the fermentation conditions some citric or itaconic acid producing strains of the *A.flavus-oryzae* group may yield kojic acid (MIALL, 1978; BUCHTA, 1982).

VII. Dihydroxyacetone

There has been an increase in the industrial production of dihydroxyacetone by microbial conversion of glycerol (Fig. 15). Dihydroxyacetone is used for tanning purposes in the cosmetics industry. Several *Acetobacter* strains are able to convert a 10% glycerol solution in a suitable medium of yeast extract 0.5%, KH_2PO_4 0.5%, and $CaCO_3$ 2.0% at 28°C in 72–96 h. The current industrial process for dihydroxyacetone production with *Acetobacter suboxydans* runs under slightly acid conditions (pH ≤ 6) to prevent further metabolizing of dihydroxyacetone by the microbes (GREEN, 1960).

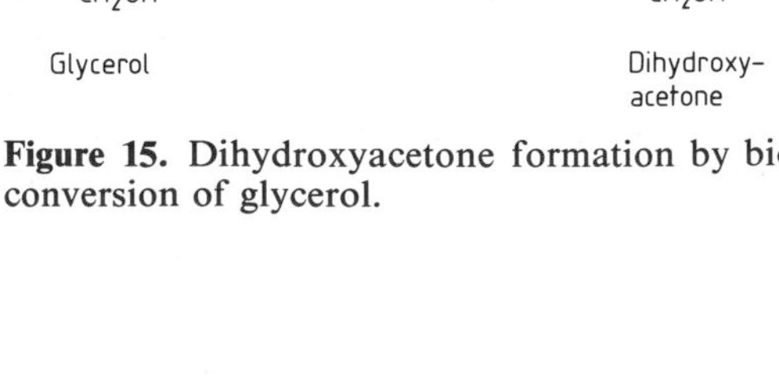

Figure 15. Dihydroxyacetone formation by bioconversion of glycerol.

Gluconobacter melanogenus IFO 3 293 may even convert a 10% glycerol solution in merely 33 h under oxygen-enriched aeration. Under these conditions the yield of dihydroxyacetone was 35.8 g/g cell mass compared to 12.2 g/g cell mass in cultures without O_2-enrichment (FLICKINGER and PERLMAN, 1977).

Cells of *Gluconobacter oxydans* subsp. *suboxydans,* which were immobilized in polyacrylamide gels, were used to convert glycerol into dihydroxyacetone in bubble column reactors. A glycerol concentration of 6% at pH 6.0 and 30 °C was found to be optimal for this conversion reaction (SONAER and ÇAĞLAR, 1981).

VIII. Isomaltulose

Isomaltulose (palatinose) is a reducing disaccharide [6-*O*-(α-D-glucopyranosyl)-D-fructofuranose]. It is being developed as a substitute for sucrose in the preparation of food, pharmaceuticals and other products. It has a similar taste, size, and structure, as well as preserving abilities similar as sucrose, but it is less sweet and non-cariogenic.

Isomaltulose can be produced by several microorganisms such as *Leuconostoc mesenteroides, Serratia plymutica, S.marcescens, Erwinia carotovora,* and *E.rhapontici* (MAUCH and SCHMIDT-BERG-LORENZ, 1964; SCHIWECK et al., 1972; BUCKE and CHEETHAM, 1979).

The corresponding enzyme is a sucrose-specific glucotransferase which only utilizes fructose derived from sucrose (Fig. 16); free

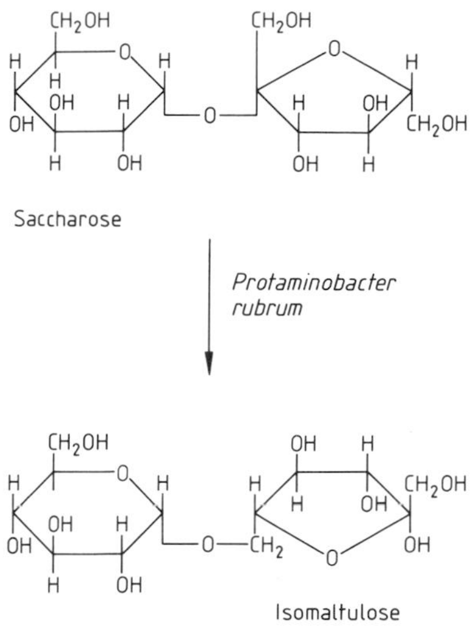

Figure 16. Isomerization of sucrose to isomaltulose.

fructose or fructose bound to other molecules can not be metabolized by this enzyme. The enzymatic conversion of sucrose into isomaltulose can be carried out by batch or continuous fermentation. Good media for fermentations are vegetable juices containing sucrose obtained as intermediary products in cane or sugar beet extraction.

The following description presents an example of a continuous fermentation with *Protaminobacter rubrum* (CBS 57 477) (CRUEGER et al., 1977):
Cells are resuspended in 10 mL of a sterile mixture of one part concentrated juice (65% dry substance) and two parts tap water plus 0.5 g/L of $(NH_4)_2HPO_4$. This suspension is used as an inoculum for the shaking machine preculture which is transferred to seed fermenters. A 3 000 L fermenter with 1 800 L of nutrient solution (concentrated liquid/sucrose/water mixture, at 25% dry substances and a purity of 98%) and 200 L of inoculum is operated at 30 °C. The aeration is set at 0.4 vvm and the stirring speed adjusted to 140 rpm. The course of fermentation is constantly observed via measurement of CO_2 content in the exhaust gas. As soon as 2.6% CO_2 has been reached the fermenter is operated with a dilution rate of 0.13 h^{-1}; the outflow contains only isomaltulose while sucrose becomes completely metabolized.

A production method of isomaltulose from sucrose solutions was described by BUCKE and CHEETHAM, 1979, in which cells of *Erwinia rhapontici* were immobilized with sodium alginate:

A 55% solution of sucrose (w/v) was converted in a column to isomaltulose at pH 7.0 and a temperature of 30 °C. The activity was 0.2 g of isomaltulose/g wet cells/hour.

The hydrogenation products of isomaltulose notably α-D-glucopyranosyl-1,6-sorbitol and α-D-glucopyranosyl-1,6-mannitol are available as a low calorie sweetener under the trade name "Palatinit®".

IX. Raffinose Conversion by α-Galactoside-galactohydrolase

A frequently applied method of sugar transformation in the sugar industry, is the hydrolysis of raffinose in beet-sugar molasses by α-galactosidase yielding sucrose and galactose. A detailed description of the procedures is given by OBARA et al. (1976/1977).

A. Introduction

Raffinose, a trisaccharide consisting of D-galactose, D-glucose, and D-fructose, is contained in sugar beets in concentrations up to 0.16%. In beet-sugar manufacturing processes raffinose retards the crystallization of sucrose even at concentrations of 0.05–1.5% resulting in lower yields of sucrose. Especially the calcium saccharate process (Steffen process) for sucrose production leads to an accumulation of raffinose after several weeks of the production campaign. The content of raffinose in these final molasses can be in the range of 7–8%. Therefore, these molasses must be discarded. The use of α-galactosidase for the hydrolysis of raffinose to sucrose and galactose made it possible to prevent the retardation of crystallization and thus, the loss of valuable molasses. About two-thirds of the hydrolyzed raffinose are recovered as sucrose; galactose is decomposed by adjusting to pH 12 and heating at 80 °C for 30 min.

B. α-Galactosidase Producing Microorganisms

α-D-Galactosidase (melibiase; α-D-galactoside-galactohydrolase, EC 3.2.1.22) is found in several microorganisms, such as *Aspergillus oryzae, A.niger, Penicillium duponti,* several strains of *Lactobacillus,* and few *Streptomyces* species.

A suitable microorganism should not produce invertase, which hydrolyzes raffinose to melibiose and fructose. A combination of α-galactosidase and invertase activities results in the undesired total hydrolysis of raffinose to galactose, glucose, and fructose (Fig. 17). Strains with a high α-galactosidase production but little ability to produce invertase are found among various species of *Streptomyces* which excrete the α-galactosidase mainly into the medium.

In 1969, SUZUKI et al. described the mold *Mortierella vinacea* var. raffinoseutilizer (ATCC 20 034) as a suitable organism for raffinose conversion. The α-galactosidase of this strain is bound to the mycelium fraction (Table 13). Therefore, it was possible to use the mycelial pellets of *M.vinacea* like an immobilized enzyme for a continuous process.

Mycelium-bound α-galactosidase is also formed by the mold *Circinella muscae* nova typica *coreanus* (ATCC 20 394) (NARITA et al., 1975). The mold *Absidia griseola* var. *iguchii* (ATCC 20 431) produces α-galactosidase without exerting any invertase activity (NARITA et al., 1976). A thermostable intracellular α-galactosidase is produced by *Bacillus stearothermophilus* (DELENTE et al., 1974).

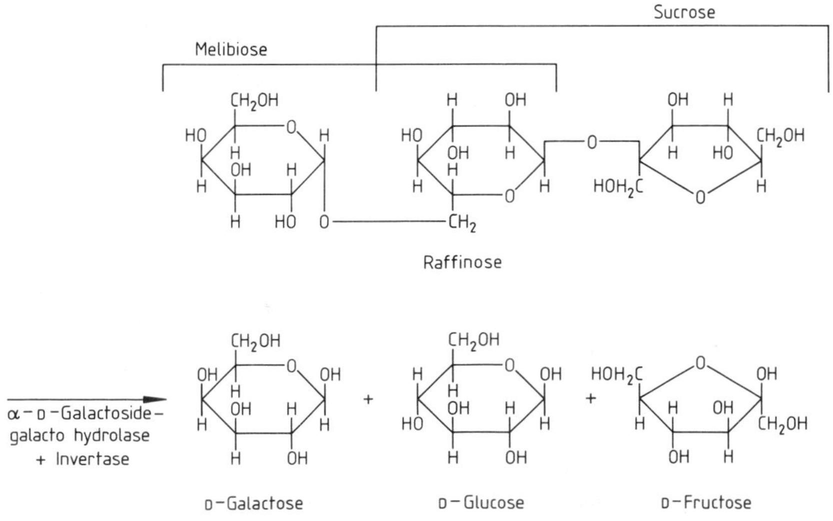

Figure 17. Total hydrolysis of raffinose by invertase and α-galactosidase.

Table 13. Location and Occurrence of α-Galactosidase

α-Galactosidase Producer	α-Galactosidase (units/mL)		
	Mycelium	Medium	Total
Streptomyces olivaceus	6	532	538
S.fradiae	61	324	385
S.roseospinus nov. sp.	181	293	474
Mortierella vinacea, var. raffinoseutilizer	28700	3300	32000

For sufficient α-galactosidase formation with *Mortierella vinacea* it was necessary to add certain carbohydrates as inducers to the medium, such as galactose, raffinose, or lactose. The effect of several sugars on α-galactosidase formation is shown in Table 14. Lactose was the most effective inducer of the enzyme but led to poor growth of *M.vinacea* when used as the sole carbon source. Therefore, in the first stage of fermentation glucose is added for mycelial growth, while lactose later on is used as an inducer for efficient α-galactosidase production.

In the case of the molds *Circinella* or *Absidia* the production of α-galactosidase is enhanced by addition of organic acids such as citric acid, lactic acid, or glycolic acid (WATANABE et al., 1975).

In 1970, α-galactosidase from *Mortierella vinacea* var. raffinoseutilizer was obtained in pure crystal form. The enzyme hydrolyzes the α-1,6-galactoside linkages of saccharides, such as raffinose, melibiose, verbascose, or stachyose. The K_m and V_{max} values of the crystalline enzyme on raffinose were 1.83 ± 0.13 mM and 19.2 μmol/min/mg protein, respectively.

Immobilized α-galactosidase has been prepared by cross-linking the enzyme of *Bacillus stearothermophilus* on nylon (REYNOLDS, 1974). For the same enzyme immobilization in hollow fiber cartridges is described by KORUS and OLSON (1977).

For technical applications in a continuous process the mycelium-bound enzyme of *M.vinacea* is used.

C. Industrial Processing

1. Production of mycelial pellets

Mortierella vinacea var. raffinoseutilizer (ATCC 20 034) is inoculated into a spore

Table 14. Influence of Several Carbohydrates on α-Galactosidase Production by *Mortierella vinacea* var. raffinoseutilizer (OBARA et al., 1976/77)

Carbohydrates	Growth in Dry Weight (mg/100 mL)	Sugar Consumption (%)	Total Activity of α-Galactosidase (units/mL)
Xylose	365	93.2	0
Arabinose	416	57.6	0
Rhamnose	122	29.5	0
Glucose	541	98.4	0
Fructose	555	97.7	0
Mannose	519	98.1	4.6
Galactose	488	95.6	1129
Lactose	353	63.7	3822
Melibiose	527	96.9	1978
Raffinose	442	98.6	2505
Soluble starch	330	70.5	0

forming medium (malt extract 2%, peptone 2%, glucose 2%, agar 2%) at 28 °C for about one week.

The resulting spore suspension is used to inoculate a seed culture broth (lactose 0.5%, glucose 0.7%, corn steep liquor 1.2%, $(NH_4)_2SO_4$ 0.1%, $(NH_4)_2CO_3$ 0.02%; KH_2PO_4 0.03%, $MgSO_4 \cdot 7\ H_2O$ 0.2%, NaCl 0.2%, cocoline as antifoam 0.08%). The culture is grown under aeration and agitation at 30 °C for 25–30 h.

The production culture (lactose 1.5%, glucose 1.0%, corn steep liquor 0.88%, $(NH_4)_2SO_4$ 0.72%, KH_2PO_4 0.3%, $MgSO_4 \cdot 7\ H_2O$ 0.2%, NaCl 0.2%, $CaCO_3$ 0.24%, cocoline 0.064%) was incubated at 30 °C for 70–80 h.

The mycelial pellets are separated from the broth, washed, and dried to result in 92% dry substance. The α-galactosidase activity of these dried pellets is about 2 000 000 units/g. The material can be stored at −15 °C without decrease of activity.

2. Hydrolysis of raffinose

For the hydrolysis of raffinose in beet sugar syrups by α-galactosidase containing mycelial pellets, SHIMIZU and KAGA (1972, 1974) developed a horizontal continuous reactor. Column and batch reactors were not suitable, the first because of the excessive pressure which lumps the mycelial pellets, the latter by impairing enzyme activity during separation steps through contact with oxygen.

The enzyme reactor according to the process of SHIMIZU and KAGA is an agitated U-shaped open vessel consisting of several chambers. The mycelial pellets are retained inside the reactor by screens (mesh diameter from 0.25–0.42 mm).

For hydrolysis of raffinose the final molasses of the Steffen process are continuously adjusted to 29–31° Brix, to pH 5.0–5.2, and a temperature of 48–52 °C. The adjusted syrup is continuously fed into the enzyme reactor. A reaction time of 1.5–2.5 h has been shown to be sufficient.

With slight modifications the described hydrolysis procedure is also suitable for beet sugar manufacturing in combination with ion exchange demineralization.

D. Economics

The efficiency of raffinose hydrolysis under laboratory conditions is shown in Table 15. 500 g of enzyme pellets with α-galactosidase activity of 210 000 units/g were used to hydrolyze a solution of 50 g raffinose · 5 H_2O in 1 350 g of dest. H_2O. Hydrolysis was achieved after 8 h at 50–54 °C and pH 5.0. The given weights refer to the weight of raffinose.

Table 15. Sugar Analysis after Hydrolysis of Raffinose by Mycelial Pellets of *Mortierella vinacea* var. raffinoseutilizer (ATCC 20 034)

	Estimated Weight (rel. to raffinose weight)
Prior to Hydrolysis	
Raffinose · 5 H_2O	100
Reducing substances (impurities)	0.25
After Hydrolysis	
Raffinose · 5 H_2O	0
Sucrose	58.6[a)]
Galactose	32.0[b)]
Glucose (from mycelia)	4.8
Fructose	≤0.54
Melibiose	≤0.32

[a] 101.7% of theoretical value
[b] 105.6% of theoretical value

Hydrolyzation of 1 ton of raffinose requires 0.03 tons of mycelial pellets (α-galactosidase activity 2 000 000 units/g), on the premise that the enzyme pellets are used for 100 days and that a 60% hydrolysis of raffinose is assumed (OBARA et al., 1976/1977). The result of raffinose hydrolysis is a 3–5% increase in the overall sugar yield. Raffinose hydrolysis in combination with

Table 16. Comparison of Processes With and Without Raffinose Hydrolysis (Obara et al., 1976/1977) at the Kitami Factory of Hokkaido Sugar Co. Ltd.

	Calcium Saccharate Process			
	Batchwise	Batchwise + Raffinose Hydrolysis	Continuous	Continuous + Raffinose Hydrolysis
Yield of sucrose (tons)	33914	35813	34223	36359
Recovery ratio of sucrose (%)	87.29	92.18	88.09	93.58
Raffinose content of molasses supplied to the calcium saccharate process (%)	6.28	1.84	7.42	2.00
Amount of discarded molasses (tons)	3855	none	4045	none
Raffinose content of discarded molasses (%)	9.39	-	9.98	-

the continuous calcium saccharate process has yielded the results presented in Table 16; this procedure has been used since 1973 by Hokkaido Sugar Co., Ltd.

X. Lactose Hydrolysis by Lactase

A. Introduction

Lactose is the main sugar found in milk and whey (~ 42%). Whey is the liquid which remains after solids have been removed from cream or milk during cheese production.

The enzyme β-D-galactosidase (β-D-galactosido galactohydrolase, EC 3.2.1.23, lactase) hydrolyzes the disaccharide lactose to glucose and galactose (Fig. 18). Acid hydrolysis can be carried out on lactose (2 h at 120 °C) but this is generally not recommended as side reactions may occur giving rise to a product with an undesirable flavor (Maillard reactions).

The presence of lactose in milk and whey is responsible for several problems which may arise:

a) 70–90% Asians, Africans, and American Negros and 2–15% white Americans and West Europeans have β-galactosidase deficiency. The effects are abdominal bloating, rumbling, and diarrhea. Children are sometimes unable to convert galactose to glucose. Galactose is accumulated in the blood, and the liver becomes enlarged. This disorder is called galactosemia. However, milk based products can be ingested by these people if lactose is hydrolyzed in the milk.

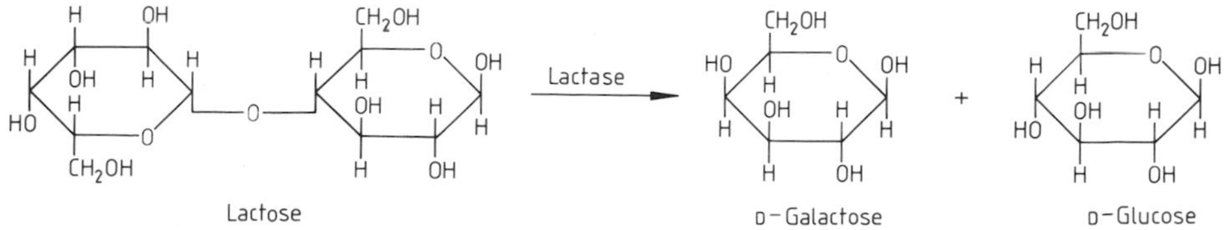

Figure 18. Hydrolysis of lactose by lactase.

b) Due to its tendency to crystallize, its low solubility, and comparably low sweetness, lactose would be impractical as a food sugar. Therefore, lactose is mostly used for animal feed, as a liquid concentrate, or dry powder.
c) It has been suggested that prehydrolysis of milk with lactase would shorten yoghurt and cheese production time by 20%, as well as increase the yields and sweetness. Significant reductions of manufacturing time, ripening time, and improved quality has already been observed with cheddar cheese.
d) About 50 million tons of whey are produced worldwide. 70% of the dry matter is lactose, 9–14% crude protein, and the ash content is about 9% (Table 17). 50% of the whey goes to waste into the municipal sewage systems. The organic material of the whey causes pollution problems. 1 000 gallons of raw whey are equivalent to the waste of 1 800 people. Solids of sweet whey are used more in animal feeds. The major portion of the by-products from cottage cheese and fresh cheese production is acidic whey.

Table 17. Content of Whey (MEYRATH and BAYER, 1979)

Components (%)	Sweet Whey	Acid Whey
Dry matter	6.7	5.6
Lactose	4.5–5.0	3.8–4.2
Ash	0.5–0.7	0.7–0.8
Crude protein	0.8–1.0	0.8–1.0
pH	4.5–6.7	3.9–4.5

New methods to enzymatically hydrolyze lactose have been developed for commercial purposes in order to increase the usage of lactose in foods and to avoid environmental and economical problems. Whey is an important potential source of lactose as it is inexpensive and a raw material from which other valuable products can be produced (e. g. ethyl alcohol or organic acids). SINGH et al. (1983) have described the production of ethanol from sweet whey with a *Kluyveromyces* species in a continuous process. Hydrolysis of the whey ultrafiltrate yields a sweet syrup which can be used as a sweetener.

Reviews on the industrial use of lactase have been written by MILLER and BRAND, 1980, BARKER and SHIRLEY, 1980, and GREENBERG and MAHONEY, 1981.

B. Microorganisms and Their Lactases

β-Galactosidase is produced by a number of microorganisms, such as:
Yeasts: *Kluyveromyces lactis, K.fragilis, Candida pseudotropicalis, Zygosaccharomyces lactis, Saccharomyces anamensis.*
Fungi: *Aspergillus oryzae, A.niger Mortierella vinacea, Penicillium multicolor, Curvularia inaequalis, Fusarium moniliforme, Mucor pusillus, M.miehei, Humicola grisea, H.lanuginosa, Sporotrichum* sp., *S.thermophile, Torula thermophila.*
Bacteria: *Bacillus coagulans, B.stearothermophilus, B.circulans, Escherichia coli, Lactobacillus bulgaricus, L.thermophila, Caldariella acidophila, Leuconostoc citrivorum, Streptomyces coelicolor.*

Those microorganisms most important for production are:

Kluyveromyces lactis
Kluyveromyces fragilis
Aspergillus niger
Aspergillus oryzae
Escherichia coli

The properties of lactases in microorganisms vary considerably. Therefore, the customer must choose the enzyme that best suits his needs (Table 18).

The intracellular yeast enzymes have a narrow pH stability range around 7.0. Therefore, their application is limited to treatment of sweet whey or milk. Contamination is supported by the low temperature optimum. The temperature stability of these enzymes lies below 40 °C. The fungal enzymes, however, are stable over a broad pH range and suitable for treatment of acid

Table 18. Properties of Commercially Available Lactases

Source	Producer	pH Optimum	pH Stability	Temperature Optimum (°C)	K_m (Lactose) (mM)	Cofactors
Aspergillus niger	Wallerstein, Rapidase	3.0–4.0	2.5–8.0	55	85	–
Aspergillus oryzae	Tokyo Tanabe, Miles Lab.	5.0	2.5–7.0	55	50	–
Kluyveromyces fragilis	Novo Ind.	6.6	6.5–7.5	30–35	14	Mn^{2+}, K^+
Kluyveromyces lactis	Gist Brocades	6.5–7.0	6.5–7.5	30–35	12–17	Mn^{2+}, Na^+
Escherichia coli	Aldrich Chem.	7.2	6.0–8.0	35		Na^+, K^+

wheys. Their optimal temperature is 55 °C, but the enzymes are thermostable up to 70 °C. The enzyme from *Aspergillus niger* is intracellular, while that of *A.oryzae* is extracellular.

During the application of lactase from *A.niger* in acid whey (55 °C at pH 3.5) there is a mutarotation of β-galactose to α-galactose. Lactase can distinguish between β- and α-galactose. α-Galactose acts as a competitive and anticompetitive inhibitor while β-galactose is a competitive one (FLASCHEL et al., 1982). Therefore, the reaction rate is not linearly correlated to the enzyme concentration. It increases more than proportionally.

The bacterial enzyme has a neutral pH optimum for hydrolysis of milk and sweet whey; however, the enzyme's temperature optimum is at 35 °C, which is critical with respect to microbial contamination.

C. Fermentation for Lactase Production

1. *Aspergillus oryzae*

The following example is a typical solid fermentation with *Aspergillus* (KIUCHI and TANAKA, 1975). 20 g of whcat bran and 12 mL of tap water are sterilized in 500 mL Erlenmeyer flasks. The solid culture is inoculated with *A.oryzae* spores and incubated for 7 days at 30 °C.

A solid culturing apparatus containing 61 kg of wheat bran and 84 L of tap water is inoculated with ten of the above flasks. The bioreactor is aerated with moist air for 3 days at 30 °C. For enzyme extraction 500 L of tap water are added, the mixture is stirred for 3 h, and then filtered. The filtrate is decolorized with activated carbon, concentrated, and precipitated with cold isopropanol.

Aspergillus oryzae lactase may be obtained from a submerse fermentation. The yields in solid fermentation are much higher, however, than in submerse fermentation.

2. *Escherichia coli*

The optimization process for β-galactosidase production by *Escherichia coli* is described by GRAY et al., 1972 and HIGGINS et al., 1978. The production medium is:

	%
$(NH_4)_2SO_4$	0.8
NaCl	0.3
Na_2HPO_4	0.53
KH_2PO_4	0.16
Triammonium citrate	0.05
Glycerol	1.5
Glucose	0.39

Antifoam agent: polypropylene glycol

The fermentation is controlled at pH 7.0 with NaOH. The incubation temperature is 37 °C. The agitation and aeration rates for a 1 000 L fermenter are:

h	agitation rpm	aeration L/min
0–14.4	300	200
14.4–end	350	500

Batch fermentations are harvested after 8–20 hours. Continuous fermentations are also possible. Yeast extract used as a complex nitrogen source reduces the enzyme titer. Glucose causes repression of β-galactosidase. The *E.coli* mutants are constitutive and up to 20% of the total protein is β-galactosidase. *E.coli* growth and β-galactosidase production in a batch and continuous fermentation is shown in Fig. 19.

The isolation and purification procedure starts with the cooled fermentation broth (Fig. 20). The cells are concentrated 6.7 fold and homogenized with a high pressure homogenizer at 550 kg/cm^2. RNA is removed by adding 0.1 M $MnSO_4$; the solution is then gently mixed for 10 min. RNA can more efficiently be removed by heating (5 min at 55 °C). After removal of the RNA through centrifugation ammonium sulfate is added for precipitation.

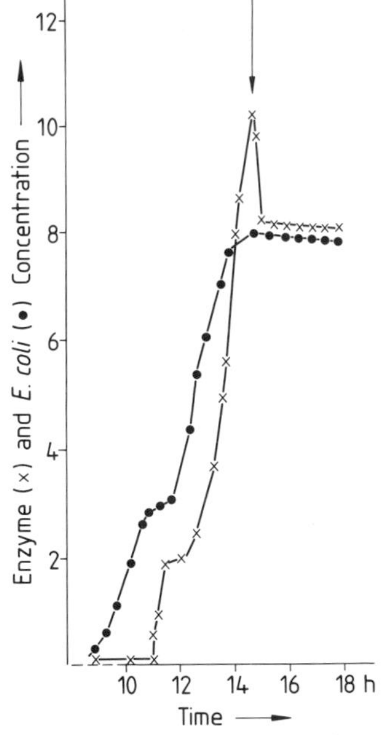

Figure 19. Batch and continuous fermentation of *Escherichia coli* for lactase production (HIGGINS et al., 1978). – Enzyme concentration (units/0.1 mL culture), bacterial concentration (mg dry wt/mL). Arrow indicates start of continuous fermentation ($D = 0.25\ h^{-1}$).

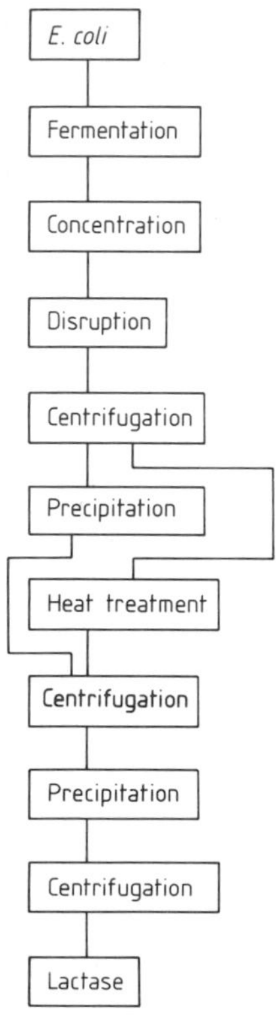

Figure 20. Isolation and purification of *E.coli* lactase.

3. *Bacillus stearothermophilus*

The lactases of *Bacillus stearothermophilus* are very thermostable. The HRI strains of *B.stearothermophilus* grow in a medium containing (mg/L): nitrilotriacetic acid 100, $MgSO_4 \cdot 7\ H_2O$ 100, $CaSO_4 \cdot 2\ H_2O$ 60, NaCl 8, KNO_3 103, $NaNO_3$ 689, Na_2HPO_4 111, $FeCl_3 \cdot 6H_2O$ 0.28, $MnSO_4 \cdot H_2O$ 2.2, $ZnSO_4 \cdot 7H_2O$ 0.5, H_3BO_3 0.5, $CuSO_4$ 0.016, $Na_2MoO_4 \cdot 2\ H_2O$ 0.025, $CaCl_2 \cdot 6\ H_2O$ 0.046, and lactose 6 400 (GRIFFITHS et al., 1979). The organisms grow while aerated at 65 °C in a batch fermenter or under continuous conditions.

The purified enzyme has an activity half life of at least 1.5 h at 55 °C, 1 h at 60 °C, and 10 min at 65 °C. The temperature dependent activity of the whole cells, immobilized cells, and the purified enzyme is shown in Fig. 21.

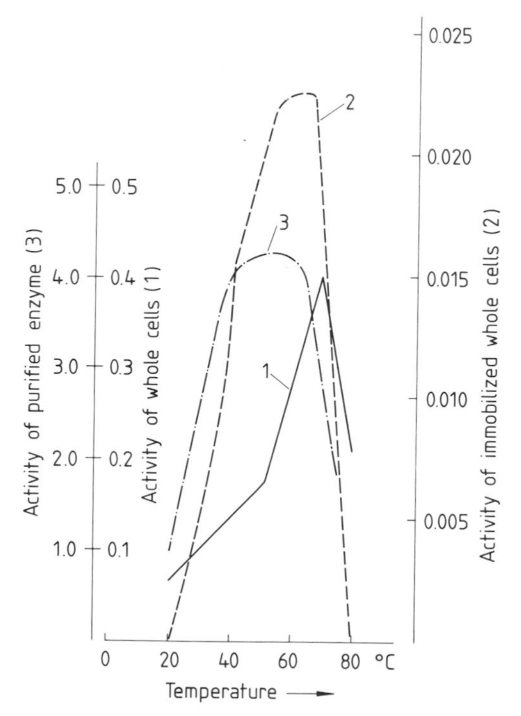

Figure 21. Comparison of activity of whole cells (1), immobilized whole cells (2), and purified lactase (3) from *Bacillus stearothermophilus* (GRIFFITHS et al., 1979).

D. Immobilization of Lactase

Nearly all of the previously described immobilization methods have been experimentally employed with lactase.

Supports used were: Agar gel, alumina, carbon, cellulose, chitin, chitosan, collagen, glass beads, ion exchange resins, nickel, nylon, phenol-formaldehyde resin, polyacrylamide, polyacrylate, polyisocyanate, polyvinylpyrrolidone, Sephadex, silk fibroin, and stainless steel (GREENBERG and MAHONEY, 1981). Immobilization is essential for reducing the cost of lactose hydrolysis as well as enabling changes of operational conditions (pH, temperature, half life). RUGH (1982) has shown that the immobilization of thermophilic lactase causes a decrease in the formation of intermediary products during lactose hydrolysis. This decrease, in addition, leads to a higher enzyme efficiency.

Publications on the immobilization of lactase have been written by PITCHER et al. (1976); PASTORE and MORISI, 1976; MARCONI and MORISI, 1979; and BARKER 1980.

As there are no extraction and purification steps involved, the immobilization of microorganisms is becoming more important than the immobilization of enzymes.

Immobilized lactases from yeasts and fungi are now commercially available. Lactase from *Kluyveromyces lactis,* entrapped in cellulose-triacetate fibers, was originally used to produce low lactose milk (10 t/day). The conversion rate of lactose to the monosaccharides in the dairy industry is 70–80%. In 50 batches, or after treatment of 10 000 L of milk, the activity decrease was less than 10%. Three companies in Italy, Spain, and Japan are using this method, which at the present time is the most advanced commercial application of the lactose hydrolysis procedure.

Aspergillus niger has been immobilized on a porous silica support with a 50% coupling efficiency. The half life of this material is 60 days. By frequent sterilization of the enzyme reactor microbial contamination has been contained. This immobilization procedure is being used to convert cheese whey into a sweet protein-rich syrup which can be used in baked goods and ice cream.

E. Economics

Today the production of low lactose milk is constantly expanding. Pollution caused from excessive quantities of whey is one of the most important factors in the lactose business. By hydrolyzing lactose in permeate whey a suitable sweetener for the food industry may be produced. Elimination of the whey lactose by enzyme treatment and additional fermentation may solve the problem. Fig. 22 shows an acid

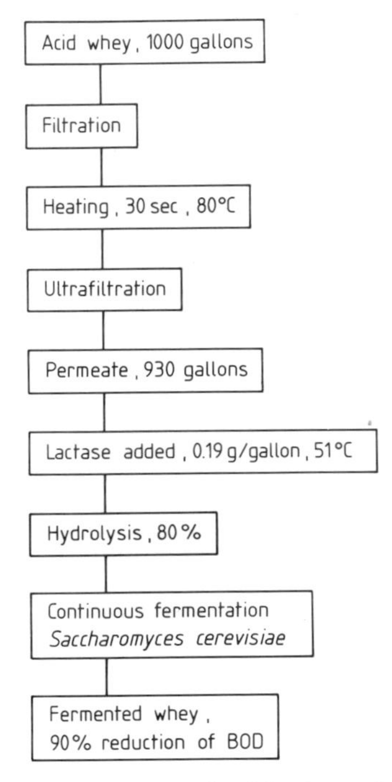

Figure 22. Treatment of acid whey with lactase and *Saccharomyces* for BOD reduction (STINEMAN et al., 1978).

whey treatment with lactase and *Saccharomyces cerevisiae*. A 90% BOD reduction was achieved (STINEMAN et al., 1978). The same results can be observed after whey treatment with immobilized lactase in a column reactor. Modern plants are able to run *Kluyveromyces fragilis* in a continuous fermenter using the Bel process. In a fermenter with a 23 m^3 working volume fermentation may run at a dilution rate of 0.25 h^{-1}, at 38 °C and pH 3.5. Strains with high lactose and ethanol tolerance have also been isolated (GAWEL and KOSIKOWSKI, 1978). Another method is a continuous lactic acid fermentation (*Lactobacillus bulgaricus*) followed by a continuous yeast fermentation (*Candida krusei*).

XI. Sucrose Conversion by Invertase

A. Introduction

Although glucose isomerization for fructose production has gained great importance, invert sugar still has its market in the production of sweets, jam, or ice cream. The hydrolysis of sucrose can be performed by acid treatment, by an ion exchange process, or by enzymatic hydrolysis. For about 10 years, large scale industrial production has used strong cationic resins in a continuous column process, where inversion of sucrose and chromatographic separation is accomplished in one step (LAUER, 1980).

Enzymatic inversion has the advantage of forming fewer by-products. But for a technical application it is disadvantageous that the free invertase of most strains show a severe substrate (sucrose) inhibition as was shown with *Candida utilis* (DICKENSHEETS et al., 1977). In addition, LOPEZ SANTIN et al. (1982) supposed a product inhibition from their experiments with free invertase (Merck) and the clay-immobilized enzyme.

B. Invertase Producing Microorganisms

Invertase (β-fructofuranosidase; β-D-fructofuranoside fructohydrolase : β-fructosidase; EC 3.2.1.26) is produced as a cell-bound or extracellular enzyme by many

Figure 23. The invertase reaction.

microorganisms (BARKER and SHIRLEY, 1980), such as:

Candida utilis
Pichia polymorpha
Saccharomyces sp.
  S.cerevisiae
  S.carlsbergensis
  S.fragilis
  S.pastorianus
Aspergillus awamori
  A.niger
Penicillium chrysogenum
  P.expansum
  P.frequentas
  P.vitale
Neurospora sp.
Actinomyces viscosus
Pullularia pullulans
Streptococcus mutans SL-1
Clostridium pasteurianum

Invertase cleaves sucrose into α-D-glucose and β-D-fructose (Fig. 23). In addition to sucrose, raffinose and stachyose are used as substrates.

In many strains invertase production is induced by sucrose. A strong inductive effect is shown by modified substrates, e. g., sucrose palmitate (REESE et al., 1969). In the case of *Pullularia pullulans* the invertase formation has been shown to increase 80-fold with this analog.

Higher glucose concentration (> 1% with *Saccharomyces cerevisiae*) during growth cause catabolite repression of invertase synthesis (ELORZA et al., 1977).

TODA (1976) found that invertase specific activity of *Saccharomyces carlsbergensis* in chemostat culture was increased to 12 times that in batch culture which was 1.2 U/mg. The continuous supply of feed medium to the fermenter, where very low concentrations of glucose were maintained, was necessary for sufficient invertase formation; maximal invertase specific activity was observed at an intermediate dilution rate of about 0.15 h^{-1} as can be seen in Fig. 24.

Furthermore, TODA (1976) succeeded in isolating a 2-deoxyglucose resistant mutant of *Saccharomyces carlsbergensis* which was resistant to catabolite repression by glucose. In chemostat fermentation the invertase specific activity of this mutant increased to 25 U/mg due to this regulatory effect.

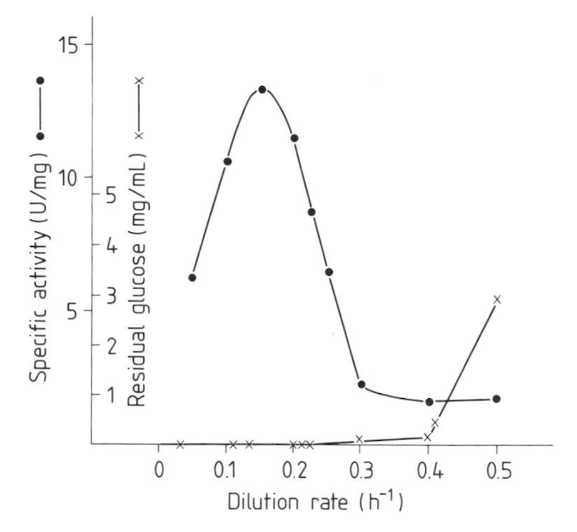

Figure 24. Specific activity of invertase (*Saccharomyces carlsbergensis*) and residual glucose concentration versus dilution rate (TODA, 1976).

C. Inversion Processes

For sucrose inversion many immobilized systems have been under investigation. TODA and SHODA (1975) reported on a continuously fed fluidized bed reactor where sucrose hydrolysis was carried out by cells of *Saccharomyces pastorianus* entrapped in agar pellets.

Active immobilized forms of invertase have been produced

- by entrapment in cellulose acetate and other fibers (DINELLI, 1972; MARCONI et al., 1974; MARCONI and MORISI, 1979).
- by radiopolymerization of acrylamide (KAWASHIMA and UMEDA, 1974).
- by cross-linking with glutaraldehyde on magnetite in the presence of sucrose (VAN LEEMPUTTEN and HORISBERGER, 1974).
- by immobilization in polysulfone hollow fiber cartridges (KORUS and OLSON, 1977).

Further immobilization techniques were described by BARKER (1980) and OOSHIMA et al. (1980).

Other suitable methods are immobilization procedures on porous supports which minimize the inhibitory effect of sucrose owing to diffusion limitations as found with yeast invertase entrapped in spun fibers of cellulose triacetate (MARCONI et al., 1974; MARCONI and MORISI, 1979). Fig. 25 shows the inversion efficiency of immobilized enzyme to decrease by an increase of the amount of entrapped invertase due to diffusional restriction of substrate molecules. The activity displayed by invertase fibers was 15–65% of that of the free enzyme depending on the porosity of fibers and on the amount of entrapped enzyme.

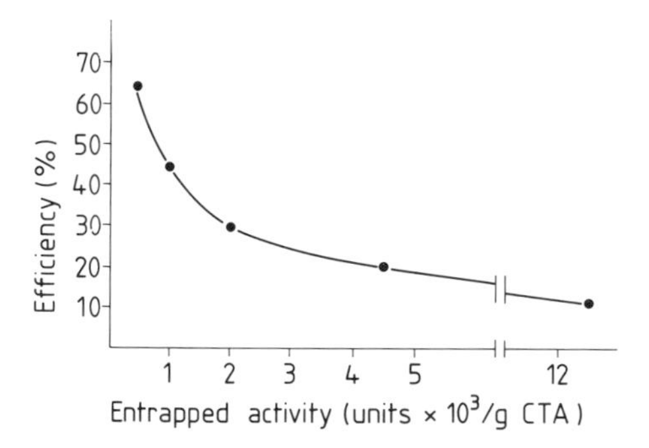

Figure 25. Activity of invertase fibers as a function of enzyme activity entrapped per g of cellulose triacetate (CTA). – 50 mL of a 20% (w/w) sucrose solution in 0.1 M sodium phosphate, at pH 4.5, were mixed with 1.0 g of fiber under shaking for 10 min at 25 °C. The activity of invertase fibers is expressed as the ratio of displayed to entrapped activity (MARCONI et al., 1974).

Under laboratory conditions, a glass column, packed with 90 g of fiber (total entrapped activity: 13 000 units), was continuously fed with a 20% sucrose solution in 0.1 M sodium phosphate, at pH 4.5 and 25 °C. After inversion to a total amount of 750 kg of sucrose the invertase fibers showed only a 10% decrease compared to their initial activity. With this technology a continuous process was developed where a 50% (w/w) sucrose solution was processed at a flow rate of about 6 L/h/kg of fiber. This process, which has a further advantage in reducing the risk of contamination due to the high sugar concentration, can compete with the traditional processes for the production of invert sugar syrups.

XII. References

AMOTZ, S., and THIESEN, N. O. (1973). Ger. Patent DP 2 317 680.

ANHEUSER-BUSCH INC. (1975). Brit. Patent 1 399 408.

ANTRIM, R. L., COLILLA, W., and SCHNYDER, B. J. (1979). Appl. Biochem. Bioeng. *2,* 97–155.

BARKER, S. A. (1976). Process Biochem. Dec., 20–21.

BARKER, S. A. (1980). In "Economic Microbiology", Vol. 5 (A. H. ROSE, ed.), pp. 331–367. Academic Press, London.

BARKER, S. A., and SHIRLEY, J. A. (1980). In "Economic Microbiology", Vol. 5 (A. H. ROSE, ed.), pp. 171–226. Academic Press, London.

BARKER, S. A., PELMORE, H., and SOMERS, P. J. (1983). Enzyme Microbiol. Technol. *5,* 121–124.

BENGTSON, B. L., and LAMM, W. R. (1974). Brit. Patent 1 368 511.

BRONN, W. K. (1976). Monatsschr. Brauerei *7,* 277.

BUCHTA, K. (1982). In "Handbuch der Biotechnologie" (P. PRÄVE, U. FAUST, W. SITTIG, and D. A. SUKATSCH, eds.), p. 354. Akademische Verlagsgesellschaft, Wiesbaden.

BUCKE, C. (1977). Top. Enzyme Ferment. Biotechnol. *1,* 147–171.

BUCKE, C., and CHEETHAM, P. S. J. (1979). Brit. Patent 7 938 563.

BUNGARD, S. J., REAGAN, R., RODGERS, P. J., and WYNCOLL, K. R. (1979). In "Immobilized Microbial Cells" (K. VENKATSUBRAMANIAN, ed.), pp. 139–146. American Chemical Society, Washington, DC.

CHEN, W.-P. (1980a). Process Biochem. June/July, 30–35.

CHEN, W.-P. (1980b). Process Biochem. August/Sept., 36–41.

CHO, Y. K., and BAILEY, J. E. (1977). Biotechnol. Bioeng. *19,* 185–198.

CLARK, L. C. (1970). U.S. Patent 3 539 455.

CPC INTERNATIONAL INC. (1975a). Brit. Patent 1 411 763.

CPC INTERNATIONAL INC. (1975b). Brit. Patent 1 411 764.

CPC INTERNATIONAL INC. (1975c). Brit. Patent 1 411 765.

CRUEGER, W., DRAHT, L., and MUNIR, M. (1977). Ger. Offenlegungsschrift 2 741 197.

D'ANGIURO, L., and CREMONESI, P. (1982). Biotechnol. Bioeng. *24,* 207–216.

DANNO, G., YOSHIMURA, S., and NATAKE, M. (1967). Agric. Biol. Chem. *31,* 284–292.

DELENTE, J., JOHNSON, J. H., KUO, M. J., O'CONNOR, R. J., and WEEKS, L. E. (1974). Biotechnol. Bioeng. *16,* 1 227–1 243.

DEMNEROVA, K., POSPISIL, S., BLUMAUEROVA, M., and KAS, J. (1979). Biotechnol. Lett. *1,* 299–304.

DICKENSHEETS, P. A., CHEN, L. F., and TSAO, G. T. (1977). Biotechnol. Bioeng. *19,* 365–375.

DINELLI, D. (1972). Process Biochem. July, 9–12.

D'SOUZA, S. F., and NADKARNI, G. B. (1980). Biotechnol. Bioeng. *22,* 2 179–2 189.

ELORZA, M. V., VILLANUEVA, J. R., and SENTANDREU, R. (1977). Biochim. Biophys. Acta *475,* 103–112.

ENFORS, S.-O. (1982). Appl. Biochem. Biotechnol. *7,* 113–119.

FLASCHEL, E., RAETZ, E., and RENKEN, A. (1982). Biotechnol. Bioeng. *24,* 2 499–2 518.

FLICKINGER, M. C., and PERLMAN, D. (1977). Appl. Environ. Microbiol. *33,* 706–712.

FULMER, E. I., DUNNING, J. W., GUYMON, J. F., and UNDERKOFLER, L. A. (1936). J. Am. Chem. Soc. *58,* 1 012–1 013.

GAWEL, J., and KOSIKOWSKI, F. V. (1978). J. Food Sci. *43,* 1 717–1 719.

GIST BROCADES (1979). Technical Data Sheet Mgi-01-01/79.02.En.05.

GONDO, S., MORISHITA, M., and OSAKI, T. (1980). Biotechnol. Bioeng. *22,* 1 287–1 291.

GONG, C.-S., CHEN, L. F., and TSAO, G. T. (1980). Biotechnol. Bioeng. *22,* 833–845.

GRAY, P. P., DUNNILL, P., and LILLY, M. D. (1972). In "Fermentation Technology Today" (G. TERUI, ed.), pp. 347–351, Society of Fermentation Technology, Tokyo.

GREEN, S. R. (1960). U.S. Patent 2 948 658.

GREENBERG, N. A., and MAHONEY, R. R. (1981). Process Biochem. Febr./March, 2–8, 49.

GRIFFITHS, M. W., MUIR, D. D., and PHILLIPS, J. D. (1979). UK Patent Application 2 022 595.

HEBEDA, R. E., and LEACH, H. W. (1975). U.S. Patent 3 922 201.

HERP, A., HORTON, D., and PIGMAN, W. (1970). In "The Carbohydrates, Vol. II A" (W. PIGMAN and D. HORTON, eds.), pp. 385–853. Academic Press, New York.

HIGGINS, J. J., LEWIS, D. J., DALY, W. H., MOSQUEIRA, F. G., DUNNILL, P., and LILLY, M. D. (1978). Biotechnol. Bioeng. *20,* 159–182.

HUGHES, R. E. (1983). Trends Biol. Sci. April, 146–147.

HUPKES, J. V., and VAN TILBERG, R. (1976). Stärke *28,* 356–360.

KARUBE, I., HIRANO, K.-I., and SUZUKI, S. (1977). Biotechnol. Bioeng. *19,* 1 233–1 238.

KATWA, L. C., and RAO, M. R. R. (1983). Biotechnol. Lett. *5,* 191–196.

KAWASHIMA, K., and UMEDA, K. (1974). Biotechnol. Bioeng. *16,* 609–621.

KELLER, H. W., REENTS, A. C., and LARAWAY, J. W. (1981). Starch *33,* 55–57.

KIERSTAN, M. (1980). Process Biochem. May, 2,3,32.

KIESLICH, K. (1976). "Microbial Transformations." Georg Thieme Verlag, Stuttgart.

KITA, D. A. (1979). U.S. Patent 4 155 812.

KITA, D. A., and HALL, K. E. (1979). Ger. Offen. (Published Patent Application) DOS 2 849 393.

KIUCHI, A., and TANAKA, Y. (1975). U.S. Patent 3 919 049.

KORUS, R. A., and OLSON, A. C. (1977). Biotechnol. Bioeng. *19,* 1–8.

KULHANEK, M. (1970). Adv. Appl. Microbiol. *12,* 11–33.

LAUER, K. (1980). Starch *32,* 11–13.
LEE, C. K., HAYES, L. E., and LONG, M. E. (1972). U.S. Patent 3 645 848.
LIEBSTER, J., LUKSIK, B., FÄRBER, G., and SVOBODA, V. (1956). Chem. Listy *50,* 395–397.
LINKO, P. (1981) In "Advances in Biotechnology", Vol. I (M. MOO-YOUNG, ed.), pp. 711–716. Proc. 6th Int. Ferm. Symp., London, Canada.
LOCKWOOD, L. B. (1954). In "Industrial Fermentations" (L. A. UNDERKOFLER, and R. J. HICKEY, eds.), pp. 1–23. Chemical Publishers, New York.
LONG, M. E. (1978). U.S. Patent Reissue 29 692.
LOPEZ SANTIN, J., SOLA, C., and LEMA, J. M. (1982). Biotechnol. Bioeng. *24,* 2 721–2 724.
MAKOVER, S., and PRUESS, D. L. (1975). U.S. Patent 3 907 639.
MARCONI, W., GULINELLI, S., and MORISI, F. (1974). Biotechnol. Bioeng. *16,* 501–511.
MARCONI, W., and MORISI, F. (1979). Appl. Biochem. Bioeng. *2,* 219–258.
MARSHALL, R. O., and KOOI, E. R. (1957). Science *125,* 648–649.
MARTIN, C. K. A., and PERLMAN, D. (1975). Biotechnol. Bioeng. *17,* 1 473–1 483.
MARTIN, C. K. A., and PERLMAN, D. (1976). Biotechnol. Bioeng. *18,* 217–237.
MAUCH, W., and SCHMIDT-BERG-LORENZ, S. (1964). Z. Zuckerind. *14,* 309–315, 375–383.
MEYRATH, J., and BAYER, K. (1979). In "Microbial Biomass" (A. H. ROSE, ed.), pp. 207–269. Academic Press, New York.
MIALL, L. M. (1978). In "Economic Microbiology", Vol. 2 (A. H. ROSE, ed.), pp. 47–119. Academic Press, London.
MILES LABORATORIES INC. (1972). Brit. Patent 1 280 396.
MILES LABORATORIES INC. (1974). Brit. Patent 1 376 787.
MILLER, J. J., and BRAND, J. C. (1980). Food Technol. Australia *32,* 144–147.
NAARDEN INTERNATIONAL N. V. (1976). Brit. Patent 1 456 262.
NARITA, S., NAGANISHI, H., IZUMI, C., YOKOUCHI, A., and YAMADA, M. (1975). U.S. Patent 3 867 256.
NARITA, S., NAGANISHI, H., YOKOUCHI, A., and KAGAYA, I. (1976). U.S. Patent 3 957 578.
NATAKE, M. (1966). Agric. Biol. Chem. *30,* 887–895.
NATAKE, M. (1968). Agric. Biol. Chem. *32,* 303–312.
NATAKE, M., and YOSHIMURA, S. (1963). Agric. Biol. Chem. *27,* 342–348.
NATAKE, M., and YOSHIMURA, S. (1964). Agric. Biol. Chem. *28,* 510–516.
NISIZAWA, K., and HASHIMOTO, Y. (1970). In "The Carbohydrates, Vol. II B" (W. PIGMAN and D. HORTON, eds.), pp. 241–300. Academic Press, New York.
NOVO (1976). Novo Enzyme Information IB 132a-GB.
OBARA, J., HASHIMOTO, S., and SUZUKI, H. (1976/1977). Sugar Technol. Rev. *4,* 209–258.
OESTERGAARD, J., and KNUDSEN, S. L. (1976). Stärke *28,* 350–356.
OOSHIMA, H., SAKIMOTO, M., and HARANO, Y. (1980). Biotechnol. Bioeng. *22,* 2 155–2 167.
PARK, Y. H., CHUNG, T. W., and HAN, M. H. (1980). Enzyme Microbiol. Technol. *2,* 227–233.
PASTORE, M., and MORISI, F. (1976). Methods Enzymol. *44,* 822–831.
PIFFERI, P. G., VACCARI, A., RICCI, G., POLI, G., and RUGGERI, O. (1982). Biotechnol. Bioeng. *24,* 2 155–2 165.
PITCHER, W. H., FORD, J. R., and WEETALL, H. H. (1976). Methods Enzymol. *44,* 792–809.
REESE, E. T., LOLA, J. E., and PARRISH, F. W. (1969). J. Bacteriol. *100,* 1 151–1 154.
REICHSTEIN, T., and GRÜSSNER, A. (1934). Helv. Chim. Acta *17,* 311–328.
REYNOLDS, R. J. (1973). Brit. Patent 1 328 980.
REYNOLDS, J. H. (1974). Biotechnol. Bioeng. *16,* 135–147.
ROELS, J. A., and VAN TILBERG, R. (1979). In "Immobilized Microbial Cells" (K. VENKATSUBRAMANIAN, ed.), pp. 147–172. American Chemical Society, Washington, DC.
RUGH, S. (1982). Appl. Biochem. Biotechnol. *7,* 27–29.
SCALLET, B. L., SHIEH, K., EHRENTAL, I., and SLAPSHAK, L. (1974). Starch *26,* 405–408.
SCHIWECK, H., STEINLE, G., and HABERL, L. (1972). Ger. Patent 2 217 628.
SHIEH, K. K. (1977). U.S. Patent 4 003 793.
SHIMIZU, J., and KAGA, T. (1972). U.S. Patent 3 664 927.
SHIMIZU, J., and KAGA, T. (1974). U.S. Patent 3 836 432.
SINGH, V., HSU, C. C., CHEN, D. C., and TZENG, C. H. (1983). Process Biochem. March/April, 13–17, 25.
SONAER, A. H., and ÇAĞLAR, M. A. (1981). 2. Eur. Congr. Biotechnol., Eastbourne, England, Abstracts of Communication, p. 157.
SONOYAMA, T., KAGEYAMA, B., and HONJO, T. (1975). U.S. Patent 3 922 194.
SONOYAMA, T., TANI, H., KAGEYAMA, B., KOBAYASHI, K., HONJO, T., and YAGI, S. (1976a). U.S. Patent 3 998 697.
SONOYAMA, T., TANI, H., KAGEYAMA, B., KOBAYASHI, K., HONJO, T., and YAGI, S. (1976b). U.S. Patent 3 959 076.

SPENCER, J. F. T., and GORIN, P. A. J. (1965). Prog. Ind. Microbiol. *7,* 177–220.

STINEMAN, T. L., EDWARDS, J. D., and GROSSKOPF, J. C. (1978). U.S. Patent 964 990.

STROSHANE, R. M., and PERLMAN, D. (1977). Biotechnol. Bioeng. *19,* 459–465.

SUEKANTE, M., TAMURA, M., and TOMIMURA, C. (1978). Agric. Biol. Chem. *42,* 909–917.

SUEKANTE, M., and IIZUKA, H. (1981) Z. Allg. Mikrobiol. *21,* 457–468.

SUZUKI, H., OZAWA, Y., OOTA, H., and YOSHIDA, H. (1969). Agric. Biol. Chem. *33,* 506–513.

SUZUKI, S., SATOH, I., and KARUBE, I. (1982). Appl. Biochem. Biotechnol. *7,* 147–155.

TAKASAKI, Y., and TANABE, O. (1962). Hakko Kyokaishi *20,* 449–455, Chem. Abstr. *60,* 806f (1964).

TAKASAKI, Y., and TANABE, O. (1963). Kogyo Gijutsuin, Hakko Kenkyusho Kenkyu Hokuku *23,* 41–97; Chem. Abstr. *60,* 14 860c (1964).

TODA, K. (1976). Biotechnol. Bioeng. *18,* 1 103–1 115.

TODA, K., and SHODA, M. (1975). Biotechnol. Bioeng. *17,* 481–497.

TRAMPER, J., LUYBEN, K. Ch. A. M., and VAN DEN TWEEL, W. J. J. (1983). Eur. J. Appl. Microbiol. Biotechnol. *17,* 13–18.

VAHERI, M., and KAUPPINEN, V. (1977). Process Biochem. July/August, 5–8.

VAN LEEMPUTTEN, E., and HORISBERGER, M. (1974). Biotechnol. Bioeng. *16,* 385–396.

WALON, R. G. P. (1977). U.S. Patent 4 009 074.

WATANABE, M., NARITA, S., KAGAYA, I, MIURA, M., SUZUKI, Y., TANAKA, S., SASAKI, M., MIYATANI, S., and TERAYAMA, K. (1975). U.S. Patent 3 894 913.

WELLS, P. A., STUBBS, J. J., LOCKWOOD, L. B., and ROE, E. T. (1937). Ind. Eng. Chem. *29,* 1 385–1 388.

WOVCHA, M. G., STEUERWALD, D. L., and BROOKS, K. E. (1983). Appl. Environ. Microbiol. *45,* 1 402–1 404.

YAMADA, K. (1977). Biotechnol. Bioeng. *19,* 1 563–1 621.

YAMANAKA, K. (1962). Agric. Biol. Chem. *26.* 167–174.

YAMANAKA, K. (1963a). Agric. Biol. Chem. *27,* 265–270.

YAMANAKA, K. (1963b). Agric. Biol. Chem. *27.* 271–278.

YAMANAKA, K. (1965). Japan. Patent 20 230/65.

YAMANAKA, K. (1968). Biochim. Biophys. Acta *151,* 670–680.

YOSHIMURA, S., DANNO, G., and NATAKE, M. (1966). Agric. Biol. Chem. *30,* 1 015–1 023.

ZITTAN, L., POULSEN, P. B., and HEMMINGSEN, S.-H. (1975). Stärke *27,* 236–241.

Index

A

Absidia coerulea alicyclic compounds hydroxylation 160
Absidia hyalospora ionone esters racemate resolution 114
Acetobacter suboxydans glucose oxidase formation 439
— oxidation of cyclane polyols 146
— shikimic acid bioconversion 174
— sorbitol dehydrogenase of 436
Acetobacter xylinum sorbitol dehydrogenase of 436
Acetylhydroxyproline use of 412
Achromobacter ampicillin formation 265
— degradation of phenols 305
— degradation of pyridinium derivatives 318
— microbial degradation of chlorinated biphenyls 312
Achromobacter liquidum cell immobilization 13
Achromobacter lyticus serine protease 407
Acinetobacter lipid content of alkane grown 366
Acinetobacter calcoaceticus aldehyde dehydrogenases 354
— alkane monooxygenase 347
Acremonium chrysogenum as antibiotic-blocked mutant 247
— penicillin N formation 251
Acronycine aromatic hydroxylation 226
Actinomycin 243 f
Actinoplanes missouriensis cell immobilization 13
— glucose isomerase of 428
Acylation of antibiotics 256 f
Adamantanones microbial reduction 142
Adriamycin 251
Aerobacter aerogenes degradation of pyridinium derivatives 318
Ajmaline bioconversions 218
Alanine formation by bioconversion 390
Alanine racemase selective inactivation 23
Alcaligenes microbial degradation of chlorinated biphenyls 312
— microbial degradation of phthalates 309
Alcaligenes faecalis alicyclic compounds hydroxylation 152
— aniline degradation 286
Alcohol dehydrogenases from alkane-assimilating microorganisms 353
Alcohol oxidations of steroids 43
Alcohols alicyclic oxidation 146
Aldehyde dehydrogenases from alkane-assimilating microorganisms 354
— substrate specificity 342
Aldrin biodegradation 182
Alicyclic compounds β-oxidation 164
— hydrolytic reactions 148
— hydroxylations 151
— microbial aromatization 167
— microbial degradation 164
— microbial transformations 127 ff
— oxidation of alcohols 146
— reduction of carbon-carbon double bonds 144
— stereochemistry of reduction 139
Aliheterocyclic compounds microbial hydroxylation 162
Aliphatic hydrocarbons 329 ff
Alkaloids microbial transformations 207 ff
— — type-reactions 209
Alkane dioic acids production from alkanes 374
Alkane monooxygenase cytochrome-linked 346
— rubredoxin-linked 347
Alkane oxidation dehydrogenation 349
— diterminal 354
— genetic control 369
— hydroperoxide mechanism 350
— hydroxylation 345
— monoterminal 350
— subterminal 357
Alkanes branched-chain biotransformations 344

— — microbial transformation 360
Alkanes long-chain biotransformations 344
— — microbial co-oxidation 361
— — terminal pathways 352
— microemulsions 334
— oxidation by methylotrophs 340
— short-chain transformation of 337 ff
— solubility 334
— uptake mechanism 333
Alkanoic acids production from hydrocarbons 373
Alkanols production from hydrocarbons 370
Alkene formation in alkane degradation 349
Alkenes biotransformations 344
— epoxidation by methylotrophs 341
— oxidation by methylotrophs 340
— oxidation of 359
Alkyl oxides production from hydrocarbons 370
Allantoin microbial hydrolysis 150
Alternaria kikuchiana aminoacylase 398
Amido acids production volume 410
Amikacin 250, 268
Amination of antibiotics 263
Amino acids aromatic biosynthesis of 174
— biotransformation 387 ff
— food use 411
— pharmaceutical use 412
— price/volume relationship 411
— production facilities 412
L-Amino acids synthesis from α-keto acids by bioconversion 392
Aminoacylase immobilization of 15
Aminoglutaric acid formation by bioconversion 399
Aminoglycosides 267 ff
— formation of analogs 252
6-Aminopenicillanic acid 241, 253
Amoxycillin 265
Amphetamine oxidative mammalian metabolites 210
Ampicillin 242, 265
Angiotensin enzymatic synthesis 405
Anilines microbial degradation 306
Ansamycin 251
Anthracene biotransformation of 288
Anthracene analog microbial reduction 139
Anthranilic acid biotransformation of 289
Antibiotics 239 ff
— acylation 256 f
— amination 263
— biotransformations of 252 ff
— deamination 263
— demethylation 264
— directed biosynthesis 241 ff
— genetic manipulation 251
— glycosidation 263
— hybrid biosynthesis 246
— hydration 264
— hydrolysis 254 f
— idiotrophs in formation of 247
— isomerization 264
— metabolic inhibitors for biotransformation 245
— methylation 264
— nucleotidylation 259
— oxidation 260 f
— phosphorylation 258
— precursors in fermentation media 241
— produced by antibiotic-blocked mutants 247
— production volume 240
— reduction 262
6-APA 253, 265
— formation of semi-synthetic penicillins 253
Apomorphine microbial transformations 219
Aporphines bioconversions 219
Arginomonas non-fermentans menthyl ester asymmetric hydrolysis 109
Aromatic compounds alkamino groups biotransformation of 285
— alkoxy substituents biotransformation of 284
— alkyl substituents biotransformation of 284
— amides biotransformation of 285
— amino substituents biotransformation of 286
— anaerobic degradation 293
— anaerobic photometabolism 293
— biotransformations 277 ff
— carboxyl substituents biotransformation of 284
— entry into the cell 284
— esters biotransformation of 285
— fatty acid side chains biotransformation of 285
— halogen substituents biotransformation of 286
— metabolism common pathways 283
— nitriles biotransformation of 285
— nitro substituents biotransformation of 286
— ring cleavage 289
— ring fission substrates 287
— side chain transformation 284
— sulfonic acid substituents biotransformation of 286
— thioalkyl substituents biotransformation of 285
— polynuclear microbial oxidation 324
Arthrobacter branched-chain alkane transformation 360
— degradation of phenols 305
— degradation of pyridinium derivatives 318
— degradation of triazine derivatives 321

— *globiformis* ephedrine degradation 234
— glucose isomerase of 428
— microbial degradation of phthalates 308
— phenolic pesticide degradation 303
— squalene transformation 122
Arthrobacter colchovorum colchicine bioconversion 228
Arthrobacter hydrocarboglutamicus microbial degradation of phthalates 308
Arthrobacter simplex cell immobilization 13
— dehydrogenase 14, 69
— fatty acid oxidation 363
— in steroid dehydrogenations 67
— in steroid reductions 49
— microbial degradation of biphenyl 312
— sterol side chain cleavage 83, 91
Ascorbic acid production 436
Ascorbic acid formation bioconversion steps 437
Aspartame stereospecific enzymatic synthesis 402
— use of 413
Aspartic acid formation by bioconversion 396
Aspartyl phenylalanine *see* Aspartame
Aspergillus alliaceus papaverine dealkylation 223
— pergolide bioconversion 215
Aspergillus awamori invertase 453
Aspergillus flavipes glaucine oxidation 221
Aspergillus fumigatus degradation of triazine derivatives 320
Aspergillus nidulans degradation of dibenzofuran 317
Aspergillus niger alicyclic compounds hydroxylation 152, 160
— degradation of furan derivatives 316
— glucose oxidase formation 439
— hydroxylation of β-ionone 112
— immobilized cells 440
— lactase 448
— microbial biotin synthesis 176
— pinene transformation 110
— prostaglandin synthon bioconversion 185
— steroid hydroxylase 63
— transformation of damascone 118
Aspergillus ochraceus in steroid hydroxylations 55
— steroid hydroxylase 63
Aspergillus oryzae aminoacylase 7
— lactase 448
Astaxanthin 115
— synthesis of 114
Atropine bioconversion 233
Azo dyes microbial degradation 311
Azotobacter phenolic pesticide degradation 303

B

Bacillus co-oxidation of long-chain alkanes 362
Bacillus alvei tropine bioconversion 233
Bacillus brevis tyrocidine formation 243
Bacillus cereus azo dyes microbial degradation 311
— nabilone transformation 198
Bacillus circulans amikacin formation 250
— as antibiotic-blocked mutant 247
— butirosin formation 248
Bacillus coagulans glucose isomerase of 428
Bacillus firmus degradation of anilines 306
Bacillus lentus in steroid dehydrogenations 67
— in steroid reductions 49
Bacillus megaterium ampicillin formation 265
— formation of 6-APA 265
— steroid hydroxylase 63
Bacillus sphaericus degradation of furan derivatives 316
— degradation of oxathiine derivatives 319
— dehydrogenase 68
— in steroid dehydrogenations 67
— L-leucine dehydrogenase 392
Bacillus stearothermophilus α-galactosidase formation 445
— cell immobilization 451
Bacillus subtilis alicyclic compounds hydrolysis of esters of 149
— menthyl ester asymmetric hydrolysis 109
— recombinant DNA technology 22
— Subtilysin formation 334
Bacteria hydrocarbon-degrading 332
Bacterium cyclooxydans dehydrogenase 68
— in steroid dehydrogenations 67
Baeyer-Villiger oxidation 101
Baeyer-Villiger reaction 164
Beauveria sulfurescens alicyclic compounds hydroxylation 152
— reduction of unsaturated cyclenones 145
Beijerinckia degradation of dibenzofuran 317
— degradation of dioxin derivatives 319
— microbial degradation of polycyclic hydrocarbons 313
Benzaldehyde biotransformation of 288
Benzene biotransformation of 288
Benzimidazole derivatives microbial degradation 316
Benzo[*a*]anthracene microbial degradation 313
Benzo[*a*]pyrene microbial degradation 313
Benzoic acid anaerobic metabolism 294
— biotransformation of 288
— ring cleavage during methane fermentation 296
Benzthiazole derivatives microbial degradation 317
Benzylpenicillin hydrolysis 252

— production by directed fermentation 241
Bicyclodecane compounds reduction of carbonyl functions 130
Bile acids use in steroid transformations 40
Biotin microbial degradation 178
— preparation by microbial transformation 323
Biotin synthesis 174
Biotransformation applications 6
— biphasic reaction systems 8
— methodology 5ff
— multiphase systems 16
— process design 8
— processes 10ff
— product isolation 24
— reaction types 6
— sequential by different microorganisms 18
— side reactions elimination of 22
— with growing cultures 10
— with immobilized cells 12
— with previously grown cells 11
— with purified enzymes 14
— with spores 11
— with water-immiscible organic solvents 17
Biphenyl compounds microbial degradation 312
Bleomycin 242
Bleomycinic acid 243
Botryosphaeria rhodina transformation of damascone 118
Botrytis cineria tomatine hydrolysis 230
Brevibacterium erythrogenes branched-chain alkane transformation 360
Brevibacterium flavum cell immobilization 13
Brevibacterium imperiale alicyclic compounds hydrolysis of esters of 150
Brevibacterium lipolyticum sterol side chain cleavage 91
Brevicarine bioconversions 217
Bromelin peptide synthesis 400
Brucine bioconversions 218
Bufanolids microbial transformation of 86
Butirosin 249
Byssochlamys fulva 128

C

Caffeic acid 285
Calonectria decora alicyclic compounds hydroxylation 160
Camphor 104
Camptothecin microbial hydroxylation 226
Candida cloacae formation of dioic acids 355
Candida guilliermondii formation of dioic acids 356
Candida lipolytica acyl-CoA synthetase 363
— alkene oxidation 359
— fatty acid degradation 363
— fatty acid oxidation 363
— lipid content of alkane grown 366
— methylcitric acid cycle 364
Candida reukaufii citronellal transformations 102
Candida rugosa formation of dioic acids 356
Candida tropicalis alcohol dehydrogenases 351
— alkane monooxygenase 346
— formation of dioic acids 354
— lipid content of alkane grown 366
Candida utilis invertase 453
Cannabidiol microbial oxidation 194
Cannabinoids bioconversions 191
— microbial nitration 198
Cannabinol microbial oxidation 194
Carbenicillin 242
Carboline alkaloids bioconversions 217
Carbomycin 247
Carbon-carbon bond scissions in steroids 44
Carbon-carbon double bonds microbial reduction 144
Carboxypeptidase A insulin bioconversion by 407
Carboxypeptidase Y enkephalin synthesis 404
— human insulin synthesis by 408
— transpeptidation of insulin chain 406
Cardenolids microbial transformation of 86
Carvone 106
Catechol meta-fission pathways 291
— ortho-fission pathways 289
— ring cleavage 290ff
— ring cleavage during methane fermentation 296
Catharanthus roseus alkaloids microbial transformations 210
Cathepsin peptide synthesis 400
Celesticetin 244
Cell immobilization methods 12
Cephalosporin formation of analogs 252
Cephalosporin C 266
Cerulenin 246, 365
Chaetomium globosum tetrahydrocannabinol transformation 198
Chiral center introduction of 7
Chlorella co-oxidation of long-chain alkanes 362
Cholesterol microbial conversion of 88
— microbial side chain cleavage 81
— use in steroid transformations 40
Chorismic acid microbial transformation 174
Chymotrypsin dynorphin synthesis 403
— enkephalin synthesis 404

— peptide synthesis 401
Cineole microbial transformation 111
Cinnamic acid ring cleavage during methane fermentation 296
Cirramycin 251
Citral microbial transformation 103
Citronellal microbial transformation 102
Citronellol 102
Cladosporium limonene transformation 107
Cladosporium cladosporioides alkanes subterminal oxidation of 358
— morphine bioconversion 228
Cladosporium resinae alcohol dehydrogenases 351
— alkane monooxygenase 347
Claviceps purpurea clavine alkaloids bioconversions 215
Clavine alkaloids bioconversions 215
Clostridium reductive dechlorination of hydrocarbons 296
Cloxacillin 242, 253
Codeine bioconversions 227
Codeinone bioconversions 227
Coenzymes regeneration 15
Colchiceinamide bioconversions 229
— homologs metabolism of 230
Colchicine alkaloids microbial transformations 228
Conessine microbial hydroxylation 231
Conjugations in steroids 47
Cornsteep in media for antibiotic production 241
Coronaridine bioconversions 218
Cortexone use in steroid transformations 35
Corticosterone as product of steroid hydroxylations 35
Cortisone production by steroid transformation 51
Corynebacterium alkane monooxygenase 346
— branched-chain alkane transformation 360
Corynebacterium cyclohexanicum alicyclic compounds aromatization 167
— alicyclic compounds hydroxylation 154
Corynebacterium equi sterol side chain cleavage 91
Corynebacterium hydrocarboclastum limonene transformation 107
Corynebacterium mediolanum in steroid alcohol dehydrogenation 49
Corynebacterium petrophilum microbial degradation of phthalates 308
Corynebacterium primorioxydans microbial biotin synthesis 174
Corynebacterium simplex cell immobilization 13
— sterol side chain cleavage 83
Cresol biotransformation of 289
Crustecdysone side chain cleavage selective 83
Cryptococcus laurentii amino caprolactamase induction of 21
— amino-ε-caprolactamase 395
Cryptococcus macerans reduction of tetralones 134
Culture collections 52
Cunninghamella bainieri alicyclic compounds hydroxylation 161
Cunninghamella blakesleeana prostaglandin degradation 189
— Rauwolfia alkaloids microbial hydroxylation 216
— reduction of unsaturated cyclenones 146
— tetrahydrocannabinol transformation 198
Cunninghamella echinulata degradation of benzimidazole 317
— degradation of thiazole derivatives 317
— degradation of triazine derivatives 321
— ergoline lergotrile bioconversions 215
— papaverine dealkylation 223
Cunninghamella elegans apomorphine bioconversion 220
— degradation of dibenzofuran 317
Curvularia falcata dioxodecaline reduction 132
Curvularia lunata cell immobilization 13
— in steroid hydroxylations 49, 57
— ketone reduction 128
— reduction of unsaturated cyclenones 146
— steroid hydroxylase 63
Cyclenones unsaturated microbial reduction 144
Cycloheptane microbial hydroxylation 152
Cycloheptanol microbial oxidation 146
Cyclohexane β-oxidation 164
— microbial hydroxylation 151f
Cyclohexanone microbial reduction 128
Cyclohexenes microbial hydroxylation 157
Cyclopentane microbial hydroxylation 151
Cyclopentanol microbial oxidation 146
Cyclopentanone microbial reduction 128
Cyclopropane microbial hydroxylation 151
Cyclosterols side chain cleavage selective 83
Cylindrocarpon radicicola in steroid lactonization 49, 70
Cysteine formation by bioconversion 397
— use of 412
Cytidine microbial degradation 301
Cytochrome c 346

D

β-Damascone microbial transformation 117
Daunomycin 247
Deamination of antibiotics 263

Decalones heterocyclic analogs reduction of 134
Dehydrogenation of steroids by dehydrogenases 67f
— by dried vegetative cells 68
— spore processes 68
— with immobilized cells 69
— with immobilized enzymes 69
Demethylation of antibiotics 264
2-Deoxystreptamine 249
Desosaminyl protylonolide 247
Diazepines biodegradation 182
Dibekacin 268
Dieldrin biodegradation 182
Dihydroxyacetone bioconversion from glycerol 442
2,5-Diketo-D-gluconate bioconversion from 2-keto-D-gluconate 438
Diosgenin use in steroid transformations 40
Dioxolane microbial reduction 146
Dipodascus albidus prostaglandin synthons racemate resolution of 184
Dipodascus uninucleatus prostaglandin synthon bioconversion 184
L-DOPA racemate resolution 393
— use of 412
2-DOS 249
Double bond introductions in steroids 44
Dynorphin enzymatic synthesis 404

E

Ellipticine bioconversions 217
Emulsifying agents microbial 334
Enkephalin enzymatic synthesis 404
Enterobacter cloacae cysteine desulfhydrase activity 397
Enzyme immobilization 15
Enzyme inactivation selective 23
Enzyme inhibitors use in sterol transformation 87
Enzymes constitutive 19
— inducible 19
— induction 21
Ephedrine microbial degradation 234
Epoxidations of steroids 43
Equilin synthesis of 45
Ergoline alkaloids bioconversions 215
Erwinia herbicola β-tyrosinase 393
Erwinia rhapontici glucotransferase 443
Erythromycin 251
Escherichia coli aspartase 396
— cell immobilization 12
— degradation of anilines 307
— degradation of triazole derivatives 317
— formation of 6-APA 265
— glutamylcysteine synthetase 403
— glutathione synthetase 403
— lactase 448
— penicillin G acylase 400
— recombinant DNA technology 22
— tryptophanase 393
Estradiol formation from prochiral intermediates 71
— microbial synthesis 168
Estradiol analogs synthons microbial reduction 169
Estradiol synthon microbial reduction 168
Eubacterium sp. sterol hydrogenation by 94

F

Fatty acid metabolism in alkane-utilizing yeasts 365
Fatty acids α-oxidation 363
— β-oxidation 362
Ferulic acid 285
— ring cleavage during methane fermentation 296
Flavin coenzymes regeneration 16
Flavobacterium azo dyes microbial degradation 311
— degradation of phenols 304
Flavobacterium dehydrogenans in steroid alcohol dehydrogenation 49
Flexibacterium phenolic pesticide degradation 303
Fluoroacetic acid enzyme inhibition 287
Fluorohydrocortisone production by steroid transformation 51
Formaldehyde oxidation 342
Formate oxidation 342
Fructose production by glucose isomerase 423
Fructose syrup production 433
— production volume 434
Fumarase selective inactivation 23
Fungi hydrocarbon-degrading 332
Fungicides microbial degradation 306, 316, 319
Furan microbial degradation of 298
Furan-2-carboxylic acid microbial hydroxylation of 298
Furan derivatives microbial degradation 316
Fusarium lini alkanes subterminal oxidation of 358
Fusarium oxysporum degradation of anilines 306
Fusarium roseum degradation of triazine derivatives 320
Fusarium solani glaucine bioconversion 220
— phenolic pesticide degradation 303
— sterol side chain cleavage 83

G

α-Galactosidase microbial sources 444
β-Galactosidase *see* Lactase
Gallic acid microbial ring cleavage 293
Genetic control of alkane oxidation 369
Genetic engineering for biodegradation of aromatic compounds 283
— for biodegradation of heterocyclic compounds 283
Genetic strain improvement mutation 20
— recombinant DNA technology 20
— selection techniques 20
Gentamicin 249, 267
Gentisic acid biotransformation of 289
— microbial ring cleavage 290, 293
Geotrichum candidum reduction of dioxolane 146
Geranic acid 103
Glaucine bioconversions 220
— oxidation by *Fusarium solani* 222
Glucoamylase immobilization of 15
Gluconic acid production from glucose 439
— production volume 441
D-Gluconic acid bioconversion from glucose 438
Gluconobacter oxydans immobilized cells 440
δ-Gluconolactone production volume 441
Glucose isomerase co-factors 433
— commercial preparations properties 433
— fermentation conditions for production of 430
— for fructose production 427
— immobilization of 15, 431
— immobilized cell systems 431
— microbial sources 427
Glucose oxidase applications 441
— immobilization 440
Glutamic acid production by fermentation 396
Glutathione 403
Glycine use of 413
Glycosidation of antibiotics 263
Gyrochiral compounds 141

H

Halocatechols dehalogenation 287
Hecogenin use in steroid transformations 40
Heliotrine bioconversion 233
Herbicide biodegradation 182
Herbicides microbial degradation 303, 306, 317f, 320
Heteroatom introductions in steroids 48
Heteroatom oxidations of steroids 46
Heterocyclic compounds biotransformations 277ff
— degradation of six-membered rings 318
— five-membered rings microbial degradation 298
— metabolism common reactions 297
— microbial transformations 127ff
— six-membered rings microbial degradation 299
Hexahydroindan derivatives by microbial degradation of sterols 92
Hexahydroindene derivatives microbial degradation 172
Histidine microbial ring cleavage 299
Homoannulation in steroids 47
Homogentisic acid microbial ring cleavage 293
— ring cleavage 290
Homoprotocatechuic acid ring cleavage 290
Hybrid biosynthesis of antibiotics 246
Hydantoinase from *Agrobacterium radiobacter* 398
Hydantoins racemate resolution 399
Hydration of antibiotics 264
Hydrocarbon compounds polycyclic microbial degradation 312
Hydrocarbons aliphatic industrial applications 370
— — microbial transformations 329ff
— *see also* Aliphatic hydrocarbons
— uptake mechanism 336
Hydrocortisone as product of steroid hydroxylations 35
— as product of steroid transformations 36, 51
Hydrolysis of antibiotics 254f
— of steroid fatty acyl esters 48
Hydroxycholestenone side chain cleavage selective 83
Hydroxylase steroid ring fission 86
— of *Streptomyces roseochromogenes* 214
Hydroxylation of steroids 42
— 11α- of steroids 55
— 11β- of steroids 57
— 16α- of steroids 57
Hydroxylation of steroids by immobilized vegetative cells 60
— by resting vegetative cells 59
— side reactions prevention of 64
— spore processes 61
Hydroxyprogesterone as product of steroid hydroxylations 35
Hydroxysterols side chain cleavage selective 83
Hydroxy tryptophan formation by bioconversion 394
Hypocrea Cf.pilulifera degradation of thiazole derivatives 317

I

Idiotrophs antibiotics of 247
Immobilization of β-tyrosinase 393
— of cells 12
Immobilized cells conversion of sterols 93
— of *Aspergillus niger* 440
— of *Gluconobacter oxydans* 440
Indenone microbial reduction 137
Indole alkaloids microbial transformations 210
Industrial pollutants microbial degradation 302
Insecticides microbial degradation 304, 316, 321
Insect molting hormones steroid precursors side chain cleavage 85
Insulin amino acid composition bovine 405
— — human 405
— — porcine 405
Insulin human formation by bioconversion 405 ff
Insulin analogs formation by bioconversion 407
Invertase immobilization 454
— microbial sources 452
α-Ionone microbial transformation 115
β-Ionone microbial transformation 112
Iremycin 251
Isoleucine formation by bioconversion 391
Isomaltulose bioconversion from sucrose 442
Isomerization of antibiotics 264
— of steroids 47
Isophorone microbial hydroxylation 157
Isoquinoline alkaloids microbial transformations 219

K

Kanamycin 249, 267 f
2-Keto-L-gulonic acid bioconversion from calcium gluconate 438
— bioconversion from sorbosone 438
Ketones microbial reduction 128
— production from hydrocarbons 372
Klebsiella pneumoniae degradation of triazine derivatives 320
Kloeckera jensenii prostaglandin synthon bioconversion 186
Kluyvera citrophila ampicillin formation 266
— formation of 6-APA 265
Kluyveromyces fragilis invertase 453
— lactase 448
Kluyveromyces lactis lactase 448
Kojic acid bioconversion from glucose 441

L

Laccase of *Polyporus versicolor* 219
— of *Polyporus* anceps 212
β-Lactams 252 ff
— hydrolysis 267
Lactase commercial preparations 449
— immobilization of 15, 45
— production from *Aspergillus oryzae* 449
— production from *Bacillus stearothermophilus* 450
— production from *Escherichia coli* 449
— properties 449
Lactobacillus pastorianus quinic acid bioconversion 174
Lactonization of steroids 70
Lactose hydrolysis by lactase 447
— whey as source of 447
Lasiodiplodia theobromae 101
— β-ionone transformation 113
— transformation of damascone 118
Laudanosine microbial dealkylation 224
Leucine formation by bioconversion 391
Leucomycin 246
Ligno-aromatic compounds ring cleavage during methane fermentation 296
Lignosulfonates microbial degradation 303
Limonene degradation pathways 105
Linalool microbial transformation 103
Lincomycin 244, 246
Lindane biodegradation of 181
Lipase *Candida cylindracea* 222
Lipids of alkane grown microbes 365
Lipo-peptide microbial production of 335
Lupanine bioconversion 233
Lysergic acid amide derivatives bioconversions 213
Lysine formation by bioconversion 394
L-Lysine chemoenzymatic synthesis 396

M

Makisterone side chain cleavage selective 83
Mammalian metabolism microbial models 209
Mandelic acid biotransformation of 288
Mannosidostreptomycin 269
Menthol 102
— racemate resolution 108
Methane fermentation degradation of aromatic compounds 295
Methane monooxygenase 339
Methane oxidation metabolic pathways 338
Methanol dehydrogenase 341
Methanotrophs properties 338
— species 338
Methicillin 242, 253

Methionine formation by bioconversion 397
— racemate resolution 398
Methylation of antibiotics 264
Methyl migration in steroids 47
Methylobacterium epoxidation by 341
— methane monooxygenase 343
Methylobacterium ethanolicum 368
Methylobacterium organophilum 368
— methane monooxygenase 339
Methylococcus methane monooxygenase 343
Methylococcus capsulatus alicyclic compounds hydroxylation 151
Methylomonas methane monooxygenase 343
Methylomonas methanica methane monooxygenase 339
Methylophilus methylotrophus methane oxidation regulation of 368
Methylosinus epoxidation by 341
— methane monooxygenase 343
Methylosinus trichosporium methane monooxygenase 339
— methane oxidation regulation of 368
Methylotrophic microbes genetics 367
Methylotrophs methane oxidation regulation of 368
Microascus trigonosporus prostaglandin degradation 191
Micrococcus microbial degradation of phthalates 308
Micrococcus agilis aminoacylase 398
Micromonospora actinomycin formation 243
Micromonospora inyoensis sisomicin formation 248
Micromonospora polytrota mycinamicin formation 249
Micromonospora purpurea gentamicin formation 248
Micromonospora sagamiensis sagamicin formation 248
Microorganisms in mixed cultures degradation of aromatic compounds 282
— degradation of heterocyclic compounds 282
Microsporum gypseum nicotine demethylation 232
Moraxella lwoffii catechol dioxygenase 291
Morphine alkaloids microbial transformations 226
Mortierella isabellina alkanes subterminal oxidation of 358
Mortierella vinacea α-galactosidase formation 445
— cell immobilization 445
Mutasynthesis antibiotics of 247 ff
Mutational biosynthesis 8
Mycobacterium diernhoferi sterol side chain cleavage 91
Mycobacterium fortuitum sterol side chain cleavage 91
Mycobacterium globiforme dehydrogenase 69
— in steroid dehydrogenations 67
Mycobacterium parafortuitum sterol side chain cleavage 91
Mycobacterium phlei microbial degradation of phthalates 308
— sterol degradation by immobilized cells of 94
— sterol side chain cleavage 83
Mycobacterium rhodochrous cannabinol degradation 194
— prostaglandin degradation 190
— tetrahydrocannabinol degradation 196
Mycobacterium smegmatis alicyclic compounds hydroxylation 157
— hexahydroindene derivatives degradation of 172
Mycobacterium vaccae alkene oxidation 360
— sterol side chain cleavage 91
Mycotorula japonica lipid content of alkane grown 366
Myrtenol 111

N

Nabilone microbial transformation 198
Nafcillin 242, 253
Naphthalene biotransformation of 288
Naphthyridine microbial hydroxylation 323
Neomycin 250, 268
Nicotinamide coenzymes regeneration 16
Nicotine microbial transformations 232
Nicotinic acid microbial degradation 300
— microbial hydroxylation of 298
Nocardia degradation of anilines 306
— degradation of pyridinium derivatives 318
Nocardia corallina sterol hydrogenation by 94
Nocardia erythropolis microbial degradation of phthalates 308
— sterol degradation by immobilized cells of 94
Nocardia globerula alicyclic compounds microbial degradation 164
Nocardia mediterranei ansamycin formation 251
— as antibiotic-blocked mutant 247
— rifamycin formation 269
Nocardia opaca sterol hydrogenation by 94
Nocardia petrophila alicyclic compounds hydroxylation 152
Nocardia restricta degradation of phenols 305
— dehydrogenase 68
— hexahydroindene derivatives degradation of 172

— steroid hydroxylase 63
— sterol side chain cleavage 83
Nocardia rhodochrous cell immobilization 13
— sterol degradation by immobilized cells of 94
Nocardia rubra sterol side chain cleavage 83
Nocardia salmonicolor alicyclic compounds hydroxylation 155
— alkene oxidation 360
— nabilone transformation 198
Norpatchoulenol 120
Norsterols side chain cleavage selective 83
Novobiocin 244, 250
Nuciferine microbial dealkylation 223
Nucleotidylation of antibiotics 259

O

Oleandomycin 251
Oleuropeic acid 104
Oszillatoria microbial degradation of biphenyl 312
Oxacillin 242
Oxidation of antibiotics 260f
Oxidosterols side chain cleavage selective 84
Oxo-isophorone microbial transformation 116
Oxytetracycline 244

P

Paecilomyces carneus hydroxylation of patchoulol 120
Papain angiotenson synthesis 405
— dynorphin synthesis 403
— enkephalin synthesis 404
— peptide synthesis 400
Papaverine microbial dealkylation 223
Paromomycin 249
Patchoulol microbial transformation 119
Pellicularia filamentosa diazepine biodegradation 182
— tetrahydrocannabinol degradation 196
Penicillin 253
— formation of analogs 252
— semi-synthetic 242
Penicillin acylase 265
— immobilization of 15
Penicillinase 253
Penicillin N 251
Penicillin V production by directed fermentation 241
Penicillium degradation of furan derivatives 316
— penicillin production 242
Penicillium adametzi naphthyridine hydroxylation 322
Penicillium chrysogenum formation of 6-APA 253
— glucose oxidase formation 439
— invertase 453
Penicillium concavorugulosum alicyclic compounds hydroxylation 152
Penicillium digitatum citronellal transformations 102
— limonene transformation 106
Penicillium italicum limonene transformation 106
Penicillium lilacinum microbial degradation of phthalates 308
Peptides biotransformation 400ff
Pergolide bioconversions 215
Peroxidase hydroxylation of L-tyrosine 394
Peroxidations of steroids 45
Pesticides biodegradation of 181
Phenanthrene biotransformation of 288
— microbial hydroxylation 161
Phenol biotransformation of 288
— ring cleavage during methane fermentation 296
Phenolic pesticides microbial degradation 302
Phenylalanine formation by bioconversion 392
— racemate resolution 393
Phenylglycine racemate resolution 398
L-Phenylglycine formation by bioconversion 393
Phloroglucinol anaerobic metabolism 294
Phospholipid microbial production of 335
Phosphorylation of antibiotics 258
Phosphotricin racemate resolution 399
Phthalates microbial degradation 308
Phthalic acid biotransformation of 289
Phthalic acid esters microbial degradation 302
Phytophthora infestans solanine hydrolysis 231
Pichia vanriji lipid content of alkane grown 366
Picolinic acid microbial hydroxylation of 298
Picromycin 246
Pinene microbial transformation 110
Platenomycin 246
Pleurotus ostreatus cell immobilization 13
Polychlorinated biphenyl compounds microbial degradation 312
Polyoxin 245
Ponasterone A side chain cleavage selective 83
Prednisolone as product of steroid transformations 36
— production by steroid transformation 51
Pristane microbial oxidation of 361
Process design biotransformations 8
Product isolation by supercritical CO_2 extraction 25

— lipophilic products 24
Progesterone 11-α-hydroxylation 7
— use in steroid transformations 35
Prostaglandin microbial degradation 189
— optically active production of 186
— racemate resolution 189
Prostaglandin synthons racemate resolution 183
Protaminobacter rubrum glucotransferase 443
Protocatechuic acid biotransformation of 289
— meta-fission pathways 291
— microbial ring cleavage 293
— ortho-fission pathways 289
— ring cleavage 290 f
Pseudomonas degradation of dioxin derivatives 319
— degradation of phenols 305
— degradation of triazine derivatives 320
— phenolic pesticide degradation 303
— pyrrolnitrin formation 245
Pseudomonas aeruginosa alcohol dehydrogenases 351
— alicyclic compounds hydroxylation 152
— — microbial degradation 164
— alkane monooxygenase 346
— alkanes subterminal oxidation of 358
— citral transformation 103
— citronellal transformations 102
— oxidation of toluene 284
Pseudomonas arvilla pyrocatechase 286
Pseudomonas cepacia alkanes subterminal oxidation of 358
Pseudomonas dacunhae aspartate decarboxylase induction of 21
— cell immobilization 13
— decarboxylation of L-aspartate 391
Pseudomonas extorquens methane oxidation regulation of 368
Pseudomonas flava isolation from eucalyptus leaves 111
— transformation of cineole 111
Pseudomonas fluorescens degradation of pyridinium derivatives 318
Pseudomonas incognita linalool transformation 103
Pseudomonas lupani lupanine bioconversion 233
Pseudomonas melanogenum ampicillin formation 266
— formation of 6-APA 265
Pseudomonas multivorans aniline degradation 286, 306
Pseudomonas oleovorans alkene oxidation 359
Pseudomonas ovalis ketone reduction 128
— quinic acid bioconversion 174
Pseudomonas pseudomallei linalool transformation 103
Pseudomonas putida aldehyde dehydrogenases 354
— alicyclic compounds aromatization 167
— alkane monooxygenase 347
— cell immobilization 13
— dealkylation of methoxy groups 285
— degradation of anilines 307
— ephedrine degradation 234
— nitration of cannabinoids 198
Pseudomonas testosteroni dehydrogenase 68
— microbial degradation of phthalates 308
— morphine alkaloid bioconversions 227
Pyridine microbial degradation 300
Pyridine alkaloids microbial transformations 232
Pyridinium derivatives microbial degradation 318
Pyridocarbazole alkaloids bioconversions 217
Pyrimidine microbial degradation 301
Pyrrol microbial degradation of 298
Pyrrolidone microbial hydrolysis 150
Pyrrolizidine alkaloids microbial transformations 233
Pyrrolnitrin 244

Q

Quinic acid microbial transformation 174
Quinoline alkaloids microbial transformations 225
Quinolizidine alkaloids microbial transformations 233
Quinomycin 244

R

Racemate resolution 7
— D,L-amino acid derivatives 388
Raffinose hydrolysis process 446
Raffinose conversion by α-galactosidase 443
Rauwolfia alkaloids bioconversions 215
Reaction specificity of enzymes 6
Reduction of antibiotics 262
— of steroids 46
Regiospecificity of enzymes 6
Reichstein's substance S *see* Substance S
Reticuline oxidative phenolic coupling 224
Rhamnolipid microbial production of 335
Rhizopus arrhizus alicyclic compounds hydroxylation 152
— sterol side chain cleavage 83

Rhizopus japonicus degradation of furan derivatives 316
— degradation of oxathiine derivatives 319
— degradation of triazine derivatives 321
— microbial degradation of chlorinated biphenyls 312
Rhizopus nigricans alicyclic compounds hydrolysis of esters of 148
— alicyclic compounds hydroxylation 154
— degradation of furan derivatives 316
— dioxodecaline reduction 132
— in steroid hydroxylations 49, 55
— steroid hydroxylase 63
— sterol side chain cleavage 83
Rhizopus peka degradation of furan derivatives 316
Rhodococcus corallinus sterol side chain cleavage 91
Rhodococcus erythropolis trehalose dimycolate formation 334
Rhodococcus rhodochrous formation of dioic acids 354
Rhodopseudomonas gelatinosa phloroglucinol photometabolism of 294
Rhodopseudomonas palustris benzoic acid photometabolism of 294
Rhodotorula flava microbial biotin synthesis 177
Rhodotorula glutinis lipid content of alkane grown 366
— L-phenylalanine ammonia lyase 393
Rhodotorula gracilis amino acid oxydase induction of 21
— lipid content of alkane grown 366
Rhodotorula minuta alicyclic compounds hydrolysis of esters of 148
— cell immobilization 13
Rhodotorula mucilaginosa ketone reduction 128
— menthyl acetate asymmetric hydrolysis 108
Ribostamycin 249
Rifamycin 251, 269 f
— formation of analogs 252
Rubredoxin 347
Rumen bacteria degradation of benzthiophen 316

S

Saccharomyces cerevisiae chorismic acid bioconversion 174
— glutamylcysteine synthetase 403
— glutathione synthetase 403
— in stereospecific reductions of steroids 71
— invertase 453
— ketone reduction 128
— recombinant DNA technology 22
— reduction of oxo-isophorone 116
Saccharomyces chevalieri in stereospecific reductions of steroids 71
Saccharomyces drosophylarum prostaglandin synthons racemate resolution of 184
Saccharomyces pastorianus cell immobilization 13, 454
— in stereospecific reductions of steroids 71
Saccharomyces uvarum in stereospecific reductions of steroids 71
Saccharomyces uvarum (carlsbergensis) invertase 453
— in steroid ketone reduction 49
Salicylic acid biotransformation of 288
— microbial degradation 325
Sanguinarine microbial reduction 219
Sapogenins steroid precursors side chain cleavage 85
Sarsasapogenin use in steroid transformations 40
Selection of microorganisms for biotransformations 9
Sepedonium ampullosporum in steroid hydroxylations 59
Sepedonium chrysospermum vinca alkaloids bioconversions 211
Septomyxa affinis in steroid dehydrogenations 67
— in steroid reductions 49
Sequential reactions in steroids 48
Serine formation by bioconversion 397
Serotonin 394
Sisomicin 249
Sitosterol conversion by *Mycobacterium fortuitum* mutants 90
— use in steroid transformations 40
Smilagenin use in steroid transformations 40
Solanine bioconversion 231
Solasodine microbial hydroxylation 231
Solvents water immiscible for biotransformations 17
Sophoroselipid microbial production of 335
Sorbitol dehydrogenase microbial sources 436
— production of 436
Sorbose production from sorbitol 436
Sorbosone bioconversion from sorbose 437
Spiramycin 246, 251
Spores use in biotransformations 11
Sporobolomyces pararoseus reduction of tetralones 134
Sporotrichum sulfurescens carboline alkaloid hydroxylation 217
Squalene microbial oxidation of 361
— microbial transformation 122
Stachylidium bicolor sterol side chain cleavage 83
Stereospecificity of enzymes 6

Steroid ring system 33
Steroid dehydrogenations by cell-free enzymes 68
Steroid hydroxylation by cell-free hydroxylases 62
— by-products 66
Steroid-related compounds microbial reduction 134
— racemate resolution 134
Steroid ring fission hydroxylase inhibition of 86
Steroid transformation heteroatom oxidations 46
— reductions 46
— alcohol oxidations 43
— analytical methods 53
— B-ring scissions 45
— carbon-carbon bond scissions 44
— conjugations 47
— dehydrogenations commercial 67
— double bond introductions 44
— epoxidations 43
— heteroatom introductions 48
— history 34
— homoannulations 47
— hydrolyses 48
— hydroxylation processes commercial 55
— hydroxylations 41
— isomerizations 47
— lactonization 70
— methyl migration 47
— microbial fermentations for 52
— peroxidations 45
— product recovery 54
— reaction types 41
— resolution of racemic mixtures 71
— sequential reactions 48
Steroids microbiological transformation 31 ff
— production volume 40
— ring system cleavage 81
Steroidal alkaloids bioconversions 230
— type-reactions 230
Sterols conversion by mutants 89
— conversion in organic solvents 93
— microbial hydrogenation of 94
— microbial transformations of 79 ff
— side chain cleavage selective 82
— side chain degradation 80
Stigmasterol use in steroid transformations 40
Streptococcus allantoicus alicyclic compounds hydrolysis of esters of 150
Streptomutin A 245
Streptomyces co-oxidation of long-chain alkanes 362
— degradation of pyridinium derivatives 318
Streptomyces albogriseolus vinca alkaloids bioconversions 211
Streptomyces albus glucose isomerase of 428
Streptomyces antibioticus formation of oleandomycin analogs 251
— oleandomycin formation 249
Streptomyces aureofaciens as antibiotic-blocked mutant 247
— tetracycline formation 245
Streptomyces cacaoi polyoxin formation 245
Streptomyces celestis as antibiotic-blocked mutant 247
— celesticetin formation 244
Streptomyces cinnamonensis vinca alkaloids bioconversions 211
Streptomyces clavuligerus as antibiotic-blocked mutant 247
Streptomyces erythraeus erythromycin formation 249
Streptomyces fradiae neomycin formation 248
Streptomyces griseoviridus griseoviridin formation 244
Streptomyces griseus apomorphine bioconversion 219
— colchicine alkaloid bioconversions 229
— streptomycin formation 246, 249
— vinca alkaloids bioconversions 211
Streptomyces halstedii in steroid hydroxylations 58
Streptomyces hydrogenans hydroxysteroid dehydrogenase induction of 21
Streptomyces hygroscopicus alicyclic compounds hydrolysis of esters of 149
— turimycin formation 252
Streptomyces kanamyceticus kanamycin formation 248
Streptomyces lavendulae tetrahydrocannabinol degradation 197
Streptomyces lincolnensis lincomycin formation 244
Streptomyces mediocidicus alicyclic compounds hydroxylation 152
Streptomyces niveus novobiocin formation 249, 250
Streptomyces olivaceus α-galactosidase formation 445
— glucose isomerase of 428
Streptomyces olivochromogenes glucose isomerase of 428
Streptomyces peucetius adriamycin formation 249, 251
— as antibiotic-blocked mutant 247
— daunomycin formation 247
Streptomyces phaeochromogenes glucose isomerase of 428
Streptomyces platensis as antibiotic-blocked mutant 247
Streptomyces punipalus vinca alkaloids bioconversions 212

Streptomyces ribosidificus ribostamycin formation 248
Streptomyces rimosus apomorphine bioconversion 220
— ergoline lergotrile bioconversions 215
— paromomycin formation 248
Streptomyces roseochromogenes in steroid hydroxylations 49, 57
— metabolism of lysergic acid dialkyl homologs 214
— steroid hydroxylase 63
Streptomyces ruber prostaglandin degradation 189
Streptomyces spectabilis colchicine alkaloid bioconversions 229
Streptomyces spheroides novobiocin formation 244
Streptomyces venezuelae as antibiotic-blocked mutant 247
— cell immobilization 13
Streptomyces verticillus bleomycin formation 243
Streptomyces violaceus violamycin formation 252
Streptomycetes actinomycin formation 243
Streptomycin 245
Strychnine bioconversions 218
Substance S in steroid hydroxylations 49
— use in steroid transformations 35f
Substance S acetate synthesis of 43
Succinate dehydrogenase selective inactivation 23
Sucrose conversion by invertase 452
— *see* Invertase
Surface active agents production from hydrocarbons 375
Surfactants microbial degradation 311
— microbial production of 335
Syncephalastrum racemosum cannabidiol degradation 194
Syringic acid ring cleavage during methane fermentation 296

T

Talaromyces wortmannii degradation of anilines 307
Terpenoids microbial transformation 97ff
Terpenoid transformations sesquiterpenoids 118
— diterpenoids 121
— enrichment cultures 99
— isolation of products 100
— monoterpenoids 101
— substrate addition 99
— transformation sites 100
— triterpenoids 121
Terpineol 106
Testololactone synthesis of 44
Tetracyclines 245
1-Tetralone microbial reduction 133
Tetrandrine microbial dealkylation 225
Thalicarpine microbial oxidation 225
Thebaine bioconversions 227
Thiophene microbial degradation of 298
Threonine formation by bioconversion 397
Thymine microbial degradation 301
Tieghemella orchidis in steroid hydroxylations 60
Tobacco flavor 114, 118
Tobramycin 268
Toluene microbial degradation 284
Tomatine bioconversion 230
Torulopsis candida formation of dioic acids 356
Torulopsis gropengiesseri alkene oxidation 360
Trametes cinnabarina morphine alkaloid bioconversions 227
Trametes sanguinea morphine alkaloid bioconversions 227
Trehaloselipid microbial production of 335
Triazine derivatives microbial degradation 317, 320
Trichoderma reesei alicyclic compounds hydrolysis of esters of 149
— menthyl acetate asymmetric hydrolysis 108
Trichoderma virgatum phenolic pesticide degradation 303
Trichospora brinkmanii prostaglandin synthon bioconversion 186
Trichothecium roseum alicyclic compounds hydroxylation 152
Tropane alkaloids microbial transformations 233
Tropine bioconversion 233
Trypsin dynorphin synthesis 403
— enkephalin synthesis 404
— peptide synthesis 400
— transpeptidation of insulin chain 406
Tryptophan formation by bioconversion 393
— use of 412
L-Tryptophan formation by bioconversion 393
Tryptophanase immobilization 393
Tryptophan oxygenase pyrrol ring cleavage 299
Turimycin 252
Tylosin 246, 251
β-Tyrosinase immobilization 393
Tyrosine formation by bioconversion 393

U

Uracil microbial degradation 301
Urocanase selective inactivation 23

V

Valine formation by bioconversion 391
L-Valine formation by bioconversion 393
Vanillin ring cleavage during methane fermentation 296
Verticillium dahliae reduction of sanguinarine 219
Verticillium theobromae sterol side chain cleavage 83
Vinblastine bioconversions 210
Vinca alkaloids microbial transformations 210
Vincristine bioconversions 210
Vindoline bioconversion 211
Violamycin 252

W

Warfarin aromatic hydroxylation 210
Whey lactose hydrolysis 452

X

Xenobiotics 301

Y

Yeasts hydrocarbon-degrading 332
Yohimbine *see* Rauwolfia alkaloids

Z

Zeaxanthin 117